PROCESS HEAT TRANSFER

PROCESS HEAT TRANSFER

BY

DONALD Q. KERN

*D. Q. Kern Associates, and
Professorial Lecturer in Chemical Engineering
Case Institute of Technology*

McGRAW-HILL BOOK COMPANY

Auckland Bogotá Guatemala Hamburg Lisbon London
Madrid Mexico New Delhi Panama Paris San Juan
São Paulo Singapore Sydney Tokyo

PROCESS HEAT TRANSFER
INTERNATIONAL EDITION 1965

Exclusive rights by McGraw-Hill Book Co-Singapore for manufacture and export. This book cannot be re-exported from the country to which it is consigned by McGraw-Hill.

10 1 2 3 4 5 6 7 8 9 20 FSP 9 5 4 3

Copyright, 1950, by the McGraw-Hill Book Company, Inc. All rights reserved. No part of this publication may be reproduced, stored in a retrieval system, or transmitted, in any form or by any means, electronic, mechanical, photocopying, recording, or otherwise, without the prior written permission of the publisher.

When ordering this title use ISBN 0-07-085353-3

To my wife
NATALIE W. KERN
for her real help

PREFACE

It is the object of this text to provide fundamental instruction in heat transfer while employing the methods and language of industry. This treatment of the subject has evolved from a course given at the Polytechnic Institute of Brooklyn over a period of years. The possibilities of collegiate instruction patterned after the requirements of the practicing process engineer were suggested and encouraged by Dr. Donald F. Othmer, Head of the Department of Chemical Engineering. The inclusion of the practical aspects of the subject as an integral part of the pedagogy was intended to serve as a supplement rather than a substitute for a strong foundation in engineering fundamentals. These points of view have been retained throughout the writing of the book.

To provide the rounded group of heat-transfer tools required in process engineering it has been necessary to present a number of empirical calculation methods which have not previously appeared in the engineering literature. Considerable thought has been given to these methods, and the author has discussed them with numerous engineers before accepting and including them in the text. It has been a further desire that all the calculations appearing in the text shall have been performed by an experienced engineer in a conventional manner. On several occasions the author has enlisted the aid of experienced colleagues, and their assistance is acknowledged in the text. In presenting several of the methods some degree of accuracy has been sacrificed to permit the broader application of fewer methods, and it is hoped that these simplifications will cause neither inconvenience nor criticism.

It became apparent in the early stages of writing this book that it could readily become too large for convenient use, and this has affected the plan of the book in several important respects. A portion of the material which is included in conventional texts is rarely if ever applied in the solution of run-of-the-mill engineering problems. Such material, as familiar and accepted as it may be, has been omitted unless it qualified as important fundamental information. Secondly, it was not possible to allocate space for making bibliographic comparisons and evaluations and at the same time present industrial practice. Where no mention has been made of a recent contribution to the literature no slight was intended. Most of the literature references cited cover methods on which the author has obtained additional information from industrial application.

PREFACE

The author has been influenced in his own professional development by the excellent books of Prof. W. H. McAdams, Dr. Alfred Schack, and others, and it is felt that their influence should be acknowledged separately in addition to their incidence in the text as bibliography.

For assistance with the manuscript indebtedness is expressed to Thomas H. Miley, John Blizard, and John A. Jost, former associates at the Foster Wheeler Corporation. For checking the numerical calculations credit is due to Krishnabhai Desai and Narendra R. Bhow, graduate students at the Polytechnic Institute. For suggestions which led to the inclusion or exclusion of certain material thanks are due Norman E. Anderson, Charles Bliss, Dr. John F. Middleton, Edward L. Pfeiffer, Oliver N. Prescott, Everett N. Sieder, Dr. George E. Tait, and to Joseph Meisler for assistance with the proof. The Tubular Exchanger Manufacturers Association has been most generous in granting permission for the reproduction of a number of the graphs contained in its Standard. Thanks are also extended to Richard L. Cawood, President, and Arthur E. Kempler, Vice-President, for their personal assistance and for the cooperation of The Patterson Foundry & Machine Company.

DONALD Q. KERN

NEW YORK, N.Y.
April, 1950

CONTENTS

PREFACE vii

INDEX TO THE PRINCIPAL APPARATUS CALCULATIONS xi

CHAPTER
1. Process Heat Transfer 1
2. Conduction 6
3. Convection 25
4. Radiation 62
5. Temperature 85
6. Counterflow: Double-pipe Exchangers 102
7. 1-2 Parallel-counterflow: Shell-and-Tube Exchangers 127
8. Flow Arrangements for Increased Heat Recovery 175
9. Gases 190
10. Streamline Flow and Free Convection 201
11. Calculations for Process Conditions 221
12. Condensation of Single Vapors 252
13. Condensation of Mixed Vapors 313
14. Evaporation 375
15. Vaporizers, Evaporators, and Reboilers 453
16. Extended Surfaces 512
17. Direct-contact Transfer: Cooling Towers 563
18. Batch and Unsteady State Processes 624
19. Furnace Calculations 674
20. Additional Applications 716
21. The Control of Temperature and Related Process Variables 765

APPENDIX OF CALCULATION DATA 791

AUTHOR INDEX 847

SUBJECT INDEX 851

INDEX TO THE PRINCIPAL APPARATUS CALCULATIONS

EXCHANGERS

Double-pipe counterflow exchanger (benzene–toluene)	113
Double-pipe series-parallel exchanger (lube oil–crude oil)	121
Tubular exchanger (kerosene–crude oil)	151
Tubular exchanger (water–water)	155
Tubular cooler (K_2PO_4 solution–water)	161
Tubular heater, unbaffled (sugar solution–steam)	167
Tubular 2-4 cooler (33.5° API oil–water)	181
Tubular exchangers in series (acetone–acetic acid)	184
Tubular gas aftercooler (ammonia gas–water)	193
Tubular gas intercooler (CO_2–water vapor–water)	346
Tubular streamline flow heater (crude oil–steam)	203
Tubular free convection heater (kerosene–steam)	207
Core tube heater (gas oil–steam)	211
Tank heater (aniline–steam)	217
Tubular exchanger (straw oil–naphtha)	231
Tubular 4-8 exchanger (lean oil–rich oil)	235
Tubular cooler (NaOH solution–water)	238
Tubular heater (alcohol–steam)	241
Tubular split-flow cooler (flue gas–water)	246
Jacketed vessel (aqueous solution–steam)	719
Tube coil (aqueous solution–steam)	723
Pipe coil cooler (slurry–water)	725
Trombone cooler (SO_2 gas–water)	729
Atmospheric cooler (jacket water–water)	736
Electric resistance heater	758

CONDENSERS (TUBULAR)

Condenser, horizontal (propanol–water)	274
Condenser, vertical (propanol–water)	277
Desuperheater-condenser, horizontal (butane–water)	285
Condenser-subcooler, vertical (pentanes–water)	290
Condenser-subcooler, horizontal (pentanes–water)	295
1-1 Reflux condenser, vertical (carbon disulfide–water)	299
Surface condenser (turbine exhaust steam–water)	308
Condenser, horizontal (hydrocarbon mixture–water)	331
Condenser, horizontal (steam, CO_2 mixture–water)	346
Condenser, horizontal (hydrocarbon mixture, gas, steam–water)	356

EVAPORATORS (TUBULAR)

Raw water evaporator	387
Power plant makeup evaporator	388

INDEX TO THE PRINCIPAL APPARATUS CALCULATIONS

Process multiple effect evaporator.................389
Heat transformer evaporator..................390
Salt water distiller......................393
Cane sugar multiple effect evaporator..............418
Paper pulp waste liquor multiple effect evaporator........427
Caustic soda multiple effect forced circulation evaporator....437
Thermocompression cane sugar evaporator............447

VAPORIZING EXCHANGERS (TUBULAR)

Vaporizer, forced circulation (butane–steam)..........464
Kettle reboiler (hydrocarbons–steam)..............475
Thermosyphon reboiler, horizontal (naphtha–gas oil)......482
Thermosyphon reboiler, vertical (butane–steam)........488

EXTENDED SURFACES

Longitudinal fin double-pipe cooler (gas oil–water)......530
Tubular longitudinal fin cooler (oxygen–water).........535
Transverse fin crossflow cooler (air–water)...........556

DIRECT CONTACT TRANSFER

Cooling tower requirement...................602
Cooling tower guarantee....................605
Cooling tower rerating.....................609
Gas cooler (nitrogen–water)..................615
Gas cooler, approximate solution (nitrogen–water).......620

RADIANT HEATERS

Tube still...........................702
Direct fired vessel......................709

CHAPTER 1

PROCESS HEAT TRANSFER

Heat Transfer. The science of thermodynamics deals with the quantitative transitions and rearrangements of energy as heat in bodies of matter. *Heat transfer* is the science which deals with the rates of exchange of heat between hot and cold bodies called the *source* and *receiver*. When a pound of water is vaporized or condensed, the energy change in either process is identical. The rates at which either process can be made to progress with an independent source or receiver, however, are inherently very different. Vaporization is generally a much more rapid phenomenon than condensation.

Heat Theories. The study of heat transfer would be greatly enhanced by a sound understanding of the nature of heat. Yet this is an advantage which is not readily available to students of heat transfer or thermodynamics because so many manifestations of heat have been discovered that no simple theory covers them all. Laws which may apply to mass transitions may be inapplicable to molecular or atomic transitions, and those which are applicable at low temperatures may not apply at high temperatures. For the purposes of engineering it is necessary to undertake the study with basic information on but a few of the many phenomena. The phases of a single substance, solid, liquid, and gaseous, are associated with its energy content. In the solid phase the molecules or atoms are close together, giving it rigidity. In the liquid phase sufficient thermal energy is present to extend the distance of adjacent molecules such that rigidity is lost. In the gas phase the presence of additional thermal energy has resulted in a relatively complete separation of the atoms or molecules so that they may wander anywhere in a confined space. It is also recognized that, whenever a change of phase occurs outside the critical region, a large amount of energy is involved in the transition.

For the same substance in its different phases the various thermal properties have different orders of magnitude. As an example, the specific heat per unit mass is very low for solids, high for liquids, and usually intermediate for gases. Similarly in any body absorbing or losing heat, special consideration must be given whether the change is one of sensible or latent heat or both. Still further, it is also known that a hot source is capable of such great subatomic excitement that it emits energy without any direct contact with the receiver, and this is the underlying principle of radiation. Each type of change exhibits its own peculiarities.

Mechanisms of Heat Transfer. There are three distinct ways in which heat may pass from a source to a receiver, although most engineering applications are combinations of two or three. These are *conduction, convection,* and *radiation.*

Conduction. Conduction is the transfer of heat through fixed material such as the stationary wall shown in Fig. 1.1. The direction of heat flow will be at right angles to the wall if the wall surfaces are isothermal and the body homogeneous and isotropic. Assume that a source of heat exists on the left face of the wall and a receiver of heat exists on the right face. It has been known and later it will be confirmed by derivation that the flow of heat per hour is proportional to the change in temperature through the wall and the area of the wall A. If t is the temperature at any point in the wall and x is the thickness of the wall in the direction of heat flow, the quantity of heat flow dQ is given by

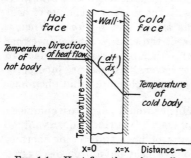

Fig. 1.1. Heat flow through a wall.

$$dQ = kA\left(-\frac{dt}{dx}\right) \quad \text{Btu/hr} \tag{1.1}$$

The term $-dt/dx$ is called the *temperature gradient* and has a negative sign if the temperature has been assumed higher at the face of the wall where $x = 0$ and lower at the face where $x = X$. In other words, the instantaneous quantity of heat transfer is proportional to the area and temperature difference dt, which drives the heat through the wall of thickness dx. The proportionality constant k is peculiar to conductive heat transfer and is known as the *thermal conductivity*. It is evaluated experimentally and is basically defined by Eq. (1.1). The thermal conductivities of solids have a wide range of numerical values depending upon whether the solid is a relatively good conductor of heat such as a metal or a poor conductor such as asbestos. The latter serve as *insulators*. Although heat conduction is usually associated with heat transfer through solids, it is also applicable with limitations to gases and liquids.

Convection. Convection is the transfer of heat between relatively hot and cold portions of a fluid by mixing. Suppose a can of liquid were placed over a hot flame. The liquid at the bottom of the can becomes heated and less dense than before owing to its thermal expansion. The liquid adjacent to the bottom is also less dense than the cold upper portion and rises through it, transferring its heat by mixing as it rises. The transfer of heat from the hot liquid at the bottom of the can to the remainder is *natural* or *free convection*. If any other agitation occurs, such as that produced by a stirrer, it is *forced convection*. This type of heat transfer may be described in an equation which imitates the form of the conduction equation and is given by

$$dQ = hA\, dt \qquad (1.2)$$

The proportionality constant h is a term which is influenced by the nature of the fluid and the nature of the agitation and must be evaluated experimentally. It is called the *heat-transfer coefficient*. When Eq. (1.2) is written in integrated form, $Q = hA\, \Delta t$, it is called Newton's law of cooling.

Radiation. Radiation involves the transfer of radiant energy from a source to a receiver. When radiation issues from a source to a receiver, part of the energy is absorbed by the receiver and part reflected by it. Based on the second law of thermodynamics Boltzmann established that the rate at which a source gives off heat is

$$dQ = \sigma\epsilon\, dA\, T^4 \qquad (1.3)$$

This is known as the fourth-power law in which T is the absolute temperature. σ is a dimensional constant, but ϵ is a factor peculiar to radiation and is called the *emissivity*. The emissivity, like the thermal conductivity k or the heat-transfer coefficient h, must also be determined experimentally.

Process Heat Transfer. *Heat transfer* has been described as the study of the rates at which heat is exchanged between heat sources and receivers usually treated independently. *Process heat transfer* deals with the rates of heat exchange as they occur in the heat-transfer equipment of the engineering and chemical processes. This approach brings to better focus the importance of the temperature difference between the source and receiver, which is, after all, the driving force whereby the transfer of heat is accomplished. A typical problem of process heat transfer is concerned with the quantities of heats to be transferred, the rates at which they may be transferred because of the natures of the bodies, the driving potential, the extent and arrangement of the surface separating the source and receiver, and the amount of mechanical energy which may be expended to facilitate the transfer. Since heat transfer involves an *exchange* in a *system*, the loss of heat by the one body will equal the heat absorbed by another *within the confines of the same system*.

In the chapters which follow studies will first be made of the three individual heat-transfer phenomena and later of the way in which their combination with a simultaneous source and receiver influences an apparatus as a whole. A large number of the examples which follow have been selected from closely related processes to permit gradual comparisons. This should not be construed as limiting the broadness of the underlying principles.

Many of the illustrations and problems in the succeeding chapters refer to liquids derived from petroleum. This is quite reasonable, since petroleum refining is a major industry, petroleum products are an important fuel for the power industry, and petroleum derivatives are the starting point for many syntheses in the chemical industry.

Petroleum is a mixture of a great many chemical compounds. Some can be isolated rather readily, and the names of common hydrocarbons present in petroleum may be identified on Fig. 7 in the Appendix. But more frequently there is no need to obtain pure compounds, since the ultimate use of a mixture of related compounds will serve as well. Thus lubricating oil is a mixture of several compounds of high molecular weight, all of which are suitable lubricants. Similarly, gasoline which will ultimately be burned will be composed of a number of volatile combustible compounds. Both of these common petroleum products were present in the crude oil when it came from the ground or were formed by subsequent reaction and separated by distillation. When dealt with in a process or marketed as mixtures, these products are called *fractions* or *cuts*. They are given common names or denote the refinery operation by which they were produced, and their specific gravities are defined by a scale established by the American Petroleum Institute and termed either degrees API or °API. The °API is related to the specific gravity by

$$°\text{API} = \frac{141.5}{\text{sp gr at } 60°\text{F}/60°\text{F}} - 131.5 \qquad (1.4)$$

Being mixtures of compounds the petroleum fractions do not boil isothermally like pure liquids but have boiling ranges. At atmospheric pressure the lowest temperature at which a liquid starts to boil is identified as the *initial boiling point*, IBP, °F. A list of the common petroleum fractions derived from crude oil is given below:

Fractions from crude oil	Approx °API	Approx IBP, °F
Light ends and gases............	114	
Gasoline........................	75	200
Naphtha........................	60	300
Kerosene.......................	45	350
Absorption oil..................	40	450
Straw oil.......................	40	500
Distillate.......................	35	550
Gas oil.........................	28	600
Lube oil........................	18–30	
Reduced crude..................		
Paraffin wax and jelly...........		
Fuel oil (residue)................	25–35	500
Asphalt.........................		

A method of defining the chemical character of petroleum and correlating the properties of mixtures was introduced by Watson, Nelson, and Murphy.[1] They observed that, when a crude oil of uniform distilling behavior is distilled into narrow cuts, the ratio of the cube root of the absolute average boiling points to the specific gravities of the cuts is a constant or

$$K = \frac{T_B^{1/3}}{s} \tag{1.5}$$

where K = characterization factor
T_B = average boiling point, °R
s = specific gravity at 60°/60°

NOMENCLATURE FOR CHAPTER 1

A Heat-transfer surface, ft^2
h Individual heat-transfer coefficient, Btu/(hr)(ft^2)(°F)
K Characterization factor
k Thermal conductivity, Btu/(hr)(ft^2)(°F/ft)
Q Heat flow, Btu/hr
s Specific gravity, dimensionless
T Temperature, °R
T_B Absolute average boiling temperature, °R
t Temperature in general, °F
x, X Distance, ft
σ A constant, Btu/(hr)(ft^2)(°R^4)
ϵ Emissivity, dimensionless

[1] Watson, K. M., E. F. Nelson, and G. B. Murphy, *Ind. Eng. Chem.*, **25,** 880 (1933 **27,** 1460 (1935).

CHAPTER 2

CONDUCTION

The Thermal Conductivity. The fundamentals of heat conduction were established over a century ago and are generally attributed to Fourier. In numerous systems involving flow such as heat flow, fluid flow, or electricity flow, it has been observed that the flow quantity is directly proportional to a driving potential and inversely proportional to the resistances applying to the system, or

$$\text{Flow} \propto \frac{\text{potential}}{\text{resistance}} \qquad (2.1)$$

In a simple hydraulic path the pressure along the path is the driving potential and the roughness of the pipe is the flow resistance. In an electric circuit the simplest applications are expressed by Ohm's law: The voltage on the circuit is the driving potential, and the difficulty with which electrons negotiate the wire is the resistance. In heat flow through a wall, flow is effected by a temperature difference between the hot and cold faces. Conversely, from Eq. (2.1) when the two faces of a wall are at different temperatures, a flow of heat and a resistance to heat flow are necessarily present. The *conductance* is the reciprocal of the resistance to heat flow and Eq. (2.1) may be expressed by

$$\text{Flow} \propto \text{conductance} \times \text{potential} \qquad (2.2)$$

To make Eq. (2.2) an equality the conductance must be evaluated in such a way that both sides will be dimensionally and numerically correct. Suppose a measured quantity of heat Q' Btu has been transmitted by a wall of unknown size in a measured time interval θ hr with a measured temperature difference Δt °F. Rewriting Eq. (2.2)

$$Q = \frac{Q'}{\theta} = \text{conductance} \times \Delta t \qquad Btu/hr \qquad (2.3)$$

and the conductance has the dimensions of Btu/(hr)(°F). The conductance is a measured property of the entire wall, although it has also been found experimentally that the flow of heat is influenced independently by the thickness and the area of the wall. If it is desired to design a wall to have certain heat-flow characteristics, the conductance obtained above is not useful, being applicable only to the experimental wall. To enable a broader use of experimental information, it has become conventional to report the conductance only when all the dimensions are referred to unit values. When the conductance is reported for a quantity of material 1 ft thick with heat-flow area 1 ft², time unit 1 hr, and temperature difference 1°F, it is called the *thermal conductivity k*. The relationship between the thermal conductivity and the conductance of an entire wall of thickness L and area A is then given by

$$\text{Conductance} = k\frac{A}{L}$$

and

$$Q = k\frac{A}{L}\Delta t \qquad (2.4)$$

where k has the dimensions resulting from the expression $QL/A\ \Delta t$ or Btu/(hr)(ft² of flow area)(°F of temperature difference)/(ft of wall thickness).[1]

Experimental Determination of k:* **Nonmetal Solids.** An apparatus for the determination of the thermal conductivity of nonmetal solids is shown in Fig. 2.1. It consists of an electrical heating plate, two identical test specimens through which heat passes, and two water jackets which remove heat. The temperatures at both faces of the specimens and at their sides are measured by thermocouples. A guard ring is provided to assure that all the measured heat input to the plate passes through the specimens with a negligible loss from their sides. The guard ring surrounds the test assembly and consists of an auxiliary heater sandwiched between pieces of the material being tested. While current enters the heating plate, the input to the

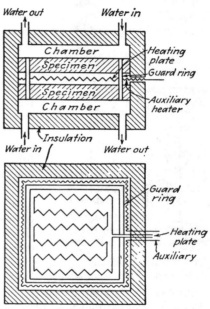

FIG. 2.1. Guarded conductivity apparatus.

---

[1] In the metric system it is usual to report the thermal conductivity as cal/(sec)(cm²)(°C/cm).

* An excellent review of experimental methods will be found in Saha and Srivastava, "Treatise on Heat," The Indian Press, Calcutta, 1935. Later references are Bates, O. K., *Ind. Eng. Chem.*, **25**, 432 (1933); **28**, 494 (1936); **33**, 375 (1941); **37**, 195 (1945). Bolland, J. L. and H. W. Melville, *Trans. Faraday Soc.*, **33**, 1316 (1937). Hutchinson, E., *Trans. Faraday Soc.*, **41**, 87 (1945).

auxiliary heater is adjusted until no temperature differences exist between the specimens and adjacent points in the guard ring. Observations are made when the heat input and the temperatures on both faces of each specimen remain steady. Since half of the measured electrical heat input to the plate flows through each specimen and the temperature difference and dimensions of the specimen are known, k can be computed directly from Eq. (2.4).

Liquids and Gases. There is greater difficulty in determining the conductivities of liquids and gases. If the heat flows through a thick layer of liquid or gas, it causes free convection and the conductivity is deceptively high. To reduce convection it is necessary to use very thin films and small temperature differences with attendant errors of measurement. A method applicable to viscous fluids consists of a bare electric wire passing through a horizontal tube filled with test liquid. The tube is immersed in a constant-temperature bath. The resistance of the wire is calibrated against its temperature. For a given rate of heat input and for the temperature of the wire obtained from resistance measurements the conductivity can be calculated

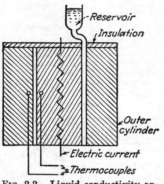

Fig. 2.2. Liquid conductivity apparatus. (*After J. F. D. Smith.*)

by suitable equations. A more exact method, however, is that of Bridgman and Smith,[1] consisting of a very thin fluid annulus between two copper cylinders immersed in a constant-temperature bath as shown in Fig. 2.2. Heat supplied to the inner cylinder by a resistance wire flows through the film to the outer cylinder, where it is removed by the bath. This apparatus, through the use of a reservoir, assures that the annulus is full of liquid and is adaptable to gases. The film is $\frac{1}{64}$ in. thick, and the temperature difference is kept very small.

Influence of Temperature and Pressure on k. The thermal conductivities of solids are greater than those of liquids, which in turn are greater than those of gases. It is easier to transmit heat through a solid than a liquid and through a liquid than a gas. Some solids, such as metals, have high thermal conductivities and are called *conductors*. Others have low conductivities and are poor conductors of heat. These are *insulators*. In experimental determinations of the type described above the thermal conductivity has been assumed independent of the temperature at any point in the test material. The reported values of k are consequently

[1] Smith, J. F. D., *Ind. Eng. Chem.*, **22**, 1246 (1930); *Trans. ASME*, **58**, 719 (1936).

the averages for the entire specimen, and the error introduced by this assumption can be estimated by an examination of Tables 2 to 5 in the Appendix. The conductivities of solids may either increase or decrease with temperature and in some instances may even reverse their rate of change from a decrease to an increase. For the most practical problems there is no need to introduce a correction for the variation of the thermal conductivity with temperature. However, the variation can usually be expressed by the simple linear equation

$$k = k_0 + \gamma t$$

where k_0 is the conductivity at 0°F and γ is a constant denoting the change in the conductivity per degree change in temperature. The conductivities of most liquids decrease with increasing temperature, although water is a notable exception. For all the common gases and vapors there is an increase with increasing temperature. Sutherland[1] deduced an equation from the kinetic theory which is applicable to the variation of the conductivity of gases with temperature

$$k = k_{32} \frac{492 + C_k}{T + C_k} \left(\frac{T}{492}\right)^{3/2}$$

where C_k = Sutherland constant
 T = absolute temperature of the gas, °R
 k_{32} = conductivity of the gas at 32°F

The influence of pressure on the conductivities of solids and liquids appears to be negligible, and the reported data on gases are too inexact owing to the effects of free convection and radiation to permit generalization. From the kinetic theory of gases it can be concluded that the influence of pressure should be small except where a very low vacuum is encountered.

Contact Resistance. One of the factors which causes error in the determination of the thermal conductivity is the nature of the bond formed between the heat source and the fluid or solid specimen which contacts it and transmits heat. If a solid receives heat by contacting a solid, it is almost impossible to exclude the presence of air or other fluid from the contact. Even when a liquid contacts a metal, the presence of minute pits or surface roughness may permanently trap infinitesimal bubbles of air, and it will be seen presently that these may cause considerable error.

Derivation of a General Conduction Equation. In Eqs. (2.1) to (2.4) a picture of heat conduction was obtained from an unqualified observation of the relation between heat flow, potential, and resistance. It is now feasible to develop an equation which will have the broadest applicability

[1] Sutherland, W., *Phil. Mag.*, **36,** 507 (1893).

and from which other equations may be deduced for special applications. Equation (2.4) may be written in differential form

$$\frac{dQ'}{d\theta} = k\, dA\, \frac{dt}{dx} \qquad (2.5)$$

In this statement k is the only property of the matter and it is assumed to be independent of the other variables. Referring to Fig. 2.3, an elemental cube of volume $dv = dx\, dy\, dz$ receives a differential quantity of heat dQ'_1 Btu through its left yz face in the time interval $d\theta$. Assume all but the left and right yz faces are insulated. In the same interval the quantity of heat dQ'_2 leaves at the right face. It is apparent that any of three effects may occur: dQ'_1 may be greater than dQ'_2 so that the elemental volume stores heat, increasing the *average* temperature of the cube; dQ'_2 may be greater than dQ'_1 so that the cube loses heat; and lastly, dQ'_1 and dQ'_2 may be equal so that the heat will simply pass through the cube without affecting the storage of heat. Taking either of the first two cases as being more general, a storage or depletion term dQ' can be defined as the difference between the heat entering and the heat leaving or

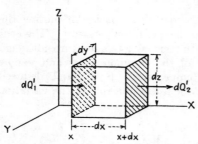

Fig. 2.3. Unidirectional heat flow.

$$dQ' = dQ'_1 - dQ'_2 \qquad (2.6)$$

According to Eq. (2.5) the heat entering on the left face may be given by

$$\frac{dQ'_1}{d\theta} = k\, dy\, dz \left(-\frac{\partial t}{\partial x}\right) \qquad (2.7)$$

The temperature gradient $-\dfrac{\partial t}{\partial x}$ may vary with both time and position in the cube. The variation of $-\dfrac{\partial t}{\partial x}$ as $f(x)$ only is $-\dfrac{\partial(\partial t/\partial x)}{\partial x}$. Over the distance dx from x to $x + dx$, if $dQ'_2 > dQ'_1$, the total change in the temperature gradient will be $-\dfrac{\partial(\partial t/\partial x)}{\partial x} dx$ or $-\dfrac{\partial^2 t}{\partial x^2} dx$. Then at x the gradient is $-\dfrac{\partial t}{\partial x}$, and at $x + dx$ the temperature gradient is

$$-\frac{\partial t}{\partial x} - \frac{\partial^2 t}{\partial x^2} dx$$

CONDUCTION

dQ_2' out of the cube at the right face and in the same form as Eq. (2.7) is given by

$$\frac{dQ_2'}{d\theta} = k\, dy\, dz \left(-\frac{\partial t}{\partial x} - \frac{\partial^2 t}{\partial x^2}\, dx\right) \quad (2.8)$$

from which

$$\frac{dQ'}{d\theta} = \frac{dQ_1'}{d\theta} - \frac{dQ_2'}{d\theta} = k\, dy\, dz \left(\frac{\partial^2 t}{\partial x^2}\right) dx \quad (2.9)$$

The cube will have changed in temperature by $-dt$ deg. The change in temperature per unit time will be $dt/d\theta$ and over the time interval $d\theta$ it is given by $(dt/d\theta)\, d\theta$ deg. Since the analysis has been based on an elemental *volume*, it is now necessary to define a volumetric specific heat, c_v Btu/(ft^3)(°F) obtained by multiplying the weight specific heat c Btu/(lb)(°F) by the density ρ. To raise the volume $dx\, dy\, dz$ by

$$\frac{dt}{d\theta}\, d\theta\ \text{°F}$$

requires a heat change in the cube of

$$\frac{dQ'}{d\theta} = c\rho\, dx\, dy\, dz\, \frac{\partial t}{\partial \theta} \quad (2.10)$$

and combining Eqs. (2.9) and (2.10)

$$c\rho\, dx\, dy\, dz\, \frac{\partial t}{\partial \theta} = k\, dy\, dz \left(\frac{\partial^2 t}{\partial x^2}\right) dx \quad (2.11)$$

from which

$$\frac{\partial t}{\partial \theta} = \frac{k}{c\rho}\left(\frac{\partial^2 t}{\partial x^2}\right) \quad (2.12)$$

which is *Fourier's general equation*, and the term $k/c\rho$ is called the thermal diffusivity, since it contains all the properties involved in the conduction of heat and has the dimensions of ft^2/hr. If the insulation is removed from the cube so that the heat travels along the X, Y, and Z axes, Eq. (2.12) becomes

$$\frac{\partial t}{\partial \theta} = \frac{k}{c\rho}\left(\frac{\partial^2 t}{\partial x^2} + \frac{\partial^2 t}{\partial y^2} + \frac{\partial^2 t}{\partial z^2}\right) \quad (2.13)$$

When the flow of heat into and out of the cube is constant as in the *steady state*, t does not vary with time, and $dt/d\theta = 0$, in Eq. (2.12). $\partial t/\partial x$ is a constant and $\partial^2 t/\partial x^2 = 0$. $dQ_1' = dQ_2'$, and Eq. (2.8) reduces to Eq. (2.5) where $dx\, dy = dA$. Substituting dQ for $dQ'/d\theta$, both terms having dimensions of Btu/hr, the *steady-state* equation is

$$dQ = k\, dA\, \frac{dt}{dx} \tag{2.14}$$

Equation (2.14) applies to many of the common engineering problems.

Thermal Conductivity from Electrical-conductivity Measurements. The relationship between the thermal and electrical conductivities of metals demonstrates an application of Fourier's derivation incorporated in Eq. (2.9) and is a useful method of determining the thermal conductivities of metals. An insulated bar of metal as shown in Fig. 2.4 has its left and right cross-sectional faces exposed to different constant-temperature baths at t_1 and t_2, respectively. By fastening electric leads to the left and right faces, respectively, a current of I amp may be passed in the direction indicated, generating heat throughout the length of the bar. The quantities of heat leaving both ends of the bar in the steady state must be equal to the amount of heat received as electrical energy, $I^2 R^\omega$, where R^ω is the resistance in ohms. From Ohm's law

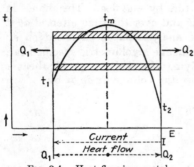

FIG. 2.4. Heat flow in a metal.

$$I = \frac{E_1 - E_2}{\sigma(L/A)} = \frac{A}{\sigma}\frac{dE}{dx}$$

where $E_1 - E_2$ is the voltage difference, σ the resistivity of the wire in ohm-ft and K, the reciprocal of the resistivity, is the electrical conductivity.

$$I = KA\frac{dE}{dx} \tag{2.15}$$

$$R^\omega = \frac{\sigma\, dx}{A} = \frac{dx}{KA} \tag{2.16}$$

Substituting Eqs. (2.15) and (2.16) for $I^2 R^\omega$,

$$dQ = I^2 R^\omega = K^2 A^2 \left(\frac{dE}{dx}\right)^2 \frac{dx}{KA} = KA\left(\frac{dE}{dx}\right)^2 dx \tag{2.17}$$

But this is the same as the heat transferred by conduction and given by

Eq. (2.9). When $t_1 = t_2$ and equating (2.9) and (2.17),

$$k \frac{d^2t}{dx^2} dx - K \left(\frac{dE}{dx}\right)^2 dx = 0 \qquad (2.18)$$

But

$$\frac{dt}{dx} = \frac{dt}{dE} \frac{dE}{dx} \qquad (2.19)$$

Differentiating,

$$\frac{d^2t}{dx^2} = \left(\frac{dE}{dx}\right)^2 \frac{d^2t}{dE^2} + \frac{dt}{dE}\left(\frac{d^2E}{dx^2}\right) \qquad (2.20)$$

If I and A are constant for the bar, then $K(dE/dx)$ is constant. Since K does not vary greatly with t or x, dE/dx is constant, $d^2E/dx^2 = 0$, and from Eq. (2.18) substituting Eq. (2.20) for d^2t/dx^2

$$k \frac{d^2t}{dE^2} - K = 0 \qquad (2.21)$$

$$\frac{d^2t}{dE^2} = \frac{K}{k} \qquad (2.22)$$

$$\frac{kt}{K} = \frac{1}{2} E^2 + C_1 E + C_2 \qquad (2.23)$$

where C_1 and C_2 are integration constants. Since there are three constants in Eq. (2.23), C_1, C_2, and k/K, three voltages and three temperatures must be measured along the bar for their evaluation. C_1 and C_2 are determined from the end temperatures and k is obtained from k/K using the more simply determined value of the electrical conductivity K.

Flow of Heat through a Wall. Equation (2.14) was obtained from the general equation when the heat flow and temperatures into and out of the two opposite faces of the partially insulated elemental cube $dx\,dy\,dz$ were constant. Upon integration of Eq. (2.14) when all of the variables but Q are independent the steady-state equation is

$$Q = \frac{kA}{L} \Delta t \qquad (2.24)$$

Given the temperatures existing on the hot and cold faces of a wall, respectively, the heat flow can be computed through the use of this equation. Since kA/L is the conductance, its reciprocal R is the resistance to heat flow, or $R = L/kA$ (hr)(°F)/Btu.

Example 2.1. Flow of Heat through a Wall. The faces of a 6-in. thick wall measuring 12 by 16 ft will be maintained at 1500 and 300°F, respectively. The wall is made of kaolin insulating brick. How much heat will escape through the wall?

Solution. The average temperature of the wall will be 900°F. From Table 2 in the

Appendix the thermal conductivity at 932°F is 0.15 Btu/(hr)(ft²)(°F/ft). Extrapolation to 900°F will not change this value appreciably.

$$Q = \frac{kA}{L} \Delta t$$

Where $\Delta t = 1500 - 300 = 1200°F$
$A = 16 \times 12 = 192$ ft²
$L = \frac{6}{12} = 0.5$ ft

$$Q = 0.15 \times \frac{192}{0.5} \times 1200 = 69{,}200 \text{ Btu/hr}$$

Flow of Heat through a Composite Wall: Resistances in Series. Equation (2.24) is of interest when a wall consists of several materials placed together in series such as in the construction of a furnace or boiler firebox. Several types of refractory brick are usually employed, since those capable of withstanding the higher inside temperatures are more fragile and expensive than those required near the outer surface, where the temperatures are considerably lower. Referring to Fig. 2.5, three different refractory materials are placed together indicated by the subscripts a, b, and c. For the entire wall

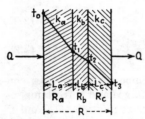

Fig. 2.5. Heat flow through a composite wall.

$$Q = \frac{\Delta t}{R} \quad (2.25)$$

The heat flow in Btu per hour through material a must overcome the resistance R_a. But in passing through material a the heat must also pass through materials b and c in series. The heat flow entering at the left face must be equal to the heat flow leaving the right face, since the steady state precludes heat storage. If R_a, R_b, and R_c are unequal, as the result of differing conductivities and thicknesses, the ratio of the temperature difference across each layer to its resistance must be the same as the ratio of the total temperature difference is to the total resistance or

$$Q = \frac{\Delta t}{R} = \frac{\Delta t_a}{R_a} = \frac{\Delta t_b}{R_b} = \frac{\Delta t_c}{R_c} \quad (2.26)$$

For any composite system using actual temperatures

$$Q = \frac{\Delta t}{R} = \frac{t_0 - t_1}{R_a} = \frac{t_1 - t_2}{R_b} = \frac{t_2 - t_3}{R_c} \quad (2.27)$$

CONDUCTION

Rearranging and substituting,

$$Q = \frac{\Delta t}{R} = \frac{t_0 - t_3}{(L_a/k_a A) + (L_b/k_b A) + (L_c/k_c A)} \qquad (2.28)$$

Example 2.2. Flow of Heat through a Composite Wall. The wall of an oven consists of three layers of brick. The inside is built of 8 in. of firebrick, $k = 0.68$ Btu/(hr)(ft²)(°F/ft), surrounded by 4 in. of insulating brick, $k = 0.15$, and an outside layer of 6 in. of building brick, $k = 0.40$. The oven operates at 1600°F and it is anticipated that the outer side of the wall can be maintained at 125°F by the circulation of air. How much heat will be lost per square foot of surface and what are the temperatures at the interfaces of the layers?

Solution:

For the firebrick, $R_a = L_a/k_a A = 8/12 \times 0.68 \times 1 = 0.98$ (hr)(°F)/(Btu)
Insulating brick, $R_b = L_b/k_b A = 4/12 \times 0.15 \times 1 = 2.22$
Building brick, $R_c = L_c/k_c A = 6/12 \times 0.40 \times 1 = \underline{1.25}$
$\qquad\qquad\qquad\qquad\qquad\qquad\qquad\qquad R = 4.45$

Heat loss/ft² of wall, $Q = \Delta t/R = (1600 - 125)/4.45 = 332$ Btu/hr
For the individual layers:

$\Delta t = QR \quad$ and $\quad \Delta t_a = QR_a$, etc.
$\Delta t_a = 332 \times 0.98 = 325°F \qquad t_1 = 1600 - 325 = 1275°F$
$\Delta t_b = 332 \times 2.22 = 738°F \qquad t_2 = 1275 - 738 = 537°F$

Example 2.3. Flow of Heat through a Composite Wall with an Air Gap. To illustrate the poor conductivity of a gas, suppose an air gap of ¼ in. were left between the insulating brick and the firebrick. How much heat would be lost through the wall if the inside and outside temperatures are kept constant?

Solution. From Table 5 in the Appendix at 572°F air has a conductivity of 0.0265 Btu/(hr)(ft²)(°F/ft), and this temperature is close to the range of the problem.

$$R_{\text{air}} = 0.25/12 \times 0.0265 = 0.79 \text{ (hr)(°F)/Btu}$$
$$R = 4.45 + 0.79 = 5.24$$
$$Q = \frac{1600 - 125}{5.24} = 281 \text{ Btu/hr}$$

It is seen that in a wall 18 in. thick a stagnant air gap only ¼ in. thick reduces the heat loss by 15 per cent.

Heat Flow through a Pipe Wall. In the passage of heat through a flat wall the area through which the heat flows is constant throughout the entire distance of the heat flow path. Referring to Fig. 2.6 showing a unit length of pipe, the area of the heat flow path through the pipe wall increases with the distance of the path from r_1 to r_2. The area at any radius r is given by $2\pi r 1$ and if the heat flows out of the cylinder the temperature gradient for the

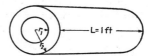

Fig. 2.6. Heat flow through a pipe wall.

incremental length dr is dt/dr. Equation (2.14) becomes

$$q = 2\pi rk \left(-\frac{dt}{dr}\right) \quad \text{Btu/(hr)(lin ft)} \tag{2.29}$$

Integrating,

$$t = -\frac{q}{2\pi k} \ln r + C_1 \tag{2.30}$$

When $r = r_i$, $t = t_i$; and when $r = r_o$, $t = t_o$; where i and o refer to the inside and outside surfaces, respectively. Then

$$q = \frac{2\pi k(t_i - t_o)}{2.3 \log r_o/r_i} \tag{2.31}$$

and if D is the diameter,

$$\frac{r_o}{r_i} = \frac{D_o}{D_i}$$

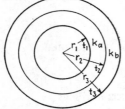

Fig. 2.7. Cylindrical resistances in series.

Referring to Fig. 2.7 where there is a composite cylindrical resistance,

$$t_1 = t_2 + \frac{2.3q}{2\pi k_a} \log \frac{D_2}{D_1} \tag{2.32}$$

$$t_2 = t_3 + \frac{2.3q}{2\pi k_b} \log \frac{D_3}{D_2} \tag{2.33}$$

Adding,

$$t_1 - t_3 = \frac{2.3q}{2\pi k_a} \log \frac{D_2}{D_1} + \frac{2.3q}{2\pi k_b} \log \frac{D_3}{D_2} \tag{2.34}$$

Example 2.4. Heat Flow through a Pipe Wall. A glass pipe has an outside diameter of 6.0 in. and an inside diameter of 5.0 in. It will be used to transport a fluid which maintains the inner surface at 200°F. It is expected that the outside of the pipe will be maintained at 175°F. What heat flow will occur?

Solution. $k = 0.63$ Btu/(hr)(ft²)(°F/ft) (see Appendix Table 2).

$$q = \frac{2\pi k(t_i - t_o)}{2.3 \log D_o/D_i} = \frac{2 \times 3.14 \times 0.63(200 - 175)}{2.3 \log 6.0/5.0} = 538 \text{ Btu/lin ft}$$

If the inside diameter of a cylinder is greater than 0.75 of the outside diameter, the mean of the two may be used. Then per foot of length

$$q = \frac{\Delta t}{R} = \frac{\Delta t}{L_a/k_a A_m} = \frac{t_1 - t_2}{\dfrac{(D_2 - D_1)/2}{\pi k_a (D_1 + D_2)/2}} \tag{2.35}$$

where $(D_2 - D_1)/2$ is the thickness of the pipe. Within the stated limitations of the ratio D_2/D_1, Eq. (2.35) will differ from Eq. (2.34) by about 1 per cent. Actually there are 1.57 ft² of external surface per linear foot and 1.31 ft² of internal surface. The heat loss per square foot is 343 Btu/hr based on the outside surface and 411 Btu/hr based on the inside surface.

Heat Loss from a Pipe. In the preceding examples it was assumed that the cold external surface could be maintained at a definite temperature. Without this assumption the examples would have been indeterminate, since both Q and Δt would be unknown and independent in a single equation. In reality the temperature assigned to the outer wall depends not only on the resistances between the hot and cold surfaces but also on the ability of the surrounding colder atmosphere to remove the heat arriving at the outer surface. Consider a pipe as shown in Fig. 2.8 covered (lagged) with rock wool insulation and carrying steam at a temperature t_s considerably above that of the atmosphere, t_a. The overall temperature difference driving heat out of the pipe is $t_s - t_a$. The resistances to heat flow taken in order are (1) the resistance of the steam to condense upon and give up heat to the inner pipe surface, a resistance which has been found experimentally to be very small so that t_s and t'_s are nearly the same; (2) the resistance of the pipe metal, which is very small except for thick-walled conduits so that t'_s and t''_s are nearly the same; (3) the resistance of the rock wool insulation; and (4) the resistance of the surrounding air to remove heat from the outer surface. The last is appreciable, although the removal of heat is effected by the natural convection of ambient air in addition to the radiation caused by the temperature difference between the outer surface and colder air. The natural convection results from warming air adjacent to the pipe, thereby lowering its density. The warm air rises and is replaced continuously by cold air. The combined effects of natural convection and radiation cannot be represented by a conventional resistance term $R_a = L_a/K_a A$, since L_a is indefinite and the conductance of the air is simultaneously supplemented by the transfer of heat by radiation. Experimentally, a temperature difference may be created between a known outer surface and the air, and the heat passing from the outer surface to the air can be determined from measurements on the fluid flowing in the pipe. Having Q, A, and Δt, the combined resistance of both effects is obtained as the quotient of $\Delta t/Q$. The flow of heat from a pipe to ambient air is usually a heat loss, and it is therefore desirable to report the data as a *unit conductance* term k/L Btu/(hr)(ft² of

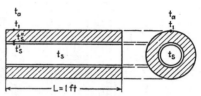

Fig. 2.8. Heat loss from an insulated pipe.

external surface)(°F of temperature difference). The unit conductance is the reciprocal of the *unit resistance* L/k instead of the reciprocal of the resistance for the entire surface L/kA. In other words, it is the conductance per square foot of heat-flow surface rather than the conductance of the total surface. The unit resistance has the dimensions

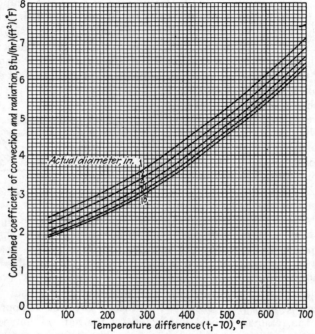

FIG. 2.9. Heat transfer by convection and radiation from horizontal pipes at temperature t_1 to air at 70°F.

(hr)(ft²)(°F)/Btu. The reciprocal of the unit resistance, h_a, has the dimensions of Btu/(hr)(ft²)(°F) and is sometimes designated the *surface coefficient* of heat transfer. Figure 2.9 shows the plot of the surface coefficient from pipes of different diameters and surface temperatures to ambient air at 70°F. It is based upon the data of Heilman,[1] which has been substantiated by the later experiments of Bailey and Lyell.[2] The four resistances in terms of the equations already discussed are

Condensation of steam:

$$q = h_s \pi D'_s (t_s - t'_s) \tag{1.2}$$

[1] Heilman, R. H., *Ind. Eng. Chem.*, **16**, 445–452 (1924).
[2] Bailey, A., and N. C. Lyell, *Engineering*, **147**, 60–62 (1939).

CONDUCTION

Pipe wall:
$$q = \frac{2\pi k_b}{2.3 \log D''_s/D_s} (t'_s - t''_s) \qquad (2.31)$$

Insulation:
$$q = \frac{2\pi k_c}{2.3 \log D_1/D''_s} (t''_s - t_1) \qquad (2.31)$$

Radiation and convection to air:
$$q = h_a \pi D_1 (t_1 - t_a) \qquad (1.2)$$

or combining
$$t_s - t_a = q \left(\frac{1}{h_s \pi D'_s} + \frac{2.3}{2\pi k_b} \log \frac{D''_s}{D'_s} + \frac{2.3}{2\pi k_c} \log \frac{D_1}{D''_s} + \frac{1}{h_a \pi D_1} \right)$$

The terms inside the parentheses are the four resistances, and of these the first two can usually be neglected. The equation then reduces to

$$q = \frac{\pi(t_s - t_a)}{\frac{2.3}{2k_c} \log \frac{D_1}{D''_s} + \frac{1}{h_a D_1}}$$

From the abscissa of Fig. 2.9 it is seen that h_a depends upon not only the temperature difference but the actual temperatures at the outside of the insulation and of the air. Its reciprocal is also one of the resistances necessary for the calculation of the total temperature difference, and therefore the surface coefficient h_a cannot be computed except by trial-and-error methods.

Example 2.5. Heat Loss from a Pipe to Air. A 2-in. steel pipe (dimensions in Table 11 in the Appendix) carries steam at 300°F. It is lagged with ½ in. of rock wool, $k = 0.033$, and the surrounding air is at 70°F. What will be the heat loss per linear foot?

Solution. Assume $t_1 = 150°F$, $t_1 - 70 = 80°F$, $h_a = 2.23$ Btu/(hr)(ft²)(°F).

$$q = \frac{3.14(300 - 70)}{\frac{2.3}{2 \times 0.033} \log \frac{3.375}{2.375} + \frac{1}{2.23 \times 3.375/12}} = 104.8 \text{ Btu/(hr)(lin ft)}$$

Check between t_s and t_1, since $\Delta t/R = \Delta t_c/R_c$.

$$q = 104.8 = \frac{2 \times 3.14 \times 0.033(300 - t_1)}{2.3 \log 3.375/2.375}$$

$$t_1 = 123.5°F \quad \textit{No check}$$

Assume $t_1 = 125°F$, $t_1 - 70 = 55°F$, $h_a = 2.10$ Btu/(hr)(ft²)(°F).

$$q = \frac{3.14(300 - 70)}{\frac{2.3}{2 \times 0.033} \log \frac{3.375}{2.375} + \frac{1}{2.10 \times 3.375/12}} = 103.2 \text{ Btu/(hr)(lin ft)}$$

Check between t_s and t_1.

$$q = 103.2 = \frac{2 \times 3.14 \times 0.033(300 - t_1)}{2.3 \log 3.375/2.375}$$

$$t_1 = 125.8°F \quad Check$$

The total heat loss q does not appear to vary significantly for the different assumed values of t_1. This is because the insulation and not the small surface coefficient affords the major resistance to heat flow. When the variation in q is considerable for different assumed temperatures of t_1, it indicates insufficient insulation.

The Maximum Heat Loss through Pipe Insulation. It would seem at first that the thicker the insulation the less the total heat loss. This is always true for flat insulation but not for curved insulation. Consider a pipe with successive layers of cylindrical insulation. As the thickness of the insulation is increased, the surface area from which heat may be removed by air increases and the total heat loss may also increase if the area increases more rapidly than the resistance. Referring to Fig. 2.10, the resistance of the insulation per linear foot of pipe is

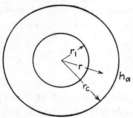

FIG. 2.10. The critical radius.

$$R_b = \frac{1}{2\pi k_b} \ln \frac{r}{r_1} \quad (2.36)$$

and the resistance of the air per linear foot of pipe, although a function of the surface and air temperatures, is given by

$$R_a = \frac{1}{h_a 2\pi r} \quad (2.37)$$

The resistance is a minimum and the heat loss a maximum when the derivative of the sum of the resistances R with respect to the radius r is set equal to zero or

$$\frac{dR}{dr} = 0 = \frac{1}{2\pi k_b} d \ln \frac{r}{r_1} + \frac{1}{h_a 2\pi} d \frac{1}{r} \quad (2.38)$$

$$= \frac{1}{2\pi k_b r} - \frac{1}{h_a 2\pi r^2}$$

At the maximum heat loss $r = r_c$, the critical radius, or

$$r_c = \frac{k_b}{h_a} \quad (2.39)$$

In other words, the maximum heat loss from a pipe occurs when the critical radius equals the ratio of the thermal conductivity of the insulation to the surface coefficient of heat transfer. The ratio has the dimension of ft. It is desirable to keep the critical radius as small as possible so that the application of insulation will result in a reduction and not an increase in the heat loss from a pipe. This is obviously accomplished by using an insulation of small conductivity so that the critical radius is less than the radius of the pipe, or $r_c < r_1$.

The Optimum Thickness of Insulation. The optimum thickness of insulation is arrived at by a purely economic approach. If a bare pipe were to carry a hot fluid, there would be a certain hourly loss of heat whose value could be determined from the cost of producing the Btu in the plant heat-generating station. The lower the heat loss the greater the thickness and initial cost of the insulation and the greater the annual fixed charges (maintenance and depreciation) which must be added to the annual heat loss. The fixed charges on pipe insulation will be about 15 to 20 per cent of the initial installed cost of the insulation. By assuming a number of thicknesses of insulation and adding the fixed charges to the value of the heat lost, a minimum cost will be obtained and the thickness corresponding to it will be the optimum economic thickness of the insulation. The form of such an analysis is shown in Fig. 2.11.

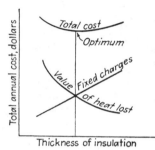

FIG. 2.11. Optimum thickness of insulation.

The most difficult part is obtaining reliable initial-installation-cost data, since they vary greatly with plant to plant and with the amount of insulating to be done at a single time.

Graphical Solutions of Conduction Problems. Thus far in the treatment of conduction, only those cases have been considered in which the heat input per square foot of surface was uniform. It was also characteristic of these cases that the heat removal per square foot of surface was also uniform. This was likewise true for the cylinder, even though the internal and external surfaces were not identical. Some of the common problems of steady-state conduction in solids involve the removal or input of heat where it is not uniform over a surface, and although the solution of such problems by mathematical analysis is often complicated it is possible to obtain close approximations graphically. The method employed here is that of Awbery and Schofield[1] and earlier investigators.

Consider the section of a metal-sheathed wall, as shown in Fig. 2.12, with hot side ABC at the uniform temperature t_1. At recurring intervals DF on the cold side DEF at the uniform temperature t_2, metal bracing

[1] Awbery, J. and F. Schofield, *Proc. Intern. Congr. Refrig.*, 5th Congr., **3**, 591–610 (1929).

ribs are attached to the outer sheath and imbedded two-thirds into the thickness of the wall. Since the sheath and metal rib both have a high thermal conductivity compared with the wall material itself, the rib and sheath may both be considered to be at very nearly the same temperature. The predominantly horizontal lines indicated on the drawing represent isothermal planes perpendicular to the plane of the drawing. Consequently there is no heat flow to be considered in the direction perpendicular to the plane of the drawing.

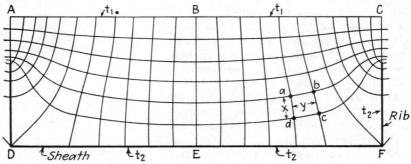

Fig. 2.12. Graphical representation of heat conduction.

Since the drawing is symmetrical about the vertical line BE, consider only the right half of the drawing bounded by $BCFE$. Assume an arbitrary number of isotherms n_x in the direction from B to E so that, if k is constant, $\Delta t = n_x \Delta t_x$. If k varies with t then $k \Delta t_x = \dfrac{1}{n_x} \int_1^2 k\, dt$. The greater the assumed number of isotherms the greater the precision of the solution. Next, consider the heat to flow from t_1 to metal at t_2 through n_l lanes emanating from BC and forming the network indicated. Now refer to any small portion of any lane, such as $abcd$ with length x, mean width y, where $y = (ab + cd)/2$, and unit depth $z = 1$ perpendicular to the drawing. The steady-heat flow into each lane is Q_l. The conduction equation is then $Q_l = k(yz)\, \Delta t_x / x$. The temperature difference from one isotherm to the next is naturally the same, and since Q_l is constant for the lane it is evident from the conduction equation that the ratio y/x must also be constant, although y and x may vary. The network of the drawing is constructed such that, for each quadrilateral, $y = x$. Where x is small it is because the isotherms are crowded together owing to the high heat removal by the rib. The heat flow per lane is then given by

$$k\frac{(t_1 - t_2)}{n_x}.$$

CONDUCTION 23

The total heat flow from BC thus requires $n_l = Qn_x/k(t_1 - t_2)$ lanes, where Q is the total heat flow. Figure 2.12 was constructed in this manner starting with six isotherms.

Although the individual portions of the network are neither squares nor rectangles their corners are at right angles in accordance with the steady-state principle that the flow of heat is always at right angles to the isotherms comprising the temperature difference. In Fig. 2.12 it is seen that 11 lanes were obtained for each half of a symmetrical section. If the isotherms were undisturbed by the rib, the portions $abcd$ would then be squares and the heat entering BC would flow normal to it and 8.3 lanes would be required. The rib is therefore equivalent to increasing the heat removal by 33 per cent. When the ribs are spaced more closely together, the fractional heat removal increases.

PROBLEMS

2.1. A furnace is enclosed by walls made (from inside out) of 8 in. of kaolin firebrick, 6 in. of kaolin insulating brick, and 7 in. of fireclay brick. What is the heat loss per square foot of wall when the inside of the furnace is maintained at 2200°F and the outside at 200°F?

2.2. A furnace wall is to consist in series of 7 in. of kaolin firebrick, 6 in. of kaolin insulating brick, and sufficient fireclay brick to reduce the heat loss to 100 Btu/(hr)(ft²) when the face temperatures are 1500 and 100°F, respectively. What thickness of fireclay brick should be used? If an effective air gap of $\frac{1}{8}$ in. can be incorporated between the fireclay and insulating brick when erecting the wall without impairing its structural support, what thickness of insulating brick will be required?

2.3. A furnace wall consists of three insulating materials in series. 32 per cent chrome brick, magnesite bricks, and low-grade refractory bricks ($k = 0.5$). The magnesite bricks cannot withstand a face temperature above 1500°F, and the low-grade bricks cannot exceed 600°F. What thickness of the wall will give a heat loss not in excess of 1500 Btu/(hr)(ft²) when the extreme face temperatures are 2500 and 200°F, respectively?

2.4. A 6-in. IPS pipe is covered with three resistances in series consisting from the inside outward of $\frac{1}{2}$ in. of kapok, 1 in. of rock wool, and $\frac{1}{2}$ in. of powdered magnesite applied as a plaster. If the inside surface is maintained at 500°F and the outside at 100°F, what is the heat loss per square foot of outside pipe surface?

2.5. A 2-in. IPS line to a refrigerated process covered with $\frac{1}{2}$ in. of kapok carries 25% NaCl brine at 0°F and at a flow rate of 30,000 lb/hr. The outer surface of the kapok will be maintained at 90°F. What is the equation for the flow of heat? Calculate the heat leakage into the pipe and the temperature rise of the fluid for a 60-ft length of pipe.

2.6. A vertical cylindrical kiln 22 ft in diameter is enclosed at the top by a hemispherical dome fabricated from an 8-in. layer of interlocking and self-supporting 32 per cent chrome bricks. Derive an expression for conduction through the dome. When the inside and outside of the hemispherical dome are maintained at 1600 and 300°F, respectively, what is the heat loss per square foot of internal dome surface? How does the total heat loss for the dome compare with the total heat loss for a flat structurally supported roof of the same material when exposed to the same difference in temperature?

24   PROCESS HEAT TRANSFER

2.7. A 4-in. steel pipe carrying 450°F steam is lagged with 1 in. of kapok surrounded by 1 in. of powdered magnesite applied as a plaster. The surrounding air is at 70°F. What is the loss of heat from the pipe per linear foot?

2.8. A 3-in. IPS main carries steam from the powerhouse to the process plants at a linear velocity of 8000 fpm. The steam is at 300 psi (gage), and the atmosphere is at 70°F. What percentage of the total heat flow is the bare-tube heat loss per 1000 ft of pipe? If the pipe is lagged with half a thickness of kapok and half a thickness of asbestos, what total thickness will reduce the insulated heat loss to 2 per cent of the bare-tube heat loss?

2.9. In a 6-in. steam line at 400°F the unit resistance for the condensation of steam at the inside pipe wall has been found experimentally to be 0.00033 (hr)(ft²)(°F)/Btu. The line is lagged with ½ in. of rock rool and ½ in. of asbestos. What is the effect of including the condensation and metal pipe-wall resistances in calculating the total heat loss per linear foot to atmospheric air at 70°F?

NOMENCLATURE FOR CHAPTER 2

A	Heat-flow area, ft²
C_1, C_2	Constants of integration
C_k	Sutherland constant
c_v	Volumetric specific heat, Btu/(ft³)(°F)
c	Specific heat at constant pressure, Btu/(lb)(°F)
D	Diameter, ft
E	Voltage or electromotive force
h_a	Surface coefficient of heat transfer, Btu/(hr)(ft²)(°F)
I	Current, amp
K	Electrical conductivity, 1/ohm-ft
k	Thermal conductivity, Btu/(hr)(ft²)(°F/ft)
L	Thickness of wall or length of pipe, ft
n_l	Number of heat-flow lanes
n_x	Number of isotherms
Q	Heat flow, Btu/hr
Q_l	Heat flow per lane, Btu/hr
Q'	Heat, Btu
q	Heat flow, Btu/(hr)(lin ft)
R	Resistance to heat flow, (hr)(°F)/Btu
R^ω	Resistance to electric flow, ohms
r	Radius, ft
t	Temperature at any point, °F
Δt	Temperature difference promoting heat flow, °F
T	Absolute temperature, °R
v	Volume, ft³
x, y, z	Coordinates of distance, ft
γ	Change in thermal conductivity per degree
θ	Time, hr
ρ	Density, lb/ft³
σ	Resistivity, ohm-ft

CHAPTER 3
CONVECTION

Introduction. Heat transfer by convection is due to fluid motion. Cold fluid adjacent to a hot surface receives heat which it imparts to the bulk of the cold fluid by mixing with it. Free or natural convection occurs when the fluid motion is not implemented by mechanical agitation. But when the fluid is mechanically agitated, the heat is transferred by forced convection. The mechanical agitation may be supplied by stirring, although in most process applications it is induced by circulating the hot and cold fluids at rapid rates on the opposite sides of pipes or tubes. Free- and forced-convection heat transfer occur at very different speeds, the latter being the more rapid and therefore the more common. Factors which promote high rates for forced convection do not necessarily have the same effect on free convection. It is the purpose of this chapter to establish a general method for obtaining the rates of heat transfer particularly in the presence of forced convection.

Film Coefficients. In the flow of heat through a pipe to air it was seen that the passage of heat into the air was not accomplished solely by conduction. Instead, it occurred partly by radiation and partly by free convection. A temperature difference existed between the pipe surface and the *average* temperature of the air. Since the distance from the pipe surface to the region of average air temperature is indefinite, the resistance cannot be computed from $R_a = L_a/k_a A$, using k for air. Instead the resistance must be determined experimentally by appropriately measuring the surface temperature of the pipe, the temperature of the air, and the heat transferred from the pipe as evidenced by the quantity of steam condensed in it. The resistance for the entire surface was then computed from

$$R_a = \frac{\Delta t_a}{Q} \quad \text{(hr)(°F)/Btu}$$

If desired, L_a can also be calculated from this value of R_a and would be the length of a fictitious conduction film of air equivalent to the combined resistance of conduction, free convection, and radiation. The length of the film is of little significance, although the concept of the fictitious film finds numerous applications. Instead it is preferable to deal directly with the reciprocal of the unit resistance h, which has an experimental origin. Because the use of the unit resistance L/k is so much more common than the use of the total surface resistance L/kA, the letter R will now be used to designate L/k (hr)(ft²)(°F)/Btu and it will simply be called the resistance.

Not all effects other than conduction are necessarily combinations of two effects. Particularly in the case of free or forced convection to liquids and, in fact, to most gases at moderate temperatures and temperature differences the influence of radiation may be neglected and the experimental resistance corresponds to forced or free convection alone as the case may be.

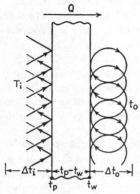

Fig. 3.1. Two convection coefficients.

Consider a pipe wall with forced convection of different magnitudes on both sides of the pipe as shown in Fig. 3.1. On the inside, heat is deposited by a hot flowing liquid, and on the outside, heat is received by a cold flowing liquid. Either resistance can be measured independently by obtaining the temperature difference between the pipe surface and the average temperature of the liquid. The heat transfer can be determined from the sensible-heat change in either fluid over the length of the pipe in which the heat transfer occurs. Designating the resistance on the inside by R_i and on the outside by R_o, the inside and outside pipe-wall temperatures by t_p and t_w, and applying an expression for the steady state,

$$Q = \frac{A_i(T_i - t_p)}{R_i} = \frac{A_o(t_w - t_o)}{R_o} \quad (3.1)$$

where T_i is the temperature of the hot fluid on the inside and t_o the temperature of the cold fluid on the outside. Replacing the resistances by their reciprocals h_i and h_o, respectively,

$$Q = h_i A_i \Delta t_i = h_o A_o \Delta t_o \quad (3.2)$$

The reciprocals of the heat-transfer resistances have the dimensions of Btu/(hr)(ft²)(°F of temperature difference) and are called *individual film coefficients* or simply *film coefficients*.

Inasmuch as the film coefficient is a measure of the heat flow for unit surface and unit temperature difference, it indicates the rate or speed with which fluids having a variety of physical properties and under varying degrees of agitation transfer heat. Other factors influence the film coefficient such as the size of the pipe and whether or not the fluid is considered to be on the inside or outside of the pipe. With so many variables, each having its own degree of influence on the rate of heat transfer (film coefficient), it is fairly understandable why a rational derivation is not available for the direct calculation of the film coefficient. On the other hand, it is impractical to run an experiment to determine the coefficient each time heat is to be added or removed from a fluid. Instead it is desirable to study some method of correlation whereby several basic experiments performed with a wide range of the variables can produce a relationship which will hold for any other combinations of the variables. The immediate problem is to establish a method of correlation and then apply it to some experimental data.

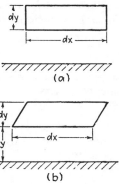

FIG. 3.2. Fluid strain.

The Viscosity. It is not possible to proceed very far in the study of convection and fluid flow without defining a property which has an important bearing upon both, *viscosity*. In order to evaluate this property by fluid dynamics two assumptions are required: (1) Where a solid-liquid interface exists, there is no slip between the solid and liquid, and (2) Newton's rule: Shear stress is proportional to the rate of shear in the direction perpendicular to motion. An unstressed particle of liquid as shown in Fig. 3.2a will assume the form in Fig. 3.2b when a film of liquid is subjected to shear.

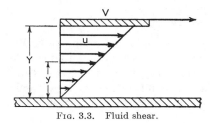

FIG. 3.3. Fluid shear.

The *rate of shear* is proportional to the velocity gradient du/dy. Applying Newton's rule, if τ is the shear stress,

$$\tau = \mu \frac{du}{dy} \qquad (3.3)$$

where μ is the proportionality constant or

$$\mu = \frac{\tau}{du/dy}$$

{In solids it results in deformation and is equivalent to the modulus of elasticity.
In liquids it results in deformation at a given rate.

To evaluate μ refer to Fig. 3.3 where shear is produced by maintaining the liquid film between a stationary plate at distance Y and a moving plate with velocity V. At any point in the film the velocity u is given by $u = Vy/Y$.

$$\text{Rate of shear} = \frac{du}{dy} = \frac{V}{Y} \tag{3.3}$$

$$\tau = \mu \frac{V}{Y} \tag{3.4}$$

where μ is called the viscosity when V and Y have unit values.

The actual force required to move the plate is τA. If F is the pound-force, L the length, and θ the time, the dimensions of the viscosity are

$$\mu = \tau \frac{Y}{V} = \frac{F}{L^2} \frac{L}{L/\theta} = \frac{F\theta}{L^2}$$

or using the pound-mass M, where $F = Mg$ and $g = L/\theta^2$, the acceleration of gravity,

$$\mu = \frac{ML}{L^2\theta^2} \frac{L}{L/\theta} = \frac{M}{L\theta}$$

When evaluated in cgs metric units μ is commonly called the *absolute* viscosity.

$$\mu = \frac{\text{gram-mass}}{\text{centimeter} \times \text{second}}$$

This unit has been named the poise after the French scientist Poiseuille. This is a large unit, and it is customary to use and speak of the centipoise, or one-hundredth poise. In engineering units its equivalent is defined by

$$\mu = \frac{\text{pound-mass}}{\text{foot} \times \text{hour}}$$

Viscosities in centipoises can be converted to engineering units on multiplying by 2.42. This unit has no name. Another unit, the kinematic viscosity, is also used because it occurs frequently in physical systems and produces straighter graphs of viscosity vs. temperature on logarithmic coordinates. The kinematic viscosity is the absolute viscosity in centipoises divided by the specific gravity.

$$\text{Kinematic viscosity} = \frac{\text{absolute viscosity}}{\text{specific gravity}}$$

The unit of kinematic viscosity is the stokes, after the English mathematician Stokes, and the hundredth of the stokes is the centistokes.

The viscosity can be determined indirectly by measuring the time of efflux from a calibrated flow device having an orifice and a controlled temperature. The commonest is the Saybolt viscometer, and the time of efflux from a standard cup into a standard receiver is measured in seconds and recorded as Saybolt Seconds Universal, SSU. Conversion factors from the time of efflux to centistokes for the Saybolt and other viscometers are given in Fig. 13.[1]

Heat Transfer between Solids and Fluids: Streamline and Turbulent Flow. The Reynolds Number. When a liquid flows in a horizontal pipe, it may flow with a random eddying motion known as *turbulent flow*, as shown in Fig. 3.4 by a plot of the local velocity in the pipe vs. the distance from its center. If the linear velocity of the liquid is decreased below some threshold value, the nature of the flow changes and the turbulence disappears. The fluid particles flow in lines along the axis of the pipe, and this is known as *streamline flow*. An experiment used for the visual determination of the type of flow consists of a glass tube

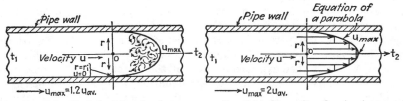

FIG. 3.4. Turbulent flow in pipes.   FIG. 3.5. Streamline flow in pipes.

through which water flows. A thin stream of ink is injected at the center of the tube, and if the ink remains at the center for a reasonable distance, it is indicative of streamline flow. Synonyms for streamline flow are viscous, laminar, and rodlike flow. Additional experimentation has indicated that streamline flow proceeds as if by the sliding of concentric thin cylinders of liquid one within the other as shown in Fig. 3.5. It appears that the distribution of the velocities of the cylinders is parabolic with the maximum at the center and approaching zero at the tube wall.

Reynolds[2] observed that the type of flow assumed by a liquid flowing in a tube was influenced by the velocity, density, and viscosity of the liquid and the diameter of the tube. When related as the quotient of $Du\rho/\mu$, called the Reynolds number, it was found that turbulent flow always existed when the numerical value of $Du\rho/\mu$ exceeded about 2300 and usually when it exceeded 2100. By definition, the transfer of heat by convection proceeds mainly as the result of mixing, and while this requirement appears to be satisfied by turbulent flow, it is not fulfilled by stream-

---

[1] Figure numbers which are not preceded by a chapter number will be found in the Appendix.

[2] Reynolds, O., "Scientific Papers of Osborne Reynolds," p. 81, Cambridge University Press, London, 1901.

line flow. Streamline flow is, in fact, a form of conduction whose study will be deferred to a later chapter. The speed with which heat is transferred to or from a given liquid to a tube is considerably less for streamline than for turbulent flow, and in industrial practice it is almost always desirable to avoid conditions such as low liquid velocity which promote streamline flow.

Dimensional Analysis. A method of correlating a number of variables into a single equation expressing an effect is known as *dimensional analysis*. Certain equations describing physical phenomena can be obtained rationally from basic laws derived from experiments. An example is the time of vibration of a pendulum from Newton's second law and the gravitational constant. Still other effects can be described by differential equations, and the course or extent of the phenomena deduced by means of the calculus. Numerous examples of this type are encountered in elementary physics. In still other types of phenomena there is insufficient information to permit the formulation of either differential equations or a clear picture by which fundamental laws may be applied. This last group must be studied experimentally, and the correlation of the observations is an empirical approach to the equation. Equations which can be obtained theoretically can also be obtained empirically, but the reverse is not true.

Bridgman[1] has presented by far the most extensive proof of the mathematical principles underlying dimensional analysis. Because it operates only upon the dimensions of the variables, it does not directly produce numerical results from the variables but instead yields a modulus by which the observed data can be combined and the relative influence of the variables established. As such, it is one of the important cornerstones of empirical study. It recognizes that any combination of a number and a dimension, such as 5 lb or 5 ft, possesses two identifying aspects, one of pure magnitude (numerical) and the other quantitative (dimensional). *Fundamental dimensions* are quantities such as length, time, and temperature which are directly measurable. Derived dimensions are those which are expressed in terms of fundamental dimensions such as velocity = length/time or density = mass/length^3. The end results of a dimensional analysis may be stated as follows: If a dependent variable having given dimensions depends upon some relationship among a group of variables, the individual variables of the group must be related in such a way that the net dimensions of the group are identical with those of the dependent variable. The independent variables may also be related in such a way that the dependent variable is defined by the sum of several

---

[1] Bridgman, P. W., "Dimensional Analysis," Yale University Press, New Haven, 1931.

CONVECTION 31

different groups of variables, each group having the net dimensions of the dependent variable. As a very simple illustration consider the *continuity equation* which is frequently written in elementary physics and thermodynamics texts in the form

$$w = \frac{ua}{v} \qquad (3.5)$$

where w = fluid flow rate, weight/time
u = fluid velocity in conduit, length/time
a = cross section of the conduit, length $\times$ length = length^2
v = specific volume, length $\times$ length $\times$ length/weight = length^3/weight

Why does Eq. (3.5) have this particular form? u, a, and v must be related so that their net dimensions are the same as those of the dependent variable w, namely, weight/time. An equation involving both pure numbers and dimensions must be correct with respect to either and both. Checking the dimensions alone, writing for the variables in Eq. (3.5) their individual dimensions,

$$\frac{\text{Weight}}{\text{Time}} = \frac{\text{length}}{\text{time}} \times \text{length}^2 \times \frac{1}{\frac{\text{length}^3}{\text{weight}}} = \frac{\text{weight}}{\text{time}} \qquad (3.6)$$

It is seen that the dimensions on the left side are identical with the net dimensions of the group only when the variables of the group are arranged in the particular manner indicated by the formula. The three independent variables above give an answer in weight/time only when arranged in a single way, ua/v. Conversely *it may be deduced that the form of an equation is determined only by its dimensions;* the form which produces dimensional equality represents the necessary relationship among the variables. Any physical equation may be written and evaluated in terms of a power series containing all the variables. If the form were not known in the illustration above and it was desired to find a relationship which must exist between the variables w, u, a, and v, it may be expressed by a power series such as

$$\phi(w, u, a, v) = \alpha w^b u^c a^d v^e + \alpha' w^{b'} u^{c'} a^{d'} v^{e'} + \cdots = 0 \qquad (3.7)$$

The factors α and α' are dimensionless proportionality constants. Since the dimensions of all the consecutive terms of the series are identical, it is not necessary to consider any of the terms beyond the first. Accordingly one can write

$$\phi'(w^b u^c a^d v^e) = 1 \qquad (3.8)$$

where ϕ' indicates the function. Arbitrarily setting $b = -1$ so that w

will not appear in the final equation raised to a fractional exponent,

$$w = \alpha u^c a^d v^e \tag{3.9}$$

Substituting dimensions,

$$\frac{\text{Weight}}{\text{Time}} = \alpha \left(\frac{\text{length}}{\text{time}}\right)^c \times (\text{length}^2)^d \times \left(\frac{\text{length}^3}{\text{weight}}\right)^e \tag{3.10}$$

If a group of independent variables will establish numerical equality with a dependent variable, the same is true of their dimensions. Equation (3.6) imposes this condition. The exponents c, d, and e may then assume such values as are necessary to effect the dimensional equality between the left and right sides. The remainder of the solution is merely to evaluate c, d, and e by simple algebra. Summing the exponents of the dimensions on both sides and recalling that an exponent of zero reduces a number to unity,

$$\Sigma \text{ length, } 0 = c + 2d + 3e$$
$$\Sigma \text{ weight, } 1 = -e$$
$$\Sigma \text{ time, } -1 = -c$$

Solving for the unknown, d is found to be $+1$. The three exponents are then $c = +1$, $d = +1$, and $e = -1$. Substituting these in Eq. (3.9),

$$w = \alpha u^{+1} a^{+1} v^{-1} = \alpha \frac{ua}{v} \tag{3.11}$$

Inasmuch as this is an exact relationship, the proportionality constant α is equal to 1.0 and

$$w = \frac{ua}{v}$$

Thus by purely algebraic means the correct form of the equation has been established. This has been an extremely elementary illustration in which all of the exponents were integers and the dependent and independent variables were expressed in only three kinds of dimensions: weight, length, and time. In systems involving mechanics and heat it is often necessary to employ other dimensions such as temperature and a derived heat unit H, the Btu or calorie.

In mechanical and chemical engineering it is customary to use a set of six dimensions: force F, heat H, length L, mass M, temperature T, and time θ. One of the important alternatives, however, hinges about the unit of force and the unit of mass. In the preceding illustration the *weight* was employed. The relationship would hold whether the pound-mass or gram-mass or the pound-force (**poundal**) or **gram-force (dyne)** were used

CONVECTION

TABLE 3.1. DIMENSIONS AND UNITS

Dimensions:* Force = F, heat = H, length = L, mass = M, temperature = T, time = θ. The force-pound is the poundal, the force-gram is the dyne

Symbol	Quantity: consistent engineering and metric units	Dimension
g	Acceleration of gravity, ft/hr², cm/sec²	L/θ^2
A	Area or surface, ft², cm²	L^2
K_H	Conversion from kinetic energy to heat	$ML^2/H\theta^2$
K_M	Conversion from force to mass	$ML/F\theta^2$
ρ	Density, lb/ft³, g/cm³	M/L^3
D	Diameter, ft, cm	L
k_d	Diffusivity (volumetric), ft²/hr, cm²/sec	L^2/θ
F	Force, force-pound (poundal), force-gram (dyne)	F
H	Heat, Btu, cal	H
L	Length, ft, cm	L
M	Mass, lb, g	M
W	Mass flow, lb/hr, g/sec	M/θ
G	Mass velocity, lb/(hr)(ft²), g/(sec)(cm²)	$M/\theta L^2$
J	Mechanical equivalent of heat, (force-lb)(ft)/Btu, (force-g)(cm)/cal	FL/H
P	Pressure, force-lb/ft², force-g/cm²	F/L^2
P_0	Power, (force-lb)(ft)/hr, (force-g)(cm)/sec	FL/θ
r	Radius, ft, cm	L
c	Specific heat, Btu/(lb)(°F), cal/(g)(°C)	H/MT
v	Specific volume, ft³/lb, cm³/g	L^3/M
τ	Stress, force-lb/ft², force-g/cm²	F/L^2
σ	Surface tension, force-lb/ft, force-g/cm	F/L
T	Temperature, °F, °C	T
k	Thermal conductivity, Btu/(hr)(ft²)(°F/ft), cal/(sec)(cm²)(°C/cm)	$H/LT\theta$
α	Thermal diffusivity, ft²/hr, cm²/sec	L^2/θ
β	Thermal coefficient of expansion, 1/°F, 1/°C	$1/T$
R_t	Thermal resistivity, (°F)(ft)(hr)/Btu, (°C)(cm)(sec)/cal	$LT\theta/H$
θ	Time, hr, sec	θ
u	Velocity, ft/hr, cm/sec	L/θ
μ_g	Viscosity (force-lb)(hr)/ft², (force-g)(sec)/cm²	$F\theta/L^2$
μ	Viscosity (abs), lb/(ft)(hr), g/(cm)(sec)	$M/L\theta$
M	Mass, lb, g	M
w_0	Work, (force-lb)(ft), (force-g)(cm)	FL

* For a system without heat changes these automatically reduce to $FLM\theta$.

as long as the weight was always treated in the same way. Consider a system in which the mass is a fundamental dimension as M, L, T, θ. From the acceleration equation, force = mass × acceleration,

$$F = ML\theta^{-2}.$$

In another set of dimensions it may be more convenient to consider force the fundamental dimension, in which case mass is expressed by

$$M = FL^{-1}\theta^2.$$

When some of the variables are commonly expressed in units of force such as pressure FL^{-2} and other variables by units of mass such as the density ML^{-3}, it is necessary to introduce a dimensional constant into the series expression before solving for the exponents. The constant relating M and F naturally has the dimensions of the gravitational acceleration constant $L\theta^{-2}$. A similar situation also arises when describing a phenomenon by which a work or kinetic energy change occurs in a system. Some of the variables may ordinarily be expressed in terms of foot-pounds (force-pound × foot) and others in terms of heat energy such as the Btu. A conversion factor which is the heat equivalent of work must be introduced to convert FL to H, or vice versa. The constant is the kinetic energy equivalent of the heat $ML^2/H\theta^2$. A number of common variables and dimensional constants are given in Table 3.1 together with their net dimensions in a six-dimension system. Typical sets of engineering and metric units are included.

Analysis for the Form of a Fluid-flow Equation. When an incompressible fluid flows in a straight horizontal uniform pipe with a constant mass rate, the pressure of the fluid decreases along the length of the pipe owing to friction. This is commonly called the *pressure drop* of the system, ΔP. The pressure drop per unit length is referred to as the *pressure gradient*, dP/dL, which has been found experimentally to be influenced by the following pipe and fluid properties: diameter D, velocity u, fluid density ρ, and viscosity μ. What relationship exists between the pressure gradient and the variables?

Solution. The pressure has dimensions of force/area, whereas the density is expressed by mass/volume so that a dimensional constant relating M to F must be included, $K_M = ML/F\theta^2$. The same result may be accomplished by including the acceleration constant g along with the variables above. While the *viscosity* is determined experimentally as a force effect and has the dimensions $F\theta/L^2$, it is a very small unit, and it is more common in the engineering sciences to use the absolute viscosity $M/L\theta$ in which the conversion from force to mass has already been made.

Using the same method of notation as before,

$$\frac{dP}{dL} \propto D, u, \rho, \mu, K_M \tag{3.12}$$

$$\frac{dP}{dL} = \alpha D^a u^b \rho^c \mu^d K_M^e \tag{3.13}$$

Substituting dimensions and arbitrarily setting the exponent of dP/dL equal to 1,

$$\frac{F}{L^3} = \alpha(L)^a \left(\frac{L}{\theta}\right)^b \left(\frac{M}{L^3}\right)^c \left(\frac{M}{L\theta}\right)^d \left(\frac{ML}{F\theta^2}\right)^e \quad (3.14)$$

Summing exponents,

$\Sigma F,\quad 1 = -e$
$\Sigma L,\quad -3 = a + b - 3c - d + e$
$\Sigma M,\quad 0 = c + d + e$
$\Sigma \theta,\quad 0 = -b - d - 2e$

Solving simultaneously,

$a = -1 - d$
$b = 2 - d$
$c = 1 - d$
$d = d$
$e = -1$

Substituting back in Eq. (3.13),

$$\frac{dP}{dL} = \alpha D^{-1-d} u^{2-d} \rho^{1-d} \mu^d K_M^{-1} = \frac{\alpha u^2 \rho}{D K_M}\left(\frac{Du\rho}{\mu}\right)^{-d} \quad (3.15)$$

where α and $-d$ must be evaluated from experimental data. A convenient term of almost universal use in engineering is the *mass velocity G*, which is identical with $u\rho$ and corresponds to the weight flow per square foot of flow area. To obtain the pressure drop from Eq. (3.15), replace dP by ΔP, dL by the length of pipe L, or ΔL, and substituting for K_M its equivalent g,

$$\Delta P = \frac{\alpha G^2 L}{D\rho g}\left(\frac{DG}{\mu}\right)^{-d} \quad (3.16)$$

where $Du\rho/\mu$ or DG/μ is the Reynolds number.

Analysis for the Form of a Forced-convection Equation. The rate of heat transfer by forced convection to an incompressible fluid traveling in turbulent flow in a pipe of uniform diameter at constant mass rate has been found to be influenced by the velocity u, density ρ, specific heat c, thermal conductivity k, viscosity μ, as well as the inside diameter of the pipe D. The velocity, viscosity, density, and diameter affect the thickness of the fluid film at the pipe wall through which the heat must first be conducted, and they also influence the extent of fluid mixing. k is the thermal conductivity of the fluid, and the specific heat reflects the variation of the average fluid temperature as a result of unit heat absorption. What relationship holds between the film coefficient or rate of heat transfer, $h_i = H/\theta L^2 T$ [such as Btu/(hr)(ft^2)(°F)], and the other variables?

Solution. It is not known whether all energy terms will be expressed mechanically or thermally by the dimensions of the variables so that the dimensional constant $K_H = ML^2/H\theta^2$ must be included. If all the dimensions combine to give only thermal quantities such as the Btu, which appears in the dimensions of h_i, the exponent of K_H in the series expression should be zero and the constant will reduce dimensionally to 1.0, a pure number.

$$h_i \propto u, \rho, c, D, k, \mu, K_H$$

$$h_i = \alpha u^a \rho^b c^d D^e k^f \mu^g K_H^i \tag{3.17}$$

$$\frac{H}{\theta L^2 T} = \alpha \left(\frac{L}{\theta}\right)^a \left(\frac{M}{L^3}\right)^b \left(\frac{H}{MT}\right)^d (L)^e \left(\frac{H}{\theta L T}\right)^f \left(\frac{M}{L\theta}\right)^g \left(\frac{ML^2}{H\theta^2}\right)^i \tag{3.18}$$

Summing exponents,

$$\Sigma H, \quad 1 = d + f - i$$
$$\Sigma L, \quad -2 = a - 3b + e - f - g + 2i$$
$$\Sigma M, \quad 0 = b - d + g + i$$
$$\Sigma T, \quad -1 = -d - f$$
$$\Sigma \theta, \quad -1 = -a - f - g - 2i$$

Solving simultaneously,

$$a = a$$
$$b = a$$
$$d = 1 - f$$
$$e = a - 1$$
$$f = f$$
$$g = 1 - f - a$$
$$i = 0$$

Substituting back,

$$h_i = \alpha u^a \rho^a c^{1-f} D^{a-1} k^f \mu^{1-f-a} K_H^0 \tag{3.19}$$

or collecting terms,

$$\frac{h_i D}{k} = \alpha \left(\frac{Du\rho}{\mu}\right)^a \left(\frac{c\mu}{k}\right)^{1-f} \tag{3.20}$$

where α, a, and $1 - f$ must be evaluated from a minimum of three sets of experimental data. Substituting the mass velocity for $u\rho$ in the above,

$$\frac{h_i D}{k} = \alpha \left(\frac{DG}{\mu}\right)^a \left(\frac{c\mu}{k}\right)^{1-f} \tag{3.21}$$

The dimensionless groups hD/k and $c\mu/k$, like the Reynolds number

CONVECTION

$Du\rho/\mu$ or DG/μ, have been assigned names to honor earlier investigators in the field of fluid mechanics and heat transfer. A list of the common groups and the names assigned to them are included in Table 3.2.

TABLE 3.2. COMMON DIMENSIONLESS GROUPS

Symbol	Name	Group
Bi	Biot number	hr/k
Fo	Fourier number	$k\theta/\rho c r^2$
Gz	Graetz number	wc/kL
Gr	Grashof number	$D^3\rho^2 g\beta\,\Delta t/\mu^2$
Nu	Nusselt number	hD/k
Pe	Peclet number	DGc/k
Pr	Prandtl number	$c\mu/k$
Re	Reynolds number	$DG/\mu,\ Du\rho/\mu$
Sc	Schmidt number	$\mu/\rho k_d$
St	Stanton number	h/cG

One of the useful aspects of dimensional analysis is its ability to provide a relationship among the variables when the information about a phenomenon is incomplete. One may have speculated that both fluid friction and forced convection are influenced by the surface tension of the fluid. The surface tension could have been included as a variable and new equations obtained, although the form of the equations would have been altered considerably. Nevertheless it would have been found that the exponents for any dimensionless groups involving the surface tension would be nearly zero when evaluated from experimental data. By the same token, the equations obtained above may be considered to be predicated on incomplete information. In either case a relationship is obtainable by dimensional analysis.

Consistent Units. In establishing the preceding formulas the dimensions were referred to in general terms such as length, time, temperature, etc., without specifying the units of the dimensions. The dimension is the basic measurable quantity, and convention has established a number of different basic units such as temperature, °F and °C; area, square foot, square inch, square meter, square centimeter; time, second or hour; etc. In order that the net dimensions of the variables may be obtained by cancellation among the fundamental and derived dimensions, all must employ the same basic measured units. Thus if several variables employ dimensions containing length such as velocity L/θ, density M/L^3, and thermal conductivity $H/\theta LT$, each must employ the same basic unit of length such as the foot. Accordingly, when substituting values of the variables into a dimensionless group, it is not permissible to signify the dimensions of some variables in feet, some in inches, and still others in centimeters. However, any unit of length is acceptable provided all the lengths involved in the variables are expressed in the same unit of length. The same rule applies to the other fundamental and derived dimensions. When a group of dimensions are expressed in this manner, they are called *consistent units*. Any group of consistent units will yield the same numerical result when the values of the variables are substituted into the dimensionless groups.

The Pi Theorem. One of the important mathematical proofs of dimensional analysis is attributed to Buckingham,[1] who deduced that the number of dimensionless groups is equal to the difference between the number of variables and the number of dimensions used to express them. Dimensional constants are also included as variables. The proof of this statement has been presented quite completely by Bridgman.[2] Designating dimensionless groups by the letters π_1, π_2, π_3, the complete physical statement of a phenomenon can be expressed by

$$\phi(\pi_1, \pi_2, \pi_3 \ldots) = 0 \qquad (3.22)$$

where the total number of π terms or dimensionless groups equals the number of variables minus the number of dimensions. In the preceding example there were, including h_i, eight variables. These were expressed in five dimensions, and the number of dimensionless groups consequently was three. There is a notable exception which must be considered, however, or this method of obtaining the number of dimensionless groups by inspection can lead to an incorrect result. When two of the variables are expressed by the same dimension such as the length and diameter of a pipe, neither is a unique variable, since the dimensions of either are indistinguishable, and to preserve the identity of both they must be combined as a dimensionless constant ratio, L/D or D/L. When treated in this manner, the equation so obtained will apply only to a system which is geometrically similar to the experimental arrangement by which the coefficients and exponents were evaluated, namely, one having the same ratio of L/D or D/L. For this reason the form of a fluid-flow equation, Eq. (3.15), was solved for the pressure gradient rather than the pressure drop directly. Although the solution for a forced-convection equation has already been obtained algebraically, it will be solved again to demonstrate the Pi theorem and the extent to which it differs from the direct algebraic solution. In general, it is desirable to solve for the dimensionless groups appearing in Table 3.2.

[1] Buckingham, E., *Phys. Rev.*, **4**, 345–376 (1914).
[2] Bridgman, *op. cit.*

Analysis for a Forced-convection Equation by the Pi Theorem.

$$\phi(\pi_1, \pi_2, \pi_3 \ldots) = 0$$
$$\pi = \phi'(h_i^a u^b \rho^c c^e D^f k^g \mu^m K_H^i) = 1 \quad (3.23)$$
$$\pi = \alpha \left(\frac{H}{\theta L^2 T}\right)^a \left(\frac{L}{\theta}\right)^b \left(\frac{M}{L^3}\right)^d \left(\frac{H}{MT}\right)^e (L)^f \left(\frac{H}{\theta LT}\right)^g \left(\frac{M}{L\theta}\right)^m \left(\frac{ML^2}{H\theta^2}\right)^i \quad (3.24)$$

Summing exponents,

$$\Sigma H, \ 0 = a + e + g - i$$
$$\Sigma L, \ 0 = -2a + b - 3d + f - g - m + 2i$$
$$\Sigma M, \ 0 = d - e + m + i$$
$$\Sigma T, \ 0 = -a - e - g$$
$$\Sigma \theta, \ 0 = -a - b - g - m - 2i$$

π_1, π_2, and π_3 may be evaluated by simple algebra. All the exponents need not be summed up in one operation, since it has been seen in Eq. (3.20) that the dimensionless groups comprising it are only composed of three or four variables each. It is requisite in summing for the three groups individually only that all the exponents be included at some time and that three summations be made equal to the difference between the eight variables and five dimensions or three groups, π_1, π_2, and π_3.

π_1. Since it is desired to establish an expression for h_i as the dependent variable, it is preferable that it be expressed raised to the first power or $a = 1$. This will assure that in the final equation the dependent variable will not be present raised to some odd fractional power. Since all the exponents need not be included in evaluating π_1, assume $b = 0$ and $e = 0$. Referring to Eq. (3.20) it will be seen that as a result of these assumptions neither the Reynolds number nor the Prandtl number will appear as solutions for π_1. When $b = 0$, $u^b = 1$ and $Du\rho/\mu = 1$; and when $e = 0$, $c^e = 1$ and $c\mu/k = 1$.

Assume $a = 1$, $b = 0$, $e = 0$. Solving the simultaneous equations above, $d = 0$, $f = +1$, $g = -1$, $m = 0$, $i = 0$,

$$\pi_1 = \phi' \left(\frac{h_i D}{k}\right)$$

π_2. Having already obtained h_i it is desirable not to have it appear again in either π_2 or π_3. This can be accomplished in solving for the next group by assuming $a = 0$. The entire Nusselt number $h_i D/k$ will then reduce to 1. If a Reynolds number is desired, because it is a useful criterion of fluid flow, assume b or $f = 1$. Lastly, if the Prandtl number is to be eliminated, assume that the exponent of c or k is zero. If the exponent of the viscosity is assumed to be zero, it will not be possible to obtain either a Reynolds number or a Prandtl number.

Assume $f = 1$, $a = 0$, $e = 0$. Solving the simultaneous equations above, $b = 1$, $d = 1$, $g = 0$, $m = -1$, $i = 0$,

$$\pi_2 = \phi'\left(\frac{Du\rho}{\mu}\right)$$

π_3. To prevent the h_i term and the velocity or density from appearing again, assume $a = 0$, $e = 1$, $f = 0$. All the exponents will have now appeared in one or more solutions.

Assume $a = 0$, $e = 1$, $f = 0$. Solving the simultaneous equations above, $b = 0$, $d = 0$, $g = -1$, $m = 1$, $i = 0$,

$$\pi_3 = \phi'\left(\frac{c\mu}{k}\right)$$

The final expression is

$$\phi\left(\frac{h_iD}{k}, \frac{Du\rho}{\mu}, \frac{c\mu}{k}\right) = 0 \qquad (3.25)$$

or

$$\begin{aligned}\frac{h_iD}{k} &= \phi_1\left(\frac{Du\rho}{\mu}\right)\phi_2\left(\frac{c\mu}{k}\right) \\ &= \alpha\left(\frac{Du\rho}{\mu}\right)^p\left(\frac{c\mu}{k}\right)^q = \alpha\left(\frac{DG}{\mu}\right)^p\left(\frac{c\mu}{k}\right)^q\end{aligned} \qquad (3.26)$$

where the proportionality constant and the exponents must be evaluated from experimental data.

Development of an Equation for Streamline Flow. Since streamline flow is a conduction phenomenon, it is subject to rational mathematical analysis. On the assumption that the distribution of velocities at any cross section is parabolic, the inside surface of the pipe is uniform, and the velocity at the wall is zero, Graetz[1] obtained for radial conduction to a fluid moving in a pipe in rodlike flow

$$\frac{t_2 - t_1}{t_p - t_1} = 1 - 8\phi\left(\frac{wc}{kL}\right) \qquad (3.27)$$

where t_1 and t_2 are the inlet and outlet temperatures of the fluid, t_p is the uniform inside pipe surface temperature, $t_p - t_1$ the temperature difference at the inlet, and $\phi(wc/kL)$ is the numerical value of an infinite series having exponents which are multiples of wc/kL. Equation (3.27) may be replaced through dimensional analysis with an empirical expression which must be evaluated from experiments. If $t_2 - t_1$, the rise in the temperature of the fluid flowing in the pipe, is considered to be influenced

---

[1] Graetz, L., *Ann. Physik*, **25**, 337 (1885). For a review of the treatment of conduction in moving fluids see T. B. Drew, *Trans. AIChE*, **26**, 32 (1931).

CONVECTION

in radial conduction by the length of path L, the rate of flow w, specific heat c, thermal conductivity k, and the temperature difference between the pipe inside surface and the fluid temperature so that $\Delta t_i = t_p - t_1$,

$$t_2 - t_1 = \alpha L^a w^b c^d k^e \, \Delta t_i^f \tag{3.28}$$

Solving by the method of dimensional analysis

$$t_2 - t_1 = \alpha \left(\frac{wc}{kL}\right)^d \Delta t_i \tag{3.29}$$

or

$$\frac{t_2 - t_1}{\Delta t_i} = \alpha \left(\frac{wc}{kL}\right)^d \tag{3.30}$$

which is similar to Eq. (3.27). Now note that neither Eq. (3.28) nor (3.29) contains h_i or the viscosity μ. But $Q = h_i A_i \, \Delta t_i$ or

$$wc(t_2 - t_1) = h_i \pi \, DL \, \Delta t_i$$

and substituting $G = \dfrac{w}{\pi D^2/4}$ in Eq. (3.30),

$$\frac{h_i L}{DG\,c} = \alpha \left(\frac{D^2 Gc}{kL}\right)^d \tag{3.31}$$

Now *synthetically* introducing the viscosity by multiplying the right term of Eq. (3.31) by $(\mu/\mu)^d$, one obtains

$$\frac{h_i D}{k} = \alpha \left[\left(\frac{DG}{\mu}\right)\left(\frac{c\mu}{k}\right)\left(\frac{D}{L}\right)\right]^{d+1} = \alpha \left(\frac{wc}{kL}\right)^{d+1} \tag{3.32}$$

This is a convenient means of representing streamline flow using the dimensionless groups incidental to turbulent flow and including the dimensionless ratio D/L. It should not be inferred, however, that because of this method of representation the heat-transfer coefficient is influenced by the viscosity even though the Reynolds number, which is the criterion of streamline flow, is inversely proportional to the viscosity. The values of μ in Eq. (3.32) actually cancel out.

The Temperature Difference between a Fluid and a Pipe Wall. Before attempting to evaluate the constants of a forced-convection equation from experimental data, an additional factor must be taken into account. When a liquid flows along the axis of a tube and absorbs or transmits sensible heat, the temperature of the liquid varies over the entire length of the pipe. In the case of heat flow through a flat wall, the temperature over the entire area of each face of the flat wall was identical and the temperature difference was simply the difference between any points on the two faces. If the temperature of the inner circumference of a pipe wall is nearly constant over its entire length, as it might be when the fluid inside the tube is heated by steam, there will be two distinct temperature differences at each end: one between the pipe wall and the inlet liquid and one at the other end between the pipe wall and the heated liquid. What is the appropriate temperature difference for use in the equation $Q = h_i A_i (t_p - t) = h_i A_i \Delta t_i$, where t_p is the constant temperature of the inside pipe wall and t is the varying temperature of the liquid inside the pipe?

Referring to Fig. 3.6, the constant temperature of the inside pipe wall is shown by the horizontal line t_p. If the specific heat is assumed con-

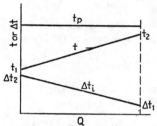

FIG. 3.6. Temperature difference between a fluid and a pipe wall.

stant for the liquid, the rise in temperature is proportional to the total heat received by the liquid in passing from the inlet temperature t_1 to the outlet temperature t_2 and if h_i is considered constant

$$dQ = h_i \, dA_i \, \Delta t_i \tag{3.33}$$

The slope of the lower line defining the temperature difference Δt_i as a function of Q is

$$\frac{d \Delta t_i}{dQ} = \frac{\Delta t_2 - \Delta t_1}{Q} \tag{3.34}$$

where $\Delta t_2 = t_p - t_1$ and $\Delta t_1 = t_p - t_2$. Eliminating dQ from Eqs. (3.33) and (3.34),

$$\frac{h_i \, dA_i}{Q} (\Delta t_2 - \Delta t_1) = \int_{\Delta t_1}^{\Delta t_2} \frac{d \Delta t_i}{\Delta t_i} \tag{3.35}$$

Integrating

$$Q = \frac{h_i A_i (\Delta t_2 - \Delta t_1)}{\ln \Delta t_2 / \Delta t_1} \tag{3.36}$$

The expression $\dfrac{\Delta t_2 - \Delta t_1}{\ln \Delta t_2 / \Delta t_1}$ is the *logarithmic mean temperature difference*, abbreviated LMTD, and the value of h_i, which has been computed from $Q = h_i A_i \Delta t_i$ when Δt_i is the logarithmic mean, is a distinct value of

h_i. If the value of Δt_i were arbitrarily taken as the arithmetic mean of Δt_2 and Δt_1, the value of h_i would have to be designated to show that it does not correspond to the logarithmic mean. This is usually accomplished by affixing the subscript a or m for arithmetic mean, as h_a or h_m. When Δt_2 and Δt_1 approach equality, the arithmetic and logarithmic means approach each other.

Experimentation and Correlation. Suppose experimental apparatus was available with known diameter and length and through which liquid could be circulated at various measurable flow rates. Suppose, furthermore, that it was equipped with suitable devices to permit measurement of the inlet and outlet liquid temperatures and the temperature of the pipe wall. The heat absorbed by the liquid in flow through the pipe would be identical with the heat passing into the pipe in directions at right angles to its longitudinal axis, or

$$Q = wc(t_2 - t_1) = h_i A_i \Delta t_i \tag{3.37}$$

From the observed values of the experiment and the calculation of Δt_i, as given in Eq. (3.36), h_i can be computed from

$$h_i = \frac{wc(t_2 - t_1)}{A_i \Delta t_i} \tag{3.38}$$

The problem encountered in industry, as compared with experiment, is not to determine h_i but to apply experimental values of h_i to obtain A_i, the heat-transfer surface. The process flow sheet ordinarily contains the heat and material balances about the various items of equipment which enter into a process. From these balances are obtained the conditions which must be fulfilled by each item if the process is to operate as a unit. Thus between two points in the process it may be required to raise the temperature of a given weight flow of liquid from t_1 to t_2 while another fluid is cooled from T_1 to T_2. The question in industrial problems is to determine how much heat-transfer surface is required to fulfill these process conditions. The clue would appear to be present in Eq. (3.38), except that not only A_i but also h_i is unknown unless it has been established by former experiments for identical conditions.

In order to prepare for the solution of industrial problems it is not practical to run experiments on all liquids and under an infinite variety of experimental conditions so as to have numerical values of h_i available. For example, h_i will differ for a single weight flow of a liquid absorbing identical quantities of heat when the numerical values of t_1 and t_2 differ, since the liquid properties are related to these temperatures. Other factors which affect h_i are those encountered in the dimensional analysis such as the velocity of the liquid and the diameter of the tube through which heat transfer occurs. It is here that the importance of the equations obtained through dimensional analysis becomes apparent. If the values of the exponents and coefficients of the dimensionless equation for extreme conditions of operation are established from experiments, the value of h_i can be calculated for any intermediate combination of velocity and pipe and liquid properties from the equation.

A typical apparatus for the determination of the heat-transfer coefficient to liquids flowing inside pipes or tubes is shown in Fig. 3.7. The principal part is the test exchanger, which consists of a test section of pipe enclosed by a concentric pipe. The annulus is usually connected

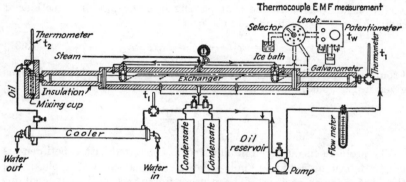

FIG. 3.7. Test apparatus for the tube transfer coefficient.

to permit the condensation of steam for liquid-heating experiments or the rapid circulation of water for liquid-cooling experiments. The auxiliary exchanger is connected to perform the opposing operation of the test section and cools when the test section is used for heating. For a heating experiment, cold liquid from the reservoir is circulated through the circuit by a centrifugal pump. A by-pass is included on the pump discharge to permit the regulation of flow through the flow meter. The liquid then passes through a temperature-measuring device, such as a calibrated thermometer or thermocouple, where t_1 is obtained. t_1 is taken at some distance in the pipe before the test section, so that the temperature-measuring device does not influence the convection eddies in the test section itself. The liquid next passes through the test section and another unheated length of pipe before it is mixed and its temperature t_2 is measured. The unheated extensions of the test pipe are referred to as *calming sections*. Next the liquid passes through a cooler where its temperature is returned to t_1. The annulus of the test section is connected to calibrated condensate drain tanks where the heat balance can be checked by measuring the steam, Dowtherm (a fluid which permits the attainment of high vapor temperatures at considerably lower pressures than are obtained with steam), or other vapor quantity. The pressure of steam can be adjusted by a pressure-reducing valve, and in the event that the heating vapor does not have a well-established temperature-pressure saturation curve, thermocouples or thermometers can also be inserted in the shell. The temperature of the *outside* surface of the heated length of the pipe is obtained by attaching a number of thermocouples about the pipe surface and obtaining their average temperature. The thermocouples may be calibrated by circulating preheated oil through the tube while the annulus outside the test pipe is kept under vacuum. The temperature at the outside of the test-pipe surface can then be calculated from the uniform temperature of the preheated oil after correcting for the temperature drop through the pipe wall and calibrating temperature against emf by means of a potentiometer. The leads from the pipe-surface thermocouples are brought out through the gaskets at the ends of the test section.

The performance of the experiment requires the selection of an initial reservoir temperature t_1 which may be attained by recirculating the liquid through the exchanger at a rapid rate until the liquid in the reservoir reaches the desired temperature. A flow rate is selected and cooling water is circulated through the cooler so that the temperature of the oil returning to the reservoir is also at t_1. When steady conditions of t_1 and t_2 have persisted for perhaps 5 min or more, the temperatures t_1 and t_2 are recorded along with the flow rate, readings of the pipe-surface thermocouples, and gain in condensate level during the test interval. With versatile apparatus using good regulating valves an experiment can usually be carried out in less than an hour.

Several important points must be safeguarded in the design of experimental apparatus if consistent results are expected. The covers at the ends of the test section cannot be connected directly to the test pipe, since they act as gatherers of heat. Should they touch the test pipe in a metal-to-metal contact, they add large quantities of heat in local sections. To prevent inaccuracies from this source the covers and the test pipe should be separated by a nonconducting packing gland. Another type of error results from the accumulation of air in the annulus, which prevents the free condensation of vapor on the test pipe. This usually is apparent if there is a lack of uniformity in the readings of the pipe-surface thermocouples and annulus thermocouples when the latter are employed. This can be overcome by providing a bleed for the removal of any trapped air. Problems incidental to the installation and calibration of thermocouples and thermometers appear elsewhere.[1] The same is true of equations to correct for fluid flow through standard orifices when the properties of the liquid vary.

Evaluation of a Forced Convection Equation from Experimental Data. As an example of correlation, data are given in Table 3.3 which were obtained by Morris and Whitman[2] on heating gas oil and straw oil with steam in a ½-in. IPS pipe with a heated length of 10.125 ft.

The viscosity data are given in Fig. 3.8 and are taken from the original publication. The thermal conductivities may be obtained from Fig. 1, and the specific heats from Fig. 4. Both are plotted in the Appendix

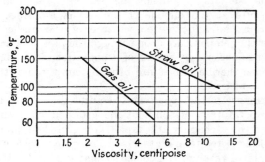

FIG. 3.8. Viscosities of test oils.

with °API as parameters. The thermal conductivity of the metal was taken by Morris and Whitman to be 35 Btu/(hr)(ft^2)(°F/ft) and constant, although this is higher than the value given in Appendix Table 3. Only columns 2, 3, 4, and 5 in Table 3.3 were observed where

t_1 = oil inlet temperature, °F
t_2 = oil outlet temperature, °F
t_w = temperature of *outer* pipe surface averaged from thermocouples
w = weight flow, lb/hr

The first step in correlating a forced convection equation is to determine that the data are in turbulent flow, otherwise an attempted correlation by Eq. (3.26) would be incorrect. In column 11 the Reynolds numbers have been calculated using the diameter and flow area of a ½-in. IPS pipe which may be found in Table 11. Fluid properties have been obtained at the mean temperature $(t_1 + t_2)/2$. The Reynolds numbers all exceed

---

[1] American Institute of Physics, "Temperature: Its Measurement and Control in Science and Industry," Reinhold Publishing Corporation, New York, 1941.

[2] Morris, F. H., and W. G. Whitman, *Ind. Eng. Chem.*, **20**, 234 (1928).

CONVECTION

TABLE 3.3. DATA OF MORRIS AND WHITMAN
Runs Heating 36.8°API Gas Oil with Steam

(1) Run No.	(2) w	(3) t_1	(4) t_2	(5) t_w	(6) Q	(7) Δt_i	(8) h_i	(9) $\frac{h_i D}{k}$	(10) G	(11) $\frac{DG}{\mu}$	(12) $\frac{c\mu}{k}$	(13) $j_H = \frac{Nu}{Pr}$	(14) $j_H = \frac{Nu}{Pr^{1/3}}$
B1	722	77.1	106.9	210.1	10,150	115.7	53.6	35.5	342,000	2,280	47.2	0.75	9.83
B2	890	77.9	109.3	209.0	13,150	112.6	71.4	46.3	421,000	2,825	46.7	0.99	12.85
B3	1056	85.6	117.6	208.9	16,100	104.0	94.5	62.3	501,000	3,710	43.3	1.44	17.75
B4	1260	89.8	121.9	208.0	19,350	98.5	120	79.5	597,000	4,620	41.4	1.92	23.0
B5	1497	91.6	123.3	207.5	22,700	96.0	144	95	709,000	5,780	40.7	2.33	27.7
B6	1802	99.1	129.2	207.2	26,200	88.5	181	120.5	854,000	7,140	38.7	3.12	35.5
B7	2164	102.3	131.7	206.9	30,900	84.7	223	147.5	1,026,000	8,840	37.7	3.91	44.0
B8	2575	106.5	134.3	206.4	35,000	80.3	266	176.5	1,220,000	10,850	36.5	4.83	53.2
B9	3265	111.5	137.1	205.0	41,100	74.2	338	223	1,548,000	14,250	35.3	6.32	68.0
B10	3902	113.9	138.2	203.8	46,600	70.6	403	266.5	1,850,000	17,350	35.1	7.60	81.4
B11	4585	116.8	139.7	203.0	51,900	66.9	474	313	2,176,000	20,950	34.1	9.18	96.5
B12	5360	122.2	142.9	202.9	54,900	62.3	538	356	2,538,000	25,550	32.9	10.82	111.2
B13	6210	124.8	144.1	202.2	59,500	59.1	615	407	2,938,000	30,000	32.7	12.43	127.5

Runs Heating 29.4°API Straw Oil with Steam

(1) Run No.	(2) w	(3) t_1	(4) t_2	(5) t_w	(6) Q	(7) Δt_i	(8) h_i	(9) $\frac{h_i D}{k}$	(10) G	(11) $\frac{DG}{\mu}$	(12) $\frac{c\mu}{k}$	(13) $j_H = \frac{Nu}{Pr}$	(14) $j_H = \frac{Nu}{Pr^{1/3}}$
C8	2900	100.0	115.4	206.3	20,700	95.8	132	87.5	1,375,000	3,210	133	0.66	17.2
C9	2920	86.7	99.3	208.0	16,800	112.7	91.1	60.4	1,387,000	2,350	179	0.34	10.7
C10	3340	101.7	117.6	206.0	24,800	92.9	163	108	1,585,000	3,820	129.5	0.83	21.4
C11	3535	100.5	115.7	205.5	25,000	94.0	162	107.5	1,675,000	3,880	133.3	0.81	21.1
C12	3725	163.0	175.1	220.1	22,600	47.9	288	191	1,767,000	10,200	57.8	3.30	49.5
C13	3810	160.5	173.6	220.5	24,900	50.0	304	201.5	1,805,000	10,150	59.3	3.39	51.6
C14	3840	109.0	124.4	205.7	27,800	85.2	199	131.5	1,820,000	4,960	115	1.15	27.1
C15	4730	112.0	127.3	205.3	34,100	81.0	257	170	2,240,000	6,430	110	1.55	35.5
C16	5240	154.9	167.6	217.7	33,150	51.9	389	257.5	2,485,000	13,150	62.9	4.08	64.7
C17	5270	150.9	164.3	217.0	34,900	54.7	389	257.5	2,500,000	12,520	65.6	3.92	63.7
C18	5280	142.3	156.8	216.7	37,500	62.0	369	244	2,510,000	11,250	72.6	3.36	58.5
C19	5320	132.3	148.1	215.0	40,600	70.2	353	234	2,520,000	9,960	81.5	2.87	54.1
C20	5620	118.8	133.1	204.7	38,200	73.5	317	210	2,660,000	8,420	100.4	2.09	45.2
C21	6720	122.2	135.9	204.3	43,900	69.2	387	256	3,185,000	10,620	95.6	2.68	56.0
C22	8240	124.1	137.0	204.4	47,900	67.3	434	287	3,905,000	12,650	93.3	3.08	63.3

2100 in substantiation of turbulent flow. Equation (3.26) is given as

$$\frac{h_i D}{k} = \alpha \left(\frac{DG}{\mu}\right)^p \left(\frac{c\mu}{k}\right)^q$$

and α, p, and q can be found algebraically by taking the data for three test points.

Algebraic Solution. This method of correlation is demonstrated by using the three points $B4$, $B12$, and $C12$ in Table 3.3 which include a large range of $h_i D/k$, DG/μ, and $c\mu/k$ as calculated presently from flow and fluid properties and tabulated in columns 9, 11, and 12.

$$\frac{h_i D}{k} = \alpha \left(\frac{DG}{\mu}\right)^p \left(\frac{c\mu}{k}\right)^q$$

$C12$: $191 = \alpha(10{,}200)^p (57.8)^q$
$B12$: $356 = \alpha(25{,}550)^p (32.9)^q$
$B4$: $79.5 = \alpha(4{,}620)^p (41.4)^q$

Taking the logarithms of both sides,

$C12$: $2.2810 = \log \alpha + 4.0086p + 1.7619q$
$B12$: $2.5514 = \log \alpha + 4.4065p + 1.5172q$
$B4$: $1.9004 = \log \alpha + 3.6646p + 1.6170q$

Eliminating the unknowns one by one gives $\alpha = 0.00682$, $p = 0.93$, and $q = 0.407$, and the final equation is

$$\frac{h_i D}{k} = 0.00892 \left(\frac{DG}{\mu}\right)^{0.93} \left(\frac{c\mu}{k}\right)^{0.407}$$

When the equation is to be used frequently, it can be simplified by fixing q as the cube root of the Prandtl number and solving for new values of α and p. The simplified equation would be

$$\frac{h_i D}{k} = 0.0089 \left(\frac{DG}{\mu}\right)^{0.965} \left(\frac{c\mu}{k}\right)^{1/3}$$

Graphical Solution. For the correlation of a large number of points the graphical method is preferable. Rewriting Eq. (3.26),

$$\frac{h_i D}{k} \left(\frac{c\mu}{k}\right)^{-q} = \alpha \left(\frac{DG}{\mu}\right)^p \qquad (3.39)$$

which is an equation of the form

$$y = \alpha x^p \qquad (3.40)$$

CONVECTION

Taking the logarithms of both sides,

$$\log y = \log \alpha + p \log x$$

which reduces on logarithmic coordinates to an equation of the form

$$y = \alpha + px \qquad (3.41)$$

On logarithmic coordinates the entire group $(h_i D/k)(c\mu/k)^{-q}$ is the ordinate y in Eq. (3.41), the Reynolds number is x, p is the slope of the data when plotted as y vs. x, and α is the value of the intercept when

$$p \log x = 0$$

which occurs when the Reynolds number is 1.0. To plot values of $j_H = (h_i D/k)(c\mu/k)^{-q}$, the exponent q must be assumed. The most satisfactory assumed value of the exponent will be the one which permits the data to be plotted with the smallest deviation from a straight line. A value of q must be assumed for the entire series of the experiments, and j_H computed accordingly. This is a more satisfactory method than an algebraic solution, particularly when a large number of tests on different oils and pipes are being correlated. If q is assumed too large, the data will scatter when plotted as y vs. x. If q is assumed too small, the data will not scatter but will give a large deviation by producing a curve. Preparatory to a graphical solution, run $B1$ in Table 3.3 will be computed completely from observed data alone. Run $B1$ consists of a test employing a 36.8°API gas oil in a ½-in. IPS pipe.

Observed test data:

Weight flow of gas oil, $w = 722$ lb/hr
Temperature of oil at pipe inlet, $t_1 = 77.1$°F
Temperature of oil at pipe outlet, $t_2 = 106.9$°F
Average temperature of outside pipe surface, $tw = 210.1$°F

Physical data and calculated results:

Heat load, Btu/hr:

Average oil temperature $= \dfrac{77.1 + 106.9}{2} = 92.0$°F

Average specific heat, $c = 0.472$ Btu/(lb)(°F)
$Q = wc(t_2 - t_1) = 722 \times 0.472(106.9 - 77.1) = 10{,}150$ Btu/hr

Temperature of pipe at inside surface, t_p:

I.D. of ½-in. IPS $= 0.62$ in.; O.D. $= 0.84$ in.
Length, 10.125 ft; surface, 1.65 ft²

Thermal conductivity of steel, 35 Btu/(hr)(ft²)(°F/ft)

Q per lin ft, $q = \dfrac{10{,}150}{10.125} = 1007$ Btu

$$t_p = t_w - \frac{2.3q}{2\pi k}\log\frac{D_2}{D_1} = 210.1 - \frac{2.3\times 1007}{2\times 3.14\times 35}\log\frac{0.84}{0.62} = 208.7°F$$

Δt_i in expression $Q = h_i A_i \Delta t_i$:

At inlet, $\Delta t_2 = 208.7 - 77.1 = 131.6°F$
At outlet, $\Delta t_1 = 208.7 - 106.9 = 101.8°F$

$$\Delta t_i = \text{LMTD} = \frac{131.6 - 101.8}{2.3\log(131.6/101.8)} = 115.7°F \qquad (3.36)$$

$$h_i = \frac{Q}{A_i\Delta t_i} = \frac{10{,}150}{1.65\times 115.7} = 53.6 \text{ Btu/(hr)(ft}^2\text{)(°F)}$$

The thermal conductivity of the oil will be considered constant at 0.078 Btu/(hr)(ft²)(°F/ft).

Nusselt number, $Nu = \dfrac{h_i D}{k} = \dfrac{53.6\times 0.62}{0.078\times 12} = 35.5$, dimensionless

Mass velocity, $G = \dfrac{w}{\pi D^2/4} = \dfrac{722}{(3.14\times 0.62^2)/(4\times 12^2)} = 342{,}000$ lb/(hr)(ft²)

The viscosity from Fig. 3.8 at 92°F is 3.22 centipoises [gram-mass/100 (cm)(sec)] or $3.22 \times 2.42 = 7.80$ lb/(ft)(hr).

Reynolds number, $Re = \dfrac{DG}{\mu} = \dfrac{0.62}{12}\times 342{,}000 \times \dfrac{1}{7.80}$
$\hspace{5cm}= 2280$, dimensionless

Prandtl number, $Pr = \dfrac{c\mu}{k} = \dfrac{0.472\times 7.80}{0.078} = 47.2$, dimensionless

Assume values of q of 1.0 and ⅓ respectively.
First trial: $j_H = Nu/Pr = 0.75$ plotted in Fig. 3.9
Second trial: $j_H = Nu/Pr^{1/3} = 9.83$ plotted in Fig. 3.10

The values of the first trial in which the ordinate is $j_H = \dfrac{h_i D}{k}\Big/\left(\dfrac{c\mu}{k}\right)$ for an assumed value of $q = 1$ are plotted in Fig. 3.9 where two distinct lines result, one for each of the oils. It is the object of a good correlation to provide but one equation for a large number of liquids and this can be accomplished by adjusting the exponent of the Prandtl number. By assuming a value of $q = \frac{1}{3}$ and plotting the ordinate $j_H = \dfrac{h_i D}{k}\Big/\left(\dfrac{c\mu}{k}\right)^{1/3}$, it is possible to obtain the single line as shown in Fig. 3.10. By drawing the best straight line through the points of Fig. 3.10, the slope can be measured in the same way as on rectangular coordinates, which in this particular case is found to be 0.90. By extrapolating the straight line

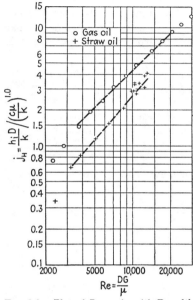

FIG. 3.9. Plot of Re vs. j_H with Prandtl exponent of 1.0.

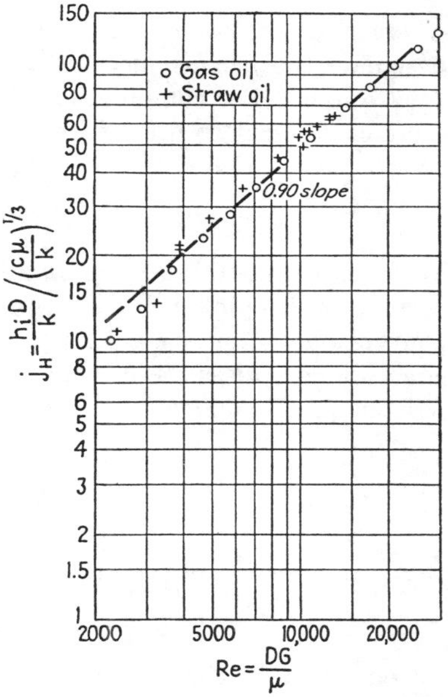

FIG. 3.10. Plot of Re vs. j_H with Prandtl exponent of $\frac{1}{3}$.

until the Reynolds number is 1.0, a value of $\alpha = 0.0115$ is obtained as the intercept. The equation for all the data is thus

$$\frac{h_i D}{k} = 0.0115 \left(\frac{DG}{\mu}\right)^{0.90} \left(\frac{c\mu}{k}\right)^{\frac{1}{3}} \tag{3.42}$$

A value of $q = 0.40$ would cause less bowing and a smaller deviation. The correlation of data need not be confined to heating or cooling runs on liquids separately. It is entirely possible to combine both types of data into a single correlation, called the isothermal heat-transfer equation, but the procedure involves an additional consideration which is deferred until Chap. 5.

Correlation of Fluid Friction in Pipes. When a fluid flows in a pipe isothermally, it undergoes a decrease in pressure. From Eq. (3.16) it is seen that for isothermal turbulent flow this pressure drop is a function of the Reynolds number and, in addition, to the roughness of the pipe

Rewriting Eq. (3.16) in dimensionless form,

$$f' = \frac{\Delta P\, g\rho D}{G^2 L} = \alpha \left(\frac{DG}{\mu}\right)^{-d}$$

where f' is one of the dimensionless factors found in the literature to designate the *friction factor* and ΔP is the pressure drop in pounds per square foot. For combination with other hydrodynamic equations it is more convenient to use a friction factor f so that

$$f = \frac{\Delta P\, 2g\rho D}{4G^2 L} = \frac{\alpha}{Re^d} \quad (3.43)$$

When experimental data are available, it is thus convenient to obtain a correlation by plotting f as a function of the Reynolds number and the conventional Fanning equation as shown in Fig. 3.11. The Fanning equation comprises the first and second terms of Eq. (3.43) and is usually written as $\Delta F = \Delta P/\rho$ where ΔF is the pressure drop expressed in *feet of liquid*, or

$$\Delta F = \frac{4fG^2 L}{2g\rho^2 D} \quad (3.44)$$

For the portion of the graph corresponding to streamline flow

$$(Re < 2100 \text{ to } 2300),$$

the equation for the pressure drop may be deduced from theoretical considerations alone and has been verified by experiment. The equation is

$$\Delta F = \frac{32\mu G}{g\rho^2 D^2} \quad (3.45)$$

By equating (3.44) and (3.45), since each applies at the transition point from streamline to turbulent flow, the equation of this line, known as the Hagen-Poiseuille equation, where f is used with Eq. (3.44), is

$$f = \frac{16}{DG/\mu} \quad (3.46)$$

To the right of the transition region in turbulent flow there are two lines, one for commercial pipes and the other for tubes. Tubes have smoother finishes than pipes and therefore give lower pressure drops when all other factors are the same. This is not so of streamline flow where the fluid at the pipe or tube wall is assumed to be stationary or nearly stationary and the pressure drop is not influenced by roughness. The equation of f in Eq. (3.44) for fluids in tubes in turbulent flow is given by

Drew, Koo, and McAdams[1] within ±5 per cent by

$$f = 0.00140 + \frac{0.125}{(DG/\mu)^{0.32}} \quad (3.47a)$$

For clean commercial iron and steel pipes an equation given by Wilson,

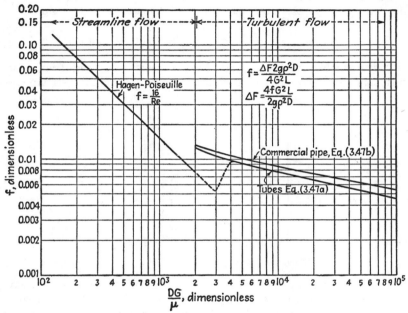

FIG. 3.11. Friction factors for flow in pipes and tubes.

McAdams, and Seltzer[2] within ±10 per cent is

$$f = 0.0035 + \frac{0.264}{(DG/\mu)^{0.42}} \quad (3.47b)$$

It can be seen that, if the transition from streamline to turbulent flow is given by $Du\rho/\mu = 2300$, approximately, then the velocity at which a fluid in a pipe changes from streamline to turbulent flow is

$$u_{crit} = \frac{2300\mu}{D\rho}$$

For water flowing in a 1-in. IPS pipe at 100°F, the viscosity is 0.72 centi-

[1] Drew, T. B., E. C. Koo, and W. H. McAdams, *Trans. AIChE*, **28**, 56–72 (1932).
[2] Wilson, R. E., W. H. McAdams, and M. Seltzer, *Ind. Eng. Chem.*, **14**, 105–119 (1922).

poise or 0.72 gram-mass × 100 (cm)(sec) or 0.72 × 2.42 = 1.74 lb/(ft)(hr), and the inside diameter of the pipe is 1.09 in., or 1.09/12 = 0.091 ft.

$$u_{crit} = \frac{2300 \times 1.74}{0.091 \times 62.3} = 707 \text{ ft/hr, or } 0.196 \text{ fps}$$

For air at 100°F the viscosity is 0.0185 × 2.42 = 0.0447 lb/(ft)(hr) and the density is approximately 0.075 lb/ft.³ Then for the same pipe

$$u_{crit} = \frac{2300 \times 0.0447}{0.091 \times 0.075} = 15{,}100 \text{ ft/hr, or } 4.19 \text{ fps}$$

The Reynolds Analogy. Both heat transfer and fluid friction in turbulent flow have been treated empirically, whereas their streamline-

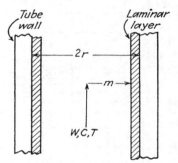

FIG. 3.12. Momentum transfer between a fluid and a boundary layer.

flow equivalents may be studied theoretically with reasonable accuracy. Turbulent flow is of greater importance to industry, yet this empiricism exists because of the lack of simple mathematics by which derivations might be obtained. As early as 1874, however, Osborne Reynolds[1] pointed out that there was probably a relationship between the transfer of heat and the fluid friction between a hot fluid and a surface.

There are several advantages which might accrue from an equation relating heat transfer and fluid friction. For the most part fluid-friction experiments are simpler to perform than are heat-transfer experiments, and the understanding of both fields could be increased by experimentation in either. The fundamentals of the mechanism of both might also be better understood if they were directly related. The analogy between the two is made possible by the fact that the transfer of heat and the transfer of fluid momentum can be related. A simplified proof follows.

Referring to Fig. 3.12 a fluid in amount W lb/hr and specific heat C flows through a tube of radius r. In the section of the pipe between L and

[1] Reynolds, *op. cit.*, pp. 81–85.

CONVECTION 55

$L + dL$ the temperature of the fluid is T, and the temperature at the inside surface of the pipe wall is t_p. Of the total fluid flow along the axis of the pipe, assume that m lb/(hr)(ft^2) impinges on the pipe wall where its velocity falls to zero and its temperature approaches the temperature of the pipe wall, t_p. Each particle of fluid which thus contacts the wall gives up its axial momentum and upon returning to the main body of the fluid has its axial momentum renewed at the expense of the energy of the main stream. The constant loss and renewal of momentum are the cause of the pressure drop. The traction or drag on unit area of the pipe wall is obtained by equating the drag on unit length of pipe wall to the product of the pressure gradient and the cross-sectional area of the pipe, which reduces to

$$\tau = \frac{f}{2} \rho u^2 \tag{3.48}$$

where τ is the drag. Since the drag is assumed to be equal to the loss of momentum of the fluid,

$$mu = \tau = \frac{f}{2} \rho u^2 \tag{3.49}$$

The rate of heat transfer between the fluid and the wall is given by

$$mC 2\pi r \, dL(T - t_p) = WC \, dT \tag{3.50}$$

or from Eqs. (3.49) and (3.50)

$$m = \frac{W \, dT}{2\pi r \, dL(T - t_p)} = \frac{\tau}{u} \tag{3.51}$$

In simple terms the last two members of Eq. (3.51) state

Heat actually given up to pipe wall
Total heat available to be given up
$$= \frac{\text{momentum lost by skin friction to pipe wall}}{\text{total momentum available}}$$
$$= \text{a constant}$$

Rewriting Eq. (3.50) to include the heat-transfer coefficient,

$$WC \, dT = h_i 2\pi r \, dL(T - t_p) \tag{3.52}$$

$$h_i = \frac{WC \, dT}{2\pi r \, dL(T - t_p)} = \frac{\tau C}{u} = \frac{f}{2} \rho u C = \frac{fCG}{2} \tag{3.53}$$

or in dimensionless form,

$$\frac{h_i}{CG} = \frac{f}{2}$$

It is interesting to note in Eq. (3.53) that an equation has been obtained for the heat-transfer coefficient which involves the friction factor and which can be determined from an experiment in which no heat was transferred. Like most derivations which require a number of assumptions, the use of Eq. (3.53) applies only to a small range of fluids, particularly permanent gases.

It was Reynolds as quoted by Stanton[1] who predicted that the coefficient of heat transfer obtained from Eq. (3.53) should be affected by the ratio of the thermal conductivity and viscosity of a specific fluid. While the presence of the thermal conductivity suggests the influence of conduction, it was observed experimentally that the entire cross section of a fluid in turbulent flow is not turbulent. Instead it has been found that a laminar layer exists near the pipe wall through which conduction must occur. Prandtl[2] and Taylor[3] independently included this layer. If it is assumed that the layer has a thickness b and that the temperature of its inner circumference is t', the heat flow per square foot of layer is given by

$$Q = \frac{k(t' - t_p)}{b} \tag{3.54}$$

where k is the mean conductivity for the layer. Assume that the transfer of heat and momentum is carried through this layer by molecular motion without disturbing its laminar flow. The inner surface of the layer moves with a velocity u' in streamline flow, and writing $(u - u')$ for u in Eq. (3.53),

$$Q = h_i(T - t') = \frac{\tau C(T - t')}{(u - u')} \tag{3.55}$$

From the definition of viscosity given by Eq. (3.4),

$$\tau = \mu \frac{u'}{b} \tag{3.56}$$

where μ is the viscosity of the fluid in the layer. From Eq. (3.54),

$$t' - t_p = \frac{Qb}{k} = \frac{\mu u' Q}{\tau k} \tag{3.57}$$

And from Eq. (3.55),

$$T - t' = \frac{Q(u - u')}{\tau C} \tag{3.58}$$

Combining Eqs. (3.57) and (3.58),

$$T - t_p = Q\left(\frac{\mu u'}{\tau k} + \frac{u}{\tau C} - \frac{u'}{\tau C}\right) \tag{3.59}$$

---

[1] Stanton, T. E., *Phil. Trans. Roy. Soc. (London)*, **A 190**, 67–88 (1897).
[2] Prandtl, L., *Physik. Z.*, **29**, 487–489 (1928).
[3] Taylor, G. I., *Brit. Adv. Comm. Aero., Rept. and Memo* 272 (1917).

The corrected value of h_i becomes

$$h_i = \frac{Q}{(T - t_p)} = \frac{\tau}{\dfrac{\mu u'}{k} + \dfrac{u}{C} - \dfrac{u'}{C}} \qquad (3.60)$$

Substituting r' for the ratio u'/u and eliminating τ by means of Eq. (3.48),

$$h_i = \frac{f\,CG}{2\,[1 - r' + r'(C\mu/k)]} \qquad (3.61)$$

or in dimensionless form using c for C and h for h_i as usually given in the literature,

$$\frac{h}{cG} = \frac{f}{2}\frac{1}{1 - r' + r'(c\mu/k)} \qquad (3.62)$$

Equation (3.61) is the Prandtl modification of the Reynolds analogy, which is sometimes called the *Prandtl analogy*. The dimensionless Prandtl group $c\mu/k$ has appeared earlier in this chapter, and when it is numerically equal to 1.0, Eq. (3.61) reduces to Eq. (3.53). This is approximately the case for permanent gases. While Eq. (3.61) is a notable extension of the Reynolds analogy, it too has definite limitations. Modern theory now presumes that the distribution of velocities no longer ends abruptly at the laminar layer but that there is instead a buffer layer within the laminar layer in which the transition occurs. Other extensions of the analogies also appear in the literature.

PROBLEMS

3.1. The heat-transfer coefficient h from a hot horizontal pipe to a gas by free convection has been found to be influenced by the specific heat c, thermal conductivity k, density ρ, viscosity μ, thermal coefficient of expansion β of the gas, pipe diameter D, gravitational constant g, and the temperature difference Δt between the surface of the pipe and the main body of the gas. Establish the form of a dimensionless equation for the heat-transfer coefficient.

3.2. The heat-transfer coefficient for the condensation of a vapor on a horizontal pipe has been found to be influenced by the diameter of the pipe D, acceleration of gravity g, the temperature difference Δt between the saturated vapor and the pipe wall, thermal conductivity k, latent heat of vaporization λ, viscosity μ, and the density ρ of the vapor. Establish a dimensionless expression for the heat-transfer coefficient.

3.3. The rate of cooling of a hot solid in still air has been found to be influenced by the specific heat c, thermal conductivity k, density ρ and viscosity μ of the gas, the length of the solid l, and the temperature difference Δt between the surface of the solid and the bulk temperature of the gas. Establish a dimensionless equation for the rate of cooling h.

3.4. When a fluid flows around a sphere, the force exerted by the fluid has been found to be a function of the viscosity μ, density ρ and velocity u of the gas, and the diameter D of the sphere. Establish an expression for the pressure drop of the fluid as a function of the Reynolds number of the gas.

3.5. The heating of gas oil and straw oil in a pipe has been found for a ½-in. IPS pipe to follow Eq. (3.42).

$$\frac{h_i D}{k} = 0.0115 \left(\frac{DG}{\mu}\right)^{0.90} \left(\frac{c\mu}{k}\right)^{\frac{1}{3}}$$

It is desired to circulate 5600 lb/hr of the 36.8°API gas oil through a 1-in. IPS pipe while its temperature is raised from 110 to 130°F. In the absence of any additional heat-transfer data, calculate the heat-transfer coefficient in the 1-in. IPS pipe. How does this compare with the value of h_i when the same quantity of gas oil flows in a ½-in. IPS pipe as calculated by the same equation? The data will be found in the illustration in the chapter.

3.6. Using Eq. (3.42), it is desired to circulate 4000 lb/hr of amyl acetate through a ¾-in. IPS pipe while its temperature is raised from 130 to 150°F. (a) From data available in the Appendix on the physical properties of amyl acetate calculate the heat-transfer coefficient. It may be necessary to extrapolate some of the data. (b) Do the same for 6000 lb/hr of ethylene glycol in the same pipe when heated from 170 to 200°F.

If only one point is given for a property, such as the thermal conductivity, and it is lower than the average temperature, its use will actually introduce a slight factor of safety.

3.7. On a ½-in. IPS pipe 10.125 ft long Morris and Whitman reported the following for heating water in a pipe while steam is circulated on the outside. Note that G', the reported mass velocity, is not in consistent units.

G', lb/(sec)(ft²)	t_1, °F	t_2, °F	t_w, °F
58.6	91.6	181.5	198.4
60.5	92.7	180.3	198.0
84.3	102.2	175.3	196.5
115	103.1	171.3	194.5
118	103.4	168.2	194.9
145	105.0	165.9	194.0
168	107.2	163.3	192.4
171	106.7	164.6	191.1
200	108.5	160.1	190.0
214	106.3	158.9	188.3
216	110.1	160.2	190.2
247	107.6	158.2	186.3

Viscosities and conductivities can be found in the Appendix. The specific heat and gravity should be taken as 1.0. Establish an equation of the form of Eq. (3.26) using all the data. (*Hint.* To save time in the selection of the exponent of the Prandtl number, take three random points such as the first, last, and an intermediate one and solve algebraically.)

3.8. On cooling a 35.8°API gas oil with water Morris and Whitman reported the following for 10.125 ft of ½-in. IPS pipe:

CONVECTION

G', lb/(sec)(ft²)	t_1, °F	t_2, °F	t_w, °F
82.6	150.3	125.5	66.0
115	138.7	118.7	68.0
164	130.5	113.9	70.0
234	141.2	124.2	79.2
253	210.0	179.8	115.4
316	197.6	173.3	122.9
334	132.3	119.4	82.1
335	191.1	168.4	116.3
413	194.4	173.0	121.6
492	132.5	122.1	89.5
562	200.2	182.4	139.4
587	188.6	171.9	127.7
672	190.0	175.1	140.7
682	191.8	176.4	139.5
739	132.6	124.6	97.2

The viscosity of the oil is 2.75 centipoises at 100°F and 1.05 centipoises at 200°F. To obtain intermediate values of the viscosity plot temperature vs. viscosity on logarithmic paper. Establish an equation of the form of Eq. (3.26) using all the data. (The hint given in Prob. 3.7 is applicable.)

3.9. On the cooling of the 29.4°API straw oil with water Morris and Whitman reported the following for 10.125 ft of ½-in. IPS pipe:

G', lb/(sec)(ft²)	t_1, °F	t_2, °F	t_w, °F
141	362.3	296.3	139.8
143	477.6	385.7	194.8
165	317.8	270.0	116.9
172	163.7	148.2	60.2
252	140.5	130.1	62.9
292	142.3	131.5	66.9
394	133.0	124.6	69.6
437	244.9	218.7	141.4
474	130.6	123.1	71.7
474	376.1	330.6	232.9
485	181.1	166.5	102.3
505	134.9	126.8	75.4
556	171.9	159.5	99.8
572	137.9	129.6	80.6
618	310.8	281.6	217.0
633	244.1	222.8	160.4
679	162.4	152.0	99.3
744	196.6	182.7	129.1
761	155.4	146.0	97.6

(a) Establish an equation of the form of Eq. (3.26) for all the data. (b) Combine this data with that of Prob. 3.8 to obtain one correlation for both oils. (c) Calculate j_H

when the exponent of the Prandtl number is ⅓, and plot together with the data for the illustration in the text on heating. What conclusion may be drawn?

3.10. Sieder and Tate[1] obtained data on the cooling of a 21°API oil flowing on the inside of copper tube having an inside diameter of 0.62 in. and 5.1 ft long:

w, lb/hr	t_1, °F	t_2, °F	t_p
1306	136.85	135.15	73.0
1330	138.0	136.2	74.0
1820	160.45	158.5	76.5
1388	160.25	157.9	75.5
231	157.75	149.5	77.0
239	157.5	148.45	78.0
457	212.8	203.2	89.0
916	205.5	200.4	86.0
905	205.0	200.0	85.5
1348	206.35	202.9	87.5
1360	207.6	204.0	87.5
1850	206.9	203.7	88.5
1860	207.0	204.0	90.0
229	141.6	134.65	82.5
885	140.35	138.05	77.0
1820	147.5	146.0	79.8
473	79.6	84.75	118.5
469	80.2	86.5	136.0
460	80.0	82.0	137.0

The temperature t_p is for the inside surface of the tube. Inasmuch as these data fall below a Reynolds number of 2100, obtain an equation of the form of Eq. (3.32). The viscosity of the oil is 24.0 centipoises at 200°F and 250 centipoises at 100°F. Intermediate viscosities can be obtained by drawing a straight line on logarithmic paper as shown in Fig. 3.8. (The hint of Problem 3.7 is applicable.)

NOMENCLATURE FOR CHAPTER 3

A Heat-transfer surface, ft²
a Fluid flow area, ft²
b Thickness of laminar layer, ft
C Specific heat of hot fluid in derivations, Btu/(lb)(°F)
c Specific heat of cold fluid, Btu/(lb)(°F)
D Inside diameter of pipe or tube, ft
F Fundamental dimension of force, force-lb
ΔF Pressure drop, ft of liquid
f Friction factor in the Fanning equation, dimensionless
f' Friction factor, dimensionless
G Mass velocity, lb/(hr)(ft²)

[1] *Ind. Eng. Chem.*, **28**, 1429–1435 (1936).

CONVECTION

g	Acceleration of gravity, ft/hr^2
H	Heat unit, Btu
h	Heat-transfer coefficient in general, Btu/(hr)(ft^2)(°F)
h_i	Heat-transfer coefficient based on the inside pipe surface, Btu/(hr)(ft^2)(°F)
h_o	Heat-transfer coefficient based on the outside pipe surface, Btu/(hr)(ft^2)(°F)
j_H	Factor for heat transfer, dimensionless
K_H	Conversion factor between kinetic energy and heat, (mass-lb)(ft)/Btu
K_M	Conversion factor between force and mass, mass-lb/force-lb
k	Thermal conductivity, Btu/(hr)(ft^2)(°F/ft)
L	Length, ft
M	Fundamental dimension of mass, mass-lb
m	Mass velocity perpendicular to inside pipe surface, lb/(hr)(ft^2)
ΔP	Pressure drop, lb/ft^2
Q	Heat flow, Btu/hr
R_i	Thermal resistance on inside of pipe, (hr)(ft^2)(°F)/Btu
R_o	Thermal resistance on outside of pipe, (hr)(ft^2)(°F)/Btu
r	Radius, ft
r'	Ratio of u'/u, dimensionless
T	Temperature of hot fluid, °F
T_i	Temperature of hot fluid inside a pipe, °F
t'	Temperature of inner surface of laminar layer, °F
t_o	Temperature of cold fluid outside a pipe, °F
t_p	Temperature at inside surface of a pipe, °F
t_w	Temperature at outside surface of a pipe, °F
t_1, t_2	Inlet and outlet cold fluid temperatures, °F
Δt	Temperature difference for heat transfer, °F
Δt_i	Temperature difference between inside pipe fluid and inside pipe wall, °F
Δt_o	Temperature difference between outside pipe fluid and outside pipe wall, °F
u	Velocity in general, ft/hr
u'	Velocity of laminar layer, ft/hr
v	Specific volume, ft^3/lb
W	Weight flow of hot fluid, lb/hr
w	Weight flow of cold fluid, lb/hr
x, y	Coordinates, ft (y is also used to indicate an ordinate)
Y	Distance, ft
α	Any of several proportionality constants, dimensionless
θ	Time, hr
μ	Viscosity, lb/(ft)(hr)
π	Dimensionless group
ρ	Density, lb/ft^3
τ	Shear stress, lb/ft^2
ϕ	Function

Superscripts

p, q	Constants

Subscripts (except as noted above)

i	Inside a pipe or tube
o	Outside a pipe or tube

CHAPTER 4

RADIATION

Introduction. All too often radiation is regarded as a phenomenon incidental only to hot, luminous bodies. In this chapter it will be seen that this is not the case and that radiation, as a third means of transferring heat, differs greatly from conduction and convection. In heat conduction through solids the mechanism consists of an energy transfer through a body whose molecules, except for vibrations, remain continuously in fixed positions. In convection the heat is first absorbed from a source by particles of fluid immediately adjacent to it and then transferred to the interior of the fluid by mixture with it. Both mechanisms require the presence of a medium to convey the heat from a source to a receiver. Radiant-heat transfer does not require an intervening medium, and heat can be transmitted by radiation across an absolute vacuum.

Wavelength and Frequency. It is convenient to mention the characteristics of radiant energy in transit before discussing the origins of radiant energy. Radiant energy is of the same nature as ordinary visible light. It is considered, in accordance with Maxwell's electromagnetic theory, to consist of an oscillating electric field accompanied by a magnetic field oscillating in phase with it. College physics texts usually treat the theory in detail.

The variation with time of the intensity of the electric field passing a given point can be represented by a sine wave having finite length from crest to crest and which is λ, the *wavelength*. The number of waves passing a given point in unit time is the *frequency* of the radiation, and the product of the frequency and wavelength is the velocity of the wave. For travel in a vacuum the velocity of propagation of radiation is very nearly 186,000 miles/sec. For travel through a medium the velocity is somewhat less, although the deviation is generally neglected.

The wavelength of radiation may be specified in any units of length, but the micron, 1×10^{-4} cm, is common. All the known waves included in the electromagnetic theory lie between the short wavelengths of cosmic rays, less than 1×10^{-6} micron, and the long wavelength of radio above 1×10^{7} microns. Of these, only waves in the region between the near and far infrared with wavelengths of $\frac{3}{4}$ to 400 microns are of importance to radiant-heat transfer as found in ordinary industrial equipment.

The Origins of Radiant Energy. Radiant energy is believed to originate within the molecules of the radiating body, the atoms of such molecules vibrating in a simple harmonic motion as linear oscillators. The emission of radiant energy is believed to represent a decrease in the amplitudes of the vibrations within the molecules, while an absorption of energy represents an increase. In its essence the quantum theory postulates that for every frequency of radiation there is a small minimum pulsation of energy which may be emitted. This is the *quantum*, and a smaller quantity cannot be emitted although many such quanta may be emitted. The total radiation of energy of a given frequency emitted by a body is an integral number of quanta at that frequency. For different frequencies, the number of quanta and thus the total energy may be different. Planck showed that the energy associated with a quantum is proportional to the frequency of the vibration or, if the velocity of all radiation is considered constant, inversely proportional to the wavelength. Thus radiant energy of a given frequency may be pictured as consisting of successive pulses of radiant energy, each pulse of which has the value of the quantum for a given frequency.

The picture of the atom proposed by Bohr is helpful to a clearer understanding of one possible origin of radiant energy. Electrons are presumed to travel about the nucleus of an atom in elliptical orbits at varying distances from the nucleus. The outermost orbital electrons possess definite energies comprising their kinetic and potential energies, by virtue of their rotation about the nucleus. The potential energy is the energy required to remove an electron from its orbit to an infinite distance from the nucleus. A given electron in an orbit at a given distance from the nucleus will have a certain energy. Should a disturbance occur such as the collision of the atom with another atom or electron, the given electron may be displaced from its orbit and may (1) return to its original orbit, (2) pass to another orbit whose electrons possess a different energy, or (3) entirely leave the system influenced by the nucleus. If the transition is from an orbit of high energy to one of lower energy, the readjustment is affected by the radiation of the excess energy.

Another origin of radiant energy may be attributed to the changes in the energies of atoms and molecules themselves without reference to their individual electrons. If two or more nuclei of the molecule are vibrating with respect to each other, a change in the amplitude or amplitudes of vibration will cause a change in energy content. A decrease in amplitude is the result of an emission of radiant energy, while an increase is the result of the absorption of radiant energy. The energy of a molecule may be changed by an alteration of its kinetic energy of translation or rotation, and this will likewise result in the emission of radiant energy. A decrease in velocity corresponds to the emission of radiant energy, while an increase corresponds to the absorption of radiant energy.

Since temperature is a measure of the average kinetic energy of molecules, the higher the temperature the higher the average kinetic energy both of translation and of vibration. It can therefore be expected that the higher the temperature the greater the quantity of radiant energy emitted from a substance. Since molecular movement ceases completely only at the absolute zero of temperature, it may be concluded that all substances will emit or absorb radiant energy provided the temperature of the substances is above absolute zero.

For radiant energy to be emitted from the interior of a solid it must penetrate the surface of the solid without being dissipated by producing other energy changes within its molecules. There is little probability that radiant energy generated in the interior of a solid will reach its surface without encountering other molecules, and therefore all radiant energy emitted from the surfaces of solid bodies is generated by energy-level changes in molecules near and on their surfaces. The quantity of radiant energy emitted by a solid body is consequently a function of the surface of the body, and conversely, radiation incident on solid bodies is absorbed at the surface. The probability that internally generated radiant energy will reach the surface is far greater for hot radiating gases than for solids, and the radiant energy emitted by a gas is a function of the gas volume rather than the surface of the gas shape. In liquids the situation is intermediate between gases and solids, and radiation may originate somewhat below the surface, depending on the nature of the liquid.

The Distribution of Radiant Energy. A body at a given temperature will emit radiation of a whole range of wavelengths and not a single wavelength. This is attributed to the existence of an infinite variety of linear oscillators. The energy emitted at each wavelength can be determined through the use of a dispersing prism and thermopiles. Such measurements on a given body will produce curves as shown in Fig. 4.1 for each given temperature. The curves are plots of the intensities of the radiant energy I_λ Btu/(hr)(ft²)(micron) against the wavelengths in microns λ as determined at numerous wavelengths and connecting points. For any given temperature each curve possesses a wavelength at which the amount of spectral energy given off is a maximum. For the same body at a lower temperature the maximum intensity of radiation is obviously less, but it is also significant that the wavelength at which the maximum occurs is longer. Since the curve for a single temperature depicts the amount of energy emitted for a single wavelength, the area under the curve must equal the sum of all the energy radiated by the body at all its wavelengths. The maximum intensity falls between ¾ and 400 microns, indicating that red heat is a far better source of energy than white heat. Were it not for this fact, the near-white incandescent lamp would require more energy for illumination and give off uncomfortable quantities of heat.

When dealing with the properties of radiation, it is necessary to differentiate between two kinds of properties: monochromatic and total. A monochromatic property, such as the maximum values of I_λ in Fig. 4.1, refers to a single wavelength. A total property indicates that it is the

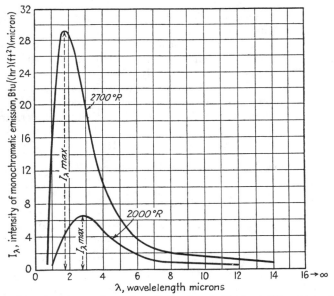

FIG. 4.1. Intensity of monochromatic radiation for a hot body at different temperatures.

algebraic sum of the monochromatic values of the property. Monochromatic radiation literally means "one color" or one wavelength, but experimentally it actually refers to a group or band of wavelengths, since wavelengths cannot be resolved individually. Monochromatic values are not important to the direct solution of engineering problems but are necessary for the derivation of basic radiation relationships.

The Emissive Power. The total quantity of radiant energy of all wavelengths emitted by a body per unit area and time is the *total emissive power* E, Btu/(hr)(ft²). If the intensity of the radiant energy at any wavelength in Fig. 4.1 is I_λ Btu/(hr)(ft²)(micron), the total emissive power is the area under the curve and may be computed by

$$E = \int_0^\infty I_\lambda \, d\lambda \tag{4.1}$$

A relationship between I_λ and λ was the subject of many investigations both experimental and mathematical during the nineteenth century. Planck was the first to recognize the quantum nature of radiant energy and developed an equation which fits the spectral energy curve of Fig. 4.1 at any temperature. It is given by

$$I_\lambda = \frac{C_1 \lambda^{-5}}{e^{C_2/\lambda T} - 1} \tag{4.2}$$

where I_λ = monochromatic intensity of emission, Btu/(hr)(ft²)(micron)
λ = wavelength, microns
C_1 and C_2 = constants with values 1.16×10^8 and 25,740
T = temperature of the body, °R

Wien postulated another law known as Wien's displacement law, which states that the product of the wavelength of the maximum value of the monochromatic intensity of emission and the absolute temperature is a constant, or

$$\lambda T = 2884 \text{ micron }°R \tag{4.3}$$

Equation (4.3) may be derived from Eq. (4.2) as follows,

$$dI_\lambda = d\left(\frac{C_1 \lambda^{-5}}{e^{C_2/\lambda T} - 1}\right) = 0 \tag{4.4}$$

$$(-5C_1\lambda^{-6})(e^{C_2/\lambda T} - 1) + C_1\lambda^{-5}(e^{C_2/\lambda T})\frac{C_2}{\lambda^2 T} = 0$$

$$\left(-5 + \frac{C_2}{\lambda T}\right)e^{C_2/\lambda T} + 5 = 0 \tag{4.5}$$

By trial and error, the first term equals -5 when $\lambda T = 2884$.

Spectral measurements of the radiation received on the earth's surface from the sun and allowing for absorption by the atmosphere indicate that the peak I_λ lies approximately at 0.25 micron, which is well in the ultraviolet. This accounts for the high ultraviolet content of the sun's rays and the predominance of blue in the visible portion of the spectrum. The location of the peak allows an estimation of the sun's temperature from Eq. (4.3) at 11,000°F.

The Incidence of Radiant Energy: The Black Body. The preceding discussion has dealt with the generation of radiant energy. What happens when radiant energy falls upon a body? In the simple case of light it may be all or partially absorbed or reflected. If the receiving medium is transparent to the radiation, it will transmit some of the energy through itself. The same effects are applicable to radiant energy, and an energy balance about a receiver on which the total incident energy is unity is

RADIATION

given by
$$a + r + \tau = 1 \tag{4.6}$$

where the *absorptivity* a is the fraction absorbed, the *reflectivity* r the fraction reflected, and the *transmissivity* τ the fraction transmitted. The majority of engineering materials are opaque substances having zero transmissivities, but there are none which completely absorb or reflect all the incident energy. The substances having nearly complete or unit absorptivities are lampblack, platinum black, and bismuth black, absorbing 0.98 to 0.99 of all incident radiation.

If an ordinary body emits radiation to another body, it will have some of the emitted energy returned to itself by reflection. When Planck developed Eq. (4.2), he assumed that none of the emitted energy was returned, this was equivalent to an assumption that bodies having zero transmissivity also had zero reflectivity. This is the concept of the perfect *black body* and for which $a = 1.0$.

Relationship between Emissivity and Absorptivity: Kirchhoff's Law. Consider a body of given size and shape placed within a hollow sphere of constant temperature, and assume that the air has been evacuated. After thermal equilibrium has been reached, the temperature of the body and that of the enclosure will be the same, inferring that the body is absorbing and radiating heat at identical rates. Let the intensity of radiation falling on the body be I Btu/(hr)(ft^2), the fraction absorbed a_1, and the total emissive power E_1 Btu/(hr)(ft^2). Then the energy emitted by the body of total surface A_1 is equal to that received, or

$$E_1 A_1 = I a_1 A_1 \tag{4.7}$$
$$E_1 = I a_1 \tag{4.8}$$

If the body is replaced by another of identical shape and equilibrium is again attained,

$$E_2 = I a_2 \tag{4.9}$$

If a third body, a black body, is introduced, then

$$E_b = I a_b \tag{4.10}$$

But by definition the absorptivity of a black body is 1.0.

$$\frac{E_1}{a_1} = \frac{E_2}{a_2} = E_b \tag{4.11}$$

or at thermal equilibrium the ratio of the total emissive power to the absorptivity for all bodies is the same. This is known as Kirchhoff's law. Since the maximum absorptivity of the black body is taken as 1.0 from Eq. (4.6), its reflectivity must be zero. Absolute values of the total emissive power are not obtainable, but

$$E_1 = a_1 E_b \qquad (4.12)$$
$$E_2 = a_2 E_b \qquad (4.13)$$
$$\frac{E_1}{E_b} = a_1 = \epsilon_1 \qquad (4.14)$$
$$\frac{E_2}{E_b} = a_2 = \epsilon_2 \qquad (4.15)$$

The use of the ratio of the actual emissive power to the black-body emissive power under identical conditions is the *emissivity* ϵ. Since it is the reference, the emissivity of the black body is unity. The emissivities of common materials cover a large range and are tabulated in Table 4.1. Emissivities are influenced by the finish or polish of the surface and increase with its temperature. Highly polished and white surfaces generally have lower values than rough or black surfaces. From Eq. (4.12) it can be seen that any body having a high emissivity as a radiator will have a high absorptivity when acting as a receiver. The usual statement is as follows: Good radiators make good absorbers.

Experimental Determination of the Emissivity. The experimental determination of the emissivities of materials is particularly difficult at high temperatures. The problems of maintaining a system free of conduction, convection, and a radiation-absorbing medium require careful analysis. A method is given here which is satisfactory for the measurement of emissivities in the range of room temperatures and might be applicable to the calculation of such problems as the loss of heat from a pipe to air by radiation alone. A hollow, opaque, internally blackened cylinder is maintained in a constant-temperature bath as shown in Fig. 4.2. A total radiation receiver is mounted by a bracket to the wall of the cylinder. The radiation receiver consists of a copper cylinder a, which is blackened on the inside and highly polished on the outside. Two extremely thin, blackened, and highly conducting copper discs b and b' are mounted in the receiver for the purpose of absorbing radiation. By mounting the discs at equal distances from the top and bottom of the small cylinder, the angles α_1 and α_2 are equal and the discs have equal areas for receiving radiation. The lower disc receives radiation from the blackened constant-temperature walls of the vessel. The upper disc receives radiation from a plate of specimen material c which is electrically maintained at a fixed temperature. The two discs are wired together by a sensitive thermocouple so that they oppose each other, and only net differences in the quantity of radiation are measured by the galvanometer. By wiring them to oppose each other, any effects within the receiver itself are also canceled. If the galvanometer deflection for the specimen nonblack body is measured and then c is replaced by a perfect black body, the ratio of the two galvanometer deflections is the emissivity of the specimen. Data obtained in this manner are the *normal total emissivity* as given in Table 4.1. They may also be used in the solution of problems having hemispherical radiation except in the presence of highly polished surfaces.

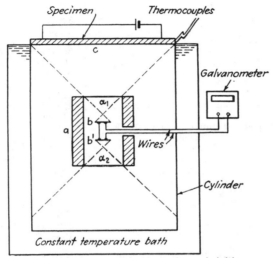

FIG. 4.2. Apparatus for measuring emissivities.

Influence of Temperature on the Emissive Power: Stefan-Boltzmann Law. If a perfect black body radiates energy, the total radiation may be determined from Planck's law. Starting with the monochromatic black-body equation

$$I_\lambda = \frac{C_1 \lambda^{-5}}{e^{C_2/\lambda T} - 1}$$

it may be applied to sum up all the energy by integration of the area under a curve in Fig. 4.1 or at a fixed temperature

$$E = \int_0^\infty \frac{C_1 \lambda^{-5}}{e^{C_2/\lambda T} - 1} \, d\lambda \tag{4.16}$$

Let $x = C_2/\lambda T$, $\lambda = C_2/Tx$, $d\lambda = (-C_2/Tx^2)\, dx$, from which

$$E_b = -\frac{C_1 T^4}{C_2^4} \int_0^\infty x^3 (e^x - 1)^{-1} \, dx \tag{4.17}$$

TABLE 4.1. THE NORMAL TOTAL EMISSIVITY OF VARIOUS SURFACES (HOTTEL)
A. Metals and Their Oxides

Surface	t, °F*	Emissivity*
Aluminum		
Highly polished plate, 98.3% pure	440–1070	0.039–0.057
Polished plate	73	0.040
Rough plate	78	0.055
Oxidized at 1110°F	390–1110	0.11–0.19
Al-surfaced roofing	100	0.216
Calorized surfaces, heated at 1110°F		
Copper	390–1110	0.18–0.19
Steel	390–1110	0.52–0.57
Brass		
Highly polished		
73.2% Cu, 26.7% Zn	476–674	0.028–0.031
62.4% Cu, 36.8% Zn, 0.4% Pb, 0.3% Al	494–710	0.033–0.037
82.9% Cu, 17.0% Zn	530	0.030
Hard rolled, polished, but direction of polishing visible	70	0.038
but somewhat attacked	73	0.043
but traces of stearin from polish left on	75	0.053
Polished	100–600	0.096–0.096
Rolled plate, natural surface	72	0.06
Rubbed with coarse emery	72	0.20
Dull plate	120–660	0.22
Oxidized by heating at 1110°F	390–1110	0.61–0.59
Chromium (see Nickel alloys for Ni-Cr steels)	100–1000	0.08–0.26
Copper		
Carefully polished electrolytic copper	176	0.018
Commercial emeried, polished, but pits remaining	66	0.030
Commercial, scraped shiny but not mirrorlike	72	0.072
Polished	242	0.023
Plate, heated long time, covered with thick oxide layer	77	0.78
Plate heated at 1110°F	390–1110	0.57–0.57
Cuprous oxide	1470–2010	0.66–0.54
Molten copper	1970–2330	0.16–0.13
Gold		
Pure, highly polished	440–1160	0.018–0.035
Iron and steel		
Metallic surfaces (or very thin oxide layer)		
Electrolytic iron, highly polished	350–440	0.052–0.064
Polished iron	800–1880	0.144–0.377
Iron freshly emeried	68	0.242
Cast iron, polished	392	0.21
Wrought iron, highly polished	100–480	0.28
Cast iron, newly turned	72	0.435
Polished steel casting	1420–1900	0.52–0.56
Ground sheet steel	1720–2010	0.55–0.61
Smooth sheet iron	1650–1900	0.55–0.60
Cast iron, turned on lathe	1620–1810	0.60–0.70

RADIATION

TABLE 4.1. THE NORMAL TOTAL EMISSIVITY OF VARIOUS SURFACES (HOTTEL).—
(*Continued*)

Surface	t, °F*	Emissivity*
Iron and steel—(*Continued*)		
Oxidized surfaces		
Iron plate, pickled, then rusted red............	68	0.612
Completely rusted........................	67	0.685
Rolled sheet steel........................	70	0.657
Oxidized iron............................	212	0.736
Cast iron, oxidized at 1100°F................	390–1110	0.64–0.78
Steel, oxidized at 1100°F...................	390–1110	0.79–0.79
Smooth oxidized electrolytic iron.............	260–980	0.78–0.82
Iron oxide...............................	930–2190	0.85–0.89
Rough ingot iron..........................	1700–2040	0.87–0.95
Sheet steel, strong, rough oxide layer........	75	0.80
Dense, shiny oxide layer..................	75	0.82
Cast plate, smooth........................	73	0.80
Rough................................	73	0.82
Cast iron, rough, strongly oxidized...........	100–480	0.95
Wrought iron, dull oxidized.................	70–680	0.94
Steel plate, rough........................	100–700	0.94–0.97
High-temperature alloy steels (see Nickel alloys)		
Molten metal		
Cast iron................................	2370–2550	0.29–0.29
Mild steel...............................	2910–3270	0.28–0.28
Lead		
Pure (99.96%), unoxidized...................	260–440	0.057–0.075
Gray oxidized...........................	75	0.281
Oxidized at 390°F.........................	390	0.63
Mercury...................................	32–212	0.09–0.12
Molybdenum filament........................	1340–4700	0.096–0.292
Monel metal, oxidized at 1110°F...............	390–1110	0.41–0.46
Nickel		
Electroplated on polished iron, then polished.....	74	0.045
Technically pure (98.9% Ni, + Mn), polished....	440–710	0.07–0.087
Electroplated on pickled iron, not polished......	68	0.11
Wire...................................	368–1844	0.096–0.186
Plate, oxidized by heating at 1110°F...........	390–1110	0.37–0.48
Nickel oxide.............................	1200–2290	0.59–0.86
Nickel alloys		
Chromnickel.............................	125–1894	0.64–0.76
Nickelin (18–32 Ni; 55–68 Cu; 20 Zn), gray oxidized	70	0.262
KA-2S alloy steel (8% Ni; 18% Cr), light silvery, rough, brown, after heating............	420–914	0.44–0.36
after 42 hr. heating at 980°F...............	420–980	0.62–0.73
NCT-3 alloy (20 Ni; 25 Cr). Brown, splotched, oxidized from service.....................	420–980	0.90–0.97
NCT-6 alloy (60 Ni; 12 Cr). Smooth, black, firm adhesive oxide coat from service........	520–1045	0.89–0.82
Platinum		
Pure, polished plate........................	440–1160	0.054–0.104
Strip...................................	1700–2960	0.12–0.17
Filament................................	80–2240	0.036–0.192
Wire...................................	440–2510	0.073–0.182

TABLE 4.1. THE NORMAL TOTAL EMISSIVITY OF VARIOUS SURFACES (HOTTEL).—
(*Continued*)

Surface	t, °F*	Emissivity*
Silver		
Polished, pure	440–1160	0.0198–0.0324
Polished	100–700	0.0221–0.0312
Steel (see Iron)		
Tantalum filament	2420–5430	0.194–0.31
Tin, bright tinned iron sheet	76	0.043 and 0.064
Tungsten		
Filament, aged	80–6000	0.032–0.35
Filament	6000	0.39
Zinc		
Commercial 99.1% pure, polished	440–620	0.045–0.053
Oxidized by heating at 750°F	750	0.11
Galvanized sheet iron, fairly bright	82	0.228
Galvanized sheet iron, gray oxidized	75	0.276
B. Refractories, Building Materials, Paints, and Miscellaneous		
Asbestos		
Board	74	0.96
Paper	100–700	0.93–0.945
Brick		
Red, rough, but no gross irregularities	70	0.93
Silica, unglazed, rough	1832	0.80
Silica, glazed, rough	2012	0.85
Grog brick, glazed	2012	0.75
(See Refractory materials)		
Carbon		
T-carbon (Gebruder Siemens) 0.9% ash	260–1160	0.81–0.79
This started with emissivity at 260°F of 0.72, but on heating changed to values given		
Carbon filament	1900–2560	0.526
Candle soot	206–520	0.952
Lampblack-water-glass coating	209–362	0.959–0.947
Same	260–440	0.957–0.952
Thin layer on iron plate	69	0.927
Thick coat	68	0.967
Lampblack, 0.003 in. or thicker	100–700	0.945
Enamel, white fused, on iron	66	0.897
Glass, smooth	72	0.937
Gypsum, 0.02 in. thick on smooth or blackened plate	70	0.903
Marble, light gray, polished	72	0.931
Oak, planed	70	0.895
Oil layers on polished nickel (lub. oil)	68	
Polished surface, alone	....	0.045
+0.001 in. oil	....	0.27
+0.002 in. oil	....	0.46
+0.005 in. oil	....	0.72
∞ thick oil layer	....	0.82

TABLE 4.1. THE NORMAL TOTAL EMISSIVITY OF VARIOUS SURFACES (HOTTEL).—(*Continued*)

Surface	t, °F*	Emissivity*
Oil layers on aluminum foil (linseed oil)		
Al foil	212	0.087†
+1 coat oil	212	0.561
+2 coats oil	212	0.574
Paints, lacquers, varnishes		
Snow-white enamel varnish on rough iron plate	73	0.906
Black shiny lacquer, sprayed on iron	76	0.875
Black shiny shellac on tinned iron sheet	70	0.821
Black matte shellac	170–295	0.91
Black lacquer	100–200	0.80–0.95
Flat black lacquer	100–200	0.96–0.98
White lacquer	100–200	0.80–0.95
Oil paints, 16 different, all colors	212	0.92–0.96
Aluminum paints and lacquers		
10% Al, 22% lacquer body, on rough or smooth surface	212	0.52
26% Al, 27% lacquer body, on rough or smooth surface	212	0.3
Other Al paints, varying age and Al content	212	0.27–0.67
Al lacquer, varnish binder, on rough plate	70	0.39
Al paint, after heating to 620°F	300–600	0.35
Paper, thin		
Pasted on tinned iron plate	66	0.924
Rough iron plate	66	0.929
Black lacquered plate	66	0.944
Plaster, rough lime	50–190	0.91
Porcelain, glazed	72	0.924
Quartz, rough, fused	70	0.932
Refractory materials, 40 different	1110–1830	
Poor radiators	....	$\left.\begin{array}{c}0.65\\0.70\end{array}\right\}$ –0.75
Good radiators	....	$\left.\begin{array}{c}0.80\\0.85\end{array}\right\}\left\{\begin{array}{c}0.85\\0.90\end{array}\right.$
Roofing paper	69	0.91
Rubber		
Hard, glossy plate	74	0.945
Soft, gray, rough (reclaimed)	76	0.859
Serpentine, polished	74	0.900
Water	32–212	0.95–0.963

NOTE. The results of many investigators have been omitted because of obvious defects in experimental method.

* When two temperatures and two emissivities are given, they correspond, first to first and second to second, and linear interpolation is permissible.

† Although this value is probably high, it is given for comparison with the data, by the same investigator, to show the effect of oil layers (see Aluminum, part A of this table).

Expanding the term in parentheses,

$$E_b = -\frac{C_1 T^4}{C_2^4} \int_0^\infty x^3(e^{-x} + e^{-2x} + e^{-3x} + e^{-4x} + \cdots)\,dx \quad (4.18)$$

Integrating each term and summing only the first four as significant,

$$E_b = \frac{C_1 T^4}{C_2^4} \times 6.44 \quad (4.19)$$

Evaluating constants,

$$E_b = 0.173 \times 10^{-8} T^4 \quad (4.20)$$

Equation (4.20) being the area under a curve in Fig. 4.1 from $\lambda = 0$ to $\lambda = \infty$ states that the total radiation from a perfect black body is proportional to the fourth power of the absolute temperature of the body. This is known as the Stefan-Boltzmann law. The constant 0.173×10^{-8} Btu/(hr)(ft^2)(°R^4) is known as the Stefan-Boltzmann constant, usually designated by σ. The equation was also deduced by Boltzmann from the second law of thermodynamics. Equation (4.20) serves as the principal relationship for the calculation of radiation phenomena and is to radiation what $Q = hA\,\Delta t$ is to convection. However, Eq. (4.20) was derived for a perfect black body. From Eq. (4.14) if a body is nonblack, the emissivity is the ratio E/E_b and E can be written $E = E_b\epsilon$. Equation (4.20) becomes

$$E = \epsilon\sigma T^4 \quad (4.21)$$

and

$$\frac{Q}{A} = \epsilon\sigma T^4 \quad (4.22)$$

Exchange of Energy between Two Large Parallel Planes. Quantitative considerations so far have dealt with the energy change when radiation occurs only from a single body, and it has been assumed that energy once radiated is no longer returned to its source. This is true only if one black body radiates to another black body with no medium between them or no absorption occurs by the medium between them. Of the gases, chlorine, hydrogen, oxygen, and nitrogen are classified as nonabsorbing. Carbon monoxide, carbon dioxide, and organic gases and vapors are absorbing to greater or lesser extents. From earlier discussions it should be conceded that radiation from a small plate proceeds outwardly in a hemisphere with the plate as center and that the radiation incident upon unit area of a body at great distance is very small. In radiation it is necessary to qualify the condition under which all the radiation from the source is completely received by the receiver. It will occur if two plates

RADIATION

or radiant planes are infinitely large, so that the amount of radiation which fans out from the edges of the source and over the edges of the receiver is insignificant. If both of the plates or planes are black bodies, the energy from the first is $E_{b1} = \sigma T_1^4$ and from the second $E_{b2} = \sigma T_2^4$. By the definition of a black body, all the energy it receives is absorbed and the net exchange per square foot between two planes maintained at constant temperatures is

$$\frac{Q}{A} = E_{b1} - E_{b2} = \sigma(T_1^4 - T_2^4) \tag{4.23}$$

$$= 0.173 \left[\left(\frac{T_1}{100}\right)^4 - \left(\frac{T_2}{100}\right)^4 \right] \tag{4.24}$$

Example 4.1. Radiation between Two Large Planes. Two very large walls are at the constant temperature of 800 and 1000°F. Assuming they are black bodies, how much heat must be removed from the colder wall to maintain a constant temperature?

Solution:

$$T_1 = 1000 + 460 = 1460°R \qquad T_2 = 800 + 460 = 1260°R$$

$$\frac{Q}{A} = 0.173\,[(14.6)^4 - (12.6)^4] = 3{,}500 \text{ Btu/(hr)(ft}^2)$$

Exchange of Energy between Two Parallel Planes of Different Emissivity. The preceding discussion applied to black bodies. If the two planes are not black bodies and have different emissivities, the net exchange of energy will be different. Some of the energy emitted from the first plane will be absorbed, and the remainder radiated back to the source. For two walls of infinite size the radiation of each wall can be traced. Thus if energy is emitted by the first wall per square foot of an amount E_1 and emissivity ϵ_1, the second wall will absorb $E_1\epsilon_2$ and reflect $1 - \epsilon_2$ of it. The first wall will then again radiate but in the amount $E_1(1 - \epsilon_2)(1 - \epsilon_1)$. The changes at the two planes are

Hot plane
Radiated: E_1
Returned: $E_1(1 - \epsilon_2)$
Radiated: $E_1(1 - \epsilon_2)(1 - \epsilon_1)$
Returned: $E_1(1 - \epsilon_2)(1 - \epsilon_1)(1 - \epsilon_2)$

Cold plane
Radiated: E_2
Returned: $E_2(1 - \epsilon_1)$
Radiated: $E_2(1 - \epsilon_1)(1 - \epsilon_2)$
Returned: $E_2(1 - \epsilon_1)(1 - \epsilon_2)(1 - \epsilon_1)$

and
$$\frac{Q}{A} = [E_1 - E_1(1 - \epsilon_2) - E_1(1 - \epsilon_1)(1 - \epsilon_2)^2$$
$$- E_1(1 - \epsilon_2)(1 - \epsilon_1)^2(1 - \epsilon_2)^2 + \cdots] - [\epsilon_1 E_2$$
$$- \epsilon_1 E_2(1 - \epsilon_1)(1 - \epsilon_2) - \epsilon_1 E_2(1 - \epsilon_1)^2(1 - \epsilon_2)^2 + \cdots] \quad (4.25)$$

E_1 is given by $\epsilon_1 \sigma T_1^4$, E_2 by $\epsilon_2 \sigma T_2^4$, and Eq. (4.25) is a series whose solution is

$$\frac{Q}{A} = \frac{\sigma}{(1/\epsilon_1) + (1/\epsilon_2) - 1}(T_1^4 - T_2^4) \quad (4.26)$$

Example 4.2. Radiation between Planes with Different Emissivities. If the two walls in Example 4.1 have emissivities of 0.6 and 0.8, respectively, what is the net exchange?

Solution:

$$\frac{Q}{A} = \frac{0.173}{(1/0.6) + (1/0.8) - 1}[(14.6)^2 - (12.6)^2] = 1825 \text{ Btu/(hr)(ft}^2)$$

For perfect black bodies the value was 3500 Btu/(hr)(ft^2).

Radiation Intercepted by a Shield. Suppose two infinite and parallel planes are separated by a third plane which is opaque to direct radiation between the two and which is extremely thin (or has infinite thermal conductivity) as shown in Fig. 4.3. The net exchange between the two initial planes is given by Eq. (4.26).

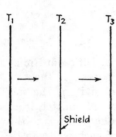

FIG. 4.3. Radiation with a shield.

$$Q = \frac{\sigma A}{(1/\epsilon_1) + (1/\epsilon_3) - 1}(T_1^4 - T_3^4) \quad (4.27)$$

If $\epsilon_1 = \epsilon_3$ but $\epsilon_1 \neq \epsilon_2$, the net exchange from 1 to 3 is given by

$$Q_1 = \frac{\sigma A}{(1/\epsilon_1) + (1/\epsilon_2) - 1}(T_1^4 - T_2^4) = \frac{\sigma A}{(1/\epsilon_2) + (1/\epsilon_3) - 1}(T_2^4 - T_3^4) \quad (4.28)$$

from which

$$T_2^4 = \tfrac{1}{2}(T_1^4 + T_3^4) \quad (4.29)$$

Then

$$Q_1 = \frac{\sigma A}{(1/\epsilon_1) + (1/\epsilon_2) - 1} \frac{1}{2}(T_1^4 - T_3^4) \quad (4.30)$$

When $\epsilon_1 = \epsilon_2$, $Q_1 = \tfrac{1}{2}Q$, and for the simple case where n shields are employed, each having the same emissivities as the initial planes,

$$Q_n = \frac{1}{n+1}Q$$

where Q is the exchange if the initial planes were not separated.

Spheres or Cylinders with Spherical or Cylindrical Enclosures. The radiation between a sphere and an enclosing sphere of radii r_1 and r_2 may be treated in the same manner as Eq. (4.26). The radiation emitted initially by the inner sphere is $E_1 A_1$, all of which falls on A_2. Of this total, however, $(1 - \epsilon_2) E_1 A_1$ is reflected of which $\left(\dfrac{r_1}{r_2}\right)^2 (1 - \epsilon_2) E_1 A_1$ falls on A_1 and $\left[1 - \left(\dfrac{r_1}{r_2}\right)^2\right] (1 - \epsilon_2) E_1 A_1$ falls on A_2. If this analysis is continued as before, the energy exchange will again be represented by a geometrical series and the net exchange between the inner and outer sphere is given by

$$Q = \frac{\sigma A_1}{\dfrac{1}{\epsilon_1} + \left(\dfrac{r_1}{r_2}\right)^2 \left(\dfrac{1}{\epsilon_2} - 1\right)} (T_1^4 - T_2^4) = \frac{\sigma A_1}{\dfrac{1}{\epsilon_1} + \dfrac{A_1}{A_2}\left(\dfrac{1}{\epsilon_2} - 1\right)} (T_1^4 - T_2^4) \quad (4.31)$$

The same relation will be seen to hold for infinitely long concentric cylinders except that A_1/A_2 is r_1/r_2 instead of r_1^2/r_2^2.

Radiation of Energy to a Completely Absorbing Receiver. In Example 2.5, calculations were made on the loss of heat from a pipe to air. When a heat source is small by comparison with the enclosure it is customary to make the simplifying assumption that none of the heat radiated from the source is reflected to it. In such cases, Eq. (4.26) reduces to

$$\frac{Q}{A_1} = \epsilon_1 \sigma (T_1^4 - T_2^4) \quad (4.32)$$

Sometimes it is convenient to represent the net effect of the radiation in the same form employed in convection; namely,

$$Q = h_r A_1 (T_1 - T_2) \quad (4.33)$$

where h_r is a fictitious film coefficient representing the rate at which the radiation passes from the surface of the radiator. The values of Q in Eqs. (4.32) and (4.33) are identical, but the value in Eq. (4.32) is related to the mechanism by which the heat was transferred. Eq. (4.33) is a statement of the heat balance as applied before in the Fourier equation to conduction and convection. Fishenden and Saunders[1] have treated a number of interesting aspects of the subject.

[1] Fishenden, M., and O. A. Saunders, "The Calculation of Heat Transmission," His Majesty's Stationery Office, London, 1932.

Example 4.3. Calculation of Radiation from a Pipe. In Example 2.5 the outside temperature of a lagged pipe carrying steam at 300°F was 125°F and the surrounding atmosphere was at 70°F. The heat loss by free convection and radiation was 103.2 Btu/(hr)(lin ft), and the combined coefficient of heat transfer was 2.10 Btu/(hr)(ft²)(°F). How much of the heat loss was due to radiation, and what was the equivalent coefficient of heat transfer for the radiation alone?

$$\text{Area/lin ft} = \pi \times \frac{3.375}{12} \times 1 = 0.88 \text{ ft}^2$$

From Table 4.1B, the emissivity is approximately 0.90.

$$T_1 = 125 + 460 = 585°R \qquad T_2 = 70 + 460 = 530°R$$
$$q = 0.90 \times 0.88 \times 0.173[(585/100)^4 - (530/100)^4] = 52.5 \text{ Btu/(hr)(lin ft)}$$
$$h_r = \frac{Q}{A(T_1 - T_2)} = \frac{52.5}{0.88(125 - 70)} = 1.08 \text{ Btu/(hr)(ft}^2)(°F)$$

Exchange of Energy between Any Source and Any Receiver. The three preceding illustrations have been extremely limited. The study of two planes was directed only to sources and receivers which were infinitely large so that every point on one plane could be connected with every point on the second and no radiation from the one "leaked" past the other and out of the system. A slightly more complex arrangement can be achieved between two concentric spheres or two concentric cylinders. In either of these all the radiation from the source may be seen to fall on the receiver. But this is very rarely the case in practical engineering problems, particularly in the design of furnaces. The receiving surface, such as the banks of tubes, is cylindrical and may partially obscure some of the surface from "seeing" the source.

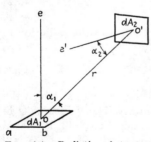

FIG. 4.4. Radiation between two plates.

In a system composed of walls and pipes running in different arrangements the geometry by which the radiation falls on the surfaces and the manner in which these surfaces reflect their energy are difficult to evaluate. The simplest elements are treated here, but many practical applications require the empirical methods of Chap. 19.

Consider the arrangement of two radiating plates at temperatures T_1 and T_2 as shown in Fig. 4.4. The two plates are not face-to-face and therefore have only an askance view of each other. The lower plate, represented isometrically in the horizontal plane, radiates in all directions upward and outward. Some of the radiation from the surface of the hot plate, dA_1 falls upon the second plate but not perpendicular to it. The second plate dA_2 will

RADIATION

reflect some of the incident energy, but only a part of it will return to the first plate. What is the net exchange of energy between the two?

The lines eo and $e'o'$ are perpendiculars to the two plates, respectively, on their mutually exposed surfaces. The length of the shortest line joining the two plates at their centers is r. When viewed from o', a foreshortened picture of dA_1 is obtained. Instead of isometrically, the plate dA_1 may be viewed from its end as in Fig. 4.5, where dA_2 is assumed perpendicular to the plane of the paper for simplicity, the line ab represents a side of dA_1, and the line $a'b$ represents the width of ab in the view obtained from o'. Since eo and oo' are mutually perpendicular to its

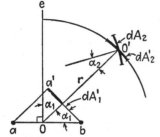 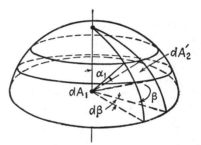

FIG. 4.5. Section view between two plates.   FIG. 4.6. Solid angle.

sides, angle aba' must equal α_1 and the side $a'b$ corresponding to dA_1' is $dA_1' = dA_1 \cos \alpha_1$. For the second plate, $dA_2' = dA_2 \cos \alpha_2$. The center of surface dA_2' may be considered to lie in the hemisphere receiving radiation from dA_1, and the quantity falling on a surface in the hemisphere consequently diminishes with the square of the radius. If both are black bodies, the radiation from plate 1 to plate 2 is proportional to the normal surfaces exposed to each other and inversely to the square of the distance between them.

$$dQ_{1-2} = \frac{I_1}{r^2} dA_1' dA_2' \qquad (4.34)$$

where I_1 is the proportionality constant dimensionally equal to the intensity of radiation. Substituting the original surfaces,

$$dQ_{1-2} = \frac{I_1}{r^2} \cos \alpha_1 \cos \alpha_2 \, dA_1 \, dA_2 \qquad (4.35)$$

An important relationship exists between the intensity I_1 and the emissive power E. In Fig. 4.6 let $d\omega_1$ be the solid angle which is by definition the intercepted area on a sphere divided by r^2. dA_1 is a small plate in the center of the isometric plane of the base. Then

$$d\omega_1 = \frac{dA_2'}{r^2} = \frac{r \sin \alpha \, d\beta \, r \, d\alpha}{r^2}$$
$$= \sin \alpha \, d\alpha \, d\beta \qquad (4.36)$$

From Eqs. (4.35), (4.35a), and (4.36),

$$\frac{dQ}{dA} = E_b = I_1 \int_0^{\pi/2} \sin \alpha \cos \alpha \, d\alpha \int_0^{2\pi} d\beta$$
$$= I_1 \pi \qquad (4.37)$$
$$I_1 = \frac{E_b}{\pi} = \frac{\sigma T^4}{\pi} \qquad (4.38)$$

Substituting Eq. (4.38) in Eq. (4.35) the net exchange between T_1 and T_2 is

$$dQ = \frac{\cos \alpha_1 \cos \alpha_2 \, dA_1 \, dA_2}{\pi r^2} \sigma(T_1^4 - T_2^4) \qquad (4.39)$$

If $\cos \alpha_1 \cos \alpha_2 \, dA_2/\pi r^2$ is written F_A, F_A is known as the configuration or geometric factor. For some systems it is very difficult to derive, but for

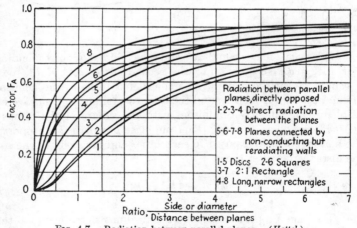

Fig. 4.7. Radiation between parallel planes. (*Hottel.*)

several basic arrangements it is fairly simple. Hottel[1] has integrated a number of cases, the commonest of which are plotted in Figs. 4.7, 4.8 and 4.9. Equation (4.39) can thus be written in integrated form

$$Q = F_A A_1 \sigma(T_1^4 - T_2^4) \qquad (4.40)$$

[1] Hottel, H. C., *Mech. Eng.*, **52**, 699 (1930).

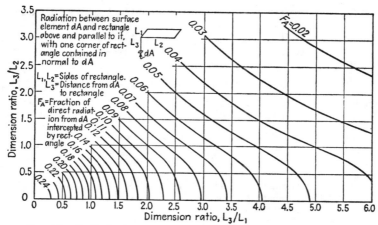

Fig. 4.8. Radiation between an element and a parallel plane. (*Hottel.*)

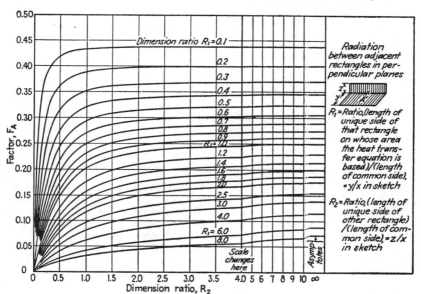

Fig. 4.9. Radiation between perpendicular planes. (*Hottel.*)

If the two surfaces are gray and therefore nonblack, from Eq. (4.26)

$$Q = \frac{F_A A_1 \sigma}{(1/\epsilon_1) + (1/\epsilon_2) - 1}(T_1^4 - T_2^4) \qquad (4.41)$$

Writing F_ϵ for the emissivity correction, Eq. (4.41) becomes

$$Q = F_A F_\epsilon A_1 \sigma (T_1^4 - T_2^4) \qquad (4.42)$$

The summary in Table 4.2 gives the values of F_A and F_ϵ for a number of the common cases derived here and elsewhere.

TABLE 4.2. VALUES OF F_A AND F_ϵ

	F_A	F_ϵ
(a) Surface A_1 small compared with the totally enclosing surface A_2	1	ϵ_1
(b) Surfaces A_1 and A_2 of parallel discs squares, 2:1 rectangles, long rectangles	Fig. 4.7	$\epsilon_1 \epsilon_2$
(c) Surface dA_1 and parallel rectangular surface A_2 with one corner of rectangle above dA_1	Fig. 4.8	$\epsilon_1 \epsilon_2$
(d) Surfaces A_1 or A_2 of perpendicular rectangles having a common side	Fig. 4.9	$\epsilon_1 \epsilon_2$
(e) Surfaces A_1 and A_2 of infinite parallel planes or surface A_1 of a completely enclosed body is small compared with A_2	1	$\dfrac{1}{\left(\dfrac{1}{\epsilon_1} + \dfrac{1}{\epsilon_2}\right) - 1}$
(f) Concentric spheres or infinite concentric cylinders with surfaces A_1 and A_2	1	$\dfrac{1}{\dfrac{1}{\epsilon_1} + \dfrac{A_1}{A_2}\left(\dfrac{1}{\epsilon_2} - 1\right)}$

Example 4.4. Radiation from a Pipe to a Duct. Calculate the radiation from a 2-in. IPS steel pipe carrying steam at 300°F and passing through the center of a 1- by 1-ft galvanized sheet-iron duct at 75°F and whose outside is insulated.

Solution. From Table 11 in the Appendix $A = 0.622$ ft² of external surface per linear foot of pipe. The emissivity of oxidized steel from Table 4.1 is $\epsilon_1 = 0.79$.

The surface of the duct is $A_2 = 4 (1 \times 1) = 4.0$ ft²/lin ft.

The surface of the pipe is not negligible by comparison with that of the duct, and (f) of Table 4.2 applies most nearly.

$\epsilon_2 = 0.276$ (oxidized zinc in Table 4.1)

$F_A = 1$ (Table 4.2)

$F_\epsilon = \dfrac{1}{\dfrac{1}{\epsilon_1} + \dfrac{A_1}{A_2}\left(\dfrac{1}{\epsilon_2} - 1\right)} = \dfrac{1}{\dfrac{1}{0.79} + \dfrac{0.622}{4.0}\left(\dfrac{1}{0.276} - 1\right)} = 0.60$ (Table 4.1)

$Q = F_A F_\epsilon A \sigma (T_1^4 - T_2^4)$ (4.42)

$= 1 \times 0.60 \times 0.622 \times 0.173 \times 10^{-8}(760^4 - 535^4)$

$= 164$ Btu/(hr)(lin ft)

RADIATION

PROBLEMS

4.1. A 2-in. IPS steel pipe carries steam at 325°F through a room at 70°F. What decrease in radiation occurs if the bare pipe is coated with 26 per cent aluminum paint?

4.2. One wall of a corridor 8 by 28 by 4 ft wide between a drying chamber and an outside wall of the building is at 200°F, and the 8- by 28-ft wall of the building will be at 40°F during normal winter weather. What will be the heat passing across the corridor in winter if the drying chamber is surfaced with unglazed silica brick and the building wall is plastered?

4.3. A chamber for heat-curing steel sheets lacquered black on both sides operates by passing the sheets vertically between two steel plates 6 ft apart. One of the plates is at 600°F, and the other, exposed to the atmosphere, is at 80°F. What heat is transferred between the walls, and what is the temperature of the lacquered sheet when equilibrium has been reached?

4.4. A 3-in. IPS lagged pipe carries steam at 400°F through a room at 70°F. The lagging consists of a $\frac{1}{2}$-in. thick layer of asbestos. The use of an overcoat of 26 per cent aluminum paint is to be investigated. What percentage savings in heat loss can be effected?

4.5. A molten organic compound is carried in the smaller of two concentric steel pipes 2- and 3-in. IPS. The annulus can be flooded with steam to prevent solidification, or the liquid can be heated somewhat and circulated without steam so that the annulus acts as an insulator. If the molten fluid is at 400°F and the outer pipe is at room temperature (80°F), what heat loss occurs from the molten material in 40 lin ft of pipe?

4.6. Calculate the radiant-heat loss from a furnace through a 2-in.-diameter peep door when the inside temperature is 1750°F and the outside temperature is 70°F. Consider the emission due to a black body.

4.7. A bare concrete pump house 10 by 20 by 10 ft high is to be heated by pipes laid in the concrete floor. Hot water is to be used as the heating medium to maintain a floor surface temperature of 78°F. The walls and ceiling are of such thickness as to maintain temperatures of 62°F on all inside surfaces in the winter. (a) What is the rate of radiation between the floor and the ceiling if the walls are considered nonconducting and reradiating? The pumps cover a negligible area of the floor. (b) What additional heat will be required if the floor area is doubled by enlarging the room to 20 by 20 ft?

4.8. A bath of molten zinc is located in the corner of the floor of a zinc dipping room 20 by 20 by 10 ft high. Zinc melts at 787°F, and the ceiling could be maintained at 90°F in the summer by conduction through the roof. (a) What heat will be radiated from a 1-ft^2 bath? (b) If the bath is moved to the center of the room, what heat will be radiated?

NOMENCLATURE FOR CHAPTER 4

A	Heat-transfer of emitting or absorbing surface, ft^2
A'	Effective surface, ft^2
a	Absorptivity, dimensionless
C_1, C_2	Constants in Planck's law
E	Emissive power, Btu/(hr)(ft^2)
F_A	Geometric factor, dimensionless
F_ϵ	Emissivity factor, dimensionless
h_r	Heat-transfer coefficient equivalent to radiation, Btu/(hr)(ft^2)(°F)

I	Intensity of radiation, Btu/(hr)(ft^2)
I_λ	Monochromative intensity of emission, Btu/(hr)(ft^2)(micron)
n	Number of radiation shields
Q	Heat flow or net heat exchange, Btu/(hr)
r	Reflectivity, dimensionless; radius, ft
T	Temperature, °R
α, β, ω	Angle, deg
ϵ	Emissivity, dimensionless
λ	Wavelength, microns
σ	Stefan-Boltzmann constant, 0.173×10^{-8} Btu/(hr)(ft^2)(°R^4)
τ	Transmissivity, dimensionless

Subscripts

b	Black body
1	Source
2	Receiver

CHAPTER 5

TEMPERATURE

The Temperature Difference. A temperature difference is the driving force by which heat is transferred from a source to a receiver. Its influence upon a heat-transfer system which includes both a source and a receiver is the immediate subject for study.

In the experimental data of Chap. 3 the temperature of the inside pipe wall t_p was calculated from the reported value of the outside pipe-wall temperature t_w. The logarithmic mean of the differences $t_p - t_1$ and $t_p - t_2$ was used to calculate Δt_i. The reported temperature of the pipe was the average of a number of thermocouples which were not actually constant over the length of the pipe. It is not ordinarily possible in industrial equipment to measure the average pipe-wall temperature. Only the inlet and outlet temperatures of the hot and cold fluids are known or can be measured, and these are referred to as the *process temperatures*.[1]

Plots of temperature vs. pipe length, t vs. L, for a system of two concentric pipes in which the annulus fluid is cooled sensibly and the pipe fluid heated sensibly are shown in Figs. 5.1 and 5.2. When the two fluids travel in opposite directions along a pipe as in Fig. 5.1, they are in *counterflow*. Figure 5.1 can be compared with Fig. 3.6 to which it is similar except that one is a plot of t vs. L and the other is a plot of t vs. Q, the heat transferred. When the fluids travel in the same direction as in Fig. 5.2, they are in *parallel* flow. The temperature of the inner pipe fluid in either case varies according to one curve as it proceeds along the length of the pipe, and the temperature of the annulus fluid varies according to

---

[1] In the remainder of this book the subscript 1 always denotes the inlet and subscript 2 the outlet. The cold terminal difference Δt_1 or Δt_c is given by $T_2 - t_1$, and the hot terminal difference Δt_2 or Δt_h by $T_1 - t_2$. There are two additional terms which are often used in industry. These are the *range* and the *approach*. By "range" is meant the actual temperature rise or fall, which for the hot fluid is $T_1 - T_2$ and for the cold fluid $t_2 - t_1$. The "approach" has two different meanings depending upon whether counterflow equipment such as concentric pipes or other types of equipment are intended. For counterflow the approach is the number of degrees between the hot-fluid inlet and cold-fluid outlet, $T_1 - t_2$, or hot-fluid outlet and cold fluid inlet, $T_2 - t_1$, whichever is smaller. Thus a close approach means that one terminal difference will be very small, a significant factor in heat transfer. The definition of approach for other types of equipment will be discussed in Chap. 7.

another. The temperature difference at any length from the origin where $L = 0$ is the vertical distance between the two curves.

Overall Coefficient of Heat Transfer. The concentric pipes in Figs. 5.1 and 5.2 bring together two streams each having a particular film coefficient and whose temperatures vary from inlet to outlet. For convenience, the method of calculating the temperature difference between the two should employ only the process temperatures, since these alone are generally known. To establish the temperature difference in this manner between some general temperature T of a hot fluid and some general temperature t of a cold fluid, it is necessary to account also for all the resistances between the two temperatures. In the case of two con-

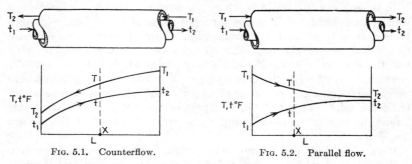

Fig. 5.1. Counterflow.   Fig. 5.2. Parallel flow.

centric pipes, the inner pipe being very thin, the resistances encountered are the pipe fluid-film resistance, the pipe-wall resistance, L_m/k_m, and the annulus fluid-film resistance. Since Q is equal to $\Delta t/\Sigma R$ as before,

$$\Sigma R = \frac{1}{h_i} + \frac{L_m}{k_m} + \frac{1}{h_o} \tag{5.1}$$

where ΣR is the overall resistance. It is customary to substitute $1/U$ for ΣR where U is called the *overall coefficient of heat transfer*. Inasmuch as a real pipe has different areas per linear foot on its inside and outside surfaces, h_i and h_o must be referred to the same heat-flow area or they will not coincide per unit length. If the outside area A of the inner pipe is used, then h_i must be multiplied by A_i/A to give the value that h_i would have if originally calculated on the basis of the larger area A instead of A_i. For a pipe with a thick wall Eq. (5.1) becomes

$$\frac{1}{U} = \frac{1}{h_i(A_i/A)} + \frac{2.3 D_o}{2 k_m} \log \frac{D_o}{D_i} + \frac{1}{h_o} = \frac{1}{h_i(D_i/D_o)} + \frac{2.3 D_o}{2 k_m} \log \frac{D_o}{D_i} + \frac{1}{h_o} \tag{5.2}$$

The integrated steady-state modification of Fourier's general equation may then be written

$$Q = UA\,\Delta t \tag{5.3}$$

where Δt is the temperature difference between the two streams for the entire surface A. Using the simplification that the thin metal pipe-wall resistance is negligible, Eq. (5.2) becomes

$$\frac{1}{U} = \frac{1}{h_i(A_i/A)} + \frac{1}{h_o} \qquad (5.4)$$

Hereafter Eq. (5.3) will be referred to simply as the *Fourier equation*. Just as h_i was obtained from $h_i = Q/A_i \Delta t_i$ in Eq. (3.2) using thermocouples, so U can be obtained from $U = Q/A \Delta t$ using process temperatures alone. In experiments involving sensible heat transfer between two fluids, Eqs. (5.2) and (5.4) can be used to obtain either individual film coefficient from the overall coefficient U only if some supplementary means are available for computing the other film coefficient. Fortunately, the condensation of steam can provide a negligible resistance, so that h_i or h_o can usually be determined individually with suitable accuracy from an experiment using either of the fluids and steam.

Equation (5.3) is of particular value in design when the individual film coefficients can both be calculated through the use of equations of the types obtained by dimensional analysis such as Eq. (3.26) or (3.32), and U can be solved for accordingly. Then Eq. (5.3) is used in computing the total area or length of path required when Q is given and Δt is calculated from the process temperatures. When the process temperatures of the two respective streams are given, the total heat transfer Q Btu/hr is also given, being computed from $Q = wc(t_2 - t_1) = WC(T_1 - T_2)$.

The Controlling Film Coefficient. When the resistance of the pipe metal is small by comparison with the sum of the resistances of both film coefficients, and it usually is, it may be neglected. If one film coefficient is small and the other very large, the small coefficient provides the major resistance and the overall coefficient of heat transfer for the apparatus is very nearly the reciprocal of the major resistance. Suppose $h_i(A_i/A) = 10$ and $h_o = 1000$ Btu/(hr)(ft²)(°F) $R_i = \frac{1}{10} = 0.1$, $R_o = \frac{1}{1000} = 0.001$, and $\Sigma R = 0.101$. A variation of 50 per cent in R_o does not materially influence Q, since a value of $h_o = 500$ will change ΣR only from 0.101 to 0.102. When a significant difference exists, the smaller coefficient is the *controlling film coefficient*.

Logarithmic Mean Temperature Difference: Counterflow. Generally both fluids undergo temperature variations which are not straight lines when temperatures are plotted against length as in Figs. 5.1 and 5.2. At every point $T - t$ between the two streams differs, yet this should lead to the same result as Eq. (3.36), in which the logarithmic mean temperature difference was obtained from a study of $T - t$ vs. Q. However, there is an advantage to a derivation based on $T - t$ vs. L, since it permits

88   PROCESS HEAT TRANSFER

identification of the temperature difference anywhere along the pipe length. Later, when more complex flow patterns are encountered, this information will be essential. Although two fluids may transfer heat in a concentric pipe apparatus in either counterflow or parallel flow, *the relative direction* of the two fluids influences the value of the temperature difference. This point cannot be overemphasized: Any flow pattern formed by two fluids must be identified with its unique temperature difference. For the derivation of the temperature difference between the two fluids of Fig. 5.1 in counterflow, the following assumptions must be made:

1. The overall coefficient of heat transfer U is constant over the entire length of path.
2. The pounds per hour of fluid flow is constant, obeying the steady-state requirement.
3. The specific heat is constant over the entire length of path.
4. There are no partial phase changes in the system, *i.e.*, vaporization or condensation. The derivation is applicable for sensible-heat changes and when vaporization or condensation is **isothermal** over the whole length of path.
5. Heat losses are negligible.

Applying the differential form of the steady-state equation,

$$dQ = U(T - t)a'' \, dL \tag{5.5}$$

where a'' is the square feet of surface per foot of pipe length or

$$a'' \, dL = dA.$$

From a differential heat balance

$$dQ = WC \, dT = wc \, dt \tag{5.6}$$

where Q is the limit as dQ varies from 0 to Q. At any point in the pipe from left to right the heat gained by the cold fluid is equal to that given up by the hot fluid. Taking a balance from $L = 0$ to $L = X$

$$WC(T - T_2) = wc(t - t_1) \tag{5.7}$$

from which

$$T = T_2 + \frac{wc}{WC}(t - t_1) \tag{5.8}$$

From Eqs. (5.5) and (5.6) substituting for T,

$$dQ = wc \, dt = U\left[T_2 + \frac{wc}{WC}(t - t_1) - t\right] a'' \, dL$$

t and L are the only variables. Collecting terms of t and L,

$$\int \frac{Ua'' \, dL}{wc} = \int \frac{dt}{T_2 - \frac{wc}{WC}t_1 + \left(\frac{wc}{WC} - 1\right)t} \tag{5.9}$$

This right-hand term is of the form

$$\int \frac{dt}{a_1 + b_1 t} = \frac{1}{b_1} \log(a_1 + b_1 t)$$

Integrating dL between 0 and L and dt between t_1 and t_2,

$$\frac{UA}{wc} = \frac{1}{\left(\frac{wc}{WC} - 1\right)} \ln \frac{T_2 - \frac{wc}{WC}t_1 + \left(\frac{wc}{WC} - 1\right)t_2}{T_2 - \frac{wc}{WC}t_1 + \left(\frac{wc}{WC} - 1\right)t_1} \tag{5.10}$$

To simplify this expression substitute for T_2 in the numerator the expression from Eq. (5.7), expand the denominator, and cancel terms

$$\frac{UA}{wc} = \frac{1}{(wc/WC) - 1} \ln \frac{T_1 - t_2}{T_2 - t_1} \tag{5.11}$$

Substitute for wc/WC the expression from Eq. (5.7)

$$\frac{UA}{wc} = \frac{1}{(T_1 - T_2)/(t_2 - t_1) - 1} \ln \frac{T_1 - t_2}{T_2 - t_1}$$

$$= \frac{t_2 - t_1}{(T_1 - t_2) - (T_2 - t_1)} \ln \frac{T_1 - t_2}{T_2 - t_1} \tag{5.12}$$

Since $wc(t_2 - t_1) = Q$ and substituting Δt_2 and Δt_1 for the hot and cold terminal temperature differences $T_1 - t_2$ and $T_2 - t_1$,

$$Q = UA \left(\frac{\Delta t_2 - \Delta t_1}{\ln \Delta t_2/\Delta t_1}\right) \tag{5.13}$$

If the difference between the two terminals $\Delta t_2 - \Delta t_1$ is written so as to be positive then the ratio of the two terminals taken in the same order is numerically greater than unity and any confusion due to negative signs is eliminated. The expression in parentheses in Eq. (5.13) again is the logarithmic mean or *log mean temperature difference* and is abbreviated LMTD. Equation (5.13) for counterflow may be written

$$Q = UA \, \Delta t = UA \times \text{LMTD} \tag{5.13a}$$

and

$$\Delta t = \text{LMTD} = \frac{(T_1 - t_2) - (T_2 - t_1)}{\ln (T_1 - t_2)/(T_2 - t_1)} = \frac{\Delta t_2 - \Delta t_1}{\ln \Delta t_2/\Delta t_1} \tag{5.14}$$

Parallel Flow. Referring to Fig. 5.2 for the case where both fluids flow in the same direction, the basic equations are essentially the same. For the steady state,
$$dQ = U(T - t)a'' \, dL$$
but
$$dQ = WC \, dT = -wc \, dt$$
since t declines in the direction of increasing values of T. Taking the heat balance between X and the left end,
$$WC(T - T_2) = wc(t_2 - t)$$
Again considering the hot terminal difference $\Delta t_2 = T_1 - t_1$ as the *greater temperature difference* in parallel flow and $\Delta t_1 = T_2 - t_2$ the lesser temperature difference, the result is
$$Q = UA \frac{(T_1 - t_1) - (T_2 - t_2)}{\ln (T_1 - t_1)/(T_2 - t_2)} = UA \frac{\Delta t_2 - \Delta t_1}{\ln \Delta t_2/\Delta t_1} \quad (5.15)$$

Relation between Parallel Flow and Counterflow. It may appear from the final form of the derivations for the two flow arrangements that there is little to choose between the two. The examples which follow demonstrate that except where one fluid is isothermal (such as condensing steam) there is a distinct thermal *disadvantage* to the use of parallel flow.

Example 5.1. Calculation of the LMTD. A hot fluid enters a concentric-pipe apparatus at a temperature of 300°F and is to be cooled to 200°F by a cold fluid entering at 100°F and heated to 150°F. Shall they be directed in parallel flow or counterflow?

Solution. It is convenient to write the temperatures in the form employed here and to realize that the log mean is always somewhat less than the arithmetic mean $(\Delta t_2 + \Delta t_1)/2$.

(a) *Counterflow:*

$$\begin{array}{ll} \text{Hot fluid} & \text{Cold fluid} \\ (T_1) \; 300 \; - \; 150 \; (t_2) = 150 \; (\Delta t_2) \\ (T_2) \; 200 \; - \; 100 \; (t_1) = \underline{100} \; (\Delta t_1) \\ & 50 \; (\Delta t_2 - \Delta t_1) \end{array}$$

$$\text{LMTD} = \frac{\Delta t_2 - \Delta t_1}{2.3 \log \Delta t_2/\Delta t_1} = \frac{50}{2.3 \log 15\%_{00}} = 123.5°F$$

(b) *Parallel flow:*

$$\begin{array}{ll} \text{Hot fluid} & \text{Cold fluid} \\ (T_1) \; 300 \; - \; 100 \; (t_1) = 200 \; (\Delta t_2) \\ (T_2) \; 200 \; - \; 150 \; (t_2) = \underline{50} \; (\Delta t_1) \\ & 150 \; (\Delta t_2 - \Delta t_1) \end{array}$$

$$\text{LMTD} = \frac{150}{2.3 \log 20\%_{0}} = 108°F$$

The LMTD for the same process temperatures in parallel flow is lower than for counterflow.

TEMPERATURE 91

Example 5.2. Calculation of the LMTD with Equal Outlet Temperatures. A hot fluid enters a concentric-pipe apparatus at 300°F and is to be cooled to 200°F by a cold fluid entering at 150°F and heated to 200°F.

(a) *Counterflow:*

$$\begin{array}{ccc} \text{Hot fluid} & \text{Cold fluid} & \\ 300 & - \quad 200 & = 100 \ (\Delta t_2) \\ 200 & - \quad 150 & = \underline{50} \ (\Delta t_1) \\ & & 50 \ (\Delta t_2 - \Delta t_1) \end{array}$$

$$\text{LMTD} = \frac{50}{2.3 \log \, ^{100}\!/_{50}} = 72$$

(b) *Parallel flow:*

$$\begin{array}{ccc} \text{Hot fluid} & \text{Cold fluid} & \\ 300 & - \quad 150 & = 150 \ (\Delta t_2) \\ 200 & - \quad 200 & = \underline{0} \ (\Delta t_1) \\ & & 150 \ (\Delta t_2 - \Delta t_1) \end{array}$$

$$\text{LMTD} = \frac{150}{2.3 \log \, ^{150}\!/_0} = 0$$

In parallel flow the lowest temperature theoretically attainable by the hot fluid is that of the outlet temperature of the cold fluid, t_2. If this temperature were attained, the LMTD would be zero. In the Fourier equation $Q = UA \, \Delta t$, since Q and U are finite, the heat-transfer surface A would have to be infinite. The last is obviously not feasible.

The inability of the hot fluid in parallel flow to fall below the outlet temperature of the cold fluid has a marked effect upon the ability of parallel-flow apparatus to *recover* heat. Suppose it is desired to recover as much heat as possible from the hot fluid in Example 5.1 by using the same quantities of hot and cold fluid as before but by assuming that more heat-transfer surface is available. In a counterflow apparatus it is possible to have the hot-fluid outlet T_2 fall to within perhaps 5 or 10° of the cold-fluid inlet t_1, say 110°F. In parallel-flow apparatus the heat transfer would be restricted by the cold-fluid outlet temperature rather than the cold-fluid inlet and the difference would be the loss in recoverable heat. Parallel flow is used for cold viscous fluids, however, since the arrangement may enable a higher value of U to be obtained.

Consider next the case where the hot terminal difference (Δt_2) in the preceding examples is not the greater temperature difference.

Example 5.3. Calculation of the LMTD When $\Delta t_c > \Delta t_h$. While a hot fluid is cooled from 300 to 200°F in counterflow, a cold fluid is heated from 100 to 275°F.

Counterflow:

$$\begin{array}{ccc} \text{Hot fluid} & \text{Cold fluid} & \\ 300 & - \quad 275 & = 25 \ (\Delta t_h) \\ 200 & - \quad 100 & = \underline{100} \ (\Delta t_c) \\ & & 75 \ (\Delta t_c - \Delta t_h) \end{array}$$

$$\text{LMTD} = \frac{75}{2.3 \log \, ^{100}\!/_{25}} = 54.3°\text{F}$$

Lastly, when one of the fluids proceeds through the apparatus isothermally (condensing steam), parallel flow and counterflow yield identical temperature differences.

Example 5.4. Calculation of the LMTD with One Isothermal Fluid. A cold fluid is heated from 100 to 275°F by steam at 300°F.

(a) *Counterflow:*

$$\begin{array}{ccc} \text{Hot fluid} & \text{Cold fluid} & \\ 300 & - \quad 275 & = 25 \\ 300 & - \quad 100 & = 200 \end{array}$$

(b) *Parallel flow:*

$$\begin{array}{ccc} \text{Hot fluid} & \text{Cold fluid} & \\ 300 & - \quad 100 & = 200 \\ 300 & - \quad 275 & = 25 \end{array}$$

These are identical.

Hereafter, unless specifically qualified, all temperature arrangements will be assumed in counterflow. Many industrial types of equipment are actually a compromise between parallel flow and counterflow and receive additional study in later chapters.

Heat Recovery in Counterflow. Very often a counterflow apparatus is available which has a given length L and therefore a fixed surface A. Two process streams are available with inlet temperatures T_1, t_1 and flow rates and specific heats W, C and w, c. What outlet temperatures will be attained in the apparatus?

This problem requires an estimate of U which can be checked by the methods of succeeding chapters for different types of counterflow heat-transfer equipment. Rewriting Eq. (5.12),

$$wc(t_2 - t_1) = UA \frac{(T_1 - t_2) - (T_2 - t_1)}{\ln (T_1 - t_2)/(T_2 - t_1)}$$

Rearranging

$$\ln \frac{T_1 - t_2}{T_2 - t_1} = \frac{UA}{wc} \left(\frac{T_1 - T_2}{t_2 - t_1} - 1 \right) \qquad (5.16)$$

Since $WC(T_1 - T_2) = wc(t_2 - t_1)$, $wc/WC = (T_1 - T_2)/(t_2 - t_1)$. This means that the ratio of the temperature ranges can be established without recourse to actual working temperatures. Calling this unique ratio R without a subscript

$$R = \frac{wc}{WC} = \frac{T_1 - T_2}{t_2 - t_1}$$

Substituting in Eq. (5.16) and removing logarithms,

$$\frac{T_1 - t_2}{T_2 - t_1} = e^{(UA/wc)(R-1)} \qquad (5.17)$$

To obtain an expression for T_2 alone.

$$t_2 = t_1 + \frac{T_1 - T_2}{R}$$

Substituting in Eq. (5.17) and solving,

$$T_2 = \frac{(1-R)T_1 + [1 - e^{(UA/wc)(R-1)}]Rt_1}{1 - Re^{(UA/wc)(R-1)}} \qquad (5.18)$$

For parallel flow it becomes

$$T_2 = \frac{[R + e^{(UA/wc)(R+1)}]T_1 + [e^{(UA/wc)(R+1)} - 1]Rt_1}{(R+1)e^{(UA/wc)(R+1)}} \qquad (5.19)$$

t_2 may be obtained from T_2 by applying the heat balance

$$WC(T_1 - T_2) = wc(t_2 - t_1)$$

The Caloric or Average Fluid Temperature. Of the four assumptions used in the derivation of Eq. (5.14) for the LMTD, the one which is subject to the largest deviation is that of a constant overall heat-transfer coefficient U. In the calculations of Chap. 3 the film coefficient was computed for the properties of the fluid at the arithmetic mean temperature between inlet and outlet, although the correctness of this calculation was not verified. In fluid-fluid heat exchange the hot fluid possesses a viscosity on entering which becomes greater as the fluid cools. The cold counterflow fluid enters with a viscosity which decreases as it is heated. There is a hot terminal $T_1 - t_2$ and a cold terminal $T_2 - t_1$, and the values of h_o and $h_i(A_i/A)$ vary over the length of the pipe to produce a larger U at the hot terminal than at the cold terminal. As a simple example, take the case of an individual transfer coefficient at the inlet and outlet as obtained from the data of Morris and Whitman through use of Eq. (3.42).

Example 5.5. Calculation of h_1 and h_2. Calculation of point B-6:

$$t_1 = 99.1° \qquad t_2 = 129.2°F$$

Inlet at 99.1°F:

$c = 0.478$ Btu/lb

$\mu = 2.95$ cp

$k = 0.078$ Btu/(hr)(ft)(°F/ft)

$\left(\dfrac{DG}{\mu}\right)^{0.9} = \left(\dfrac{0.622 \times 854{,}000}{12 \times 2.95 \times 2.42}\right)^{0.9} = 2570$

$\left(\dfrac{c\mu}{k}\right)^{1/3} = \left(\dfrac{0.478 \times 2.95 \times 2.42}{0.078}\right)^{1/3} = 3.52$

$h_i = 0.078 \times \dfrac{12}{0.622} \times 0.0115 \times 2570 \times 3.52 = 156$ at inlet

94   PROCESS HEAT TRANSFER

Outlet at 129.2°F:

$c = 0.495$

$\mu = 2.20$

$k = 0.078$

$\left(\dfrac{DG}{\mu}\right)^{0.9} = \left(\dfrac{0.622 \times 854{,}000}{12 \times 2.20 \times 2.42}\right)^{0.9} = 3390$

$\left(\dfrac{c\mu}{k}\right)^{1/3} = \left(\dfrac{0.495 \times 2.20 \times 2.42}{0.078}\right)^{1/3} = 3.23$

$h_2 = 0.078 \times \dfrac{12}{0.622} \times 0.0115 \times 3390 \times 3.23 = 190$ at outlet

At the arithmetic mean (114.3°F) $h_a = 174.5$, which is only 3.6 per cent in error of the experimental value of 181 but the variations against h_i at the arithmetic mean are $\left(\dfrac{156 - 174.5}{174.5}\right) 100 = -10.6$ per cent and $\left(\dfrac{190 - 174.5}{174.5}\right) 100 = +8.9$ per cent.

From the above it is seen that under actual conditions the variation of U may be even greater than that of h_i alone, since the outside film coefficient h_o will vary at the same time and in the same direction as h_i. The variation of U can be taken into account by numerical integration of dQ, the heat transferred over incremental lengths of the pipe $a''dL = dA$, and using the average values of U from point to point in the differential equation $dQ = U_{av}\, dA\, \Delta t$. The summation from point to point then gives $Q = UA\, \Delta t$ very closely. This is a time-consuming method, and the increase in the accuracy of the result does not warrant the effort. Colburn[1] has undertaken the solution of problems with varying values of U by assuming the variation of U to be linear with temperature and by deriving an expression for the true temperature difference accordingly. The ratio of the LMTD for constant U and the true temperature difference for varying U is then used as the basis for establishing a single overall coefficient which is the *true* mean rather than the arithmetic mean.

Assume:
1. The variation of U is given by the expression $U = a'(1 + b't)$
2. Constant weight flow
3. Constant specific heat
4. No partial phase changes

Over the whole transfer path

$$Q = WC(T_1 - T_2) = wc(t_2 - t_1)$$

Since $R = wc/WC = (T_1 - T_2)/(t_2 - t_1)$ or generalized as in Fig. 5.1,

$$R = \dfrac{(T - T_2)}{(t - t_1)}$$

The heat balance for the differential area dA is given by

$$dQ = U(T - t)\, dA = wc\, dt$$

[1] Colburn, A. P., *Ind. Eng. Chem.*, **35**, 873–877 (1933).

where U is the average value for the increment or

$$\frac{dt}{U(T-t)} = \frac{dA}{wc}$$

Since $U = a'(1 + b't)$, substitute for U.

$$\frac{dt}{a'(1+b't)(T-t)} = \frac{dA}{wc}$$

From the heat balance obtain the expression for T in terms of t and separate into parts.

$$\frac{1}{a'(R-1-b'T_2+b'Rt_1)} \int_{t_1}^{t_2} \left[\frac{(R-1)\,dt}{T_2 - Rt_1 + (R-1)t} - \frac{b'\,dt}{1+b't} \right] = \int \frac{dA}{wc} \quad (5.20)$$

Integrating,

$$\frac{1}{a'(R-1-b'T_2+b'Rt_1)} \left[\ln \frac{T_2 - Rt_1 + (R-1)t_2}{T_2 - Rt_1 + (R-1)t_1} - \ln \frac{1+b't_2}{1+b't_1} \right] = \frac{A}{wc} \quad (5.21)$$

Using the subscript 1 to indicate the cold terminal and 2 the hot terminal as heretofore

$$U_1 = a'(1 + b't_1) \qquad U_2 = a'(1 + b't_2)$$

As before,

$$\Delta t_1 = T_2 - t_1 \qquad \Delta t_2 = T_1 - t_2$$

and factoring Eq. (5.21)

$$\frac{t_2 - t_1}{U_1 \Delta t_2 - U_2 \Delta t_1} \ln \frac{U_1 \Delta t_2}{U_2 \Delta t_1} = \frac{A}{wc} \quad (5.22)$$

Combining with $Q = wc(t_2 - t_1)$,

$$\frac{Q}{A} = \frac{U_1 \Delta t_2 - U_2 \Delta t_1}{\ln U_1 \Delta t_2 / U_2 \Delta t_1} \quad (5.23)$$

Equation (5.23) is a modification of Eq. (5.13) which accounts for the variation of U by replacing it with U_1 and U_2, where $A = 0$ and $A = A$, respectively. This is still unsatisfactory, however, since it requires twice calculating both individual film coefficients to obtain U_1 and U_2. Colburn chose to obtain a *single* overall coefficient, U_x, at which all the surface can be regarded to be transferring heat at the LMTD. U_x is then defined by

$$\frac{Q}{A} = \frac{U_1 \Delta t_2 - U_2 \Delta t_1}{\ln U_1 \Delta t_2 / U_2 \Delta t_1} = U_x \left(\frac{\Delta t_2 - \Delta t_1}{\ln \Delta t_2 / \Delta t_1} \right) \quad (5.24)$$

Substituting $U_x = a'(1 + b't_c)$,

$$U_x = a'(1 + b't_c) = \frac{\dfrac{a'(1 + b't_1)\,\Delta t_2 - a'(1 + b't_2)\,\Delta t_1}{\ln\left[(1 + b't_1)\,\Delta t_2/[(1 + b't_2)\,\Delta t_1]\right]}}{\dfrac{\Delta t_2 - \Delta t_1}{\ln \Delta t_2/\Delta t_1}} \quad (5.25)$$

U_x will now be identified by finding t_c, the *temperature* of the properties at which h_i and h_o are computed and at which such a value of U_x exists. Let F_c be a fraction. By multiplying the temperature rise of the *controlling* (film) *stream* by F_c and adding the resulting fractional rise to the lower terminal temperature of the stream, a temperature is obtained at which to evaluate heat transfer properties and calculate h_i, h_o, and U_x.

$$F_c = \frac{t_c - t_1}{t_2 - t_1} \quad (5.26)$$

t_c is the *caloric temperature* of the cold stream. By definition, let

$$K_c = \frac{t_2 - t_1}{1/b' + t_1} = \frac{U_2 - U_1}{U_1} \qquad r = \frac{\Delta t_1}{\Delta t_2} = \frac{\Delta t_c}{\Delta t_h}$$

and substituting the equivalents in Eq. (5.25),

$$\frac{1 + b't_2}{1 + b't_1} = K_c + 1 \qquad \frac{1 + b't_c}{1 + b't_1} = K_c F_c + 1$$

from which

$$F_c = \frac{(1/K_c) + [r/(r-1)]}{1 + \dfrac{\ln(K_c + 1)}{\ln r}} - \frac{1}{K_c} \quad (5.27)$$

Equation (5.27) has been plotted in Fig. 17 in the Appendix with

$$K_c = \frac{U_2 - U_1}{U_1} = \frac{U_h - U_c}{U_c}$$

as the parameter, where c and h refer to the cold and hot terminals, respectively. The caloric fraction F_c can be obtained from Fig. 17 by computing K_c from U_h and U_c and $\Delta t_c/\Delta t_h$ for the process conditions. The caloric temperature of the hot fluid T_c is

$$T_c = T_2 + F_c(T_1 - T_2) \quad (5.28)$$

and for the cold fluid

$$t_c = t_1 + F_c(t_2 - t_1) \quad (5.29)$$

Colburn has correlated in the insert of Fig. 17 the values of K_c where the *controlling* film is that of a petroleum cut. A correlation of this type can be made in any industry which deals with a particular group of fluids by obtaining a' and b' from properties and eliminates the calculation of U_h and U_c. If an apparatus transfers heat between two petroleum cuts, the cut giving the largest value of K_c is controlling and can be used directly to establish F_c for both streams from the figure. Thus, whenever there is a sizeable difference between U_h and U_c *the LMTD is not the true temperature difference* for counterflow. The LMTD may be retained, however, if a suitable value of U is employed to compensate for its use in Eq. (5.13).

Example 5.6. Calculation of the Caloric Temperature. A 20°API crude oil is cooled from 300 to 200°F by heating cold 60°API gasoline from 80 to 120°F in a counterflow apparatus. At what fluid temperatures should U be evaluated?

Solution:

	Shell	Tubes		
	20°API crude:	60°API gasoline		
242.5	Caloric temp.	97		
250	Mean	100		
300	Higher temp.	120	180	Δt_2
200	Lower temp.	80	120	Δt_1
100	Diff	40		

Crude, $T_1 - T_2 = 300 - 200 = 100°F$, $K_c = 0.68$ from Fig. 17 insert
Gasoline, $t_2 - t_1 = 120 - 80 = 40°F$, $K_c \leq 0.10$

The larger value of K_c corresponds to the controlling heat-transfer coefficient which is assumed to establish the variation of U with temperature. Then

$$\frac{\Delta t_c}{\Delta t_h} = \frac{200 - 80}{300 - 120} = 0.667$$
$$F_c = 0.425 \text{ from Fig. 17}$$

Caloric temperature of crude, $T_c = 200 + 0.425(300 - 200) = 242.5°F$.
Caloric temperature of gasoline, $t_c = 80 + 0.425(120 - 80) = 97.0°F$.

It should be noted that there can be but one caloric mean and that the factor F_c applies to both streams but is determined by the controlling stream.

The Pipe-wall Temperature. The temperature of the pipe wall can be computed from the caloric temperatures when both h_i and h_o are known. Referring to Fig. 5.3 it is customary to neglect the temperature difference across the pipe metal $t_w - t_p$ and to consider the entire pipe to be at the temperature of the outside surface of the wall t_w. If the outside caloric temperature is T_c and the inside caloric temperature t_c and $1/R_{io} = h_{io} = h_i(A_i/A) = h_i \times (ID/OD)$, where the subscript *io* refers to the value of the coefficient inside the pipe referred to the outside surface of the pipe.

$$Q = \frac{\Delta t}{\Sigma R} = \frac{T_c - t_c}{R_o + R_{io}} = \frac{t_w - t_c}{R_{io}} \qquad (5.30)$$

Replacing the resistances in the last two terms by film coefficients,

$$\frac{T_c - t_c}{1/h_o + 1/h_{io}} = \frac{t_w - t_c}{1/h_{io}}$$

Solving for t_w

$$t_w = t_c + \frac{h_o}{h_{io} + h_o}(T_c - t_c) \qquad (5.31)$$

and

$$t_w = T_c - \frac{h_{io}}{h_{io} + h_o}(T_c - t_c) \qquad (5.32)$$

When the hot fluid is inside the pipe these become

$$t_w = t_c + \frac{h_{io}}{h_{io} + h_o}(T_c - t_c) \qquad (5.31a)$$

and

$$t_w = T_c - \frac{h_o}{h_{io} + h_o}(T_c - t_c) \qquad (5.32a)$$

Isothermal Representation of Heating and Cooling. In streamline flow when a fluid flows isothermally, the velocity distribution is assumed to be parabolic. When a given quantity of liquid is heated as it travels along a pipe, the viscosity near the pipe wall is lower than that of the bulk of the fluid. The fluid near the wall travels at a faster velocity than it would in isothermal flow and modifies the parabolic velocity

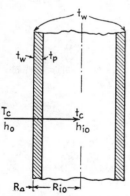

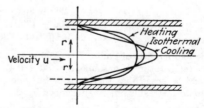

FIG. 5.3. Pipe-wall temperature.

FIG. 5.4. Heating, cooling, and isothermal streamline flow.

distribution as indicated by the heating curve in Fig. 5.4. If the liquid is cooled, the reverse occurs: The fluid near the wall flows at a lower velocity than in isothermal flow, producing the velocity distribution indicated for cooling. For the liquid to flow more rapidly at the wall during heating some of the liquid near the center axis of the pipe must flow outward toward the wall to maintain the increased velocity. This is a radial velocity component which actually modifies the nature of the streamline flow. If data for heating an oil in a given temperature range are plotted as in Fig. 3.10 together with data for cooling the oil in the same temperature range, two families of points are obtained. The data on heating give higher heat-transfer coefficients than those on cooling. Colburn[1] undertook to convert both heating and cooling data to a single isothermal line. He was able to employ a basic equation of the form of Eq. (3.32) by multiplying the right-hand term by the dimensionless ratio $(\mu/\mu_f)^{r'}$ where μ is the viscosity at the caloric temperature and μ_f is the viscosity at an arbitrary film temperature defined for streamline flow by

$$t_f = t_{av} + \tfrac{1}{4}(t_w - t_{av}) \tag{5.33}$$

and for turbulent flow by

$$t_f = t_{av} + \tfrac{1}{2}(t_w - t_{av}) \tag{5.34}$$

Sieder and Tate[2] undertook the correlation of a large quantity of data in tubes, rather than pipes, and obtained a dimensionless factor $(\mu/\mu_w)^{r''}$ where μ_w is the viscosity at the tube-wall temperature t_w. Using the Sieder and Tate correction, Eq. (3.32) for streamline flow becomes

$$\frac{h_i D}{k} = \alpha \left[\left(\frac{DG}{\mu}\right) \left(\frac{c\mu}{k}\right) \left(\frac{D}{L}\right) \right]^{d+1} \left(\frac{\mu}{\mu_w}\right)^{r''} \tag{5.35}$$

Equation (3.26) for turbulent flow becomes

$$\frac{h_i D}{k} = \alpha \left(\frac{DG}{\mu}\right)^p \left(\frac{c\mu}{k}\right)^q \left(\frac{\mu}{\mu_w}\right)^{r'''} \tag{5.36}$$

By incorporating the correlation factor for heating and cooling in this manner a single curve is obtained for both heating and cooling, since the value of μ/μ_w is greater than 1.0 for liquid heating and lower than 1.0 for liquid cooling. Inasmuch as the viscosities of gases increase rather than decrease with higher temperature, the deviations from the isothermal velocity distribution are the reverse of liquids.

PROBLEMS

5.1. For a concentric-pipe heat-transfer apparatus having a 1-in. IPS inner pipe the film coefficient h_i has been computed to be 10.0 Btu/(hr)(ft²)(°F). By suitable calculation three different fluids, when circulated through the annulus, will have film coefficients of 10.0, 20.0, and 200, respectively. Neglecting the resistance of the pipe, how does the value of the annulus coefficient affect the value of the overall coefficient?

5.2. (a) For values of $h_i = 100$ and $h_o = 500$, what error results in the calculated value of U for a concentric-pipe heat-transfer apparatus having a 2-in. IPS inner pipe when the metal resistance is obtained from $R_m = L_m/k_m A$ instead of $(2.3/2\pi k_m)$

[1] Colburn, A. P., *Trans. AIChE*, **29**, 174–210 (1933).
[2] Sieder, E. N., and G. E. Tate, *Ind. Eng. Chem.*, **28**, 1429–1436 (1936).

log D_2/D_1 and when the pipe resistance is omitted entirely. (b) What are the errors when the coefficients are $h_i = 10$ and $h_o = 50$?

5.3. Calculate the LMTD for counterflow in the following cases in which the hot fluid is cooled from 200 to 100°F and the cold fluid, heated through an equal range in each case, is (a) 90 to 140°F, (b) 80 to 130°F, and (c) 60 to 110°F. Observe the nature of the deviation of the LMTD from the arithmetic means of the two terminal differences in each case.

5.4. A hot fluid is cooled from 245 to 225°F in each case. Compare the advantage of counterflow over parallel flow in the size of the LMTD when a cold fluid is to be heated from (a) 135 to 220°F, (b) 125 to 210°F, and (c) 50 to 135°F.

5.5. 10,000 lb/hr of cold benzene is heated under pressure from 100°F by cooling 9000 lb/hr of nitrobenzene at a temperature of 220°F. Heat transfer will occur in a concentric-pipe apparatus having a 1¼-in. IPS inner pipe 240 ft long. Tests on similar equipment transferring heat between the same liquids indicates that a value of $U = 120$ based on the outside surface of the inner pipe may be expected. (a) What outlet temperatures may be expected in counterflow? (b) What outlet temperatures may be expected in parallel flow? (c) If part of the concentric pipe is removed leaving only 160 lin ft, what outlet temperatures may be expected in counterflow?

5.6. Benzene is to be heated in a concentric-pipe apparatus having a 1¼-in. IPS inner pipe from 100 to 140°F by 8000 lb/hr of nitrobenzene having an initial temperature of 180°F. A value of $U = 100$ may be expected based on the outside surface of the pipe. How much cold benzene can be heated in 160 lin ft of concentric pipe (a) in counterflow, (b) in parallel flow? (*Hint.* Trial-and-error solution.)

5.7. Aniline is to be cooled from 200 to 150°F in a concentric-pipe apparatus having 70 ft² of external pipe surface by 8600 lb/hr of toluene entering at 100°F. A value of $U = 75$ may be anticipated. How much hot aniline can be cooled in counterflow?

5.8. In a counterflow concentric-pipe apparatus a liquid is cooled from 250 to 200°F by heating another from 100 to 225°F. The value of U_1, at the cold terminal, is calculated to be 50.0 from the properties at the cold terminal, and U_2 at the hot terminal is calculated to be 60.0. At what fluid temperatures should U be computed to express the overall heat transfer for the entire apparatus?

5.9. In a counterflow concentric-pipe apparatus a liquid is cooled from 250 to 150°F by heating another from 125 to 150°F. The value of U_1 at the cold terminal is 52 and at the hot terminal U_2 is 58. At what liquid temperatures should U for the overall transfer be computed?

5.10. The calculation of the caloric temperatures can be accomplished directly by evaluating a' and b' in $U = a'(1 + b't)$ for a given temperature range. If the hot liquid in Prob. 5.8 always provides the controlling film coefficient, what are the numerical values of the constants a' and b'?

5.11. A 40°APJ kerosene is cooled from 400 to 200°F by heating 34°API crude oil from 100 to 200°F. Between what caloric temperatures is the heat transferred, and how do these deviate from the mean?

5.12. A 35° API distillate used as a heating oil is cooled from 400 to 300°F by fresh 35°API distillate heated from 200 to 300°F. Between what caloric temperatures is the heat transferred, and how do these deviate from the mean?

NOMENCLATURE FOR CHAPTER 5

A Heat-transfer surface or outside surface of pipes, ft²
a'' External pipe surface per foot of length, ft
a', b' Constants in the equation $U = a'(1 + b't)$

TEMPERATURE

a_1, b_1	Constants
C	Specific heat of hot fluid in derivations, Btu/(lb)(°F)
c	Specific heat of cold fluid, Btu/(lb)(°F)
D	Inside diameter of pipe, ft
F_c	Caloric fraction, dimensionless
G	Mass velocity, lb/(hr)(ft^2)
h	Heat-transfer coefficient, Btu/(hr)(ft^2)(°F)
h_i, h_o	Inside and outside film coefficients, Btu/(hr)(ft^2)(°F)
h_{io}	$h_i A_i/A$, inside film coefficient referred to outside surface, Btu/(hr)(ft^2)(°F)
K_c	Caloric factor, dimensionless
k	Thermal conductivity, Btu/(hr)(ft^2)(°F/ft)
L	Length, ft
LMTD	Log mean temperature difference, °F
Q	Heat flow, Btu/hr
R	Ratio of $wc/WC = (T_1 - T_2)/(t_2 - t_1)$, dimensionless
ΣR	Overall resistance to heat flow, (hr)(ft^2)(°F)/Btu
r	Ratio of cold to hot terminal differences, dimensionless
T	Temperature of the hot fluid, °F
T_c	Caloric temperature of hot fluid, °F
T_1, T_2	Hot-fluid inlet and outlet temperatures, respectively, °F
t	Temperature of the cold fluid, °F
t_c	Caloric temperature of cold fluid, °F
t_f	Film temperature in Eqs. (5.33) and (5.34), °F
t_p	Inside pipe-wall temperature, °F
t_w	Outside pipe-wall temperature, °F
t_1, t_2	Cold-fluid inlet and outlet temperatures, respectively, °F
Δt	Temperature difference at a point or mean over an area, °F
$\Delta t_c, \Delta t_h$	Temperature difference at the cold and hot terminals, respectively, °F
Δt_l	Logarithmic mean of $t_p - t_1$ and $t_p - t_2$, °F
U	Overall coefficient of heat transfer in general, Btu/(hr)(ft^2)(°F)
U_c, U_h	Overall coefficient of heat transfer at cold and hot terminals, Btu/(hr)(ft^2)(°F)
U_1, U_2	
U_x	Value of U at t_c, Btu/(hr)(ft^2)(°F)
W	Weight flow of hot fluid, lb/hr
w	Weight flow of cold fluid, lb/hr
α	Proportionality constant, dimensionless
μ	Viscosity at mean or caloric temperature, lb/(ft)(hr)
μ_f, μ_w	Viscosity at the film and pipe-wall temperatures, respectively, lb/(ft)(hr)

Superscripts

d, p, q, r', r'', r'''	Constants

Subscripts (except as noted above)

i	Inside a pipe or tube
o	Outside a pipe or tube
io	Value based on inside of a pipe or tube referred to the outside of the tube.

CHAPTER 6

COUNTERFLOW: DOUBLE PIPE EXCHANGERS

Definitions. Heat-transfer equipment is defined by the function it fulfills in a process. *Exchangers* recover heat between two process streams. Steam and cooling water are utilities and are not considered in the same sense as recoverable process streams. *Heaters* are used primarily to heat process fluids, and steam is usually employed for this purpose, although in oil refineries hot recirculated oil serves the same purpose. *Coolers* are employed to cool process fluids, water being the main cooling medium. *Condensers* are coolers whose primary purpose is the removal of latent instead of sensible heat. The purpose of *reboilers* is to supply the heat requirements of a distillation process as latent heat. *Evaporators* are employed for the concentration of a solution by the evaporation of water. If any other fluid is vaporized besides water, the unit is a *vaporizer*.

Double Pipe Exchangers. For the derivations in Chap. 5, a concentric-pipe heat-transfer apparatus was employed. The industrial counter-

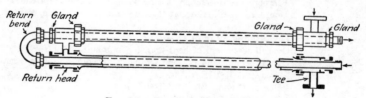

FIG. 6.1. Double pipe exchanger.

part of this apparatus is the double pipe exchanger shown in Fig. 6.1. The principal parts are two sets of concentric pipes, two connecting Tees, and a return head and a return bend. The inner pipe is supported within the outer pipe by packing glands, and the fluid enters the inner pipe through a threaded connection located outside the exchanger section proper. The Tees have nozzles or screwed connections attached to them to permit the entry and exit of the annulus fluid which crosses from one leg to the other through the return head. The two lengths of inner pipe are connected by a return bend which is usually exposed and does not provide effective heat-transfer surface. When arranged in two legs as in Fig. 6.1, the unit is a *hairpin*.

The double pipe exchanger is extremely useful because it can be assembled in any pipe-fitting shop from standard parts and provides inexpensive heat-transfer surface. The standard sizes of Tees and return heads are given in Table 6.1.

TABLE 6.1. DOUBLE PIPE EXCHANGER FITTINGS

Outer pipe, IPS	Inner pipe, IPS
2	1¼
2½	1¼
3	2
4	3

Double pipe exchangers are usually assembled in 12-, 15-, or 20-ft effective lengths, the effective length being the distance in *each* leg over which heat transfer occurs and excludes inner pipe protruding beyond the exchanger section. When hairpins are employed in excess of 20 ft in length corresponding to 40 effective linear feet or more of double pipe, the inner pipe tends to sag and touch the outer pipe, thereby causing a poor flow distribution in the annulus. The principal disadvantage to the use of double pipe exchangers lies in the small amount of heat-transfer surface contained in a single hairpin. When used with distillation equipment on an industrial process a very large number are required. These require considerable space, and each double pipe exchanger introduces no fewer than 14 points at which leakage might occur. The time and expense required for dismantling and periodically cleaning are prohibitive compared with other types of equipment. However, the double pipe exchanger is of greatest use where the total required heat-transfer surface is small, 100 to 200 ft² or less.

Film Coefficients for Fluids in Pipes and Tubes. Equation (3.42) was obtained for heating several oils in a pipe based on the data of Morris and Whitman. Sieder and Tate[1] made a later correlation of both heating and cooling a number of fluids, principally petroleum fractions, in horizontal and vertical tubes and arrived at an equation for streamline flow where $DG/\mu < 2100$ in the form of Eq. (5.35).

$$\frac{h_i D}{k} = 1.86 \left[\left(\frac{DG}{\mu}\right)\left(\frac{c\mu}{k}\right)\left(\frac{D}{L}\right)\right]^{\frac{1}{3}} \left(\frac{\mu}{\mu_w}\right)^{0.14} = 1.86 \left(\frac{4}{\pi}\frac{wc}{kL}\right)^{\frac{1}{3}} \left(\frac{\mu}{\mu_w}\right)^{0.14} \quad (6.1)$$

where L is the total length of the heat-transfer path before mixing occurs. Equation (6.1) gave maximum mean deviations of approximately ±12 per cent from $Re = 100$ to $Re = 2100$ except for water. Beyond the transition range, the data may be extended to turbulent flow in the form of Eq. (5.36).

$$\frac{h_i D}{k} = 0.027 \left(\frac{DG}{\mu}\right)^{0.8} \left(\frac{c\mu}{k}\right)^{\frac{1}{3}} \left(\frac{\mu}{\mu_w}\right)^{0.14} \quad (6.2)$$

[1] Sieder, E. N., and G. E. Tate, *Ind. Eng. Chem.*, **28**, 1429–1436 (1936).

Equation (6.2) gave maximum mean deviations of approximately $+15$ and -10 per cent for the Reynolds numbers above 10,000. While Eqs. (6.1) and (6.2) were obtained for tubes, they will also be used indiscriminately for pipes. Pipes are rougher than tubes and produce more turbulence for equal Reynolds numbers. Coefficients calculated from tube-data correlations are actually lower and safer than corresponding calculations based on pipe data and there are no pipe correlations in the literature so extensive as tube correlations. Equations (6.1) and (6.2) are applicable for organic liquids, aqueous solutions, and gases. They are not conservative for water, and additional data for water will be given later. In

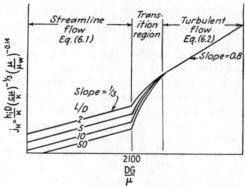

Fig. 6.2. Tube-side heat-transfer curve.

order to permit the graphical representation of both equations on a single pair of coordinates, refer to Fig. 6.2. Using the ordinate

$$j_H = \left(\frac{h_i D}{k}\right)\left(\frac{c\mu}{k}\right)^{-\frac{1}{3}}\left(\frac{\mu}{\mu_w}\right)^{-0.14}$$

and the abscissa (DG/μ) only Eq. (6.2) can be shown. By using D/L or L/D as a parameter, Eq. (6.1) can also be included. The transition region joins the two. Working plots of Eqs. (6.1) and (6.2) are given in Fig. 24 of the Appendix together with a line of slope 0.14 to facilitate the solution of the ratio $\phi = (\mu/\mu_w)^{0.14}$.

Fluids Flowing in Annuli: The Equivalent Diameter. When a fluid flows in a conduit having other than a circular cross section, such as an annulus, it is convenient to express heat-transfer coefficients and friction factors by the same types of equations and curves used for pipes and tubes. To permit this type of representation for annulus heat transfer it has been found advantageous to employ an *equivalent diameter* D_e. The equivalent diameter is four times the hydraulic radius, and the hydraulic radius is, in turn, the radius of a pipe equivalent to the annulus cross section. The hydraulic radius is obtained as the ratio of the flow area to the wetted perimeter. For a fluid flowing in an annulus as shown in Fig. 6.3 the flow area is evidently $(\pi/4)(D_2^2 - D_1^2)$ but the wetted perimeters for heat transfer and pressure drops are different. For heat transfer the wetted perimeter is the outer circumference of the inner pipe with diameter D_1, and for heat transfer in annuli

$$D_e = 4r_h = \frac{4 \times \text{flow area}}{\text{wetted perimeter}} = \frac{4\pi(D_2^2 - D_1^2)}{4\pi D_1} = \frac{D_2^2 - D_1^2}{D_1} \quad (6.3)$$

In pressure-drop calculations the friction not only results from the resistance of the outer pipe but is also affected by the outer surface of the inner pipe. The total wetted perimeter is $\pi(D_2 + D_1)$, and for the pressure drop in annuli

$$D_e' = \frac{4 \times \text{flow area}}{\text{frictional wetted perimeter}} = \frac{4\pi(D_2^2 - D_1^2)}{4\pi(D_2 + D_1)} = D_2 - D_1 \quad (6.4)$$

This leads to the anomalous result that the Reynolds numbers for the same flow conditions, w, G, and μ, are different for heat transfer and pressure drop since D_e might be above 2100 while D_e' is below 2100. Actually both Reynolds numbers should be considered only approximations, since the sharp distinction between streamline and turbulent flow at the Reynolds number of 2100 is not completely valid in annuli.

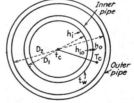

FIG. 6.3. Annulus diameters and location of coefficients.

Film Coefficients for Fluids in Annuli. When the equivalent diameter from Eq. (6.3) is substituted for D, Fig. 24 in the Appendix may be considered a plot of $D_e G/\mu$ vs. $(h_o D_e/k)(c\mu/k)^{-\frac{1}{3}}(\mu/\mu_w)^{-0.14}$. h_o is the outside or annulus coefficient and is obtained in the same manner as h_i by multiplication of the ordinate. Even though D differs from D_e, h_o is effective at the outside diameter of the inner pipe. In double pipe exchangers it is customary to use the outside surface of the inner pipe as the reference surface in $Q = UA \, \Delta t$, and since h_i has been determined for A_i and not A, it must be corrected. h_i is based on the area corresponding to the inside diameter where the surface per foot of length is $\pi \times \text{ID}$. On the outside of the pipe the surface per foot of length is $\pi \times \text{OD}$; and again letting h_{io} be the value of h_i referred to the outside diameter,

$$h_{io} = h_i \frac{A_i}{A} = h_i \frac{\text{ID}}{\text{OD}} \quad (6.5)$$

Fouling Factors. The overall coefficient of heat transfer *required* to fulfill the process conditions may be determined from the Fourier equation when the surface A is known and Q and Δt are calculated from the process conditions. Then $U = Q/A\,\Delta t$. If the surface is not known U can be obtained independently of the Fourier equation from the two film coefficients. Neglecting the pipe-wall resistance,

$$\frac{1}{U} = R_{io} + R_o = \frac{1}{h_{io}} + \frac{1}{h_o} \qquad (6.6)$$

or

$$U = \frac{h_{io}h_o}{h_{io} + h_o} \qquad (6.7)$$

The locations of the coefficients and temperatures are shown in Fig. 6.3. When U has been obtained from values of h_{io} and h_o and Q and Δt are calculated from the process conditions, the surface A required for the process can be computed. The calculation of A is known as *design*.

When heat-transfer apparatus has been in service from some time, however, dirt and scale deposit on the inside and outside of the pipe, adding two more resistances than were included in the calculation of U by Eq. (6.6). The additional resistances reduce the original value of U, and the required amount of heat is no longer transferred by the original surface A; T_2 rises above and t_2 falls below the desired outlet temperatures, although h_i and h_o remain substantially constant. To overcome this eventuality, it is customary in designing equipment to anticipate the deposition of dirt and scale by introducing a resistance R_d called the *dirt, scale,* or *fouling factor,* or resistance. Let R_{di} be the dirt factor for the inner pipe fluid at its inside diameter and R_{do} the dirt factor for the annulus fluid at the outside diameter of the inner pipe. These may be considered very thin for dirt but may be appreciably thick for scale, which has a higher thermal conductivity than dirt. The resistances are shown in Fig. 6.4. The value of U obtained in Eq. (6.7) only from $1/h_{io}$ and $1/h_o$ may be considered the *clean overall coefficient* designated by U_C to show that dirt has not been taken into account. The coefficient which includes the dirt resistance is called the *design* or *dirty overall coefficient* U_D. The value of A corresponding to U_D rather than U_C provides the basis on which equipment is ultimately built. The relationship between the two overall coefficients U_C and U_D is

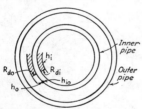

Fig. 6.4. Location of fouling factors and heat-transfer coefficients.

$$\frac{1}{U_D} = \frac{1}{U_C} + R_{di} + R_{do} \qquad (6.8)$$

COUNTERFLOW

or setting

$$R_{di}{}^* + R_{do} = R_d \qquad (6.9)$$

$$\frac{1}{U_D} = \frac{1}{U_C} + R_d \qquad (6.10)$$

Thus suppose that for a double pipe exchanger, h_{io} and h_o have been computed to be 300 and 100 respectively, then

$$\frac{1}{U_C} = \frac{1}{h_{io}} + \frac{1}{h_o} = 0.0033 + 0.01 = 0.0133,$$

or $U_C = 1/0.0133 = 75.0$ Btu/(hr)(ft^2)(°F). From experience, let us say, it has been found that a thermal dirt resistance of $R_{di} = 0.001$ (hr)(ft^2)(°F)/Btu will deposit annually inside the pipe and $R_{do} = 0.0015$ will deposit on the outside of the pipe. For what overall coefficient should the surface be calculated so that the apparatus need be cleaned only once a year? Then $R_d = R_{di} + R_{do} = 0.0025$, and

$$\frac{1}{U_D} = \frac{1}{U_C} + R_d = \frac{1}{75.0} + 0.0025 = 0.0158 \text{ (hr)(ft}^2\text{)(°F)/Btu}$$

or

$$U_D = \frac{1}{0.0158} = 63.3 \text{ Btu/(hr)(ft}^2\text{)(°F)}$$

The Fourier equation for surface on which dirt will be deposited becomes

$$Q = U_D A \, \Delta t \qquad (6.11)$$

If it is desired to obtain A, then h_{io} and h_o must first be calculated from equations such as Eq. (6.1) and (6.2) which are independent of the extent of the surface but dependent upon its form, such as the diameter and fluid flow area. With these, U_C is obtained from Eq. (6.6) and U_D is obtained from U_C using Eq. (6.10). Sometimes, however, it is desirable to study the rates at which dirt accumulates on a known surface A. U_C will remain constant if the scale or dirt deposit does not alter the mass velocity by constricting the fluid flow area. U_D and Δt will obviously change as the dirt accumulates because the temperatures of the fluids will vary from the time the surface is freshly placed in service until it becomes fouled. If Δt is calculated from *observed* temperatures instead of process temperatures then Eq. (6.11) may be used to determine R_d

---

* Actually R_{di} should be referred to the outside diameter as $R_{dio} = R_i(A/A_i)$. When a cylinder is very thin compared with its diameter, as a layer of dirt, its resistance is nearly the same as that through a flat wall. As shown by Eqs. (2.34) and (2.35), failure to correct to the outside will introduce a negligible error, usually well below 1 per cent. For thick scale, however, the error may be appreciable.

for a given fouling period. Then from Eq. (6.10)

$$R_d = \frac{1}{U_D} - \frac{1}{U_C} \tag{6.12}$$

which can also be written

$$R_d = \frac{U_C - U_D}{U_C U_D} \tag{6.13}$$

When R_d (deposited) > R_d (allowed), as after a period of service, the apparatus no longer delivers a quantity of heat equal to the process requirements and must be cleaned.

Numerical values of the dirt or fouling factors for a variety of process services are provided in Appendix Table 12. The tabulated fouling factors are intended to protect the exchanger from delivering less than the required process heat load for a period of about a year to a year and a half. Actually the purpose of the tabulated fouling factors should be considered from another point of view. In designing a process plant containing many heat exchangers but without alternate or spare pieces of heat-transfer equipment, the process must be discontinued and the equipment cleaned as soon as the first exchanger becomes fouled. It is impractical to shut down every time one exchanger or another is fouled, and by using the tabulated fouling factors, it can be arranged so that all the exchangers in the process become dirty at the same time regardless of service. At that time all can be dismantled and cleaned during a single shutdown. The tabulated values may differ from those encountered by experience in particular services. If too frequent cleaning is necessary, a greater value of R_d should be kept in mind for future design.

It is to be expected that heat-transfer equipment will transfer more heat than the process requirements when newly placed in service and that it will deteriorate through operation, as a result of dirt, until it just fulfills the process requirements. The calculation of the temperatures delivered initially by a clean exchanger whose surface has been designed for U_D but which is operating without dirt and which is consequently oversurfaced is not difficult. Referring to Eqs. (5.18) and (5.19) use U_C for U and the actual surface of the exchanger A (which is based on U_D). This calculation is also useful in checking whether or not a clean exchanger will be able to deliver the process heat requirements when it becomes dirty.

Pressure Drop in Pipes and Pipe Annuli. The *pressure-drop* allowance in an exchanger is the static fluid pressure which may be expended to drive the fluid through the exchanger. The pump selected for the circulation of a process fluid is one which develops sufficient head at the desired capacity to overcome the frictional losses caused by connecting piping, fittings, control regulators, and the pressure drop in the exchanger itself. To this head must be added the static pressure at the end of the line such as the elevation or pressure of the final receiving vessel. Once a definite pressure drop allowance has been designated for an exchanger as a part of a pumping circuit, it should always be utilized as completely as possible in the exchanger, since it will otherwise be blown off or expanded through a pressure reducer. Since in Eq. (3.44)

$$\Delta F \propto G^2 \quad \left(\text{nearly, since } f \text{ varies somewhat with } \frac{DG}{\mu}\right)$$

and in Eq. (6.2) for turbulent flow

$$h_i \propto G^{0.8} \quad \text{(nearly)}$$

the best use of available pressure is to increase the mass velocity which also increases h_i and lessens the size and cost of the apparatus. It is customary to allow a pressure drop of 5 to 10 psi for an exchanger or battery of exchangers fulfilling a single process service except where the flow is by gravity. For each pumped stream 10 psi is fairly standard. For gravity flow the allowable pressure drop is determined by the elevation of the storage vessel above the final outlet z in feet of fluid. The feet of fluid may be converted to pounds per square inch by multiplying z by $\rho/144$.

The pressure drop in pipes can be computed from the Fanning equation [Eq. (3.44)], using an appropriate value of f from Eq. (3.46) or Eq. (3.47b), depending upon the type of flow. For the pressure drop in fluids flowing in annuli, replace D in the Reynolds number by D'_e to obtain f. The Fanning equation may then be modified to give

$$\Delta F = \frac{4fG^2L}{2g\rho^2 D'_e} \quad (6.14)$$

Where several double pipe exchangers are connected in series, annulus to annulus and pipe to pipe as in Fig. 6.5, the length in Eq. (3.44) or (6.14) is the total for the entire path.

The pressure drop computed by Eq. (3.44) or (6.14) does not include the pressure drop encountered when the fluid enters or leaves exchangers. For the inner pipes of double pipe exchangers connected in series, the entrance loss is usually negligible, but for annuli it may be significant. The allowance of a pressure drop of one velocity head, $V^2/2g'$, per hairpin will ordinarily suffice. Suppose water flows in an annulus with a mass velocity of 720,000 lb/(hr)(ft²). Since $\rho = 62.5$ lb/ft³ (approximately),

$$V = \frac{G}{3600\rho} = \frac{720,000}{3600 \times 62.5} = 3.2 \text{ fps}$$

The pressure drop per hairpin will be $3.2^2/(2 \times 32.2) = 0.159$ ft of water or 0.07 psi. Unless the velocity is well above 3 fps, the entrance and exit losses may be neglected. Values of $V^2/2g'$ are plotted directly against the mass velocity for a fluid with a specific gravity of 1.0 in Fig. 27 in the Appendix.

The Calculation of a Double Pipe Exchanger. All the equations developed previously will be combined to outline the solution of a double pipe exchanger. The calculation consists simply of computing h_o and h_{io} to obtain U_C. Allowing a reasonable fouling resistance, a value of U_D is calculated from which the surface can be found with the use of the Fourier equation $Q = U_D A \, \Delta t$.

Usually the first problem is to determine which fluid should be placed in the annulus and which in the inner pipe. This is expedited by establishing the relative sizes of the flow areas for both streams. For equal allowable pressure drops on both the hot and cold streams, the decision rests in the arrangement producing the most nearly equal mass velocities and pressure drops. For the standard arrangements of double pipes the flow areas are given in Table 6.2.

TABLE 6.2. FLOW AREAS AND EQUIVALENT DIAMETERS IN DOUBLE PIPE EXCHANGERS

Exchanger, IPS	Flow area, in.²		Annulus, in.	
	Annulus	Pipe	d_e	d'_e
2 × 1¼	1.19	1.50	0.915	0.40
2½ × 1¼	2.63	1.50	2.02	0.81
3 × 2	2.93	3.35	1.57	0.69
4 × 3	3.14	7.38	1.14	0.53

In the outline below, hot- and cold-fluid temperatures are represented by upper and lower case letters, respectively. All fluid properties are indicated by lower case letters to eliminate the requirement for new nomenclature.

Process conditions required:

Hot fluid: T_1, T_2, W, c, s or ρ, μ, k, ΔP, R_{do} or R_{di}
Cold fluid: t_1, t_2, w, c, s or ρ, μ, k, ΔP, R_{di} or R_{do}

The diameter of the pipes must be given or assumed.

COUNTERFLOW

A convenient order of calculation follows:

(1) From T_1, T_2, t_1, t_2 check the heat balance, Q, using c at T_{mean} and t_{mean}.

$$Q = WC(T_1 - T_2) = wc(t_2 - t_1)$$

Radiation losses from the exchanger are usually insignificant compared with the heat load transferred in the exchanger.

(2) LMTD, assuming counterflow. (5.14)

(3) T_c and t_c: If the liquid is neither a petroleum fraction nor a hydrocarbon the caloric temperatures cannot be determined through the use of Fig. 17 and Eqs. (5.28) and (5.29). Instead, the calculation of U_C must be performed for the hot and cold terminals giving U_h and U_c from which one may obtain K_c. F_c is then gotten from Fig. 17 or Eq. (5.27). If neither of the liquids is very viscous at the cold terminal, say not more than 1.0 centipoise, if the temperature ranges do not exceed 50 to 100°F, and if the temperature difference is less than 50°F, the arithmetic means of T_1 and T_2 and t_1 and t_2 may be used in place of T_c and t_c for evaluating the physical properties. For nonviscous fluids $\phi = (\mu/\mu_w)^{0.14}$ may be taken as 1.0 as assumed below.

Inner pipe:

(4) Flow area, $a_p = \pi D^2/4$, ft².
(5) Mass velocity, $G_p = w/a_p$, lb/(hr)(ft²).
(6) Obtain μ at T_c or t_c depending upon which flows through the inner pipe. μ, lb/(ft)(hr) = centipoise × 2.42.
From D ft, G_p lb/(hr)(ft²), μ lb/(ft)(hr) obtain the Reynolds number, $Re_p = DG_p/\mu$.
(7) From Fig. 24 in which $j_H = (h_i D/k)(c\mu/k)^{-1/3}(\mu/\mu_w)^{-0.14}$ vs. DG_p/μ obtain j_H.
(8) From c Btu/(lb)(°F), μ lb/(ft)(hr), k Btu/(hr)(ft²)(°F/ft), all obtained at T_c or t_c compute $(c\mu/k)^{1/3}$.
(9) To obtain h_i multiply j_H by $(k/D)(c\mu/k)^{1/3}$ ($\phi = 1.0$) or

$$\frac{h_i D}{k}\left(\frac{c\mu}{k}\right)^{-1/3}\left(\frac{\mu}{\mu_w}\right)^{-0.14} \frac{k}{D}\left(\frac{c\mu}{k}\right)^{1/3} \times 1.0 = h_i \text{ Btu/(hr)(ft}^2\text{)(°F)} \quad (6.15a)$$

(10) Convert h_i to h_{io}; $h_{io} = h_i(A_i/A) = h_i \times \text{ID/OD}$. (6.5)

Annulus:

(4') Flow area, $a_a = \pi(D_2^2 - D_1^2)/4$, ft²

Equivalent diameter $D_e = \dfrac{4 \times \text{flow area}}{\text{wetted perimeter}} = \dfrac{D_2^2 - D_1^2}{D_1}$ ft (6.3)

(5') Mass velocity, $G_a = w/a_a$, lb/(hr)(ft²)
(6') Obtain μ at T_c or t_c, lb/(ft)(hr) = centipoise × 2.42. From D_e ft, G_a lb/(hr)(ft²), μ lb/(ft)(hr) obtain the Reynolds number,

$$Re_a = \frac{D_e G_a}{\mu}$$

(7') From Fig. 24 in which $j_H = (h_o D_e/k)(c\mu/k)^{-1/3}(\mu/\mu_w)^{-0.14}$ vs. $D_e G_a/\mu$ obtain j_H.
(8') From c, μ, and k, all obtained at T_c or t_c compute $(c\mu/k)^{1/3}$.
(9') To obtain h_o multiply j_H by $(k/D_e)(c\mu/k)^{1/3}$ ($\phi = 1.0$) or

$$\frac{h_o D_e}{k}\left(\frac{c\mu}{k}\right)^{-1/3}\left(\frac{\mu}{\mu_w}\right)^{-0.14} \frac{k}{D_e}\left(\frac{c\mu}{k}\right)^{1/3} \times 1.0 = h_o \text{ Btu/(hr)(ft²)(°F)} \quad (6.15b)$$

Overall coefficients:

(11) Compute $U_C = h_{io}h_o/(h_{io} + h_o)$, Btu/(hr)(ft²)(°F). (6.7)
(12) Compute U_D from $1/U_D = 1/U_C + R_d$. (6.10)
(13) Compute A from $Q = U_D A \Delta t$ which may be translated into length. If the length should not correspond to an integral number of hairpins, a change in the dirt factor will result. The recalculated dirt factor should equal or exceed the required dirt factor by using the next larger integral number of hairpins.

Calculation of ΔP. This requires a knowledge of the total length of path satisfying the heat-transfer requirements.

Inner pipe:

(1) For Re_p in (6) above obtain f from Eq. (3.46) or (3.47b).
(2) $\Delta F_p = 4fG^2L/2g\rho^2 D$, ft. (3.45)
$\Delta F_p \rho/144 = \Delta P_p$, psi.

Annulus:

(1') Obtain $D'_e = \dfrac{4\pi(D_2^2 - D_1^2)}{4\pi(D_2 + D_1)} = (D_2 - D_1)$. (6.4)

Compute the frictional Reynolds number, $Re'_a = D'_e G_a/\mu$. For Re'_a obtain f from Eq. (3.46) or (3.47b).
(2') $\Delta F_a = 4fG^2L/2g\rho^2 D'_e$, ft. (6.14)
(3') Entrance and exit losses, one velocity head per hairpin:

$$\Delta F_l = \frac{V^2}{2g'} \text{ ft/hairpin}$$

$(\Delta F_a + \Delta F_l)\rho/144 = \Delta P_a$, psi.

There is an advantage if both fluids are computed side by side, and the use of the outline in this manner will be demonstrated in Example 6.1.

Example 6.1. Double Pipe Benzene-Toluene Exchanger. It is desired to heat 9820 lb/hr of cold benzene from 80 to 120°F using hot toluene which is cooled from 160 to 100°F. The specific gravities at 68°F are 0.88 and 0.87, respectively. The other fluid properties will be found in the Appendix. A fouling factor of 0.001 should be provided for each stream, and the allowable pressure drop on each stream is 10.0 psi.

A number of 20-ft hairpins of 2- by 1¼-in. IPS pipe are available. How many hairpins are required?

Solution:

(1) Heat balance:

Benzene, $t_{av} = \dfrac{80 + 120}{2} = 100°F \qquad c = 0.425$ Btu/(lb)(°F) \qquad (Fig. 2)

$Q = 9820 \times 0.425(120 - 80) = 167{,}000$ Btu/hr

Toluene, $T_{av} = \dfrac{160 + 100}{2} = 130°F \qquad c = 0.44$ Btu/(lb)(°F) \qquad (Fig. 2)

$W = \dfrac{167{,}000}{0.44(160 - 100)} = 6330$ lb/hr

(2) LMTD, (see the method of Chap. 3):

Hot fluid		Cold fluid	Diff.	
160	Higher temp	120	40	Δt_2
100	Lower temp	80	20	Δt_1
			20	$\Delta t_2 - \Delta t_1$

$$\text{LMTD} = \dfrac{\Delta t_2 - \Delta t_1}{2.3 \log \Delta t_2 / \Delta t_1} = \dfrac{20}{2.3 \log {}^{40}\!/_{20}} = 28.8°F \qquad (5.14)$$

(3) Caloric temperatures: A check of both streams will show that neither is viscous at the cold terminal (the viscosities less than 1 centipoise) and the temperature ranges and temperature difference are moderate. The coefficients may accordingly be evaluated from properties at the arithmetic mean, and the value of $(\mu/\mu_w)^{0.14}$ may be assumed equal to 1.0.

$T_{av} = \frac{1}{2}(160 + 100) = 130°F \qquad t_{av} = \frac{1}{2}(120 + 80) = 100°F$

Proceed now to the inner pipe. A check of Table 6.2 indicates that the flow area of the inner pipe is greater than that of the annulus. Place the larger stream, benzene in the inner pipe.

Hot fluid: annulus, toluene
(4') Flow area,
$D_2 = 2.067/12 = 0.1725$ ft
$D_1 = 1.66/12 = 0.138$ ft
$a_a = \pi(D_2^2 - D_1^2)/4$
$\quad = \pi(0.1725^2 - 0.138^2)/4 = 0.00826$ ft²
Equiv diam, $D_e = (D_2^2 - D_1^2)/D_1$ ft
[Eq (6.3)]
$D_e = (0.1725^2 - 0.138^2)/0.138$
$\qquad = 0.0762$ ft

Cold fluid: inner pipe, benzene
(4) $D = 1.38/12 = 0.115$ ft
Flow area, $a_p = \pi D^2/4$
$\quad = \pi \times 0.115^2/4 = 0.0104$ ft²

Hot fluid: annulus, toluene

(5') Mass vel, $G_a = W/a_a$
   $= 6330/0.00826 = 767,000$ lb/(hr)(ft^2)
(6') At 130°F, $\mu = 0.41$ cp   [Fig. 14]
   $= 0.41 \times 2.42 = 0.99$ lb/(ft)(hr)

Reynolds no., $Re_a = \dfrac{D_e G_a}{\mu}$
   $= 0.0762 \times 767,000/0.99 = 59,000$

(7') $j_H = 167$ [Fig. 24]
(8') At 130°F, $c = 0.44$ Btu/(lb)(°F)   [Fig. 2]
$k = 0.085$ Btu/(hr)(ft^2)(°F/ft)   [Table 4]
$\left(\dfrac{c\mu}{k}\right)^{1/3} = \left(\dfrac{0.44 \times 0.99}{0.085}\right)^{1/3} = 1.725$

(9') $h_o = j_H \dfrac{k}{D_e}\left(\dfrac{c\mu}{k}\right)^{1/3}\left(\dfrac{\mu}{\mu_w}\right)^{0.14}$
   [Eq. (6.15b)]
   $= 167 \times \dfrac{0.085}{0.0762} \times 1.725 \times 1.0$
   $= 323$ Btu/(hr)(ft^2)(°F)

Cold fluid: inner pipe, benzene

(5) Mass vel, $G_p = w/a_p$
   $= 9820/0.0104 = 943,000$ lb/(hr)(ft^2)
(6) At 100°F, $\mu = 0.50$ cp   [Fig. 14]
   $= 0.50 \times 2.42 = 1.21$ lb/(ft)(hr)

Reynolds no., $Re_p = \dfrac{DG_p}{\mu}$
   $= 0.115 \times 943,000/1.21 = 89,500$

(7) $j_H = 236$ [Fig. 24]
(8) At 100°F, $c = 0.425$ Btu/(lb)(°F)   [Fig. 2]
$k = 0.091$ Btu/(hr)(ft^2)(°F/ft)   [Table 4]
$\left(\dfrac{c\mu}{k}\right)^{1/3} = \left(\dfrac{0.425 \times 1.21}{0.091}\right)^{1/3} = 1.78$

(9) $h_i = j_H \dfrac{k}{D}\left(\dfrac{c\mu}{k}\right)^{1/3}\left(\dfrac{\mu}{\mu_w}\right)^{0.14}$
   [Eq. (6.15a)]
   $= 236 \times \dfrac{0.091}{0.115} \times 1.78 \times 1.0$
   $= 333$ Btu/(hr)(ft^2)(°F)

(10) Correct h_i to the surface at the OD
$h_{io} = h_i \times \dfrac{\text{ID}}{\text{OD}}$ [Eq. (6.5)]
   $= 333 \times \dfrac{1.38}{1.66} = 276$

Now proceed to the annulus.

(11) Clean overall coefficient, U_C:
$$U_C = \dfrac{h_{io}h_o}{h_{io} + h_o} = \dfrac{276 \times 323}{276 + 323} = 149 \text{ Btu/(hr)(ft}^2\text{)(°F)} \tag{6.7}$$

(12) Design overall coefficient, U_D:
$$\dfrac{1}{U_D} = \dfrac{1}{U_C} + R_d \tag{6.10}$$
$R_d = 0.002$ (required by problem)
$\dfrac{1}{U_D} = \dfrac{1}{149} + 0.002$
$U_D = 115$ Btu/(hr)(ft^2)(°F)

Summary

323	h outside	276
U_C	149	
U_D	115	

(13) Required surface:

$$Q = U_D A\, \Delta t \qquad A = \dfrac{Q}{U_D\, \Delta t}$$

$$\text{Surface} = \dfrac{167,000}{115 \times 28.8} = 50.5 \text{ ft}^2$$

From Table 11 for 1¼-in. IPS standard pipe there are 0.435 ft² of external surface per foot length.

$$\text{Required length} = \frac{50.5}{0.435} = 116 \text{ lin ft}$$

This may be fulfilled by connecting three 20-ft hairpins in series.

(14) The surface supplied will actually be 120 × 0.435 = 52.2 ft². The dirt factor will accordingly be greater than required. The actual design coefficient is

$$U_D = \frac{167{,}000}{52.2 \times 28.8} = 111 \text{ Btu/(hr)(ft}^2)(°\text{F})$$

$$R_d = \frac{U_C - U_D}{U_C U_D} = \frac{149 - 111}{149 \times 111} = 0.0023 \text{ (hr)(ft}^2)(°\text{F})/\text{Btu} \qquad (6.13)$$

Pressure Drop

(1') D'_e for pressure drop differs from D_e for heat transfer.
$D'_e = (D_2 - D_1)$ [Eq. (6.4)]
$= (0.1725 - 0.138) = 0.0345$ ft

$$Re'_a = \frac{D'_e G_a}{\mu}$$

$= 0.0345 \times 767{,}000/0.99 = 26{,}800$

$$f = 0.0035 + \frac{0.264}{26{,}800^{0.42}} = 0.0071$$

[Eq. (3.47b)]

$s = 0.87$, $\rho = 62.5 \times 0.87 = 54.3$
[Table 6]

(2') $\Delta F_a = \dfrac{4fG_a^2 L}{2g\rho^2 D'_e}$

$= \dfrac{4 \times 0.0071 \times 767{,}000^2 \times 120}{2 \times 4.18 \times 10^8 \times 54.3^2 \times 0.0345}$

$= 23.5$ ft

(3') $V = \dfrac{G}{3600\rho} = \dfrac{767{,}000}{3600 \times 54.3} = 3.92$ fps

$F_l = 3\left(\dfrac{V^2}{2g'}\right) = 3 \times \dfrac{3.92^2}{2 \times 32.2} = 0.7$ ft

$\Delta P_a = \dfrac{(23.5 + 0.7)54.3}{144} = 9.2$ psi

Allowable $\Delta P_a = 10.0$ psi

(1) For $Re_p = 89{,}500$ in (6) above
$f = 0.0035 + \dfrac{0.264}{(DG/\mu)^{0.42}}$ [Eq. (3.47b)]

$= 0.0035 + \dfrac{0.264}{89{,}500^{0.42}} = 0.0057$

$s = 0.88$, $\rho = 62.5 \times 0.88 = 55.0$
[Table 6]

(2) $\Delta F_p = \dfrac{4fG_p^2 L}{2g\rho^2 D}$

$= \dfrac{4 \times 0.0057 \times 943{,}000^2 \times 120}{2 \times 4.18 \times 10^8 \times 55.0^2 \times 0.115}$

$= 8.3$ ft

$\Delta P_p = \dfrac{8.3 \times 55.0}{144} = 3.2$ psi

Allowable $\Delta P_p = 10.0$ psi

A check of U_h and U_c gives 161 and 138, respectively, and $K_c = 0.17$. From Fig. 17 for $\Delta t_c / \Delta t_h = {}^{20}\!/_{40} = 0.5$, $F_c = 0.43$, whereas in the solution above the arithmetic mean temperatures were used. The arithmetic mean assumes $F_c = 0.50$. However, since the ranges are small for both fluids, the error is too small to be significant. If the ranges of the fluids or their viscosities were large, the error might be considerable for $F_c = 0.43$.

Double Pipe Exchangers in Series-parallel Arrangements. Referring to Example 6.1, it is seen that a calculated pressure drop of 9.2 psi is obtained against an allowable pressure drop of 10.0 psi. Suppose, however, that the calculated pressure drop were 15 or 20 psi and exceeded the available head. How then might the heat load be transferred with the available pressure head? One possibility is the use of a by-pass so that only three-quarters or two-thirds of the fluid flows through the exchanger and the remainder through the by-pass. This does not provide an ideal solution, since the reduced flow causes several unfavorable changes in the design. (1) The reduced flow through the exchanger reduces the mass velocity G_a and the film coefficient h_o. Since both of the coefficients are nearly alike, 323 vs. 276, any sizable reduction in G_a alone decreases U_C by nearly $G_a^{0.8}$. (2) If less liquid circulates through the annulus, it has to be cooled over a longer range than from 160 to 100°F so that, upon

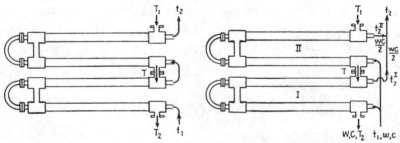

FIG. 6.5. Double pipe exchangers in series.   FIG. 6.6. Series—parallel arrangement.

mixing with the by-pass fluid, the process outlet temperature of 100°F results. As an example, the portion circulating through the annulus might have to be cooled over the range from 160 to 85°F depending upon the percentage by-passed. The outlet temperature of 85°F is closer to the inner pipe inlet of 80°F than originally, and the new cold-terminal difference Δt_1 of only 5°F greatly decreases the LMTD. The two effects, decreased U_C and LMTD, increase the required number of hairpins greatly even though the heat load is constant. Reversing the location of the streams by placing the benzene in the annulus does not provide a solution in this case, since the benzene stream is larger than the toluene stream. The possibility of reversing the location of the streams should always be examined first whenever the allowable pressure drop cannot be met.

A solution is still possible, however, even when all the above have failed. When two double pipe exchangers are connected in series, the arrangement is shown in Fig. 6.5. Suppose that the stream which is too large to be accommodated in several exchangers in series is divided in half and each half traverses but one exchanger through the inner pipes in

Fig. 6.6. Dividing a stream in half while keeping the flow area constant produces about *one-eighth* of the series pressure drop, since G and L will be half and the product of G^2L in Eq. (6.14) will be one-eighth. While the film coefficient will also be reduced, the unfavorable temperature difference of by-passing can be circumvented. Where there is a substantial unbalance between the weight flow of the two streams because one operates over a long range and the other over a very short range, the large stream may be divided in three, four, or more parallel streams. In larger services each parallel stream may also flow through several exchangers in series in each parallel bank. The term "parallel streams" should not be confused with "parallel flow." The former refers to the division of the flow of one fluid, while the latter refers to the direction of flow between two fluids.

The True Temperature Difference for Series-parallel Arrangements. The LMTD calculated from T_1, T_2, t_1, and t_2 for the series arrangement will not be the same for a series-parallel arrangement. Half of the pipe fluid enters the upper exchanger II in Fig. 6.6 where the annulus fluid is hot, and half enters the lower exchanger I in which the annulus fluid has already been partially cooled. While exchangers in series do not transfer equal quantities of heat, the series-parallel relationship is even more adverse, the lower exchanger accounting for relatively less of the total heat transfer. If the true temperature difference is called Δt, it will not be identical with the LMTD for the process conditions although both of the exchangers operate in counterflow.

Consider the two exchangers in Fig. 6.6 designated by I and II. The intermediate temperature is T, and the outlets of the parallel streams are designated by t_2^{II} and t_2^{I}. Their mixed temperature is t_2.

For exchanger I, containing half the surface,

$$Q_I = WC(T - T_2) = \frac{UA}{2} \times \text{LMTD}_I \qquad (6.16)$$

and

$$\text{LMTD}_I = \frac{(T - t_2^I) - (T_2 - t_1)}{\ln (T - t_2^I)/(T_2 - t_1)} \qquad (6.17)$$

Substituting in Eq. (6.16),

$$\frac{UA}{2WC} = \frac{(T - T_2)}{(T - t_2^I) - (T_2 - t_1)} \ln \frac{T - t_2^I}{T_2 - t_1}$$

Rearranging,

$$\frac{UA}{2WC} = \frac{(T - T_2)}{(T - T_2) - (t_2^I - t_1)} \ln \frac{(T - t_2^I)}{(T_2 - t_1)}$$

$$= \frac{1}{1 - (t_2^I - t_1)/(T - T_2)} \ln \frac{T - t_2^I}{T_2 - t_1} \qquad (6.18)$$

Let
$$R^{\mathrm{I}} = \frac{(T - T_2)}{(t_2^{\mathrm{I}} - t_1)} = \frac{wc}{2WC}$$
$$\frac{UA}{2WC} = \frac{R^{\mathrm{I}}}{R^{\mathrm{I}} - 1} \ln \frac{T - t_2^{\mathrm{I}}}{T_2 - t_1} \tag{6.19}$$

Similarly for exchanger II
$$Q_{\mathrm{II}} = WC(T_1 - T) = \frac{UA}{2} \times \mathrm{LMTD}_{\mathrm{II}} \tag{6.20}$$
$$\mathrm{LMTD}_{\mathrm{II}} = \frac{(T_1 - t_2^{\mathrm{II}}) - (T - t_1)}{\ln (T_1 - t_2^{\mathrm{II}})/(T - t_1)} \tag{6.21}$$

Let
$$R^{\mathrm{II}} = \frac{T_1 - T}{t_2^{\mathrm{II}} - t_1} = \frac{wc}{2WC}$$
$$\frac{UA}{2WC} = \frac{R^{\mathrm{II}}}{R^{\mathrm{II}} - 1} \ln \frac{T_1 - t_2^{\mathrm{II}}}{T - t_1} \tag{6.22}$$

Since c and C were assumed constant,
$$R^{\mathrm{I}} = R^{\mathrm{II}} = R' = \frac{wc}{2WC} \tag{6.23}$$

Let
$$S^{\mathrm{I}} = \frac{t_2^{\mathrm{I}} - t_1}{T - t_1} \qquad M^{\mathrm{I}} = \frac{T - T_2}{T - t_1}$$
$$M^{\mathrm{I}} = R'S^{\mathrm{I}}$$

Similarly let
$$S^{\mathrm{II}} = \frac{t_2^{\mathrm{II}} - t_1}{T_1 - t_1} \qquad M^{\mathrm{II}} = \frac{T_1 - T}{T_1 - t_1}$$

R' and S are ratios which recur frequently in obtaining the true temperature difference Δt from the LMTD. S is the ratio of the cold fluid range to the maximum temperature span, the latter being the difference between both inlet temperatures, T_1 and t_1.

But
$$M^{\mathrm{II}} = R'S^{\mathrm{II}}$$
$$1 - S^{\mathrm{I}} = \frac{T - t_2^{\mathrm{I}}}{T - t_1} = \frac{T - t_1}{T - t_1} - \frac{t_2^{\mathrm{I}} - t_1}{T - t_1}$$
$$\frac{T - t_2^{\mathrm{I}}}{T_2 - t_1} = \frac{1 - S^{\mathrm{I}}}{1 - R'S^{\mathrm{I}}}$$

and from Eq. (6.19)
$$\frac{UA}{2WC} = \frac{R'}{R' - 1} \ln \frac{1 - S^{\mathrm{I}}}{1 - R'S^{\mathrm{I}}} \tag{6.24}$$

and from Eq. (6.22)
$$\frac{UA}{2WC} = \frac{R'}{R' - 1} \ln \frac{1 - S^{\mathrm{II}}}{1 - R'S^{\mathrm{II}}} \tag{6.25}$$

and equating Eqs. (6.24) and (6.25),

$$\frac{1 - S^{\mathrm{I}}}{1 - R'S^{\mathrm{I}}} = \frac{1 - S^{\mathrm{II}}}{1 - R'S^{\mathrm{II}}}$$

Therefore
$$S^{\mathrm{I}} = S^{\mathrm{II}}$$
$$M^{\mathrm{I}} = M^{\mathrm{II}}$$

Adding Eqs. (6.24) and (6.25),

$$\frac{UA}{WC} = \frac{2R'}{R' - 1} \ln \frac{1 - S^{\mathrm{I}}}{1 - R'S^{\mathrm{I}}} = \frac{2R'}{R' - 1} \ln \frac{T - t_2^{\mathrm{I}}}{T_2 - t_1} \quad (6.26)$$

in which T is the only unknown, and since $M^{\mathrm{I}} = M^{\mathrm{II}}$,

$$\frac{T_1 - T}{T_1 - t_1} = \frac{T - T_2}{T - t_1}$$
$$T^2 - 2t_1 T + t_1(T_1 + T_2) - T_1 T_2 = 0 \quad (6.27)$$

Equation (6.27) is a quadratic whose solution is

$$T = \frac{2t_1 \pm \sqrt{4t_1^2 - 4t_1[(T_1 + T_2) - 4T_1 T_2]}}{2}$$
$$= t_1 \pm \sqrt{(T_1 - t_1)(T_2 - t_1)} \quad (6.28)$$

The minus sign applies when the heating medium is in the pipes. The plus sign applies when the cooling medium is in the pipes.

Substituting for T in Eq. (6.26),

$$\frac{UA}{WC} = \frac{2R'}{R' - 1} \ln \left[\frac{(R' - 1)(T_1 - t_1) + \sqrt{(T_1 - t_1)(T_2 - t_1)}}{R' \sqrt{(T_1 - t_1)(T_2 - t_1)}} \right]$$
$$= \frac{2R'}{R' - 1} \ln \left[\left(\frac{R' - 1}{R'}\right) \left(\frac{T_1 - t_1}{T_2 - t_1}\right)^{\frac{1}{2}} + \frac{1}{R'} \right] \quad (6.29)$$

Δt is the single value for the entire series-parallel arrangement; thus

$$Q = UA \, \Delta t = WC(T_1 - T_2) \quad (6.30)$$
$$\Delta t = \frac{Q}{UA} = \frac{WC}{UA} (T_1 - T_2) \quad (6.31)$$

It is convenient in this derivation to employ a definition for the true temperature difference in terms of the maximum temperature span $T_1 - t_1$:

$$\Delta t = \gamma(T_1 - t_1) \quad (6.32)$$

Equating (6.31) and (6.32),

$$\frac{WC}{UA}(T_1 - T_2) = \gamma(T_1 - t_1)$$
$$\gamma = \frac{WC(T_1 - T_2)}{UA(T_1 - t_1)}$$

Since $M = (T_1 - T_2)/(T_1 - t_1)$, define $P' = (T_2 - t_1)/(T_1 - t_1)$ and $UA/WC = M/\gamma$; then

$$P' + M = 1 \quad \text{or} \quad M = 1 - P'$$

Substituting in Eq. (6.29),

$$\frac{UA}{WC} = \frac{2R'}{R' - 1} \ln \left[\left(\frac{R' - 1}{R'} \right) \left(\frac{1}{P'} \right)^{\frac{1}{2}} + \frac{1}{R'} \right] \quad (6.33)$$

or

$$\frac{1 - P'}{\gamma} = 2 \left(\frac{R'}{R' - 1} \right) \ln \left[\left(\frac{R' - 1}{R'} \right) \left(\frac{1}{P'} \right)^{\frac{1}{2}} + \frac{1}{R'} \right] \quad (6.34)$$

If developed in a generalized manner it can be shown that, for *one series hot stream* and *n parallel cold streams*, Eq. (6.34) becomes

$$\frac{1 - P'}{\gamma} = 2.3 \frac{nR'}{R' - 1} \log \left[\left(\frac{R' - 1}{R'} \right) \left(\frac{1}{P'} \right)^{1/n} + \frac{1}{R'} \right] \quad (6.35a)$$

where

$$R' = \frac{T_1 - T_2}{n(t_2 - t_1)}$$

For *one series cold stream* and *n parallel hot streams*,

$$\frac{1 - P''}{\gamma} = 2.3 \frac{n}{1 - R''} \log \left[(1 - R'') \left(\frac{1}{P''} \right)^{1/n} + R'' \right] \quad (6.35b)$$

where

$$P'' = \frac{T_1 - t_2}{T_1 - t_1} \quad \text{and} \quad R'' = \frac{n(T_1 - T_2)}{t_2 - t_1}$$

Example 6.2. Calculation of the True Temperature Difference. A bank of double pipe exchangers operates with the hot fluid in series from 300 to 200°F and the cold fluid in six parallel streams from 190 to 220°F. What is the true temperature difference Δt?

$$P' = \frac{T_2 - t_1}{T_1 - t_1} = \frac{200 - 190}{300 - 190} = 0.091 \quad R' = \frac{T_1 - T_2}{n(t_2 - t_1)} = \frac{300 - 200}{6(220 - 190)} = 0.558$$

Substituting in Eq. (6.35a) and solving, $\gamma = 0.242$.

$$\Delta t = 0.242(300 - 190) = 26.6°F \quad (6.32)$$

The LMTD would be 33.7°F, and an error of 27 per cent would be introduced by its use.

Exchangers with a Viscosity Correction, ϕ. For heating or cooling fluids, the use of Fig. 24 with an assumed value of $(\mu/\mu_w)^{0.14} = 1.0$ also assumes a negligible deviation of fluid properties from isothermal flow. For nonviscous fluids the deviation from isothermal flow **during heating**

or cooling does not introduce an appreciable error in the calculation of the heat-transfer coefficient. When the pipe-wall temperature differs appreciably from the caloric temperature of the controlling fluid and the controlling fluid is viscous, the actual value of $\phi = (\mu/\mu_w)^{0.14}$ must be taken into account. To include the correction, t_w may be determined by Eq. (5.31) or by (5.32) from uncorrected values of h_o/ϕ_a and h_{io}/ϕ_p, which are then corrected accordingly by multiplication by ϕ_a and ϕ_p respectively. The corrected coefficients where $\phi \neq 1.0$ are

$$h_o = \left(\frac{h_o}{\phi_a}\right)\phi_a \qquad (6.36)$$

$$h_{io} = \left(\frac{h_{io}}{\phi_p}\right)\phi_p \qquad (6.37)$$

Similarly for two resistances in series employing the viscosity corrections for deviation from the isothermal the clean overall coefficient is again

$$U_C = \frac{h_{io}h_o}{h_{io} + h_o} \qquad (6.38)$$

Example 6.3. Double Pipe Lube Oil–Crude Oil Exchanger. 6,900 lb/hr of a 26°API lube oil must be cooled from 450 to 350°F by 72,500 lb/hr of 34°API mid-continent crude oil. The crude oil will be heated from 300 to 310°F.

A fouling factor of 0.003 should be provided for each stream, and the allowable pressure drop on each stream will be 10 psi.

A number of 20-ft hairpins of 3- by 2-in. IPS pipe are available. How many must be used, and how shall they be arranged? The viscosity of the crude oil may be obtained from Fig. 14. For the lube oil, viscosities are 1.4 centipoises at 500°F, 3.0 at 400°F, and 7.7 at 300°F. These viscosities are great enough to introduce an error if $(\mu/\mu_w)^{0.14} = 1$ is assumed.

Solution:

(1) Heat Balance:
  Lube oil, $Q = 6900 \times 0.62(450 - 350) = 427{,}000$ Btu/hr
  Crude oil, $Q = 72{,}500 \times 0.585(310 - 300) = 427{,}000$ Btu/hr

(2) Δt:

Hot Fluid		Cold Fluid	Diff.	
450	Higher temp	310	140	Δt_2
350	Lower temp	300	50	Δt_1
			90	$\Delta t_2 - \Delta t_1$

It will be impossible to put the 72,500 lb/hr of crude into a single pipe or annulus, since the flow area of each is too small. Assume, as a trial, that it will be employed in two parallel streams.

$$\Delta t = 87.5°\text{F} \qquad (6.35a)$$

(3) Caloric temperatures:
$$\frac{\Delta t_c}{\Delta t_h} = \frac{50}{140} = 0.357 \qquad K_c = 0.43 \qquad F_c = 0.395 \qquad \text{(Fig. 17)}$$
$$T_c = 350 \times 0.395(450 - 350) = 389.5°\text{F} \qquad (5.28)$$
$$t_c = 300 \times 0.395(310 - 300) = 304°\text{F} \qquad (5.29)$$

Proceed now to the inner pipe.

Hot fluid: annulus, lube oil
(4′) Flow area, $D_2 = 3.068/12 = 0.256$ ft
$D_1 = 2.38/12 = 0.199$ ft
$a_a = \pi(D_2^2 - D_1^2)/4$
$= \pi(0.256^2 - 0.199^2)/4 = 0.0206$ ft^2
Equiv diam, $D_e = (D_2^2 - D_1^2)/D_1$
[Eq. (6.3)]
$= (0.256^2 - 0.199^2)/0.199 = 0.13$ ft
(5′) Mass vel, $G_a = W/a_a$
$= 6900/0.0206 = 335{,}000$ lb/(hr)(ft^2)

(6′) At 389.5°F, $\mu = 3.0$ cp
$= 3.0 \times 2.42 = 7.25$ lb/(ft)(hr)
[Fig. 14]
$Re_a = D_e G_a/\mu$ [Eq. (3.6)]
$= 0.13 \times 335{,}000/7.25 = 6{,}000$
If only 2 hairpins in series are required, L/D will be $2 \times 40/0.13 = 614$. Use $L/D = 600$.
(7′) $j_H = 20.5$ [Fig. 24]
(8′) At $T_c = 389.5°$F, $c = 0.615$ Btu/(lb)(°F) [Fig. 4]
$k = 0.067$ Btu/(hr)(ft^2)(°F/ft) [Fig. 1]
$\left(\dfrac{c\mu}{k}\right)^{1/3} = \left(\dfrac{0.615 \times 7.25}{0.067}\right)^{1/3} = 4.05$
(9′) $h_o = j_H \dfrac{k}{D_e}\left(\dfrac{c\mu}{k}\right)^{1/3}\phi_a$ [Eq. (6.15)]
$\dfrac{h_o}{\phi_a} = \dfrac{20.5 \times 0.067 \times 4.05}{0.13}$
$= 42.7$ Btu/(hr)(ft^2)(°F)
$t_w = t_c + \dfrac{h_o/\phi_a}{h_{io}/\phi_p + h_o/\phi_a}(T_c - t_c)$
[Eq. (5.31)]
$= 304 + \dfrac{42.7}{297 + 42.7}(389.5 - 304)$
$= 314°$F
$\mu_w = 6.6 \times 2.42 = 16.0$ lb/(ft)(hr)
[Fig. 14]
$\phi_a = (\mu/\mu_w)^{0.14}$
$= (7.25/16.0)^{0.14} = 0.90$ [Fig. 24]
$h_o = \dfrac{h_o}{\phi_a}\phi_a$ [Eq. (6.36)]
$= 42.7 \times 0.90 = 38.4$

Cold fluid: inner pipe, crude oil
(4) Flow area, $D = 2.067/12 = 0.172$ ft
$a_p = \pi D^2/4$
$= \pi \times 0.172^2/4 = 0.0233$ ft^2
Since two parallel streams have been assumed, $w/2$ lb/hr will flow in each pipe.

(5) Mass vel, $G_p = w/a_p$
$= \dfrac{72{,}500}{2 \times 0.0233}$
$= 1{,}560{,}000$ lb/(hr)(ft^2)

(6) At 304°F, $\mu = 0.83$ cp
$= 0.83 \times 2.42 = 2.01$ lb/(ft)(hr)
[Fig. 14]
$Re_p = DG_p/\mu$
$= 0.172 \times 1{,}560{,}000/2.01 = 133{,}500$

(7) $j_H = 320$ [Fig. 24]
(8) At $t_c = 304°$F, $c = 0.585$ Btu/(lb)(°F)
[Fig. 4]
$k = 0.073$ Btu/(hr)(ft^2)(°F/ft) [Fig. 1]
$\left(\dfrac{c\mu}{k}\right)^{1/3} = \left(\dfrac{0.585 \times 2.01}{0.073}\right)^{1/3} = 2.52$
(9) $h_i = j_H \dfrac{k}{D}\left(\dfrac{c\mu}{k}\right)^{1/3}\phi_p$ [Eq. (6.15a)]
$\dfrac{h_i}{\phi_p} = \dfrac{320 \times 0.073 \times 2.52}{0.172}$
$= 342$ Btu/(hr)(ft^2)(°F)
(10) $\dfrac{h_{io}}{\phi_p} = \dfrac{h_i}{\phi_p} \times \dfrac{\text{ID}}{\text{OD}}$
$= 342 \times 2.067/2.38 = 297$
Now proceed from (4′) to (9′) to obtain t_w.
$\mu_w = 0.77 \times 2.42 = 1.86$ [Fig. 14]
$\phi_p = (\mu/\mu_w)^{0.14}$
$= (2.01/1.86)^{0.14} = 1.0$ nearly
[Fig. 24]
$h_{io} = \dfrac{h_{io}}{\phi_p}\phi_p$ [Eq. (6.37)]
$= 297 \times 1.0 = 297$

COUNTERFLOW

(11) Clean overall coefficient, U_C:

$$U_C = \frac{h_{io}h_o}{h_{io}+h_o} = \frac{297 \times 38.4}{297 + 38.4} = 34.0 \text{ Btu/(hr)(ft}^2\text{)(°F)} \tag{6.38}$$

(12) Design overall coefficient, U_D:

$$\frac{1}{U_D} = \frac{1}{U_C} + R_d \tag{6.10}$$

$R_d = 0.003 + 0.003 = 0.006 \text{ (hr)(ft}^2\text{)(°F)/Btu}$
$U_D = 28.2$

Summary

38.4	h outside	297
U_C	34.0	
U_D	28.2	

(13) Surface:

$$A = \frac{Q}{U_D \, \Delta t} = \frac{427{,}000}{28.2 \times 87.5} = 173 \text{ ft}^2$$

External surface/lin ft, $a'' = 0.622$ ft (Table 11)

Required length $= \dfrac{173}{0.622} = 278$ lin ft

This is equivalent to more than six 20-ft hairpins or 240 lin feet. Since two parallel streams are employed, use eight hairpins or 320 lin. feet. The hairpins should have the annuli connected in series and the tubes in two parallel banks of four exchangers.

The corrected U_D will be $U_D = Q/A \, \Delta t = 427{,}000/320 \times 0.622 \times 87.5 = 24.5$.
The corrected dirt factor will be $R_d = 1/U_D - 1/U_C = 1/24.5 - 1/34.0 = 0.0114$.

Pressure Drop

(1') $D'_e = (D_2 - D_1)$ [Eq. (6.4)]
$= (0.256 - 0.198) = 0.058$ ft
$Re'_a = D'_e G_a/\mu$
$= 0.058 \times 335{,}000/7.25 = 2680$
$f = 0.0035 + \dfrac{0.264}{2680^{0.42}} = 0.0132$
[Eq. (3.47b)]
$s = 0.775, \rho = 62.5 \times 0.775 = 48.4$
[Fig. 6]

(2') $\Delta F_a = \dfrac{4fG_a^2 L_a}{2g\rho^2 D'_e}$

$= \dfrac{4 \times 0.0132 \times 335{,}000^2 \times 320}{2 \times 4.18 \times 10^8 \times 48.4^2 \times 0.058}$
$= 16.7$ ft

(3') $V = \dfrac{G_a}{3600\rho} = \dfrac{335{,}000}{3600 \times 48.4} = 1.9$ fps

$\Delta F_l = 8\left(\dfrac{V^2}{2g'}\right) = 8\left(\dfrac{1.9^2}{2 \times 32.2}\right) = 0.45$ ft

$\Delta P_a = \dfrac{(16.7 + 0.45) \times 48.4}{144} = 5.8$ psi

Allowable $\Delta P_a = 10.0$ psi

(1) For $Re_p = 133{,}500$ in (6) above
$f = 0.0035 + \dfrac{0.264}{133{,}500^{0.42}} = 0.005375$
[Eq. (3.47b)]
$s = 0.76, \rho = 62.5 \times 0.76 = 47.5$
[Fig. 6]
Halves of the tube fluid will flow through only four exchangers.

(2) $\Delta F_p = \dfrac{4fG_p^2 L_p}{2g\rho^2 D}$

$= \dfrac{4 \times 0.005375 \times 1{,}560{,}000^2 \times 160}{2 \times 4.18 \times 10^8 \times 47.5^2 \times 0.172}$
$= 25.7$ ft

$\Delta P_p = \dfrac{25.7 \times 47.5}{144} = 8.5$ psi

Allowable $\Delta P_p = 10.0$ psi
If the flow had not been divided, the pressure drop would be nearly eight times as great, or about 60 psi.

PROBLEMS

6.1. What is the fouling factor when (a) $U_C = 30$ and $U_D = 20$, (b) $U_C = 60$ and $U_D = 50$, and (c) $U_C = 110$ and $U_D = 100$? Which do you consider reasonable to specify between two moderately clean streams?

6.2. A double pipe exchanger was oversized because no data were available on the rate at which dirt accumulated. The exchanger was originally designed to cool 13,000 lb/hr of 100 per cent acetic acid from 250 to 150°F by heating 19,000 lb/hr of butyl alcohol from 100 to 157°F. A design coefficient $U_D = 85$ was employed, but during initial operation a hot-liquid outlet temperature of 117°F was obtained. It rose during operation at the average rate of 3°F per month. What dirt factor should have been specified for a 6-month cleaning cycle?

6.3. O-xylene coming from storage at 100°F is to be heated to 150°F by cooling 18,000 lb/hr of butyl alcohol from 170 to 140°F. Available for the purpose are five 20-ft hairpin double pipe exchangers with annuli and pipes each connected in series. The exchangers are 3- by 2-in. IPS. What is (a) the dirt factor, (b) the pressure drops? (c) If the hot and cold streams in (a) are reversed with respect to the annulus and inner pipe, how does this justify or refute your initial decision where to place the hot stream?

6.4. 10,000 lb/hr of 57°API gasoline is cooled from 150 to 130°F by heating 42°API kerosene from 70 to 100°F. Pressure drops of 10 psi are allowable with a minimum dirt factor of 0.004. (a) How many 2½- by 1¼-in. IPS hairpins 20 ft long are required? (b) How shall they be arranged? (c) What is the final fouling factor?

6.5. 12,000 lb/hr of 26°API lube oil (see Example 6.3 in text for viscosities) is to be cooled from 450 to 350°F by heating 42°API kerosene from 325 to 375°F. A pressure drop of 10 psi is permissible on both streams, and a minimum dirt factor of 0.004 should be provided. (a) How many 20-ft hairpins of 2½- by 1¼-in. IPS double pipe are required? (b) How shall they be arranged, and (c) what is the final dirt factor?

6.6. 7,000 lb/hr of aniline is to be heated from 100 to 150°F by cooling 10,000 lb/hr of toluene with an initial temperature of 185°F in 2- by 1-in. IPS double pipe hairpin exchangers 15 ft long. Pressure drops of 10 psi are allowable, and a dirt factor of 0.005 is required. (a) How many hairpin sections are required? (b) How shall they be arranged? (c) What is the final dirt factor?

6.7. 24,000 lb/hr of 35°API distillate is cooled from 400 to 300°F by 50,000 lb/hr of 34°API crude oil heated from an inlet temperature of 250°F. Pressure drops of 10 psi are allowable, and a dirt factor of 0.006 is required. Using 20-ft hairpins of 4- by 3-in. IPS (a) how many are required, (b) how shall they be arranged, and (c) what is the final fouling factor?

6.8. A liquid is cooled from 350 to 300°F by another which is heated from 290 to 315°F. How does the true temperature difference deviate from the LMTD if (a) the hot fluid is in series and the cold fluid flows in two parallel counterflow paths, (b) the hot fluid is in series and the cold fluid flows in three parallel-flow–counterflow paths, (c) The cold-fluid range in (a) and (b) is changed to 275 to 300°F.

6.9. A fluid is cooled from 300 to 275°F by heating a cold fluid from 100 to 290°F. If the hot fluid is in series, how is the true temperature difference affected by dividing the hot stream into (a) two parallel streams and (b) into three parallel streams?

6.10. 6330 lb/hr of toluene is cooled from 160 to 100°F by heating amyl acetate from 90 to 100°F using 15-ft hairpins. The exchangers are 2- by 1¼-in. IPS. Allowing 10 psi pressure drops and providing a minimum dirt factor of 0.004 (a) how many hairpins are required, (b) how shall they be arranged, and (c) what is the final dirt factor?

COUNTERFLOW

6.11. 13,000 lb/hr of 26°API gas oil (see Example 6.3 in text for viscosities) is cooled from 450 to 350°F by heating 57°API gasoline under pressure from 220 to 230°F in as many 3- by 2-in. IPS double pipe 20-ft hairpins as are required. Pressure drops of 10 psi are permitted along with a minimum dirt factor of 0.004. (a) How many hairpins are required? (b) How shall they be arranged? (c) What is the final dirt factor?

6.12. 100,000 lb/hr of nitrobenzene is to be cooled from 325 to 275°F by benzene heated from 100 to 300°F. Twenty-foot hairpins of 4- by 3-in. IPS double pipe will be employed, and pressure drops of 10 psi are permissible. A minimum dirt factor of 0.004 is required. (a) How many hairpins are required? (b) How shall they be arranged? (c) What is the final dirt factor?

NOMENCLATURE FOR CHAPTER 6

A	Heat-transfer surface, ft²
a	Flow area, ft²
a''	External surface per linear foot of pipe, ft
C	Specific heat of hot fluid in derivations, Btu/(lb)(°F)
C_1	A constant
c	Specific heat of cold fluid in derivations or either fluid in calculations Btu/(lb)(°F)
D	Inside diameter, ft
D_1, D_2	For annuli D_1 is the outside diameter of inner pipe, D_2 is the inside diameter of the outer pipe, ft
D_e, D_e'	Equivalent diameter for heat-transfer and pressure drop, ft
d_e, d_e'	Equivalent diameter for heat-transfer and pressure drop, in.
D_o	Outside diameter, ft
F_c	Caloric fraction, dimensionless
ΔF	Pressure drop, ft
f	Friction factor, dimensionless
G	Mass velocity, lb/(hr)(ft²)
g	Acceleration of gravity 4.18×10^8 ft/hr²
g'	Acceleration of gravity 32.2 ft/sec²
h, h_i, h_o	Heat-transfer coefficient in general, for inside fluid, and for outside fluid, respectively, Btu/(hr)(ft²)(°F)
h_{io}	Value of h_i when referred to the pipe outside diameter, Btu/(hr)(ft²)(°F)
ID	Inside diameter, ft or in.
j_H	Heat-transfer factor, dimensionless
K_c	Caloric factor, dimensionless
k	Thermal conductivity, Btu/(hr)(ft²)(°F/ft)
L	Pipe length or length of path, ft
M	Temperature group $(T_1 - T_2)/(T_1 - t_1)$, dimensionless
n	Number of parallel streams
OD	Outside diameter, ft or in.
P'	Temperature group $(T_2 - t_1)/(T_1 - t_1)$, dimensionless
P''	Temperature group $(T_1 - t_2)/(T_1 - t_1)$, dimensionless
ΔP	Pressure drop, psi
Q	Heat flow, Btu/hr
R	Temperature group $(T_1 - T_2)/(t_2 - t_1)$, dimensionless
R'	Temperature group $(T_1 - T_2)/n(t_2 - t_1)$, dimensionless
R''	Temperature group $n(T_1 - T_2)/(t_2 - t_1)$, dimensionless

R_d, R_i, R_o	Combined dirt factor, inside dirt factor, outside dirt factor, (hr)(ft²)(°F)/Btu
Re, Re'	Reynolds number for heat transfer and pressure drop, dimensionless
r_h	Hydraulic radius, ft
S	Temperature group $(t_2 - t_1)/(T_1 - t_1)$, dimensionless
s	Specific gravity, dimensionless
T, T_1, T_2	Hot-fluid temperature in general, inlet and outlet of hot fluid, °F
T_c	Caloric temperature of hot fluid, °F
t, t_1, t_2	Cold-fluid temperature in general, inlet and outlet of cold fluid, °F
t_c	Caloric temperature of cold fluid, °F
Δt	True or effective temperature difference in $Q = U_D A \, \Delta t$
$\Delta t_c, \Delta t_h$	Cold- and hot-terminal temperature differences, °F
U, U_C, U_D	Overall coefficient of heat transfer, clean coefficient, design coefficient, Btu/(hr)(ft²)(°F)
V	Velocity, fps
W	Weight flow of hot fluid, lb/hr
w	Weight flow of cold fluid, lb/hr
z	Height, ft
γ	A constant, dimensionless
μ	Viscosity at the caloric temperature, centipoises $\times$ 2.42 = lb/(ft)(hr)
μ_w	Viscosity at the pipe-wall temperature, centipoises $\times$ 2.42 = lb/(ft)(hr)
ρ	Density, lb/ft³
ϕ	$(\mu/\mu_w)^{0.14}$

Subscripts and Superscripts

a	Annulus
l	Loss
p	Pipe
I	First of two exchangers
II	Second of two exchangers

CHAPTER 7

1-2 PARALLEL-COUNTERFLOW: SHELL-AND-TUBE EXCHANGERS

INTRODUCTION

The Tubular Element. The fulfillment of many industrial services requires the use of a large number of double pipe hairpins. These consume considerable ground area and also entail a large number of points at which leakage may occur. Where large heat-transfer surfaces are required, they can best be obtained by means of shell-and-tube equipment.

Shell-and-tube equipment involves expanding a tube into a tube sheet and forming a seal which does not leak under reasonable operating con-

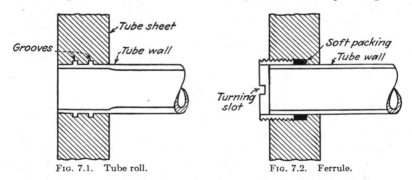

FIG. 7.1. Tube roll.   FIG. 7.2. Ferrule.

ditions. A simple and common example of an expanded tube is shown in Fig. 7.1. A tube hole is drilled in a tube sheet with a slightly greater diameter than the outside diameter of the tube, and two or more grooves are cut in the wall of the hole. The tube is placed inside the tube hole, and a tube roller is inserted into the end of the tube. The roller is a rotating mandril having a slight taper. It is capable of exceeding the elastic limit of the tube metal and transforming it into a semiplastic condition so that it flows into the grooves and forms an extremely tight seal. Tube rolling is a skill, since a tube may be damaged by rolling to paper thinness and leaving a seal with little structural strength.

In some industrial uses it is desirable to install tubes in a tube sheet so that they can be removed easily as shown in Fig. 7.2. The tubes are

128   PROCESS HEAT TRANSFER

actually packed in the tube sheet by means of ferrules using a soft metal packing ring.

Heat-exchanger Tubes. Heat-exchanger tubes are also referred to as condenser tubes and should not be confused with steel pipes or other types of pipes which are extruded to iron pipe sizes. The outside diameter of heat exchanger or condenser tubes is the actual outside diameter in inches within a very strict tolerance. Heat-exchanger tubes are available in a variety of metals which include steel, copper, admiralty, Muntz metal, brass, 70-30 copper-nickel, aluminum bronze, aluminum, and the stainless steels. They are obtainable in a number of different wall thicknesses defined by the Birmingham wire gage, which is usually referred to as the BWG or *gage* of the tube. The sizes of tubes which are generally available are listed in Table 10 of the Appendix of which the ¾ outside diam-

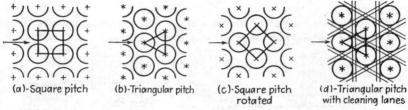

FIG. 7.3.   Common tube layouts for exchangers.

eter and 1 in. OD are most common in heat-exchanger design. The data of Table 10 have been arranged in a manner which will be most useful in heat-transfer calculations.

Tube Pitch. Tube holes cannot be drilled very close together, since too small a width of metal between adjacent tubes structurally weakens the tube sheet. The shortest distance between two adjacent tube holes is the *clearance* or *ligament*, and these are now fairly standard. Tubes are laid out on either square or triangular patterns as shown in Fig. 7.3a and b. The advantage of square pitch is that the tubes are accessible for external cleaning and cause a lower pressure drop when fluid flows in the direction indicated in Fig. 7.3a. The *tube pitch* P_T is the shortest center-to-center distance between adjacent tubes. The common pitches for square layouts are ¾ in. OD on 1-in. square pitch and 1 in. OD on 1¼-in. square pitch. For triangular layouts these are ¾ in. OD on 15/16-in. triangular pitch, ¾ in. OD on 1-in. triangular pitch, and 1-in. OD on 1¼-in. triangular pitch. In Fig. 7.3c the square-pitch layout has been rotated 45°, yet it is essentially the same as in Fig. 7.3a. In Fig. 7.3d a mechanically cleanable modification of triangular pitch is shown. If the tubes are spread wide enough, it is possible to allow the cleaning lanes indicated.

Shells. Shells are fabricated from steel pipe with nominal IPS diameters up to 12 in. as given in Table 11. Above 12 and including 24 in. the actual outside diameter and the nominal pipe diameter are the same. The standard wall thickness for shells with inside diameters from 12 to 24 in. inclusive is $\frac{3}{8}$ in., which is satisfactory for shell-side operating pressures up to 300 psi. Greater wall thicknesses may be obtained for greater pressures. Shells above 24 in. in diameter are fabricated by rolling steel plate.

Stationary Tube-sheet Exchangers. The simplest type of exchanger is the *fixed* or *stationary tube-sheet* exchanger of which the one shown in Fig. 7.4 is an example. The essential parts are a shell (1), equipped with two nozzles and having tube sheets (2) at both ends, which also serve as flanges for the attachment of the two channels (3) and their respective

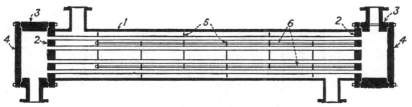

Fig. 7.4. Fixed-head tubular exchanger.

channel covers (4). The tubes are expanded into both tube sheets and are equipped with transverse baffles (5) on the shell side. The calculation of the effective heat-transfer surface is frequently based on the distance between the inside faces of the tube sheets instead of the overall tube length.

Baffles. It is apparent that higher heat-transfer coefficients result when a liquid is maintained in a state of turbulence. To induce turbulence outside the tubes it is customary to employ baffles which cause the liquid to flow through the shell at right angles to the axes of the tubes. This causes considerable turbulence even when a small quantity of liquid flows through the shell. The center-to-center distance between baffles is called the *baffle pitch* or *baffle spacing*. Since the baffles may be spaced close together or far apart, the mass velocity is not entirely dependent upon the diameter of the shell. The baffle spacing is usually not greater than a distance equal to the inside diameter of the shell or closer than a distance equal to one-fifth the inside diameter of the shell. The baffles are held securely by means of baffle spacers (6) as shown in Fig. 7.4, which consist of through-bolts screwed into the tube sheet and a number of smaller lengths of pipe which form shoulders between adjacent baffles. An enlarged detail is shown in Fig. 7.5.

There are several types of baffles which are employed in heat exchangers, but by far the most common are segmental baffles as shown in Fig. 7.6. Segmental baffles are drilled plates with heights which are generally 75 per cent of the inside diameter of the shell. These are known

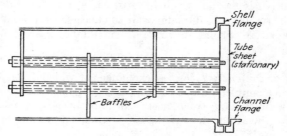

Fig. 7.5. Baffle spacer detail (enlarged).

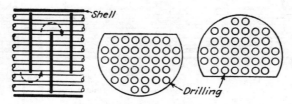

Fig. 7.6. Segmental baffle detail.

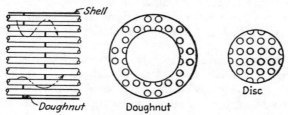

Fig. 7.7. Disc and doughnut baffle.

as 25 *per cent cut baffles* and will be used throughout this book although other fractional baffle cuts are also employed in industry. An excellent review of the influence of the baffle cut on the heat-transfer coefficient has been presented by Donohue.[1] They may be arranged, as shown, for "up-and-down" flow or may be rotated 90° to provide "side-to-side" flow, the latter being desirable when a mixture of liquid and gas flows

[1] Donohue, D. A., *Ind. Eng. Chem.*, **41**, 2499–2510 (1949).

through the shell. The baffle pitch and not the 25 per cent cut of the baffles, as shown later, determines the effective velocity of the shell fluid.

Other types of baffles are the *disc* and *doughnut* of Fig. 7.7 and the *orifice* baffle in Fig. 7.8. Although additional types are sometimes employed, they are not of general importance.

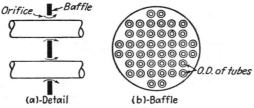

FIG. 7.8. Orifice baffle.

Fixed-tube-sheet Exchanger with Integral Channels. Another of several variations of the fixed-tube-sheet exchanger is shown in Fig. 7.9, in which the tube sheets are inserted into the shell, forming channels which are integral parts of the shell. In using stationary tube-sheet exchangers it is often necessary to provide for differential thermal expansion between the tubes and the shell during operation, or thermal stresses will develop across the tube sheet. This can be accomplished by the use of an *expansion joint* on the shell, of which a number of types of flexible joints are available.

FIG. 7.9. Fixed-tube-sheet exchanger with integral channels. (*Patterson Foundry & Machine Co.*)

Fixed-tube-sheet 1-2 Exchanger. Exchangers of the type shown in Figs. 7.4 and 7.9 may be considered to operate in counterflow, notwithstanding the fact that the shell fluid flows across the outside of the tubes. From a practical standpoint it is very difficult to obtain a high velocity when one of the fluid flows through all the tubes in a single pass. This can be circumvented, however, by modifying the design so that the tube fluid is carried through fractions of the tubes consecutively. An example of a *two-pass* fixed-tube-sheet exchanger is shown in Fig. 7.10, in which all the tube fluid flows through the two halves of the tubes successively.

The exchanger in which the shell-side fluid flows in one shell pass and the tube fluid in two or more passes is the 1-2 *exchanger*. A single channel is employed with a *partition* to permit the entry and exit of the tube fluid from the same channel. At the opposite end of the exchanger a bonnet is provided to permit the tube fluid to cross from the first to the second pass. As with all fixed-tube-sheet exchangers, the outsides of the tubes

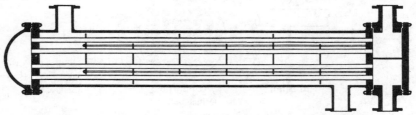

Fig. 7.10. Fixed-head 1-2 exchanger. (*Patterson Foundry & Machine Co.*)

are inaccessible for inspection or mechanical cleaning. The insides of the tubes can be cleaned in place by removing only the channel cover and using a rotary cleaner or a wire brush. Expansion problems are extremely critical in 1-2 fixed-tube-sheet exchangers, since both passes, as well as the shell itself, tend to expand differently and cause stress on the stationary tube sheets.

Removable-bundle Exchangers. In Fig. 7.11 is shown a counterpart of the 1-2 exchanger having a tube bundle which is removable from the

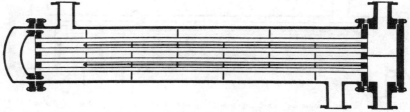

Fig. 7.11. Pull-through floating-head 1-2 exchanger. (*Patterson Foundry & Machine Co.*)

shell. It consists of a stationary tube sheet, which is clamped between the single channel flange and a shell flange. At the opposite end of the bundle the tubes are expanded into a freely riding *floating tube sheet* or *floating head*. A *floating-head cover* is bolted to the tube sheet, and the entire bundle can be withdrawn from the channel end. The shell is closed by a shell bonnet. The floating head illustrated eliminates the differential expansion problem in most cases and is called a *pull-through floating head*.

The disadvantage to the use of a pull-through floating head is one of simple geometry. To secure the floating-head cover it is necessary to bolt it to the tube sheet, and the bolt circle requires the use of space where it would be possible to insert a great number of tubes. The bolting not only reduces the number of tubes which might be placed in the tube bundle but also provides an undesirable flow channel between the bundle and the shell. These objections are overcome in the more conventional

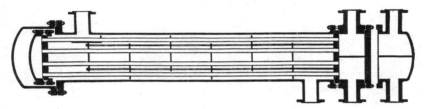

FIG. 7.12. Floating-head 1-2 exchanger. (*Patterson Foundry & Machine Co.*)

split-ring floating-head 1-2 exchanger shown in Fig. 7.12. Although it is relatively expensive to manufacture, it does have a great number of mechanical advantages. It differs from the pull-through type by the use of a split-ring assembly at the floating tube sheet and an oversized shell cover which accommodates it. The detail of a split ring is shown in Fig. 7.13. The floating tube sheet is clamped between the floating-head cover and a clamp ring placed in back of the tube sheet which is split in half to permit dismantling. Different manufacturers have different

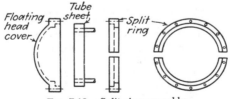

FIG. 7.13. Split-ring assembly.

modifications of the design shown here, but they all accomplish the purpose of providing increased surface over the pull-through floating head in the same size shell. Cast channels with nonremovable channel covers are also employed as shown in Fig. 7.12.

Tube-sheet Layouts and Tube Counts. A typical example of the layout of tubes for an exchanger with a split-ring floating head is shown in Fig. 7.14. The actual layout is for a $13\frac{1}{4}$ in. ID shell with 1 in. OD tubes on $1\frac{1}{4}$-in. triangular pitch arranged for six tube passes. The partition arrangement is also shown for the channel and floating-head cover along with the orientation of the passes. Tubes are not usually laid out symmetrically in the tube sheet. Extra entry space is usually allowed in the shell by omitting tubes directly under the inlet nozzle so as to minimize the contraction effect of the fluid entering the shell. When tubes are

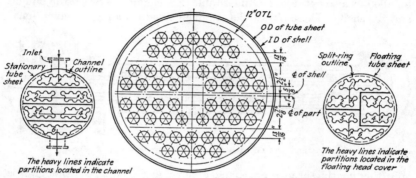

FIG. 7.14. Tube-sheet layout for a $13\frac{1}{4}$ in. ID shell employing 1 in. OD tubes on $1\frac{1}{4}$-in. triangular pitch with six tube passes.

laid out with minimum space allowances between partitions and adjoining tubes and within a diameter free of obstruction called the *outer tube limit*, the number of tubes in the layout is the *tube count*. It is not always possible to have an equal number of tubes in each pass, although in large exchangers the unbalance should not be more than about 5 per cent. In Appendix Table 9 the tube counts for $\frac{3}{4}$ and 1 in. OD tubes are given for one pass shells and one, two, four, six, and eight tube pass arrangements.

TABLE 7.1. TUBE COUNT ENTRY ALLOWANCES

Shell ID, in.	Nozzle, in.
Less than 12	2
$12-17\frac{1}{4}$	3
$19\frac{1}{4}-21\frac{1}{4}$	4
$23\frac{1}{4}-29$	6
31 $\ \ $ –37	8
Over 39	10

These tube counts include a free entrance path below the inlet nozzle equal to the cross-sectional area of the nozzles shown in Table 7.1. When a larger inlet nozzle is used, extra entry space can be obtained by flaring the inlet nozzle at its base or removing the tubes which ordinarily lie close to the inlet nozzle.

Packed Floating Head. Another modification of the floating-head 1-2 exchanger is the packed floating-head exchanger shown in Fig. 7.15.

This exchanger has an extension on the floating tube sheet which is confined by means of a packing gland. Although entirely satisfactory for shells up to 36 in. ID, the larger packing glands are not recommended for higher pressures or services causing vibration.

U-bend Exchangers. The 1-2 exchanger shown in Fig. 7.16 consists of tubes which are bent in the form of a U and rolled into the tube sheet.

Fig. 7.15. Packed floating head 1-2 exchanger. (*Patterson Foundry & Machine Co.*)

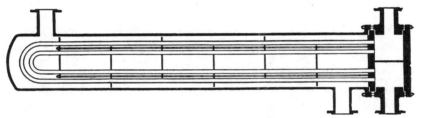

Fig. 7.16. U-bend 1-2 exchanger. (*Patterson Foundry & Machine Co.*)

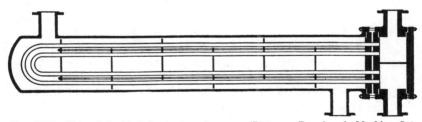

Fig. 7.17. U-bend double-tube-sheet exchanger. (*Patterson Foundry & Machine Co.*)

The tubes can expand freely, eliminating the need for a floating tube sheet, floating-head cover, shell flange, and removable shell cover. Baffles may be installed in the conventional manner on square or triangular pitch. The smallest diameter U-bend which can be turned without deforming the outside diameter of the tube at the bend has a diameter of three to four times the outside diameter of the tubing. This means that it will usually be necessary to omit some tubes at the center of the bundle, depending upon the layout.

An interesting modification of the U-bend exchanger is shown in Fig. 7.17. It employs a double stationary tube sheet and is used when the leakage of one fluid stream into the other at the tube roll can cause serious corrosion damage. By using two tube sheets with an air gap between them, either fluid leaking through its adjoining tube sheet will escape to the atmosphere. In this way neither of the streams can contaminate the other as a result of leakage except when a tube itself corrodes. Even tube failure can be prevented by applying a pressure shock test to the tubes periodically.

THE CALCULATION OF SHELL-AND-TUBE EXCHANGERS

Shell-side Film Coefficients. The heat-transfer coefficients outside tube bundles are referred to as shell-side coefficients. When the tube bundle employs baffles directing the shell-side fluid across the tubes from top to bottom or side to side, the heat-transfer coefficient is higher than for undisturbed flow along the axes of the tubes. The higher transfer coefficients result from the increased turbulence. In square pitch, as seen in Fig. 7.18, the velocity of the fluid undergoes continuous fluctuation because of the constricted area between adjacent tubes compared with the flow area between successive rows. In triangular pitch even greater turbulence is encountered because the fluid flowing between adjacent tubes at high velocity impinges directly on the succeeding row. This would indicate that, when the pressure drop and cleanability are of little consequence, triangular pitch is superior for the attainment of high shell-side film coefficients. This is actually the case, and under comparable conditions of flow and tube size the coefficients for triangular pitch are roughly 25 per cent greater than for square pitch.

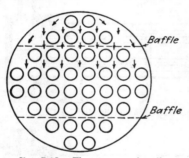

Fig. 7.18. Flow across a bundle.

Several factors not treated in preceding chapters influence the rate of heat transfer on the shell side. Suppose the length of a bundle is divided by six baffles. All the fluid travels across the bundle seven times. If ten baffles are installed in the same length of bundle, it would require that the bundle be crossed a total of eleven times, the closer spacing causing the greater turbulence. In addition to the effects of the baffle spacing the shell-side coefficient is also affected by the type of pitch, tube size, clearance, and fluid-flow characteristics. Furthermore, there is no true flow area by which the shell-side mass velocity can be computed, since the flow area varies *across* the diameter of the bundle with the different number of tube clearances in each longitudinal row of tubes. *The correlation obtained for fluids flowing in tubes is obviously not applicable to fluids flowing over tube bundles with segmental baffles,* and this is indeed borne out by experiment. However, in establishing a method of correlation the form of the heat-transfer factor $j_H = (hD/k)(c\mu/k)^{-\frac{1}{3}}(\mu/\mu_w)^{-0.14}$ vs. DG/μ has been retained, in agreement with the suggestion of McAdams,[1] but using fictitious values for the equivalent diameter D_e and the mass velocity G_s as discussed below.

Figure 28 in the Appendix is a correlation of industrial data which gives satisfactory results for the hydrocarbons, organic compounds, water, aqueous solutions, and gases when the bundle employs baffles with acceptable clearances between baffles and tubes and between baffles and shells.[2] It is not the mean curve through the data but a safe curve such that the deviation of the test points from the curve ranges from 0 to approximately 20 per cent high. Inasmuch as the line expressing the equation possesses a curvature, it cannot be evaluated in the simple form of Eq. (3.42), since the proportionality constant and the exponent of the Reynolds number actually vary. For values of Re from 2000 to 1,000,000, however, the data are closely represented by the equation

$$\frac{h_o D_e}{k} = 0.36 \left(\frac{D_e G_s}{\mu}\right)^{0.55} \left(\frac{c\mu}{k}\right)^{\frac{1}{3}} \left(\frac{\mu}{\mu_w}\right)^{0.14}$$

where h_o, D_e and G_s are as defined below. Calculations using Fig. 28 agree very well with the methods of Colburn[3] and Short[4] and the test data of Breidenbach[5] and O'Connell on a number of commercial heat exchangers. It will be observed in Fig. 28 that there is no discontinuity at a Reynolds number of 2100 such as occurs for fluids in tubes. The different equivalent diameters used in the correlation of shell and tube data precludes comparison between fluids flowing in tubes and across tubes on the basis of the Reynolds number alone. All the data in Fig. 28 refer to turbulent flow.

Shell-side Mass Velocity. The linear and mass velocities of the fluid change continuously across the bundle, since the width of the shell and the number of tubes vary from zero at the top and bottom to maxima at the center of the shell. The width of the flow area in the correlation

---

[1] McAdams, W. H., "Heat Transmission," 2d ed., p. 217, McGraw-Hill Book Company, Inc., New York, 1942.

[2] For mechanical details and standards see *Standards of the Tubular Exchanger Manufacturers' Association*, New York (1949).

[3] Colburn, A. P., *Trans. AIChE*, **29**, 174–210 (1933).

[4] Short, B. E., *Univ. Texas Pub.* 3819 (1938).

[5] Breidenbach, E. P., and H. E. O'Connell, *Trans. AIChE*, **42**, 761–776 (1946).

represented by Fig. 28 was taken at the hypothetical tube row possessing the maximum flow area and corresponding to the center of the shell. The length of the flow area was taken equal to the baffle spacing B. The tube pitch is the sum of the tube diameter and the clearance C'. If the inside diameter of the shell is divided by the tube pitch, it gives a fictitious, but not necessarily integral, number of tubes which may be assumed to exist at the center of the shell. Actually in most layouts there is no row of tubes through the center but instead two equal maximum rows on either side of it having fewer tubes than computed for the center. These deviations are neglected. For each tube or fraction there are considered to be $C' \times 1$ in.[2] of crossflow area per inch of baffle space. The shell-side or bundle crossflow area a_s is then given by

$$a_s = \frac{\text{ID} \times C'B}{P_T \times 144} \quad \text{ft}^2 \tag{7.1}$$

and as before, the mass velocity is

$$G_s = \frac{W}{a_s} \quad \text{lb/(hr)(ft}^2) \tag{7.2}$$

Shell-side Equivalent Diameter. By definition, the hydraulic radius corresponds to the area of a circle equivalent to the area of a noncircular flow channel and consequently in a plane at right angles to the direction of flow. The hydraulic radius employed for correlating shell-side coefficients for bundles having baffles is not the true hydraulic radius. The direction of flow in the shell is partly along and partly at right angles to the long axes of the tubes of the bundle. The flow area at right angles to the long axes is variable from tube row to tube row. A hydraulic radius based upon the flow area across any one row could not distinguish between square and triangular pitch. In order to obtain a simple correlation combining both the size and closeness of the tubes and their type of pitch, excellent agreement is obtained if the hydraulic radius is calculated *along* instead of across the long axes of the tubes. The equivalent diameter for the shell is then taken as four times the hydraulic radius obtained for the pattern as layed out on the tube sheet. Referring to Fig. 7.19, where the crosshatch covers the free area[1] for *square pitch*

$$D_e = \frac{4 \times \text{free area}}{\text{wetted perimeter}} \quad \text{ft} \tag{7.3}$$

or

$$d_e = \frac{4 \times (P_T^2 - \pi d_0^2/4)}{\pi d_0} \quad \text{in.} \tag{7.4}$$

---

[1] The expression free area is used to avoid confusion with the free-flow area, an actual entity in the hydraulic radius.

where P_T is the tube pitch in inches and d_0 the tube outside diameter in inches. For triangular pitch as shown in Fig. 7.19 the wetted perimeter of the element corresponds to half a tube.

$$d_e = \frac{4 \times (\tfrac{1}{2}P_T \times 0.86 P_T - \tfrac{1}{2}\pi d_0^2/4)}{\tfrac{1}{2}\pi d_0} \quad \text{in.} \quad (7.5)$$

The equivalent diameters for the common arrangements are included in Fig. 28.

It would appear that this method of evaluating the hydraulic radius and equivalent diameter does not distinguish between the relative percentage of right-angle flow to axial flow, and this is correct. It is possible, using the same shell, to have equal mass velocities, equivalent diameters, and

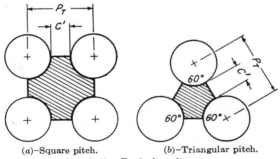

(a)—Square pitch.   (b)—Triangular pitch.
Fig. 7.19.   Equivalent diameter.

Reynolds numbers using a large quantity of fluid and a large baffle pitch or a small quantity of fluid and a small baffle pitch, although the proportions of right-angle flow to axial flow differ. Apparently, where the range of baffle pitch is restricted between the inside diameter and one-fifth the inside diameter of the shell, the significance of the error is not too great to permit correlation.

Example 7.1. Compute the shell-side equivalent diameter for a ¾ in. OD tube on 1-in. square pitch. From Eq. (7.4)

$$d_e = \frac{4(1^2 - 3.14 \times 0.75^2/4)}{3.14 \times 0.75} = 0.95 \text{ in.}$$

$$D_e = \frac{0.95}{12} = 0.079 \text{ ft}$$

The True Temperature Difference Δt in a 1-2 Exchanger. A typical plot of temperature vs. length for an exchanger having one shell pass and two tube passes is shown in Fig. 7.20 for the nozzle arrangement indicated. Relative to the shell fluid, one tube pass is in counterflow and the other in parallel flow. Greater temperature differences have been found, in Chap. 5, to result when the process streams are in counterflow and lesser differences for parallel flow. The 1-2 exchanger is a combination of both, and the LMTD for counterflow or parallel flow alone cannot be the true temperature difference for a parallel flow–counterflow arrangement. Instead it is necessary to develop a new equation for calculation of the effective or true temperature difference Δt to replace the counterflow LMTD. The method employed here is a modification of the derivation of Underwood[1] and is presented in the final form proposed by Nagle[2] and Bowman, Mueller, and Nagle.[3]

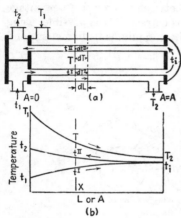

FIG. 7.20. Temperature relations in a 1-2 exchanger.

The temperature of the shell fluid may undergo either of two variations as it proceeds from the inlet to outlet, crossing the tube bundle several times in its progress: (1) So much turbulence is induced that the shell fluid is completely mixed at any length X from the inlet nozzle, or (2) so little turbulence is induced that there is a selective temperature atmosphere about the tubes of each tube pass individually. The baffles and turbulent nature of the flow components across the bundle appear to eliminate (2) so that (1) is taken as the first of the assumptions for the derivation of the true temperature difference in a 1-2 exchanger. The assumptions are

1. The shell fluid temperature is an average isothermal temperature at any cross section.
2. There is an equal amount of heating surface in each pass.
3. The overall coefficient of heat transfer is constant.
4. The rate of flow of each fluid is constant.
5. The specific heat of each fluid is constant.
6. There are no phase changes of evaporation or condensation in a part of the exchanger.
7. Heat losses are negligible.

The overall heat balance, where Δt is the true temperature difference, is

$$Q = UA\,\Delta t = WC(T_1 - T_2) = wc(t_2 - t_1) \tag{7.6}$$

[1] Underwood, A. J. V., *J. Inst. Petroleum Technol.*, **20**, 145–158 (1934).

[2] Nagle, W. M., *Ind. Eng. Chem.*, **25**, 604–608 (1933).

[3] Bowman, R. A., A. C. Mueller, and W. M. Nagle. *Trans. ASME.*, **62**, 283–294 (1940).

1-2 PARALLEL-COUNTERFLOW

From which

$$\Delta t = \left(\frac{T_1 - T_2}{UA/WC}\right)_{\text{true}} = \left(\frac{t_2 - t_1}{UA/wc}\right)_{\text{true}} \quad (7.7)$$

In Fig. 7.20a let T be the temperature of the shell fluid at any cross section of the shell $L = X$ between $L = 0$ and $L = L$. Let t^I and t^{II} represent temperatures in the first and second tube passes, respectively, and at the same cross section as T. Let a'' be the external surface per foot of length. In the incremental surface $dA = a''dL$ the shell temperature changes by $-dT$. Over the area dA

$$-WC\, dT = U \frac{dA}{2} (T - t^I) + U \frac{dA}{2} (T - t^{II}) \quad (7.8)$$

$$-WC\, dT = U\, dA \left(T - \frac{t^I + t^{II}}{2}\right) \quad (7.9)$$

$$-\int \frac{U\, dA}{WC} = \int \frac{dT}{T - (t^I + t^{II})/2} \quad (7.10)$$

But in this equation T, t^I, and t^{II} are dependent variables. The heat balance from $L = X$ to the hot-fluid inlet is

$$WC(T - T_2) = wc(t^{II} - t^I) \quad (7.11)$$

and the heat balance per pass

$$wc\, dt^I = U \frac{dA}{2} (T - t^I) \quad (7.12)$$

$$wc\, dt^{II} = -U \frac{dA}{2} (T - t^{II}) \quad (7.13)$$

Dividing Eq. (7.13) by Eq. (7.12),

$$\frac{dt^{II}}{dt^I} = -\frac{T - t^{II}}{T - t^I} \quad (7.14)$$

To eliminate t^{II} and dt^{II} from Eq. (7.11) and (7.13)

$$t^{II} = \frac{WC}{wc} (T_2 - T) + t^I \quad (7.15)$$

Differentiating Eq. (7.15), with the hot fluid inlet T_1 constant,

$$dt^{II} = -\frac{WC}{wc}\, dT + dt^I \quad (7.16)$$

Substituting in Eq. (7.14) and rearranging,

$$\frac{WC}{wc} \frac{dT}{dt^I} = 1 + \frac{T - t^I - (WC/wc)(T_2 - T)}{T - t^I} \quad (7.17)$$

The number of variables in Eq. (7.15) has been reduced from three (T, t^I, t^{II}) to two (T and t^I). For a solution it is necessary to eliminate either T or t^I. Simplifying by the use of parameters as in the case of the double pipe exchanger let

$$R = \frac{T_1 - T_2}{t_2 - t_1} = \frac{wc}{WC} \quad \text{and} \quad S = \frac{t_2 - t_1}{T_1 - t_1}$$

Rearranging Eq. (7.8),

$$WC \frac{dT}{dA} + \frac{U}{2}(T - t^I) + \frac{U}{2}(T - t^{II}) = 0 \tag{7.18}$$

Simplifying and substituting $WC = wc/R$,

$$\frac{dT}{dA} + \frac{URT}{wc} - \frac{UR}{2wc}(t^I + t^{II}) = 0 \tag{7.19}$$

Differentiating with respect to A,

$$\frac{d^2T}{dA^2} + \frac{UR}{wc}\frac{dT}{dA} - \frac{UR}{2wc}\left(\frac{dt^I}{dA} + \frac{dt^{II}}{dA}\right) = 0 \tag{7.20}$$

Substituting Eqs. (7.12) and (7.13),

$$\frac{d^2T}{dA^2} + \frac{UR}{wc}\frac{dT}{dA} - \frac{U^2R}{(2wc)^2}(t^{II} - t^I) = 0 \tag{7.21}$$

Since the heat change is sensible, a direct proportionality exists between the percentage of the temperature rise (or fall) and Q.

$$\frac{T - T_2}{T_1 - T_2} = \frac{t^I - t^{II}}{t_1 - t_2} \tag{7.22}$$

or

$$t^{II} - t^I = \frac{T - T_2}{R} \tag{7.23}$$

$$\frac{d^2T}{dA^2} + \frac{UR}{wc}\frac{dT}{dA} - \frac{U^2T}{(2wc)^2} = -\frac{U^2T_2}{(2wc)^2} \tag{7.24}$$

Differentiating again with respect to A,

$$\frac{d^3T}{dA^3} + \frac{UR}{wc}\frac{d^2T}{dA^2} - \frac{U^2}{(2wc)^2}\frac{dT}{dA} = 0 \tag{7.25}$$

The solution of this equation will be found in any standard differential-equations text. The equation is

$$T = K_1 + K_2 e^{-(UA/2wc)(R+\sqrt{R^2+1})} + K_3 e^{-(UA/2wc)(R-\sqrt{R^2+1})} \tag{7.26}$$

When $T = T_2$, A will have increased from 0 to A, and from the solution

1-2 PARALLEL-COUNTERFLOW

of Eq. (7.24) $K_1 = T_2$ so that Eq. (7.26) becomes

$$-K_2 e^{-(UA/2wc)(R+\sqrt{R^2+1})} = K_3 e^{-(UA/2wc)(R-\sqrt{R^2+1})} \quad (7.27)$$

Taking logarithms of both sides and simplifying,

$$\frac{UA}{wc} = \frac{1}{\sqrt{R^2+1}} \ln\left(-\frac{K_2}{K_3}\right) \quad (7.28)$$

Differentiate Eq. (7.26):

$$\frac{dT}{dA} = -K_2 \frac{U}{2wc}(R+\sqrt{R^2+1})e^{-(UA/2wc)(R+\sqrt{R^2+1})}$$
$$- K_3 \frac{U}{2wc}(R-\sqrt{R^2+1})e^{-(UA/2wc)(R-\sqrt{R^2+1})} \quad (7.29)$$

Substituting the value of dT/dA from Eq. (7.19) and since at $A = 0$, $t^{\mathrm{I}} = t_1$, $t^{\mathrm{II}} = t_2$, and $T = T_1$, $t^{\mathrm{I}} + t^{\mathrm{II}} = t_1 + t_2$.

$$R(t_1 + t_2) - 2RT_1 = -K_2(R+\sqrt{R^2+1})$$
$$- K_3(R-\sqrt{R^2+1}) \quad (7.30)$$

From Eq. (7.26) at $A = 0$ and $T = T_1$ and $K_1 = T_2$

$$T_1 - T_2 = K_2 + K_3 \quad (7.31)$$

Multiplying both sides of Eq. (7.31) by $(R+\sqrt{R^2+1})$,

$$(R+\sqrt{R^2+1})(T_1 - T_2) = K_2(R+\sqrt{R^2+1})$$
$$+ K_3(R+\sqrt{R^2+1}) \quad (7.32)$$

Adding Eqs. (7.31) and (7.32) and solving for K_3,

$$K_3 = \frac{R(t_1+t_2) + (T_1-T_2)(R+\sqrt{R^2+1}) - 2RT_1}{2\sqrt{R^2+1}} \quad (7.33)$$

Returning to Eq. (7.31),

$$-K_2 = K_3 - (T_1 - T_2) =$$
$$\frac{(R+\sqrt{R^2+1})(T_1-T_2) - 2\sqrt{R^2+1}\,(T_1-T_2) - 2RT_1 + R(t_1+t_2)}{2\sqrt{R^2+1}}$$
$$(7.34)$$

Since $R = (T_1 - T_2)/(t_2 - t_1)$,

$$-\frac{K_2}{K_3} = \frac{(R-\sqrt{R^2+1})(t_1-t_2) - (T_1-t_1) - (T_1-t_2)}{(R+\sqrt{R^2+1})(t_1-t_2) - (T_1-t_1) - (T_1-t_2)} \quad (7.35)$$

Dividing by $T_1 - t_1$ and substituting $S = (t_2 - t_1)/(T_1 - t_1)$ and $1 - S = (T_1 - t_2)/(T_1 - t_1)$,

$$-\frac{K_2}{K_3} = \frac{2 - S(R + 1 - \sqrt{R^2 + 1})}{2 - S(R + 1 + \sqrt{R^2 + 1})} \quad (7.36)$$

Substituting in Eq. (7.28),

$$\left(\frac{UA}{wc}\right)_{\text{true}} = \frac{1}{\sqrt{R^2 + 1}} \ln \frac{2 - S(R + 1 - \sqrt{R^2 + 1})}{2 - S(R + 1 + \sqrt{R^2 + 1})} \quad (7.37)$$

Equation (7.37) is the relationship for the true temperature difference for 1-2 parallel flow–counterflow. How does this compare with the LMTD for counterflow employing the same process temperatures? For counterflow

$$Q = wc(t_2 - t_1) = UA \frac{(T_1 - t_2) - (T_2 - t_1)}{\ln (T_1 - t_2)/(T_2 - t_1)} \quad (7.38)$$

from which

$$\left(\frac{UA}{wc}\right)_{\text{counterflow}} = \frac{t_2 - t_1}{\dfrac{(T_1 - t_2) - (T_2 - t_1)}{\ln (T_1 - t_2)/(T_2 - t_1)}} = \frac{\ln (1 - S)/(1 - RS)}{R - 1} \quad (7.39)$$

The ratio of the true temperature difference to the LMTD is

$$\frac{t_2 - t_1}{(UA/wc)_{\text{true}}} \bigg/ \frac{t_2 - t_1}{(UA/wc)_{\text{counterflow}}} = \frac{(UA/wc)_{\text{counterflow}}}{(UA/wc)_{\text{true}}} \quad (7.40)$$

Calling the fractional ratio of the true temperature difference to the LMTD F_T,

$$F_T = \frac{\sqrt{R^2 + 1} \ln (1 - S)/(1 - RS)}{(R - 1) \ln \dfrac{2 - S(R + 1 - \sqrt{R^2 + 1})}{2 - S(R + 1 + \sqrt{R^2 + 1})}} \quad (7.41)$$

The Fourier equation for a 1-2 exchanger can now be written:

$$Q = UA \, \Delta t = UA F_T (\text{LMTD}) \quad (7.42)$$

To reduce the necessity of solving Eq. (7.37) or (7.41), correction factors F_T for the LMTD have been plotted in Fig. 18 in the Appendix as a function of S with R as the parameter. When a value of S and R is close to the vertical portion of a curve, it is difficult to read the figure and F_T should be computed from Eq. (7.41) directly. When an exchanger has one shell pass and four, six, eight, or more even-numbered tube passes such as a 1-4, 1-6, or 1-8 exchanger, Eq. (7.10) becomes for a 1-4 exchanger

$$-\int \frac{U \, dA}{WC} = \int \frac{dT}{T - (t^{\text{I}} + t^{\text{II}} + t^{\text{III}} + t^{\text{IV}})/4}$$

for a 1-6 exchanger

$$-\int \frac{U\,dA}{WC} = \int \frac{dT}{T - (t^{\mathrm{I}} + t^{\mathrm{II}} + t^{\mathrm{III}} + t^{\mathrm{IV}} + t^{\mathrm{V}} + t^{\mathrm{VI}})/6} \cdots$$

It can be shown that the values of F_T for a 1-2 and 1-8 exchanger are less than 2 per cent apart in the extreme case and generally considerably less. It is therefore customary to describe any exchanger having one shell pass and *two or more even-numbered* tube passes in parallel flow–counterflow as a 1-2 exchanger and to use the value of F_T obtained from Eq. (7.41). The reason F_T will be less than 1.0 is naturally due to the fact that the tube passes in parallel with the shell fluid do not contribute so effective a temperature difference as those in counterflow with it.

There is an important limitation to the use of Fig. 18. Although any exchanger having a value of F_T above zero will theoretically operate, it is not practically true. The failure to fulfill in practice all the assumptions employed in the derivation, assumptions 1, 3, and 7 in particular, may cause serious discrepancies in the calculation of Δt. As a result of the discrepancies if the actual value of t_i in Fig. 7.20a at the end of the parallel pass is required to approach T_2 more closely than the derived value of t_i, it

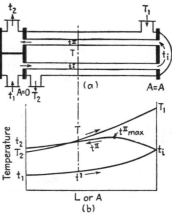

FIG. 7.21. Temperature relations in a 1-2 exchanger with conventional nozzle arrangement.

may impose a violation of the rule of parallel flow; namely, the outlet of the one stream t_i may not attain the outlet of the other, T_2, without infinite surface. Accordingly it is not advisable or practical to use a 1-2 exchanger whenever the correction factor F_T is computed to be less than 0.75. Instead some other arrangement is required which more closely resembles counterflow.

The temperature relationship for the case where the orientation of the shell nozzles has been reversed is shown in Fig. 7.21 for the same inlet and outlet temperatures plotted in Fig. 7.20. Underwood[1] has shown that the values of F_T for both are identical.[2] Since a 1-2 exchanger is a combination of counterflow and parallel-flow passes, it may be expected that the outlet of one process stream cannot approach the inlet of the

---

[1] Underwood, *op. cit.*

[2] The values of t_i, however, differ for both cases.

other very closely. In fact it is customary in parallel flow–counterflow equipment to call $T_2 - t_2$ the *approach*, and if $t_2 > T_2$, then $t_2 - T_2$ is called the *temperature cross*.

It is useful to investigate several typical process temperatures and to note the influence of different approaches and crosses upon the value of

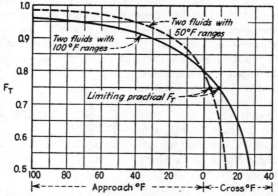

FIG. 7.22. Influence of approach temperature on F_T with fluids having equal ranges in a 1-2 exchanger.

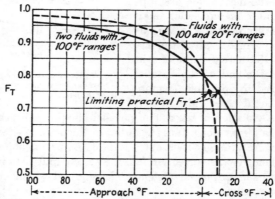

FIG. 7.23. Influence of approach temperature on F_T with fluids having unequal ranges in a 1-2 exchanger.

F_T. For a given service the reduction of F_T below unity in Eq. (7.42) is compensated for by increasing the surface. Thus if the process temperatures are fixed it may be inadvisable to employ a parallel flow–counterflow exchanger as against a counterflow exchanger, since it increases the cost of the equipment beyond the value of its mechanical advantages. In Fig. 7.22 two pairs of fluids each with equal *ranges* of 100 and 50°F

1-2 PARALLEL-COUNTERFLOW

are studied. The operating temperatures of the cold fluid are fixed, while the hot-fluid temperatures are variable thereby changing the approach in each case. Note the conditions under which F_T shrinks rapidly, particularly the approach at the practical minimum $F_T = 0.75$ and the influence of the relationship between T_2 and t_2. The calculation of several points is demonstrated.

Example 7.2. Calculation of F_T for Fluids with Equal Ranges

Point:
(a) 50° approach
(T_1) 350   200 (t_2)
(T_2) 250   100 (t_1)
$\overline{100}$ $\overline{100}$

(b) Zero approach
(T_1) 300   200 (t_2)
(T_2) 200   100 (t_1)
$\overline{100}$ $\overline{100}$

(c) 20° cross
(T_1) 280   200 (t_2)
(T_2) 180   100 (t_1)
$\overline{100}$ $\overline{100}$

$R = \dfrac{T_1 - T_2}{t_2 - t_1} = \dfrac{100}{100} = 1.0$ $R = 1.0$ $R = 1.0$

$S = \dfrac{t_2 - t_1}{T_1 - t_1} = \dfrac{100}{350 - 100} = 0.40$ $S = 0.50$ $S = 0.555$

$F_T = 0.925$ (Fig. 18) $F_T = 0.80$ $F_T = 0.64$

In Fig. 7.23 are shown the results of the calculation when one fluid has a range five times as great as the other.

Shell-side Pressure Drop. The pressure drop through the shell of an exchanger is proportional to the number of times the fluid crosses the bundle between baffles. It is also proportional to the distance across the bundle each time it is crossed. Using a modification of Eq. (3.44) a correlation has been obtained using the product of the distance across the bundle, taken as the inside diameter of the shell in feet D_s and the number of times the bundle is crossed $N + 1$, where N is the number of baffles. If L is the tube length in feet,

Number of crosses, $N + 1 =$ tube length, in./baffle space, in.

$$= 12 \times L/B \quad (7.43)$$

If the tube length is 16'0'' and the baffles are spaces 18 in. apart, there will be 11 crosses or 10 baffles. There should always be an odd number of crosses if both shell nozzles are on opposite sides of the shell and an even number if both shell nozzles are on the same side of the shell. With close baffle spacings at convenient intervals such as 6 in. and under, one baffle may be omitted if the number of crosses is not an integer. The equivalent diameter used for calculating the pressure drop is the same as for heat transfer, the additional friction of the shell itself being neglected. The isothermal equation for the pressure drop of a fluid being heated or cooled and including entrance and exit losses is

$$\Delta P_s = \frac{fG_s^2 D_s (N + 1)}{2g\rho D_e \phi_s} = \frac{fG_s^2 D_s (N + 1)}{5.22 \times 10^{10} D_e s \phi_s} \quad \text{psf} \quad (7.44)$$

where s is the specific gravity of the fluid. Equation (7.44) gives the pressure drop in pounds per square foot. The common engineering unit is pounds per square inch. To permit the direct solution for ΔP_s in psi *dimensional* shell-side friction factors, square foot per square inch, have been plotted in Fig. 29. To obtain the pressure drop in consistent units by Eq. (7.44) multiply f in Fig. 29 by 144.

Tube-side Pressure Drop. Equation (3.44) may be used to obtain the pressure drop in tubes, but it applies principally to an isothermal fluid. Sieder and Tate have correlated friction factors for fluids being heated or cooled in tubes. They are plotted in dimensional form in Fig. 26 and are used in the equation

$$\Delta P_t = \frac{fG_t^2 Ln}{5.22 \times 10^{10} D_e s \phi_t} \qquad \text{psf} \qquad (7.45)$$

where n is the number of tube passes, L the tube length, and Ln is the total length of path in feet. The deviations are not given, but the curve has been accepted by the Tubular Exchanger Manufacturers Association. In flowing from one pass into the next at the channel and floating head the fluid changes direction abruptly by 180°, although the flow area provided in the channel and floating-head cover should not be less than the combined flow area of all the tubes in a single pass. The change of direction introduces an additional pressure drop ΔP_r, called the return loss and accounted for by allowing four velocity heads per pass. The velocity head $V^2/2g'$ has been plotted in Fig. 27 against the mass velocity for a fluid with a specific gravity of 1, and the return losses for any fluid will be

$$\Delta P_r = \frac{4n}{s} \frac{V^2}{2g'} \qquad \text{psi} \qquad (7.46)$$

where V = velocity, fps
$\quad s$ = specific gravity
$\quad g'$ = acceleration of gravity, ft/sec²

The total tube-side pressure drop ΔP_T will be

$$\Delta P_T = \Delta P_t + \Delta P_r \qquad \text{psi} \qquad (7.47)$$

The Analysis of Performance in an Existing 1-2 Exchanger. When all the pertinent equations are used to calculate the suitability of an existing exchanger for given process conditions, it is known as *rating* an exchanger. There are three significant points in determining the suitability of an existing exchanger for a new service.

1. What clean coefficient U_C can be "performed" by the two fluids as the result of their flow and individual film coefficients h_{io} and h_o?

2. From the heat balance $Q = WC(T_1 - T_2) = wc(t_2 - t_1)$, known surface A, and the true temperature difference for the process temperatures a value of the design or dirty coefficient, U_D is obtained. U_C must exceed U_D sufficiently so that the dirt factor, which is a measure of the excess surface, will permit operation of the exchanger for a reasonable period of service.

3. The allowable pressure drops for the two streams may not be exceeded.

When these are fulfilled, an existing exchanger is suitable for the process conditions for which it has been rated. In starting a calculation the first point which arises is to determine whether the hot or cold fluid should be placed in the shell. There is no fast rule. One stream may be large and the other small, and the baffle spacing may be such that in one instance the shell-side flow area a_s will be larger. Fortunately any selection can be checked by switching the two streams and seeing which arrangement gives the larger value of U_C without exceeding the allowable pressure drop. Particularly in preparation for later methods there is some advantage, however, in starting calculations with the tube side, and it may be well to establish the habit. The detailed steps in the rating of an exchanger are outlined below. The subscripts s and t are used to distinguish between the shell and tubes, and for the outline the hot fluid has been assumed to be in the shell. By always placing the hot fluid on the left the usual method of computing the LMTD may be retained.

The Calculation of an Existing 1-2 Exchanger. Process conditions required

Hot fluid: T_1, T_2, W, c, s, μ, k, R_d, ΔP
Cold fluid: t_1, t_2, w, c, s, μ, k, R_d, ΔP

For the exchanger the following data must be known:

Shell side	Tube side
ID	Number and length
Baffle space	OD, BWG, and pitch
Passes	Passes

(1) Heat balance, $Q = WC(T_1 - T_2) = wc(t_2 - t_1)$
(2) True temperature difference Δt:

LMTD, $\qquad R = \dfrac{T_1 - T_2}{t_2 - t_1}, \qquad S = \dfrac{t_2 - t_1}{T_1 - t_1}$ \hfill (5.14)

$\Delta t = \text{LMTD} \times F_T$ (F_T from Fig. 18) \hfill (7.42)

(3) Caloric temperature T_c and t_c:[1] (5.28), (5.29)

Hot fluid: shell side	Cold fluid: tube side

(4′) Flow area, $a_s = \text{ID} \times C'B/144P_T$, ft² [Eq. (7.1)]

(4) Flow area, a_t:
Flow area per tube a_t' from Table 10, in.²
$$a_t = \frac{\text{No. of tubes} \times \text{flow area/tube}}{\text{No. of passes}}$$
$$= N_t a_t'/144n, \text{ ft}^2 \qquad \text{Eq. (7.48)}$$

(5′) Mass vel, $G_s = W/a_s$, lb/(hr)(ft²) [Eq. (7.2)]

(5) Mass vel, $G_t = w/a_t$, lb/(hr)(ft²)

(6′) Obtain D_e from Fig. 28 or compute from Eq. (7.4).
Obtain μ at T_c, lb/(ft)(hr) = cp × 2.42
$Re_s = D_e G_s/\mu$

(6) Obtain D from Table 10, ft.
Obtain μ at t_c, lb/(ft)(hr) = cp × 2.42
$Re_t = DG_t/\mu$

(7′) Obtain j_H from Fig. 28.

(7) Obtain j_H from Fig. 24.

(8′) At T_c obtain c, Btu/(lb)(°F) and k, Btu/(hr)(ft²)(°F/ft).
Compute* $(c\mu/k)^{1/3}$.

(8) At t_c obtain c, Btu/(lb)(°F) and k, Btu/(hr)(ft²)(°F/ft).
Compute* $(c\mu/k)^{1/3}$.

(9′) $h_o = j_H \dfrac{k}{D}\left(\dfrac{c\mu}{k}\right)^{1/3} \phi_s$ [Eq. (6.15b)]

(9) $h_i = j_H \dfrac{k}{D}\left(\dfrac{c\mu}{k}\right)^{1/3} \phi_t$ [Eq. (6.15a)]

(10′) Tube-wall temp, t_w

$$t_w = t_c + \frac{h_o/\phi_s}{h_{io}/\phi_t + h_o/\phi_s}(T_c - t_c)$$
[Eq. (5.31)]

(10) $\dfrac{h_{io}}{\phi_t} = \dfrac{h_i}{\phi_t} \times \dfrac{\text{ID}}{\text{OD}}$ [Eq. (6.5)]

(11′) Obtain μ_w and $\phi_s = (\mu/\mu_w)^{0.14}$. [Fig. 24]

(11) Obtain t_w from (10′).
Obtain μ_w and $\phi_t = (\mu/\mu_w)^{0.14}$. [Fig. 24]

(12′) Corrected coefficient, $h_o = \dfrac{h_o \phi_s}{\phi_s}$ [Eq. (6.36)]

(12) Corrected coefficient, $h_{io} = \dfrac{h_{io}\phi_t}{\phi_t}$ [Eq. (6.37)]

(13) Clean overall coefficient U_C:
$$U_C = \frac{h_{io}h_o}{h_{io} + h_o} \qquad (6.38)$$

(14) Design overall coefficient U_D: Obtain external surface/lin ft a'' from Appendix Table 10.
Heat-transfer surface, $A = a''LN_t$, ft²
$$U_D = \frac{Q}{A\,\Delta t} \quad \text{Btu/(hr)(ft}^2)(°F)$$

(15) Dirt factor R_d:
$$R_d = \frac{U_C - U_D}{U_C U_D} \quad \text{(hr)(ft}^2)(°F)/\text{Btu} \qquad (6.13)$$

If R_d equals or exceeds the required dirt factor, proceed under the pressure drop.

[1] The use of caloric temperatures is in partial contradiction of the derivation for the 1-2 parallel flow–counterflow temperature difference in which U was assumed constant. The use of caloric temperatures presumes that a linear variation of U with t can be accounted for as the product of $U_{\text{caloric}} \Delta t$ where Δt is the true parallel flow–counterflow temperature difference when U is constant.

* A convenient graph of $k(c\mu/k)^{1/3}$ vs. μ for petroleum fractions is given in Fig. 16.

1-2 PARALLEL-COUNTERFLOW

Pressure Drop

(1') For Re_s in (6') obtain f, ft^2/in.^{2} [Fig. 29]

(2') No. of crosses, $N + 1 = 12L/B$ [Eq. (7.43)]

(3') $\Delta P_s = \dfrac{fG_s^2 D_s(N+1)}{5.22 \times 10^{10} D_e s \phi_s}$ psi [Eq. (7.44)]

(1) For Re_t in (6) obtain f, ft^2/in.^{2} [Fig. 26]

(2) $\Delta P_t = \dfrac{fG_t^2 L n}{5.22 \times 10^{10} D s \phi_t}$ [Eq. (7.45)]

(3) $\Delta P_r = \dfrac{4n}{s} \dfrac{V^2}{2g'} \dfrac{62.5}{144}$ psi [Eq. (7.46)]

$\Delta P_T = \Delta P_t + \Delta P_r$ psi [Eq. (7.47)]

Example 7.3. Calculation of a Kerosene–Crude Oil Exchanger. 43,800 lb/hr of a 42°API kerosene leaves the bottom of a distilling column at 390°F and will be cooled to 200°F by 149,000 lb/hr of 34°API Mid-continent crude coming from storage at 100°F and heated to 170°F. A 10 psi pressure drop is permissible on both streams, and in accordance with Table 12, a combined dirt factor of 0.003 should be provided.

Available for this service is a 21¼ in. ID exchanger having 158 1 in. OD, 13 BWG tubes 16'0" long and laid out on 1¼-in. square pitch. The bundle is arranged for four passes, and baffles are spaced 5 in. apart.

Will the exchanger be suitable; *i.e.*, what is the dirt factor?

Solution:

Exchanger:

    Shell side   *Tube side*
    ID = 21¼ in.   Number and length = 158, 16'0"
    Baffle space = 5 in.   OD, BWG, pitch = 1 in., 13 BWG, 1¼-in. square
    Passes = 1   Passes = 4

(1) Heat balance:
    Kerosene, $Q = 43{,}800 \times 0.605(390 - 200) = 5{,}100{,}000$ Btu/hr
    Mid-continent crude, $Q = 149{,}000 \times 0.49(170 - 100) = 5{,}100{,}000$ Btu/hr

(2) Δt:

Hot Fluid		Cold Fluid	Diff.	
390	Higher Temp	170	220	
200	Lower Temp	100	100	
190	Differences	70	120	$(\Delta t_2 - \Delta t_1)$
$(T_1 - T_2)$		$(t_2 - t_1)$		

LMTD = 152.5°F   (5.14)
$R = {}^{190}\!/_{70} = 2.71$
$S = \dfrac{70}{390 - 100} = 0.241$
$F_T = 0.905$ (Fig. 18)
$\Delta t = 0.905 \times 152.5 = 138°F$ (7.42)

(3) T_c and t_c:

$\dfrac{\Delta t_c}{\Delta t_h} = 0.455$ (Fig. 17)

$K_c = 0.20$ (crude oil controlling)
$F_c = 0.42$
$T_c = 200 + 0.42 \times 190 = 280°F$ (5.28)
$t_c = 100 + 0.42 \times 70 = 129°F$ (5.29)

Since the flow area of both the shell and tube sides will be nearly equal, assume the larger stream to flow in the tubes and start calculation on the tube side.

Hot fluid: shell side, kerosene

(4') Flow area, $a_s = ID \times C'B/144P_T$ [Eq. (7.1)]
$= 21.25 \times 0.25 \times 5/144 \times 1.25$
$= 0.1475$ ft²

(5') Mass vel, $G_s = W/a_s$ [Eq. (7.2)]
$= 43{,}800/0.1475 = 297{,}000$ lb/(hr)(ft²)

(6') $Re_s = D_e G_s/\mu$ [Eq. (7.3)]
At $T_c = 280°F$, $\mu = 0.40 \times 2.42$
$= 0.97$ lb/(ft)(hr) [Fig. 14]
$D_e = 0.99/12 = 0.0825$ ft [Fig. 28]
$Re_s = 0.0825 \times 297{,}000/0.97 = 25{,}300$
(7') $j_H = 93$ [Fig. 28]

(8') At $T_c = 280°F$,
$c = 0.59$ Btu/(lb)(°F) [Fig. 4]
$k = 0.0765$ Btu/(hr)(ft²)(°F/ft) [Fig. 1]
$(c\mu/k)^{1/3} = (0.59 \times 0.97/0.0765)^{1/3} = 1.95$

(9') $h_o = j_H \dfrac{k}{D_e}\left(\dfrac{c\mu}{k}\right)^{1/3} \phi_s$ [Eq. (6.15b)]

$\dfrac{h_o}{\phi_s} = 93 \times \dfrac{0.0765}{0.0825} \times 1.95 = 169$

(10') Tube-wall temperature:

$t_w = t_c + \dfrac{h_o/\phi_s}{h_{io}/\phi_t + h_o/\phi_s}(T_c - t_c)$

[Eq. (5.31)]

$= 129 + \dfrac{169}{109 + 169}(280 - 129)$

$= 221°F$

(11') At $t_w = 221°F$, $\mu_w = 0.56 \times 2.42$
$= 1.36$ lb/(ft)(hr) [Fig. 14]
$\phi_s = (\mu/\mu_w)^{0.14} = (0.97/1.36)^{0.14}$
$= 0.96$ [Fig. 24 insert]

(12') Corrected coefficient, $h_o = \dfrac{h_o}{\phi_s}\phi_s$
[Eq. (6.36)]
$= 169 \times 0.96 = 162$ Btu/(hr)(ft²)(°F)

Cold fluid: tube side, crude oil

(4) Flow area, $a'_t = 0.515$ in.² [Table 10]
$a_t = N_t a'_t/144n$ [Eq. (7.48)]
$= 158 \times 0.515/144 \times 4 = 0.141$ ft²

(5) Mass vel, $G_t = w/a_t$
$= 149{,}000/0.141 = 1{,}060{,}000$
lb/(hr)(ft²)

(6) $Re_t = DG_t/\mu$
At $t_c = 129°F$, $\mu = 3.6 \times 2.42$
$= 8.7$ lb/(ft)(hr) [Fig. 14]
$D = 0.81/12 = 0.0675$ ft [Table 10]
$Re_t = 0.0675 \times 1{,}060{,}000/8.7 = 8{,}220$
(7) $L/D = 16/0.0675 = 237$
$j_H = 31$ [Fig. 24]

(8) At $t_c = 129°F$,
$c = 0.49$ Btu/(lb)(°F) [Fig. 4]
$k = 0.077$ Btu/(hr)(ft²)(°F/ft) [Fig. 1]
$(c\mu/k)^{1/3} = (0.49 \times 8.7/0.077)^{1/3} = 3.81$

(9) $h_i = j_H \left(\dfrac{k}{D}\right)\left(\dfrac{c\mu}{k}\right)^{1/3}\phi_t$

[Eq. (6.15a)]

$\dfrac{h_i}{\phi_t} = 31 \times \dfrac{0.077}{0.0675} \times 3.81 = 135$

(10) $\dfrac{h_{io}}{\phi_t} = \dfrac{h_i}{\phi_t} \times \dfrac{ID}{OD} = 135 \times \dfrac{0.81}{1.0} = 109$
[Eq. (6.5)]

(11) At $t_w = 221°F$, $\mu_w = 1.5 \times 2.42$
$= 3.63$ lb/(ft)(hr) [Fig. 14]
$\phi_t = (\mu/\mu_w)^{0.14} = (8.7/3.63)^{0.14}$
$= 1.11$ [Fig. 24 insert]

(12) Corrected coefficient, $h_{io} = \dfrac{h_{io}}{\phi_t}\phi_t$
[Eq. (6.37)]
$= 109 \times 1.11 = 121$ Btu/(hr)(ft²)(°F)

1-2 PARALLEL-COUNTERFLOW

(13') Clean overall coefficient U_C:

$$U_C = \frac{h_{io}h_o}{h_{io} + h_o} = \frac{121 \times 162}{121 + 162} = 69.3 \text{ Btu/(hr)(ft}^2)(°F) \quad (6.38)$$

(14) Design overall coefficient U_D:

$$a'' = 0.2618 \text{ ft}^2/\text{lin ft} \quad \text{(Table 10)}$$
$$\text{Total surface, } A = 158 \times 16'0'' \times 0.2618 = 662 \text{ ft}^2$$
$$U_D = \frac{Q}{A\Delta t} = \frac{5,100,000}{662 \times 138} = 55.8 \text{ Btu/(hr)(ft}^2)(°F)$$

(15) Dirt factor R_d:

$$R_d = \frac{U_C - U_D}{U_C U_D} = \frac{69.3 - 55.8}{69.3 \times 55.8} = 0.00348 \text{ (hr)(ft}^2)(°F)/\text{Btu} \quad (6.13)$$

Summary

162	h outside	121
U_C	69.3	
U_D	55.8	
R_d Calculated	0.00348	
R_d Required	0.00300	

Pressure Drop

(1') For $Re_s = 25,300$,
$f = 0.00175 \text{ ft}^2/\text{in.}^2$ [Fig. 29]
$s = 0.73$ [Fig. 6]
$D_s = 21.25/12 = 1.77$ ft

(2') No. of crosses, $N + 1 = 12L/B$ [Eq.(7.43)]
$= 12 \times 16/5 = 39$

(3') $\Delta P_s = \dfrac{fG_s^2 D_s(N + 1)}{5.22 \times 10^{10} D_e s \phi_s}$ [Eq. (7.44)]

$= \dfrac{0.00175 \times 297,000^2 \times 1.77 \times 39}{5.22 \times 10^{10} \times 0.0825 \times 0.73 \times 0.96}$

$= 3.5$ psi

Allowable $\Delta P_s = 10.0$ psi

(1) For $Re_t = 8220$,
$f = 0.000285 \text{ ft}^2/\text{in}^2$ [Fig. 26]
$s = 0.83$ [Fig. 6]

(2) $\Delta P_t = \dfrac{fG_t^2 Ln}{5.22 \times 10^{10} Ds\phi_t}$ [Eq. (7.45)]

$= \dfrac{0.000285 \times 1,060,000^2 \times 16 \times 4}{5.22 \times 10^{10} \times 0.0675 \times 0.83 \times 1.11}$

$= 6.3$ psi

(3) $G_t = 1,060,000$, $\dfrac{V^2}{2g'} = 0.15$ [Fig. 27]

$\Delta P_r = \dfrac{4n}{s} \dfrac{V^2}{2g'}$ [Eq. (7.46)]

$= \dfrac{4 \times 4}{0.83} \times 0.15 = 2.9$ psi

(4) $\Delta P_T = \Delta P_t + \Delta P_r$
$= 6.3 + 2.9 = 9.2$ psi [Eq. (7.47)]
Allowable $\Delta P_T = 10.0$ psi

It is seen that a dirt factor of 0.00348 will be obtained although only 0.003 will be required to provide reasonable maintenance period. The pressure drops have not been exceeded and the exchanger will be satisfactory for the service.

Exchangers Using Water. Cooling operations using water in tubular equipment are very common. Despite its abundance, the heat-transfer characteristics of water separate it from all other fluids. It is corrosive to steel, particularly when the tube-wall temperature is high and dissolved air is present, and many industrial plants use nonferrous tubes exclusively for heat-transfer services involving water. The commonest nonferrous tubes are admiralty, red brass, and copper, although in certain localities there is a preference for Muntz metal, aluminum bronze, and aluminum. Since shells are usually fabricated of steel, water is best handled in the tubes. When water flows in the tubes, there is no serious problem of corrosion of the channel or floating-head cover, since these parts are often made of cast iron or cast steel. Castings are relatively passive to water, and large corrosion allowances above structural requirements can be provided inexpensively by making the castings heavier. Tube sheets may be made of heavy steel plate with a corrosion allowance of about ⅛ in. above the required structural thickness or fabricated of naval brass or aluminum without a corrosion allowance.

When water travels slowly through a tube, dirt and slime resulting from microorganic action adhere to the tubes which would be carried away if there were greater turbulence. As a standard practice, the use of cooling water at velocities less than 3 fps should be avoided, although in certain localities minimum velocities as high as 4 fps are required for continued operation. Still another factor of considerable importance is the deposition of mineral scale. When water of average mineral and air content is brought to a temperature in excess of 120°F, it is found that tube action becomes excessive, and for this reason an outlet water temperature above 120°F should be avoided.

Cooling water is rarely abundant or without cost. One of the serious problems facing the chemical and power industries today results from the gradual deficiency of surface and subsurface water in areas of industrial concentration. This can be partially overcome through the use of cooling towers (Chap. 17), which reuse the cooling water and reduce the requirement to only 2 per cent of the amount of water required in once-through use. River water may provide part of the solution to a deficiency of ground water, but it is costly and presupposes the proximity of a river. River water must usually be strained by moving screens and pumped considerable distances, and in some localities the water from rivers servicing congested industrial areas requires cooling in cooling towers before it can be used.

Many sizable municipalities have legislated against the use of public water supplies for large cooling purposes other than for make-up in cooling towers or spray-pond systems. Where available, municipal water may average about 1 cent per 1000 gal., although it has the advantage of being generally available at from 30 to 60 psi pressure which is adequate for most process needs including the pressure drops in heat exchangers. Where a cooling tower is used, the cost of the water is determined by the cost of fresh water, pumping power, fan power, and write-off on the original investment.

The shell-side heat-transfer curve (Fig. 28) correlates very well for the flow of water across tube bundles. The high thermal conductivity of water results in relatively high film coefficients compared with organic fluids. The use of the tube-side curve (Fig. 24), however, gives coefficients which are generally high. In its place the data of Eagle and Ferguson[1] for water alone are given in Fig. 25 and are recommended whenever water flows in tubes. Since this graph deals only with water, it has been possible to plot film coefficients vs. velocity in feet per second with temperature as the parameter. The data have been plotted with the $\frac{3}{4}$ in., 16 BWG tube as the base, and the correction factor obtained from the insert in Fig. 25 should be applied when any other inside diameter is used.

In a water-to-water exchanger with individual film coefficients ranging from 500 to 1500 for both the shell and tube, the selection of the required dirt factor merits serious judgment. As an example, if film coefficients of 1000 are obtained on the shell and tube sides, the combined resistance is 0.002, or $U_C = 500$. If a fouling factor of 0.004 is required, the fouling factor becomes the controlling resistance. When the fouling factor is 0.004, U_D must be less than 1/0.004 or 250. Whenever high coefficients exist on both sides of the exchanger, the use of an unnecessarily large fouling factor should be avoided.

The following heat-recovery problem occurs in powerhouses. Although it involves a moderate-size exchanger, the heat recovery is equivalent to nearly 1500 lb/hr of steam, which represents a sizable economy in the course of a year.

Example 7.4. Calculation of a Distilled-water–Raw-water Exchanger. 175,000 lb/hr of distilled water enters an exchanger at 93°F and leaves at 85°F. The heat will be transferred to 280,000 lb/hr of raw water coming from supply at 75°F and leaving the exchanger at 80°F. A 10 psi pressure drop may be expended on both streams while providing a fouling factor of 0.0005 for distilled water and 0.0015 for raw water when the tube velocity exceeds 6 fps.

Available for this service is a $15\frac{1}{4}$ in. ID exchanger having 160 $\frac{3}{4}$ in. OD, 18 BWG tubes 16'0" long and laid out on $1\frac{5}{16}$-in. triangular pitch. The bundle is arranged for two passes, and baffles are spaced 12 in. apart.

Will the exchanger be suitable?

[1] Eagle, A., and R. M. Ferguson, *Pro. Roy. Soc.*, **A127**, 540–566 (1930).

Solution:

Exchanger:

Shell side	Tube side
ID = 15¼ in.	Number and length = 160, 16'0''
Baffle space = 12 in.	OD, BWG, pitch = ¾ in., 18 BWG, 15/16 in. tri.
Passes = 1	Passes = 2

(1) Heat balance:
Distilled water, $Q = 175{,}000 \times 1(93 - 85) = 1{,}400{,}000$ Btu/hr
Raw water, $Q = 280{,}000 \times 1(80 - 75) = 1{,}400{,}000$ Btu/hr

(2) Δt:

	Hot Fluid		Cold Fluid	Diff.
	93	Higher Temp	80	13
	85	Lower Temp	75	10
	8	Differences	5	3

$$\text{LMTD} = 11.4°\text{F} \qquad (5.14)$$
$$R = \frac{8}{5} = 1.6 \qquad S = \frac{5}{93 - 75} = 0.278$$
$$F_T = 0.945 \qquad \text{(Fig. 18)}$$
$$\Delta t = 0.945 \times 11.4 = 10.75°\text{F} \qquad (7.42)$$

(3) T_c and t_c:
The average temperatures T_a and t_a of 89 and 77.5°F will be satisfactory for the short ranges and ϕ_s and ϕ_t taken as 1.0. Try hot fluid in shell as a trial, since it is the smaller of the two.

Hot fluid: shell side, distilled water

(4') $a_s = \text{ID} \times C'B/144P_T$ [Eq. (7.1)]
$= 15.25 \times 0.1875 \times 12/144 \times 0.9375 = 0.254$ ft²

(5') $G_s = W/a_s$ [Eq. (7.2)]
$= 175{,}000/0.254$
$= 690{,}000$ lb/(hr)(ft²)

(6') At $T_a = 89°\text{F}$, $\mu = 0.81 \times 2.42$
$= 1.96$ lb/(ft)(hr)
[Fig. 14]
$D_e = 0.55/12 = 0.0458$ ft [Fig. 28]
$Re_s = D_e G_s/\mu$ [Eq. (7.3)]
$= 0.0458 \times 690{,}000/1.96 = 16{,}200$

(7') $j_H = 73$ [Fig. 28]

(8') At $T_a = 89°\text{F}$, $c = 1.0$ Btu/(lb)(°F)
$k = 0.36$ Btu/(hr)(ft²)(°F/ft) [Table 4]
$(c\mu/k)^{\frac{1}{3}} = (1.0 \times 1.96/0.36)^{\frac{1}{3}} = 1.76$

Cold fluid: tube side, raw water

(4) $a'_t = 0.334$ in.² [Table 10]
$a_t = N_t a'_t/144n$ [Eq. (7.48)]
$= 160 \times 0.334/144 \times 2 = 0.186$ ft²

(5) $G_t = w/a_t$
$= 280{,}000/0.186$
$= 1{,}505{,}000$ lb/(hr)(ft²)
Vel, $V = G_t/3600\rho$
$= 1{,}505{,}000/3600 \times 62.5 = 6.70$ fps

(6) At $t_a = 77.5°\text{F}$, $\mu = 0.92 \times 2.42$
$= 2.23$ lb/(ft)(hr)
[Fig. 14]
$D = 0.65/12 = 0.054$ ft (Re_t is for pressure drop only) [Table 10]
$Re_t = DG_t/\mu$
$= 0.054 \times 1{,}505{,}000/2.23 = 36{,}400$

1-2 PARALLEL-COUNTERFLOW

Hot fluid: shell side, distilled water

(9') $h_o = j_H \dfrac{k}{D_e}\left(\dfrac{c\mu}{k}\right)^{1/3} \times 1$ [Eq. (6.15b)]

$= 73 \times 0.36 \times 1.76/0.0458 = 1010$

(10') (11') (12') The small difference in the average temperatures eliminates the need for a tube-wall correction, and $\phi_s = 1$.

Cold fluid: tube side, raw water

(9) $h_i = 1350 \times 0.99 = 1335$ [Fig. 25]
$h_{io} = h_i \times \text{ID/OD} = 1335 \times 0.65/0.75$
$= 1155$ [Eq. (6.5)]

(13) Clean overall coefficient U_C:

$$U_C = \frac{h_{io}h_o}{h_{io}+h_o} = \frac{1155 \times 1010}{1155 + 1010} = 537 \text{ Btu/(hr)(ft}^2\text{)(°F)} \qquad (6.38)$$

When both film coefficients are high the thermal resistance of the tube metal is not necessarily insignificant as assumed in the derivation of Eq. (6.38). For a steel 18 BWG tube $R_m = 0.00017$ and for copper $R_m = 0.000017$.

(14) Design overall coefficient U_D:

External surface/ft, $a'' = 0.1963 \text{ ft}^2/\text{ft}$
$A = 160 \times 16'0'' \times 0.1963 = 502 \text{ ft}^2$

$$U_D = \frac{Q}{A\Delta t} = \frac{1{,}400{,}000}{502 \times 10.75} = 259 \qquad (5.3)$$

(15) Dirt factor R_d:

$$R_d = \frac{U_C - U_D}{U_C \times U_D} = \frac{537 - 259}{537 \times 259} = 0.0020 \text{(hr)(ft}^2\text{)(°F)/Btu} \qquad (6.13)$$

Summary

1010	h outside	1155
U_C	537	
U_D	259	
R_d Calculated 0.0020		
R_d Required 0.0020		

Pressure Drop

(1') For $Re_s = 16{,}200, f = 0.0019 \text{ ft}^2/\text{in.}^2$ [Fig. 29]

(2') No. of crosses, $N + 1 = 12L/B$ [Eq. (7.43)]
$= 12 \times 16/12$
$= 16$

$D_s = 15.25/12 = 1.27 \text{ ft}$

(1) For $Re_t = 36{,}400, f = 0.00019 \text{ ft}^2/\text{in.}^2$ [Fig. 26]

(2) $\Delta P_t = \dfrac{fG_t^2 Ln}{5.22 \times 10^{10} Ds\phi_t}$ [Eq. (7.45)]

$= \dfrac{0.00019 \times 1{,}505{,}000^2 \times 16 \times 2}{5.22 \times 10^{10} \times 0.054 \times 1.0 \times 1.0}$

$= 4.9 \text{ psi}$

Pressure Drop

(3') $\Delta P_s = \dfrac{fG_s^2 D_s (N+1)}{5.22 \times 10^{10} D_e s \phi_t}$ [Eq. (7.44)]

$= \dfrac{0.0019 \times 690{,}000^2 \times 1.27 \times 16}{5.22 \times 10^{10} \times 0.0458 \times 1.0 \times 1.0}$

$= 7.7$ psi

Allowable $\Delta P_s = 10.0$ psi

(3) ΔP_r: $G_t = 1{,}505{,}000$, $V^2/2g' = 0.33$ [Fig. 27]

$\Delta P_r = (4n/s)(V^2/2g')$ [Eq. (7.46)]

$= \dfrac{4 \times 2}{1} \times 0.33 = 2.6$ psi

(4) $\Delta P_T = P_t + P_r$ [Eq. (7.47)]

$= 4.9 + 2.6 = 7.5$ psi

Allowable $\Delta P_T = 10.0$ psi

It is seen that the overall coefficient for this problem is five times that of the oil-to-oil exchange of Example 7.3, the principal difference being due to the excellent thermal properties of water. The exchanger is satisfactory for the service.

Optimum Outlet-water Temperature. In using water as the cooling medium for a given duty it is possible to circulate a large quantity with a small temperature range or a small quantity with a large temperature range. The temperature range of the water naturally affects the LMTD. If a large quantity is used, t_2 will be farther from T_1 and less surface is required as a result of the larger LMTD. Although this will reduce the original investment and fixed charges, since depreciation and maintenance will also be smaller, the operating cost will be increased owing to the greater quantity of water. It is apparent that there must be an optimum between the two conditions: much water and small surface or little water and large surface.

In the following it is assumed that the line pressure on the water is sufficient to overcome the pressure drop in the exchanger and that the cost of the water is related only to the amount used. It is also assumed that the cooler operates in true counterflow so that $\Delta t = $ LMTD. If the approach is small or there is a temperature cross, the derivation below requires an estimate of F_T by which the LMTD is multiplied.

The total annual cost of the exchanger to the plant will be the sum of the annual cost of water and the fixed charges, which include maintenance and depreciation.

If C_T is the total annual cost,

$C_T = $ (water cost/lb)(lb/hr)(annual hr)
$\qquad\qquad\qquad\qquad\qquad\qquad +$ (annual fixed charges/ft^2)(ft^2)

$$Q = wc(t_2 - t_1) = UA(\text{LMTD}) \qquad (7.49)$$

Substituting the heat-balance terms in Eq. (7.49), where $w = Q/[c(t_2 - t_1)]$ and the surface $A = Q/U(\text{LMTD})$,

$$C_T = \dfrac{Q\theta C_W}{c(t_2 - t_1)} + \dfrac{C_F Q}{U(\text{LMTD})}$$

1-2 PARALLEL-COUNTERFLOW

where θ = annual operating hours
C_W = water cost/lb
C_F = annual fixed charges/ft^2

Assuming U is constant

$$\text{LMTD} = \frac{\Delta t_2 - \Delta t_1}{\ln \Delta t_2/\Delta t_1}$$

Keeping all factors constant except the water-outlet temperature and consequently Δt_2,

$$C_T = \frac{Q\theta C_W}{c(t_2 - t_1)} + \frac{C_F Q}{U\left[\dfrac{T_1 - t_2 - \Delta t_1}{\ln (T_1 - t_2)/\Delta t_1}\right]} \tag{7.50}$$

The optimum condition will occur when the total annual cost is a minimum, thus when $dC_T/dt_2 = 0$.

Differentiating and equating the respective parts,

$$\frac{U\theta C_W}{C_F c}\left(\frac{T_1 - t_2 - \Delta t_1}{t_2 - t_1}\right)^2 = \ln \frac{T_1 - t_2}{\Delta t_1} - \left[1 - \frac{1}{(T_1 - t_2)/\Delta t_1}\right] \tag{7.51}$$

Equation (7.51) has been plotted by Colburn and is reproduced in Fig. 7.24.

Example 7.5. Calculation of the Optimum Outlet-water Temperature. A viscous fluid is cooled from 175 to 150°F with water available at 85°F. What is the optimum-outlet water temperature?

$$175 - x = \Delta t_2$$
$$150 - 85 = \Delta t_1 = 65$$

It will first be necessary to assume a value of U. Since the material is viscous, assume $U = 15$. To evaluate the group $U\theta C_W/C_F c$:

θ = 8000 operating hours annually
C_W computed at $0.01/1000 gal. = 0.01/8300, dollars/lb

For annual charges assume 20 per cent repair and maintenance and 10 per cent depreciation. At a unit cost of $4 per square foot the annual fixed charge is

$$\$4 \times 0.30 = \$1.20$$

The specific heat of water is taken as 1.0.

$$\frac{U\theta C_W}{C_F c} = \frac{15 \times 8000}{1.20 \times 1.0}\left(\frac{0.01}{8300}\right) = 0.1205$$

$$\frac{T_1 - T_2}{\Delta t_1} = \frac{175 - 150}{150 - 85} = 0.39$$

From Figure 7.24,

$$\frac{\Delta t_2}{\Delta t_1} = 0.96$$

$$\Delta t_2 = T_1 - t_2 = 0.96 \times 65 = 62.3°F$$
$$t_2 = 175 - 62.3 = 112.7°F$$

When the value of U is high or there is a large hot-fluid range, the optimum outlet-water temperature may be considerably above the upper limit of 120°F. This is not completely correct, since the maintenance cost will probably rise considerably above 20 per cent of the initial cost when the temperature rises above 120°F. Usually information is not available on the increase in maintenance cost with increased water-outlet

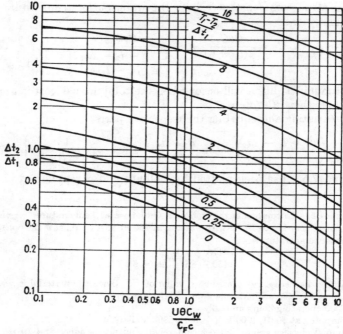

Fig. 7.24. Optimum outlet water temperature. (*Perry, Chemical Engineers' Handbook*, McGraw-Hill Book Company, Inc., New York, 1950.)

temperature, since such data entail not only destructive tests but records kept over a long period of time.

Solution Exchangers. One of the commonest classes of exchangers embraces the cooling or heating of solutions for which there is a paucity of physical data. This is understandable, since property vs. temperature plots are required not only for each combination of solute and solvent but for different concentrations as well. Some of the data available in the literature and other studies permit the formulation of rules for estimating the heat-transfer properties of solutions when the rules are used with considerable caution. They are given as follows:

1-2 PARALLEL-COUNTERFLOW

Thermal conductivity:

Solutions of *organic liquids:* use the weighted conductivity.

Solutions of organic liquids and water: use 0.9 times the weighted conductivity.

Solutions of salts and water circulated through the shell: use 0.9 times the conductivity of water up to concentrations of 30 per cent.

Solutions of salts and water circulating through the tubes and not exceeding 30 per cent: use Fig. 24 with a conductivity of 0.8 that of water.

Colloidal dispersions: use 0.9 times the conductivity of the dispersion liquid.

Emulsions: use 0.9 times the conductivity of the liquid surrounding the droplets.

Specific heat:

Organic solutions: use the weighted specific heat.

Organic solutions in water: use the weighted specific heat.

Fusable salts in water: use the weighted specific heat where the specific heat of the salt is for the crystalline state.

Viscosity:

Organic liquids in organics: use the reciprocal of the sum of the terms, (weight fraction/viscosity) for each component.

Organic liquids in water: use the reciprocal of the sum of the terms, (weight fraction/viscosity) for each component.

Salts in water where the concentration does not exceed 30 per cent and where it is known that a sirup-type of solution does not result: use a viscosity twice that of water. A solution of sodium hydroxide in water under even very low concentrations should be considered sirupy and can not be estimated.

Wherever laboratory tests are available or data can be obtained, they will be preferable to any of the foregoing rules. The following demonstrates the solution of a problem involving an aqueous solution:

Example 7.6. Calculation of a Phosphate Solution Cooler. 20,160 lb/hr of a 30% K_3PO_4 solution, specific gravity at 120°F = 1.30, is to be cooled from 150 to 90°F using well water from 68 to 90°F. Pressure drops of 10 psi are allowable on both streams, and a total dirt factor of 0.002 is required.

Available for this service is a 10.02 in. ID 1-2 exchanger having 52 ¾ in. OD, 16 BWG tubes 16'0" long laid out on 1-in. square pitch. The bundle is arranged for two passes, and the baffles are spaced 2 in. apart.

Will the exchanger be suitable?

162 PROCESS HEAT TRANSFER

Solution:
Exchanger:

Shell side	Tube side
ID = 10.02 in.	Number and length = 52, 16'0''
Baffle space = 2 in.	OD, BWG, pitch = ¾ in., 16 BWG, 1 in. square
Passes = 1	Passes = 2

(1) Heat balance:
 Specific heat of phosphate solution = $0.3 \times 0.19 + 0.7 \times 1 = 0.757$ Btu/(lb)(°F)
 30% K_3PO_4 solution, $Q = 20{,}160 \times 0.757(150 - 90) = 915{,}000$ Btu/hr
 Water, $Q = 41{,}600 \times 1.0(90 - 68) = 915{,}000$ Btu/hr

(2) Δt:

	Hot Fluid		Cold Fluid	Diff.
	150	Higher Temp	90	60
	90	Lower Temp	68	22
	60	Differences	22	38

LMTD = 37.9°F (5.14)

$R = \dfrac{60}{22} = 2.73 \qquad S = \dfrac{22}{150 - 68} = 0.268$

$F_T = 0.81$ (Fig. 18)
$\Delta t = 0.81 \times 37.9 = 30.7°F$ (7.42)

(3) T_c and t_c: The average temperatures T_a and t_a of 120 and 79°F will be satisfactory.

Hot fluid: shell side, phosphate solution

(4') $a_s = \text{ID} \times C'B/144P_T$ [Eq. (7.1)]
 $= 10.02 \times 0.25 \times 2/144 \times 1$
 $= 0.0347$ ft²

(5') $G_s = w/a_s$
 $= 20{,}160/0.0347$
 $= 578{,}000$ lb/(hr)(ft²)

(6') At $T_a = 120°F$,
 $\mu = 2\mu_\text{water} = 1.20 \times 2.42$
 $= 2.90$ lb/(ft)(hr) [Fig. 14]
 $D_e = 0.95/12 = 0.079$ ft [Fig. 28]
 $Re_s = D_e G_s/\mu$ [Eq. (7.3)]
 $= 0.079 \times 578{,}000/2.90 = 15{,}750$

(7') $j_H = 71$ [Fig. 28]

(8') At $T_a = 120°F$, $k = 0.9 k_\text{water}$
 $= 0.9 \times 0.37$
 $= 0.33$ Btu/(hr)(ft²)(°F/ft)
 $(c\mu/k)^{1/3} = (0.757 \times 2.90/0.33)^{1/3} = 1.88$

(9') $h_o = j_H \dfrac{k}{D_e}\left(\dfrac{c\mu}{k}\right)^{1/3} \times 1$ [Eq. (6.15b)]
 $= 71 \times 0.33 \times 1.88/0.079$
 $= 558$ Btu/(hr)(ft²)(°F)

(10') (11') (12') ϕ_s and $\phi_t = 1$

Cold fluid: tube side, water

(4) $a_t' = 0.302$ in.² [Table 10]
 $a_t = N_t a_t'/144n$
 $= 52 \times 0.302/144 \times 2 = 0.0545$ ft²

(5) $G_t = w/a_t$
 $= 41{,}600/0.0545$
 $= 762{,}000$ lb/(hr)(ft²)
 $V = G_t/3600\rho = 762{,}000/3600 \times 62.5$
 $= 3.40$ fps

(6) At $t_a = 79°F$, $\mu = 0.91 \times 2.42$
 $= 2.20$ lb/(ft)(hr) [Table 14]
 $D = 0.62/12 = 0.0517$ ft [Table 10]
 (Re_t is for pressure drop only)
 $Re_t = DG_t/\mu$
 $= 0.0517 \times 762{,}000/2.20 = 17{,}900$

(9) $h_i = 800 \times 1.0 = 800$ [Fig. 25]
 $h_{io} = h_i \times \text{ID/OD} = 800 \times 0.62/0.75$
 $= 662$ Btu/(hr)(ft²)(°F) [Eq. (6.5)]

1-2 PARALLEL-COUNTERFLOW

(13) Clean overall coefficient U_C:

$$U_C = \frac{h_{io}h_o}{h_{io} + h_o} = \frac{662 \times 558}{662 + 558} = 303 \text{ Btu/(hr)(ft}^2\text{)(°F)} \qquad (6.38)$$

(14) Design overall coefficient U_D:

External surface/ft, $a'' = 0.1963$ ft (Table 10)
$A = 52 \times 16'0'' \times 0.1963 = 163 \text{ ft}^2$

$$U_D = \frac{Q}{A\Delta t} = \frac{915{,}000}{163 \times 30.7} = 183 \text{ Btu/(hr)(ft}^2\text{)(°F)}$$

(15) Dirt factor R_d:

$$R_d = \frac{U_C - U_D}{U_C U_D} = \frac{303 - 183}{303 \times 183} = 0.00216 \text{ (hr)(ft}^2\text{)(°F)/Btu} \qquad (6.13)$$

Summary

558	h outside	662
U_C		303
U_D		183
R_d Calculated		0.00216
R_d Required		0.0020

Pressure Drop

(1′) For $Re_s = 15{,}750$, $f = 0.0019 \text{ ft}^2/\text{in.}^2$ [Fig. 29]

(2′) No. of crosses, $N + 1 = 12L/B$ [Eq. (7.43)]
$= 12 \times 16\frac{1}{2}$
$= 96$

$D_s = 10.02/12 = 0.833$ ft

(3′) $\Delta P_s = \dfrac{fG_s^2 D_s(N+1)}{5.22 \times 10^{10} D_e s \phi_s}$ [Eq. (7.44)]

$= \dfrac{0.0019 \times 578{,}000^2 \times 0.833 \times 96}{5.22 \times 10^{10} \times 0.079 \times 1.30 \times 1.0}$
$= 9.5$ psi

Allowable $\Delta P_s = 10.0$ psi

(1) For $Re_t = 17{,}900$, $f = 0.00023 \text{ ft}^2/\text{in.}^2$ [Fig. 26]

(2) $\Delta P_t = \dfrac{fG_t^2 Ln}{5.22 \times 10^{10} Ds\phi_t}$ [Eq. (7.45)]

$= \dfrac{0.00023 \times 762{,}000^2 \times 16 \times 2}{5.22 \times 10^{10} \times 0.0517 \times 1.0 \times 1.0}$
$= 1.6$ psi

(3) $G_t = 762{,}000$, $\dfrac{V^2}{2g'} = 0.08$ [Fig. 27]

$\Delta P_r = \dfrac{4n}{s}\dfrac{V^2}{2g'}$ [Eq. (7.46)]

$= \dfrac{4 \times 2}{1} \times 0.08 = 0.7$ psi

$\Delta P_T = \Delta P_t + \Delta P_r$ [Eq. (7.47)]
$= 1.6 + 0.7 = 2.3$ psi
Allowable $\Delta P_T = 10.0$ psi

The exchanger is satisfactory for the service.

Steam as a Heating Medium. Thus far none of the heat-transfer services studied has employed steam although it is by far the commonest heating medium. Steam as a heating medium introduces several difficulties: (1) Hot steam condensate is fairly corrosive, and care must be exercised to prevent condensate from accumulating within an exchanger where continuous contact with metal will cause damage. (2) The condensate line must be connected with discretion. Suppose exhaust steam at 5 psig and 228°F is used to heat a cold fluid entering at a temperature of 100°F. The tube wall will be at a temperature between the two but nearer that of the steam, say 180°F, which corresponds to a saturation pressure of only 7.5 psia for the condensate at the tube wall. Although the steam entered at 5 psig, the pressure on the steam side may drop locally to a pressure below that of the atmosphere, so that the condensate will *not* run out of the heater. Instead it will remain and build up in the exchanger until it blocks off all the surface available for heat transfer. Without surface, the steam will not continue to condense and will retain its inlet pressure long enough to blow out some or all of the accumulated condensate so as to reexpose surface, depending upon the design. The heating operation will become cyclical and to overcome this difficulty and attain uniform flow, it may be necessary to employ a trap or suction for which piping arrangements will be discussed in Chap. 21.

The heat-transfer coefficients associated with the condensation of steam are very high compared with any which have been studied so far. It is customary to adopt a conventional and conservative value for the film coefficient, since it is never the controlling film, rather than obtain one by calculation. In this book in all heating services employing relatively air-free steam a value of 1500 Btu/(hr)(ft^2)(°F) will be used for the condensation of steam without regard to its location. Thus $h_i = h_o = h_{io} = 1500$.

It is advantageous in heating to connect the steam to the tubes of the heater rather than the shell. In this way, since the condensate may be corrosive, the action can be confined to the tube side alone, whereas if the steam is introduced into the shell, both may be damaged. When steam flows through the tubes of a 1-2 exchanger, there is no need for more than two tube passes. Since steam is an isothermally condensing fluid, the true temperature difference Δt and the LMTD are identical.

When using superheated steam as a heating medium, except in desuperheaters, it is customary to disregard the temperature range of desuperheating and consider all the heat to be delivered at the saturation temperature corresponding to the operating pressure. A more intensive analysis of the condensation of steam will be undertaken in the chapters dealing with condensation.

Pressure Drop for Steam. When steam is employed in two passes on the tube side, the allowable pressure drop should be very small, less than 1.0 psi, particularly if there is a gravity return of condensate to the boiler. In a gravity-return system, condensate flows back to the boiler because of the difference in static head between a vertical column of steam and a vertical column of condensate. The pressure drop including entrance and end losses through an exchanger can be calculated by taking one-half of the pressure drop for steam as calculated in the usual manner by Eq. (7.45) for the inlet vapor conditions. The mass velocity is calculated from the inlet steam rate and the flow area of the first pass (which need not be equal to that of the second pass). The Reynolds number is based on the mass velocity and the viscosity of steam as found in Fig. 15. The specific gravity used with Eq. (7.45) is the density of the steam obtained from Table 7 for the inlet pressure divided by the density of water taken as 62.5 lb/ft^3.

Quite apparently this calculation is an approximation. It is conservative inasmuch as the pressure drop per foot of length decreases successively with the square of the mass velocity while the approximation assumes a value more nearly the mean of the inlet and outlet.

The Optimum Use of Exhaust and Process Steam. Some plants obtain power from noncondensing turbines or engines. In such places there may be an abundance of exhaust steam at low pressures from 5 to 25 psig which is considered a by-product of the power cycles in the plant. While there are arbitrary aspects to the method of estimating the cost of exhaust steam, it will be anywhere from one-quarter to one-eighth of the cost of process or live steam. Although it possesses a high latent heat, exhaust steam is of limited process value, since the saturation temperature is usually about 215 to 230°F. If a liquid is to be heated to 250 or 275°F, it is necessary to use process steam at 100 to 200 psi developed at the powerhouse specially for process purposes.

When a fluid is to be heated to a temperature close to or above that of exhaust steam, all the heating can be done in a single shell using only process steam. As an alternative the heat load can be divided into two shells, one utilizing as much exhaust steam as possible and the other using as little process steam as possible. This leads to an optimum: If the outlet temperature of the cold fluid in the first exchanger is made to approach the exhaust steam temperature too closely, a small Δt and large first heater will result. On the other hand, if the approach is not close, the operating cost of the higher process steam requirement in the second heater increases so that the initial cost of two shells may not be justified.

In the following analysis it is assumed that the pressure drop, pumping cost, and overall coefficient are identical in a single and double heater arrangement. It is assumed also that the fixed charges per square foot of surface is constant, although this too is not strictly true. The cost equation is taken as the sum of steam and fixed charges and because steam condenses isothermally, $\Delta t = \text{LMTD}$.

$$C_T = wc(t - t_1)\theta C_E + A_1 C_F + wc(t_2 - t)\theta C_P + A_2 C_F \qquad (7.52)$$

where C_T = total annual cost, dollars
C_F = annual fixed charges, dollars/ft^2
C_E = cost of exhaust steam, dollars/Btu
C_P = cost of process steam, dollars/Btu
T_E = temperature of exhaust steam, °F
T_P = temperature of process steam, °F
t = intermediate temperature between shells
θ = total annual operating hours

$$A_1 = \frac{Q_1}{U\,\Delta t_1} = \frac{wc}{U}\ln\frac{T_E - t_1}{T_E - t} \quad \text{and} \quad A_2 = \frac{Q_2}{U\,\Delta t_2} = \frac{wc}{U}\ln\frac{T_P - t}{T_P - t_2}$$

Substituting, differentiating Eq. (7.52) with respect to t, and setting equal to zero

$$(T_P - t)(T_E - t) = \frac{C_F(T_P - T_E)}{(C_P - C_E)U\theta} \qquad (7.53)$$

Example 7.7. The Optimum Use of Exhaust and Process Steam. Exhaust steam at 5 psi ($\simeq$ 228°F) and process steam at 85 psi ($\simeq$ 328°F) are available to heat a liquid from 150 to 250°F. Exhaust steam is available at 5 cents per per 1000 lb, and process steam at 30 cents per 1000 lb. From experience an overall rate of 50 Btu/(hr)(ft^2)(°F) may be expected. The assumption may be checked later. Use annual fixed charges of $1.20 per square foot, 8000 annual hours, latent heats of 960.1 Btu/lb for exhaust, and 888.8 Btu/lb for process steam.

Solution:

$$(328 - t)(228 - t) = \frac{1.20(328 - 228)}{(0.30/1000 \times 888.8 - 0.05/1000 \times 960)50 \times 8000} \qquad (7.53)$$
$$t = 218°\text{F}$$

1-2 Exchangers without Baffles. Not all 1-2 exchangers have 25 per cent cut segmental baffles. When it is desired that a fluid pass through the shell with an extremely small pressure drop, it is possible to depart from the use of segmental baffles and use only support plates. These will usually be half-circle, 50 per cent cut plates which provide rigidity and prevent the tubes from sagging. Successive support plates overlap at the shell diameter so that the entire bundle can be supported by two

1-2 PARALLEL-COUNTERFLOW

half circles which support one or two rows of tubes in common. These may be spaced farther apart than the outside diameter of the shell, but when they are employed, the shell fluid is considered to flow along the axis instead of across the tubes. When the shell fluid flows *along* the tubes or the baffles are cut more than 25 per cent, Fig. 28 no longer applies. The flow is then analogous to the annulus of a double pipe exchanger and can be treated in a similar manner, using an equivalent diameter based on the distribution of flow area and the wetted perimeter for the entire shell. The calculation of the shell-side pressure drop will also be similar to that for an annulus.

Example 7.8. Calculation of a Sugar-solution Heater without Baffles. 200,000 lb/hr of a 20 per cent sugar solution ($s = 1.08$) is to be heated from 100 to 122°F using steam at 5 psi pressure.

Available for this service is a 12 in. ID 1-2 exchanger without baffles having 76 ¾ in. OD, 16 BWG tubes 16'0" long laid out on a 1-in. square pitch. The bundle is arranged for two passes.

Can the exchanger provide a 0.003 dirt factor without exceeding a 10.0 psi solution pressure drop?

Solution:

Exchanger:

Shell side	Tube side
ID = 12 in.	Number and length = 76, 16'0"
Baffle space = half circles	OD, BWG, pitch = ¾ in., 16 BWG, 1 in. square
Passes = 1	Passes = 2

(1) Heat balance:
Specific heat of 20 per cent sugar at 111°F = $0.2 \times 0.30 + 0.8 \times 1$
$= 0.86$ Btu/(lb)(°F)
Sugar solution, $Q = 200{,}000 \times 0.86(122 - 100) = 3{,}790{,}000$ Btu/hr
Steam $Q = 3950 \times 960.1 = 3{,}790{,}000$ Btu/hr   (Table 7)

(2) Δt:

Hot Fluid		Cold Fluid	Diff.
228	Higher Temp	122	106
228	Lower Temp	100	128
0	Differences	22	22

When $R = 0$, $\Delta t = $ LMTD $= 116.5$°F.   (5.14)

(3) T_c and t_c: The steam coefficient will be very great compared with that for the sugar solution, and the tube wall will be considerably nearer 228°F than the caloric temperature of the fluid. Obtain F_c from U_1 and U_2. Failure to correct for wall effects, however, will keep the heater calculation on the safe side. Use 111°F as the average, t_a.

Hot fluid: tube side, steam

(4) $a'_t = 0.302$ in^2 [Table 10]
$a_t = N_t a'_t / 144n$ [Eq. (7.48)]
$= 76 \times 0.302 / 144 \times 2 = 0.0797$ ft^2

(5) G_t (for pressure drop only) $= W/a_t$
$= 3950/0.0797 = 49{,}500$ lb/(hr)(ft^2)

(6) At $T_a = 228°$F,
$\mu_{\text{steam}} = 0.0128 \times 2.42$
$= 0.031$ lb/(ft)(hr)   [Fig. 15]
$D = 0.62/12 = 0.0517$ ft   [Table 10]
$Re_t = DG_t/\mu$ [Eq. (3.6)]
$= 0.0517 \times 49{,}500 / 0.031 = 82{,}500$
Re_t is for pressure drop.

(9) Condensation of steam:
$h_{io} = 1500$ Btu/(hr)(ft^2)(°F)
(10) $t_w{}^*$:
$t_w = t_c + \dfrac{h_{io}}{h_{io} + h_o}(T_a - t_a)$ [Eq. (5.31a)]
$= 111 + \dfrac{1500}{1500 + 278}(228 - 111)$
$= 210°$F

Cold fluid: shell side, sugar solution

(4′) a_s = (area of shell) − (area of tubes)
$= \tfrac{1}{144}(\pi 12^2/4 - 76 \times \pi \times 0.75^2/4)$
$= 0.55$ ft^2

(5′) $G_s = w/a_s$ [Eq. (7.2)]
$= 200{,}000/0.55$
$= 364{,}000$ lb/(hr)(ft^2)

(6′) At $t_a = 111°$F, $\mu = 2\mu_{\text{water}}$
$= 1.30 \times 2.42 = 3.14$ lb/(ft)(hr)
    [Fig. 14]
$D_e = 4a_s/(\text{wetted perimeter})$ [Eq. (6.3)]
$= 4 \times 0.55/(76 \times \pi \times 0.75/12)$
$= 0.148$ ft
$Re_s = D_e G_s/\mu$ [Eq. (7.3)]
$= 0.148 \times 364{,}000/3.14 = 17{,}100$
(7′) From Fig. 24 (tube-side data)
$j_H = 61.5$
(8′) At $t_a = 111°$F,
$k = 0.9 \times 0.37$
$= 0.333$ Btu/(hr)(ft^2)(°F/ft)
$(c\mu/k)^{1/3} = (0.86 \times 3.14/0.333)^{1/3} = 2.0$

(9′) $h_o = j_H \dfrac{k}{D_e}\left(\dfrac{c\mu}{k}\right)^{1/3}\phi_s$ [Eq. (6.15b)]
$\dfrac{h_o}{\phi_s} = 61.5 \times 0.333 \times 2.0/0.148 = 278$

(11′) At $t_w = 210°$F, $\mu_w = 2\mu_{\text{water}}$
$= 0.51 \times 2.42 = 1.26$ lb/(ft)(hr)
    [Fig. 14]
$\phi_s = (\mu/\mu_w)^{0.14} = (3.14/1.26)^{0.14} = 1.12$

(12′) Corrected coefficient, $h_o = \dfrac{h_o \phi_s}{\phi_s}$
    [Eq. (6.36)]
$= 278 \times 1.12 = 311$ Btu/(hr)(ft^2)(°F)

(13) Clean overall coefficient U_C:

$$U_C = \frac{h_{io} h_o}{h_{io} + h_o} = \frac{1500 \times 311}{1500 + 311} = 257 \text{ Btu/(hr)(ft}^2\text{)(°F)} \qquad (6.38)$$

(14) Design overall coefficient U_D:

$a'' = 0.1963$ ft^2/lin ft   (Table 10)
$A = 76 \times 16'0'' \times 0.1963 = 238$ ft^2

$$U_D = \frac{Q}{A\,\Delta t} = \frac{3{,}790{,}000}{238 \times 116.5} = 137 \text{ Btu/(hr)(ft}^2\text{)(°F)}$$

(15) Dirt factor R_d:

$$R_d = \frac{U_C - U_D}{U_C U_D} = \frac{257 - 137}{257 \times 137} = 0.0034 \text{ (hr)(ft}^2\text{)(°F)/Btu} \qquad (6.13)$$

* Note h_{io} in the numerator.

Summary

1500	h outside	311
U_C		257
U_d		137
R_d Calculated		0.0034
R_d Required		0.003

Pressure Drop

(1) Specific vol of steam from Table 7:
$v = 20.0$ ft^3/lb

$s = \dfrac{1/20.0}{62.5} = 0.00080$

$Re_t = 82{,}500, f = 0.000155$ ft^2/in.^{2} [Fig. 26]

$\Delta P_t = \dfrac{1}{2} \times \dfrac{fG_t^2 Ln}{5.22 \times 10^{10} Ds\phi_t}$ [Eq. (7.45)]

$= \dfrac{1}{2} \times \dfrac{0.000155 \times 49{,}500^2 \times 16 \times 2}{5.22 \times 10^{10} \times 0.0517 \times 0.0008 \times 1.0}$

$= 2.8$ psi

This is a relatively high pressure drop for steam with a gravity condensate return. The exchanger is satisfactory.

(1') $D_{e'} = 4 \times$ flow area/frictional wetted perimeter [Eq. (6.4)]
$= 4 \times 0.55/(76 \times 3.14 \times 0.75/12 + 3.14 \times 1\frac{2}{12}) = 0.122$ ft

$Re_s' = D_{e'}G_s/\mu$ [Eq. (7.3)]
$= 0.122 \times 364{,}000/3.14 = 14{,}100$

f(from Fig. 26 for tube side)
$= 0.00025$ ft^2/in.^{2}

(2') $\Delta P_s = \dfrac{fG_s^2 Ln}{5.22 \times 10^{10} D_{e'} s \phi_s}$ [Eq. (7.45)]

$= \dfrac{0.00025 \times 364{,}000^2 \times 16 \times 1}{5.22 \times 10^{10} \times 0.122 \times 1.08 \times 1.12}$

$= 0.07$ psi

Heat Recovery in a 1-2 Exchanger. When an exchanger is clean, the hot-fluid outlet temperature is lower than the process outlet temperature and the cold-fluid outlet temperature is higher than the process outlet temperature. For counterflow it was possible to obtain the value of T_2 and t_2 for a clean exchanger from Eq. (5.18), starting with

$$wc(t_2 - t_1) = UA \times \text{LMTD}$$

For a 1-2 exchanger the outlet temperatures can be obtained starting with the expression $wc(t_2 - t_1) = UAF_T \times \text{LMTD}$, where the LMTD is defined in terms of parameters R and S by Eq. (7.39) and F_T is defined by Eq. (7.41).

Recognizing that F_T can be eliminated when UA/wc in Eq. (7.37) is plotted against S, Ten Broeck[1] developed the graph shown in Fig. 7.25. In an existing 1-2 exchanger A and wc are known. U can be computed

[1] Ten Broeck, H., *Ind. Eng. Chem.*, **30**, 1041–1042 (1938).

from the flow quantities and temperatures, and R can be evaluated from wc/WC. This permits S to be read directly from the graph. Since $S = (t_2 - t_1)/(T_1 - t_1)$ and T_1 and t_1 are known, it is then possible to obtain t_2 and from the heat balance $wc(t_2 - t_1) = WC(T_1 - T_2)$. The line designated *threshold* represents the initial points at which a temperature cross occurs. Values on this line correspond to $T_2 = t_2$.

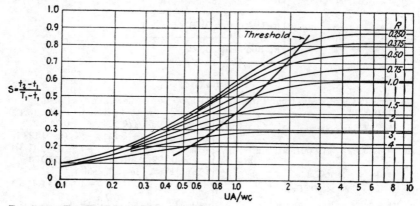

Fig. 7.25. Ten Broeck chart for determining t_2 when T_1 and t_1 are known in a 1-2 exchanger. (*Industrial & Engineering Chemistry*.)

Example 7.9. Outlet Temperatures for a Clean 1-2 Exchanger. In Example 7.3, kerosene-crude oil exchanger, what will the outlet temperatures be when the exchanger is freshly placed in service?

Solution:

$$U_C = 69.3 \quad A = 662 \quad w = 149{,}000 \quad c = 0.49$$
$$W = 43{,}800 \quad C = 0.60$$

$$\frac{UA}{wc} = \frac{69.3 \times 662}{149{,}000 \times 0.49} = 0.63$$

$$R = \frac{wc}{WC} = \frac{149{,}000 \times 0.49}{43{,}800 \times 0.60} = 2.78$$

From Figure 7.25,

$$S = \frac{t_2 - t_1}{T_1 - t_1} = 0.265$$
$$t_2 = t_1 + 0.265(T_1 - t_1) = 100 + 0.265(390 - 100) = 177°F$$
$$T_2 = T_1 - R(t_2 - t_1) = 390 - 2.78(177 - 100) = 176°F$$

The Efficiency of an Exchanger. In the design of many types of apparatus it is frequently desirable to establish a standard of maximum performance. The efficiency is then defined as the fractional performance of an apparatus delivering less than the standard. Dodge[1] gives the definition of the efficiency of an exchanger as the ratio of the quantity

---

[1] Dodge, B. F., "Chemical Engineering Thermodynamics," McGraw-Hill Book Company, Inc., New York, 1944.

of heat removed from a fluid to the maximum which might have been removed. Using the usual nomenclature,

$$e = \frac{wc(t_2 - t_1)}{wc(T_1 - t_1)} = \frac{t_2 - t_1}{T_1 - t_1} \tag{7.54}$$

which is identical with the temperature group S and presumed that $t_2 = T_1$. Depending upon whether the hot or cold terminal approaches zero, the efficiency may also be expressed by

$$e = \frac{WC(T_1 - T_2)}{WC(T_1 - t_1)} \tag{7.55}$$

Although there is merit to this definition from the standpoint of thermodynamics there is a lack of realism in an efficiency definition which involves a terminal difference and a temperature difference of zero. It is the same as defining the efficiency as the ratio of the heat transferred by a real exchanger to an exchanger with infinite surface.

In process heat transfer there is another definition which is useful. The process temperatures are capable of providing a maximum temperature difference if arranged in counterflow. There appears to be some value in regarding the efficiency of an exchanger as the ratio of the temperature difference attained by any other exchanger to that for true counterflow. This is identical with F_T, which proportionately influences the surface requirements. It will be seen in the next and later chapters that other flow arrangements besides 1-2 parallel flow-counterflow can be attained in tubular equipment and by which the value of F_T may be increased for given process temperatures. These obviously entail flow patterns which approach true counterflow more closely than the 1-2 exchanger.

PROBLEMS

7.1. A 1-2 exchanger is to be used for heating 50,000 lb/hr of methyl ethyl ketone from 100 to 200°F using hot amyl alcohol available at 250°F. (a) What minimum quantity of amyl alcohol is required to deliver the desired heat load in a 1-2 exchanger? (b) If the amyl alcohol is available at 275°F, how does this affect the total required quantity?

7.2. A 1-2 exchanger has one shell and two tube passes. The passes do not have equal surfaces. Instead X per cent of the tubes are in the first pass and $(1 - X)$ per cent are in the second, but if the tube-side film coefficient is not controlling, the assumption of constant U is justifiable. (a) Develop an expression for the true temperature difference when X per cent of the tubes are in the colder of the two tube passes. (b) What is the true temperature difference when the hot fluid is cooled from 435 to 225°F by a noncontrolling cooling medium in the tubes which is heated from 100 to 150°F when 60 per cent of the tubes are in the colder tube pass and (c) when 40 per cent of the tubes are in the colder pass? How do these compare with the 1-2 true temperature difference with equal surfaces in each pass?

7.3. A double pipe exchanger has been designed for the nozzle arrangement shown in Fig. 7.26. If the hot stream is cooled from 275 to 205°F while the cold stream enters at 125°F and is heated to $t_2 = 190$°F, what is the true temperature difference? (*Hint.* Establish an equation for the temperature difference with the nozzle arrangement shown and sufficient to allow a numerical trial-and-error solution.) How does this compare with the LMTD for counterflow?

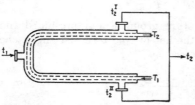

Fig. 7.26. Illustration for Prob. 7.3.

7.4. 43,800 lb/hr of 42°API kerosene between 390 and 200°F is used to heat 149,000 lb/hr of 34°API Mid-continent crude from 100 to 170°F in a 662-ft² exchanger (Example 7.3). The clean coefficient is 69.3 Btu/(hr)(ft²)(°F). When the 1-2 exchanger is clean, what outlet temperatures will be obtained? Calculate the outlet temperatures directly from F_T. How does the total heat load compare with that which could be delivered by a true counterflow exchanger assuming that the same U could be obtained?

7.5. It is necessary on a new installation to preheat 149,000 lb/hr of 34°API crude oil from 170°F to a temperature of 285°F, corresponding to that of the feed plate of a fractionating tower. There is a utility 33°API gas oil line running near the tower at 530°F of relatively unlimited quantity. Because the pumping cost for cold gas oil is prohibitive, the temperature of the gas oil from the heat exchanger, returning to the line, should not be less than 300°F.

Available on the site is a 25 in. ID 1-2 exchanger containing 252 tubes 1 in. OD. 13 BWG, 16'0" long arranged on a six-pass 1¼-in. triangular pitch layout. The shell baffles are spaced at 5-in. centers. A pumping head of 10 psi is allowable on the gas oil line and 15 psi on the feed line. Will the exchanger be acceptable if cleaned, and if so, what will the fouling factor be? For the gas oil the viscosities are 0.4 centipoise at 530°F and 0.7 centipoise at 300°F. For the crude oil the viscosities are 0.9 centipoise at 285°F and 2.1 centipoise at 170°F. (Interpolate by plotting °F vs. centipoise on logarithmic paper.)

7.6. 96,000 lb/hr of 35°API absorption oil in being cooled from 400 to 200°F is used to heat 35°API distillate from 100 to 200°F. Available for the service is a 29 in. ID 1-2 exchanger having 338 tubes 1 in. OD, 14 BWG, 16'0" long on 1¼-in. triangular pitch. Baffles are spaced 10 in. apart, and the bundle is arranged for four tube passes. What arrangement gives the more nearly balanced pressure drops, and what is the dirt factor? The viscosity of the absorption oil is 2.6 centipoise at 100°F and 1.15 centipoise at 210°F. (Plot on logarithmic paper °F vs. viscosity in centipoise, and extrapolate as a straight line.) The viscosity of the distillate is 3.1 centipoise at 100°F and 1.3 centipoise at 210°F.

7.7. 43,200 lb/hr of 35°API distillate is cooled from 250 to 120°F using cooling water from 85 to 120°F. Available for the service is a 19¼ in. ID 1-2 exchanger having 204 tubes ¾ in. OD, 16 BWG, 16'0" long on 1-in. square pitch. Baffles are spaced 5 in. apart, and the bundle is arranged for four passes. What arrangement gives the more nearly balanced pressure drops, and what is the dirt factor? What is the optimum outlet-water temperature? (Viscosities of the distillate are given in Prob. 7.6.)

7.8. 75,000 lb/hr of ethylene glycol is heated from 100 to 200°F using steam at 250°F. Available for the service is a 17¼ in. ID 1-2 exchanger having 224 tubes

1-2 PARALLEL-COUNTERFLOW 173

¾ in. OD, 14 BWG, 16'0" long on $1\frac{5}{16}$-in. triangular pitch. Baffles are spaced 7 in. apart, and there are two tube passes to accommodate the steam. What are the pressure drops, and what is the dirt factor?

7.9. 100,000 lb/hr of 20 per cent potassium iodide solution is to be heated from 80 to 200°F using steam at 15 psig. Available for the service is a 10 in. ID 1-2 exchanger without baffles having 50 tubes ¾ in. OD, 16 BWG, 16'0" long arranged for two passes on $1\frac{5}{16}$-in. triangular pitch. What are the pressure drops and the dirt factor?

7.10. 78,359 lb/hr of isobutane (118°API) is cooled from 203 to 180°F by heating butane (111.5°API) from 154 to 177°F. Available for the service is a $17\frac{1}{4}$ in. ID 1-2 exchanger having 178 tubes ¾ in. OD, 14 BWG, 12'0" long on 1-in. triangular pitch. Baffles are spaced 6 in. apart, and the bundle is arranged for four passes. What are the pressure drops and the dirt factor?

7.11. A 1-2 exchanger recovers heat from 10,000 lb/hr of boiler blowdown at 135 psig by heating raw water from 70 to 96°F. Raw water flows in the tubes. Available for the service is a 10.02 in. ID 1-2 exchanger having 52 tubes ¾ in. OD, 16 BWG, 8'0" long. Baffles are spaced 2 in. apart, and the bundle is arranged for two tube passes. What are the pressure drops and fouling factors?

7.12. 60,000 lb/hr of a 25% NaCl solution is cooled from 150 to 100°F using water with an inlet temperature of 80°F. What outlet water temperature may be used? Available for the service is a $21\frac{1}{4}$ in. ID 1-2 exchanger having 302 tubes ¾ in. OD, 14 BWG, 16'0" long. Baffles are spaced 5 in. apart, and the bundle is arranged for two passes. What are the pressure drops and fouling factor?

NOMENCLATURE FOR CHAPTER 7

A	Heat-transfer surface, ft^2
a	Flow area, ft^2
a''	External surface per linear foot, ft
B	Baffle spacing, in.
C	Specific heat of hot fluid, in derivations, Btu/(lb)(°F)
C'	Clearance between tubes, in.
c	Specific heat of fluid, Btu/(lb)(°F)
C_E	Cost of exhaust steam, dollars/Btu
C_F	Annual fixed charges, dollars/ft^2
C_P	Cost of process steam, dollars/Btu
C_T	Total annual cost, dollars/year
C_W	Cost of water, dollars/lb
D	Inside diameter of tubes, ft
d_o	Outside diameter of tubes, in.
D_e, D'_e	Equivalent diameter for heat transfer and pressure drop, ft
d_e, d'_e	Equivalent diameter for heat transfer and pressure drop, in.
D_s	Inside diameter of shell, ft
e	Efficiency, dimensionless
F_c	Caloric fraction, dimensionless
F_T	Temperature difference factor, $\Delta t = F_T \times$ LMTD, dimensionless
f	Friction factor, dimensionless; for ΔP in psi, ft^2/in^2
G	Mass velocity, lb/(hr)(ft^2)
g	Acceleration of gravity, ft/hr^2
g'	Acceleration of gravity, ft/sec^2
h, h_i, h_o	Heat-transfer coefficient in general, for inside fluid, and for outside fluid, respectively, Btu/(hr)(ft^2)(°F)

h_{io}	Value of h_i when referred to the tube outside diameter, Btu/(hr)(ft²)(°F)
ID	Inside diameter, in.
j_H	Factor for heat transfer, dimensionless
K_c	Caloric constant, dimensionless
K_1, K_2	Numerical constants
k	Thermal conductivity, Btu/(hr)(ft²)(°F/ft)
L	Tube length, ft
LMTD	Log mean temperature difference, °F
N	Number of shell-side baffles
N_t	Number of tubes
n	Number of tube passes
P_T	Tube pitch, in.
$\Delta P_T, \Delta P_t, \Delta P_r$	Total, tube side and return pressure drop, respectively, psi
Q	Heat flow, Btu/hr
R	Temperature group, $(T_1 - T_2)/(t_2 - t_1)$, dimensionless
R_d, R_i, R_o	Combined, inside and outside dirt factors, respectively, (hr)(ft²)(°F)/Btu
Re, Re'	Reynolds number for heat transfer and pressure drop, dimensionless
S	Temperature group, $(t_2 - t_1)/(T_1 - t_1)$, dimensionless
s	Specific gravity
T, T_1, T_2	Temperature in general, inlet and outlet of hot fluid, °F
T_E, T_P	Saturation temperatures of exhaust and pressure steam, °F
T_a	Average temperature of hot fluid, °F
T_c	Caloric temperature of hot fluid, °F
t, t_1, t_2	Temperature in general or outlet of the first of two heaters, inlet and outlet of cold fluid, °F
$t^{\mathrm{I}}, t^{\mathrm{II}}$	Temperatures in first and second passes, °F
t_a	Average temperature of cold fluid, °F
t_i	Temperature at end of first pass, °F
t_c	Caloric temperature of cold fluid, °F
t_w	Tube wall temperature, °F
Δt	True temperature difference in $Q = U_D A \Delta t$, °F
$\Delta t_c, \Delta t_h$	Temperature differences at the cold and hot terminals, °F
U, U_C, U_D	Overall coefficient of heat transfer, clean coefficient, design coefficient, Btu/(hr)(ft²)(°F)
V	Velocity, fps
v	Specific volume, ft³/lb
W	Weight flow in general, weight flow of hot fluid, lb/hr
w	Weight flow of cold fluid, lb/hr
X	Length, ft
z	Height, ft
ϕ	The viscosity ratio $(\mu/\mu_w)^{0.14}$
μ	Viscosity, centipoises $\times 2.42$ = lb/(ft)(hr)
μ_w	Viscosity at tube-wall temperature, centipoises $\times 2.42$ = lb/(ft)(hr)
ρ	Density, lb/ft³

Subscripts (except as noted above)

s	Shell
t	Tubes

CHAPTER 8

FLOW ARRANGEMENTS FOR INCREASED HEAT RECOVERY

The Lack of Heat Recovery in Exchangers. The important limitation of the 1-2 exchangers treated in Chap. 7 lies in their inherent inability to provide effective heat recovery. The advantages of 1-2 exchangers have already been discussed. When a temperature cross occurs in a 1-2 exchanger, the value of F_T drops sharply, and the small extent to which the shell-outlet temperature can fall below the tube-outlet temperature eliminates them from consideration for high heat recoveries. Assume conditions in which the shell fluid is reduced from 200 to 140°F while the tube fluid rises from 80 to 160°F. All the heat in the hot fluid from 140 to 80°F is necessarily lost in a 1-2 exchanger because of the close approach required between the tube fluid at the end of the parallel pass and the shell fluid outlet T_2, as in Figs. 7.20 and 7.21.

This chapter deals with shell-and-tube equipment and the methods by which the temperature cross of two fluid streams $t_2 - T_2$ can be increased with an accompanying increase in heat recovery. Consider an exchanger similar to a 1-2 exchanger except that it is equipped

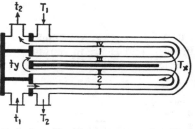

FIG. 8.1. Pass arrangements in a 2-4 exchanger.

with a longitudinal baffle (heavy line) as shown in Fig. 8.1. In this exchanger the fluid enters the shell through one of the nozzles adjacent to the tube sheet and travels the length of the shell before reversing its direction about the longitudinal baffle and returning to an exit nozzle also adjacent to the tube sheet. Assume that the tube bundle contains four or more passes with equal surface in each pass. Such an exchanger is a *2-4 exchanger*. A generalized sketch of temperature vs. length for a 2-4 exchanger is shown in Fig. 8.2a. In a 1-2 exchanger operating with the identical temperatures and shown in Fig. 8.2b, it is seen that a cross exists so that hot fluid leaving the shell at 140°F is forced to pass over tubes carrying heated cold fluid having a temperature of 160°F. Thus the shell fluid may be cooled at some point to a temperature lower than its outlet, and the tube fluid may be heated to a temperature above its outlet. When the two fluids are near their outlets, the shell fluid, being cooled, is actually heated and the tube fluid is actually cooled. In exchangers this is called *reheating*.

The True Temperature Difference Δt in a 2-4 Exchanger. In the 2-4 exchanger the longitudinal baffle reduces reheating as shown in Fig. 8.2a so that the 140°F shell-

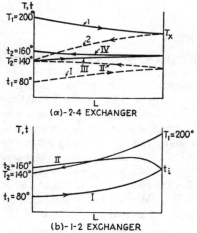

FIG. 8.2. Temperature relations in 1-2 and 2-4 exchangers.

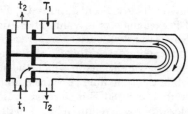

FIG. 8.3. The 2-2 true counterflow exchanger.

outlet fluid need never come in contact with the 160°F tube-outlet fluid. Passes I and II are in contact only with 2 and passes III and IV are in contact only with 1. If there are two shell passes and but two tube passes, they can be arranged in true counterflow as shown in Fig. 8.3. However, where the shell contains two shell passes and the tube bundle contains four or more tube passes, the flow pattern differs from any encountered heretofore. The derivation of a factor F_T for the 2-4 exchanger can be readily established.

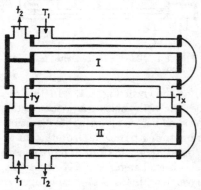

FIG. 8.4. Two 1-2 exchangers connected in series.

In the 2-4 exchanger it is assumed that there is no fluid leakage between the longitudinal shell baffle and the shell and that no heat is transferred across it, although this may lead to an error of 10 to 15 per cent when a large temperature difference exists between the average temperatures of the shell fluid in the two shell passes. The assumptions for the 1-2 exchanger also apply. Referring to the 2-4 exchanger in Fig. 8.1,

FLOW ARRANGEMENTS FOR INCREASED HEAT RECOVERY

the shell fluid temperature is T_x where it reverses direction after the first shell pass and the tube temperature is t_y where it reverses direction after the second tube pass. The entire 2-4 exchanger may then be considered the same as two 1-2 exchangers in series as shown in Fig. 8.4 with intermediate temperatures T_x and t_y. Calling these exchangers I and II in Fig. 8.4, the heat balances are, respectively,

I. $$WC(T_1 - T_x) = wc(t_2 - t_y) \tag{8.1}$$

and

II. $$WC(T_x - T_2) = wc(t_y - t_1) \tag{8.2}$$

The quantities of heat transferred in I and II are obviously not the same. Equation (7.37) may be written for each of the exchangers:

I. $$\frac{UA}{wc} = \frac{1}{2\sqrt{R^2+1}} \ln \frac{T_1 + T_x - t_y - t_2 + (T_1 - T_x)\sqrt{R^2+1}}{T_1 + T_x - t_y - t_2 - (T_1 - T_x)\sqrt{R^2+1}} \tag{8.3}$$

II. $$\frac{UA}{wc} = \frac{1}{2\sqrt{R^2+1}} \ln \frac{T_x + T_2 - t_1 - t_y + (T_x - T_2)\sqrt{R^2+1}}{T_x + T_2 - t_1 - t_y - (T_x - T_2)\sqrt{R^2+1}} \tag{8.4}$$

By algebraically eliminating T_x and t_y in Eqs. (8.3) and (8.4) through the use of S and the heat balances in Eqs. (8.1) and (8.2), F_T is given by

$$F_T = \frac{[\sqrt{R^2+1}/2(R-1)] \ln(1-S)/(1-RS)}{\ln \dfrac{2/S - 1 - R + (2/S)\sqrt{(1-S)(1-RS)} + \sqrt{R^2+1}}{2/S - 1 - R + (2/S)\sqrt{(1-S)(1-RS)} - \sqrt{R^2+1}}} \tag{8.5}$$

Equation (8.5) has been plotted in Fig. 19 and will be used for 2-4 exchangers in the Fourier equation $Q = UA\ \Delta t = UAF_T \times \text{LMTD}$. A comparison has been made between the values of F_T in 1-2 and 2-4 exchangers as shown in Fig. 8.5, where both exchangers employ fluids operating with identical temperature ranges. The advantage of the 2-4 flow arrangement is apparent from the large temperature crosses permissible. In Fig. 8.6 the hot fluid at each point has a 100° range with varying approach and the cold fluid a 20° fixed range from 180 to 200°F. With a 5° cross the 1-2 exchanger has a value of $F_T = 0.70$ compared with $F_T = 0.945$ for a 2-4 exchanger. It is general that the greater the number of shell passes in an exchanger the greater the cross or the greater the heat recovery which may be obtained. Mechanically, however, it is impractical to design single items of heat-transfer equipment with removable bundles having more than two shell passes, although it is seen that the 2-4 exchanger is thermally identical with two 1-2 exchangers in series. Greater crosses than those possible in a 2-4 exchanger can be achieved by using three 1-2 exchangers in series (3-6 arrangement) or two 2-4 exchangers in series (4–8 arrangement). Values of F_T for arrangements up to six shell passes and twelve tube passes are plotted in Figs. 18 through 23. A Ten Broeck chart for heat-recovery calculations in a 2–4 exchanger is given in Fig. 8.7 and is used in the same manner as Fig. 7.25.

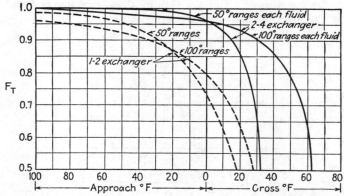

FIG. 8.5. Comparison of efficiency of 2–4 and 1–2 exchangers with equal fluid temperature ranges.

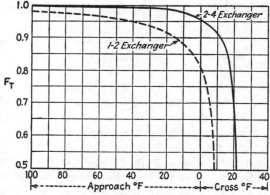

FIG. 8.6. Influence of approach temperature on F_T for unequal fluid temperature ranges.

Exchangers having odd numbers of tube passes have not been treated here because they may create mechanical problems in stationary tube-sheet exchangers and are not often employed. Fischer[1] has calculated and plotted the values of F_T for several odd-numbered tube arrangements. Naturally for a maximum value of F_T the odd tube-pass arrangements must be piped so that the majority of the tube passes are in counterflow with the shell fluid rather than in parallel flow.

[1] Fischer, F. K., *Ind. Eng. Chem.*, **30**, 377–383 (1938).

FLOW ARRANGEMENTS FOR INCREASED HEAT RECOVERY

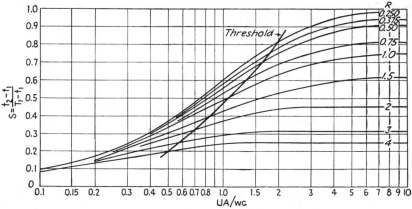

FIG. 8.7. Ten Broeck chart for determining t_2 in a 2-4 exchanger. (*Industrial and Engineering Chemistry.*)

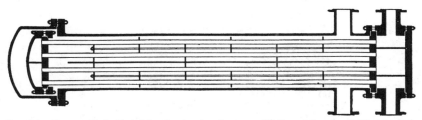

FIG. 8.8. Removable baffle 2-4 floating-head exchanger. (*Patterson Foundry & Machine Co.*)

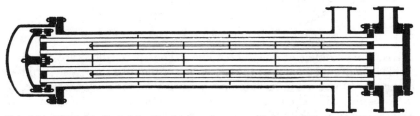

FIG. 8.9. Welded baffle 2-4 floating-head exchanger. (*Patterson Foundry & Machine Co.*)

2-4 Tubular Exchangers. Two methods by which the 2-4 temperature arrangement is achieved in tubular exchangers are shown in Figs. 8.8 and 8.9. The exchanger in Fig. 8.8 is similar to an ordinary 1-2 exchanger except that both of the shell nozzles are adjacent to the stationary tube sheet. Through the use of split segmental baffles a *removable* longitudinal baffle is inserted in the tube bundle. Usually some type of seal is provided between the longitudinal baffle and the shell, since any appreciable leakage between the two shell passes invalidates the calculated value of F_T for the 2-4 exchanger.

A more expensive but more positive form of the 2-4 exchanger is shown in Fig. 8.9. In this exchanger the baffle is welded to the shell. This requires cutting smaller inside diameter shells in half and welding the baffle from the outside of the shell. In larger shells the baffle is installed internally. Furthermore, in order that the bundle be removable it is necessary to make the floating tube sheet in two halves which are joined by a single floating-head cover and a backing piece to prevent the two

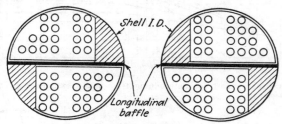

Fig. 8.10. Vertically cut segmental baffles.

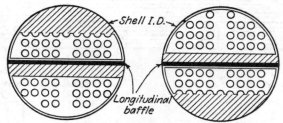

Fig. 8.11. Horizontally cut segmental baffles.

halves from leaking at the center line. The space allowance for the welded longitudinal baffle at the center line of the tube sheet also permits fewer tubes to be accommodated in the shell than for a 1-2 exchanger.

To permit the introduction of the longitudinal baffle, the segmental baffles may possess either of the two forms shown in Fig. 8.10 and 8.11. Those in Fig. 8.10 are *vertically cut* baffles and are similar to the segmental baffles found in 1-2 exchangers. The crosshatch section is the flow area. The flow distribution is nearly identical with that which exists in 1-2 exchangers with but one-half the flow area per inch of baffle spacing. Consequently the mass velocity for any given weight flow of fluid, W or w, will be twice that of a 1-2 exchanger of the same inside diameter and baffle spacing. The baffles shown in Fig. 8.11 are *horizontally cut* baffles in which the cut-out portions are equal to those of ordinary segmental

FLOW ARRANGEMENTS FOR INCREASED HEAT RECOVERY

baffles. The mass velocities in this case are the same per inch of baffle spacing as for 1-2 exchangers for a given weight flow although the fluid traverses a depth but half the inside diameter of the shell. Horizontally cut baffles are seldom used, since most fluids involved in large temperature crosses necessitating 2-4 exchangers also have long temperature ranges and relatively small flow quantities. The reduced flow area and increased mass velocity and film coefficient afforded by vertically cut baffles are usually desirable.

The Calculation of a 2-4 Exchanger. A 2-4 exchanger can be used when the process temperatures give a correction factor F_T of less than 0.75 for a 1-2 exchanger. If the factor F_T obtained from Fig. 19 for a 2-4 exchanger exceeds 0.90 with a removable longitudinal baffle or 0.85 with a welded longitudinal baffle, a single 2-4 exchanger will be adequate. If the value of F_T is below these limits, it will then be necessary to use a larger number of shell passes until an arrangement is found in which F_T approximates these values.

The calculation of a 2-4 exchanger differs in only three minor respects from the calculation of a 1-2 exchanger as outlined in Chap. 7. (1) F_T will be read from Fig. 19, (2) the flow area for vertically cut baffles will be half the values computed from Eq. (7.44), and (3) the number of crosses used for computing the pressure drop will be double, since one set of baffles is above and one is below the longitudinal baffle.

Example 8.1. Calculation of a 2-4 Oil Cooler. A 33.5°API oil has a viscosity of 1.0 centipoise at 180°F and 2.0 centipoise at 100°F. 49,600 lb/hr of oil leaves a distilling column at 358°F and is to be used in an absorption process at 100°F. Cooling will be achieved by water from 90 to 120°F. Pressure-drop allowances of 10 psi may be used on both streams along with a combined dirt factor of 0.004.

Available for this service from a discontinued operation is a 35 in. ID 2-4 exchanger having 454 1 in. OD, 11 BWG tubes 12'0" long and laid out on 1¼-in. square pitch. The bundle is arranged for six tube passes, and vertically cut baffles are spaced 7 in. apart. The longitudinal baffle is welded to the shell.

Is it necessary to use a 2-4 exchanger? Will the available exchanger fulfill the requirements?

Solution:

Exchanger:

Shell side	Tube side
ID = 35 in.	Number and length = 454, 12'0"
Baffle space = 7 in. (vert. cut)	OD, BWG, pitch = 1 in., 11 BWG, 1¼-in. square
Passes = 2	Passes = 6

(1) Heat balance:
 Oil, $Q = 49,600 \times 0.545(358 - 100) = 6,980,000$ Btu
 Water, $Q = 233,000 \times 1.0(120 - 90) = 6,980,000$ Btu

182   PROCESS HEAT TRANSFER

(2) Δt:

	Hot Fluid		Cold Fluid	Diff.
	358	Higher Temp	120	238
	100	Lower Temp	90	10
	258	Differences	30	228

$\quad$ LMTD = 72.0°F $\hfill$ (5.14)

$\quad R = \dfrac{258}{30} = 8.6 \qquad S = \dfrac{30}{358 - 90} = 0.112$

R and S do not intersect on Fig. 18, making a 2-4 exchanger imperative.
From Fig. 19 for a 2-4 exchanger, $F_T = 0.93$
$\quad \Delta t = F_T \times \text{LMTD} = 0.93 \times 72.0 = 66.9°\text{F}$ $\hfill$ (7.42)

(3) T_c and t_c:

$\quad \dfrac{\Delta t_c}{\Delta t_h} = \dfrac{10}{238} = 0.042 \qquad K_c = 0.47 \qquad F_c = 0.25$ $\hfill$ (Fig. 17)

$\quad T_c = 100 + 0.25(258) = 165°\text{F}$ $\hfill$ (5.28)
$\quad t_c = 90 + 0.25(30) = 98°\text{F}$ $\hfill$ (5.29)

Water will flow in the tubes to prevent corrosion of the shell.

Hot fluid: shell side, 33.5°API oil

(4′) $a_s = \frac{1}{2}(\text{ID} \times C'B)/144P_T$ (modified for a 2-4 exchanger) [Eq. (7.1)]
$\quad = \frac{1}{2} \times 35 \times 0.25 \times 7/144 \times 1.25 = 0.17 \text{ ft}^2$

(5′) $G_s = w/a_s$ $\hfill$ [Eq. (7.2)]
$\quad = 49,600/0.17$
$\quad = 292,000 \text{ lb/(hr)(ft}^2)$

(6′) At $T_c = 165°\text{F}$,
By obtaining X and Y in Fig. 14 from original data, $\mu = 1.12$ cp
$\quad = 1.12 \times 2.42 = 2.71 \text{ lb/(ft)(hr)}$
$D_e = 0.99/12 = 0.0825 \text{ ft}$ $\hfill$ [Fig. 28]
$Re_s = D_e G_s/\mu$ $\hfill$ [Eq. (7.3)]
$\quad = 0.0825 \times 292,000/2.71 = 8900$

(7′) $j_H = 52.5$ $\hfill$ [Fig. 28]

(8′) At $T_c = 165°\text{F}$, when $\mu = 1.12$ cp
$k(c\mu/k)^{1/3} = 0.20 \text{ Btu/(hr)(ft}^2\text{)(°F)}$
$\hfill$ [Fig. 16]

(9′) $h_o = j_H \dfrac{k}{D_e} (c\mu/k)^{1/3} \phi_s$ $\hfill$ [Eq. (6.15b)]

$\dfrac{h_o}{\phi_s} = 52.5 \times 0.20/0.0825 = 127$

Cold fluid: tube side, water

(4) $a'_t = 0.455 \text{ in.}^2$ $\hfill$ [Table 10]
$a_t = N_t a'_t / 144n$
$\quad = 454 \times 0.455/144 \times 6 = 0.239 \text{ ft}^2$

(5) $G_t = w/a_t$
$\quad = 233,000/0.239$
$\quad = 975,000 \text{ lb/(hr)(ft}^2)$
$V = G_t/3600\rho = 975,000/3600 \times 62.5$
$\quad = 4.33 \text{ fps}$

(6) At $t_c = 98°\text{F}$,
$\mu = 0.73 \times 2.42 = 1.77 \text{ lb/(ft)(hr)}$
$\hfill$ [Fig. 14]
$D = 0.76/12 = 0.0633 \text{ ft}$ $\hfill$ [Table 10]
(Re_t is for pressure drop only)
$Re_t = DG_t/\mu$
$\quad = 0.0633 \times 975,000/1.77 = 34,900$

(9) $h_i = 1010 \times 0.96$
$\quad = 970 \text{ Btu/(hr)(ft}^2\text{)(°F)}$ $\hfill$ [Fig. 25]

FLOW ARRANGEMENTS FOR INCREASED HEAT RECOVERY

Hot fluid: shell side, 33.5° API oil
(10') Tube-wall temp:

$$t_w = t_c + \frac{h_o}{h_{io} + h_o}(T_c - t_c) \quad \text{[Eq. (5.31)]}$$

$$= 98 + \frac{127}{737 + 127}(165 - 98) = 108°F$$

(11') At t_w,
$\mu_w = 1.95 \times 2.42 = 4.72$ lb/(ft)(hr) [Fig. 14]
$\phi_s = (\mu/\mu_w)^{0.14} = (2.71/4.72)^{0.14}$
$\quad = 0.92$ [Fig. 24 insert]
(12') Corrected coefficient h_o:
$h_o = 127 \times 0.92 = 117$ Btu/(hr)(ft²)(°F)
  [Eq. (6.36)]

(13') Clean overall coefficient U_C:

$$U_C = \frac{h_{io}h_o}{h_{io} + h_o} = \frac{737 \times 117}{737 + 117} = 101 \text{ Btu/(hr)(ft}^2\text{)(°F)} \quad (6.38)$$

(14') Design overall coefficient U_D:

$$a'' = 0.2618 \text{ ft}^2/\text{lin ft} \quad \text{(Table 10)}$$
Total surface, $A = 454 \times 12'0'' \times 0.2618 = 1425$ ft²
$$U_D = \frac{Q}{A\,\Delta t} = \frac{6{,}980{,}000}{1425 \times 66.9} = 73.3 \text{ Btu/(hr)(ft}^2\text{)(°F)}$$

(15) Dirt factor R_d:

$$R_d = \frac{U_C - U_D}{U_C U_D} = \frac{101 - 73.3}{101 \times 73.3} = 0.0038 \text{ (hr)(ft}^2\text{)(°F)/Btu} \quad (6.13)$$

Cold fluid: tube side, water
(10) $h_{io} = h_i \times \text{ID/OD} = 970 \times 0.76/1.0 = 737$ [Eq. (6.5)]

(11) Correction unnecessary for water.

Summary

117	h outside	737
U_C	101	
U_D	73.3	
R_d Calculated 0.0038		
R_d Required 0.0040		

Pressure Drop

(1') For $Re_s = 8900, f = 0.00215$ ft²/in.² [Fig. 29]

(2') No. of crosses, $N + 1 = 12L/B$ [Eq. (7.43)]
$\quad = 12 \times 12/7 = 20.1$
Say 21 per pass or 42 for bundle
$D_s = {}^{35}\!/_{12} = 2.92$ ft
$s = 0.82$ [Fig. 6]

(1) For $Re_t = 34{,}900$,
$f = 0.000195$ ft²/in.² [Fig. 26]

(2) $\Delta P_t = \dfrac{fG_t^2 Ln}{5.22 \times 10^{10} Ds\phi_t}$ [Eq. (7.45)]

$\quad = \dfrac{0.000195 \times 975{,}000^2 \times 12 \times 6}{5.22 \times 10^{10} \times 0.0633 \times 1.0 \times 1.0}$

$\quad = 4.0$ psi

Pressure Drop

(3') $\Delta P_s = \dfrac{fG_s^2 D_s(N+1)}{5.22 \times 10^{10} D_e s \phi_s}$ [Eq. (7.44)]

$= \dfrac{0.00215 \times 292{,}000^2 \times 2.92 \times 42}{5.22 \times 10^{10} \times 0.0825 \times 0.82 \times 0.92}$

$= 7.0$ psi

Allowable $\Delta P_s = 10.0$ psi

(3) $G_t = 975{,}000$, $\dfrac{V^2}{2g'} = 0.13$ [Fig. 27]

$\Delta P_r = \dfrac{4n}{s}\dfrac{V^2}{2g'}$ [Eq. (7.46)]

$= \dfrac{4 \times 6}{1} \times 0.13 = 3.2$ psi

(4) $\Delta P_T = \Delta P_t + \Delta P_r$ [Eq. (7.47)]
   $= 4.0 + 3.2 = 7.2$ psi

Allowable $\Delta P_T = 10.0$ psi

The exchanger will have a slightly low dirt factor but is otherwise satisfactory.

Exchangers in Series. In plants where large numbers of exchangers are used, certain size standards (total number of tubes, pass arrangements, baffle spacing) are established for 1-2 exchangers so that a majority of future services can be fulfilled by an arrangement of a number of standard exchangers in series or parallel. Although this may sometimes cause awkwardness because of the impossibility of utilizing the equipment most efficiently, it has the great advantage of reducing the type and number of replacement parts, tubes, and tools. In these plants when a process has become obsolete, it is customary to find a number of exchangers of identical size available for other uses. If the tube bundle is simply retubed, the exchanger will frequently be as serviceable as when new. When two exchangers are connected in series on *both* the shell and tube sides, they form a temperature arrangement which has been shown to be identical with the 2-4 exchanger. When a temperature cross involves a correction factor for an arrangement which approximates true counterflow more closely than that possible in a 1-2 exchanger, it can be met by a series arrangement of a number of 1-2 exchangers. The 2-4, 3-6, 4-8, etc., arrangements are all based upon shells and channels being connected in series. Any arrangement which is an even multiple of two shell passes such as 2-4, 4-8, etc., may be fulfilled by a number of 1-2 exchangers or half as many 2-4 exchangers.

The calculation for process conditions requiring more than one shell pass follows the method used for 1-2 exchangers except that the entire group of exchangers is treated as a unit.

Example 8.2. Calculation of an Acetone–Acetic Acid Exchanger. Acetone ($s = 0.79$) at 250°F is to be sent to storage at 100°F and at a rate of 60,000 lb/hr. The heat will be received by 185,000 lb/hr of 100 per cent acetic acid ($s = 1.07$) coming from storage at 90°F and heated to 150°F. Pressure drops of 10.0 psi are available for both fluids, and a combined dirt factor of 0.004 should be provided.

Available for the service are a large number of 1-2 exchangers having 21¼ in. ID shells with 270 tubes ¾ in. OD, 14 BWG, 16'0" long and laid out on 1-in. square

FLOW ARRANGEMENTS FOR INCREASED HEAT RECOVERY

pitch. The bundles are arranged for two tube passes with segmental baffles spaced 5 in. apart.

How many of the 1-2 exchangers should be installed in series?

Solution:

Exchanger:

Shell side	Tube side
ID = 21¼ in.	Number and length = 270, 16'0''
Baffle space = 5 in.	OD, BWG, pitch = ¾ in., 14 BWG, 1 in. square
Passes = 1	Passes = 2

(1) Heat balance:
Acetone, $Q = 60{,}000 \times 0.57(250 - 100) = 5{,}130{,}000$ Btu/hr (Fig. 2)
Acetic acid, $Q = 168{,}000 \times 0.51(150 - 90) = 5{,}130{,}000$ Btu/hr (Fig. 2)

(2) Δt:

	Hot Fluid		Cold Fluid	Diff.
	250	Higher Temp	150	100
	100	Lower Temp	90	10
	150	Difference	60	90

LMTD = 39.1°F (5.14)
$R = {}^{150}\!/_{60} = 2.5$; $S = 60/(250 - 90) = 0.375$ (7.18)
F_T: 1-2 exchanger, F_T = not possible (Fig. 18)
2-4 exchanger, $F_T = 0.67$ (too low) (Fig. 19)
3-6 exchanger, $F_T = 0.88$ (Fig. 20)
$\Delta t = 0.88 \times 39.1 = 34.4°F$ (7.42)

To permit the transfer of heat with the temperatures given by the process, a minimum of three shell passes is required. If the sum of the surfaces in three shells is insufficient, a greater number will be required.

(3) T_c and t_c: These liquids are not viscous, and the viscosity correction will be negligible, $\phi_s = \phi_t = 1$. Average temperatures may be used.

Hot fluid: shell side, acetone	Cold fluid: tube side, acetic acid
(4') $a_s = \text{ID} \times C'B/144P_T$ [Eq. (7.1)]	(4) $a'_t = 0.268$ in.² [Table 10]
$= 21.25 \times 0.25 \times 5/144 \times 1.0$	$a_t = N_t a'_t/144n$ [Eq. (7.48)]
$= 0.185$ ft²	$= 270 \times 0.268/144 \times 2 = 0.251$ ft²
(5') $G_s = W/a_s$ [Eq. (7.2)]	(5) $G_t = w/a_t$
$= 60{,}000/0.185$	$= 168{,}000/0.251$
$= 324{,}000$ lb/(hr)(ft²)	$= 670{,}000$ lb/(hr)(ft²)
(6') At $T_a = 175°F$,	(6) At $t_a = 120°F$,
$\mu = 0.20 \times 2.42 = 0.49$ lb/(ft)(hr)	$\mu = 0.85 \times 2.42 = 2.06$ lb/(ft)(hr)
[Fig. 14]	[Fig. 14]
$D_e = 0.95/12 = 0.079$ ft [Fig. 28]	$D = 0.584/12 = 0.0487$ ft
$Re_s = D_e G_s/\mu$ [Eq. (7.3)]	$Re_t = DG_t/\mu$
$= 0.079 \times 324{,}000/0.49 = 52{,}200$	$= 0.0487 \times 670{,}000/2.06 = 15{,}800$
(7') $j_H = 137$ [Fig. 28]	(7) $j_H = 55$ [Fig. 24]

Hot fluid: shell side, acetone

(8') At $T_a = 175°F$,
$c = 0.63$ Btu/(lb)(°F) [Fig. 2]
$k = 0.095$ Btu/(hr)(ft²)(°F/ft) [Table 4]

$(c\mu/k)^{1/3} = (0.63 \times 0.49/0.095)^{1/3} = 1.47$

(9') $h_o = j_H \dfrac{k}{D_e}\left(\dfrac{c\mu}{k}\right)^{1/3} \times 1$ [Eq. (6.15b)]

$= 137 \times 0.095 \times 1.47/0.079$
$= 242$ Btu/(hr)(ft²)(°F)

(10') (11') (12') The viscosity correction will be very small.

Cold fluid: tube side, acetic acid

(8) At $t_a = 120°F$,
$c = 0.51$ Btu/(lb)(°F) [Fig. 2]
$k = 0.098$ Btu/(hr)(ft²)(°F/ft) [Table 4]

$(c\mu/k)^{1/3} = (0.51 \times 2.06/0.098)^{1/3} = 2.21$

(9) $h_i = j_H \dfrac{k}{D}\left(\dfrac{c\mu}{k}\right)^{1/3} \times 1$ [Eq. (6.15a)]

$= 56 \times 0.098 \times 2.21/0.0487$
$= 249$ Btu/(hr)(ft²)(°F)

(10) $h_{io} = h_i \times ID/OD$ [Eq. (6.5)]

$= 249 \times \dfrac{0.584}{0.75} = 194$

(13) Clean overall coefficient U_C:

$$U_C = \frac{h_{io}h_o}{h_{io} + h_o} = \frac{242 \times 194}{242 + 194} = 107.5 \text{ Btu/(hr)(ft}^2\text{)(°F)} \qquad (6.38)$$

(14) Design overall coefficient U_D:

$$a'' = 0.1963 \text{ ft}^2/\text{lin ft} \qquad \text{(Table 10)}$$

Total surface, $A = 3(270 \times 16'0'' \times 0.1963) = 2540$ ft²

$$U_D = \frac{Q}{A\,\Delta t} = \frac{5{,}130{,}000}{2540 \times 34.4} = 58.8 \text{ (Btu/(hr)(ft}^2\text{)(°F)}$$

(15) Dirt factor R_d:

$$R_d = \frac{U_C - U_D}{U_C U_D} = \frac{107.5 - 58.8}{107.5 \times 58.8} = 0.0077 \text{ (hr)(ft}^2\text{)(°F)/Btu} \qquad (6.13)$$

Summary

242	h outside	194
U_C	107.5	
U_D	58.8	
R_d Calculated	0.0077	
R_d Required	0.0040	

Pressure Drop

(1') For $Re_s = 52{,}200$,
$f = 0.00155$ ft²/in.² [Fig. 29]

(2') No. of crosses, $N + 1 = 12L/B$ [Eq. (7.43)]

$= 16 \times 12/5 = 39$

Total for 3 exchangers $= 39 \times 3 = 117$

$D_s = 21.25/12 = 1.78$ ft
$s = 0.79$ [Table 6]

(1) For $Re_t = 158{,}000$,
$f = 0.00024$ ft²/in.² [Fig. 26]
$s = 1.07$ [Table 6]

(2) $\Delta P_t = \dfrac{fG_t^2 L n}{5.22 \times 10^{10} Ds\phi_t}$ [Eq. (7.45)]

$= \dfrac{0.00024 \times 670{,}000^2 \times 16 \times 2 \times 3}{5.22 \times 10^{10} \times 0.0487 \times 1.07 \times 1.0}$

$= 3.8$ psi

Pressure Drop

(3') $\Delta P_s = \dfrac{fG_s^2 D_s (N+1)}{5.22 \times 10^{10} D_e s \phi_s}$

$= \dfrac{0.00155 \times 324{,}000^2 \times 1.78 \times 117}{5.22 \times 10^{10} \times 0.079 \times 0.79 \times 1.0}$

$= 10.4$ psi

Allowable $\Delta P_s = 10.0$ psi

(3) $G_t = 670{,}000$,
$V^2/2g' = 0.063$ [Fig. 27]
$\Delta P_r = 3(4n/s)(V^2/2g')$ [Eq. (7.46)]

$= \dfrac{3 \times 4 \times 2}{1.07} \times 0.063 = 1.4$ psi

(4) $\Delta P_T = \Delta P_t + \Delta P_r$ [Eq. (7.47)]
$= 3.8 + 1.4 = 5.2$ psi

Allowable $\Delta P_T = 10.0$ psi

The three exchangers are more than adequate for heat transfer, even though the pressure drop is insignificantly high. Fewer exchangers cannot be used.

The 1-1 True Counterflow Exchanger. There are instances in which the temperature cross is so great that the only solution lies in the use of

FIG. 8.12. The 1-1 floating-head exchanger. (*Patterson Foundry & Machine Co.*)

true counterflow. In fixed-tube-sheet equipment this is readily accomplished, but in floating-head equipment it is slightly more difficult. It can be achieved as shown in Fig. 8.12. An extension forming the tube fluid-outlet nozzle protrudes through the shell bonnet by means of a packing gland. This type of exchanger is also used when there is a great quantity of tube-side fluid.

PROBLEMS

8.1. 33,114 lb/hr of n-butyl alcohol at 210°F is to be cooled to 105°F using water from 95 to 115°F. Available for the purpose is a 19¼ in. ID, two-pass shell exchanger with 204 tubes ¾ in. OD, 16 BWG, 16'0" long on 1-in. square pitch arranged for four passes. Vertically cut baffles are spaced 5 in. apart. Pressure drops of 10 psi are allowable. What is the dirt factor?

8.2. 62,000 lb/hr of hot 26°API quench oil having viscosities equivalent to a 42°API kerosene (Fig. 14) is cooled from 425 to 304°F by heating 27,200 lb/hr of 35°API distillate (Fig. 14) with an inlet temperature of 100°F. Available for the service are 15¼ in. ID 1-2 exchangers with 108 tubes ¾ in. OD, 16 BWG, 16'0" long arranged for six passes. Baffles are spaced 6 in. apart. Total pressure drops of 10 psi are allowable. How many 1-2 exchangers are required, and what dirt factor do they yield?

8.3. 36,000 lb/hr of ethyl acetate is cooled from 190 to 100°F using water from 80 to 120°F. Available for the service is a 23¼ in. ID, two-pass shell exchanger with

188   *PROCESS HEAT TRANSFER*

292 ¾ in. OD, 14 BWG, 12'0" long tubes arranged for eight passes. Vertically cut baffles are spaced 8 in. apart. What pressure drops and dirt factor will be obtained?

8.4. 55,400 lb/hr of nitrobenzene coming from a reactor under pressure at 365°F is to be cooled to 150°F by preheating benzene under pressure from 100° to the highest outlet temperature attainable. Available for the service are two 25 in. ID, two-pass shell exchangers with 356 tubes ¾ in. OD, 13 BWG, 16'0" long arranged for six tube passes. Vertically cut baffles are spaced 8 in. apart. What are the pressure drops and the dirt factor?

NOMENCLATURE FOR CHAPTER 8

A	Heat-transfer surface, ft²
a''	External surface per lin ft, ft
a	Flow area, ft²
B	Baffle spacing, in.
C'	Clearance between tubes, in.
C	Specific heat of hot fluid in derivations, Btu/(lb)(°F)
c	Specific heat of fluid, Btu/(lb)(°F)
D	Inside diameter of tubes, ft
D_e	Equivalent diameter for heat transfer and pressure drop, ft
D_s	Inside diameter of shell, ft
d	Inside diameter of tubes, in.
d_e	Equivalent diameter for heat transfer and pressure drop, in.
F_c	Caloric fraction, dimensionless
F_T	Temperature difference factor, $\Delta t = F_T \times \text{LMTD}$, dimensionless
f	Friction factor, ft²/in.²
G	Mass velocity, lb/(hr)(ft²)
g'	Acceleration of gravity, ft/sec²
h_i, h_o	Heat-transfer coefficient for inside and outside fluids, respectively, Btu/(hr)(ft)(°²F)
h_{io}	Value of h_i when referred to the tube outside diameter, Btu/(hr)(ft²)(°F)
ID	Inside diameter, in.
j_H	Factor for heat transfer, dimensionless
K_c	Caloric constant, dimensionless
k	Thermal conductivity, Btu/(hr)(ft²)(°F/ft)
L	Tube length, ft
LMTD	Log mean temperature difference, °F
N	Number of shell-side baffles
N_t	Number of tubes
n	Number of tube passes
OD	Outside diameter of tube, in.
ΔP	Pressure drop, psi
$\Delta P_T, \Delta P_t, \Delta P_r$	Total, tube and return pressure drop, psi
P_T	Tube pitch, in.
R	Temperature group $(T_1 - T_2)/(t_2 - t_1)$, dimensionless
R_d	Combined dirt factor, (hr)(ft²)(°F)/Btu
Re	Reynolds number, dimensionless
S	Temperature group, $(t_2 - t_1)/(T_1 - t_1)$, dimensionless
s	Specific gravity, dimensionless
T, T_1, T_2	Temperature in general, inlet and outlet of hot fluid, °F

T_a	Average temperature of hot fluid, °F
T_c	Caloric temperature of hot fluid, °F
T_x	Temperature of shell fluid between first and second passes, °F
t_1, t_2	Inlet and outlet temperatures of cold fluid, °F
t_a	Average temperature of hot fluid, °F
t_c	Caloric temperature of cold fluid, °F
t_i	Temperature at end of first pass, °F
t_w	Tube-wall temperature, °F
t_y	Temperature of tube fluid between second and third passes, °F
Δt	True temperature difference in $Q = U_D A \, \Delta t$, °F
$\Delta t_c, \Delta t_h$	Temperature differences at the cold and hot terminals, °F
U_C, U_D	Clean and design overall coefficients of heat transfer, Btu/(hr)(ft^2)(°F)
V	Velocity, fps
W	Weight flow of hot fluid, lb/hr
w	Weight flow of cold fluid, lb/hr
μ	Viscosity, centipoises $\times$ 2.42 = lb/(ft)(hr)
μ_w	Viscosity at tube wall-temperature, centipoises $\times$ 2.42 = lb/(ft)(hr)
ρ	Density, lb/ft^3
ϕ	$(\mu/\mu_w)^{0.14}$

Subscripts (except as noted above)

s	Shell side
t	Tube side

CHAPTER 9

GASES

Introduction. The calculation for the heating or cooling of a gas differs in only minor respects from the calculation for liquid-liquid systems. The relationships between gas film coefficients and allowable pressure drops are critically dependent upon the operating pressure of the system whereas for incompressable fluids the operating pressure is not important. The values of film coefficients for gases are generally lower than those obtained for liquids at equal mass velocities, the difference being inherent in the properties of the gases.

Properties of Gases. The properties of gases are compared with those of liquids to emphasize the major differences between them. The viscosities of gases range from about 0.015 to 0.025 centipoise, or about one-tenth to one-fifth the values obtained for the least viscous liquids. Gas viscosities increase with temperature in contrast with liquids and Reynolds numbers are correspondingly large even when mass velocities are small. The thermal conductivities of gases, with the exception of hydrogen, are about one-fifth the values usually obtained for organic liquids and about one-fifteenth of the values for water and aqueous solutions. The specific heats of organic gases and vapors are only slightly lower than those of organic liquids. With the exception of hydrogen the specific heats of inorganic gases and light hydrocarbon vapors range from 0.2 to 0.5 Btu/(lb)(°F). Although the specific heat, viscosity, and thermal conductivity of a gas increase with temperature, the Prandtl number $c\mu/k$ shows little dependence upon temperature except near the critical. The value of $c\mu/k$ calculated at any single temperature serves sufficiently well for the solution of problems involving the same gas at another temperature within reasonable proximity. The values of $c\mu/k$ are given in Table 9.1 for the common gases.

While most of the viscosity, specific-heat, and conductivity data on gases are tabulated for atmospheric pressure, corrections to other pressures may be made by established methods. Viscosities may be corrected by means of the correlation of Comings and Egly[1] given in Fig. 13b or by employing the method of Othmer and Josefowitz.[2] Specific heats may be

---
[1] Comings, E. W., and R. S. Egly, *Ind. Eng. Chem.*, **32**, 714–718 (1940).
[2] Othmer, D. F., and S. Josefowitz, *Ind. Eng. Chem.*, **38**, 111–116 (1946).

corrected by the method of Watson and Smith.[1] These corrections will not be significant, however, unless the pressure of the gas is great. Except at high vacuums the conductivities of gases are not affected by the pressure. Calculation of the density or specific volume of a gas by the perfect gas law is suitable for moderate pressures but may be in error at high pressures. If actual compression data are available, their use is preferable at high pressures, or the perfect gas law may be replaced by a

TABLE 9.1. PRANDTL NUMBERS FOR GASES AT 1 ATM AND 212°F

Gas	$\dfrac{c\mu}{k}$
Air	0.74
Ammonia	0.78
Carbon dioxide	0.80
Carbon monoxide	0.74
Ethylene	0.83
Hydrogen	0.74
Hydrogen sulfide	0.77
Methane	0.79
Nitrogen	0.74
Oxygen	0.74
Steam	0.78
Sulfur dioxide	0.80

more acceptable equation of state such as that of Van der Waals or Beattie-Bridgman.

Pressure Drop. Equations (7.44) and (7.45) and friction factors obtained from Figs. 29 and 26 can be used for the calculation of the pressure drop on the shell or tube sides of gas heaters or coolers when the average value of the inlet and outlet specific gravities of the gas *relative to water* is used. It is apparent in the case of any gas that the specific gravity varies considerably with its *operating pressure*. The specific gravity of air in an exchanger operated at 150 psia is very nearly ten times the specific gravity when operated at atmospheric pressure, and for a given mass velocity the pressure drop will only be one-tenth as great. Moving in the other direction, air at 7½ psia has a density one-half that of the atmosphere, and the pressure drop for a given mass velocity becomes greater as the operating pressure decreases, an unfavorable consideration in vacuum processes. When a gas is operated at high pressure, however, a relatively large mass velocity can be used without obtaining a pressure drop of an impractical order. When a gas is operated under vacuum, a pressure drop of 0.5 psi may represent a very large portion of the entire head available to move the gas through the exchanger.

[1] Watson, K. M., and R. L. Smith, *Natl. Petroleum News*, July, 1936.

Film Coefficients. Film coefficients can be evaluated with reasonable accuracy through the use of the usual shell-and-tube equations or Figs. 28 and 24. No correction need be made for the viscosity ratio ϕ unless the temperature range is exceedingly great. As mentioned above, the low viscosities of gases result in large Reynolds numbers even where very small mass velocities are used. Large values of j_H are the result, but the correspondingly low thermal conductivity gives film coefficients below those obtained for liquids at equal mass velocities or at equal values of j_H. The commonest applications of gas cooling under pressure are found in the aftercooling and intercooling of gases which have undergone adiabatic or polytropic compression in single- and multistage compressors. Gas-to-liquid heat transfer is also used for the recovery of waste heat from near-atmospheric combustion gases such as *economizers*, but these usually employ a modification known as *extended surfaces*, which will be treated in Chap. 16. When gases are heated, steam is the usual heating medium.

Many of the data reported in the literature for turbulent-flow heat transfer make use of the following transformation:

$$\frac{hD}{k}\left(\frac{c\mu}{k}\right)^{-\frac{1}{3}} \frac{\mu}{DG} \frac{c}{c} = \frac{h}{cG}\left(\frac{c\mu}{k}\right)^{\frac{2}{3}} \tag{9.1}$$

A new factor j_h is then defined as

$$j_h = \frac{h}{cG}\left(\frac{c\mu}{k}\right)^{\frac{2}{3}} \tag{9.2}$$

and is plotted as a function of Re. All the convection equations given heretofore can be plotted as j_h vs. Re in place of j_H vs. Re by simply dividing the values of j_H as given in Figs. 28 and 24 by their respective values of Re. Alternate equations are obtained thereby.

There is considerable merit to the use of j_h in preference to j_H when correlating gases. Using j_H, $c\mu/k$ is a constant, but k in Eq. (6.15a) and (6.15b) must be obtained at the bulk temperature to obtain h, and conductivity data are sparse for gases. Using j_h, $c\mu/k$ is a constant, but only c is required in Eq. (9.1) to obtain h, and this will have been required for the heat load as well.

When gases enter an adiabatic compressor, their isotherms follow the equation $pv^\gamma = $ constant, where p is the absolute pressure of the gas, v its specific volume, and γ the ratio of the specific heats of the gas at constant pressure to constant volume. Applying the perfect gas law, the variation of absolute pressure with *absolute temperature* becomes

$$\left(\frac{T_2}{T_1}\right)_{abs} = \left(\frac{p_2}{p_1}\right)^{(\gamma-1)/\gamma} \tag{9.3}$$

GASES   193

Gases to be cooled or heated at moderate pressures are usually placed in the shell of shell-and-tube equipment to localize corrosion resulting from the cooling water or steam condensate. At higher pressures, however, it is customary to place the gas in the tubes where the pressure is effective only upon the tubes.

Example 9.1. Calculation of an Ammonia Compressor Aftercooler. Dry ammonia gas at 83 psia and at a rate of 9872 lb/hr is discharged from a compressor at 245°F and is to be fed to a reactor at 95°F using cooling water from 85 to 95°F. A pressure drop of 2.0 psi is allowable on the gas and 10.0 psi on the water.

Available for the service is a 23¼ in. ID 1-2 exchanger having 364 ¾ in., 16 BWG tubes 8'0" long and laid out on 1 5⁄16-in. triangular pitch. The bundle is arranged for eight passes, and baffles are spaced 12 in. apart.

What will the dirt factor and pressure drops be?

Solution:

Exchanger:

Shell side	*Tube side*
ID = 23¼	Number and length = 364, 8'0"
Baffle space = 12 in.	OD, BWG, pitch = ¾ in., 16 BWG, 1 5⁄16-in. tri.
Passes = 1	Passes = 8

(1) Heat balance:
  Ammonia gas, $Q = 9872 \times 0.53(245 - 95) = 785{,}000$ Btu/hr   (Fig. 3)
  Water, $Q = 78{,}500 \times 1(95 - 85) = 785{,}000$ Btu/hr

(2) Δt:

	Hot Fluid		Cold Fluid	Diff.
	245	Higher Temp	95	150
	95	Lower Temp	85	10
	150	Differences	10	140

LMTD = 51.8°F   (5.14)

$R = \dfrac{150}{10} = 15 \quad S = \dfrac{10}{245 - 85} = 0.0625$

$F_T = 0.837$ (Fig. 18)
$\Delta t = 0.837 \times 51.8 = 43.4°F$ (7.42)

(3) T_c and t_c: The viscosities will vary too little to require correction. Water will flow in the tubes to prevent corrosion of the shell.

Hot fluid: shell side, ammonia at 83 psia | *Cold fluid: tube side, water*

(4') $a_s = \text{ID} \times C'B/144P_T$ [Eq. (7.1)] | (4) $a_t' = 0.302$ in.^{2} [Table 10]
  $= 23.25 \times 0.1875 \times 12/144$ | $a_t = N_t a_t'/144n$
  $\qquad \times 0.937$ | $\qquad = 364 \times 0.302/144 \times 8 = 0.0954$ ft^2
  $= 0.388$ ft^2 |

Hot fluid: shell side, ammonia at 83 psia

(5′) $G_s = W/a_s$ [Eq. (7.2)]
$= 9872/0.388$
$= 25,400$ lb/(hr)(ft²)

(6′) At $T_a = 170°F$,
$\mu = 0.012 \times 2.42 = 0.029$ lb/(ft)(hr) [Fig. 15]
$D_e = 0.55/12 = 0.0458$ ft
$Re_s = D_e G_s/\mu$ [Eq. (7.3)]
$= 0.0458 \times 25,400/0.029 = 40,200$

(7′) $j_H = 118$ [Fig. 28]

(8′) At $T_a = 170°F$,
$k = 0.017$ Btu/(hr)(ft²)(°F/ft) [Table 5]
$(c\mu/k)^{\frac{1}{3}} = (0.53 \times 0.029/0.017)^{\frac{1}{3}} = 0.97$

(9′) $h_o^- = j_H \dfrac{k}{D_e}\left(\dfrac{c\mu}{k}\right)^{\frac{1}{3}} \times 1$ [Eq. (6.15b)]
$= 118 \times 0.017 \times 0.97/0.0458$
$= 42.3$ Btu/(hr)(ft²)(°F)

(10′) (11′) (12′) Viscosity correction is unnecessary.

Cold fluid: tube side, water

(5) $G_t = w/a_t$
$= 78,500/0.0954$
$= 823,000$ lb/(hr)(ft²)
$V = G_t/3600\rho = 823,000/3600 \times 62.5$
$= 3.65$ fps

(6) At $t_a = 90°F$,
$\mu = 0.82 \times 2.42 = 1.99$ lb/(ft)(hr) [Fig. 14]
$D = 0.62/12 = 0.0517$ ft [Table 10]
$Re_t = DG_t/\mu$ (Re_t is for pressure drop only)
$= 0.0517 \times 823,000/1.99 = 21,400$

(9) $h_i = 900$ Btu/(hr)(ft²)(°F) [Fig. 25]

(10) $h_{io} = h_i \times \text{ID/OD}$ [Eq. (6.5)]
$= 900 \times 0.62/0.75 = 744$

(13) Clean overall coefficient U_C:

$$U_C = \frac{h_{io} h_o}{h_{io} + h_o} = \frac{744 \times 42.3}{744 + 42.3} = 40.1 \text{ Btu/(hr)(ft²)(°F)} \qquad (6.38)$$

(14) Design overall coefficient U_D:

$a'' = 0.1963$ ft²/lin ft (Table 10)
Total surface, $A = 364 \times 8'0'' \times 0.1963 = 572$ ft²

$$U_D = \frac{Q}{A \, \Delta t} = \frac{785,000}{572 \times 43.4} = 31.7 \text{ Btu/(hr)(ft²)(°F)}$$

(15) Dirt factor R_d:

$$R_d = \frac{U_C - U_D}{U_C U_D} = \frac{40.1 - 31.7}{40.1 \times 31.7} = 0.0070 \text{ (hr)(ft²)(°F)/Btu} \qquad (6.13)$$

Summary

42.3	h outside	744
U_C	40.1	
U_D	31.7	
R_d Calculated 0.0070		
R_d Required	x	

GASES

Pressure Drop

(1') For $Re_s = 40{,}200$,
$f = 0.00162$ ft^2/in.^{2} [Fig. 29]

(2') No. of crosses, $N + 1 = 12L/B$ [Eq. (7.43)]
$= 12 \times 8/12 = 8$

$s = \dfrac{\rho_{gas}}{\rho_{water}}$

$pv = \dfrac{w'}{(MW)} \times 1545 T_{abs}$

$\rho_{gas} = \dfrac{w'}{v} = \dfrac{p(MW)}{1545 T_{abs}}$

$= \dfrac{(83 \times 144)(17.1)}{1545(460 + 170)} = 0.209$ lb/ft^3

$s = \dfrac{0.209}{62.5} = 0.00335$

$D_s = 23.25/12 = 1.94$ ft

(3') $\Delta P_s = \dfrac{fG_s^2 D_s (N+1)}{5.22 \times 10^{10} D_e s \phi_s}$

$= \dfrac{0.00162 \times 25{,}400^2 \times 1.94 \times 8}{5.22 \times 10^{10} \times 0.0458 \times 0.00335} \times 1.0$

$= 2.0$ psi

Allowable $\Delta P_s = 2.0$ psi

(1) For $Re_t = 21{,}400$,
$f = 0.000225$ ft^2/in.^{2} [Fig. 26]

(2) $\Delta P_t = \dfrac{fG_t^2 L n}{5.22 \times 10^{10} D s \phi_t}$ [Eq. (7.45)]

$= \dfrac{0.000225 \times 823{,}000^2 \times 8 \times 8}{5.22 \times 10^{10} \times 0.0517 \times 1.0} \times 1.0$

$= 3.6$ psi

(3) $G_t = 823{,}000$, $V^2/2g' = 0.090$ [Fig. 27]

$\Delta P_r = (4n/s)(V^2/2g')$ [Eq. (7.46)]

$= \dfrac{4 \times 8}{1.0} \times 0.090 = 2.9$ psi

(4) $\Delta P_T = \Delta P_t + \Delta P_r$ [Eq. (7.47)]
$= 3.6 + 2.9 = 6.5$ psi

Allowable $\Delta P_T = 10.0$ psi

The ability to meet the allowable pressure drop hinges closely upon the density of the gas. If the gas had been air at the same pressure, the density and pressure drop would have been $0.209 \times 29/17 = 0.357$ lb/ft^3 and 1.2 psi. Similarly an exchanger can be used for gases at vacuum pressure only when a very small mass velocity is employed. The latter results in very low transfer rates for vacuum services, values of U_D being as low as 2 to 10 Btu/(hr)(ft^2)(°F).

Air Compressor Intercoolers. In the compression of air for utility purposes it is common to subject atmospheric air to four or more stages of compression. The allowable pressure drop in the intercoolers following the initial stages of compression is extremely critical. Assuming that the compressors operate with compression ratios of roughly $2\frac{1}{4}:1$ or $2\frac{1}{2}:1$, a pressure drop of 1 psi in the first-stage intercooler represents a reduction in the total pressure delivered after the fourth stage of $1 \times 2.5 \times 2.5 \times 2.5 = 13.1$ psi and nearly 80 psi after the sixth stage. In addition, the presence of moisture in the inlet air makes it impossible to compute the heat load as a simple sensible-heat change. Suppose saturated air is taken into the compressor at 95°F (during a summer shower), compressed, and then cooled back to 95°F between each stage of compression. The air and water vapor both occupy the same total volume. The compression of a saturated gas raises the dew point above its initial dew point. Cooling the compressed gas back to its original temperature requires that it be cooled below its dew point. This can occur only if water condenses out of the compressed gas during cooling. The reasoning follows: The original water vapor in the saturated atmospheric air was all that could exist in the original volume of air at 95°F. Its total weight can be obtained from the specific volume of steam, in cubic feet per pound, for 95°F as found in Table 7. After compression and cooling back to 95°F the total volume of air is reduced but the specific volume of steam at 95°F is unchanged. The same specific volume can be maintained in the reduced gas volume only if some of the water condenses out of the gas.

Example 9.2. Calculation of the Heat Load for an Air Intercooler. 4670 cfm of air saturated at 95°F enters a four-stage adiabatic compressor, having a compression ratio of 2.33:1, at atmospheric pressure. (a) How much heat must be removed in the first-stage intercooler? (b) How much heat must be removed in the second-stage intercooler?

Solution:

(a) Inlet 4670 cfm:
Saturation partial pressure of water at 95°F = 0.8153 psi (Table 7)
Saturation specific volume of water at 95°F = 404.3 ft³/lb (Table 7)
The air and water both occupy the same volume at their respective partial pressures.
Lb water/hr entering = 4670 × 60/404.3 = 692 lb

First stage:

After 2.33 compression ratio

$$p_2 = 14.7 \times 2.33 = 34.2 \text{ psi}$$

$$\left(\frac{T_2}{T_1}\right)_{abs} = \left(\frac{p_2}{p_1}\right)^{(\gamma-1)/\gamma} \quad (\gamma = 1.40 \text{ for air*})$$

$$\frac{T_{2abs}}{460 + 95} = (2.33)^{(1.4-1)/1.4}$$

$$T_{2abs} = 705°R \text{ or } 245°F$$

Intercooler:

Final gas volume = 4670 × 60 × 14.7/34.2 = 120,000 ft³/hr
Water remaining in air = 120,000/404.3 = 297 lb/hr
Condensation in intercooler = 692 − 297 = 395 lb/hr
Specific volume of atmospheric air = (359/29)(555/492) 14.7/(14.7 − 0.8153)
= 14.8 ft³/lb
Air in inlet gas = 4670 × 60/14.8 = 18,900 lb/hr

Heat load (245 to 95°F):

Sensible heat:
Q_{air} = 18,900 × 0.25(245 − 95) =   708,000 Btu/hr
Q_{water} = 692 × 0.45(245 − 95) =   46,700 Btu/hr

Latent heat:
Q_{water} = 395 × 1040.1   =   411,000 Btu/hr
                          Total 1,165,700 Btu/hr

* The correction of γ for the presence of water vapor is usually omitted.

GASES 197

If condensation had not been accounted for, an error of 33 per cent would have resulted. It should also be noted that over half of the water condenses in the first-stage intercooler.

(b) *Second stage:* $p_3 = 34.2 \times 2.33 = 79.8$ psi.
Final gas volume $= 4670 \times 60 \times 14.7/79.8 = 51,500$ ft^3/hr
Pounds of water remaining in air $= 51,500/404.3 = 127.5$ lb/hr
Condensation in intercooler $= 297 - 127.5 = 169.5$ lb/hr

Heat load (245° to 95°F):

Sensible heat:

$Q_{air} = 18,900 \times 0.25(245 - 95) = 708,000$ Btu/hr
$Q_{water} = 297 \times 0.44(245 - 95) = 19,600$ Btu/hr

Latent heat:

$Q_{water} = 169.5 \times 1040.1 \qquad = \underline{170,700}$ Btu/hr
$\qquad\qquad\qquad\qquad\qquad\qquad\quad 898,300$ Btu/hr

Example 9.3. Calculation of the Dew Point after Compression. The dew point and saturation temperature of the saturated inlet air are the same. After the first-stage compression the dew point is raised. What is the dew point when air saturated at 95°F and 14.7 psi is compressed to 34.2 psi?

Solution:

At inlet:

Mols air $= 18,900/29 = 652$
Mols water $= 692/18 = \underline{38.4}$
$\qquad\qquad\qquad\qquad\quad 690.4$

After compression:

Partial pressure of water vapor $= (38.4/690.4)34.2 = 1.90$ psi
From Table 7 equivalent to 1.90 psi, dew point $= 124°F$

In other words, the gas and water vapor are cooled sensibly from 245 to 124°F in the first stage intercooler and the water vapor will start to condense at 124°F.

The Calculation of Coolers for Wet Gases. The calculation of the heat load and film coefficients in the intercoolers of adiabatic compression systems starting with initially dry gases offers no particular difficulty. The heat load is the sensible-heat requirement to cool the gas back between stages. The film coefficient is that of the dry gas.

Coolers which are required to cool wet gases present a number of additional problems. If the wet gas is to be cooled below its dew point, two zones will appear: (1) from the inlet temperature to the dew point in which both the gas and the vapor are cooled sensibly and (2) from the dew point to the outlet temperature, in which the gas and vapor are cooled and part of the vapor condenses. The first zone can be calculated rather simply as a dry gas, but the calculation of the second zone is extremely lengthy. A relatively accurate calculation will be demonstrated as an example of condensation in Chap. 13, where it will be seen that both the condensation and gas film coefficients are closely related. The film coefficient for the mixture varies considerably from the dew point to the outlet temperature as the concentration of the condensable vapor diminishes. It is also seen that, in any wet gas cooling service if the tube-wall temperature is below the dew point of the gas, even though the gas is not cooled below its dew point, the tube wall will be wet with condensate. As droplets of condensate fall from the tube, they will reflash into the gas and some fraction may drain from the cooler if the temperature should fall below the dew point. However, the film of liquid on the tube actually introduces a resistance film through which the heat must be transferred. If the condensable vapor is water, the resistance can be omitted because of the high conductivity of the film. If it is a vapor whose condensate is a viscous fluid, it may be necessary to calculate the mean resistance of the film from methods given in Chap. 13 based on the properties of the condensate. It is also well to consider that gases which are not particularly corrosive when containing a small concentration of water vapor may be corrosive when dissolved in condensate water at the cold tube wall. Desuperheaters, which are simply gas coolers, frequently operate with part of the surface moist although no actual condensate drains from the system.

The performance of commercial intercoolers for permanent gases saturated at atmospheric pressure with water at 100°F or less can be predicted rapidly by empirical rules. These rules are as follows: (1) Calculate the entire heat load of sensible cooling and condensation as if transferred at the dry gas rate, and (2) use the value of $\Delta t = F_T \times \text{LMTD}$ obtained from the inlet and outlet temperatures of the gas to and from the cooler and the temperatures of the water. These rules are actually the combination of a safe and an unsafe generalization which tend to offset each other. The combined film coefficient for condensation and gas cooling below the dew point is greater than that given by (1). The true temperature difference is less than that calculated by (2), since the log mean for the portion of the heat load delivered from the dew point to the outlet is less than that calculated by the rule.

PROBLEMS

9.1. 3500 cfm of dry nitrogen at 17 psig and 280°F is cooled to 100°F by water with an inlet temperature of 85°F. Available for the service is a 31 in. ID 1-2 exchanger having 600 ¾ in. OD, 16 BWG tubes 12'0" long arranged for eight passes on 1-in. triangular pitch. The baffle spacing is 24 in. center to center.

Pressure drops of 2.0 psi for the gas and 10.0 for the water should not be exceeded, and a minimum dirt factor of 0.01 should be provided. Will the cooler work?

9.2. 17,500 lb/hr of oxygen at atmospheric pressure is cooled from 300 to 100°F by water from 85 to 100°. Available for the service is a 31 in. ID 1-2 exchanger containing 600 ¾ in. OD, 16 BWG tubes 12'0" long arranged for eight passes on 1-in. triangular pitch. Baffles are spaced 24 in. apart.

What are the dirt factor and the pressure drops?

GASES 199

9.3. 5000 cfm of saturated air at 100°F enters the first stage of a compressor having a 2.45:1 compression ratio. The air is at atmospheric pressure. (a) How much heat must be removed after each of the four stages, assuming a 2.0 psi pressure drop in each intercooler. (b) Available for the first-stage intercooler is a 29 in. ID exchanger having 508 tubes ¾ OD in., 14 BWG by 12′0″ long arranged for eight passes on 1-in. triangular pitch. The baffle spacing is 24 in. Using cooling water with an 85°F inlet, what are the pressure drops and dirt factor?

9.4. For the second-stage intercooler of Example 9.2 in the text the following 1-2 exchanger is available: 21¼ in. ID containing 294 ¾ in. OD, 14 BWG tubes 12′0″ long arranged for eight passes on $1\frac{5}{16}$-in. triangular pitch. Baffles are spaced 20 in. apart. What are the dirt factor and the pressure drops?

NOMENCLATURE FOR CHAPTER 9

A	Heat-transfer surface, ft²
a	Flow area, ft²
a''	External surface per linear foot, ft
B	Baffle spacing, in.
C'	Clearance between tubes, in.
c	Specific heat of cold fluid, Btu/(lb)(°F)
D	Inside diameter of tubes, ft
D_e	Equivalent diameter, ft
F_T	Temperature difference factor, $\Delta t = F_T \times \text{LMTD}$, dimensionless
f	Friction factor, ft²/in.²
G	Mass velocity, lb/(hr)(ft²)
g'	Acceleration of gravity, ft/sec²
h, h_i, h_o	Heat-transfer coefficient in general, for inside fluid, and for outside fluids, respectively, Btu/(hr)(ft²)(°F)
h_{io}	Value of h_i when referred to the tube OD, Btu/(hr)(ft²)(°F)
j_H	Factor for heat transfer $(hD/k)(c\mu/k)^{1/3}$, dimensionless
j_h	Factor for heat transfer $(h/cG)(c\mu/k)^{2/3}$, dimensionless
k	Thermal conductivity, Btu/(hr)(ft²)(°F/ft)
L	Tube length, ft
LMTD	Log mean temperature difference, °F
N	Number of baffles
N_t	Number of tubes
n	Number of tube passes
P_T	Tube pitch, in.
ΔP	Pressure drop, psi
$\Delta P_T, \Delta P_t, \Delta P_r$	Total, tube side and return drop, psi
p	Pressure, psia
R	Temperature group $(T_1 - T_2)/(t_2 - t_1)$, dimensionless
R_d	Combined dirt factor, (hr)(ft²)(°F/Btu)
Re	Reynolds number, dimensionless
S	Temperature group $(t_2 - t_1)/(T_1 - t_1)$, dimensionless
s	Specific gravity, dimensionless
T_{abs}	Absolute temperature, °R
T_a	Average temperature of hot fluid, °F
T_c	Caloric temperature of hot fluid, °F
T_1, T_2	Inlet and outlet temperature of hot fluid, °F
t_a	Average temperature of cold fluid, °F

t_c	Caloric temperature of cold fluid, °F
t_1, t_2	Inlet and outlet temperature of cold fluid, °F
Δt	True temperature difference in $Q = U_D A \, \Delta t$, °F
U_C, U_D	Clean and design overall coefficient of heat transfer, Btu/(hr)(ft²)(°F)
V	Velocity, fps
v	Specific volume, ft³/lb
W	Weight flow of hot fluid, lb/hr
w	Weight flow of cold fluid, lb/hr
w'	Weight, lb
γ	Ratio of specific heats of a gas, dimensionless
μ	Viscosity, centipoises $\times$ 2.42 = lb/(ft)(hr)
μ_w	Viscosity at the tube wall, centipoises $\times$ 2.42 = lb/(ft)(hr)
ρ	Density, lb/ft³
ϕ	$(\mu/\mu_w)^{0.14}$

Subscripts (except as noted above)

s	Shell side
t	Tube side

CHAPTER 10

STREAMLINE FLOW AND FREE CONVECTION

Streamline Flow in the Tubes of Exchangers. From the Fourier equation alone for a single tube $Q = wc(t_2 - t_1) = h_i(\pi DL)\,\Delta t$. When the inside tube-wall temperature t_p is constant, the temperature difference Δt in streamline flow may be replaced by the arithmetic mean of the hot and cold terminal temperature differences $\Delta t_a = [(t_p - t_1) + (t_p - t_2)]/2$. Solving for $h_i D/k$,

$$\frac{h_i D}{k} = \left(\frac{2}{\pi}\frac{wc}{kL}\right)\frac{(t_2 - t_1)}{(t_p - t_1) + (t_p - t_2)} \tag{10.1}$$

It is interesting to note that the highest outlet temperature attainable in a tube is the constant temperature of the hot tube wall t_p. For this case Eq. (10.1) reduces to

$$\frac{h_i D}{k} = \left(\frac{2}{\pi}\frac{wc}{kL}\right) \tag{10.2}$$

No observed average value of h_i can exceed that given by Eq. (10.2), and it is a useful tool by which erroneous observations may be rejected.

Graetz obtained Eq. (3.27) from purely theoretical considerations on the assumption of a parabolic distribution of velocities as the fluid flows in a tube in laminar flow. He did not include any corrections for modifications of the parabolic distribution during heating and cooling. Sieder and Tate evaluated the equivalent empirical equation [Eq. (3.32)] and obtained Eq. (6.1), which may be credited with correcting for the modifications of the velocity distribution during heating and cooling.

Streamline flow in tubes may be construed as a conduction effect, and it is also subject to the simultaneous occurrence of free convection as well. Free convection is only significant in nonviscous fluids. Fluids flow in streamline flow due to three conditions: (1) The fluid is viscous; (2) the fluid is not viscous, but the quantity is small for the flow area provided; and (3) the flow quantity and viscosity are intermediate but combine to give streamline flow. Only when free convection is suppressed owing to the high average viscosity of the liquid, say several centipoises and over, or when the temperature difference is small does Eq. (6.1) give h_i within the stated deviation as pure conduction. For cases falling under (2) or (3) above, the value of h_i from Eq. (6.1) may be conservative, the true value of h_i being as much as 300 per cent greater due to the influence of free convection.

In the Graetz derivation the value of L in the Graetz number wc/kL or in the ratio D/L is assumed to be the length of the path over which the fluid moves with a conduction temperature gradient at right angles to the long axis of the tube. Naturally, if mixing occurs at any point in the heat-transfer tube, the distance from the inlet to the point of mixing must be regarded as the length of the path over which the temperature gradient is effective whether or not it corresponds to the total tube length of the exchanger, nL. Boussinesq[1] has advanced the theory that under ideal conditions streamline flow is not established until the liquids has traveled a length of approximately 15 tube diameters. In multipass heat-transfer equipment it is sometimes possible to consider the fluid in the tubes mixed at the end of each pass. Internal mixing is desirable, just as convection is desirable and also because it restricts the length of streamline path to the length of each pass, L.

The shorter the unmixed length the greater the value of h_i, although it is not always safe to assume that mixing occurs at the end of each tube pass of an exchanger. In modern multipass exchangers, the flow areas in the floating head and channel are usually designed to be identical with or slightly greater than the flow area of the tubes in each passs. In this way it is possible to eliminate excessive return pressure drops. If no turbulence or mixing is induced at the ends of each pass, nL is the total length of path instead of L, which leads to the calculation of safe values of h_i should mixing actually occur.

Frequently it will be found that the Reynolds number based on the viscosity at $t_a = (t_1 + t_2)/2$ is less than 2100 but near the outlet the Reynolds number based on the viscosity at t_2 is greater than 2100. Equation (6.1) is not applicable in the transitional or turbulent flow range. In calculations on multipass exchangers, the point at which $Re = 2100$ must then be obtained by trial and error, and the path beyond it excluded from calculation as streamline flow. If a portion of the tube is in the transitional range, it can best be computed by means of Fig. 24.

For a two-tube pass exchanger the maximum error in h_i calculated from Eq. (6.1) between a mixing or nonmixing assumption between passes is $(2/1)^{1/3} = 1.26$ or 26 per cent, since $h_i \propto 1/L^{1/3}$. For an eight-tube pass exchanger the error is $(8/1)^{1/3} = 2.0$ or 100 per cent. Ordinarily a decision should not be arrived at without consulting the design of the exchanger and noting whether or not provision for mixing between passes has been included.

[1] Boussinesq, J., *Compt. rend.*, **113**, 9 (1891).

STREAMLINE FLOW AND FREE CONVECTION

Example 10.1. Crude Oil Heater: Streamline Flow. A line carries 16,000 lb/hr of 34°API crude oil. It enters the tubes at 95°F and is heated to 145°F using steam at 250°F. Consider the fluid *mixed* between passes.

The viscosities of the crude oil are

°F	μ cp
250	1.15
200	1.7
150	2.8
125	3.8
100	5.2

Available for temporary service is a horizontal 1-2 exchanger having a 15¼ in. ID shell with 86 1 in. OD, 16 BWG tubes 12'0" long laid out on 1¼-in. triangular pitch. The bundle is arranged for two tube passes, and baffles are spaced 15 in. apart. Since the heater is for temporary use, no dirt factor will be included.

Will the exchanger fulfill the requirement?

Solution:

Exchanger:

    Shell side   *Tube side*
    ID = 15¼ in.   Number and length = 86, 12'0"
  Baffle space = 15 in.   OD, BWG, pitch = 1 in., 16 BWG, 1¼-in. tri.
    Passes = 1   Passes = 2

(1) Heat balance:
    Crude, $Q = 16{,}000 \times 0.485(145 - 95) = 388{,}000$ Btu/hr
    Steam, $Q = 410 \times 945.5 = 388{,}000$ Btu/hr

(2) Δt:

Hot Fluid		Cold Fluid	Diff.
250	Higher Temp	145	105
250	Lower Temp	95	155
0	Differences	50	50

$\Delta t = \text{LMTD} = 129°\text{F}$ (true counterflow)   (5.14)

(3) T_c and t_c: This fluid is in streamline flow throughout the heater [see (6), p. 204]. For streamline flow the arithmetic mean should be used. For the first pass

$$t_a = \frac{t_1 + t_i}{2} = \frac{95 + 125}{2} = 110°\text{F (approx.)}$$

On the assumption that the fluids are mixed between passes, each pass must be solved independently. Since only two passes are present in this exchanger, it is simply a matter of assuming the temperature at the end of the first pass. More than half the heat load must be transferred in the first pass; therefore assume t_i at the end of the first pass is 125°F and $t_i - t_1$ is 30°F.

Hot fluid: shell side steam	Cold fluid: tube side, crude oil

Cold fluid side:

(4) $a_t' = 0.594$ in.²   [Table 10]
$a_t = N_t a_t'/144n$ [Eq. (7.48)]
$= 86 \times 0.594/144 \times 2 = 0.177$ ft²

(5) $G_t = w/a_t$
$= 16,000/0.177$
$= 90,400$ lb/(hr)(ft²)

(6) $D = 0.87/12 = 0.0725$ ft   [Table 10]
$Re_t = DG_t/\mu$
At $t_2 = 145°F$ (the outlet temperature)
$\mu = 2.95 \times 2.42 = 7.15$ lb/(ft)(hr)
$Re_t = 0.0725 \times 90,400/7.15 = 915$
At $t_a = 110°F$,
$\mu = 4.8 \times 2.42 = 11.6$ lb/(ft)(hr)
$Re_t = 0.0725 \times 90,400/11.6 = 565$

(7) $h_i = 1.86 \dfrac{k}{D} \left(\dfrac{DG}{\mu} \dfrac{c\mu}{k} \dfrac{D}{L} \right)^{1/3} \phi_t$

                                           [Eq. (6.1)]

(8) $\dfrac{c\mu}{k} = \dfrac{0.485 \times 11.6}{0.0775} = 72.5$

$\dfrac{D}{L} = \dfrac{0.0725}{12} = 0.0060$

(9') $h_o = 1500$

(9) $\dfrac{h_i}{\phi_t} = 1.86 \times \dfrac{0.0775}{0.0725}$
$(565 \times 72.5 \times 0.0060)^{1/3} = 12.4$

(10') $t_p = t_a + \dfrac{h_o}{h_i + h_o}(T_c - t_a)$

$= 110 + \dfrac{1500}{12.4 + 1500}(250 - 110)$

$= 249°F$

(10) At $t_p = 249°F$,
$\mu = 1.20 \times 2.42 = 2.9$ lb/(ft)(hr)
$\phi_t = \left(\dfrac{\mu}{\mu_w} \right)^{0.14}$
$= \left(\dfrac{11.6}{2.9} \right)^{0.14} = 1.20$

(11) $h_i = \dfrac{h_i}{\phi_t} \phi_t = 12.4 \times 1.20$
$= 14.9$ Btu/(hr)(ft²)(°F)

$$\Delta t = t_p - t_a = 249 - 110 = 139°F$$
$$t_i - t_1 = \dfrac{h_i A_i \Delta t}{wc}$$

Internal surface per foot of length $= 0.228$ ft

$$A_i = \dfrac{86}{2} \times 12'0'' \times 0.228 = 117.5 \text{ ft}^2$$

$$t_i - t_1 = \dfrac{14.9 \times 117.5 \times 139}{16,000 \times 0.485} = 31.4°F$$

Assumed value of $t_i - t_1 = 30.0°F$ (close enough for check)

The oil now enters the second pass at 126.4°F and leaves at 152 instead of 145°F, **indicating that the heater is oversurfaced if no dirt factor is required.**

If four or more passes are present, the calculation is carried out in the same manner as shown above with a new assumed temperature at the end of each pass. If the calculated outlet from the last pass equals or exceeds t_2, the heater will operate satisfactorily.

When a dirt factor is to be provided, obtain the initial outlet temperature t_2 delivered by the oversized heater, when freshly placed in service, from the heat balance and Eq. (5.18), using U_C for U. U_C is obtained from U_D and R_d by Eq. (6.10) instead of from h_{io} and h_o, since in this case it is desired to calculate the value of h_{io} which will produce the initial value of the outlet temperature. Having calculated t_2, solve for each pass until the outlet temperature corresponds to the initial value of t_2 before dirt accumulates.

Free Convection in Tubes. Streamline flow is calculated by equations employing weight flow or mass velocity as one of the variables. However, if a horizontal tube surrounded by condensing steam carries cold

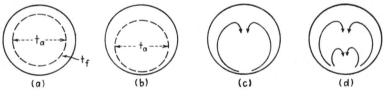

Fig. 10.1. Free convection in tubes.

liquid and its flow is suddenly halted, the liquid in the tube continues to heat. According to Eq. (6.1) the film coefficient should be zero if the mass velocity is zero. Where the heat is transferred through movements within the liquid itself without forced circulation, it occurs by free or natural convection. Some of the factors influencing free convection in liquids can be observed very readily in the laboratory owing to changes in the index of refraction which accompany changes in density. When a fluid is heated in a glass vessel on a hot plate, the convection currents are visible. Liquid at the bottom of the vessel and adjacent to the heat source is heated by conduction. The heat absorbed reduces the density of the bottom layer of liquid so that it rises and the colder liquid tends to settle.

In a *horizontal* tube the process is somewhat more orderly. Starting with a stationary liquid, heat is applied from the outside, raising the temperature of an outer layer of liquid as shown in Fig. 10.1a. The cold central core is heavier than the liquid adjacent to the wall and settles toward the bottom of the tube somewhat as shown in Fig. 10.1b. The rate of settling is retarded by the temperature-viscosity relationship between the hot fluid at the wall and the cooled core of liquid. As the free-convection currents are established, they develop and mix with the bulk of liquid presumably as in Fig. 10.1c, and if the tube is large, mixing may be accelerated as in Fig. 10.1d. The film coefficient for free convection is a function of the inside diameter of the tube D, density of the liquid ρ, coefficient of expansion β, gravitational constant g, thermal conductivity k, viscosity μ, and lastly the temperature difference Δt_a between the hot tube wall and the bulk of the fluid.

$$h_i = f(D, \rho, \beta, c, g, k, \mu, \Delta t_a)$$

Solving by dimensional analysis

$$\frac{h_i D}{k} = C_1 \left(\frac{D^3 \rho^2 g \beta \Delta t_a}{\mu^2}\right)^a \left(\frac{c\mu}{k}\right)^b \tag{10.3}$$

where $D^3 \rho^2 g \beta \Delta t_a / \mu^2$ is the Grashof number.

Combined Free Convection and Streamline Flow in Horizontal Tubes. Just as there is a transition region and not a single point separating streamline and turbulent flow, there must also be some transition region between free convection to a fluid standing still and streamline flow. At low linear velocities both are undoubtedly operative. Equation (6.1) was correlated from data obtained on small tubes with moderately viscous fluids and under moderate temperature differences such that the Grashof numbers were relatively small. Kern and Othmer[1] investigated this region in *horizontal* tubes under large temperature differences and tube diameters and evaluated free convection as a correction to the Sieder-Tate equation. Their final equation is

$$\frac{h_i D}{k}\left(\frac{\mu}{\mu_w}\right)^{-0.14} = 1.86\left[\left(\frac{DG}{\mu}\right)\left(\frac{c\mu}{k}\right)\left(\frac{D}{L}\right)\right]^{1/3} \frac{2.25(1 + 0.010 Gr_a^{1/3})}{\log Re} \tag{10.4}$$

where Gr_a is the Grashof number evaluated from properties taken at the average fluid temperature $t_a = (t_1 + t_2)/2$. h_i as ordinarily calculated by Eq. (6.1) can be corrected for free convection by multiplying by

$$\psi = \frac{2.25(1 + 0.010 Gr_a^{1/3})}{\log Re} \tag{10.5}$$

Inspection of Eq. (10.4) indicates that the influence of the free-convection currents are dissipated in the transition and turbulent region. In view of the delicate nature of free-convection currents this is quite plausible. The two factors which ordinarily influence free convection most are a low viscosity and large temperature difference. Martinelli *et al.*[2] studied

[1] Kern, D. Q., and D. F. Othmer, *Trans. AIChE*, **39**, 517–555 (1943).
[2] Martinelli, R. C., C. J. Southwell, G. Alves, H. L. Craig, E. B. Weinberg, N. E. Lansing, and L. M. K. Boelter, *Trans. AIChE*, **38**, 943 (1942).

STREAMLINE FLOW AND FREE CONVECTION

the influence of free convection in upward and downward flow in vertical tubes. They found a slight increase in the coefficient when heating water in upward flow compared with downward flow. Their final correlation is rather complicated, although they also obtained a correlation of free convection involving the Reynolds number.

Example 10.2. Kerosene Heater: Streamline Flow and Free Convection. A line carries 16,000 lb/hr of 40°API light distillate or heavy kerosene with corrosive contaminants. It enters the tubes at 95°F and is heated to 145°F using steam at 250°F. Consider the liquid *unmixed* between passes.

The viscosities of the kerosene are

°F	μ, cp
250	0.60
200	0.85
150	1.30
125	1.70
100	2.10

Available for the service is a horizontal 1-2 exchanger, having a $15\frac{1}{4}$ in. ID shell with 86 1 in. OD, 16 BWG tubes 12'0'' long laid out on $1\frac{1}{4}$-in. triangular pitch. The bundle is arranged for two tube passes, and baffles are spaced 15 in. apart (same as Example 10.1).

What is the true dirt factor?

Solution:

Exchanger:

Shell side
ID = $15\frac{1}{4}$ in.
Baffle space = 15 in.
Passes = 1

Tube side
Number and length = 86, 12'0''
OD, BWG, pitch = 1 in., 16 BWG, $1\frac{1}{4}$-in. tri.
Passes = 2

(1) Heat balance:
Kerosene, $Q = 16,000 \times 0.50(145 - 95) = 400,000$ Btu/hr
Steam, $Q = 4230 \times 945.5 = 400,000$ Btu/hr

(2) Δt:

	Hot Fluid	Cold Fluid	Diff.
Higher Temp	250	145	105
Lower Temp	250	95	155
Differences	0	50	50

$\Delta t = \text{LMTD} = 129°\text{F}$ (true counterflow) (5.14)

(3) T_c and t_c: This fluid is in streamline flow throughout the heater [see (6), p. 208]. For streamline flow the arithmetic mean is

$$t_a = \frac{t_1 + t_2}{2} = \frac{95 + 145}{2} = 120°\text{F}$$

Hot fluid: shell side, steam

Cold fluid: tube side, 40°API kerosene

(4) $a_t' = 0.594$ in.²   [Table 10]
$a_t = N_t a_t'/144n$ [Eq. (7.48)]
  $= 86 \times 0.594/144 \times 2 = 0.177$ ft²

(5) $G_t = w/a_t$
  $= 16,000/0.177$
  $= 90,400$ lb/(hr)(ft²)

(6) $D = 0.87/12 = 0.0725$ ft   [Table 10]
$Re_t = DG_t/\mu$
At $t_2 = 145°$F,
$\mu = 1.36 \times 2.42 = 3.29$ lb/(ft)(hr)
  [Fig. 14]
$Re_t = 0.0725 \times 90,400/3.29 = 1990$
At $t_a = 120°$F,
$\mu = 1.75 \times 2.42 = 4.23$ lb/(ft)(hr)
$Re_t = 0.0725 \times 90,400/4.23 = 1,550$
Streamline flow, fluid unmixed between passes
$Ln/D = 12 \times 2/0.0725 = 331$

(7) $j_H = 3.10$ [Fig. 24]

(8) At $\mu = 1.75$ cp and 40°API,
$k(c\mu/k)^{1/3} = 0.24$ Btu/(hr)(ft²)(°F/ft)
  [Fig. 16]

(9) $\dfrac{h_i}{\phi_t} = j_H \dfrac{k}{D}\left(\dfrac{c\mu}{k}\right)^{1/3}$ [Eq. (6.15a)]
$h_i/\phi_t = 3.10 \times 0.24/0.0725 = 10.25$

(10) $\dfrac{h_{io}}{\phi_t} = \dfrac{h_i}{\phi_t} \times \text{ID/OD}$ [Eq. (6.5)]
  $= 10.25 \times 0.87/1.0 = 8.91$

(11) At $t_w = 249°$F,
$\mu_w = 0.60 \times 2.42 = 1.45$ lb/(ft)(hr)
$\phi_t = (\mu/\mu_w)^{0.14}$
  $= (4.23/1.45)^{0.14} = 1.16$

(9') Condensation of steam:
$h_o = 1500$

(10') t_w:
$t_w = t_a + \dfrac{h_o}{h_{io} + h_o}(T_a - t_a)$ [Eq. (5.31)]
  $= 120 + \dfrac{1500}{8.9 + 1500}(250 - 120)$
  $= 249°$F

(12) $h_{io} = \dfrac{h_{io}}{\phi_t}\phi_t$ [Eq. (6.37)]
  $= 8.91 \times 1.16$
  $= 10.3$ Btu/(hr)(ft²)(°F)
$\Delta t_a = t_w - t_a = 249 - 120 = 129°$F
Since the kerosene has a viscosity of only 1.75 cp at the caloric temperature and $\Delta t_a = 129°$F, free convection should be investigated.
$\psi = \dfrac{2.25(1 + 0.01 Gr_a^{1/3})}{\log Re}$ [Eq. (10.5)]
Grashof number, $Gr_a = D^3\rho^2 g\beta\, \Delta t_a/\mu^2$
$s = 0.80$,
$\rho = 0.8 \times 62.5 = 50.0$ lb/ft³   [Fig. 6]
$\beta = \dfrac{1}{°\text{F}} = \dfrac{1/\rho_2 - 1/\rho_1}{(t_2 - t_1)(1/\rho_{av})}$
  $= \dfrac{s_1^2 - s_2^2}{2(t_2 - t_1)s_1 s_2}$

STREAMLINE FLOW AND FREE CONVECTION

Hot fluid: shell side, steam	Cold fluid: tube side, 40° API kerosene
	At 95°F, $s_1 = 0.810$ [Fig. 6]
	At 145°F, $s_2 = 0.792$
	$\beta = 0.000451/°F$
	$Gr_a = 0.0725^3 \times 50.0^2 \times 0.00045 \times 4.18$
	$\qquad\qquad \times 10^8 \times 129/4.23^2$
	$\qquad = 1{,}300{,}000$
	$\psi = \dfrac{2.25(1 + 0.01 \times 1{,}300{,}000^{1/3})}{\log 1550} = 1.47$
	Corrected $h_{io} = 10.3 \times 1.47 = 15.1$

(13) Clean overall coefficient U_C:

$$U_C = \frac{h_{io}h_o}{h_{io} + h_o} = \frac{15.1 \times 1500}{15.1 + 1500} = 14.9 \text{ Btu/(hr)(ft}^2\text{)(°F)} \qquad (6.38)$$

(14) Design overall coefficient U_D:

$$a'' = 0.2618 \text{ ft}^2/\text{lin ft} \qquad \text{(Table 10)}$$
Total surface, $A = 86 \times 12'0'' \times 0.2618 = 270 \text{ ft}^2$
$$U_D = \frac{Q}{A\,\Delta t} = \frac{400{,}000}{270 \times 129} = 11.5 \text{ Btu/(hr)(ft}^2\text{)(°F)}$$

(15) Dirt factor:

$$R_d = \frac{U_C - U_D}{U_C U_D} = \frac{14.9 - 11.5}{14.9 \times 11.5} = 0.0198 \text{ (hr)(ft}^2\text{)(°F)/Btu} \qquad (6.13)$$

Summary: Figures in parentheses are uncorrected for free convection

Summary

1500	h outside	(10.3) 15.1
U_C	(10.2) 14.9	
U_D	11.5	
R_d Calculated	0.0198	

If the correction for free convection were not included, it would have given the impression that the unit would not work, since U_C would be smaller than U_D. Corrected, however, it is very safe. If the viscosity were 3 or 4 centipoises and the Reynolds number remained the same because of increased weight flow, the exchanger would not be suitable.

The pressure drop may be computed as heretofore except that ϕ_t for pressure drop in streamline flow is $(\mu/\mu_w)^{0.25}$.

The Use of Core Tubes. Employing the horizontal exchanger of the preceding example, suppose that 50,000 lb/hr of a 28°API gas oil was to be heated from 105 to 130°F using steam at 250°F. Would the exchanger be satisfactory?

Solving as before, the value of the viscosity correction ϕ_t is 1.18 but the free-convection correction ψ is negligible, and U_C would be less than U_D.

Cores, dummy tubes with one end sealed (by clamping in a vise), **may** be placed within the tubes of an exchanger as shown in Fig. 10.2. They constrict the cross-sectional area by forming an annulus which replaces the inside diameter of the tube with a smaller equivalent diameter and which increases the mass velocity of the tube fluid. Although the use of a core decreases the effective tube diameter and increases the mass

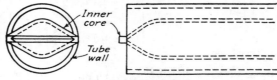

Fig. 10.2. Core detail.

velocity, it does not alter the Reynolds number from its value before introduction of the core. Using a core,

$$a_s = \frac{\pi}{4}(D_2^2 - D_1^2)$$

where D_1 is the outside diameter of the core and D_2 is the inside diameter of the tube. Wetted perimeter for heat transfer[1] = πD_2.

$$D_e = 4\frac{\pi}{4}\frac{(D_2^2 - D_1^2)}{\pi D_2} = \frac{D_2^2 - D_1^2}{D_2}$$

$$G = \frac{w}{(\pi/4)(D_2^2 - D_1^2)}$$

$$Re_t = \frac{D_e G}{\mu} = \frac{D_2^2 - D_1^2}{D_2 \mu}\frac{w}{(\pi/4)(D_2^2 - D_1^2)} = \frac{4w}{\pi D_2 \mu}$$

Without a core the result is the same.

$$Re_t = \frac{DG}{\mu} = \frac{D_2}{\mu}\frac{w}{(\pi/4)D_2^2} = \frac{4w}{\pi D_2 \mu}$$

Since the Reynolds number remains the same, the advantage of the cores is manifested by the smaller value of D_e to D in Eq. (6.1). However, the use of cores eliminates the possibility of free convection as correlated by Eq. (10.4), since the cores disrupt the natural-convection currents. Cores are most advantageously employed only when the fluid is viscous and free convection is precluded.

Any size of core may be employed, and it is customary to use 18 BWG or lighter heat-exchanger tubes for the purpose. If the tube is sealed by clamping in a vise, the core must be of such size that one-half its perimeter is greater than the inside diameter of the tube into which it is to be inserted. If the clamped width is less, the core is secured by welding

[1] Unlike the annulus of a double pipe exchanger, the wetted perimeter is the circumference at the inside diameter of the outer tube rather than the outer diameter of the inner tube.

STREAMLINE FLOW AND FREE CONVECTION 211

small rods at the clamped ends, which are larger than one-half the perimeter of the tube. The cores are inserted with the sealed end at the inlet of each pass, so that in multipass exchangers half are inserted at the channel and half at the opposite end of the bundle.

Since cores produce considerable inlet and exit turbulence, it is fair to assume that substantial mixing occurs between passes. In such cases the unmixed length of path may be taken as the tube length.

The equivalent diameter for pressure drop is smaller than that for heat transfer because the wetted perimeter is the sum of the circumferences of the inside of the tube plus the outside of the core. Ordinarily, if there is a large quantity of liquid flowing but the viscosity is high, the pressure drop will be quite large. If the flow is scanty and the viscosity is small, the pressure drop will usually be negligibly small despite the use of a core.

Example 10.3. Gas Oil Heater Using Cores. In an effort to make the exchanger in Example 10.2 suitable for the flow of 50,000 lb/hr of 28°API gas oil, the use of cores will be investigated. The oil will be heated from 105 to 130°F using steam at 250°F. Comparisons will be made with Example 10.2.

The correct core cannot always be selected for the first trial, and it is usually necessary to assume several core sizes and carry out the calculation.

Viscosities of the oil in the low temperature range are

°F	μ, cp
250	2.0
200	3.1
150	5.0
125	6.3
100	8.2

Solution:

Exchanger:

    Shell side   *Tube side*
      ID = $15\frac{1}{4}$ in.   Number and length = 86, 12'0''
  Baffle space = 15 in.   OD, BWG, pitch = 1 in., 16 BWG, $1\frac{1}{4}$-in. tri.
    Passes = 1   Passes = 2

Assume that $\frac{1}{2}$-in., 18 BWG cores are used.

(1) Heat balance:
   Gas oil, $Q = 50,000 \times 0.47(130 - 105) = 587,000$ Btu/hr
   Steam, $Q = 6220 \times 945.5 = 587,000$ Btu/hr

(2) Δt:

Hot Fluid		Cold Fluid	Diff.
250	Higher Temp	130	120
250	Lower Temp	105	145
0	Differences	25	25

Δt = LMTD = 132.5°F (true counterflow)   (5.14)

(3) T_c and t_c:

$$t_a = \frac{(t_1 + t_2)}{2} = \frac{130 + 105}{2} = 117.5°F$$

Hot fluid: shell side, steam

(4') $a_s = \text{ID} \times C'B/144P_T$ [Eq. (7.1)]
$= 15.25 \times 0.25 \times 15/144 \times 1.25$
$= 0.318 \text{ ft}^2$

(5') $G_s = W/a_s$ [Eq. (7.2)]
$= 6220/0.318$
$= 19,600 \text{ lb/(hr)(ft}^2)$

(6') At $T_a = 250°F$,
$\mu = 0.013 \times 2.42 = 0.0314 \text{ lb/(ft)(hr)}$
                                   [Fig. 15]
$D_e = 0.72/12 = 0.060 \text{ ft}$ [Fig. 29]
$Re_s = D_e G_s/\mu$ (for pressure drop)
                                   [Eq. (7.3)]
$= 0.06 \times 19,600/0.0314 = 37,400$

(9') Condensation of steam:
$h_o = 1500$

(10') t_w:
$t_w = t_a + \dfrac{h_o}{h_{io} + h_o}(T_a - t_a)$ [Eq. (5.31)]
$= 117.5 + \dfrac{1500}{19.5 + 1500}(250 - 117.5)$
$= 249°F$

Cold fluid: tube side, crude oil

(4) $a'_t = \dfrac{\pi}{4}(d_2^2 - d_1^2)$
$= \dfrac{\pi}{4}(0.87^2 - 0.50^2) = 0.40 \text{ in.}^2$

(d_2 and d_1 are the annulus diameters, in.)
$a_t = N_t a'_t/144n$ [Eq. (7.48)]
$= 86 \times 0.40/144 \times 2 = 0.119 \text{ ft}^2$

(5) $G_t = w/a_t$
$= 50,000/0.119$
$= 420,000 \text{ lb/(hr)(ft}^2)$

(6) $d_e = \dfrac{d_2^2 - d_1^2}{d_2} = \dfrac{0.87^2 - 0.50^2}{0.87}$
$= 0.582 \text{ in.}$
$D_e = 0.582/12 = 0.0485 \text{ ft}$
At $t_a = 117.5°F$,
$\mu = 6.9 \times 2.42 = 16.7 \text{ lb/(ft)(hr)}$
$Re_t = D_e G_t/\mu$
$= 0.0485 \times 420,000/16.7 = 1,220$
Assume mixing between passes L/D_e
$\approx 12/0.0485 = 247$

(7) $j_H = 3.10$
(8) At $\mu = 6.9$ cp and 28°API
$k(c\mu/k)^{1/3} = 0.35 \text{ Btu/(hr)(ft}^2)(°F/ft)$
                                     [Fig. 16]

(9) $\dfrac{h_i}{\phi_t} = j_H \dfrac{k}{D_e}\left(\dfrac{c\mu}{k}\right)^{1/3}$ [Eq. (6.15a)]
$= 3.10 \times 0.35/0.0485 = 22.4$

(10) $\dfrac{h_{io}}{\phi_t} = \dfrac{h_i}{\phi_t} \times \dfrac{\text{ID}}{\text{OD}}$ [Eq. (6.5)]
$= 22.4 \times 0.87/1.0 = 19.5$

(11) At $t_w = 249°F$,
$\mu_w = 2.0 \times 2.42 = 4.84 \text{ lb/(ft)(hr)}$
                                     [Fig. 14]
$\phi_t = (\mu/\mu_w)^{0.14}$
$= (16.7/4.84)^{0.14} = 1.18$

(12) $h_{io} = \dfrac{h_{io}}{\phi_t}\phi_t$ [Eq. (6.37)]
$= 19.5 \times 1.18$
$= 23.0 \text{ Btu/(hr)(ft}^2)(°F)$

(13) Clean overall coefficient U_C:

$$U_C = \frac{h_{io}h_o}{h_{io} + h_o} = \frac{23.0 \times 1500}{23.0 + 1500} = 22.6 \text{ Btu/(hr)(ft}^2)(°F) \quad (6.38)$$

STREAMLINE FLOW AND FREE CONVECTION

(14) Design overall coefficient U_D:

$$A = 270 \text{ ft}^2$$
$$U_D = \frac{Q}{A\,\Delta t} = \frac{587{,}000}{270 \times 132.5} = 16.4 \text{ Btu/(hr)(ft}^2\text{)(°F)}$$

(15) Dirt factor R_d:

$$R_d = \frac{U_C - U_D}{U_C U_D} = \frac{22.6 - 16.4}{22.6 \times 16.4} = 0.0172 \text{ (hr)(ft}^2\text{)(°F)/Btu} \quad (6.13)$$

Summary

1500	h outside	19.4
U_C		22.6
U_D		16.4
R_d Calculated 0.0172		

Note that the cores make the heater operable.

Pressure Drop

(1′) For $Re_s = 37{,}400$,
$f = 0.0016$ ft^2/in.. [Fig. 29]

(1) $d'_e = (d_2 - d_1)$ [Eq. (6.4)]
$= (0.87 - 0.50) = 0.37$ in.
$D'_{e_t} = 0.37/12 = 0.0309$ ft
$Re'_{e_t} = D'_{e_t} G_t/\mu$
$= 0.0309 \times 420{,}000/16.7 = 777$
$f = 0.00066$ ft^2/in.^{2} [Fig. 26]

(2′) No. of crosses, $N + 1 = 12L/B$
   [Eq. (7.43)]
$= 12 \times 12/15 = 10$
$v = 13.82$ ft^3/lb   (Table 7)
$s = \dfrac{1}{13.82 \times 62.5} = 0.00116$
$D_s = 15.25/12 = 1.27$ ft

(2) For streamline-flow pressure drop:
$\phi_t = (16.7/4.84)^{0.25} = 1.35$
$s = 0.85$ [Fig. 6]
$$\Delta P_t = \frac{fG_t^2 L n}{5.22 \times 10^{10} D'_{e_t} s \phi_t} \quad \text{[Eq. (7.45)]}$$

(3′) $\Delta P_s = \dfrac{1}{2} \dfrac{fG_s^2 (D_s)(N+1)}{5.22 \times 10^{10} D_e s \phi_s}$
   [Eq. (7.52)]
$= \dfrac{1}{2} \times \dfrac{0.0016 \times 19{,}600^2 \times 1.27 \times 10}{5.22 \times 10^{10} \times 0.060 \times 0.00116 \times 1.0}$
$= 1.4$ psi

(3) $= \dfrac{0.00066 \times 420{,}000^2 \times 12 \times 2}{5.22 \times 10^{10} \times 0.0309 \times 0.85 \times 1.35}$
$= 1.5$ psi   [Eq. (7.46)]
$\Delta P_r =$ negligible

The choice of a core was satisfactory. A large dirt factor should be used because of the inaccuracies of the calculation. It would also be advisable to investigate the use of a ⅝ or ¾ in. OD core, although either of these cores will cause a considerably greater pressure drop.

Streamline Flow in Shell. When an exchanger is selected to fulfill process conditions from among existing exchangers, streamline flow in tubes should be avoided if possible. By comparison with turbulent flow, streamline flow requires the use of a larger surface for the delivery of equal heat. Sometimes streamline flow in tubes is unavoidable as when there is a great unbalance between both the sizes of the two streams and the shell-side and tube-side flow areas. If the flow area of the shell is greater than that of the tubes and one of the fluid quantities is much larger than the other, it may be imperative in meeting the allowable pressure drop to place the smaller stream in the tubes. Still other instances of streamline flow arise when a liquid is placed in the tubes for corrosion reasons whereas from the standpoint of fluid flow it logically belongs in the shell.

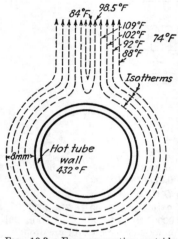

Fig. 10.3. Free convection outside tubes. (*After Ray.*)

Streamline flow in shells, on the other hand, is fortunately a rare problem. It occurs at a Reynolds number well below 10, as plotted in Fig. 28, due to the constantly changing flow area across the long axis of the bundle. The shell is consequently an excellent place in which to locate a scanty stream. The fouling factors of Table 12 have been predicated upon turbulent flow, and for low tube velocities they should be increased perhaps 50 to 100 per cent to provide additional protection.

Free Convection outside Tubes and Pipes. The mechanism of free convection outside a horizontal cylindrical shape differs greatly from that within it. On the outside of a pipe the convection currents are not restrained as they are within it, and heated fluid is usually free to rise through greater heights of cold fluid thereby increasing the convection. The atmosphere about a pipe has been explored by Ray,[1] and the isotherms are shown schematically in Fig. 10.3. Cold air from the relatively ambient atmosphere travels in toward the hot tube whence it is heated and rises. Numerous investigators have established the influence of the Grashof and Prandtl numbers on the correlation of free convection. Unfortunately most of the experimental information has been obtained on apparatuses such as single tubes and wires rather than industrial equipment. Accordingly, however, the film coefficient for free convection to

[1] Ray, B. B., *Proc. Indian Assoc. Cultivation Sci.*, **5**, 95 (1920).

gases from horizontal cylinders can be represented by

$$\frac{h_c D}{k_f} = \alpha \left[\left(\frac{D_o^3 \rho_f^2 g \beta \, \Delta t}{\mu_f^2} \right) \left(\frac{c_f \mu_f}{k_f} \right) \right]^{0.25}$$

where h_c is the free convection coefficient and all the properties are evaluated at the fictitious film temperature t_f taken as the mean of the temperature of the heating surface and the bulk temperature of the fluid being heated. Thus

$$t_f = \frac{t_w + t_a}{2} \tag{10.6}$$

For the most part it is difficult to obtain good data for the various types and sizes of equipment used in industry. This is due in part to the interference and complexities of free-convection heating elements such as banks of tubes and the inability to control an atmosphere of fluid to the extent necessary for good experimental results.

Free-convection correlations from the outside surfaces of different shapes which are of direct engineering value fall largely into two classes: free convection about single tubes or pipes and free convection about vessels and walls. McAdams[1] gives an excellent review of the work done heretofore in this field. It is apparent that free-convection currents are not only influenced by the position of the surface but also by its proximity to other surfaces. A hot horizontal surface sets up currents which differ greatly from those set up by a vertical surface. McAdams has summarized these in a simplified dimensional form[2] for free convection to air.

Horizontal pipes: $\quad h_c = 0.50 \left(\dfrac{\Delta t}{d_o} \right)^{0.25}$ \hfill (10.7)

Long vertical pipes: $\quad h_c = 0.4 \left(\dfrac{\Delta t}{d_o} \right)^{0.25}$ \hfill (10.8)

Vertical plates less than 2 ft high: $\quad h_c = 0.28 \left(\dfrac{\Delta t}{z} \right)^{0.25}$ \hfill (10.9)

Vertical plates more than 2 ft high: $\quad h_c = 0.3 \, \Delta t^{0.25}$ \hfill (10.10)

Horizontal plates:
Facing upward: $\quad h_c = 0.38 \, \Delta t^{0.25}$ \hfill (10.11)
Facing downward: $\quad h_c = 0.2 \, \Delta t^{0.25}$ \hfill (10.12)

where Δt is the temperature difference between hot surface and cold fluid in °F, d_o is the outside diameter in inches, and z is the height in feet.

[1] McAdams, W. H., "Heat Transmission," 2d ed., pp. 237–246, McGraw-Hill Book Company, Inc., New York, 1942.
[2] Perry, J. H., "Chemical Engineers' Handbook," 3rd Ed., p. 474, McGraw-Hill Book Company, Inc., New York, 1950.

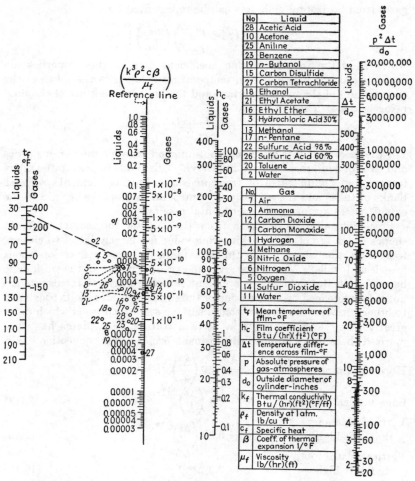

Fig. 10.4. Free convection outside horizontal pipes and tubes. [*T. H. Chilton, A. P. Colburn, R. P. Generaux,* and *H. C. Vernon, Trans. ASME Petroleum Mech. Eng.,* **55**, 5 (1933).]

For single horizontal pipes the dimensionless expression will apply except that α will vary between 0.47 and 0.53 between small and large pipes. Thus

$$\frac{h_c D_o}{k_f} = 0.47 \left[\left(\frac{D_o^3 \rho_f^2 g \beta \, \Delta t}{\mu_f^2} \right) \left(\frac{c \mu_f}{k_f} \right) \right]^{0.25} \quad (10.13)$$

Chilton, Colburn, Generaux, and Vernon[1] have developed an alignment chart which gives conservative coefficients for a single pipe but which has been used by the author and others without noticeable error for the calculations of free convection outside *banks* of tubes. The dimensional equation, plotted in Fig. 10.4 for gases and liquids, is

$$h_c = 116 \left[\left(\frac{k_f^3 \rho_f^2 c_f \beta}{\mu_f'} \right) \left(\frac{\Delta t}{d_o} \right) \right]^{0.25} \qquad (10.14)$$

where μ_f' is in centipoises. Of the four axes in the alignment chart one is the reference line of values of $k_f^3 \rho_f^2 c_f \beta / \mu_f$ which allows its use for fluids other than those whose key numbers are given in Fig. 10.4. The use of the chart for tube bundles requires that the pipes or tubes not be located too closely to the bottom of the vessel or too closely among themselves so that they interfere with the natural-convection currents. The tubes should not be located closer than several diameters from the bottom of the vessel, nor should the clearance between tubes be less than a diameter. Notwithstanding the data available, free-convection design is not very accurate, and reasonable factors of safety such as large fouling factors are recommended.

The use of the several types of free-convection correlations for fluids outside tubes is demonstrated by the typical problem which follows.

Example 10.4. Calculation of a Heating Bundle for an Aniline Storage Tank. A horizontal outdoor storage tank 5'0" ID by 12'0" long and presumed cylindrical is to

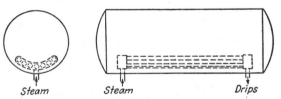

FIG. 10.5. Heating a fluid in a tank by free convection.

be used for aniline storage at a constant temperature of 100°F even though the prevailing atmospheric temperature should fall to 0°F. The tank is not insulated but is shielded from the wind. Heat will be supplied by means of exhaust steam through a bundle connected through the bottom of the tank as shown in Fig. 10.5 and consisting of 6'0" lengths of 1-in. IPS pipe. The use of a temperature controller (Chap. 21) to throttle the steam would make the entire operation automatic.

How many lengths of pipe should be provided?

Solution. The first part of the problem is to determine how much heat will be lost from the tank to the atmosphere. This gives the heat load for which the coil must be designed.

[1] Chilton, T. H., A. P. Colburn, R. P. Generaux, and H. C. Vernon, *Trans. A.S.M.E., Petroleum Mech. Eng.*, **55**, 5 (1933).

From Eq. (10.7) to (10.12) the convection coefficient to air will be somewhere between $h_c = 0.2 \Delta t^{0.25}$ and $h_c = 0.3 \Delta t^{0.25}$ for the entire tank surface. Since the equation giving the higher value of h_c is safer.

Convection loss:
$$h_c = 0.3 \Delta t^{0.25}$$

Neglecting the temperature drop through the tank metal
$$\Delta t = 100 - 0 = 100°F$$
$$h_c = 0.3 \times 100^{0.25} = 0.95 \text{ Btu/(hr)(ft}^2)(°F)$$

Radiation rate, $h_r = \dfrac{0.173 \, \epsilon[(T_{1,\text{abs}}/100)^4 - (T_{2,\text{abs}}/100)^4]}{T_{1,\text{abs}} - T_{2,\text{abs}}}$ (4.32)

Assuming an emissivity of about 0.8,
$$h_r = \frac{0.173 \times 0.8(5.6^4 - 4.6^4)}{100} = 0.75 \text{ Btu/(hr)(ft}^2)(°F)$$

Combined loss:
$$h_c + h_r = 0.95 + 0.75 = 1.70 \text{ Btu/(hr)(ft}^2)(°F)$$

Total tank area $= \dfrac{2\pi \times 5^2}{4} + \pi \times 5 \times 12 = 227.8 \text{ ft}^2$

Total heat loss, $Q = (h_c + h_r) A \, \Delta t = 1.70 \times 227.8 \times (100 - 0) = 38{,}800$ Btu/hr

This heat must be supplied by the pipe bundle.
Assuming exhaust steam to be at 212°F,

$$\frac{\Delta t}{d_o} = \frac{212 - 100}{1.32} = 85$$

$$t_f = \frac{212 + 100}{2} = 156°F \text{ (nearly)}$$

From Figure 10.4, $h_c = 48$ Btu/(hr)(ft^2)(°F).

$$U_C = \frac{h_{io} h_o}{h_{io} + h_o} = \frac{1500 \times 48}{1500 + 48} = 46.5 \text{ Btu/(hr)(ft}^2)(°F) \quad (6.38)$$

Assume a dirt factor of 0.02 (hr)(ft^2)(°F)/Btu

$$U_D = \frac{U_C \times 1/R_d}{U_C + 1/R_d} = \frac{46.5 \times 50}{46.5 + 50} = 24.1 \text{ Btu/(hr)(ft}^2)(°F)$$

Total surface, $A = \dfrac{Q}{U_D \, \Delta t}$
$$= 38{,}800/24.1 \times (212 - 100) = 14.4 \text{ ft}^2$$
Area/pipe $= 0.344 \times 6 = 2.06 \text{ ft}^2$ (Table 11)
Number of pipes $= 14.4/2.06 = 7$

These may be arranged in series or parallel.

PROBLEMS

10.1. 9000 lb/hr of 34°API crude oil at 100°F (see Example 10.1 for viscosities) enter the tubes of a 17¼ in. ID 1-2 exchanger containing 118 tubes 1 in. OD, 16 BWG, 12'0" long arranged for two passes in 1¼-in. triangular pitch. It is desired to attain an outlet temperature of 150°F using steam at 250°F. What is the actual outlet temperature?

STREAMLINE FLOW AND FREE CONVECTION

10.2. 37,000 lb/hr of an organic compound whose properties are closely approximated by the 28°API gas oil in the Appendix is to be heated by steam in the tubes of a clean exchanger because the shell baffles are half-circle support plates. The oil is to be heated from 100 to 200°F by steam at 325°F.

Available for the service is a 27 in. ID 1-2 exchanger containing 334 1 in. OD, 16 BWG, 16'0" long tubes arranged for two passes on 1¼-in. triangular pitch.

(a) Determine the dirt factor if the oil is not mixed between passes? (b) What is the actual outlet temperature of the oil if the fluid is mixed between passes?

10.3. 22,000 lb/hr of a flushing oil closely resembling 28°API gas oil is to be heated from 125 to 175°F to improve filtration. Available as a heating medium is another line corresponding to a 35°API distillate at 280°F. The temperature range of the distillate will be only 5°F so that the exchanger is essentially in counterflow.

Available for the service is a 21¼ in. ID 1-2 exchanger containing 240 ¾ in., 16 BWG tubes 8'0" long. Tubes are arranged for six passes on 1-in. square pitch. Baffles are spaced 8 in. apart. Because of contaminants the flushing oil will flow in the tubes.

Investigate all the possibilities, including the use of cores, to make the exchanger operable. Pressure drops are 10 psi, and a combined dirt factor is 0.015 (not to be confused with 0.0015).

10.4. A vertical cylindrical storage tank 6'0" ID and 8'0" tall is filled to 80 per cent of its height with ethylene glycol at 150°F. The surrounding atmosphere may fall to 0°F in the winter, but it is desired to maintain the temperature within the tank by using a rectangular bank of 1 in. IPS pipes heated with exhaust steam at 225°F How much surface is required? How should the pipes be arranged?

NOMENCLATURE FOR CHAPTER 10

A	Heat transfer surface, ft^2
a''	External surface per linear foot, ft
a	Flow area, ft^2
B	Baffle spacing, in.
C_1, C_2	Constants, dimensionless
D	Inside diameter of tubes or pipes, ft
D_o	Outside diameter of tubes or pipes, ft
D_e, D'_e	Equivalent diameter for heat transfer and pressure drop, ft
d	Inside diameter of tubes, in.
d_o	Outside diameter of tubes, in.
f	Friction factor, ft^2/in.^{2}
G	Mass velocity, lb/(hr)(ft^2)
Gr	Grashof number, dimensionless
Gr_a	Grashof number at the average or bulk temperature, dimensionless
g	Acceleration of gravity, ft/hr^2
h, h_i, h_o	Heat-transfer coefficient in general, for inside fluid, and for outside fluid, respectively, Btu/(hr)(ft^2)(°F)
h_{io}	Value of h_i when referred to the tube outside diameter
h_c	Heat-transfer coefficient for free convection, Btu/(hr)(ft^2)(°F)
h_r	Heat-transfer coefficient for radiation, Btu/(hr)(ft^2)(°F)
j_H	Factor for heat transfer, dimensionless
k	Thermal conductivity, Btu/(hr)(ft^2)(°F/ft)
L	Tube length or length of unmixed path, ft
n	Number of tube passes

ΔP	Pressure drop in general, psi
$\Delta P_T, \Delta P_r, \Delta P_t$	Total, tube, and return pressure drop, respectively, psi
P_T	Tube pitch, in.
R_d	Combined dirt factor, $(hr)(ft^2)(°F)/Btu$
Re, Re'	Reynolds number for heat transfer and pressure drop, dimensionless
s	Specific gravity, dimensionless
T_a, T_c	Average and caloric temperature of hot fluid, °F
t_1, t_2	Inlet and outlet temperatures of cold fluid, °F
t_a, t_c	Average and caloric temperatures of cold fluid, °F
t_f	Film temperature $\frac{1}{2}(t_w + t_a)$, °F
t_i	Temperature at end of first pass, °F
t_p, t_w	Inside and outside tube wall temperatures, °F
Δt	Temperature difference for heat transfer, °F
Δt_a	Temperature difference between tube wall and average fluid temperature, °F
U_C, U_D	Clean and design overall coefficients, $Btu/(hr)(ft^2)(°F)$
v	Specific volume, ft^3/lb
W, w	Weight flow of hot and cold fluid, lb/hr
z	Height, ft
α	Constant
β	Coefficient of thermal expansion, 1/°F
μ	Viscosity, centipoises $\times 2.42 = lb/(ft)(hr)$
μ_w	Viscosity at tube wall temperature, centipoises $\times 2.42 = lb/(ft)(hr)$
μ_f	Viscosity at film temperature, centipoises
ρ	Density, lb/ft^3
ϕ	$(\mu/\mu_w)^{0.14}$. For pressure drop in streamline flow, $(\mu/\mu_w)^{0.25}$
ψ	Free convection correction, Eq. (10.5), dimensionless

Subscripts (except as noted above)

f	Evaluated at the film temperature
s	Shell side
t	Tube side

CHAPTER 11

CALCULATIONS FOR PROCESS CONDITIONS

Optimum Process Conditions. The experience obtained from the calculation of existing tubular exchangers will presently be applied to cases in which only the process conditions are given. Before undertaking these calculations an investigation will be made to determine whether or not several related pieces of equipment can avail themselves of the overall process temperatures in an optimum manner. This is an economic affair similar to that for the optimum outlet-water temperature

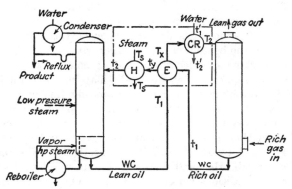

FIG. 11.1. Typical process employing heat recovery.

and the optimum use of exhaust steam discussed in Chap. 7. Often an exchanger operates in series with a cooler and a heater as shown in Fig. 11.1. The cooler regulates the final temperature of the hot fluid, and the heater adjusts the final temperature of the cold fluid[1] to the requirements of the next step in the process. The quantity of heat recovered in the exchanger alone greatly influences its size and cost, since the true temperature difference in the exchanger approaches zero as the heat recovery approaches 100 per cent. On the other hand, whatever

---
[1] This is actually the way to compensate and take advantage of the overperformance of an exchanger when it is clean if it has been designed for a dirty coefficient U_D. Unless equipped with a by-pass to reduce the flow through the exchanger, the temperature ranges of both fluids will exceed the process conditions. This is offset by throttling the steam and water in the heater and cooler and accordingly reducing the utility costs.

heat is not recovered in the exchanger must be removed or added through the use of additional steam in the heater and additional water in the cooler, which raises the initial costs of the two as well as the operating cost of the process. This arrangement suggests the presence of an optimum distribution of temperatures so that the fixed and the operating charges will combine to give a minimum.

Among a number of examples of heat recovery available in the chemical and power industries, that of Fig. 11.1 is typical. It shows the flow sheet of a simple vapor-recovery system such as is employed in the stripping of gasoline from natural gas, an absorption and distillation process, although the absorber and distilling column are not a part of the present analysis. Only the pertinent heat-flow lines are dash enclosed. The natural gas coming from the earth is laden with gasoline vapors which command a higher price when separated from the natural gas and condensed. The *rich gas* enters the absorber where it contacts an absorbent, usually a nonviscous oil, in which the gasoline vapors selectively dissolve. The outlet gas of reduced gasoline-vapor concentration is *lean gas*. The absorbent leaves the bottom of the absorber with the dissolved vapors as *rich oil*. It must next be fed to a distilling column where the gasoline and the oil are separated by steam distillation. The oil leaving the bottom of the distilling column is substantially free of the solute and therefore *lean oil*. Absorption is favored by low temperatures, while distillation requires higher temperatures so that the gasoline can be vaporized from the oil. The exchanger, heater, and cooler are represented in Fig. 11.1 by E, H, and CR, respectively. The temperatures to and from the absorber and distilling column will be considered fixed by equilibrium conditions whose definitions are beyond the scope of this study. The temperatures of the steam and water will also be fixed. The nucleus of the problem lies in determining the exchanger outlet temperatures T_x or t_y, either of which fixes the other, so that the total annual cost of the three heat-transfer items will be a minimum.

If C_E, C_H, and C_{CR} are the annual fixed charges per square foot for the exchanger, heater, and cooler, and if A_E, A_H, and A_{CR} are their surfaces, the five elements of cost are

1. Fixed charges of exchanger, $C_E A_E$
2. Fixed charges of heater, $C_H A_H$
3. Cost of steam, C_S, dollars/Btu
4. Fixed charges of cooler, $C_{CR} A_{CR}$
5. Cost of water, C_W, dollars/Btu

The surface of each unit, $A = Q/U\,\Delta t$, is dependent upon its true temperature difference. In each piece of equipment, however, the true temperature difference depends upon either T_x or t_y. To obtain an expression for the minimum cost it will be necessary to differentiate the total annual cost with respect to either T_x or t_y and solve for T_x or t_y after setting the derivative equal to zero. For simplicity assume that only true counterflow exchangers are employed so that the true temperature difference and LMTD of all units are identical and that all the *unit* fixed charges C_E, C_H, C_{CR} are the same and independent of the total number of square feet of each unit. The equation for the total annual cost C_T is obtained after eliminating t_y through the use of the dimensionless group R.

$$C_T = \frac{C_E WC(T_1 - T_x)}{U_E[(1-R)T_1 + (R-1)T_x]} \ln \frac{(1-R)T_1 - t_1 + RT_x}{T_x - t_1}$$

$$+ \frac{C_H wc}{U_H} \ln \frac{T_s - t_1 - RT_1 + RT_x}{T_s - t_2} + C_s \theta wc(t_2 - t_1 - RT_1 + RT_x)$$

$$+ \frac{C_{CR}WC(T_x - T_2)}{U_{CR}[(T_x - t_2') - (T_2 - t_1')]} \ln \frac{T_x - t_2'}{T_2 - t_1'} + C_W \theta WC(T_x - T_2) \qquad (11.1)$$

where U_E, U_H, and U_{CR} are the overall coefficients, θ is the total number of annual operating hours, and the other temperatures are indicated in Fig. 11.1. When differentiated with respect to T_x and set equal to zero, the equation for the optimum value of T_x is

$$0 = C_s \theta + C_W \theta + \frac{C_H}{U_H} \frac{1}{[R(T_x - T_1) + (T_s - t_1)]}$$

$$+ \frac{C_{CR}}{U_{CR}} \frac{(t_1' - t_2')}{[(T_x - T_2) + (t_1' - t_2')]} \ln \frac{(T_x - t_2')}{(T_2 - t_1')}$$

$$+ \frac{C_{CR}}{U_{CR}} \frac{(T_x - T_2)}{[(T_x - T_2) + (t_1' - t_2')]} \frac{1}{(T_x - t_2')}$$

$$- \frac{C_E}{U_E} \frac{(T_1 - t_1)}{[(T_1 - t_1) + R(T_x - T_1)]} \frac{1}{(T_x - T_1)} \qquad (11.2)$$

T_x can be obtained by a trial-and-error calculation.

Actually where the disposition of the temperatures justifies obtaining the optimum, the operation will be too large to permit the use of true counterflow in all of the equipment. When the flow pattern in the exchanger alone deviates from true counterflow and Δt is given by Eq. (7.37), the problem is somewhat more difficult to solve and less direct. However, if F_T is placed in the denominator of the last term in Eq. (11.2), a simplified trial-and-error solution can be obtained for a system using a 1-2, 2-4, etc., exchanger. Equation (11.2) is not particularly useful, however, unless extensive data are available on the installed costs and fixed charges of exchangers, since the cost per square foot of surface also varies with the size of the exchanger. Sieder[1] has shown in Fig. 11.2 how the

[1] Sieder, E. N., *Chem. Met. Eng.*, **46**, 322–325 (1939).

relative cost (per square foot) of surface decreases from small to large exchangers. The fixed charges are usually arrived at as a percentage (30 per cent) of the initial cost per square foot, and this varies so much

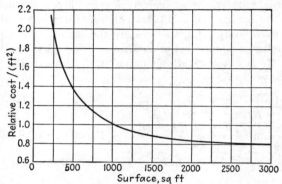

Fig. 11.2. Cost of tubular surface vs. size of exchanger. (*Sieder, Chemical Engineering.*)

with the size of the final exchanger that several successive trials are required to establish the proper range of individual costs.

Other obstacles are also encountered in the solution of a general equation for the optimum. If a typical problem is solved for a system with a 1-2 exchanger and the calculated value of T_x comes out to be less than the practical limit of $F_T = 0.75 - 0.80$ in the 1-2 exchanger, the entire calculation, though valid, must be repeated using a different exchanger. The task of obtaining the optimum process temperatures can be achieved more readily by graphical means. As the temperature T_x is varied, there are two opposed costs as follows: As T_x is increased, the cost of the utilities increases but the initial cost and fixed charges on the exchanger decrease. By assuming several values of T_x the required sizes of all the equipment can be computed for each assumption. From the overall heat balances the operating costs of the utilities can also be obtained. The total annual costs are then the sum of the two costs as plotted in Fig. 11.3 with the optimum corresponding to the point indicated.

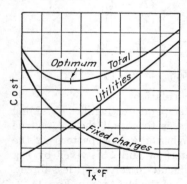

Fig. 11.3. Optimum recovery temperature.

The Optimum Exchanger. The factors which are favorable for the attainment of higher film coefficients for fluids in exchangers also increase their pressure drops. For turbulent flow in tubes as given by Eq. (6.2) the film coefficient varies with $G_t^{0.8}$, whereas the pressure drop in Eq. (7.45) varies with G_t^2. This means that if the tube-side coefficient is the controlling coefficient and if the design of the exchanger can be altered so as to increase the tube-side mass velocity, the overall coefficients U_C, U_D also increase and the size of the exchanger can be reduced accordingly. Offsetting this advantage is an increase in the pressure drop and pumping power costs for the greater mass velocity. As in the calculation of the optimum process conditions, there is an optimum exchanger capable of fulfilling process conditions with a minimum annual cost. The achievement of this design, however, requires an exchanger capable of providing the optimum fluid-flow velocities on the shell as well as the tube sides. This would frequently entail the use of an odd number of tube passes or an odd tube length, which is inconsistent with industrial practices. McAdams[1] has given an excellent résumé of the equations and approximations required for establishing the optimum exchanger to which the reader is referred.

Rating an Exchanger. In Chaps. 7 through 10 the calculations were carried out on existing exchangers. In the application of the Fourier equation to an existing exchanger, Q was determined from the heat balance; A from the number, outside diameter, and length of the tubes; and Δt from $F_T \times$ LMTD, permitting the solution for U_D. From the fluid-flow conditions, h_o, h_{io}, U_C, and the pressure drops were calculated. The criterion of performance R_d was then obtained from U_D and U_C. When there is no available exchanger and only the process conditions are known, calculation may proceed in an orderly way by assuming the existence of an exchanger and testing it as in previous examples for a suitable dirt factor and pressure drops.

To prevent the loss of considerable time, rational methods of assuming an exchanger should be developed. By returning to the component parts of the Fourier equation $Q = U_D A \, \Delta t$, the heat load Q is seen to be fixed by the process conditions, while Δt is obtained by assuming a fluid-flow pattern. Thus if it is desired to set up a 1-2 exchanger, and if the process temperatures give $F_T > 0.75 - 0.80$, the remaining unknowns are U_D and A. The design or dirty coefficient U_D is in turn related by a reasonable dirt factor to U_C, which reflects the heat-transfer characteristics of the two fluids. Previous examples in the text suggest that different numerical film coefficients may be expected within definite ranges in well-designed exchangers for different classes of fluids. It is also apparent that, except where both coefficients are approximately

[1] McAdams, W. H., "Heat Transmission," 2d ed., pp. 363–367, McGraw-Hill Book Company, Inc., New York, 1942.

equal, the lower film coefficient determines the range of U_C and U_D. In the light of this experience if a trial value of U_D is assumed and then substituted into the Fourier equation to supplement calculated values of Q and Δt, it permits the trial calculation of A. To facilitate the use of reasonable trial values of U_D approximate overall coefficients have been presented in Appendix Table 8 for common pairs of fluids.

When the value of A is combined with tube length and pitch preferences, the tube counts in Appendix Table 9 become a catalogue of all the possible exchanger shells, from which only one will usually best fulfill the process conditions. Having decided which fluid will tentatively flow in the tubes, the trial number of tube passes can be approximated by a consideration of the quantity of fluid flowing in the tubes and the number of tubes corresponding to the trial value of A. The trial mass velocity should fall somewhere between 750,000 and 1,000,000 lb/(hr)(ft^2) for fluids with allowable tube-side pressure drops of 10 psi.

If the trial number of tube passes has been assumed incorrectly, a change in the total number of passes changes the total surface which is contained in a given shell, since the number of tubes for a given shell diameter varies with the number of tube-pass partitions. The assumed number of tube passes for the trial surface is satisfactory if it gives a value of h_i above U_D and a pressure drop not exceeding the allowable pressure drop for the fluid. One may next proceed to the shell side by assuming a trial baffle spacing which can be varied, if in error, over a wide range without altering h_i, A, or Δt as computed previously for the tube side. It is always advantageous therefore to compute the tube side first to validate the use of a particular shell.

In calculating an exchanger the best exchanger is the smallest one with a standard layout which just fulfills the dirt-factor and pressure-drop requirements. There are only a few limitations to be considered. It is still assumed that there is no advantage in using less than the allowable pressure drop and that, in accordance with Fig. 28, 25 per cent cut segmental baffles will be employed within the minimum and maximum spacing. The extremes of the spacing range are

Maximum spacing, $B = $ ID of shell, in. (11.3)

Minimum spacing, $B = \dfrac{\text{ID of shell}}{5}$, or 2 in., whichever is larger (11.4)

These limitations stem from the fact that at wider spacings the flow tends to be axial rather than across the bundle and at closer spacings there is excessive leakage between the baffles and the shell. Owing to the convention of placing the inlet and outlet nozzles on opposite sides of the shell, the end baffles may not exactly conform to the chosen spacing for an even number of baffles and an odd number of crosses. When using a close convenient spacing, if an odd number of baffles is indicated, the pressure-drop and heat-transfer coefficients can be calculated for the chosen spacing, although one baffle will be omitted by respacing the extreme end baffles.

The different combinations of the number of tube passes and the baffle pitch which can be employed in a given shell permit the variation of the mass velocities and film coefficients over broad limits. The number of tube passes can be varied from two to eight and in larger shells to sixteen. As indicated above, the shell-side mass velocity can be altered fivefold between the minimum and maximum baffle spacings. It is desirable to bear in mind the latitude this permits in the event that the first trial baffle spacing and tube passes have been wide of fulfilling the process conditions. In 1-2 exchangers the least performance is obtained with two tube passes and the maximum baffle space. For the tube side in turbulent flow

$$h_i \propto G_t^{0.8}$$
$$\Delta P_t \propto G_t^2 n L$$

where nL is the total length of path. Going to eight tube passes in the same shell inside diameter the changes incurred are

$$\frac{h_{i(8\ \text{passes})}}{h_{i(2\ \text{passes})}} = \left(\frac{8}{2}\right)^{0.8} = \frac{3}{1}$$

but

$$\frac{\Delta P_{t(8\ \text{passes})}}{\Delta P_{t(2\ \text{passes})}} = \frac{8^2 \times 8 \times 1}{2^2 \times 2 \times 1} = \frac{64}{1}$$

or although the heat-transfer coefficient can be increased threefold, the pressure drop must be increased 64 times to accomplish it. For streamline flow the expenditure of the larger amount of pumping energy will increase the tube-side coefficient only by

$$\frac{h_{i(8\ \text{passes})}}{h_{i(2\ \text{passes})}} = \left(\frac{8}{2}\right)^{\frac{1}{3}} = \frac{1.58}{1}$$

provided the fluid is in streamline flow in both cases. The shell side may be represented approximately by

$$h_o \propto G_s^{0.5}$$
$$\Delta P_s \propto G_s^2(N+1)$$

where N is the number of baffles and $N+1$ the number of bundle crosses. The changes for the shell side between minimum and maximum baffle spacings are

$$\frac{h_{o,\text{min}}}{h_{o,\text{max}}} = \left(\frac{5}{1}\right)^{0.5} = \frac{2.23}{1}$$

but
$$\frac{\Delta P_{s,\min}}{\Delta P_{s,\max}} = \frac{5^2 \times 5}{1^2 \times 1} = \frac{125}{1}$$

Offsetting this, however, is the fact that the shell side yields a higher order of film coefficients for the smaller of the two streams if there is a great difference in their weight flow rates.

Through the use of the overall coefficients suggested in Table 8 and a judicious analysis of the summary of the first trial assumptions, it should be possible to obtain the most suitable exchanger on the second trial.

Tube Standards. There are numerous advantages in standardizing the outside diameter, gage, and length of the tubes used in a plant.

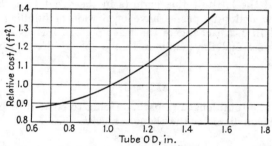

Fig. 11.4. Cost of tubular surface vs. tube outside diameter. (*Sieder, Chemical Engineering.*)

Standardization reduces the number of sizes and lengths which must be carried in storage for the replacement of tubes which develop leaks. It also reduces the number of installing and cleaning tools required for maintenance. Pitch standards have already been discussed in Chap. 7, but the selection of the tube diameter has an economic aspect. Obviously the smaller the diameter of the tube the smaller the shell required for a given surface, the greater the value of h_i, and the smaller the first cost. The nature of the variation in cost, taken from Sieder, is shown in Fig. 11.4. The difference is small between the use of ¾ and 1 in. OD tubes. The cost per square foot rises greatly, however, as the diameters increase above 1 in. OD. Similarly the longer the tubes the smaller the shell diameter for a given surface, and from Fig. 11.5 the cost variation between the use of 12-, 16-, and 20-ft tubes is not very great although 8-ft tubes sharply increase the first cost.

The lower cost of surface obtained from the smaller outside diameter and greater length tubes is offset by the fact that maintenance and particularly cleaning are costlier with long tubes of small outside diameter. If the tubes are too small, under ¾ in. OD, there are too many to be cleaned, and there is less facility in handling and cleaning the smaller tubes. If the tubes are too long, it is difficult to remove the bundle and plot space must be allocated not only for the exchanger itself but also for the withdrawal of the bundle. Long tubes are also very difficult to replace, especially where the baffles are closely spaced. It is difficult to obtain comparative maintenance data per square foot as a function of tube diameter or length, since few industrial users appear to employ an assortment of tubes or have kept cost data. It may be significant

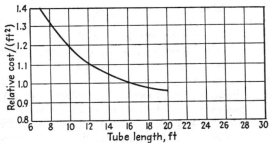

FIG. 11.5. Cost of tubular surface vs. tube length. (*Sieder, Chemical Engineering.*)

that for exchanger work with fluids of ordinary fouling characteristics the ¾ and 1 in. OD tubes are employed most frequently. For chemical evaporators, boilers, and fouling services, larger tubes are customary.

The Calculation and Design of an Exchanger. The outline for design follows:

Given:

Process conditions:

Hot fluid: T_1, T_2, W, C, s, μ, k, R_d, ΔP
Cold fluid: t_1, t_2, w, c, s, μ, k, R_d, ΔP

The tube length, outside diameter, and pitch will be specified by plant practice or may be determined from the suggestions in Chap. 7.

(1) Heat balance, $Q = WC(T_1 - T_2) = wc(t_2 - t_1)$
(2) True temperature difference, Δt:
LMTD:
$$R = \frac{T_1 - T_2}{t_2 - t_1} \qquad S = \frac{t_2 - t_1}{T_1 - t_1} \qquad (5.14)$$
F_T from Figs. 18 through 23
$\Delta t = \text{LMTD} \times F_T \qquad (7.42)$
(3) Caloric temperatures, T_c and t_c \hfill (5.28) and (5.29)

Trial 1:

For the exchanger:

(a) *Assume* a tentative value of U_D with the aid of Table 8, and compute the surface from $A = Q/U_D \, \Delta t$. It is always better to assume U_D too high than too low, as this practice ensures arriving at the minimum surface. Determine the corresponding number of tubes using Table 10.

(b) *Assume* a plausable number of tube passes for the pressure drop allowed, and select an exchanger for the nearest number of tubes from the tube counts of Table 9.

(c) Correct the tentative U_D to the surface corresponding to the actual number of tubes which may be contained in the shell.

The performance calculation for the film coefficients should start with the tube side. If the tube-side film coefficient is relatively greater than U_D and the pressure-drop allowance is reasonably fulfilled and not exceeded, the calculation can proceed to the shell side. Whenever the number of tube passes is altered, the surface in the shell is also altered, changing the value of A and U_D. For the remainder of the calculation shown here it is assumed that the cold fluid flows in the tubes as it does in a majority but not necessarily all cases.

Hot fluid: shell side	Cold fluid: tube side
(4′) Assume a plausible baffle spacing for the pressure drop allowed. Flow area, $a_s = \text{ID} \times C'B/144P_T$ ft² [Eq. (7.1)]	(4) Flow area, a_t: Flow area per tube a_t' From Table 10 $a_t = N_t a_t'/144n$, ft² [Eq. (7.48)] ID will be obtained from Table 10.
(5′) Mass vel, $G_s = W/a_s$ lb/(hr)(ft²) [Eq. (7.2)]	(5) Mass vel, $G_t = w/a_t$ lb/(hr)(ft²)
(6′) $Re_s = D_e G_s/\mu$ [Eq. (7.3)] Obtain D_e from Fig. 28 or compute from Eq (7.4). Obtain μ at T_c	(6) $Re_t = DG_t/\mu$ Obtain D from Table 10. Obtain μ at t_c.
(7′) j_H from Fig. 28	(7) j_H from Fig. 24
(8′) At T_c obtain c Btu/(lb)(°F) and k Btu/(hr)(ft²)(°F/ft) $(c\mu/k)^{1/3}$	(8) At t_c obtain c Btu/(lb)(°F) and k Btu/(hr)(ft²)(°F/ft) $(c\mu/k)^{1/3}$
(9′) $h_o = j_H \dfrac{k}{D_e}\left(\dfrac{c\mu}{k}\right)^{1/3} \phi_s$ [Eq. (6.15)]	(9) $h_i = j_H \dfrac{k}{D}\left(\dfrac{c\mu}{k}\right)^{1/3} \phi_t$ [Eq. (6.15a)]
(10′) Tube-wall temp, t_w $t_w = t_c + \dfrac{h_o}{h_{io} + h_o}(T_c - t_c)$ Eq. [5.31]	(10) $\dfrac{h_{io}}{\phi_t} = \dfrac{h_i}{\phi_t}\dfrac{\text{ID}}{\text{OD}}$ [Eq. (6.5)]
(11′) Obtain μ_w and $\phi_s = (\mu/\mu_w)^{0.14}$ [Fig. 24]	(11) Obtain μ_w and $\phi_t = (\mu/\mu_w)^{0.14}$ [Fig. 24]
(12′) Corrected coefficient, $h_o = \dfrac{h_o \phi_s}{\phi_s}$ [Eq. (6.36)]	(12) Corrected coefficient. $h_{io} = \dfrac{h_{io}\,\phi_t}{\phi_t}$ [Eq. (6.37)]
Check pressure drop. If unsatisfactory, assume a new baffle spacing.	Check pressure drop. If unsatisfactory, assume a new pass arrangement.

CALCULATIONS FOR PROCESS CONDITIONS

Pressure Drop

(1') For Re_s in (6') obtain f [Fig. 29] | (1) For Re_t in (6) obtain f [Fig. 26]

(2') No. of crosses, $N + 1 = 12L/B$ [Eq. (7.43)]

$$\Delta P_s = \frac{fG_s^2 D_s (N+1)}{5.22 \times 10^{10} D_e s \phi_s} \quad [\text{Eq. (7.44)}]$$

(2) $\Delta P_t = \dfrac{fG_t^2 L n}{5.22 \times 10^{10} D s \phi_t}$ [Eq. (7.45)]

(3) $\Delta P_r \equiv \dfrac{4n}{s} \dfrac{V^2}{2g'}$ [Eq. (7.46)]

(4) $\Delta P_T = \Delta P_t + \Delta P_r$ [Eq. (7.47)]

If both sides are satisfactory for film coefficients and pressure drop, the trial may be concluded.

(13) Clean overall coefficient U_C:

$$U_C = \frac{h_{io} h_o}{h_{io} + h_o} \qquad (6.38)$$

(14) Dirt factor R_d: U_D has been obtained in (c) above.

$$R_d = \frac{U_C - U_D}{U_C U_D} \qquad (6.13)$$

The calculation of a number of exchangers for typical sensible-heat-transfer conditions are given in this chapter. Each presents a different aspect of design. Together they should provide the perspective necessary for meeting a variety of applications encountered in modern industry. Since the method of approach involves trial-and-error calculations, the analyses and comments included in each solution should reduce the time required for subsequent calculations.

Example 11.1. Calculation of a Straw Oil-Naphtha Exchanger. 29,800 lb/hr of a 35°API light oil at 340°F is used to preheat 103,000 lb/hr of 48°API naphtha from 200 to 230°F. The viscosity of the oil is 5.0 centipoises at 100°F and 2.3 centipoises at 210°F. The viscosity of the naphtha is 1.3 centipoises at 100°F and 0.54 centipoise at 210°F. Pressure drops of 10 psi are allowable.

Because the oil may tend to deposit residues, allow a combined dirt factor of 0.005 and use square pitch. Plant practice employs ¾ in. OD, 16 BWG tubes 16'0" long wherever possible.

Solution:

(1) Heat balance:
    Straw oil, $Q = 29{,}800 \times 0.58(340 - 240) = 1{,}730{,}000$ Btu/hr
    Naphtha, $Q = 103{,}000 \times 0.56(230 - 200) = 1{,}730{,}000$ Btu/hr

(2) Δt:

	Hot Fluid		Cold Fluid	Diff.
	340	Higher Temp	230	110
	240	Lower Temp	200	40
	100	Differences	30	70

LMTD = 69.3°F (5.14)

$R = \dfrac{100}{30} = 3.3 \qquad S = \dfrac{30}{340 - 200} = 0.214 \qquad F_T = 0.885$ (Fig. 18)

A 1-2 exchanger will be satisfactory.

$\Delta t = 0.885 \times 69.3 = 61.4°F$ (7.42)

(3) T_c and t_c:

$\dfrac{\Delta t_c}{\Delta t_h} = 0.364 \qquad K_c = 0.23$ (straw oil controlling) (Fig. 17)

$F_c = 0.405$
$T_c = 240 + 0.405 \times 100 = 280.5°F$ (5.28)
$t_c = 200 + 0.405 \times 30 = 212°F$ (5.29)

Trial:

(a) Assume $U_D = 70$: From Table 8 a value of U_D between 60 and 75 should be the maximum expected. It is always better to assume U_D too high than too low so that the final exchanger will just fulfill the requirements. Place the small stream in the shell.

$A = \dfrac{Q}{U_D \Delta t} = \dfrac{1{,}730{,}000}{70 \times 61.4} = 403 \text{ ft}^2$

$a'' = 0.1963 \text{ ft}^2/\text{lin ft}$ (Table 10)

Number of tubes $= \dfrac{403}{16'0'' \times 0.1963} = 129$ (Table 10)

(b) Assume two tube passes: The quantity of tube-side fluid is very large for the small heat load and moderately large Δt and will otherwise cause difficulty in meeting the allowable ΔP if too many tube passes are employed.

From the tube counts (Table 9): 129 tubes, two passes, ¾ in. OD on 1-in. square pitch

Nearest count: 124 tubes in a 15¼ in. ID shell

(c) Corrected coefficient U_D:

$A = 124 \times 16'0'' \times 0.1963 = 390 \text{ ft}^2$ (Table 10)

$U_D = \dfrac{Q}{A \Delta t} = \dfrac{1{,}730{,}000}{390 \times 61.4} = 72.3$

Hot fluid: shell side, straw oil

(4′) Flow area, a_s: Since the minimum baffle space will provide the greatest value of h_o, assume
$B = \text{ID}/5 = 15.25/5$, say 3.5 in.
[Eq. (11.4)]
$a_s = \text{ID} \times C'B/144P_T$ [Eq. (7.1)]
$= 15.25 \times 0.25 \times 3.5/144 \times 1$
$= 0.0927 \text{ ft}^2$

(5′) Mass vel, $G_s = W/a_s$ [Eq. (7.2)]
$= 29{,}800/0.0927 = 321{,}000 \text{ lb}/(\text{hr})(\text{ft}^2)$

(6′) At $T_c = 280.5°F$,
$\mu = 1.5 \times 2.42 = 3.63 \text{ lb}/(\text{ft})(\text{hr})$
[Fig. 14]

$D_e = 0.95/12 = 0.0792 \text{ ft}$ [Fig. 28]
$Re_s = D_e G_s/\mu$
$= 0.0792 \times 321{,}000/3.63 = 7000$

Cold fluid: tube side, naphtha

(4) Flow area, $a_t' = 0.302 \text{ in.}^2$
[Table 10]
$a_t = N_t a_t'/144n$ [Eq. (7.48)]
$= 124 \times 0.302/144 \times 2 = 0.130 \text{ ft}^2$

(5) Mass vel, $G_t = w/a_t$
$= 103{,}000/0.130 = 793{,}000 \text{ lb}/(\text{hr})(\text{ft}^2)$

(6) At $t_c = 212°F$,
$\mu = 0.54 \times 2.42 = 1.31 \text{ lb}/(\text{ft})(\text{hr})$
$D = 0.62/12 = 0.0517 \text{ ft}$ [Table 10]
$Re_t = DG_t/\mu$
$= 0.0517 \times 793{,}000/1.31 = 31{,}300$

CALCULATIONS FOR PROCESS CONDITIONS

Hot fluid: shell side, straw oil

(7') $j_H = 46$

(8') For $\mu = 1.5$ cp and 35°API   [Fig. 16]

$k(c\mu/k)^{1/3} = 0.224$ Btu/(hr)(ft²)(°F/ft)

(9') $h_o = j_H \dfrac{k}{D_s}\left(\dfrac{c\mu}{k}\right)^{1/3} \phi_s$ [Eq. (6.15b)]

$\dfrac{h_o}{\phi_s} = \dfrac{46 \times 0.224}{0.0792} = 130$

(10'), (11'), (12') Omit the viscosity correction for the trial or $\phi_s = 1.0$.

$h_o = \dfrac{h_o}{\phi_s} = 130$ Btu/(hr)(ft²)(°F)

Cold fluid: tube side, naphtha

(7) $j_H = 102$ [Fig. 24]

(8) For $\mu = 0.54$ cp and 48°API   [Fig. 16]

$k(c\mu/k)^{1/3} = 0.167$ Btu/(hr)(ft²)(°F/ft)

(9) $h_i = j_H \dfrac{k}{D}\left(\dfrac{c\mu}{k}\right)^{1/3} \phi_t$ [Eq. (6.15a)]

$\dfrac{h_i}{\phi_t} = 102 \times 0.167/0.0517 = 329$

(10) $\dfrac{h_{io}}{\phi_t} = \dfrac{h_i}{\phi_t} \times \dfrac{\text{ID}}{\text{OD}} = 329 \times \dfrac{0.62}{0.75} = 272$ [Eq. (6.5)]

(11), (12) Omit the viscosity correction for the trial or $\phi_t = 1.0$.

$h_{io} = \dfrac{h_{io}}{\phi_t} = 272$ Btu/(hr)(ft²)(°F)

Proceed with the pressure drop calculation.

Pressure Drop

(1') For $Re_s = 7000$,

$f = 0.00225$ ft²/in.²   [Fig. 29]

$s = 0.76$ [Fig. 6]

(2') No. of crosses, $N + 1 = 12L/B$ [Eq. (7.43)]

$= 12 \times 16/3.5 = 55$

$s = 0.76$ [Fig. 6]

$D_s = 15.25/12 = 1.27$ ft

(3') $\Delta P_s = \dfrac{fG_s^2 D_s(N+1)}{5.22 \times 10^{10} D_e s \phi_s}$ [Eq. (7.44)]

$= \dfrac{0.00225 \times 321{,}000^2 \times 1.27 \times 55}{5.22 \times 10^{10} \times 0.0792 \times 0.76 \times 1.0}$

$= 5.2$ psi

(1) For $Re_t = 31{,}300$,

$f = 0.0002$ ft²/in.²   [Fig. 26]

$s = 0.72$ [Fig. 6]

(2) $\Delta P_t = \dfrac{fG_t^2 Ln}{5.22 \times 10^{10} Ds\phi_t}$ [Eq. (7.45)]

$= \dfrac{0.00020 \times 792{,}000^2 \times 16 \times 2}{5.22 \times 10^{10} \times 0.0517 \times 0.72 \times 1.0}$

$= 2.1$ psi

(3) and (4) may be omitted for the trial. Now proceed to the shell side.

(13) Clean overall coefficient U_C:

$$U_C = \dfrac{h_{io}h_o}{h_{io} + h_o} = \dfrac{272 \times 130}{272 + 130} = 88.2 \text{ Btu/(hr)(ft}^2\text{)(°F)} \quad (6.38)$$

(14) Dirt factor R_d: U_D from (c) is 72.3.

$$R_d = \dfrac{U_C - U_D}{U_C U_D} = \dfrac{88.2 - 72.3}{88.2 \times 72.3} = 0.0025 \text{ (hr)(ft}^2\text{)(°F)/Btu} \quad (6.13)$$

Summary

130	h outside	272
U_C	88.2	
U_D	72.3	
R_d Calculated	0.0025	
R_d Required	0.0050	
5.2	Calculated ΔP	2.1
10.0	Allowable ΔP	10.0

Discussion. The first trial is disqualified because of failure to meet the required dirt factor. What conclusions may be drawn so that the next trial will produce the satisfactory exchanger? Could any advantage be gained by reversing the streams? Obviously the film coefficient for the distillate, which is controlling, would drop considerably if the streams were reversed. Could four passes be used for the tubes? Doubling the number of tube passes would approximately double the mass velocity and give *eight* times the tube-side pressure drop thereby exceeding the allowable ΔP. All the assumptions above have been reasonable. The exchanger is simply a little too small, or in other words, the value assumed for U_D must be reduced. It will be necessary to proceed anew.

Trial 2: Assume $U_D = 60$, two tube passes and the minimum shell baffle space. Proceeding as above and carrying the viscosity correction and pressure drops to completion, the new summary is given using a 17¼ in. ID shell with 166 tubes on two passes and a 3.5-in. baffle space.

Summary

115.5	h outside	213
U_C	74.8	
U_D	54.2	
R_d Calculated	0.005	
R_d Required	0.005	
4.7	Calculated ΔP	2.1
10.0	Allowable ΔP	10.0

The final exchanger will be

Shell side
ID = 17¼ in.
Baffle space = 3.5 in.
Passes = 1

Tube side
Number and length = 166, 16'0''
OD, BWG, pitch = ¾ in., 16 BWG, 1-in. square
Passes = 2

CALCULATIONS FOR PROCESS CONDITIONS 235

Example 11.2. Calculation of a Lean-oil–Rich-oil Exchanger. 84,348 lb/hr of a 35°API lean absorption oil in a process identical with Fig. 11.1 leaves a stripping column to transfer its heat to 86,357 lb/hr of rich oil leaving the absorber at 100°F with a gravity of closely 36°API at 60°F. The range for the lean oil will be from 350 to 160°F, and the outlet temperature of the rich oil will be 295°F. The viscosity of the oil is 2.6 centipoises at 100°F and 1.15 centipoises at 210°F. Pressure drops of 10 psi are available, and in accordance with Table 12, a combined dirt factor of 0.004 should be allowed.

Plant practice again employs ¾ in. OD, 16 BWG tubes 16'0" long and laid out on square pitch.

Solution:

(1) Heat balance:
Lean oil, $Q = 84{,}438 \times 0.56(350 - 160) = 8{,}950{,}000$ Btu/hr
Rich oil, $Q = 86{,}357 \times 0.53(295 - 100) = 8{,}950{,}000$ Btu/hr

(2) Δt:

	Hot Fluid		Cold Fluid	Diff.
	350	Higher Temp	295	55
	160	Lower Temp	100	60
	190	Differences	195	5

LMTD = 57.5°F (5.14)

$R = \dfrac{190}{195} = 0.975 \qquad S = \dfrac{195}{350 - 100} = 0.78$ (7.18)

1-2 exchanger, F_T = inoperable (Fig. 18)
2-4 exchanger, F_T = inoperable (Fig. 19)
3-6 exchanger, F_T = 0.725 (Fig. 20)
4-8 exchanger, F_T = 0.875 (Fig. 21)

A 4-8 exchanger arrangement will be required. This may be met by four 1-2 exchangers in series or two 2-4 exchangers in series. The latter will be used.

$\Delta t = 0.875 \times 57.5 = 50.3°F$ (7.42)

(3) T_c and t_c:
$\dfrac{\Delta t_c}{\Delta t_h} = 1.09$
$K_c = 0.32$ (Fig. 17)
$F_c = 0.48$
$T_c = 160 + 0.48 \times 190 = 251°F$ (5.28)
$t_c = 100 + 0.48 \times 195 = 193.5°F$ (5.29)

Trial:

(a) Assume $U_D = 50$: Although both oils and quantities are almost identical, the temperature range of the cold rich oil and correspondingly greater viscosity will make the rich oil controlling. For this reason, allowable pressure drops being equal, the cold fluid should be placed in the shell. The coefficients will be lower than those of Table 8, since the pressure drops are more difficult to meet in a 4-8 exchanger and mass velocities must accordingly be kept down. In Example 11.1 with a similar controlling oil the value of U_C was approximately 75 with a minimum baffle spacing which did not completely take advantage of the allowable pressure drop. The assumption of $U_D = 50$ is a compromise between medium and heavy organics and will probably be high but will help to establish the correct unit on the following trial.

$$A = \frac{Q}{U_D \Delta t} = \frac{8,950,000}{50 \times 50.3} = 3560 \text{ ft}^2$$

Use two 2-4 exchangers in series with removable longitudinal baffles.
Number of tubes per shell, $N_t = 3560/2 \times 16'0'' \times 0.1963 = 567$ (Table 10)

(b) Assume six tube passes: From previous problems a mass velocity of 700,000 gave satisfactory tube-side pressure drops. Since the number of tube passes in two units will be greater, a maximum of about 450,000 should be employed.

Six passes (flow area/tube, $a_t' = 0.302$ in.²)
$G = w/a_t = w144n/N_t a_t' = 84.438 \times 144 \times 6/567 \times 0.302 = 426,000 \text{ lb/(hr)(ft}^2)$
From the tube counts (Table 9): 567 tubes, six passes, ¾ in. OD on 1-in. square pitch
Nearest count: 580 tubes in a 31 in. ID shell

(c) Corrected coefficient U_D:
$a'' = 0.1963$ ft²/lin ft (Table 10)
$A = 2 \times 580 \times 16'0'' \times 0.1963 = 3640 \text{ ft}^2$

$$U_D = \frac{Q}{A \Delta t} = \frac{8,950,000}{3640 \times 50.3} = 49.0$$

Hot fluid: tube side, lean oil

(4) Flow area, a_t:
$a_t' = 0.302$ in.² [Table 10]
$a_t = N_t a_t'/144n$ [Eq. (7.48)]
$= 580 \times 0.302/144 \times 6 = 0.203$ ft²

(5) Mass vel, $G_t = W/a_t$
$= 84,438/0.203 = 416,000 \text{ lb/(hr)(ft}^2)$
(6) At $T_c = 251°F$, $\mu = 0.88 \times 2.42$
$= 2.13 \text{ lb/(ft)(hr)}$ [Fig. 14]
$D = 0.62/12 = 0.0517$ ft [Table 10]
$Re_t = DG_t/\mu$
$= 0.0517 \times 416,000/2.13 = 10,100$

(7) $j_H = 36.5$ [Fig. 24]
(8) For $\mu = 0.88$ cp and 35°API
$k(c\mu/k)^{\frac{1}{3}} = 0.185 \text{ Btu/(hr)(ft}^2)(°F/ft)$ [Fig. 16]

(9) $h_i = j_H \frac{k}{D}\left(\frac{c\mu}{k}\right)^{\frac{1}{3}} \phi_t$ [Eq. (6.15a)]

$\frac{h_i}{\phi_t} = 36.5 \times \frac{0.185}{0.0517} = 130$

(10) $h_{io} = h_i \times \frac{\text{ID}}{\text{OD}}$ [Eq. (6.5)]
$= 130 \times 0.62/0.75 = 107$
(11) (12) Omit the viscosity correction for the trial, $\phi_t = 1.0$.

$h_{io} = \frac{h_{io}}{\phi_t} = 107 \text{ Btu/(hr)(ft}^2)(°F)$

Cold fluid: shell side, rich oil

(4') Flow area, a_s: Since the quantity of fluid is large, any baffle spacing may be arbitrarily assumed.
Assume $B = 12$ in.
$a_s = \text{ID} \times C'B/144P_T$ [Eq. (7.1)]
$= \frac{1}{2}(31 \times 0.25 \times 12/144 \times 1.0)$
$= 0.323$ ft² (2-4 exchanger)

(5') Mass vel, $G_s = w/a_s$ [Eq. (7.2)]
$= 86,357/0.323 = 267,000 \text{ lb/(hr)(ft}^2)$
(6') At $t_c = 193.5°F$,
$\mu = 1.30 \times 2.42 = 3.15 \text{ lb/(ft)(hr)}$
  [Fig. 14]
$D_e = 0.95/12 = 0.0792$ ft [Fig. 28]
$Re_s = D_e G_s/\mu$ [Eq. (7.3)]
$= 0.0792 \times 267,000/3.15 = 6,720$

(7') $j_H = 45$ [Fig. 28]
(8') For $\mu = 1.30$ cp and 35°API
$k(c\mu/k)^{\frac{1}{3}} = 0.213 \text{ Btu/(hr)(ft}^2)(°F/ft)$ [Fig. 16]

(9') $h_o = j_H \frac{k}{D_e}\left(\frac{c\mu}{k}\right)^{\frac{1}{3}} \phi_s$ [Eq. (6.15)]

$\frac{h_o}{\phi_s} = 45 \times \frac{0.213}{0.0792} = 121$

(10') Omit the viscosity correction for the trial, $\phi_s = 1.0$.

$h_o = \frac{h_o}{\phi_t} = 121 \text{ Btu/(hr)(ft}^2)(°F)$

CALCULATIONS FOR PROCESS CONDITIONS

Pressure Drop

(1) For $Re_t = 10,100$,
$f = 0.00027$ ft^2/in.^{2} [Fig. 26]
$s = 0.77$ [Fig. 6]

(2) $\Delta P_t = \dfrac{fG_t^2 Ln}{5.22 \times 10^{10} Ds\phi_t}$ [Eq. (7.45)]

$= \dfrac{0.00027 \times 416{,}000^2 \times 16 \times 6 \times 2}{5.22 \times 10^{10} \times 0.0517 \times 0.77 \times 1.0}$
$= 4.3$ psi

(3) $G_t = 416{,}000$, $V^2/2g' = 0.024$ [Fig. 27]

$\Delta P_r = \dfrac{4n}{s}\dfrac{V^2}{2g'} = \dfrac{4 \times 2 \times 6}{0.77} \times 0.024$

$= 1.5$ psi [Eq. (7.46)]
$\Delta P_T = \Delta P_t + \Delta P_R = 4.3 + 1.5 = 5.8$ psi [Eq. (7.47)]

The pressure drop suggests the possibility of using eight passes but a rapid check shows the pressure drop for eight passes would exceed 10 psi.
Now proceed to the shell side.

(1′) For $Re_s = 6720$,
$f = 0.0023$ ft^2/in.^{2} [Fig. 29]
$s = 0.79$ [Fig. 6]

(2′) No. of crosses, $N + 1 = 12L/B$ [Eq. (7.42)]
$= 2 \times 2 \times 12 \times 16/12 = 64$
$D_s = 31/12 = 2.58$ ft

(3′) $\Delta P_s = \dfrac{fG_s^2 D_s(N+1)}{5.22 \times 10^{10} D_e s\phi_s}$ [Eq. (7.44)]

$= \dfrac{0.0023 \times 267{,}000^2 \times 2.58 \times 64}{5.22 \times 10^{10} \times 0.0792 \times 0.79 \times 1.0}$
$= 8.3$ psi

(13) Clean overall coefficient U_C:

$$U_C = \dfrac{h_{io}h_o}{h_{io} + h_o} = \dfrac{107 \times 121}{107 + 121} = 56.8 \text{ Btu/(hr)(ft}^2)(°F) \quad (6.38)$$

(14) Dirt factor R_d: U_D from (c) is 49.0

$$R_d = \dfrac{U_C - U_D}{U_C U_D} = \dfrac{56.8 - 49.0}{56.8 \times 49.0} = 0.0028 \text{ (hr)(ft}^2)(°F)/\text{Btu} \quad (6.13)$$

Summary

107	h outside	121
U_C	56.8	
U_D	49.0	
R_d Calculated 0.0028		
R_d Required 0.0040		
5.8	Calculated ΔP	8.3
10.0	Allowable ΔP	10.0

Discussion. The initial assumptions have provided an exchanger which very nearly meets all the requirements. Eight-pass units would meet the heat-transfer requirement but would give a tube-side pressure drop of 14 psi. The trial exchanger will be somewhat less suitable when the value of ϕ_t is also taken into account. If the minimum dirt factor of 0.0040 is to be taken literally, it will be necessary to try the next size shell.

Trial 2:

Assume a 33 in. ID shell with six[1] tube passes and baffles spaced 12-in. apart, since the pressure drop increases with the diameter of the shell for a given mass velocity. The summary for the conditions are

Summary

94	h outside	118
U_C	52.3	
U_D	42.0	
R_d Calculated	0.0047	
R_d Required	0.004	
4.4	Calculated ΔP	7.9
10.0	Allowable ΔP	10.0

The two final exchangers in series will be

Shell side
ID = 33 in.
Baffle space = 12 in.
Passes = 2

Tube side
Number and length = 676, 16'0''
OD, BWG, pitch = ¾ in., 16 BWG, 1-in. square
Passes = 6

Example 11.3. Calculation of a Caustic Solution Cooler. 100,000 lb/hr of 15°Bé caustic solution (11 per cent sodium hydroxide, $s = 1.115$) leaves a dissolver at 190°F and is to be cooled to 120°F using water at 80°F. Use a combined dirt factor of 0.002 and pressure drops of 10 psi.

The viscosity of the 11 per cent sodium hydroxide may be approximated by the methods of Chap. 7, but is sirupy, and actual data should be used if possible. The viscosity at 100°F is 1.4 centipoises and at 210°F is 0.43 centipoises. For the specific heat assume the dry salt to have a value of 0.25 Btu/lb, giving a specific heat for the solution at the mean of 0.88.

Plant practice permits the use of triangular pitch with 1 in. OD tubes for solutions in which the scale may be boiled out.

Solution:

(1) Heat balance:
  Caustic, $Q = 100,000 \times 0.88(190 - 120) = 6,160,000$ Btu/hr
  Water, $Q = 154,000 \times 1(120 - 80) = 6,160,000$ Btu/hr

[1] An eight-pass exchanger would give a pressure drop of 10.8 psi.

CALCULATIONS FOR PROCESS CONDITIONS

(2) Δt:

	Hot Fluid		Cold Fluid	Diff.
	190	Higher Temp	120	70
	120	Lower Temp	80	40
	70	Differences	40	30

LMTD = 53.3°F   Eq. (5.14)

$R = \dfrac{70}{40} = 1.75 \qquad S = \dfrac{40}{190 - 80} = 0.364$

$F_T = 0.815$ (Fig. 18)

A 1-2 exchanger will be satisfactory.

$\Delta t = 0.815 \times 53.3 = 43.5°F$ Eq. (7.42)

(3) T_c and t_c: The average temperatures T_a and t_a will be satisfactory because of the closeness of the ranges and the low viscosities.

Trial:

(a) Assume $U_D = 250$. From Table 8 this value is about the minimum for a 0.001 dirt factor and should be suitable for a trial when the required dirt factor is 0.0020.

$A = \dfrac{6{,}160{,}000}{250 \times 43.5} = 567 \text{ ft}^2$

$a'' = 0.2618 \text{ ft}^2/\text{lin ft}$ (Table 10)

Number of tubes, $N_t = \dfrac{567}{16'0'' \times 0.2618} = 136$

(b) Assume four tube passes: For two tube passes $a_t = 0.258$ and $G_t = 598{,}000$ corresponding to a water velocity of only 2.65 fps.

From the tube counts (Table 9): 136 tubes, 4 passes, 1 in. OD on 1¼-in. triangular pitch

Nearest count: 140 tubes in a 19¼" ID shell

(c) Corrected coefficient U_D:

$A = 140 \times 16'0'' \times 0.2618 = 586 \text{ ft}^2$

$U_D = \dfrac{Q}{A \, \Delta t} = \dfrac{6{,}160{,}000}{586 \times 43.5} = 242$

Hot fluid: shell side, caustic

(4') From the previous problems a mass velocity of about 500,000 gave a reasonable pressure drop. By trial this corresponds to about a 7-in. baffle spacing.

$a_s = \text{ID} \times C'B/144P_T$ [Eq. (7.1)]
$= 19.25 \times 0.25 \times 7/144 \times 1.25$
$= 0.1875 \text{ ft}^2$

Cold fluid: tube side, water

(4) Flow area, $a_t' = 0.546$ in.² [Table 10]
$a_t = N_t a_t'/144n$ [Eq. (7.48)]
$= 140 \times 0.546/144 \times 4 = 0.133 \text{ ft}^2$

Hot fluid: shell side, caustic

(5′) $G_s = W/a_s$ [Eq. (7.2)]
    = 100,000/0.1875
    = **533,000** lb/(hr)(ft²)

(6′) At $T_a = 155°F$,
 $\mu = 0.76 \times 2.42 = 1.84$ lb/(ft)(hr)
 $D_e = 0.72/12 = 0.06$ ft   [Fig. 28]
 $Re_s = D_e G_s/\mu$ [Eq. (7.3)]
    = $0.06 \times 533{,}000/1.84 = 17{,}400$
(7′) $j_H = 75$ [Fig. 28]
(8′) At 155°F, $k = 0.9$ (k_{water})   [Table 4]
    = 0.9×0.38
    = 0.342 Btu/(hr)(ft²)(°F/ft)
$(c\mu/k)^{1/3} = (0.88 \times 1.84/0.342)^{1/3} = 1.68$

(9′) $h_o = j_H \dfrac{k}{D_e}\left(\dfrac{c\mu}{k}\right)^{1/3} \phi_s$ [Eq. (6.15b)]

$\dfrac{h_o}{\phi_s} = 75 \times 0.342 \times 1.68/0.06 = 717$

(10′) (11′) (12′) $\phi_s = 1$ (low viscosity)

$h_o = \dfrac{h_o}{\phi_s} = 717$ Btu/(hr)(ft²)(°F)

Cold fluid: tube side, water

(5) $G_t = w/a_t$
    = 154,000/0.133
    = 1,160,000 lb/(hr)(ft²)
Vel, $V = G_t/3600\rho$
    = $1{,}160{,}000/3600 \times 62.5 = 5.16$ fps

(6) At $t_a = 100°F$,
 $\mu = 0.72 \times 2.42 = 1.74$ lb/(ft)(hr)   [Fig. 14]
 $D = 0.834/12 = 0.0695$ ft   [Table 10]
 (Re_t is for pressure drop only)
 $Re_t = DG_t/\mu$
    = $0.0695 \times 1{,}160{,}000/1.74 = 46{,}300$

(9) $h_i = 1240 \times 0.94 = 1165$ [Fig. 25]
(10) $h_{io} = h_i \times \text{ID/OD}$ [Eq. (6.5)]
    = $1165 \times 0.834/1.0$
    = 972 Btu/(hr)(ft²)(°F)

Pressure Drop

(1′) For $Re_s = 17{,}400$,
$f = 0.0019$ ft²/in.²   [Fig. 29]

(2′) No. of crosses, $N + 1 = 12L/B$ [Eq. (7.43)]
    = $12 \times 16/7 = 28$
$D_s = 19.25/12 = 1.60$ ft

(3′) $\Delta P_s = \dfrac{fG_s^2 D_s (N+1)}{5.22 \times 10^{10} D_e s \phi_s}$ [Eq. (7.44)]

    = $\dfrac{0.0019 \times 533{,}000^2 \times 1.60 \times 28}{5.22 \times 10^{10} \times 0.06 \times 1.115 \times 1.0}$
    = 7.0 psi

(1) For $Re_t = 46{,}300$,
$f = 0.00018$ ft²/in.²   [Fig. 26]

(2) $\Delta P_t = \dfrac{fG_t^2 Ln}{5.22 \times 10^{10} Ds\phi_t}$ [Eq. (7.45)]

    = $\dfrac{0.00018 \times 1{,}160{,}000^2 \times 16 \times 4}{5.22 \times 10^{10} \times 0.0695 \times 1.0 \times 1.0}$
    = 4.3 psi   [Eq. (7.46)]

(3) $\Delta P_r = \dfrac{4n}{s}\dfrac{V^2}{2g'}$ [Fig. 27]

    = $\dfrac{4 \times 4}{1} \times 0.18 = 2.9$ psi

(4) $\Delta P_T = \Delta P_t + \Delta P_r$ [Eq. (7.47)]
    = $4.3 + 2.9 = 7.2$ psi
Now proceed to the shell side.

(13) Clean overall coefficient U_C:

$$U_C = \dfrac{h_{io} h_o}{h_{io} + h_o} = \dfrac{972 \times 717}{972 + 717} = 413 \text{ Btu/(hr)(ft}^2\text{)(°F)} \qquad (6.38)$$

(14) Dirt factor R_d: U_D from (c) is 242.

$$R_d = \dfrac{U_C - U_D}{U_C U_D} = \dfrac{413 - 242}{413 \times 242} = 0.0017 \text{ (hr)(ft}^2\text{)(°F)/Btu} \qquad (6.13)$$

CALCULATIONS FOR PROCESS CONDITIONS

Summary

717	h outside	972
U_C	413	
U_D	242	
R_d Calculated	0.0017	
R_d Required	0.0020	
7.0	Calculated ΔP	7.2
10.0	Allowable ΔP	10.0

Discussion. Adjustment of the baffle space to use the full 10 psi will still not permit the exchanger to make the 0.002 dirt factor. The value of U_D has been assumed too high. Try the next size shell.

Trial 2:

Try a 21¼ in. ID shell with four tube passes and a 6-in. baffle space. This corresponds to 170 tubes.

Summary

720	h outside	840
U_C	390	
U_D	200	
R_d Calculated	0.0024	
R_d Required	0.002	
9.8	Calculated ΔP	4.9
10.0	Allowable ΔP	10.0

The use of six tube passes exceeds the allowable tube side pressure drop. The final exchanger will be

    Shell side   *Tube side*
        ID = 21¼ in.   Number and length = 170, 16'0''
Baffle space = 6 in.   OD, BWG, pitch = 1 in., 14 BWG, 1¼-in. tri.
    Passes = 1   Passes = 4

Example 11.4. Calculation of an Alcohol Heater. 115,000 lb/hr of absolute alcohol (100 per cent ethyl alcohol, $s = 0.78$) is to be heated under pressure from a stor-

242   PROCESS HEAT TRANSFER

age temperature of 80 to 200°F using steam at 225°F.   A dirt factor of 0.002 is required along with an allowable alcohol pressure drop of 10 psi.

Plant practice is established using 1-in. OD tubes, 14 BWG, 12′0″ long.   Triangular pitch is satisfactory for clean services.

Solution:

(1) Heat balance:
Alcohol, $Q = 115{,}000 \times 0.72(200 - 80) = 9{,}950{,}000$ Btu/hr
Steam, $Q = 10{,}350 \times 962 = 9{,}950{,}000$ Btu/hr

(2) Δt: LMTD (true counterflow):

	Hot Fluid		Cold Fluid	Diff.
225	Higher Temp		200	25
225	Lower Temp		80	145
0	Differences		120	120

LMTD = 68.3°F   (5.14)

(3) T_c and t_c: Use T_a and t_a because of the low alcohol viscosity.

Trial.

(a) Assume $U_D = 200$: From Table 8 values of U_D from 200 to 700 may be expected when a dirt factor of 0.001 is employed.   Since the dirt factor required is 0.002, the very maximum value of U_D would be 500 corresponding to the dirt alone.

$$A = \frac{Q}{U_D \, \Delta t} = \frac{9{,}950{,}000}{200 \times 68.3} = 728 \text{ ft}^2$$

$a'' = 0.2618$ ft²/lin ft   (Table 10)

Number of tubes, $N_t = \dfrac{728}{12'0'' \times 0.2618} = 232$

(b) Assume two tube passes: Only one or two passes are required for steam heaters. From the tube counts (Table 9): 232 tubes, two passes, 1 in. OD on 1¼-in. triangular pitch

Nearest count: 232 tubes in a 23¼ in. ID shell

(c) Corrected coefficient U_D:
$A = 232 \times 12'0'' \times 0.2618 = 728$ ft²
$U_D = \dfrac{Q}{A \, \Delta t} = \dfrac{9{,}950{,}000}{728 \times 68.3} = 200$

Hot fluid: tube side, steam

(4) Flow area, $a_t' = 0.546$ in.² [Table 10]
$a_t = N_t a_t'/144n$ [Eq. (7.48)]
$= 232 \times 0.546/144 \times 2 = 0.44$ ft²

Cold fluid: shell side, alcohol

(4′) To obtain a mass velocity between 400,000 and 500,000 use a 7-in. baffle space.
$a_s = \text{ID} \times C'B/144P_T$ [Eq. (7.1)]
$= 23.25 \times 0.25 \times 7/144 \times 1.25$ ft²
$= 0.226$

CALCULATIONS FOR PROCESS CONDITIONS

Hot fluid: tube side, steam

(5) $G_t = W/a_t$
$= 10,350/0.44$
$= 23,500 \text{ lb}/(\text{hr})(\text{ft}^2)$

(6) At 225°F,
$\mu = 0.013 \times 2.42 = 0.0314 \text{ lb}/(\text{ft})(\text{hr})$ [Fig. 15]
$D = 0.834/12 = 0.0695 \text{ ft}$
$Re_t = DG_t/\mu$ (for pressure drop only)
$= 0.0695 \times 23,500/0.0314$
$= 52,000$

(9) $h_{io} = 1500 \text{ Btu}/(\text{hr})(\text{ft}^2)(°\text{F})$

Cold fluid: shell side, alcohol

(5') $G_s = w/a_s$ [Eq. (7.2)]
$= 115,000/0.226$
$= 508,000 \text{ lb}/(\text{hr})(\text{ft}^2)$

(6') At $t_a = 140°\text{F}$,
$\mu = 0.60 \times 2.42 = 1.45 \text{ lb}/(\text{ft})(\text{hr})$ [Fig. 14]
$D_e = 0.72/12 = 0.06 \text{ ft}$ [Fig. 28]
$Re_s = D_e G_s/\mu$ [Eq. (7.3)]
$= 0.06 \times 508,000/1.45 = 21,000$

(7') $j_H = 83$

(8') At 140°F,
$k = 0.085 \text{ Btu}/(\text{hr})(\text{ft}^2)(°\text{F}/\text{ft})$ [Table 4]
$(c\mu/k)^{1/3} = (0.72 \times 1.45/0.085)^{1/3} = 2.31$

(9') $h_o = j_H \dfrac{k}{D_e}\left(\dfrac{c\mu}{k}\right)^{1/3}\phi_s$ [Eq. (6.15b)]

$\dfrac{h_o}{\phi_s} = 83 \times 0.085 \times 2.31/0.06 = 270$

(10') $\phi_s = 1.0$,
$h_o = \dfrac{h_o}{\phi_s} = 270 \text{ Btu}/(\text{hr})(\text{ft}^2)(°\text{F})$

Pressure Drop

(1) For $Re_t = 52,000$,
$f = 0.000175 \text{ ft}^2/\text{in.}^2$ [Fig. 26]
From Table 7 the specific volume is approximately 21 ft³/lb.
$\rho = \frac{1}{21} = 0.0477 \text{ lb}/\text{ft}^3$
$s = 0.0477/62.5 = 0.00076$

(2) $\Delta P_t = \dfrac{1}{2}\dfrac{fG_t^2 Ln}{5.22 \times 10^{10} Ds\phi_t}$ [Eq. 7.53]

$= \dfrac{1}{2} \times \dfrac{0.000175 \times 23,500^2 \times 12 \times 2}{5.22 \times 10^{10} \times 0.0695 \times 0.00076 \times 1.0}$

$= 0.42 \text{ psi}$

(3) ΔP_r: Negligible because of partial condensation at end of first pass.

(1') For $Re_s = 21,000$,
$f = 0.0018 \text{ ft}^2/\text{in.}^2$ [Fig. 29]
$s = 0.78$

(2') No. of crosses, $N + 1 = 12L/B$ [Eq. (7.43)]
$= 12 \times 12/7 = 21$
$D_s = 23.25/12 = 1.94 \text{ ft}$

(3') $\Delta P_s = \dfrac{fG_s^2 D_s (N+1)}{5.22 \times 10^{10} D_e s \phi_s}$ [Eq. (7.44)]

$= \dfrac{0.0018 \times 508,000^2 \times 1.94 \times 21}{5.22 \times 10^{10} \times 0.06 \times 0.78 \times 1.0}$

$= 7.8 \text{ psi}$

(13) Clean overall coefficient U_C:

$$U_C = \dfrac{h_{io} h_o}{h_{io} + h_o} = \dfrac{1500 \times 270}{1500 + 270} = 229 \text{ Btu}/(\text{hr})(\text{ft}^2)(°\text{F}) \qquad (6.38)$$

(14) Dirt factor R_d: U_D from (c) is 200.

$$R_d = \dfrac{U_C - U_D}{U_C U_D} = \dfrac{229 - 200}{229 \times 200} = 0.000633 \text{ (hr)}(\text{ft}^2)(°\text{F})/\text{Btu} \qquad (6.13)$$

Summary

1500	h outside	270
U_C	229	
U_D	200	
R_d Calculated 0.000633		
R_d Required 0.002		
0.42	Calculated P	7.8
Neg.	Allowable P	10.0

Discussion. This is clearly an instance in which U_D was assumed too high. It is now a question of how much too high. With the aid of the summary it is apparent that in a larger shell a clean overall coefficient of about 200 may be expected. To permit a dirt factor of 0.002 the new U_D should be

$$\frac{1}{U_D} = \frac{1}{U_C} + R_d = \frac{1}{200} + 0.002$$
$$U_D = 143$$

Trial 2:

$$A = \frac{Q}{U_D \Delta t} = \frac{9{,}950{,}000}{143 \times 68.3} = 1020 \text{ ft}^2$$

No. of tubes $= \dfrac{1020}{12'0'' \times 0.2618} = 325$

Nearest count: 334 tubes in a 27 in. ID shell (Table 9).

The same baffle pitch should be retained, since the pressure drop increases with the inside diameter.

Summary

1500	h outside	250
U_C	214	
U_D	138.5	
R_d Calculated 0.0025		
R_d Required 0.002		
0.23	Calculated ΔP	7.1
Neg.	Allowable ΔP	10.0

If a 25-in. exchanger had been used, the dirt factor would be less than 0.002 and a 6-in. baffle space would give a pressure drop exceeding 10 psi. The final exchanger is

CALCULATIONS FOR PROCESS CONDITIONS

Shell side   Tube side
ID = 27 in.   Number and length = 334, 12'0''
Baffle space = 7 in.   OD, BWG, pitch = 1 in., 14 BWG, 1¼-in. tri.
Passes = 1   Passes = 2

Split Flow. Sometimes it is not possible to meet the pressure-drop requirements in a 1-2 or 2-4 exchanger. Instances will occur when (1) the true temperature difference or U_D is very great and a small exchanger is indicated for the quantity of heat to be transferred, (2) one fluid stream has a very small temperature range compared with the other, or (3) the allowable pressure drop is small. In gases and vapors the last is the more critical because of the low density of the gas or vapor. In liquids an excellent example of (2) is found in the quenching of steel, where it is customary to cool a large volume of circulated quench oil over a small range. It is also characteristic of a number of near-constant temperature operations such as the removal of heat from exothermic reactions by continuous recirculation of the reacting fluids through an external 1-2 cooler.

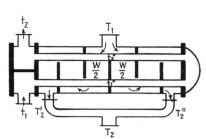

FIG. 11.6. Split-flow exchanger.

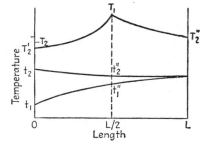

FIG. 11.7. Temperature relations in split flow.

The failure to meet the allowable pressure drop by the conventional methods in a 1-2 exchanger should be construed as an indication that fluid flow and not heat transfer is the controlling factor. Reducing the length of the tubes and increasing the diameter of the shell provides one means of reducing the pressure drop, but other means are available. By locating the shell inlet nozzle at the center instead of at the end of the shell and using two outlet nozzles as shown in Fig. 11.6, the shell-side pressure drop will be but one-eighth as great as in a conventional 1-2 exchanger of the same shell diameter. The reduction is due to halving the mass velocity and halving the length of the shell-side path. This type of flow is known as *split flow*.

As seen in Fig. 11.7 the temperature relations in a split-flow exchanger are **not** in true counterflow or identical with a 1-2 exchanger, being discontinuous at the shell mid-point. A direct solution of the equation[1] for the true temperature difference is somewhat tedious, since the values of t_1'' and t_2'' are related to the actual temperature differences and heat transferred in both parts of the exchanger on either side of the shell inlet. If $T_2 > t_2$, it is satisfactory to obtain Δt by multiplying the LMTD by the value of F_T obtained for a 1-2 exchanger. For services in which there is a temperature cross the actual split-flow equation should be used.

Another type of flow giving even lower pressure drops is *divided flow*, which is usually reserved for low-pressure gases, condensers, and reboilers. It will be discussed in Chap. 12.

Example 11.5. Calculation of a Flue Gas Cooler. 10,500 cfm of flue gas (mol. wt. = 30) at 2 psig and 250°F is to be cooled to 125°F with an allowable pressure drop of 1.0 psi. Cooling will be effected by water from 80 to 100°F and with an allowable pressure drop of 10 psi. An overall dirt factor of 0.005 should be provided with a reasonable minimum water velocity.

Plant practice uses 1 in. OD, 14 BWG tubes on square pitch for all services, and because it is sometimes difficult to meet the pressure drop in gas coolers, no tube length is specified.

Solution:

(1) Heat balance:
 Gas: 10,500 cfm of flue gas at 250°F

$$\text{Total gas} = \frac{10{,}500 \times 60 \times 30}{359 \times (711/492) \times (14.7/16.7)} = 41{,}300 \text{ lb/hr}$$

Gas, $Q = 41{,}300 \times 0.25(250 - 125) = 1{,}290{,}000$ Btu/hr
Water, $Q = 64{,}500 \times 1(100 - 80) = 1{,}290{,}000$ Btu/hr

(2) Δt:

	Hot Fluid		Cold Fluid	Diff.
	250	Higher Temp	100	150
	125	Lower Temp	80	45
	125	Differences	20	105

[1] Unpublished notes, D. Q. Kern and C. L. Carpenter. The equation in terms of terminal temperatures for a split-flow exchanger with two tube passes is

$$\Delta t = \frac{(T_1 - T_2)\lambda/2}{2.3 \log \left\{ \frac{\frac{(T_1 - T_2)2\lambda n}{n+1} e^{(T_1 - T_2)\left(\frac{\lambda}{2} - 1\right)/2\Delta t} - (T_1 - T_2)(\lambda + 2) + t_1 + t_2 - \frac{4T_2 + 2(n-1)T_1}{n+1}}{(T_1 - T_2)(\lambda - 2) + t_1 + t_2 - \frac{4T_2 + 2(n-1)T_1}{n+1}} \right\}}$$

where $n = Q_B/Q_A$

$$\lambda = \frac{\sqrt{4R^2 + 1}}{R}$$

and

Q_B/Q_A = ratio of heat transfer in each half

CALCULATIONS FOR PROCESS CONDITIONS

$\text{LMTD} = 87.4°F$ (5.14)

$R = \dfrac{125}{20} = 6.25 \qquad S = \dfrac{20}{250 - 80} = 0.118 \qquad F_T = 0.935$ (Fig. 18)

A conventional 1-2 exchanger will be satisfactory.

$\Delta t = 0.935 \times 87.4 = 81.6°F$ (7.42)

(3) T_c and t_c: The average temperatures T_a and t_a will be satisfactory because of the small variations in the individual viscosities.

Trial:

(a) Assume $U_D = 15$: From the examples of Chap. 9, at atmospheric pressure and 2 psi allowable pressure drop a coefficient of about 20 might be anticipated. Since the allowable pressure drop in this example is only 1.0 psi, the trial value of U_D must be reduced accordingly. Assume 12'0" tubes to increase the shell cross section.

$A = \dfrac{Q}{U_D \, \Delta t} = \dfrac{1{,}290{,}000}{15 \times 81.6} = 1055 \text{ ft}^2$

$a'' = 0.2618 \text{ ft}^2/\text{lin ft}$

Number of tubes $= \dfrac{1055}{12'0'' \times 0.2618} = 336$

(b) Assume eight tube passes: Because of the low design coefficient gas exchangers are large for the amount of cooling medium required.

From the tube counts (Table 9): 336 tubes, eight passes, 1 in. OD on 1¼-in. square pitch

Nearest count: 358 tubes in a 31 in. ID shell

(c) Corrected coefficient U_D:

$A = 358 \times 12'0'' \times 0.2618 = 1125 \text{ ft}^2$

$U_D = \dfrac{Q}{A \, \Delta t} = \dfrac{1{,}290{,}000}{1125 \times 81.6} = 14.0$

When solved in a manner identical with the preceding examples and using the smallest integral number of bundle crosses (five) corresponding to a 28.8-in. spacing the summary is

Summary

24.0	h outside	392
U_C	22.7	
U_D	14.0	
R_D Calculated	0.027	
R_d Required	0.005	
5.2	Calculated ΔP	1.0
1.0	Allowable ΔP	10.0

Discussion. The exchanger selected as a solution to the requirements combines two conditions which have not been met previously: The dirt factor is considerably greater than necessary, and the pressure drop is five times greater than the allowable. Had 8-ft tubes been used in place of the 12-ft tubes for $U_D = 15$, the shell inside diameter would have been 37 in. The baffles could have been spaced 32 in. apart to provide three bundle crosses, but the resulting pressure drop would be 1.7 psi. This would be unsatisfactory, since gases require large inlet connections and the flow distribution on the first and third bundle crosses would be poor and the conditions of allowable pressure drop would still not be met. The solution is found in a split-flow arrangement.

Trial 2. Split flow:

(a) Assume $U_D = 15$. Referring to the summary of the first trial it is evident that, if the pressure drop is to be met, the mass velocity must be reduced so that the new gas film coefficient will be considerably below the value of 24.0 obtained for ordinary flow.

(b) Assume 12 tube passes. The low water coefficient of 392 corresponds to a velocity of only 1.7 fps which is extremely low for corrosion and dirt even where good water may be employed. Since the size of the shell will not be altered appreciably, having a large inside diameter, it is justifiable that 12 passes be employed. Fewer tube passes would be needed if tube cores were inserted in the tubes. These may be calculated in the manner of Example 10.3. When using more than 8 passes in large shells, the tube count for 8 passes should be reduced by 5 per cent for 12 passes and 10 per cent for 16 passes. For smaller shells it is advisable to avoid the use of 12 and 16 passes. Using the same shell as in Trial 1 for 12 passes, the new tube count will be $358 \times 0.95 = 340$ tubes.

(c) Corrected coefficient U_D:
$A = 340 \times 12'0'' \times 0.2618 = 1070$ ft^2
$$U_D = \frac{Q}{A\,\Delta t} = \frac{1,240,000}{1070 \times 81.6} = 14.8$$

Hot fluid: shell side, flue gas

(4') Flow area,
$a_s = \text{ID} \times C'B/144P_T$ [Eq. (7.1)]
There must be an odd number of crosses in each half of the shell and the largest spacing is 31 in.
72 in./31 in. = 2 crosses; say 3 crosses (odd)

Actual spacing $= \dfrac{12 \times 12}{2 \times 3} = 24$ in.

$a_s = 31 \times 0.25 \times 24/144 \times 1.25 = 1.03$ ft^2

(5') Mass vel, split flow:
$G_s = \frac{1}{2}W/a_s$ [Eq. (7.2)]
$= \frac{1}{2} \times 41,300/1.03$
$= 20,000$ lb/(hr)(ft^2)

(6') At $T_a = 187.5°$F,
$\mu = 0.0206 \times 2.42 = 0.050$ lb/(ft)(hr)
[Fig. 15]
$D_e = 0.99/12 = 0.0825$ [Fig. 28]
$Re_s = D_e G_s/\mu$ [Eq. (7.3)]
$= 0.0825 \times 20,000/0.05 = 33,000$

(7') $j_H = 105$ [Fig. 28]
(8') At 187.5°F,
$k = 0.015$ Btu/(hr)(ft^2)(°F/ft)
(Table 5)
$(c\mu/k)^{\frac{1}{3}} = (0.25 \times 0.050/0.015)^{\frac{1}{3}} = 0.94$

Cold fluid: tube side, water

(4) Flow area,
$a'_t = 0.546$ in.^{2} [Table 10]
$a_t = N_t a'_t/144n$
$= 340 \times 0.546/144 \times 12 = 0.107$ ft^2

(5) $G_t = w/a_t$
$= 64,500/0.107$
$= 602,000$ lb/(hr)(ft^2)
Vel, $V = G_t/3600\rho$
$= 602,000/3600 \times 62.5 = 2.68$ fps
(6) At $t_a = 90°$F,
$\mu = 0.81 \times 2.42 = 1.96$ lb/(ft)(hr)
[Fig. 14]
$D = 0.834/12 = 0.0695$ ft
$Re_t = DG_t/\mu$
$= 0.0695 \times 602,000/1.96 = 21,300$
(Re_t is for pressure drop only)

CALCULATIONS FOR PROCESS CONDITIONS

Hot fluid: shell side, flue gas

(9') $h_o = j_H \dfrac{k}{D_e}\left(\dfrac{c\mu}{k}\right)^{1/3} \phi_s$ [Eq. (6.15)]

$\dfrac{h_o}{\phi_s} = 105 \times 0.015 \times 0.94/0.0825$
$= 17.9$

(10') (11') (12') $\phi_s = 1.0$

$h_o = \dfrac{h_o}{\phi_s} = 17.9$ Btu/(hr)(ft^2)(°F)

Cold fluid: tube side, water

(9) $h_i = 710 \times 0.94 = 667$ [Fig. 25]
(10) $h_{io} = h_i \times$ ID/OD
$= 667 \times 0.83/1.0$
$= 557$ Btu/(hr)(ft^2)(°F)

Pressure Drop

(1') For $Re_s = 33,000$,
$f = 0.0017$ ft^2/in.^{2} [Fig. 29]
(2') No. of crosses, $N + 1 = 3$ [Eq. (7.43)]
$D_s = 31/12 = 2.58$ ft
$s = 0.0012$

(3') $\Delta P_s = \dfrac{fG_s^2 D_s(N+1)}{5.22 \times 10^{10} D_e s \phi_s}$ [Eq. (7.44)]

$= \dfrac{0.00167 \times 20{,}000^2 \times 2.58 \times 3}{5.22 \times 10^{10} \times 0.0825 \times 0.0012 \times 1.0}$
$= 1.0$ psi

(1) For $Re_t = 21{,}300$,
$f = 0.00012$ ft^2/in.^{2} [Fig. 26]
(2) $\Delta P_t = \dfrac{fG_t^2 Ln}{5.22 \times 10^{10} Ds\phi_t}$ [Eq. (7.45)]

$= \dfrac{0.00022 \times 602{,}000^2 \times 12 \times 12}{5.22 \times 10^{10} \times 0.0695 \times 1.0 \times 1.0}$
$= 3.1$ psi

(3) $\Delta P_r = 4n/s(V^2/2g')$ [Fig. 27]
$= (4 \times 12/1)0.052 = 2.5$ psi

(4) $\Delta P_T = \Delta P_t + \Delta P_r$ [Eq. (7.47)]
$= 3.1 + 2.5 = 5.6$ psi

(13) Clean overall coefficient U_C:

$$U_C = \dfrac{h_{io}h_o}{h_{io} + h_o} = \dfrac{557 \times 17.9}{557 + 17.9} = 17.3 \text{ Btu/(hr)(ft}^2\text{)(°F)} \tag{6.38}$$

(14) Dirt factor R_d: U_D from (c) = 14.8

$$R_d = \dfrac{U_C - U_D}{U_C U_D} = \dfrac{17.3 - 14.8}{17.3 \times 14.8} = 0.0098 \text{ (hr)(ft}^2\text{)(°F)/Btu} \tag{6.13}$$

Summary

17.9	h outside	557
U_C	17.3	
U_D	14.8	
R_d Calculated 0.0098		
R_d Required 0.005		
1.0	Calculated ΔP	5.6
1.0	Allowable ΔP	10.0

A 16-pass unit would also be suitable but is not warranted. The final exchanger is

Shell side	Tube side
ID = 31 in.	Number and length = 340, 12'0"
Baffle space = 24 in.	OD, BWG, pitch = 1 in., 14 BWG, 1¼-in. square
Passes = split flow	Passes = 12

PROBLEMS

For the following process conditions determine the size and arrangement of exchanger to fulfill the conditions allowing pressure drops of 10 psi each stream and a combined dirt factor of 0.004. Employ 1-2 exchangers wherever possible.

11.1. 60,000 lb/hr of 42°API kerosene is cooled from 400 to 225°F by heating 35°API distillate from 100 to 200°F.
Use ¾ in. OD, 16 BWG tubes, 16'0" long on $^{15}\!/_{16}$-in. triangular pitch.

11.2. 120,000 lb/hr of aniline is cooled from 275 to 200°F by heating 100,000 lb/hr of benzene from 100 to 200°F.
Use ¾ in. OD, 14 BWG tubes, 16'0" long on 1-in. square pitch.

11.3. 84,000 lb/hr of 42°API kerosene is cooled from 300 to 100°F using water from 85 to 120°F. Calculate the requirement using the 1-2 exchangers in series.
Use ¾ in. OD, 16 BWG tubes, 16'0" long on 1-in. triangular pitch.

11.4. 22,000 lb/hr of 35°API distillate is heated from 200 to 300°F by 28°API gas oil from an inlet temperature of 500°F.
Use 1 in. OD, 14 BWG tubes, 12'0" long on 1¼-in. triangular pitch.

11.5. 68,000 lb/hr of 56°API gasoline is cooled from 200 to 100°F using water at 85°F.
Use 1 in. OD, 14 BWG tubes, 12'0" long on 1¼-in. square pitch.

11.6. 32,000 lb/hr of oxygen at 5 psig is cooled from 300 to 150°F using water at 85°F. (Allowable pressure drop for oxygen 2.0 psi.)
Use ¾ in. OD, 16 BWG tubes, 12'0" long on 1-in. square pitch.

NOMENCLATURE FOR CHAPTER 11

A	Heat-transfer surface, ft²
A_{CR}, A_E, A_H	Heat-transfer surface of cooler, exchanger, and heater, ft²
a	Flow area, ft²
a''	External surface per linear foot, ft
B	Baffle spacing, in.
C	Specific heat of hot fluid in derivations, Btu/(lb)(°F)
C_S	Cost of steam, dollars/Btu
C_T	Total annual cost, dollars
C_W	Cost of water, dollars/Btu
C_{CR}, C_E, C_H	Fixed charges of cooler, exchanger, and heater, respectively, dollars/(ft²)(year)
C'	Clearance between tubes, in.
c	Specific heat of fluid, Btu/(lb)(°F)
D	Inside diameter of tubes, ft
D_s	Inside diameter of shell, ft
D_e, D'_e	Equivalent diameter for heat transfer and pressure drop, ft
d	Inside diameter of tubes, in.
d_e, d'_e	Equivalent diameter for heat transfer and pressure drop, in.
F_c	Caloric fraction, dimensionless
F_T	Temperature-difference factor, $\Delta t = F_T \times \text{LMTD}$, dimensionles

CALCULATIONS FOR PROCESS CONDITIONS 251

f	Friction factor, $ft^2/in.^2$
G	Mass velocity, $lb/(hr)(ft^2)$
g	Acceleration of gravity, ft/hr^2
g'	Acceleration of gravity, ft/sec^2
h, h_i, h_o	Heat-transfer coefficient in general, for inside fluid, and for outside fluid, respectively, $Btu/(hr)(ft^2)(°F)$
h_{io}	Value of h_i when referred to the tube outside diameter, $Btu/(hr)(ft^2)(°F)$
j_H	Factor for heat transfer, dimensionless
K_c	Caloric constant, dimensionless
k	Thermal conductivity, $Btu/(hr)(ft^2)(°F/ft)$
L	Tube length, ft
LMTD	Log mean temperature difference, °F
N	Number of shell-side baffles
N_t	Number of tubes
n	Number of tube passes
P_T	Tube pitch, in.
$\Delta P_T, \Delta P_t, \Delta P_r$	Total, tube and return pressure drop, psi
Q	Heat flow, Btu/hr
R	Temperature group $(T_1 - T_2)/(t_2 - t_1)$, dimensionless
R_d	Dirt factor, $(hr)(ft^2)(°F)/Btu$
Re, Re'	Reynolds number for heat transfer and pressure drop, dimensionless
S	Temperature group $(t_2 - t_1)/(T_1 - t_1)$, dimensionless
s	Specific gravity, dimensionless
T_1, T_2	Inlet and outlet temperature of hot fluid, °F
T_a	Average temperature of hot fluid, °F
T_c	Caloric temperature of hot fluid, °F
T_s	Temperature of steam, °F
T_x	Optimum exchanger hot-fluid outlet temperature, °F
t_y	Optimum exchanger cold-fluid outlet temperature, °F
t_a	Average temperature of cold fluid, °F
t_c	Caloric temperature of cold fluid, °F
t_w	Tube-wall temperature, °F
t_1', t_2'	Inlet and outlet water temperatures, °F
t_1'', t_2''	Temperatures in a split-flow exchanger, °F
Δt	True temperature difference in $Q = U_D A \Delta t$, °F
$\Delta t_c, \Delta t_h$	Cold and hot terminal temperature difference, °F
U_C, U_D	Clean and design overall coefficient of heat transfer, $Btu/(hr)(ft^2)(°F)$
U_{CR}, U_E, U_H	Overall coefficient of heat transfer for cooler, exchanger, and heater, respectively, $Btu/(hr)(ft^2)(°F)$
V	Velocity, fps
W	Weight flow of hot fluid, lb/hr
w	Weight flow of cold fluid, lb/hr
ϕ	Viscosity ratio, $(\mu/\mu_w)^{0.14}$
μ	Viscosity, centipoises $\times 2.42 = lb/(ft)(hr)$
μ_w	Viscosity at tube-wall temperature, centipoises $\times 2.42 = lb/(ft)(hr)$
θ	Annual operating hours

Subscripts (except as noted above)

s	Shell
t	Tube

CHAPTER 12

CONDENSATION OF SINGLE VAPORS

Introduction. A fluid may exist as a gas, vapor, or liquid. The change from liquid to vapor is *vaporization*, and the change from vapor to liquid is *condensation*. The quantities of heat involved in the vaporization or condensation of a pound of fluid are identical. For a pure fluid compound at a given pressure the change from liquid to vapor or vapor to liquid occurs at but one temperature which is the *saturation* or *equilibrium* temperature. Since vapor-liquid heat-transfer changes usually occur at constant or nearly constant pressure in industry, the vaporization or condensation of a single compound normally occurs isothermally. When a vapor is removed upon formation from further contact with a liquid, the addition of heat to the vapor causes *superheat*, during which it behaves like a gas. If a mixture of vapors instead of a pure vapor is condensed at constant pressure, the change does not take place isothermally in most instances. The general treatment of vapor mixtures differs in certain respects from single compounds and will be studied in the next chapter with the aid of the phase rule of J. Willard Gibbs.

Condensation occurs at very different rates of heat transfer by either of the two distinct physical mechanisms, which will be discussed presently, *dropwise* or *filmwise* condensation. The condensing film coefficient is influenced by the texture of the surface on which condensation occurs and also by whether the condensing surface is mounted vertically or horizontally. In spite of these apparent complications condensation, like streamline flow, lends itself to direct mathematical study.

Dropwise and Filmwise Condensation. When a saturated pure vapor comes into contact with a cold surface such as a tube, it condenses and may form liquid droplets on the surface of the tube. These droplets may not exhibit an affinity for the surface and instead of coating the tube fall from it, leaving bare metal on which successive droplets of condensate may form. When condensation occurs by this mechanism, it is called *dropwise* condensation. Usually, however, a distinct film may appear as the vapor condenses and coats the tube. Additional vapor is then required to condense into the liquid film rather than form directly on the bare surface. This is *film* or *filmwise* condensation. The two mechanisms are distinct and independent of the quantity of vapor condensing per square foot of surface. *Filmwise* condensation is therefore not a transition from *dropwise* condensation because of the rapidity at which condensate forms on the tube. Due to the resistance of the condensate film to the heat passing through it the heat-transfer coefficients for dropwise condensation are four to eight times those for filmwise condensation. Steam is the only pure vapor known to condense in a dropwise manner, and special conditions are required for its occurrence. These are described by Drew, Nagle, and Smith[1] and principally result from the presence of dirt on the surface or the use of a contaminant which adheres to the surface. Materials have been identified by Nagle[2] which promote the dropwise condensation of steam although these also introduce an impurity into the steam. Dropwise condensation also occurs when several materials condense simultaneously as a mixture and where the condensate mixture is not miscible, as in the case of a hydrocarbon and steam. However, during various periods in the normal operation of a steam condenser the mechanism may initially be filmwise condensation, shift to dropwise condensation, and at some later time revert to film condensation. Because of the lack of control it is not customary in calculations to take advantage of the high coefficients which have been obtained in dropwise-condensation experiments. This chapter consequently deals with the calculation of condensers for various conditions and is based solely upon film-condensation heat-transfer coefficients.

It is fortunate that the phenomenon of film condensation lends itself to mathematical analysis, and the nature of condensation on a cold surface may be considered one of self-diffusion. The saturation pressure of vapor in the vapor body is greater than the saturation pressure of the cold condensate in contact with the cold surface. This pressure difference provides the potential for driving vapor out of the vapor body at a great rate. Compared with the small resistance to heat transfer by diffusion from the vapor into the condensate, the film of condensate on the cold tube wall contributes the controlling resistance. It is the slowness with which the heat of condensation passes through this film that determines the condensing coefficient. The ultimate form of an equation for the condensing coefficient may be obtained from dimensional analysis where the average condensing coefficient $\bar{h}$ is a function of the properties of the condensate film, k, ρ, g, μ, and L, Δt, and λ, the last being the latent heat of vaporization. Nusselt theoretically derived the relationships for the mechanism of film condensation, and the results he obtained are in excellent agreement with experiments.

Process Applications. In chemical industry it is a common practice to separate a liquid mixture by distilling off the compounds which have

[1] Drew, T. B., W. M. Nagle, and W. Q. Smith, *Trans. AIChE*, **31**, 605–621 (1935).
[2] Nagle, W. M., U.S. Patent 1,995,361.

254   PROCESS HEAT TRANSFER

lower boiling points in the pure condition from those having higher boiling points. In a solution of several compounds each exerts a partial pressure and the most volatile cannot be boiled off from the rest without carrying some of the heavier or higher boiling compounds along with it. The proportion of heavier compounds carried off when a solution starts to boil is less than existed in the original solution before boiling commenced. If the vapor coming off initially is condensed, it has a lower boiling point than the original solution, indicating the increase in the proportion of the more volatile compounds. By successively boiling off

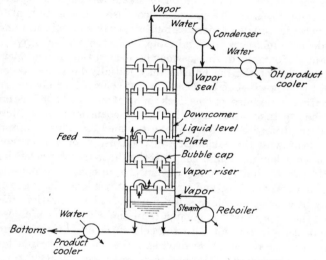

Fig. 12.1. Distilling column with auxiliaries.

only part of a liquid mixture, condensing the vapor formed, and boiling off only a part of the condensate, it is possible to obtain a nearly pure quantity of the most volatile compound by numerous repetitions of the procedure. Thus the separation by distillation is accomplished by partial vaporization and subsequent condensation.

In distillation it is customary to obtain a number of partial vaporizations and condensations by directly contacting a vapor and a liquid cooling medium in a continuous distilling column. The bubble cap distilling column shown in Fig. 12.1 is representative of modern practice and derives its name from a series of inverted slotted caps which are placed over vapor risers on each *plate* of the column. Vapor from below a plate enters the risers and is broken into bubbles as it passes through the slotted bottoms of the bubble caps and thence through the layer of liquid maintained by the *downcomer* on each plate. The *feed*, which is usually a liquid, is a mixture of more and less volatile compounds and enters the distilling column at the feed plate where the volatile compounds are partially vaporized by the rising vapors as the feed travels across the plate. The remainder of the liquid on the plate is less volatile than the feed and overflows to the plate below through the downcomer. The boiling points of the liquids on each of the lower plates is consequently higher.

To vaporize a portion of the feed, vapor from below the feed plate must exchange heat with the liquid on the feed plate, thereby driving the more volatile compounds to the plate above the feed. By supplying heat at the bottom of the column where the increased concentration of the least volatile compounds represents the highest boiling temperature in the system, a thermal gradient is established plate by plate between the bottom of the column and the top. Heat supplied at the bottom by vaporization in a *reboiler* is transmitted to the top of the column plate-by-plate due to the temperature differences corresponding to the differences in boiling points between plates. Continuous distillation requires the presence of liquid at all times on the plates, so that vapors of the less volatile compounds in the feed may be condensed and carried downward. To accomplish this, some volatile liquid from the condenser, which represents one plate above the top plate and which is therefore colder, is introduced onto the top plate and flows downward in the column. The volatile liquid which is poured back into the column from the condenser is the *reflux*. The quantity of volatile components removed from the system at the top and having the same composition as the reflux is called the *distillate* or *overhead* product. The heavier compounds removed at the bottom are variously called *waste* or *residue* or, if they are of value, *bottoms product*. The quantitative aspects of the heat balance are treated in Chap. 14.

It is the condensing temperature in the condenser which determines the operating pressure of the distilling column, since the saturation temperature of a vapor varies with its pressure. The overhead product must condense in the condenser at a temperature sufficiently high so that its latent heat can be removed by cooling water. The size of the condenser is dependent upon the difference between the condensing temperature and the range of the cooling water. If the condensing temperature is very close to the cooling water range at atmospheric pressure, the distillation pressure must be elevated to permit the attainment of a larger Δt.

In the power industry the term *surface condenser* is reserved for tubular equipment which condenses steam from the exhaust of turbines and engines. Since a turbine is primarily designed to obtain mechanical work from heat, the maximum conversion is obtained in the turbine by maintaining a low turbine-discharge temperature. If the turbine were to discharge to the atmosphere, the lowest attainable steam temperature would be 212°F, but if the steam were to discharge into a condenser under vacuum, it would be possible to operate at discharge temperatures of 75°F and lower and to convert the enthalpy difference from 212 to 75°F into useful work.

Condensation on Surfaces—Nusselt's Theory. In condensation on a vertical surface a film of condensate is formed as shown in Fig. 12.2 and further condensation and heat transfer to the surface occurs by conduction through the film which is assumed to be in laminar flow downward.

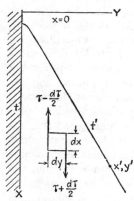

Fig. 12.2. The vertical condensate film.

The thickness of this film greatly influences the rate of condensation, since the heat accompanying the removal of vapors from the vapor phase encounters the condensate film as a resistance which may be quite large. The thickness of the film is a function of the velocity of drainage which varies with the deviation of the surface from a vertical position. For a vertical surface the thickness of the film cumulatively increases from top to bottom. For this reason the condensing coefficient for a vapor condensing on a vertical surface decreases from top to bottom, and for the attainment of a large condensing coefficient the height of the surface should not be very great. The velocity of drainage for equal quantities of condensate is also a function of the viscosity of the condensate: The lower the viscosity the thinner the film. For all liquids the viscosity decreases as the temperature increases, and the condensing coefficient consequently increases with the condensate temperature. The derivations given in this chapter through Eq. (12.34) are those of Nusselt.[1] The following assumptions are involved:

1. The heat delivered by the vapor is latent heat only.
2. The drainage of the condensate film from the surface is by laminar flow only, and the heat is transferred through the film by conduction.
3. The thickness of the film at any point is a function of the mean velocity of flow and of the amount of condensate passing at that point.
4. The velocity of the individual layers of the film is a function of the relation between frictional shearing force and the weight of the film (see Chap. 3).

[1] Nusselt, W., *Z. Ver. deut. Ing.*, **60**, 541 (1916).

CONDENSATION OF SINGLE VAPORS

5. The quantity of condensate is proportional to the quantity of heat transferred, which is in turn related to the thickness of the film and of the temperature difference between the vapor and the surface.
6. The condensate film is so thin that the temperature gradient through it is linear.
7. The physical properties of the condensate are taken at the mean film temperature.
8. The surface is assumed to be relatively smooth and clean.
9. The temperature of the surface of the solid is constant.
10. The curvature of the film is neglected.

Condensation. *Vertical Surfaces.* In Fig. 12.2 the rate at which heat passes from the vapor through the liquid condensate film and into the cooling surface per unit area is given by

$$\frac{Q}{A} = \frac{k(t' - t)}{y'} = \lambda W' = h(t' - t) \qquad (12.1)$$

where λ is the latent heat of vaporization, W' the pounds of condensate formed per hour per square foot, and y' is the thickness of the condensate film at the generalized point in the figure whose coordinates are x', y'. The other symbols have their conventional meaning.

The rate at which the vapor condenses is then given by

$$W' = \frac{k(t' - t)}{\lambda y'} \qquad (12.2)$$

The liquid flows downward over the vertical surface with a velocity u varying from zero at the tube-film interface and increasing outward to the condensate-vapor interface. The velocity also increases vertically as the condensate flows downward.

Consider a small cube of unit depth $dz = 1$, defined by $dx\, dy\, 1$ in the moving condensate film of Fig. 12.2. On the side near the cold vertical surface there is a tangential force acting upward and tending to support the cube. On the side away from the cold vertical surface there is a tangential force acting downward due to the more rapid movement of the liquid downward as the distance from the surface is increased. If the resultant force upward through the cube is designated by τ, then the respective forces are $\tau - d\tau/2$ and $\tau + d\tau/2$. The differential tangential force must be offset by gravity acting down.

$$\rho\, dx\, dy\, 1 = \left(\tau - \frac{d\tau}{dy}\frac{dy}{2}\right) - \left(\tau + \frac{d\tau}{dy}\frac{dy}{2}\right) = -d\tau \qquad (12.3)$$

On unit area, $dx\, dz = 1$

$$\rho = -\frac{d\tau}{dy}$$

From the basic definition of viscosity in Chap. 3 the tangential stress is defined by Eq. (3.3), using (lb-force)(hr)/ft² as the dimensions for the viscosity. Since it is customary to use the dimensions (lb-mass)/(ft)(hr) for the dimensions, Eq. (3.3) becomes

$$\tau = \frac{\mu}{g}\frac{du}{dy} \tag{12.4}$$

$$\frac{d\tau}{dy} = \frac{\mu}{g}\frac{d^2u}{dy^2} = -\rho \tag{12.5}$$

Take ρ/μ as constant.

$$\frac{d^2u}{dy^2} = -\frac{\rho g}{\mu} \tag{12.6}$$

$$u = -\frac{\rho g y^2}{2\mu} + C_1 y + C_2' \tag{12.7}$$

The constants C_1 and C_2 must now be evaluated. Since the liquid adheres to the wall, u must equal zero at $y = 0$, making C_2 equal zero.

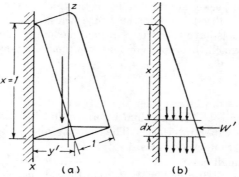

Fig. 12.3. Vertical condensate film flow.

At the outer boundary of the film (condensate-vapor interface) there is no tangential stress and from $\tau = \mu\, du/dy$ when $y = y'$,

$$\left(\frac{du}{dy}\right)_{y=y'} = 0 = -\frac{\rho g y'}{\mu} + C_1 \qquad C_1 = \frac{\rho g y'}{\mu} \tag{12.8}$$

$$u = \frac{\rho g}{\mu}\left(y y' - \frac{y^2}{2}\right)$$

At a distance x from the top of the condensing surface the average velocity downward $\bar{u}$ is given by

$$\bar{u} = \frac{1}{y'}\int_0^{y'} u\, dy = \frac{\rho g}{3\mu} y'^2 \tag{12.9}$$

CONDENSATION OF SINGLE VAPORS

When the value of x from the top of a vertical wall is taken to be unity as shown in Fig. 12.3a, the quantity of downward flow across a horizontal plane of area $1y'$ of the condensate is

$$1y'\bar{u}\rho$$

At $x + dx$ there is a gain in the amount of downward flow of condensate as shown in Fig. 12.3b. Using the value for $\bar{u}$ from Eq. (12.9), multiplying by $\rho y'$, and differentiating with respect to x to obtain the increase from x to $x + dx$,

$$d(\rho\bar{u}y') = d\left(\frac{\rho^2 g}{3\mu} y'^3\right) = \frac{\rho^2 g y'^2}{\mu} dy' \tag{12.10}$$

And this increase must come from condensation out of the vapor and into the condensate film

$$d(\rho\bar{u}y') = W'1\,dx$$

where W' is the condensate flow out of the vapor and "normal" to the falling condensate layer per unit area as in Fig. 12.3b. From Eq. (12.2), however, W' was defined in terms of the heat transfer as

$$W' = \frac{k(t' - t)}{\lambda y'}$$

Substituting for W' in Eq. (12.10) the value from Eq. (12.2),

$$\frac{k}{\lambda y'}(t' - t)\,dx = \frac{\rho^2 g y'^2}{\mu} dy' \tag{12.11}$$

$$(t' - t)\,dx = \frac{\rho^2 g \lambda y'^3}{k\mu} dy' \tag{12.12}$$

For a limited range set $t' - t$, ρ, λ, μ, and k constant and integrate. When $y' = 0$, $x = 0$

$$y'^4 \frac{\rho^2 g \lambda}{4\mu k} = (t' - t)x \tag{12.13}$$

$$y' = \left[\frac{4\mu k}{\rho^2 \lambda g}(t' - t)x\right]^{1/4} \tag{12.14}$$

The heat-transfer coefficient across the condensate layer at the distance x from the origin per unit of interfacial area is given from Eq. (12.1) by

$$h_x = \frac{Q_x/A_x}{t' - t} = \frac{k}{y'} \tag{12.15}$$

Substitute y' from Eq. (12.14):

$$h_x = \left[\frac{k^3 \rho^2 \lambda g}{4\mu(t' - t)}\right]^{1/4} \frac{1}{x^{1/4}} \tag{12.16}$$

The total heat through the condensate layer from 0 to x is Q_x

$$Q_x = \int_0^x h_x(t'-t)\,dx = \int_0^x \left[\frac{k^3\rho^2\lambda g}{4\mu(t'-t)}\right]^{1/4}(t'-t)\frac{dx}{x^{1/4}}$$

$$= \frac{4^{3/4}}{3}\left(\frac{k^3\rho^2\lambda g}{\mu}\right)^{1/4}[(t'-t)x]^{3/4} \tag{12.17}$$

If the average coefficient between the two points is $\bar{h}$,

$$\bar{h} = \frac{(Q_x)_{x=L}}{(t'-t)L} = \frac{4^{3/4}}{3}\frac{\left(\dfrac{k_f^3\rho_f^2\lambda g}{\mu_f}\right)^{1/4}[(t'-t)L]^{3/4}}{(t'-t)L}$$

$$\bar{h} = 0.943\left(\frac{k_f^3\rho_f^2\lambda g}{\mu_f L\,\Delta t_f}\right)^{1/4} \tag{12.18}$$

where k_f, ρ_f, and μ_f are evaluated at the film temperature t_f and where the film temperature is

$$t_f = \tfrac{1}{2}(t'+t) = \tfrac{1}{2}(T_v + t_w) \tag{12.19}$$

and

$$\Delta t_f = t_f - t_w$$

In the above, as in the derivation which follows, the stress caused by the passage of the saturated vapor over the condensate-vapor interface has been neglected. It can be included, although it is not of practical consequence. The variation of the film thickness and local heat-transfer coefficient are shown in Fig. 12.4. The shapes of the curves follow the thickness and consequently the resistance of the condensate film.

Inclined Surfaces. Consider a cube making the angle α as shown in Fig. 12.5. The gravity component acting in a plane parallel to the surface is $\rho \sin \alpha$, and Eq. (12.3) becomes

$$\rho \sin \alpha \, dy\, dx\, 1 = \left(\tau - \frac{d\tau}{dy}\frac{dy}{2}\right) - \left(\tau + \frac{d\tau}{dy}\frac{dy}{2}\right) = -d\tau \tag{12.20}$$

On unit area $dx\,dz = 1$

$$\rho \sin \alpha = -\frac{d\tau}{dy} \tag{12.21}$$

Equation (12.6) becomes

$$\frac{d^2u}{dy^2} = -\frac{g\rho}{\mu}\sin \alpha \tag{12.22}$$

Equation (12.7) becomes

$$u = -\frac{g\rho}{2\mu}y^2 \sin \alpha + C_1 y + C_2 \tag{12.23}$$

At the start of condensation on the tube where $y = 0$ and there is no velocity along the tube, $u = 0$ and $C_2 = 0$

$$C_1 = \frac{g\rho y'}{\mu}\sin \alpha$$

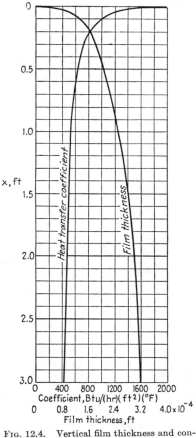

Fig. 12.4. Vertical film thickness and condensing coefficients for a descending film. (*After Nusselt.*)

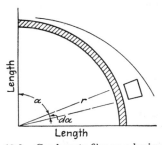

Fig. 12.5. Film on an inclined surface

Fig. 12.6. Condensate film on a horizontal tube.

and Eq. (12.9) becomes

$$\bar{u} = \frac{g\rho y'^2}{3\mu} \sin \alpha \qquad (12.24)$$

and Eq. (12.18) becomes

$$\bar{h} = 0.943 \left(\frac{k_f^3 \rho_f^2 \lambda g \sin \alpha}{\mu_f L \, \Delta t_f} \right)^{\frac{1}{4}} \qquad (12.25)$$

Horizontal Tubular Surfaces. Consider a cube of unit length at radius r making an angle α with the vertical as seen in Fig. 12.6. The mass flow of vapor into the condensate film through the area $r \, d\alpha$ and with a film thickness y' is given by the conduction equation.

$$W' = \frac{k(t' - t)r\, d\alpha}{\lambda y'}$$

The condensation must give rise to an increase in the inclined falling film. For a differential amount of condensation the increase through the condensing area $r\, d\alpha$ is $d(\rho \bar{u} y')$ and for Eq. (12.10)

$$d(\rho \bar{u} y') = \frac{\rho^2 g}{3\mu} d(y'^3 \sin \alpha) = W'\, dx$$

Substituting for W', Eq. (12.11) becomes

$$\frac{k(t' - t)r\, d\alpha}{\lambda y'} = \frac{\rho^2 g}{3\mu} d(y'^3 \sin \alpha)$$

$$\frac{3\mu k(t' - t)r\, d\alpha}{\rho^2 g \lambda} = y'\, d(y'^3 \sin \alpha)$$

Let

$$m = \frac{3\mu k(t' - t)r}{\rho^2 g \lambda}$$

$$m\, d\alpha = y'\, d(y'^3 \sin \alpha)$$

Differentiating

$$m\, d\alpha = y'(3y'^2 \sin \alpha\, dy' + y'^3 \cos \alpha\, d\alpha)$$
$$= 3y'^3 \sin \alpha\, dy' + y'^4 \cos \alpha\, d\alpha \qquad (12.26)$$

In Eq. (12.26) the term $3y'^3\, dy'$ appears, but $d(y'^4) = 4y'^3\, dy'$ and

$$3y'^3\, dy' = \tfrac{3}{4}\, dy'^4.$$

Rearranging Eq. (12.26) and substituting,

$$d\alpha = \frac{3}{4m} \sin \alpha\, dy'^4 + \frac{y'^4}{m} \cos \alpha\, d\alpha$$

Let $y'^4/m = \psi^4 = z$

$$d\alpha = \tfrac{3}{4} \sin \alpha\, dz + z \cos \alpha\, d\alpha \qquad (12.27)$$

$$\tfrac{3}{4} \sin \alpha \frac{dz}{d\alpha} + z \cos \alpha - 1 = 0 \qquad (12.28)$$

Equation (12.28) is a linear differential equation whose solution is

$$z = \frac{1}{\sin^{4/3} \alpha} \left(\frac{4}{3} \int \sin^{1/3} \alpha\, d\alpha + C_3 \right) \qquad (12.29)$$

When $\alpha = 0$, $C_3 = 0$,

$$z = \frac{4}{3} \frac{1}{\sin^{4/3} \alpha} \int \sin^{1/3} \alpha\, d\alpha \qquad (12.30)$$

CONDENSATION OF SINGLE VAPORS 263

The value of this integral for different values of α may be determined by graphical methods. From the substitution in Eq. (12.27)

$$y' = \psi m^{1/4} = \psi \left[\frac{3\mu k(t' - t)r}{\rho^2 g \lambda} \right]^{1/4} \quad (12.31)$$

As shown in Eq. (12.15), $h_x = k/y'$.

The thickness of the film actually decreases slightly as α increases from 0 to 5°, and then it increases steadily and breaks into drops. The local-heat transfer coefficient at any point is then

$$h_\alpha = \frac{k}{y'} = \frac{1}{\psi} \left[\frac{k^3 \rho^2 \lambda g}{3\mu(t' - t)r} \right]^{1/4} \quad (12.32)$$

The average heat-transfer coefficient $\bar{h}_\alpha$ of the segment between angles α_1 and α_2 is

$$\bar{h}_\alpha \Big]_{\alpha_1}^{\alpha_2} = \frac{k}{m^{1/4}(\alpha_2 - \alpha_1)} \int_{\alpha_1}^{\alpha_2} \frac{d\alpha}{\psi} \quad (12.33)$$

Employing graphical methods as before, where D_o is the outside diameter of the tube, the average heat-transfer coefficients are found to be

$$\bar{h}_\alpha \Big]_{0°}^{90°} = 0.860 \left(\frac{k^3 \rho^2 \lambda g}{\mu D_o \Delta t_f} \right)^{1/4}$$

$$\bar{h}_\alpha \Big]_{90°}^{180°} = 0.589 \left(\frac{k^3 \rho^2 \lambda g}{\mu D_o \Delta t_f} \right)^{1/4}$$

From 0 to 180° which is one-half the tube, the other half being symmetrical,

$$\bar{h} = 0.725 \left(\frac{k_f^3 \rho_f^2 \lambda g}{\mu_f D_o \Delta t_f} \right)^{1/4} \quad (12.34)$$

The variation of the film thickness and heat-transfer coefficient for steam on a horizontal tube is shown in Fig. 12.7. As in the preceding case it is governed by the resistance of the condensate film to conduction.

Development of Equations for Calculations. McAdams[1] found from the correlation of the data of several investigations that observed condensing coefficients for steam on vertical tubes were 75 per cent greater than the theoretical coefficients calculated by Eq. (12.18). The values calculated from Eq. (12.18) agree, however, for a condensate in streamline flow with the values calculated from Eq. (6.1) for ordinary streamline flow.

When a liquid descends vertically on the outside of a tube, it is certainly

[1] McAdams, *op. cit.*, p. 264.

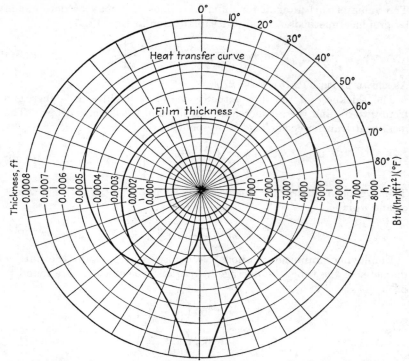

Fig. 12.7. Plot of the heat-transfer coefficient and film thickness of water on a horizontal tube. (*After Nusselt.*)

in streamline flow at the top of the tube, where the accumulation of condensate is small. If a relatively large amount of vapor is condensed on the tube, it is possible that at some point below the top the film will change to turbulent flow. This may be estimated from the diameter and length of the tube, the viscosity of the condensate, and the quantity being condensed. Referring to the tube as shown in Fig. 12.8, the crosshatched area outside the tube represents condensate film as seen at any point looking down. This is similar to the flow in the annulus of a double pipe exchanger except that the outer surface of the film is not formed by a concentric pipe. In the case of the double pipe exchanger the equivalent diameter was taken as four times the hydraulic radius.

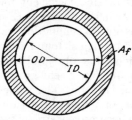

Fig. 12.8. Vertical descending film.

CONDENSATION OF SINGLE VAPORS

Then
$$D_e = 4r_h = 4 \times \frac{\text{free flow area}}{\text{wetted perimeter}}$$
and
$$Re = \frac{D_e G}{\mu}$$

For *vertical* tubes let
  A_f = cross-sectional area (shaded)
  P = wetted perimeter per tube
  $D_e = 4 \times A_f/P$

Letting the loading per tube be $w' = W/N_t$, where N_t is the number of tubes,
$$G = w'/A_f \quad \text{lb/(hr)(ft}^2\text{)}$$
$$Re = D_e G/\mu = (4A_f/P)(w'/A_f)/\mu = 4w'/\mu P \quad (12.35)$$

Calling the condensate loading per linear foot G',
$$G' = \frac{w'}{P}, \quad \text{lb/(hr)(lin ft)} \quad (12.36)$$

Eq. (12.35) becomes
$$Re = \frac{4G'}{\mu}$$

The total heat load is given by $Q = \lambda w'$.
$$\bar{h} = \frac{Q}{A \, \Delta t_f} = \frac{\lambda w'}{PL \, \Delta t_f} = \frac{\lambda}{L} \frac{G'}{\Delta t_f} \quad (12.37)$$

Substituting in Eq. (12.18),
$$\bar{h} = 0.943 \left(\frac{k_f^3 \rho_f^2 g}{\mu_f} \frac{\bar{h}}{G'} \right)^{1/4} \quad (12.38)$$

Multiplying the right term by $(4\mu/4\mu)^{1/4}$,
$$\bar{h}^{3/4} = 0.943 \left(\frac{4k_f^3 \rho_f^2 g}{\mu_f^2} \frac{\mu_f}{4G'} \right)^{1/4}$$
$$\bar{h} \left(\frac{\mu_f^2}{k_f^3 \rho_f^2 g} \right)^{1/3} = 1.47 \left(\frac{4G'}{\mu_f} \right)^{-1/3} \quad (12.39)$$

For *horizontal* tubes Eq. (12.39) becomes
$$\bar{h} \left(\frac{\mu_f^2}{k_f^3 \rho_f^2 g} \right)^{1/3} = 1.51 \left(\frac{4G''}{\mu_f} \right)^{-1/3} \quad (12.40)$$

where the loading for a single horizontal tube is
$$G'' = \frac{W}{LN_t} \quad (12.41)$$

Using the corresponding loading as given by either Eq. (12.36) or (12.41) as the case may be, Eqs. (12.39) and (12.40) may both be represented by

$$\bar{h}\left(\frac{\mu_f^2}{k_f^3\rho_f^2 g}\right)^{1/3} = 1.5 \left(\frac{4G'}{\mu_f}\right)^{-1/3} = 1.5 \left(\frac{4G''}{\mu_f}\right)^{-1/3} \qquad (12.42)$$

Equations (12.39) and (12.40) were obtained for condensation on single tubes. In a vertical-tube bundle the presence of one or more tubes does not alter the assumptions on which the derivation was predicated. However, on horizontal tubes in tube bundles it has been found that the splashing of the condensate as it drips over successive rows of tubes causes G'' to be more nearly inversely proportional to $N_t^{2/3}$ rather than N_t so that it is preferable to use a fictitious value for horizontal tubes

$$G'' = \frac{W}{LN_t^{2/3}} \qquad \text{lb/(hr)(lin ft)} \qquad (12.43)$$

Figure 12.9 is a line chart of solutions of Eq. (12.42) prepared for convenience. Its use requires that the film be in streamline flow corresponding to an average Reynolds number of about 1800 to 2100 for the flow gradient assumed by the condensate. For steam at atmospheric pressure Eq. (12.42) reduces to the equations given by McAdams:[1] For horizontal tubes

$$\bar{h} = \frac{3100}{D_o^{1/4} \Delta t_f^{1/3}} \qquad (12.44a)$$

and for vertical tubes

$$\bar{h} = \frac{4000}{L^{1/4} \Delta t_f^{1/3}} \qquad (12.44b)$$

where Δt_f ranges from 10 to 150°F.

It is frequently desirable to apply Eqs. (12.39), (12.40), and (12.42) to the calculation of condensers which are modifications of the 1-2 exchanger with condensation in the shell. Such condensers have baffled tube bundles. The baffles do not affect the condensing film coefficients in horizontal condensers, since the coefficients are independent of the vapor mass velocity, but they do influence the accumulation of condensate on the tubes of vertical condensers. Moreover, in condensers with multipass tubes the tube-wall temperature is different at every point in each pass, whereas the surface temperature was assumed constant in the derivations. A correction for the latter cannot be accounted for in the calculations except by the treatment of small surface increments of each pass individually. The error introduced by using the mean tube-wall temperature as being effective over all the surface is apparently too small to justify the lengthier calculation. Since the baffle holes are ordinarily

[1] *Ibid.*, p. 270.

CONDENSATION OF SINGLE VAPORS

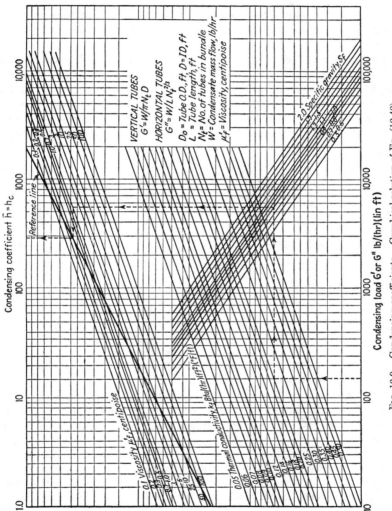

Fig. 12.9. Condensing coefficients. Graphical solution of Eq. (12.42).

about $\frac{1}{32}$ in. greater in diameter than the tube outside diameter, the baffles in vertical condensers prevent the condensate film from reaching a thickness greater than $\frac{1}{64}$ in. before impinging on the baffle. This, however, is a favorable limitation except at high tube loadings where the condensate film might otherwise grow sufficiently to change to turbulent flow.

Comparison between Horizontal and Vertical Condensation. The value of the condensing film coefficient for a given quantity of vapor on a given surface is significantly affected by the position of the condenser. In a vertical tube about 60 per cent of the vapor condenses in the upper half of the tube. By combining Eqs. (12.18) and (12.34) the ratio of the theoretical horizontal and vertical condensing coefficients is given by

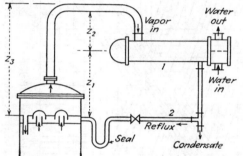

Fig. 12.10. Condenser with gravity return of reflux.

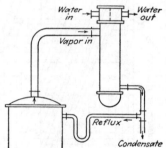

Fig. 12.11. Vertical condenser with condensation in shell and gravity return of reflux.

$(0.725/0.943)(L/D)^{1/4}$. For a ¾ in. OD tube 16′0″ long, the horizontal coefficient should be 3.07 times as great as the vertical coefficient, provided that the condensate film is in streamline flow throughout. Ordinarily, however, the advantage is not quite so great due to the other modifications which prevail such as a transition of the vertical film into turbulent flow.

For the condensation of exhaust steam from turbines with vacuum discharge the condenser surface is usually very great, from 10,000 to 60,000 ft^2 per shell, and the economics are such that tubes up to 26 ft long are employed. These large condensers are designed with overall transfer coefficients as high as 800 Btu/(hr)(ft^2)(°F) as treated later. Condensers for such services are universally installed in a horizontal position to facilitate the distribution of the vapor and the removal of the condensate.

When a condenser is employed on a distillation column, several specific factors must be taken into consideration. A typical arrangement of such

CONDENSATION OF SINGLE VAPORS

a condenser is shown in Fig. 12.10, in which the reflux is returned to the column by gravity. The condensate leg 1-2 of height z_1, must supply sufficient hydrostatic head to return the condensate to the column through the seal. In Fig. 12.11 a vertical condenser is employed for the same service, but it is not well suited for gravity return of the condensate, since it must be elevated considerably above the column, which in many cases is very tall by itself. Maintenance and structural support for the vertical condenser may be costly and considerably more difficult. On the other hand if it is desired not only to condense the overhead vapor but also to subcool the condensate, the vertical condenser is admirably suited. *Subcooling* is the operation of cooling the condensate below its saturation temperature, and this is done very frequently when the overhead product is a volatile liquid to be sent to storage. By subcooling it is possible to avoid large evaporation losses during initial storage. The combination of condensation and subcooling in a single unit eliminates the need for a separate overhead product cooler, as shown in Fig. 12.1.

Condensation inside Tubes. *Horizontal Condensers.* The equations developed so far give excellent results when applied to condensation outside tubes, although the deviations in commercial condensers have not been reported except in isolated instances. Often, however, the condensate is corrosive, or it is desired to recover the latent heat from the vapor by using it to preheat the feed to a column. In such cases it may be preferable to condense the vapor in the tubes rather than on the tubes wherein the original derivation is no longer applicable. Within the tubes of a single-pass horizontal condenser each tube condenses an equal amount of vapor and there is no change in the coefficient due to the splashing of condensate from one tube row to another. As the condensate flows along the inside bottoms of the tubes, however, it builds up a thicker condensate resistance film than that anticipated in the derivation. Little is available of a theoretical nature to permit a rational analysis, but it has been found that the film coefficient may be safely computed by Eq. (12.40) when G'', which is theoretically W/LN_t, is replaced by the fictitious loading

$$G'' = \frac{W}{0.5LN_t} \qquad \text{lb/(hr)(lin ft)} \qquad (12.45)$$

Equation (12.45) is especially useful when condensation occurs in the inner pipe of a double pipe exchanger. For condensation in the tubes of a condenser with multipass tubes it is preferable to compute the average film coefficient for each pass. The condensate formed in the first pass is carried through the second pass by one or more of the lower tubes in the pass, which may flow full of condensate and therefore expose no surface for condensation. The calculation is then carried out by trial and error to determine the true tube loading in each pass.

Vertical Condensers. Condensation inside vertical tubes follows essentially the same mechanism as condensation outside vertical tubes if the interference of the shell baffles is neglected. Since the condensate film has the ability to grow continuously in its descent down the inside or outside of tubes, it may change from streamline to turbulent flow at some height between the top and the bottom. The local condensing coefficient decreases continuously from the top downward until at some point the film changes to turbulent flow. After the transition to turbulent flow the coefficient increases in accordance with the usual behavior of forced convection. By semiempirical means Colburn[1] has combined the effect

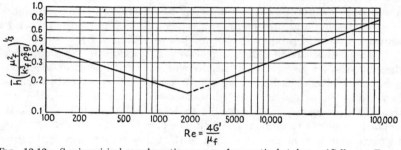

Fig. 12.12. Semiempirical condensation curve for vertical tubes. (*Colburn, Transactions of American Institute of Chemical Engineers.*)

of streamline flow in the upper portion of the tube with that of turbulent flow below the point where $4G'/\mu_f = 2100$. This required the selection of a heat-transfer factor for forced convection such that h at the transition point was roughly the same for both Nusselt condensation and turbulent flow. He then obtained the mean coefficient for the *entire height* of the tube by weighting the average coefficient for the upper portion of the tube and h for forced convection in the lower portion of the tube. The mean coefficients of condensation for the entire tube when $4G'/\mu_f > 2100$ are given in Fig. 12.12. This plot contains the values recommended by McAdams. The distance from the top of the tube at which the change from laminar to turbulent flow occurs may be obtained from Nusselt's semiempirical derivation based on the transition occurring at $4G'/\mu_f = 1400$ for steam and employing Eqs. (12.8) and (12.14). If x_c is the distance in feet from the top,

$$x_c = \frac{2668\lambda\mu_f^{5/3}}{\rho^{2/3}kg^{1/3}(T_v - t_w)} \quad \text{ft} \qquad (12.46)$$

[1] Colburn, A. P., *Trans. AIChE*, **30**, 187–193 (1934).

Where there is evidence that the transition occurs at a higher value than $4G'/\mu_f = 1400$, as for organic vapors, the value of x_c should be multiplied by the ratio of the corrected value of $4G'/\mu_f$ divided by 1400.

Naturally if a vertical condenser is to operate with condensation inside the tubes, it will have but one tube pass as shown in Fig. 12.13.

Condenser Calculations. Condensers are better classified by what goes on inside them than by their process locations or services. Often in addition to condensation they may also perform vapor desuperheating or condensate subcooling so that a separate shell need not be employed for the sensible-heat transfer. A convenient classification in which each class is indicative of a different modification of the calculation is given below:

1. Single vapors (the vapor of a pure compound or a constant boiling mixture)
   a. Saturated vapor: Total or partial condensation outside tubes
   b. Superheated vapor: Desuperheating and condensation outside tubes
   c. Saturated vapor: Condensation and subcooling outside tubes
   d. Condensation inside tubes: Desuperheating, condensing, subcooling
   e. Condensation of steam
2. Vapor mixtures (Chap. 13). The application of the phase rule.
   a. Binary mixture
   b. Vapor mixture with long condensing range
   c. Vapor mixtures forming immiscible condensates
   d. Single vapor or vapors with noncondensable gas
   e. Vapor mixtures and noncondensable gases forming immiscible condensates

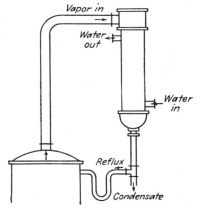

FIG. 12.13. Vertical condenser with condensation in the tubes and gravity return of reflux.

For cases 1a through 1d the majority of condensers are modifications of the 1-2 exchanger and may be referred to as 1-2 condensers. The use of a 1-2 exchanger as a condenser usually requires provision for a larger entrance space so that the vapor is not subjected to too great a pressure drop on entering the shell. This may be accomplished in either of three ways: The vapor may be introduced through a vapor belt as shown in Fig. 12.14 or by means of a flared shell inlet nozzle as shown in Fig. 12.15. A third method is the elimination of some of the tubes from the bundle which lie close to the inlet nozzle.

Although condensation reduces the volume of the vapor, it occurs at constant pressure except for the frictional pressure drop between the inlet and the outlet. In a horizontal condenser using conventional seg mental baffles it is imperative that they be arranged for side-to-side flow and not up-and-down flow as shown in Fig. 7.6. This involves merely rotating the bundle 90° before bolting the channel to the shell flange. Failure to provide side-to-side flow causes pools of condensate to form between each pair of baffles whose cutout areas are on the top of the shell, thereby impeding the passage of vapor.

The Allowable Pressure Drop for a Condensing Vapor. In the original Nusselt assumptions the condensing coefficient was considered to be independent of the vapor velocity across the bundle of the condenser and to depend only upon G' or G'', the loading as pounds of condensate per hour per linear foot. It is customary for the sake of good vapor distribution to have the vapor travel across the bundle as rapidly as the pressure-drop considerations will allow and to space the baffles accordingly. To account for the shrinkage in the total pounds of vapor as it travels the

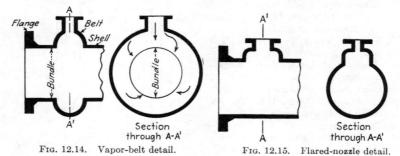

FIG. 12.14. Vapor-belt detail.   FIG. 12.15. Flared-nozzle detail.

length of the bundle, the baffle spacing may be *staggered* to give a nearly constant vapor mass velocity. At the inlet the baffles are spaced far apart but are spaced more closely toward the outlet. The use of staggered pitch is not universally accepted, since it reduces the adaptability of the equipment to other services in the event that the original process is discontinued. With uniform baffle pitch the condenser may be readily adapted to gas-liquid and liquid-liquid heat exchange.

In distillation the allowable pressure drop for a vapor coming off a column is usually very small if the condenser is installed for the gravity return of the reflux to the column. Referring to Fig. 12.10 the hydraulic circuit consists of the weight of the overhead vapor line $z_3 - z_2$, condenser pressure drop ΔP_c, the weight of the condensate leg z_1, and the pressure drop in the condensate return line, which is usually neglected. The equation is given closely in pounds per square inch by

$$\frac{\rho_v z_1}{144} + \Delta P_c = \frac{\rho_l z_1}{144}$$

where ρ_v = density of vapor, lb/ft³
ρ_l = density of liquid, lb/ft³
ΔP_c = pressure drop in condenser, psi

Expansion and contraction losses from and to the tower have also been neglected. For all practical purposes the minimum available hydrostatic head must exceed $\rho_l z_1 = 144 \Delta P_c$. To this must usually be added an additional factor to permit the use of a flow-control valve in the reflux line. For a liquid with a specific gravity of 1.0 the minimum elevation of the condensate line above the level of the top plate for a 2.0 psi allowable pressure drop in the condenser is $(2.0/14.7) \times 34 = 4.62$ ft, and for liquids of lower specific gravity the elevation of the condenser will be proportionately greater. For this reason the condenser is frequently installed at the ground level and a condensate return pump is substituted for gravity return of reflux. This is especially true at higher pressures, where hydraulic control is more difficult and serious surging may occur at the top of the column. To prevent surging the condensate leaving the condenser first enters a condensate accumulator and then goes to the pump. With pumped return a vapor pressure drop of 5 psi is often allowed in the condenser. For gravity systems the allowable vapor pressure drop in the condenser will be usually 1 to 2 psi.

In the condensation of a pure saturated vapor, the vapor enters the condenser at its saturation temperature and leaves as a liquid. The pressure drop is obviously less than that which would be calculated for a gas at the inlet specific gravity of the vapor and greater than that which would be computed using the outlet specific gravity of the condensate. The mass velocity of inlet vapor and outlet liquid are, however, the same. In the absence of extensive correlations reasonably good results are obtained when the pressure drop is calculated for a mass velocity using the total weight flow and the average specific gravity between inlet and outlet. This method can be further simplified, as in the condensation of steam, by taking one-half the conventional pressure drop computed entirely on inlet conditions. Thus for condensation in the shell,

$$\Delta P_s = \frac{1}{2} \frac{fG_s^2 D_s(N+1)}{5.22 \times 10^{10} D_e s} \tag{12.47}$$

where s is the specific gravity of vapor. For condensation in tubes

$$\Delta P_t = \frac{1}{2} \frac{fG_t^2 L n}{5.22 \times 10^{10} D_e s} \tag{12.48}$$

where s is the specific gravity of the vapor. No contraction or expansion losses need be considered. Both Eqs. (12.47) and (12.48) are on the safe side, since the mass velocity of the vapor decreases nearly linearly in the presence of a large Δt from inlet to outlet whereas the pressure drop decreases as the square of the velocity.

When circumstances make it difficult to meet the allowable drop through a 1-2 condenser, a lower pressure drop can be obtained by resorting to split flow as shown in Fig. 12.16. Still another arrangement

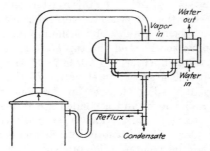

Fig. 12.16. Condenser with split flow and gravity return of reflux.

whose characteristics will be discussed later is the double-flow or *divided-flow* condenser shown in Fig. 12.17. This condenser consists of a conventional bundle with a removable longitudinal baffle and transverse support plate as well as smaller baffles to induce side-to-side flow of the vapor and condensate. The support plate, in addition to supporting all

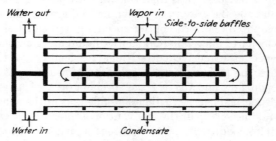

Fig. 12.17. Divided-flow 1-2 condenser.

the tubes, also serves to divide the flow. To prevent the condensate and vapor in the upper half of the shell from passing between the longitudinal baffle and the inside diameter of the shell, it is customary to provide sealing strips, which force the vapor and condensate to travel the half lengths of the divided flow condenser.

Example 12.1. Calculation of a Horizontal n-Propanol Condenser. A horizontal 1-2 condenser is required for the condensation of 60,000 lb/hr of substantially pure n-propanol (propyl alcohol) coming from the top of a distilling column operating at

CONDENSATION OF SINGLE VAPORS 275

15 psig, at which pressure it boils at 244°F. Water at 85°F will be used as the cooling medium. A dirt factor of 0.003 is required with allowable pressure drops of 2.0 psi for the vapor and 10.0 psi for the water.

Because of the location of the condenser, assume that 8'0" tubes are used. Tubes are to be ¾ in. OD, 16 BWG on 1 5⁄16-in. triangular pitch.

Solution:

(1) Heat balance:
 n-Propanol, $Q = 60{,}000 \times 285 = 17{,}100{,}000$ Btu/hr
 Water, $Q = 488{,}000 \times 1(120 - 85) = 17{,}100{,}000$ Btu/hr

(2) Δt:

	Hot Fluid		Cold Fluid	Diff.
	244	Higher Temp	120	124
	244	Lower Temp	85	159
	0	Differences	35	35

$\Delta t = \text{LMTD} = 141°\text{F}$ \hfill (5.14)

The exchanger is in true counterflow, since the shell-side fluid is isothermal.

(3) T_c and t_c: The influence of the tube-wall temperature is included in the condensing film coefficient. The mean $t_a = 102.5°\text{F}$ can be used for t_c.

Trial:

(a) Assume $U_D = 100$: Condensing film coefficients will generally range from 150 to 300. Assuming a film coefficient of 1000 for water U_C will range from 130 to 230.

$A = \dfrac{Q}{U_D \, \Delta t} = \dfrac{17{,}100{,}000}{100 \times 141} = 1213 \text{ ft}^2$

Number of tubes $= \dfrac{1213}{8'0'' \times 0.1963} = 773$

(b) Assume four tube passes: The quantity of water is large, but the condenser will have a large number of tubes, making a two-pass assumption inadvisable.

From the tube counts (Table 9): 773 tubes, four passes, ¾ in. OD on 1 5⁄16-in. triangular pitch

Nearest count: 766 tubes in a 31 in. ID shell

(c) Corrected coefficient U_D:
$A = 766 \times 8'0'' \times 0.1963 = 1205 \text{ ft}^2$

$U_D = \dfrac{Q}{A \, \Delta t} = \dfrac{17{,}100{,}000}{1205 \times 141} = 101$

Hot fluid: shell side, n-propanol

(4') Assume maximum baffle space. This will be 32½, 31, and 32½ in. equal to 96 in. or 2 baffles and 3 crosses for side-to-side flow.

$a_s = \text{ID} \times C'B/144P_T$ \hfill [Eq. (7.1)]
$ = 31 \times 0.1875 \times 31/144 \times 0.937$
$ = 1.34 \text{ ft}^2$

Cold fluid: tube side, water

(4) Flow area, $a_t' = 0.302 \text{ in.}^2$ [Table 10;
$a_t = N_t a_t'/144n$ \hfill [Eq. (7.48)]
$ = 766 \times 0.302/144 \times 4 = 0.402 \text{ ft}^2$

Hot fluid: shell side, n-propanol

(5') $G_s = W/a_s$ (for pressure drop only) [Eq. (7.2)]
$= 60,000/1.34$
$= 44,700$ lb/(hr)(ft^2)

Loading, $G'' = W/LN_t^{2/3}$ [Eq. (12.43)]
$= 60,000/8 \times 766^{2/3}$
$= 89.3$ (lb)/(hr)(lin ft)

Assume $\bar{h} = h_o = 200$
From (10) $h_{io} = 1075$

$t_w = t_a + \dfrac{h_o}{h_{io} + h_o}(T_v - t_a)$ [Eq. (5.31)]

$= 102.5 + 200/1275(244 - 102.5)$
$= 125°F$
$t_f = (T_v + t_w)/2$ [Eq. (12.19)]
$= (244 + 125)/2 = 184.5°F$
$k_f = 0.094$ Btu/(hr)(ft^2)(°F/ft) [Table 4]
$s_f = 0.80$ [Table 6]
$\mu_f = 0.62$ cp [Fig. 14]
From Fig. 12.9 or Eq. (12.42)
$\bar{h} = h_o = 172$ Btu/(hr)(ft^2)(°F)

Cold fluid: tube side, water

(5) $G_t = w/a_t$
$= 488,000/0.402$
$= 1,210,000$ lb/(hr)(ft^2)

Vel, $V = G_t/3600\rho$
$= 1,210,000/3600 \times 62.5 = 5.42$ fps

(6) At $t_a = 102.5°F$,
$\mu = 0.72 \times 2.42 = 1.74$ lb/(ft)(hr)
$D = 0.62/12 = 0.0517$ ft [Table 10]
$Re_t = DG_t/\mu$ (for pressure drop only)
$= 0.0517 \times 1,210,000/1.74 = 36,200$
(9) $h_i = 1300$ [Fig. 25]
(10) $h_{io} = h_i \times ID/OD$ [Eq. (6.5)]
$= 1300 \times 0.62/0.75$
$= 1075$ Btu/(hr)(ft^2)(°F)

Based on $\bar{h} = 172$ instead of the assumed 200 a new value of t_w and t_f could be obtained to give a more exact value of $\bar{h}$ based on fluid properties at a value of t_f more nearly correct. It is not necessary in this example because the condensate properties will not change materially.

Pressure Drop

(1') At $T_v = 244°F$
$\mu_{\text{vapor}} = 0.010 \times 2.42$
$= 0.0242$ lb/(ft)(hr) [Fig. 15]
$D_e = 0.55/12 = 0.0458$ ft [Fig. 28]
$Re_s = D_e G_s/\mu$ [Eq. (7.3)]
$= 0.0458 \times 44,700/0.0242 = 84,600$
$f = 0.00141$ ft^2/in.^{2} [Fig. 29]

(2') No. of crosses, $N + 1 = 3$

Mol. wt. = 60.1

Density, $\rho = \dfrac{60.1}{359(704/492)(14.7/29.7)}$
$= 0.238$ lb/ft^3

$s = 0.238/62.5 = 0.00381$
$D_s = 31/12 = 2.58$ ft

(3') $\Delta P_s = \dfrac{1}{2} \dfrac{fG_s^2 D_s(N+1)}{5.22 \times 10^{10} D_e s}$
[Eq. (12.47)]

$= \dfrac{1}{2} \dfrac{0.0014 \times 44,700^2 \times 2.58 \times 3}{5.22 \times 10^{10} \times 0.0458 \times 0.00381}$
$= 1.2$ psi

(1) For $Re_t = 36,200$,
$f = 0.00019$ ft^2/in.^{2} [Fig. 26]

(2) $\Delta P_t = \dfrac{fG_t^2 Ln}{5.22 \times 10^{10} Ds\phi_t}$ [Eq. (7.45)]

$= \dfrac{0.00019 \times 1,210,000^2 \times 8 \times 4}{5.22 \times 10^{10} \times 0.0517 \times 1.0 \times 1.0}$
$= 3.3$ psi

(3) $\Delta P_r = (4n/s)(V^2/2g')$ [Eq. (7.46)]
$= (4 \times 4/1)0.20 = 3.2$ psi
(4) $\Delta P_T = \Delta P_t + \Delta P_r$ [Eq. (7.47)]
$= 3.3 + 3.2 = 6.5$ psi

CONDENSATION OF SINGLE VAPORS

(13) Clean overall coefficient U_C:

$$U_C = \frac{h_{io}h_o}{h_{io} + h_o} = \frac{1075 \times 172}{1075 + 172} = 148.5 \text{ Btu/(hr)(ft}^2\text{)(°F)} \quad (6.38)$$

(14) Dirt factor R_d:[1] U_D from $(c) = 101$

$$R_d = \frac{U_C - U_D}{U_C U_D} = \frac{148.5 - 101}{148.5 \times 101} = 0.0032 \text{ (hr)(ft}^2\text{)(°F)/Btu} \quad (6.13)$$

Summary

172	h outside	1075
U_C	148.5	
U_D	101	
R_d Calculated	0.0032	
R_d Required	0.003	
1.2	Calculated ΔP	6.5
2.0	Allowable ΔP	10.0

The first trial exchanger is satisfactory and will be

Shell side
ID = 31 in.
Baffle space = 31 in. (approx.)
Passes = 1

Tube side
Number and length = 766, 8'0''
OD, BWG, pitch = ¾ in., 16 BWG, 1 5⁄16-in. tri.
Passes = 4

It is interesting at this point to compare a vertical condenser with the horizontal condenser which fulfilled the process conditions of Example 12.1. The horizontal and vertical condensing film coefficients are both affected by W and N_t, and the best basis for comparison is obtained when the number of tubes in both condensers is the same. To this end a vertical condenser will be assumed which employs the same tube count as that of the preceding example except that the tube length may be 12 or 16 ft as needed to account for the lower coefficients obtained in vertical condensation.

Example 12.2. Design of a Vertical n-Propanol Condenser. Process conditions will be taken from Example 12.1. To prevent corrosion of the shell the water will flow in the tubes.

Solution:

(1) Heat balance: same as Example 12.1, $Q = 17,100,000$ Btu/hr
(2) Δt: same as Example 12.1, $\Delta t = 141°F$
(3) T_c and t_c: same as Example 12.1

[1] In condensation calculations the omission of the resistance of the tube metal may introduce a significant error and should be checked.

Trial:

(a) Assume $U_D = 70$. The equation for the condensing film coefficient gives greater values for horizontal tubes than for vertical tubes. It will consequently be necessary to reduce the value of U_D.

$$A = \frac{Q}{U_D \, \Delta t} = \frac{17{,}100{,}000}{70 \times 141} = 1730 \text{ ft}^2$$

Nearest common tube length:

$$\frac{1730}{766 \times 0.1963} = 11.5 \text{ ft} \qquad (\text{use } 12'0'')$$

(b) The layout of Example 12.1, using ¾ in. OD tubes on 15⁄16-in. triangular pitch and four passes will be retained for comparison.

(c) Corrected coefficient U_D:

$$A = 766 \times 12'0'' \times 0.1963 = 1805 \text{ ft}^2$$
$$U_D = \frac{Q}{A \, \Delta t} = \frac{17{,}100{,}000}{1805 \times 141} = 67.2$$

Hot fluid: shell side, n-propanol

(4′) See (1′) under Pressure Drop
$D_o = 0.75/12 = 0.0625$ ft
(5′) Loading, $G' = W/3.14 N_t D_o$
 [Eq. (12.36)]
$= 60{,}000/3.14 \times 766 \times 0.0625$
$= 399$ lb/(hr)(lin ft)
Assume $\bar{h} = h_o = 100$

$t_w = t_a + \dfrac{h_o}{h_{io} + h_o}(T_v - t_a)$ [Eq. (5.31)]
$= 102.5 + (100/1175)(244 - 102.5)$
$= 114.5°\text{F}$
$t_f = \frac{1}{2}(T_v + t_w)$ [Eq. (12.19)]
$= (244 + 114.5)/2 = 179°\text{F}$
$k_f = 0.0945$ Btu/(hr)(ft²)(°F/ft)
 [Table 4]
$s_f = 0.76$
$\mu'_f = 0.65$ cp $(4G'/\mu = 1025)$ [Fig. 14]
From Fig. 12.9 or Eq. (12.41),
$\bar{h} = h_o = 102$ Btu/(hr)(ft²)(°F)

Cold fluid: tube side, water

(4)–(10) Same as Example 12.1
$h_{io} = 1075$

Pressure Drop

(1′) It will be necessary to arrange the 12-ft bundle into a minimum number of bundle crosses or $N + 1 = 5$. The spacing will be
$B = 144/5 = 29$ in.
$a_s = \text{ID} \times C'B/144P_T$ [Eq. (7.1)]
$= 31 \times 0.1875 \times 29/144 \times 0.937$
$= 1.25$ sq ft

(1) Same as Example 12.1 except for the tube length.

Pressure Drop

$G_s = W/a_s$ [Eq. (7.2)]
$= 60{,}000/1.25 = 48{,}000$ lb/(hr)(ft²)
At $T_v = 244°F$, μ_{vapor} [Fig. 15]
$= 0.010 \times 2.42 = 0.0242$ lb/(ft)(hr)
$D_e = 0.55/12 = 0.0458$ ft [Fig. 28]
$Re_s = D_e G_s/\mu$ [Eq. (7.3)]
$= 0.0458 \times 48{,}000/0.0242 = 91{,}000$
$f = 0.0014$ ft²/in.²
(2') No. of crosses,
$N + 1 = 5$ [Eq. (7.43)]
$s = 0.00381$ [Example 12.1]
$D_s = 3\frac{1}{12} = 2.58$ ft

(3') $\Delta P_s = \dfrac{1}{2}\dfrac{fG_s^2 D_s(N+1)}{5.22 \times 10^{10} D_e s}$
 [Eq. (12.47)]
$= \dfrac{1}{2} \times \dfrac{0.00142 \times 48{,}000^2 \times 2.58 \times 5}{5.22 \times 10^{10} \times 0.0458 \times 0.00381}$
$= 2.3$ psi

The pressure drop is high, and if it cannot be compensated for by elevating the condenser, it will be necessary to use the half-circle support baffles as in Example 7.8.

(2) $\Delta P_t = \dfrac{fG_t^2 L n}{5.22 \times 10^{10} D s \phi_t}$
$= \dfrac{0.00019 \times 1{,}210{,}000^2 \times 12 \times 4}{5.22 \times 10^{10} \times 0.0517 \times 1.0 \times 1.0}$
$= 5.0$ psi

(3) ΔP_r as in Example 12.1 = 3.2 psi
(4) $\Delta P_T = \Delta P_t + \Delta P_r$
$= 5.0 + 3.2 = 8.2$ psi

(13) Clean overall coefficient U_C:

$$U_C = \frac{h_{io} h_o}{h_{io} + h_o} = \frac{1075 \times 102}{1075 + 102} = 93.2 \text{ Btu/(hr)(ft}^2\text{)(°F)} \tag{6.38}$$

(14) Dirt factor R_d: U_D from (c) = 67.2

$$R_d = \frac{U_C - U_D}{U_C U_D} = \frac{93.2 - 67.2}{93.2 \times 67.2} = 0.00415 \tag{6.13}$$

Summary

102	h outside	1075
U_C	93.2	
U_D	67.2	
R_d Calculated 0.00415		
R_d Required 0.0030		
2.3	Calculated ΔP	8.2
2.0	Allowable ΔP	10.0

Discussion. The condenser is somewhat safe from a heat-transfer standpoint but exceeds the allowable pressure drop, although not seriously. The advantage of horizontal condensation may be observed from the U_C of 148.5 in the horizontal condenser as compared with 93.2 in a vertical condenser in an identical service. The vertical condenser has advantages, however, when the condensate is to be subcooled.

The final vertical condenser is

Shell side	*Tube side*
ID = 31 in.	Number and length = 766, 12'0''
Baffle space = 29 in.	OD, BWG, pitch = ¾ in., 16, $^{15}\!/_{16}$-in. tri.
Passes = 1	Passes = 4

Partial Condensers and the Balanced Pressure Drop. Sometimes it is desirable to condense only a portion of the vapor in a condenser such as may be needed only for reflux. Such a condenser is a *partial condenser* although the term *dephlegmator* was formerly used. The calculation of a partial condenser does not alter the method of computing the condensing film coefficient. The calculation of the pressure drop for a partial condenser is obtained with sufficient accuracy from the average of the pressure drops based on inlet and outlet conditions. The pressure drop for the inlet conditions has already been treated and for the outlet conditions it is obtained for practical purposes by computing the mass velocity and Reynolds number from the weight flow of vapor still in the vapor phase at the outlet.

A more precise calculation of the pressure drop involves an additional factor. During partial condensation in a horizontal condenser the vapor stream travels in the upper portion of the shell and the layer of condensate travels in parallel with it on the bottom of the shell. Both must traverse the length of the exchanger with the *same* pressure drop, since the terminal pressures for both are identical. This is called the condition of *balanced pressure drops*, and it applies whenever two fluids flow in parallel in the same channel. For condensation inside horizontal tubes, particularly in tube bundles, the pressure drop can be computed by trial and error by assuming the *average* segments of the pipe in which the condensate and vapor flow. The equivalent diameter can then be calculated for each portion from the free-flow area and the wetted perimeter. The mass velocity, Reynolds number, and pressure drops can similarly be calculated. If the pressure drops for the assumed division of the segments do not check, a new assumption must be made and the calculation repeated. The principle of balanced pressure drops is particularly important when it is desired to subcool the condensate to a very specific temperature and the area for sensible-heat transfer must be determined closely. When dealing with the shell side of 1-2 horizontal condensers, it is extremely difficult to predict accurately the effective flow area at the bottom of the shell unless it can be determined by planimeter from the actual layout, although the principle of balanced pressure drop must also apply.

Influence of Impurities upon Condensation. In distillation operations the volatile component is always only partially separated from the less volatile components and the overhead product is never 100 per cent pure. Instead it may contain from a trace to a substantial concentration of the heavier components, and it does not condense isothermally except when the overhead is a constant boiling mixture or a mixture forming two immiscible liquids. When the temperature range over which the condensation of a mixture occurs is small, perhaps not exceeding 10 to 20°F, it may be treated as a pure compound with the true temperature difference being the LMTD for 1-1 condensers or $F_T \times$ LMTD for 1-2 condensers. The use of the conventional LMTD in either case assumes that the heat load removed from the vapor per degree of temperature decrease is uniform. Particularly where close approaches to the temperature of the cooling medium are involved, this may lead to serious error as demonstrated in Chap. 13. For the majority of services the assumption does not cause a serious error.

Another type of impurity causing a deviation from isothermal condensation is the presence of a trace of noncondensable gas, such as air, mixed with the vapor. A noncondensable gas is actually a superheated gas which is not cooled to its saturation temperature while the vapor itself is condensed. A common example is the presence of air in the condensation of steam. The presence of only 1 per cent of air by volume may reduce the condensing coefficient of steam by 50 per cent. The mechanism of the condensation becomes one of diffusion of the vapor through the air, the latter serving as a resistance to the transfer of heat. Under conditions of superatmospheric pressure there is little hazard that air will leak into a system except for the small quantities which may have been dissolved in the feed water before it was vaporized. In vacuum operation the possibility of air leakage into the system requires that provision be made for its continuous removal.

Condensation of a Superheated Vapor. The condensation of a superheated vapor differs from that of a saturated vapor in so far as there is sensible heat to be removed. The superheat may be the result of an additional heat absorption by a dry saturated vapor after removal from contact with the liquid from which it was formed or from passing a saturated vapor through a pressure-reducing valve. For the condensation of superheated steam McAdams[1] cites a number of investigators who have found that the entire superheated and condensing heat loads may be

[1] McAdams, *op. cit.*, p. 279.

considered transferred by the temperature difference between the steam saturation temperature and the tube wall temperature, or

$$Q = \bar{h}A(T_s - t_w) \qquad (12.49)$$

where $\bar{h}$ is the condensing film coefficient and T_s is the saturation temperature of the superheated steam corresponding to its inlet pressure. This is actually a compromise rule. It simplifies the calculation by balancing the greater temperature difference that should be used for the desuperheating alone against a film coefficient which may be somewhat lower than the condensing coefficient. In the light of less conclusive evidence one may speculate on what possibility occurs in a desuperheater-condenser. Consider a horizontal true counterflow or 1-1 condenser with the vapor in the shell. When the superheated vapor enters at the hot end of the condenser, the temperature of the tube wall may be lower than the saturation temperature of the vapor. The superheated vapor contacting the tube wall condenses at its saturation temperature and may possibly even be subcooled. As the condensate drips from the tubes, it probably exchanges heat by reflashing into the hot superheated vapor at a great transfer rate, thereby effecting most of the desuperheating. Here, then, the desuperheating is probably controlled by the rate of subcooling, which should be relatively high for a film on a cold tube wall provided the condensate does not drain too rapidly to be subcooled.

It is also possible to have a reverse condition. If the cooling medium has been heated so that the tube-wall temperature near the hot terminal is greater than the saturation temperature, the tube wall will be dry toward the hot end and desuperheating will occur only as if the superheated steam were a dry gas. In this case it is necessary to divide the unit into a desuperheating and a condensing zone, and this method will be treated presently.

In the condensation of vapors other than steam it must be noted that a different relationship exists between the relative heat content of the superheat and the latent heat of evaporation. When steam at atmospheric pressure and superheated 100°F is condensed, the desuperheat represents slightly less than 5 per cent of the total heat load. If an organic vapor whose boiling point is in the same range, such as n-heptane, is superheated 100°F and then condensed, the desuperheat represents over 25 per cent of the heat load. Furthermore, water vapor has a very low density compared with organic vapors, and the mechanics of reflashing and diffusion in the vapor phase is probably more active for steam than for organic vapors. In condensers with multiple tube passes, condensation does not occur where the tube-wall temperature is higher than the saturation temperature of the vapor, so that it is possible to have only part of the surface wetted.

The Weighted Clean Coefficient and Temperature Difference in Desuperheater-Condensers. In establishing the true temperature difference for counterflow and 1-2 exchangers it was assumed that no partial phase change occurred in the exchanger. The desuperheater-condenser is a distinct violation of this assumption, requiring the development of a new method of computing the true temperature difference. In Fig. 12.18 the temperatures during the condensation of a pure vapor are shown. In Fig. 12.19 are shown the temperatures during the desuperheating and condensation of a superheated vapor. It is convenient to consider the

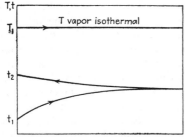

FIG. 12.18. Temperature distribution vs. tube length during isothermal condensation in a 1-2 condenser.

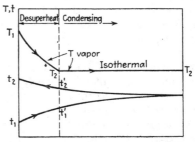

FIG. 12.19. Temperature distribution vs. tube length in a 1-2 desuperheater-condenser.

condenser as being subdivided into two zones *in series*, one for desuperheating and the other for condensation. The length L in the liquid condensate loading, $G'' = W/LN_t^{2/3}$, for a horizontal desuperheater-condenser is then the length of tube corresponding only to condensation and not the overall tube length. It will be seen that this requires an estimate of the length of tube corresponding only to condensation before $\bar{h}$ can be obtained. This difficulty in calculation does not arise in vertical units.

The temperature relations in the condensing zone of Fig. 12.19 are identical with those of a 1-2 exchanger, but those of the desuperheating zone differ from any type encountered previously. For a two-tube pass 1-2 desuperheater-condenser with desuperheating occurring at the channel end, the temperature t_1' may be obtained from differential equations for the heat transfer over the two tube passes of the desuperheating zone. Then t_1' is given by[1]

$$t_1' = \frac{1}{2}\left(\frac{W\lambda}{wc} + 2T_1 + t_1 - t_2\right)$$
$$\pm \frac{1}{2}\sqrt{\left(t_2 - t_1 - 2T_1 - \frac{W\lambda}{wc}\right)^2 - 4\left[(2T_1 - t_2)t_1 + \left(\frac{WC}{wc}\right)T_1\right]}$$

[1] Unpublished notes, D. Q. Kern and C. L. Carpenter.

By a heat balance t_2' can also be found. For four or more tube passes the calculation of the temperature difference is considerably more involved but may be simplified semiempirically. For desuperheating and condensation in a counterflow apparatus no problem arises. In a 1-2 condenser if the condensation is isothermal or the condensing range is small, and if there is no temperature cross between the condensing range and the range of the cooling medium, it is possible, with small error, to consider the entire heat load delivered in counterflow regardless of the nozzle orientation. Temperature crosses in parallel flow-counterflow desuperheater-condensers should always be avoided if at all possible. Then the temperature rise of the water during condensation corresponds to the condensing heat load. This method is not so unsafe in the desuperheating zone as it appears if the tube-wall temperature at the hot terminal of the condenser is lower than the saturation temperature of the vapor. The calculation of the surface required for desuperheating by using a dry gas film coefficient without regard to possible desuperheating by reflashing in the desuperheating zone more than compensates for the error in the temperature difference encountered in real problems. If d and c indicate the desuperheating and condensing zones, respectively, and q is the heat transferred in each, the two zones may be computed by $q_d = U_d A_d (\Delta t)_d$ and $q_c = U_c A_c (\Delta t)_c$ where U_d and U_c are the clean overall coefficients and A_d and A_c the corresponding surfaces.

The use of the two zones permits the calculation of individual values of U_d and U_c, for each zone, but in the preceding examples the practice was established to judge the suitability of an exchanger by the size of the dirt factor, which was obtained from the difference between the clean and design overall coefficients. In each zone of a desuperheater-condenser there is a separate clean coefficient, and each is effective over an independent surface. The two overall coefficients may be replaced by a single value, the *weighted overall clean coefficient*, which is obtained from

$$U_C \text{ (weighted)} = \frac{\Sigma U_c A_c}{\Sigma A_c} = \frac{U_c A_c + U_d A_d}{A_c + A_d} \qquad (12.50)$$

where U_C is the *weighted clean overall coefficient*.

To calculate a value of U_D from the Fourier equation $Q = U_D A \, \Delta t$ it is necessary first to obtain a single value for Δt. This may be weighted in terms of the heat transferred, respectively, by the temperature differences for condensation and desuperheating. Since $U_D A = Q/\Delta t$,

$$\Delta t \text{ (weighted)} = \frac{Q}{\sum \frac{q}{\Delta t}} = \frac{Q}{\frac{q_c}{(\Delta t)_c} + \frac{q_d}{(\Delta t)_d}} \qquad (12.51)$$

CONDENSATION OF SINGLE VAPORS

where Δt is the weighted temperature difference and $(\Delta t)_c$ and $(\Delta t)_d$ are LMTD in counterflow. The assumptions in the weighting method should be completely understood before attempting the solution of problems. They serve as a consistent means, if not entirely precise, for comparing the performance and design of desuperheater-condensers.

In the calculation of desuperheater-condensers for the fulfillment of process conditions, it is difficult to establish a fast rule for assuming a reasonable value of U_D for the trial calculation. It is not only dependent upon the film coefficients of the superheated and saturated vapor but to an equal extent upon the distribution of the heat load between condensation and superheat which may have any value. By weighting U_D from the individual heat loads and anticipated individual design coefficients it permits a reasonably good approximation. Thus for a trial calculation

$$U_D \text{ (weighted)} = \frac{Q}{\dfrac{q_c}{(U_D)_c} + \dfrac{q_d}{(U_D)_d}} \qquad (12.52)$$

Example 12.3. Calculation of a Butane Desuperheater-condenser. 27,958 lb/hr of isobutane with small amounts of n-butane issues from a gas reactor at 200°F and 85 psig. The gas becomes saturated at 130°F and is completely condensed at 125°F. Cooling is by well water from 65 to 100°F.

A minimum combined dirt factor of 0.003 is required along with allowable pressure drops of 2.0 psi for the butanes and 10.0 psi for the water.

Available for the service is a 23¼ in. ID horizontal 1-2 gas exchanger with 352 ¾ in. OD, 16 BWG tubes 16'0" long laid out on 1-in. triangular pitch for four passes. The baffles are spaced 12 in. apart.

Solution:

Exchanger:

Shell side	Tube side
ID = 23¼ in.	Number and length = 352, 16'0"
Baffle space = 12 in.	OD, BWG, pitch = ¾ in., 16 BWG, 1 in.-tri
Passes = 1	Passes = 4

(**1**) Heat balance:
 Butanes: Desuperheating range, $q_d = 27{,}958 \times 0.44(200 - 130)$
 $= 860{,}000$ Btu/hr (Fig. 5 or 9)

 Condensing range (From Fig. 9):
  Enthalpy of n-butane vapor at 100 psia and 130°F = 309 Btu/lb
  Enthalpy of n-butane liquid at 100 psia and 125°F = 170 Btu/lb
  $q_c = 27{,}958(309 - 170) = 3{,}880{,}000$ Btu/hr
  $Q = 860{,}000 + 3{,}880{,}000 = 4{,}740{,}000$ Btu/hr

 Water, $Q = 135{,}500 \times 1(100 - 65) = 4{,}740{,}000$ Btu/hr
 $\Delta t_{\text{water}} = 3{,}880{,}000/135{,}500 = 28.7°$F

(2) Δt weighted:

Desuperheat $(\Delta t)_d$

Hot Fluid		Cold Fluid	Diff.
200	Higher Temp	100	100
130	Lower Temp	93.7	36.3
70	Differences	6.3	63.7

Condensation $(\Delta t)_c$

Hot Fluid		Cold Fluid	Diff.
130	Higher Temp	93.7	36.3
125	Lower Temp	65.0	60.0
5	Differences	28.7	23.7

$(\Delta t)_d = \text{LMTD} = 63.0°F$ (5.14)

$\dfrac{q_d}{(\Delta t)_d} = \dfrac{860{,}000}{63.0} = 13{,}650$

$(\Delta t)_c = \text{LMTD} = 47.0°F$ (5.14)

$\dfrac{q_c}{(\Delta t)_c} = \dfrac{3{,}880{,}000}{47} = 82{,}500$

$$\text{Weighted } \Delta t = \frac{Q}{\Sigma q/\Delta t} = \frac{4{,}740{,}000}{13{,}650 + 82{,}500} = 49.3°F \quad (12.51)$$

(3) T_c and t_c: Average values will be satisfactory.

Hot fluid: shell side, butanes

(4') $a_s = \text{ID} \times C'B/144P_T$ [Eq. (7.1)]
 $= 23.25 \times 0.25 \times 12/144 \times 1.0$
 $= 0.484 \text{ ft}^2$

Desuperheating (dry gas):

(5') $G_s = W/a_t$ [Eq. (7.2)]
 $= 27{,}958/0.484$
 $= 57{,}800 \text{ lb/(hr)(ft}^2)$

(6') At $T_a = 165°F$, the mean for the superheated gas,
 $\mu = 0.01 \times 2.42 = 0.0242 \text{ lb/(ft)(hr)}$
 [Fig. 15]
$D_e = 0.73/12 = 0.0608 \text{ ft}$ [Fig. 28]
$Re_s = D_e G_s/\mu$
 $= 0.0608 \times 57{,}800/0.0242 = 145{,}000$

(7') $j_H = 239$ [Fig. 28]

(8') At 165°F,
 $k = 0.012 \text{ Btu/(hr)(ft}^2)(°F)$ [Table 5]
$(c\mu/k)^{1/3} = (0.44 \times 0.0242/0.012)^{1/3}$
 $= 0.96$

(9') $h_o = j_H \times \dfrac{k}{D_e}\left(\dfrac{c\mu}{k}\right)^{1/3}$ [Eq. (6.15b)]
 $= 239 \times 0.012 \times 0.96/0.0608$
 $= 45.2 \text{ Btu/(hr)(ft}^2)(°F)$

Cold fluid: tube side, water

(4) $a'_t = 0.302 \text{ in.}^2$ [Table 10]
 $a_t = N_t a'_t/144n$
 $= 352 \times 0.302/144 \times 4 = 0.185 \text{ ft}^2$

(5) $G_t = w/a_t$
 $= 135{,}500/0.185$
 $= 732{,}000 \text{ lb/(hr)(ft}^2)$

Vel, $V = G_t/3600\rho$
 $= 732{,}000/3600 \times 62.5 = 3.25 \text{ fps}$

(6) At $t_a = 82.5°F$,
 $\mu = 0.87 \times 2.42 = 2.11 \text{ lb/(ft)(hr)}$
 [Fig. 14]
$D = 0.62/12 = 0.0517 \text{ ft}$
$Re_t = DG_t/\mu$ (for pressure drop only)
 $= 0.0517 \times 732{,}000/2.11 = 18{,}000$

(9) $h_i = 800$ [Fig. 25]
$h_{io} = h_i \times \text{ID/OD}$ (6.5)
 $= 800 \times 0.62/0.75$
 $= 662 \text{ Btu/(hr)(ft}^2)(°F)$

Clean overall coefficient U_d, desuperheat:

$$U_d = \frac{h_{io}h_o}{h_{io} + h_o} = \frac{662 \times 45.2}{662 + 45.2} = 42.3 \quad (6.38)$$

CONDENSATION OF SINGLE VAPORS

Clean surface required for desuperheating:

$$A_d = \frac{q_d}{U_d(\Delta t)_d} = \frac{860{,}000}{42.3 \times 63.0} = 323 \text{ ft}^2$$

Condensation

(5′) Assume condensation occurs over 60% of the tube length.

$L_c = 16'0'' \times 0.60 = 9.6$ ft

$G'' = W/LN_t^{2/3} = \dfrac{27{,}958}{9.6 \times 352^{2/3}}$

$= 58.3 \text{ lb/(hr)(lin ft)}$ [Eq. (12.43)]

Assume $\bar{h} = h_o = 200$;

From (9), $h_{io} = 662$

Average temp of condensation:
$T_v = (130 + 125)/2 = 127.5°\text{F}$

Average temp during condensation
$t_a = 82.5°\text{F}$

$t_w = t_a + \dfrac{h_o}{h_{io} + h_o}(T_v - t_a)$ [Eq. (5.31)]

$= 82.5 + \dfrac{200}{662 + 200}(127.5 - 82.5)$

$= 93°\text{F}$

$t_f = (T_v + t_w)/2 = (127.5 + 93)/2$
$= 110°\text{F}$ [Eq. (12.19)]

k_f is not given in Table 4 for butane or isobutane, but from the values for pentane and hexane it should be about 0.076 at 86°F and 0.074 at 140°F. A value of 0.075 will be used.

$s_f = 0.55$ (110°API)   [Fig. 6]
$\mu_f = 0.14$ cp   [Fig. 14]
$\bar{h} = h_o = 207$ [Fig. 12.9]

The assumption of $\bar{h} = 200$ is satisfactory.

Clean overall coefficient U_c, condensation:

$$U_c = \frac{h_{io}h_o}{h_{io} + h_o} = \frac{662 \times 207}{662 + 207} = 158 \text{ Btu/(hr)(ft}^2)(°\text{F}) \tag{6.38}$$

Clean surface required for condensation:

$$A_c = \frac{q_c}{U_c(\Delta t)_c} = \frac{3{,}880{,}000}{158 \times 47} = 523 \text{ ft}^2$$

Total clean surface required A_C:

$$A_C = A_d + A_c = 323 + 523 = 846 \text{ ft}^2$$

Check of assumed condensing length L_c:

$$\frac{A_c}{A_c + A_d} \times 100 = \frac{523}{846} \times 100 = 62\%$$

Assumed length $= 60\%$ (satisfactory)

(13) Weighted clean overall coefficient:

$$U_C = \frac{\Sigma UA_C}{\Sigma A_C} = \frac{42.3 \times 323 + 158 \times 523}{846} = 114 \text{ Btu/(hr)(ft}^2\text{)(°F)} \quad (12.50)$$

(14) Design overall coefficient U_D:

$$a'' = 0.1963 \text{ ft}^2/\text{lin ft} \quad \text{(Table 10)}$$
$$\text{Total surface} = 352 \times 16'0'' \times 0.1963 = 1105 \text{ ft}^2$$
$$U_D = \frac{Q}{A \, \Delta t} = \frac{4,740,000}{1105 \times 49.3} = 87.2$$

(15) Dirt factor R_d:

$$R_d = \frac{U_C - U_D}{U_C U_D} = \frac{114 - 87.2}{114 \times 87.2} = 0.0027 \quad (6.13)$$

Pressure Drop

(1′) Desuperheat:
For $Re_s = 145{,}000$, $f = 0.0013$ ft²/in.²   [Fig. 29]

(2′) No. of crosses
($L_d = 16'0'' \times 0.40 = 6.4$ ft)
$12L/B = 6.4 \times 12/12 = 6$ [Eq. (7.43)]
Mol. wt. = 58.1
Density, $\rho = \dfrac{58.1}{359(625/492)(14.7/99.7)}$
$= 0.863$ lb/ft³
$s = 0.863/62.5 = 0.0138$
$D_s = 23.25/12 = 1.94$ ft

(3′) $\Delta P_s = \dfrac{fG_s^2 D_s(N+1)}{5.22 \times 10^{10} D_e s \phi_s}$

[Eq. (7.44)]
$= \dfrac{0.0013 \times 57{,}800^2 \times 1.94 \times 6}{5.22 \times 10^{10} \times 0.0608 \times 0.0138 \times 1.0}$
$= 1.1$ psi

(1′) Condensation:
Use of the same Reynolds number will be satisfactory. $s = 0.0146$

(2′) No. of crosses ($L_c = 9.6$ ft)
$N + 1 = 12 \times 9.6/12 = 10$ [Eq. (7.43)]

(3′) $\Delta P_s = \dfrac{1}{2} \dfrac{fG_s^2 D_s(N+1)}{5.22 \times 10^{10} D_e s}$

[Eq. (12.47)]
$= \dfrac{1}{2} \times \dfrac{0.0013 \times 57{,}800^2 \times 1.94 \times 10}{5.22 \times 10^{10} \times 0.0608 \times 0.0146}$
$= 0.90$ psi

(4′) $\Delta P_s = 1.1 + 0.9 = 2.0$ psi (total)

(1) For $Re_t = 17{,}900$, $f = 00023$ ft²/in.   [Fig. 26]

(2) $\Delta P_t = \dfrac{fG_t^2 Ln}{5.22 \times 10^{10} D s \phi_t}$ [Eq. (7.45)]

$= \dfrac{0.00023 \times 730{,}000^2 \times 16 \times 4}{5.22 \times 10^{10} \times 0.0517 \times 1.0 \times 1.0}$
$= 3.0$ psi

(3) $\Delta P_r = (4n/s)(V^2/2g')$ [Eq. (7.46)]
[Fig. 27]
$= (4 \times 4/1)0.075 = 1.2$ psi

(4) $\Delta P_T = \Delta P_t + \Delta P_r$
$= 3.0 + 1.2 = 4.2$ psi

CONDENSATION OF SINGLE VAPORS

Summary

45.2/207	h outside	662
U_C	114	
U_D	87.2	
R_d Calculated 0.0027		
R_d Required 0.0030		
2.0	Calculated ΔP	4.2
2.0	Allowable ΔP	10.0

The Vertical Condenser-subcooler. It is often desirable to subcool a vapor to a temperature lower than the saturation temperature of the vapor. This occurs in distillation when the overhead product is volatile and it is desired to send it to storage at a lower temperature to prevent excessive evaporation losses. Vertical condensers are excellent for use as condenser-subcoolers whether they be of the 1-2 type as shown in Fig. 12.11 with condensation in the shell or the 1-1 type in Fig. 12.13 with condensation inside or outside the tubes.

If a saturated vapor passes into the shell of a vertical condenser, it is possible to divide it into two distinct zones operating in series, the upper for condensation and the lower for subcooling. This is accomplished by means of a *loop seal* as shown in Fig. 12.20. The object of the loop seal is to prevent the condensate from draining from the exchanger at so rapid a rate that it emerges without subcooling. Since the shell will flow full of liquid in the subcooling zone, when a loop seal is used, the film coefficients for subcooling can be computed through the use of Fig. 28 in the same manner employed heretofore for liquids. The same is true for the tube side of a 1-1 exchanger using tube-side heat transfer data, Fig. 24, for calculating coefficients.

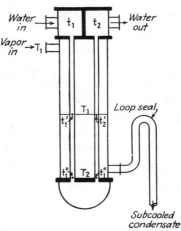

FIG. 12.20. Loop seal on vertical condenser-subcooler.

In establishing the true temperature difference for the condenser-subcooler in Fig. 12.20 there is no margin of safety such as occurs in the desuperheater-condenser. In the subcooling zone the temperatures are identical with a 1-2 exchanger having water inlet and outlet temperatures t_1' and t_2', whereas in the condensing zone the pattern again differs. For a precise calculation of the true temperature difference it is necessary to estimate t_1' and t_2' and to solve for the two condensing passes separately. This entails a lengthy trial-and-error calculation. For four or more tube passes the solution is even lengthier.

The cooling medium is heated over a greater temperature range in the first condensing pass than in the second, so that $T_1 - t_1$ is greater than $T_1 - t_2$. The temperature t_1'' at the bottom of the subcooling zone is actually higher than the mean of t_1 and t_2.

As in the case of the desuperheater-condenser, the calculation of the effective temperature difference may be simplified by means of a similar assumption. Avoid process conditions which involve a cross between the cooling medium outlet temperature and the subcooled condensate outlet temperatures. Then for 1-2 condensers with any even number of tube passes consider the cooling medium to be in counterflow and calculate the weighted temperature difference accordingly.

Example 12.4. Calculation of a Vertical Condenser-subcooler. 21,000 lb/hr of mixed n-pentane and i-pentane comes from a distilling column at 130°F and 25 psia and condenses completely at 125°F. The condensate is to be subcooled from 125 to 100°F for storage. Cooling will be effected by water from 80 to 100°F. A shell-side pressure drop of 2.0 psi is permissible for the vapor and 10.0 psi for water. A minimum fouling factor of 0.003 should be provided.

Available for the service is a 25 in. ID vertical 1-2 condenser containing 370 ¾ in. OD, 16 BWG, 16'0" tubes laid out on 1-in. square pitch. Shell baffles are 12 in. apart, and the bundle is arranged for four tube passes.

Will the condenser be a satisfactory vertical condenser-subcooler?

Solution:

Exchanger:

Shell side
    ID = 25 in.
    Baffle space = 12 in.
    Passes = 1

Tube side
    Number and length = 370, 16'0"
    OD, BWG, pitch = ¾ in., 16 BWG, 1 in.-square
    Passes = 4

(1) Heat balance:

Condensing range (130 to 125°F): (Data from Fig. 9).
    Enthalpy of n-pentane vapor at 25 psia and 130°F = 315 Btu/lb
    Enthalpy of n-pentane liquid at 25 psia and 125°F = 170 Btu/lb
    $q_c = 21{,}000(315 - 170) = 3{,}040{,}000$ Btu/hr

CONDENSATION OF SINGLE VAPORS

Subcooling range (125 to 100°F):
$q_s = 21{,}000 \times 0.57(125 - 100) = 300{,}000$ Btu/hr
$Q = \Sigma q = 3{,}040{,}000 + 300{,}000 = 3{,}340{,}000$ Btu/hr
Water, $Q = 167{,}000 \times 1(100 - 80) = 3{,}340{,}000$ Btu/hr
During condensation $\Delta t_{\text{water}} = 3{,}040{,}000/167{,}000 = 18.2°F$

(2) Δt weighted:

Condensing, $(\Delta t)_c = 36.4$; $q_c/(\Delta t)_c = 3{,}040{,}000/36.4 = 83{,}500$ Btu/(hr)(°F)
Subcooling, $(\Delta t)_s = 30.2$; $q_s/(\Delta t)_s = 300{,}000/30.2 = \underline{9{,}930}$
$93{,}430$ Btu/(hr)(°F)

$$\Delta t = Q \Big/ \sum \frac{q}{\Delta t} = \frac{3{,}340{,}000}{93{,}430} = 35.8°F \qquad (12.51)$$

(3) T_c and t_c: Average values will be satisfactory.

Hot fluid: shell side, pentanes
(4′) Condensation:
$D_o = 0.75/12 = 0.0625$ ft
$G' = W/\pi N_t D_o$ [Eq. (12.42)]
$ = 21{,}000/3.14 \times 370 \times 0.0625$
$ = 290$ lb/(hr)(lin ft)
Assume $\bar{h} = h_o = 125$
Average temp of condensing vapor:
$T_v = (130 + 125)/2 = 127.5°F$
$t_w = t_a + \dfrac{h_o}{h_{io} + h_o}(T_v - t_a)$ [Eq. (5.31)]
$ = 90 + \dfrac{125}{777 + 125}(127.5 - 90) = 95°F$
$t_f = (T_v + t_w)/2 = (127.5 + 95)/2$
$ = 111°F$ [Eq. (12.19)]
$k_f = 0.077$ Btu/(hr)(ft²)(°F/ft)
 [Table 4]
$s_f = 0.60$ (92°API) [Fig. 6]
$\mu_f' = 0.19$ cp [Fig. 14]
$\bar{h} = h_o = 120$ (vs. 125 assumed. t_f will not be changed materially by recalculation.)

Cold fluid: tube side, water
(4) $a_t' = 0.302$ in.² [Table 10]
$a_t = N_t a_t'/144 n$ [Eq. (7.48)]
$ = 370 \times 0.302/144 \times 4 = 0.194$ ft²
(5) $G_t = w/a_t$
$ = 167{,}000/0.194$
$ = 860{,}000$ lb/(hr)(ft²)
Vel, $V = G_t/3600\rho = 860{,}000/3600$
$\times 62.5$
$ = 3.84$ fps
(9) $h_i = 940$ [Fig. 25]
$h_{io} = h_i \times \text{ID}/\text{OD}$ [Eq. (6.5)]
$\phantom{h_{io}} = 940 \times 0.62/0.75$
$\phantom{h_{io}} = 777$ Btu/(hr)(ft²)(°F)

Clean overall coefficient for condensation U_c:

$$U_c = \frac{h_{io} h_o}{h_{io} + h_o} = \frac{777 \times 120}{777 + 120} = 104 \text{ Btu/(hr)(ft²)(°F)} \qquad (6.38)$$

Clean surface required for condensation A_c:

$$A_c = \frac{q_c}{U_c(\Delta t)_c} = \frac{3{,}040{,}000}{104 \times 36.4} = 803 \text{ ft}^2$$

(4') Subcooling:
$a_s = \text{ID} \times C'B/144P_T$ [Eq. (7.1)]
$= 25 \times 0.25 \times 12/144 \times 1.0 = 0.521 \text{ ft}^2$
(5') $G_s = W/a_s$ [Eq. (7.2)]
$= 21{,}000/0.521$
$= 40{,}300 \text{ lb}/(\text{hr})(\text{ft}^2)$
(6') At $T_a = 112.5°\text{F}$,
$\mu = 0.19 \times 2.42 = 0.46 \text{ lb}/(\text{ft})(\text{hr})$ [Fig. 14]
$D_e = 0.95/12 = 0.0792 \text{ ft}$
$Re_s = D_e G_s/\mu$ [Eq. (7.3)]
$= 0.0792 \times 40{,}300/0.46 = 6950$
(7') $j_H = 46.5$
(8') At 112.5°F,
$k = 0.077 \text{ Btu}/(\text{hr})(\text{ft}^2)(°\text{F}/\text{ft})$ [Table 4]
$(c\mu/k)^{1/3} = (0.57 \times 0.46/0.077)^{1/3} = 1.51$

(9') $h_o = j_H \dfrac{k}{D_e} \left(\dfrac{c\mu}{k}\right)^{1/3}$ [Eq. (6.15b)]
$= 46.5 \times 0.077 \times 1.51/0.0792$
$= 68.0 \text{ Btu}/(\text{hr})(\text{ft}^2)(°\text{F})$

(6) At 90°F, $\mu = 0.82 \times 2.42 = 1.98$ [Fig. 14]
$D = 0.62/12 = 0.0517 \text{ ft}$
$Re_i = DG_t/\mu$ (for pressure drop only)
$= 0.0517 \times 860{,}000/1.98 = 22{,}500$

(9) $h_{io} = 777 \text{ Btu}/(\text{hr})(\text{ft}^2)(°\text{F})$

Clean overall coefficient for subcooling U_s:

$$U_s = \frac{h_{io}h_o}{h_{io} + h_o} = \frac{777 \times 68.0}{777 + 68.0} = 62.5 \text{ Btu}/(\text{hr})(\text{ft}^2)(°\text{F}) \qquad (6.38)$$

Clean surface required for subcooling A_s:

$$A_s = \frac{q_s}{U_s(\Delta t)_s} = \frac{300{,}000}{62.5 \times 30.2} = 159 \text{ ft}^2$$

Total clean surface required A_C:

$$A_C = A_c + A_s = 803 + 159 = 962 \text{ ft}^2$$

(13) Weighted overall clean coefficient U_C:

$$U_C = \frac{\Sigma UA_C}{\Sigma A_C} = \frac{104 \times 803 + 62.5 \times 159}{962} = 97.1 \qquad (12.50)$$

(14) Design overall coefficient U_D:

$$a'' = 0.1963 \text{ ft}^2/\text{lin ft} \qquad \text{(Table 10)}$$
$$\text{Total surface} = 370 \times 16'0'' \times 0.1963 = 1160 \text{ ft}^2$$
$$U_D = \frac{Q}{A\,\Delta t} = \frac{3{,}340{,}000}{1160 \times 35.8} = 80.5$$

(15) Dirt factor R_d:

$$R_d = \frac{U_C - U_D}{U_C U_D} = \frac{97.1 - 80.5}{97.1 \times 80.5} = 0.0021 \text{ (hr)}(\text{ft}^2)(°\text{F})/\text{Btu} \qquad (6.13)$$

CONDENSATION OF SINGLE VAPORS

Pressure Drop

Height of zones:
Condensing: $L_c = LA_c/A_C$
$= 16 \times 803/962 = 13.4$ ft
(1′) Condensation:

At $T'_v = 127.5°F$,
$\mu = 0.0068 \times 2.42 = 0.0165$ lb/(ft)(hr)
[Fig. 15]
$Re_s = D_e G_s/\mu$
$= 0.0792 \times 40{,}300/0.0165 = 193{,}000$
$f = 0.0012$ ft^2/in.^{2} [Fig. 29]
Mol. wt. $= 72.2$
$\rho_{\text{vapor}} = \dfrac{72.2}{359(590/492)(14.7/25)}$
$= 0.284$ lb/ft^3
$s = 0.284/62.5 = 0.00454$
(2′) No. of crosses, $N + 1 = 12L/B$
$= 12 \times 13.4/12 = 13.4$, say 14
$D_s = 25/12 = 2.08$ ft
(3′) $\Delta P_s = \dfrac{1}{2} \dfrac{fG_s^2 D_s(N+1)}{5.22 \cdot 10^{10} D_e s}$
[Eq. (12.47)]
$= \dfrac{1}{2} \times \dfrac{0.0012 \times 40{,}300^2 \times 2.08 \times 14}{5.22 \times 10^{10} \times 0.0792 \times 0.00454}$
$= 1.6$ psi
(4′) ΔP of subcooling is negligible.

(1) For $Re_t = 22{,}500, f = 0.00022$ ft^2/in.^{2}
[Fig. 26]
(2) $\Delta P_t = \dfrac{fG_t^2 Ln}{5.22 \times 10^{10} D s \phi_t}$ [Eq. (7.45)]
$= \dfrac{0.00022 \times 860{,}000^2 \times 16 \times 4}{5.22 \times 10^{10} \times 0.0517 \times 1.0 \times 1.0}$
$= 3.9$ psi
(3) $\Delta P_r = (4n/s)(V^2/2g')$ [Fig. 27]
$= (4 \times 4/1)0.10 = 1.6$ psi
[Eq. (7.46)]
(4) $\Delta P_T = \Delta P_t + \Delta P_r$ [Eq. (7.47)]
$= 3.9 + 1.6 = 5.5$ psi

Summary

12⅝8	h inside	777
U_C	97.1	
U_D	80.5	
R_d Calculated	0.0021	
R_d Required	0.003	
1.6	Calculated ΔP	5.5
2.0	Allowable ΔP	10.0

The dirt factor is too small to warrant installation of the unit.

The Horizontal Condenser-subcooler.[1] The horizontal condenser can also be equipped with a loop seal as shown in Fig. 12.21 to provide subcooling surface. This can also be accomplished by means of a *dam* baffle, as shown in Fig. 12.22. The loop seal has the advantage of external adjustability. In either case the flow of vapor is predominantly the same as in a condenser. The condensing and subcooling zones are in parallel instead of in series as in the vertical unit. This requires that the condensing vapor and the condensate both travel the length of the exchanger with a balanced pressure drop, and since the specific gravity of the condensate is so much greater than that of the vapor, the flow area required for the condensate must indeed be very small. The calculation of a horizontal condenser-subcooler brings up the problem of balancing

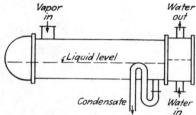

FIG. 12.21. Loop seal on horizontal condenser-subcooler.

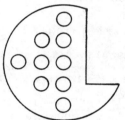

FIG. 12.22. Dam baffle.

the pressure drops and at the same time balancing the performance for the sensible and condensing heat loads to correspond to the assumed proportion of the bundle submerged for subcooling.

The calculation of the weighted clean overall coefficient to be employed here presupposes the presence of the two zones in parallel. It is further assumed that the surface for subcooling is not more than 50 per cent of the total surface. If subcooling represents more than 50 per cent, it would often be preferable to use a separate shell for subcooling alone where a higher condensate velocity can be attained. When the condensation surface requirements are more than 50 per cent of the total surface requirement, the vapor may be assumed to travel as in an ordinary condenser and its film coefficient is computed by Eq. (12.43), except that N_t in G'' is the number of tubes which are not submerged. The value of L used in calculating the vapor loading is the overall tube length. For simplicity, subcooling may be assumed to occur at a heat transfer coefficient corresponding to free convection although all the correlations for

[1] The method of computing this type of apparatus is arbitrary, although it gives overall coefficients which have proved satisfactory in a large number of cases.

CONDENSATION OF SINGLE VAPORS

free convection in Chap. 10 were for heating only. For light materials, such as organic solvents and petroleum fractions like kerosene and lighter, it is safe to assume a subcooling film coefficient of about 50, and for heavier condensates, such as aniline, straw oil, etc., it is about 25 or lower.

The calculation of the weighted temperature difference is also modified. In the orientation of the tube passes in a horizontal 1-2 condenser-subcooler the coldest pass is always placed in contact with the flooded portion of the shell, and it is a safe practice to consider the temperature rise of the cooling medium proportional to the transferred heat load. Thus if the subcooling heat load is 25 per cent of the total heat load, the cooling medium will rise the first 25 per cent of its temperature range while traveling through the subcooling passes.

The pressure drop in the shell should be computed on the basis of balancing the pressure drops in the two zones, but this requires a number of additional assumptions which cannot be entirely justified. Instead, for any given submergence, it is possible to approximate the pressure drop by calculating the average pressure drop for the vapors alone, using a segmental flow area equivalent to the vapor space and assuming that the pressure drop so calculated is identical with that of the condensate.

Example 12.5. Calculation of a Horizontal Condenser-subcooler. Using the same process conditions and condenser as in Example 12.4, which were unsatisfactory, what dirt factor would be achieved if the condenser-subcooler was operated horizontally?

It is understood that the baffle pitch will have to be somewhat greater if the unit is operated horizontally, since flooding the horizontal shell reduces the vapor flow area and therefore increases the pressure drop. But the baffle pitch is not regarded as influencing the transfer coefficients in the horizontal condenser-subcooler, and this example may be considered an illustration of the effectiveness of horizontal vs. vertical surface in condenser-subcoolers.

Solution:

Exchanger: Same as Example 12.4.

(1) Heat balance: Same as Example 12.4.
(2) Δt weighted:
Temperature rise of water during condensation = 18.2°F

Condensation $(\Delta t)_c$

Hot Fluid		Cold Fluid	Diff.
130	Higher Temp	100	30
125	Lower Temp	81.8	43.2

$(\Delta t)_c$ = LMTD = 36.4°F   [Eq. (5.14)]

Subcooling $(\Delta t)_s$

Hot Fluid		Cold Fluid	Diff.
125	Higher Temp	81.8	43.2
100	Lower Temp	80.	20.

$(\Delta t)_s$ = LMTD = 30.2°F   [Eq. (5.14)]

Condensing, $(\Delta t)_c = 36.4$ $\dfrac{q_c}{(\Delta t)_c} = \dfrac{3{,}040{,}000}{36.4} = 83{,}500$ Btu/(hr)(°F)

Subcooling, $(\Delta t)_s = 30.2$ $\dfrac{q_s}{(\Delta t)_s} = \dfrac{300{,}000}{30.2} = 9{,}930$

$\hspace{6cm} \overline{93{,}430}$ Btu/(hr)(°F)

$$\Delta t = Q \Big/ \sum \dfrac{q}{\Delta t} = \dfrac{3{,}340{,}000}{93{,}430} = 35.8°\text{F} \hspace{2cm} (12.51)$$

(3) T_c and t_c: Average values will be satisfactory.

Trial:

Assume shell is flooded to a height of $0.3D_s$. Originally subcooling represented $^{159}\!/_{962} = 16.5$ per cent of the total surface. For the horizontal condenser $\bar{h}$ will be much higher than for the vertical, hence about 25 per cent of the surface will be required for subcooling.

From mathematical tables determine the area of shell cross section which is flooded a'_s to obtain the number of submerged tubes.

For $0.3D_s$, $C_1 = 0.198$ in the formula
$a'_s = C_1 D_s^2 = 0.198 \times 25^2 = 124$ in.^{2}

Number of submerged tubes $= 370 \times \dfrac{124}{(\pi/4) \times 25^2} = 93$ (approx.)

Number of tubes for condensation $= 370 - 93 = 277$

Flooded surface $= (^{93}\!/_{370})100 = 25\%$

Condensation:

$$G'' = \dfrac{W}{LN_t^{2/3}} = \dfrac{21{,}000}{16 \times 277^{2/3}} = 30.9 \hspace{2cm} (12.43)$$

Assume the same film temperature as before.

$\bar{h} = h_o = 251$ \hfill (Fig. 12.9)

$h_{io} = 777$ from Example 12.4

Clean overall coefficient for condensation U_c:

$$U_c = \dfrac{h_{io}h_o}{h_{io} + h_o} = \dfrac{777 \times 251}{777 + 251} = 190 \text{ Btu/(hr)(ft}^2)(°\text{F}) \hspace{2cm} (6.38)$$

Clean surface required for condensation A_c:

$$A_c = \dfrac{q_c}{U_c(\Delta t)_c} = \dfrac{3{,}040{,}000}{190 \times 36.4} = 440 \text{ ft}^2$$

Subcooling: (Free-convection rate assume $h = 50$):

Clean overall coefficient for subcooling U_s:

$$U_s = \dfrac{h_{io}h_o}{h_{io} + h_o} = \dfrac{777 \times 50}{777 + 50} = 47.0 \text{ Btu/(hr)(ft}^2)(°\text{F}) \hspace{2cm} (6.9)$$

Clean surface required for subcooling A_s:

$$A_s = \dfrac{q_s}{U_s(\Delta t)_s} = \dfrac{300{,}000}{47.0 \times 30.2} = 211 \text{ ft}^2$$

Total clean surface required A_C:

$$A_C = A_c + A_s = 440 + 211 = 651 \text{ ft}^2$$

CONDENSATION OF SINGLE VAPORS

(13) Weighted clean overall coefficient U_C:

$$U_C = \frac{\Sigma UA}{\Sigma A_C} = \frac{190 \times 440 + 47.0 \times 211}{651} = 144 \text{ Btu/(hr)(ft}^2)(°F) \quad (12.50)$$

(14) Design overall coefficient U_D:

$$U_D = \frac{Q}{A\,\Delta t} = \frac{3{,}340{,}000}{1160 \times 35.8} = 80.5 \text{ Btu/(hr)(ft}^2)(°F)$$

(15) Dirt factor R_d:

$$R_d = \frac{U_C - U_D}{U_C U_D} = \frac{144 - 80.5}{144 \times 80.5} = 0.0054 \text{(hr)(ft}^2)(°F)/\text{Btu} \quad (6.13)$$

Pressure Drop

It will be necessary to spread the baffles to a spacing of 18 in. to compensate for the reduction in crossflow area due to the flooded subcooling zone. The tube-side pressure drop will be the same as before. Assume bundle flooded to $0.3D_s$.

$$a_s = 0.7 \times \text{ID} \times C' \frac{B}{144\,P_T} = 0.7 \times 25 \times 0.25 \times \frac{18}{144 \times 1.0} = 0.547 \text{ ft}^2 \quad (7.1)$$

$$G_s = \frac{W}{a_s} = \frac{21{,}000}{0.547} = 38{,}400 \text{ lb/(hr)(ft}^2) \quad (7.2)$$

$$Re_s = \frac{D_e G_s}{\mu} = 0.0792 \times \frac{38{,}400}{0.0165} = 185{,}000 \quad (7.3)$$

$$f = 0.00121 \text{ ft}^2/\text{in.}^2 \quad \text{(Fig. 29)}$$

Number of crosses, $N + 1 = \frac{12L}{B} = 12 \times \frac{16}{18} = 11 \quad (7.43)$

$D_s = 2.08$ ft

$$\Delta P_s = \frac{1}{2} \frac{fG_s^2 D_s(N+1)}{5.22 \times 10^{10} D_e s} = \frac{1}{2} \frac{0.00121 \times 38{,}400^2 \times 2.08 \times 11}{5.22 \times 10^{10} \times 0.0792 \times 0.00454} = 1.0 \text{ psi} \quad (12.47)$$

Summary

$251\!\!/\!\!50$	h inside	777
U_C	144	
U_D	80.5	
R_d Calculated 0.0054		
R_d Required 0.003		
1.0	Calculated ΔP	5.5
2.0	Allowable ΔP	10.0

Remarks. The horizontal condenser-subcooler should be equipped with a loop seal or dam baffles approximately one-third the height of the shell.

It is possible to draw some general conclusions from the summaries of Examples 12.4 and 12.5 which shed light on the relative orders of magnitude between horizontal and vertical condenser-subcoolers. The vertical condenser has the advantage of well-defined zones, but it is restricted by the height available at the top of a column when employed with a column and the smaller size of the condensing film coefficient. On the other hand, the horizontal condenser-subcooler, though less adaptable to close calculations, gives considerably higher overall clean coefficients. The majority of the condenser-subcoolers used in industry are consequently of the horizontal type.

Vertical Refluxing inside Tubes. It is often necessary to treat a solid material with a volatile boiling liquid or to maintain a liquid mixture at

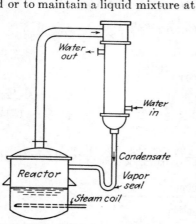

Fig. 12.23. Reflux-type condenser.

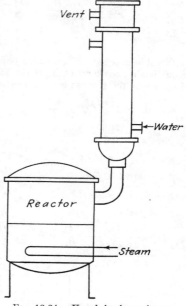

Fig. 12.24. Knock-back condenser.

its boiling point so as to complete a chemical reaction. To maintain a constant pressure the volatile liquid is boiled overhead in the reactor and returned at a continuous rate as shown in Fig. 12.23. The rate at which the volatile liquid is boiled overhead is called the *boil-up rate* and is usually expressed in pounds per hour per square foot of liquid surface. A condenser operated in a closed circuit in this manner is a *reflux-type* condenser. When the vapor enters the bundle at the bottom as in Fig. 12.24 it is a special case of the reflux type known as a *knock-back* condenser. The reflux type of condenser is not intended to produce subcooling, and the condensate drains back freely into the treating vessel. Very often the boil-up rate is so great that the condensing film may be

CONDENSATION OF SINGLE VAPORS 299

partially in turbulent flow, whence the film coefficient is calculated from Colburn's relationship as given in Fig. 12.12.

Greater headroom is required for a knock-back condenser, although it has some advantage over a horizontal condenser due to its excellent drainage. If the Reynolds number for a vertical condenser, $4G'/\mu_f$, is less than 1800 to 2100 corresponding to the transition point in Fig. 12.12, all the condensation will occur in streamline flow as computed by Eq. (12.42). If the Reynolds number is greater than about 1800, the upper surface where the condensate film is in streamline flow may be computed separately from that below it in turbulent flow. Knock-back condensers differ from reflux-type condensers, since the vapor enters at the bottom and may not, in fact, even reach the top.

Example 12.6. Calculation of Vertical Reflux-type CS_2 Condenser. A solid material is treated in a vessel with carbon disulfide and traces of corrosive sulfur compounds at 25 psig corresponding to a boiling point of 176°F and a total boil-up of 30,000 lb/hr. A negligible pressure drop will be required for the tube side, and 10 psi is available for the water, which enters at 85°F. A combined dirt factor of 0.003 will be necessary

Available for the purpose is 17¼ in. ID 1-1 exchanger having 177 ¾ in. OD, 16 BWG tubes 16'0" long. They are laid out for 1-in. square pitch, and the baffle spacing is 6 in.

Will the exchanger be satisfactory?

Solution:

Exchanger:

Shell side	Tube side
ID = 17¼ in.	Number and length = 177, 16'0"
Baffle space = 6 in.	OD, BWG, pitch = ¾ in., 16 BWG, 1-in. square
Passes = 1	Passes = 1

(1) Heat balance:
Carbon disulfide, $Q = 30,000 \times 140 = 4,200,000$ Btu/hr (Fig. 12)
Water, $Q = 120,000 \times 1(120 - 85) = 4,200,000$ Btu/hr
(2) Δt: $\Delta t =$ LMTD $= 72.1°F$ (5.14)
(3) T_c and t_c: The use of average temperatures will be satisfactory.

In this particular problem it will be advisable to compute the shell side first, since the water rate is needed to establish the tube-wall temperature for condensation.

Hot fluid: tube side, carbon disulfide
Assume $h_{io} = 300$
$t_w = t_a + \dfrac{h_{io}}{h_{io} + h_o}(T_v - t_a)$
$= 102.5 + {}^{300}\!/_{1086}(176 - 102.5)$
$= 122.5°F$
$t_f = (122.5 + 176)/2 = 149°F$
  [Eq. (12.19)]

Cold fluid: shell side, water
(4') $a_s = $ ID $\times C'B/144P_T$ [Eq. (7.1)]
$= 17.25 \times 0.25 \times 6/144 \times 1.0$
$= 0.18$ ft^2
(5') $G_s = w/a_s$ [Eq. (7.2)]
$= 120,000/0.18$
$= 667,000$ lb/(hr)(ft^2)

Hot fluid: tube side, carbon disulfide
$\mu_f = 0.28 \times 2.42 = 0.68$ lb/(ft)(hr) [Fig. 24]
$k_f = 0.09$ Btu/(hr)(ft²)(°F/ft) [Fig. 14]
$s_f = 1.26$; $\rho_f = 62.5 \times 1.26 = 78.8$ lb/ft³
$D = 0.62/12 = 0.0517$ ft
$G' = W/\pi N_t D$ [Eq. (12.42)]
$\quad = 30{,}000/3.14 \times 177 \times 0.0517$
$\quad = 1045$ lb/(hr)(lin ft)
$Re_t = 4G'/\mu_f$ [Eq. (12.37)]
$\quad = 4 \times 1045/0.68 = 6150$
The film will be in turbulent flow.
$\bar{h}(\mu_f^2/k_f^3\rho_f^2 g)^{1/3} = 0.251$ [Fig. 12.12]
$h_i = \bar{h} = 0.251 \left(\dfrac{k_f^3 \rho_f^2 g}{\mu_f^2}\right)^{1/3}$
$\quad = 0.251 \left(\dfrac{0.093^3 \times 78.8^2 \times 4.17 \times 10^8}{0.68^2}\right)^{1/3}$
$\quad = 400$
$h_{io} = 400 \times 0.62/0.75$
$\quad = 331$ Btu/(hr)(ft²)(°F) [Eq. (6.5)]

Cold fluid: shell side, water
(6') At $t_a = 102.5$°F,
$\mu = 0.70 \times 2.42 = 1.70$ lb/(ft)(hr) [Fig. 14]
$D_e = 0.95/12 = 0.0792$ ft [Fig. 28]
$Re_s = D_e G_s/\mu$ [Eq. (7.3)]
$\quad = 0.0792 \times 667{,}000/1.70 = 31{,}000$
(7') $j_H = 103$ [Fig. 28]
(8') At $t_a = 102.5$°F,
$k = 0.36$ Btu/(hr)(ft²)(°F/ft)
$(c\mu/k)^{1/3} = (1 \times 1.70/0.36)^{1/3} = 1.68$
(9') $h_o = j_H \dfrac{k}{D_e}\left(\dfrac{c\mu}{k}\right)^{1/3}$ [Eq. (6.15b)]
$\quad = 103 \times 0.36 \times 1.68/0.0792$
$\quad = 786$ Btu/(hr)(ft²)(°F)

Clean overall coefficient U_C:

$$U_C = \frac{h_{io} h_o}{h_{io} + h_o} = \frac{331 \times 786}{331 + 786} = 233 \text{ Btu/(hr)(ft²)(°F)} \qquad (6.38)$$

Design overall coefficient U_D:

Total surface $= 177 \times 16'0'' \times 0.1963 = 556$ ft²

$$U_D = \frac{Q}{A\,\Delta t} = \frac{4{,}200{,}000}{556 \times 72.1} = 105 \text{ Btu/(hr)(ft²)(°F)}$$

Dirt factor R_d:

$$R_d = \frac{U_C - U_D}{U_C U_D} = \frac{233 - 105}{233 \times 105} = 0.00522 \text{ (hr)(ft²)(°F)/Btu} \qquad (6.13)$$

Pressure Drop

(1) Flow area: $a_t' = 0.302$ in.² [Table 10]
$a_t = N_t a_t'/144n$ [Eq. (7.48)]
$\quad = 177 \times 0.302/144 \times 1 = 0.372$ ft²
(2) $G_t = W/a_t$ [Eq. (6.3)]
$\quad = 30{,}000/0.372$
$\quad = 80{,}500$ lb/(hr)(ft²)
At inlet,
$\mu = 0.012 \times 2.42 = 0.029$ lb/(ft)(hr) [Fig. 15]
$D = 0.62/12 = 0.0517$ ft
$Re = 0.0517 \times 80{,}500/0.029 = 143{,}000$
$f = 0.000138$ ft²/in.² [Fig. 26]
Mol. wt. $= 76.1$
$\rho = \dfrac{76.1}{359(636/492)(14.7/39.7)} = 0.443$ lb/ft³

(1') For $Re_s = 31{,}000$
$f = 0.0017$ ft²/in.² [Fig. 29]
(2') No. of crosses, $N + 1 = 12L/B$ [Eq. (7.43)]
$\quad = 12 \times 16/6 = 32$
$D_s = 17.25/12 = 1.44$ ft

CONDENSATION OF SINGLE VAPORS

Pressure Drop

$s = 0.443/62.5 = 0.0071$

(3) $\Delta P_t = \dfrac{1}{2} \dfrac{fG_t^2 L n}{5.22 \times 10^{10} D s}$

[Eq. (12.48)]

$= \dfrac{0.000138 \times 80{,}500^2 \times 16}{5.22 \times 10^{10} \times 0.0517 \times 0.0071}$

$= 0.4$ psi

(3') $\Delta P_s = \dfrac{fG_s^2 D_s (N+1)}{5.22 \times 10^{10} D_e s \phi_s}$

[Eq. (7.44)]

$= \dfrac{0.0017 \times 667{,}000^2 \times 1.44 \times 32}{5.22 \times 10^{10} \times 0.0792 \times 1.0}$

$\times 1 0$

$= 8.4$ psi

Summary

331	h outside	786
U_C	233	
U_D	105	
R_d Calculated	0.00522	
R_d Required	0.004	
0.4	Calculated ΔP	8.4
neg.	Allowable ΔP	10.0

The unit is satisfactory.

THE CONDENSATION OF STEAM

The Surface Condenser. Any saturated vapor can be condensed by a direct spray of cold water under appropriate conditions of temperature and pressure provided contamination of the condensate by water is not objectionable. Steam, on the other hand, as generated in power stations, is an extremely pure form of water substantially free of the impurities causing scale. The term *surface condenser* is reserved for tubular apparatus employed for the condensation of steam.

In the design and operation of a steam turbine or engine the exhaust temperature of the heat is kept as low as possible so that a maximum change in enthalpy occurs during the conversion of heat into work. This is a natural deduction of the Carnot cycle. The exhaust temperature is limited only by the coldness and abundance of the cooling medium and the allowance of a reasonable temperature difference. The economics and optimum energy distribution of power cycles is beyond consideration here, but with cooling water available at 70°F or thereabouts a turbine

302   PROCESS HEAT TRANSFER

will exhaust at 75°F, which corresponds to a vacuum saturation pressure. Consequently, such apparatus is usually designed for vacuum operation on the steam side.

The surface condenser, as a development of the power rather than chemical industry, is handled differently from preceding illustrations, and the purpose here is to indicate the exceptions. Equations (12.44a) and (12.44b) for the condensation of steam at atmospheric pressures have

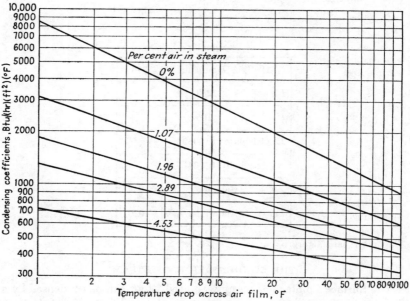

Fig. 12.25. Influence of air on the condensing coefficient of 230°F steam. (*Data of Othmer.*)

already been presented. These equations give steam-condensing coefficients at atmospheric pressure and are of little use in estimating the condensation rate under a vacuum of 1 to 1½ in. Hg abs. Furthermore, air dissolved in the boiler feed water, despite deaeration, tends to accumulate in the condenser, where it impedes heat transfer. Othmer[1] has shown that, when as little as 1 per cent of air by volume is mixed with the inlet steam, the condensing coefficient falls from 2000 to 1100 with a temperature difference between the steam and cooling medium of 20°F. When the air concentration is 2 per cent by volume, the condensing coefficient falls from 2000 to 750 at the same temperature difference. It is also found that during steady-state transfer the air tends to surround the

[1] Othmer, D. F., *Ind. Eng. Chem.*, **21**, 576 (1929).

CONDENSATION OF SINGLE VAPORS

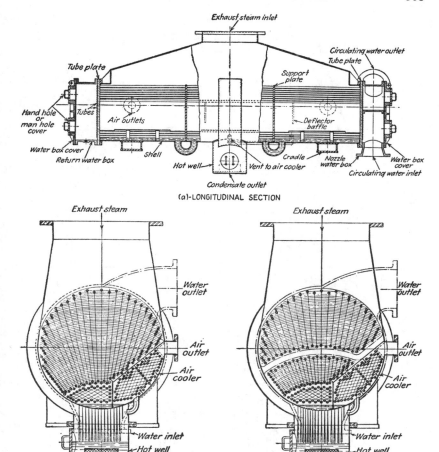

FIG. 12.26. Surface condenser. (*Foster Wheeler Corporation.*)

tube surface, setting up a resistance through which the steam must diffuse. Othmer's data are reproduced in Fig. 12.25.

Surface condensers are usually much larger than other types of tubular equipment, some containing over 60,000 ft^2 of condensing surface. The attainment of a very small steam-side pressure drop directly affects the pressure of the condenser exhaust and the cycle efficiency for a given cooling-water inlet temperature. A typical small surface condenser is shown in Fig. 12.26. To permit a low pressure drop and a deep penetra-

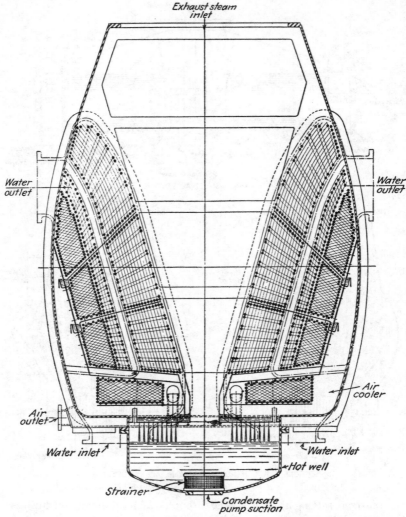

Fig. 12.27. Large dual-bank surface condenser. (*Foster Wheeler Corporation.*)

tion of the steam into the bundle tubes are laid out for *crossflow*, using *radial* pitch as shown in Fig. 12.26b. Another important consideration is the removal of air, since an accumulation of air increases the total pressure in the condenser and raises the condensation temperature. Only when surface condensers contain 15,000 ft² or less are they apt to have cylindrical bodies. When they contain surface in excess of 15,000 ft^2, they are made in the boxlike manner shown in the elevation of Fig. 12.27. In large condensers the steam is not required to pass through the entire tube bundle but instead through a divided portion of the tube bundle.

The principal parts of surface condensers, in addition to the shell and water boxes or channels, are a large exhaust steam inlet port, shell-side air-removal outlets, and a hot well equipped with condensate and air outlets. The condenser shown in Fig. 12.26 is arranged horizontally for two passes or double flow on the water side as are the majority of such condensers. The tubes are usually 22 to 26 ft in length with support plates through which all the tubes pass, since the steam travels in crossflow. The lower portion of the bundle is separated to permit the cooling of the air-vapor mixture which leads to the vacuum apparatus. This is usually an air jet pump as discussed in Chap. 14. Since the steam condenses isothermally, the LMTD in crossflow is identical with that for counterflow. The method of calculation outlined in the remainder of this chapter is in agreement with the recommended practices of the Heat Exchange Institute, whose member companies include the larger manufacturers of surface condensers.

Definitions. There are a number of definitions which apply to surface condensers. The more important are given here: The *steam load* is expressed in pounds per hour of steam having an assumed heat rejection to the cooling water of 950 Btu/lb. When used in connection with the exhaust of a steam engine the rejection is taken as 1000 Btu/lb. The *condenser vacuum* is the difference between atmospheric pressure and the pressure measured at the steam inlet and is expressed in in. Hg at a temperature of 32°F. The *absolute pressure* in a condenser is the difference between the barometric pressure and condenser vacuum and is expressed in in. Hg abs. The *heat head* is the difference between the inlet circulating water temperature and the temperature corresponding to the absolute pressure at the steam inlet to the condenser, T_S. The *temperature rise* refers to the circulating-water temperatures $t_2 - t_1$, and the terminal difference is defined as $T_S - t_2$.

Since the condensate is also under vacuum, it collects in a hot well at the bottom of the condenser and requires pumping out. The *condensate depression* is the difference between the temperature of the condensate in the hot well and the temperature corresponding to the absolute pressure of the steam at the inlet to the condenser. It is the actual number of degrees the condensate is subcooled, and must be maintained within close limits since subcooling reduces the saturation pressure and hence the suction pressure on the intake of the condensate pump, reducing its performance. It is also customary to compute the total surface in a surface condenser using the tube length between the facing sides of the tube sheets instead of the overall tube length. This is equivalent to the *effective surface* described in Chap. 7.

There are several conventions employed in surface condensers which are rarely violated. Surface condensers are seldom designed to operate at an absolute pressure below 0.7 in. Hg abs with a terminal difference of less than 5°F, with a dissolved oxygen content in the condensate of less than 0.03 cm³/liter, or with a steam loading [not to be confused with G'

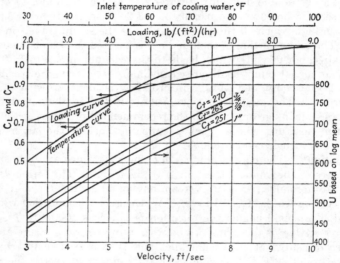

Fig. 12.28. Overall coefficients in surface condensers. (*Heat Exchange Institute.*)

or G'', lb/(hr)(lin ft)] exceeding 8 lb/(hr)(ft²). Water velocities of less than 3 fps are not used. The quantity of air which leaks into the system must be estimated for the design of the air pump whether a mechanical pump or a jet-type ejector is used. The Heat Exchange Institute gives a graph of *overall* clean heat-transfer coefficients to be used as shown in Fig. 12.28 and with the maximum loading of 8 lb/(hr)(ft²) of steam condensed when the inlet water is at 70°F. The clean overall coefficient for different tube diameters and a loading of 8 lb/(hr)(ft²) can also be obtained by

$$U_c = C_t \sqrt{V} \qquad (12.53)$$

where C_t is a constant as shown in Fig. 12.28 for each tube outside diameter and V is the water velocity in feet per second. Dirt factors are defined as a percentage of the clean overall coefficients called the *cleanli-*

ness factor. Thus a cleanliness factor of 85 per cent means that the design overall coefficient will be 85 per cent of the clean overall coefficient. The design overall coefficient U_D for a loading of 8 lb/(hr)(ft²) and inlet water temperature of 70° is

$$U_D = C_{cl}C_t \sqrt{V} \qquad (12.54)$$

And the coefficient for any other loading and temperature is

$$U_D = C_L C_T C_{cl} C_t \sqrt{V} \qquad (12.55)$$

where C_L is the loading correction equal to $\sqrt[4]{\text{loading}/8}$ and C_T is the temperature correction.

Surface Condenser Calculation. The condenser pressure-drop calculation is far more complex than for previous types of equipment and hinges very greatly upon the geometric design of the surface condenser. There is no published correlation between the pressure drop of condensing steam in crossflow and radial pitch, and the calculations performed here determine the surface required for the heat load without layout refinements. The calculations, however, permit the evaluation of a condenser for conditions other than those for which it was designed.

The pressure drop on the tube side is conventionally determined by a formula of the Williams and Hazen type using a constant of 130, so that

$$\Delta P_t = 0.0067 \frac{V^{1.84}}{d^{1.16}} \qquad (12.56)$$

where V = water velocity, fps
d = tube ID, in.

It differs somewhat on the unsafe side from Eq. (7.45) at velocities of 6 fps and less but agrees very closely at velocities approaching 10 fps. The return losses suggested by the Heat Exchange Institute correspond to less than the four velocity heads given by Eq. (7.46). The relationships for the condenser are

$$Q = w1(t_2 - t_1) = 500G_o(t_2 - t_1)$$

where G_o are the gallons per minute and multiplication by 500 gives the flow in pounds per hour.

$$\Delta t = \text{LMTD} = \frac{(T_s - t_1) - (T_s - t_2)}{\ln (T_s - t_1)/(T_s - t_2)} = \frac{t_2 - t_1}{\ln (T_s - t_1)/(T_s - t_2)}$$

$$A = \frac{Q}{U \Delta t} = \frac{500G_o(t_2 - t_1)}{U \frac{(t_2 - t_1)}{\ln (T_s - t_1)/(T_s - t_2)}} = \frac{500G_o}{V} \ln \frac{T_s - t_1}{T_s - t_2} \qquad (12.57)$$

$$t_2 = T_s - \frac{T_s - t_1}{\text{antilog } UA/(2.3 \times 500G_o)} \qquad (12.58)$$

The outlet temperature of the water t_2 depends upon gallons per minute circulated G_o or the velocity V. But U is also dependent upon V. Equation (12.58) can be solved directly for t_2, since

$$A = La''N_t$$

where a'' are the square feet of external surface per foot of tube length and

$$\frac{UA}{2.3 \times 500 G_o} = \frac{U}{2.3 \times 500} \frac{La''N_t n}{7.5 \times 60 V(a_t'/144)N_t} = \frac{0.00279 ULna''}{Va_t'}$$

and

$$t_2 = T_s - \frac{T_s - t_1}{\text{antilog } 0.00279 ULna''/Va_t'} \quad (12.59)$$

Example 12.7. Calculation of a Surface Condenser. A steam turbine exhausts 250,000 lb/hr of steam at a pressure of 1.5 in. Hg. Cooling water is available at 70°F. A preference has been designated for the use of ⅞ in. OD, 18 BWG tubes and a cleanliness factor of 85 per cent.

To permit estimation of the cost of the condenser on the basis of dollars/ft², how much surface will be required and what will be the operating quantity and range of the cooling water?

Solution. Assume the maximum loading of 8 lb/(hr)(ft²) and a water velocity of 7.5 fps.

$$C_{Cl} = 0.85 \quad C_T = 1.0 \quad C_L = 1.0$$
$$U_D = C_{Cl} C_T C_L C_t \sqrt{V} \quad (12.55)$$
$$= 0.85 \times 1.0 \times 1.0 \times 263 \sqrt{7.5} = 612$$
$$A = \frac{250{,}000}{8} = 31{,}250 \text{ ft}^2$$

The necessary outlet terminal water temperature will be

$$t_2 = T_s - \frac{T_s - t_1}{\text{antilog } 0.000279 \times ULna''/Va_t'} \quad (12.59)$$

For a ⅞ in. OD, 18 BWG tube, $a'' = 0.229$ ft²/ft, ID = 0.777 in., and $a_t' = 0.475$ in.² (computed from Table 10).

$$T_s \approx 1.5 \text{ in. Hg} = 91.72°F$$

Since this will be a relatively large condenser, assume a maximum tube length of 26'0'' and two tube passes.

$$t_2 = 91.72 - \frac{91.72 - 70}{\text{antilog } (0.000279 \times 612 \times 26 \times 2 \times 0.229)/(7.5 \times 0.475)} = 85.90°F$$

The circulation rate is

$$G_o = \frac{250{,}000 \times 950}{(85.90 - 70) \times 500} = 29{,}800 \text{ gpm}$$

Individual Film Coefficients for Steam Condensers. The method of calculating surface condensers has been based on the use of overall coefficients instead of individual coefficients. Individual coefficients are difficult to obtain from direct experimentation but can be computed from the overall coefficients. Unlike condensers employing organic vapors, the condensing coefficient for steam is considerably greater than that for the water, and under these circumstances the tube metal resistance is a sufficiently sizable portion of the total resistance to require inclusion in the calculation. For a clean tube the overall coefficient of heat transfer may be equated thus:

$$\frac{1}{U_C} = R_c + R_m + R_w$$

and for a dirty tube:

$$\frac{1}{U_D} = R_c + R_m + R_d + R_w$$

where R_c is the resistance of the condensing vapor, R_m is the resistance of the tube metal, and R_w the resistance of the water in the tubes, all based on the tube outside diameter. Wilson[1] has shown that the sum of the first three resistances for a series of tests is essentially constant and that R_w is the controlling resistance. Furthermore, the film coefficient for water in tubes is proportional to the 0.8 power of the velocity through the tubes, R_w can then be replaced by $1/a_1 V^{0.8}$, and the equations above can be rewritten with a_0 and a_1 constants.

$$\frac{1}{U} = a_0 + \frac{1}{a_1 V^{0.8}} \tag{12.60}$$

A plot of $1/U$ vs. $1/V^{0.8}$ on rectangular coordinates should yield a straight line permitting the evaluation of both constants a_0 and a_1. a_0 is actually the intercept for an infinite water velocity, and when the metal and dirt resistances are subtracted, the true value of the condensing coefficient can be obtained.

$$R_c = \frac{1}{h_c} = a_0 - R_m - R_d$$

PROBLEMS

12.1. 62,000 lb/hr of pure ethyl alcohol at 2.0 psig is to be condensed by water from 85 to 120°F. A dirt factor of 0.003 should be provided. A pressure drop of 2 psi is allowable for the vapor and 10.0 for the water.

Calculate the required size of a 1-2 horizontal condenser using 1 in. OD, 14 BWG tubes 16'0" long on 1¼-in. triangular pitch.

12.2. Using the data of Prob. 12.1 calculate the size of a 1-2 vertical condenser required to fulfill the conditions.

12.3. Using the data of Prob. 12.1 calculate the size of a 1-1 vertical condenser with condensation in the tubes.

[1] Wilson, R. E., *Trans. ASME*, **37**, 47 (1915).

310   PROCESS HEAT TRANSFER

12.4. 24,000 lb/hr of nearly pure methyl ethyl ketone vapor at 2 psig (boiling point 180°F) is to be condensed and cooled to 160°F by water from 85 to 120°F. Pressure drops of 2.0 psi for the vapor and 10.0 for the water are permissible.

Available for the service is a 25 in. ID 1-2 horizontal condenser with 468 ¾ in. OD, 16 BWG tubes, 16'0" long arranged for four passes on $1\frac{5}{16}$-in. triangular pitch. Baffles are spaced 25" apart.

(a) What is the true temperature difference?
(b) What are the dirt factor and the pressure drops?

12.5. For the data of Prob. 12.4 calculate the true temperature difference and size of a horizontal 1-2 condenser-subcooler required to fulfill the conditions using ¾ in. OD, 16 BWG tubes 12'0" long on $1\frac{5}{16}$-in. triangular pitch with a dirt factor of 0.003.

12.6. 50,000 lb/hr of ethyl acetate at 35 psig (boiling point 248°F) enters a horizontal 1-2 desuperheater-condenser at 300°F and leaves at 248°F. Cooling is effected by water from 85 to 120°F.

Available for the service is a 27 in. ID 1-2 condenser containing 432 ¾ in. OD, 16 BWG tubes 12'0" long arranged for four passes on 1-in. square pitch. Baffles are spaced 24 in. apart.

Calculate the true temperature difference and the dirt factor and pressure drops.

12.7. 57,000 lb/hr of nearly pure hexane enters the shell of a vertical 1-2 condenser at 5 psig and 220°F. The condensing range is from 177 to 170°F, at which temperature it goes to storage. Cooling water is used between 90 and 120°F.

Available for the service is a 31 in. ID condenser containing 650 ¾ in. OD, 16 BWG tubes 16'0" long on four-pass layout using 1-in. triangular pitch. Baffles are spaced 18 in. apart.

Calculate the true temperature difference, the dirt factor, and the pressure drops.

12.8. 59,000 lb/hr of a mixture of light hydrocarbons, principally propane, enters a 1-2 horizontal condenser at the initial condensing temperature of 135°F at 275 psig. The condensing range is from 135 to 115°F, at which temperature 49,000 lb/hr is condensed by cooling water from 90 to 110. The remainder of the condensation occurs with refrigerated water.

Available for the service is a 37 in. ID 1-2 horizontal condenser with 1100 ¾ in. OD, 16 BWG tubes 16'0" long arranged for four passes on $1\frac{5}{16}$-in. triangular pitch. Baffles are spaced 36 in. apart.

What is the dirt factor, and what are the pressure drops?

NOMENCLATURE FOR CHAPTER 12

A	Heat-transfer surface, ft²
A_c, A_d, A_s	Heat-transfer surface for condensation, desuperheating, and subcooling, respectively, ft²
A_f	Cross-sectional area of a film, ft²
A_C	Total clean heat-transfer surface, ft²
a	Flow area, ft²
a''	External surface per linear foot, ft
a'_s	Submerged cross section of shell, ft²
a'_t	Flow area per tube, in.²
a_0, a_1	Constants
B	Baffle spacing, in.
C	Specific heat of hot fluid in derivations, Btu/(lb)(°F)
C_{Cl}, C_T, C_L	Cleanliness, temperature, and loading factors, dimensionless
C_t	Tube factor, dimensionless

CONDENSATION OF SINGLE VAPORS

C'	Clearance between tubes, in.
C_1, C_2, C_3	Constants
c	Specific heat of cold fluid, Btu/(lb)(°F)
D	Inside diameter of tubes, ft
D_o	Outside diameter of tubes, ft
D_e	Equivalent diameter for heat transfer and pressure drop, ft
d	Inside diameter of tubes, in.
d_o	Outside diameter of tubes, in.
f	Friction factor, ft²/in.²
G	Mass velocity, lb/(hr)(ft²)
G'	Condensate loading for vertical tubes, lb/(hr)(ft)
G''	Condensate loading for horizontal tubes, lb/(hr)(ft)
G_0	Circulating water rate, gpm
g	Acceleration of gravity, ft/hr²
g'	Acceleration of gravity, ft/sec²
h, h_i, h_o	Heat-transfer coefficient in general, for inside fluid, and for outside fluid, respectively, Btu/(hr)(ft²)(°F)
h_{io}	Value of h_i when referred to the tube OD, Btu/(hr)(ft²)(°F)
h_x	Condensing film coefficient at a distance x from the top of the tube Btu/(hr)(ft²)(°F)
h_α	Condensing film coefficient at angle $\alpha°$, Btu/(hr)(ft²)(°F)
$\bar{h}$	Average value of the condensing film coefficient between two points, Btu/(hr)(ft²)(°F)
j_H	Factor for heat transfer, dimensionless
k	Thermal conductivity, Btu/(hr)(ft²)(°F/ft)
L	Tube length, ft
L_c	Tube length exposed to condensation, ft
LMTD	Log mean temperature difference, °F
m	A constant
N	Number of shell-side baffles
n	Number of tube passes
N_t	Number of tubes effective for condensation
P	Perimeter, ft
P_T	Tube pitch, in.
ΔP	Pressure drop in general, psi
$\Delta P_T, \Delta P_t, \Delta P_r$	Total, tube, return pressure drop, respectively, psi
ΔP_c	Pressure head of condensate, psi
Q	Heat flow, Btu/hr
Q_x	Heat flow at distance x from top of tube, Btu/hr
q_c, q_d, q_s	Heat flow for condensation, desuperheating and subcooling respectively, Btu/hr
R_d	Combined dirt factor, (hr)(ft²)(°F)/Btu
R_c, R_m, R_w	Resistance of condensate film, tube metal, and water, respectively, (hr)(ft²)(°F)/Btu
r	Radius of tube, ft
r_h	Hydraulic radius, r_h = flow area/wetted perimeter, ft
Re	Reynolds number, dimensionless
s	Specific gravity, dimensionless
T, T_1, T_2	Temperature in general, inlet and outlet of hot fluid, respectively, °F

T_a	Average temperature of hot fluid, °F
T_c	Caloric temperature of hot fluid, °F
T_S	Steam temperature, °F
t, t_1, t_2	Cold-fluid temperature in general, inlet, and outlet, respectively, °F
t_a	Average temperature of cold fluid, °F
t_c	Caloric temperature of cold fluid, °F
t_f, t_w	Temperature of film and tube wall respectively, °F
t'	Temperature at outer surface of condensate film, °F
Δt	True temperature difference in $Q = U_D A \, \Delta t$, °F
$(\Delta t)_c, (\Delta t)_d, (\Delta t)_s$	True or fictitious temperature difference for condensation, desuperheating, and subcooling, respectively, °F
Δt_f	$(T_v - t_w)/2$, °F
U, U_C, U_D	Overall coefficient of heat transfer, clean coefficient, and design coefficient, respectively, Btu/(hr)(ft²)(°F)
U_c, U_d, U_s	Clean overall coefficient for condensation, desuperheating, and subcooling respectively, Btu/(hr)(ft²)(°F)
u	Film velocity along axis of tube, ft/hr
$\bar{u}$	Average film velocity, ft/hr
V	Velocity, fps
W	Weight flow in general, weight flow of hot fluid, lb/hr
W'	Rate of condensation, lb/(hr)(ft²)
w	Weight flow of cold fluid, lb/hr
w'	Rate of condensation per tube W/N_t, lb/(hr)(tube)
x, x'	Length of film, ft
x_c	Distance from top of tube at which change from streamline to turbulent flow occurs, ft
y, y'	Width of film, ft
z	Distance, ft; height of a hydrostatic leg, ft; a synthetic function
α	Angle of tube, deg
λ	Latent heat of condensation or vaporization, Btu/lb
μ	Viscosity, centipoises × 2.42 = lb/(ft)(hr)
μ'	Viscosity, centipoises
μ_w	Viscosity at tube-wall temperature, centipoises × 2.42 = lb/(ft)(hr)
ρ	Density, lb/ft³
τ	Tangential stress, lb/ft²
ϕ	Viscosity ratio, $(\mu/\mu_w)^{0.14}$, dimensionless
ψ	A function

Subscripts (except otherwise noted)

c	Condensate, condensing
f	Film or film temperature
l	Liquid
s	Shell side
v	Vapor
t	Tube side

CHAPTER 13

CONDENSATION OF MIXED VAPORS

Introduction. In the preceding chapter it was assumed that the condensing vapor consisted of a pure or substantially pure compound which condensed isothermally. If the vapor was mixed with another compound having a slightly different boiling point, the mixture condensed over a small condensing range. It was further assumed that, where there was a condensing range, the latent heat of condensation was transferred to the cooling medium uniformly over the entire condensing range.

However, consider a mixture of two diverse fluids having a condensing range of 100°F. In order to effect the first 10 per cent reduction in the vapor temperature it may be necessary to remove 50 per cent of the total heat load from the vapor mixture, since the less volatile component condenses more rapidly as the vapor temperature is reduced. Previously, in using the LMTD for condensing vapors in counterflow it was assumed that during the first 10 per cent reduction in the vapor temperature only 10 per cent of the heat was removed. The substitution of the logarithmic mean for the true temperature difference based on assumed uniform condensing characteristics for the vapor mixture may lead to a conservative or an unsafe value of Δt and the selection of the wrong condenser. The problems imposed by the condensation of a vapor mixture do not end there. Depending upon the nature of the mixture, the average condensing coefficient $\bar{h}$ may not remain constant throughout the condensing range and may vary greatly with the *composition* of the vapor mixture as the less volatile components are condensed. The latter is particularly true of a mixture consisting of a vapor and a noncondensable gas, such as steam and air.

The Phase Rule.[1] The different types of vapor mixtures may be studied qualitatively through the use of the phase rule of J. Willard Gibbs. Phases are defined as homogeneous amounts of matter in the solid,[2] liquid, or gaseous form which are distinguishable from each other by the presence of an interface between any two phases. A vapor is the coexistence of a gas and a liquid and, like any boiling liquid and its vapor, consists of two phases.

[1] The treatment of the phase rule is simplified here for the particular applications in this book.
[2] In nonthermodynamic texts these are frequently referred to as states.

There are many organic chemical compounds which are not soluble in water, forming *immiscible* mixtures with it. If a cylindrical vessel is filled to the top with a mixture of pentane and water, an interface appears between the water on bottom ($s = 1.0$) and the pentane on top ($s = 0.8$) and the *system* contains two liquid phases. While agitation may be used to disperse the water into the pentane layer as droplets, it does not affect the number of phases, since only the nature of the interface is of consequence and not its shape. On the other hand, a mixture of two gases readily forms a homogeneous mixture in a vessel, and as in all gas mixtures, all the gaseous compounds form but one phase. An interesting relationship is obtained if a mixture of water and pentane is brought to a boil in an enclosed vessel: There is a mixture of pentane and water in the gas phase and two liquid phases, so that the bounded matter or *system* consists of three phases.

Suppose a mixture of two mutually soluble or miscible compounds is olaced in an insulated vessel at its boiling point as shown in Fig. 13.1.

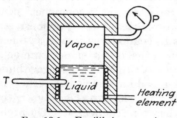

Fig. 13.1. Equilibrium vessel.

If the vapor and liquid are in *equilibrium,* no heat can escape and the system continues to vaporize and condense indefinitely. The boiling point of the system is related to the total pressure of the system as well as the ratio of the two compounds in the original liquid. Assume next that it is desired to change the pressure on the system. Does the liquid still continue to boil at the same temperature if the enthalpy remains constant? Do the chemical compositions of the vapor and liquid phases remain the same? If these questions can be answered, it is possible without running experiments to determine whether or not a given mixture condenses isothermally. If condensation is isothermal, the true temperature difference may be taken as being identical with the LMTD. If condensation is not isothermal, then other methods must be applied for the calculation of the true temperature difference, the commonest of which will be developed later in this chapter.

Gibbs[1] formulated a rule to determine the number of independent variables pertaining to a system in equilibrium which must be specified for the number of phases to be fixed permanently. Thus if a system consists of several phases in equilibrium, what conditions in the original system can be varied while keeping the kind (compositions) and number

[1] Gibbs, J. W., *Trans. Conn. Acad. Arts Sci.*, **III,** 108–248 (1876).

of phases from changing? Changes in the relative sizes of the phases are of no consequence. The numbers of variables, such as the temperature, pressure, and compositions of the system, which the process engineer is *free* to fix so as to fix the number and kinds of phases permanently are the *degrees of freedom* (of selection). Sometimes it is possible to fix the number of phases in a system by simply fixing the pressure. This is true of a pure boiling fluid such as water. If it is desired to boil water by establishing a gas and a liquid phase, can it be done under any combination of pressure and temperature the engineer desires to fix? If the pressure is regarded as a degree of freedom, since it may be freely and independently chosen, and it is fixed at 14.7 psia, it suffices to fix all the conditions for the existence of the two phases. But the temperature may not be freely and independently chosen, since it is not an independent variable but in the equilibrium system it is a property *dependent* upon the saturation (pressure) curve of the liquid. It is consequently not possible to have the water present in two phases at a pressure of 14.7 psia at any other temperature than 212°F.

The boiling of a single compound is a relatively simple matter. If several compounds are mixed together, it is considerably more difficult to determine how the compositions, pressure, and temperature serve to fix permanently the number and kinds of phases in the original system. Gibbs's phase rule allows the rapid determination of the number of degrees of freedom in systems of greater complexity. The phase rule is written

$$F = C - P + 2 \tag{13.1}$$

where F is the number of degrees of freedom, C the number of individual chemical compounds or chemical substances in the system, and P is the number of phases. Nothing in the phase rule determines how many phases will result from the admixture of any particular group of chemical components. The number of phases which a group of chemical compounds may form in a boiling system must be known beforehand from a knowledge of the miscibility or immiscibility of the components as liquids plus the addition of one gas phase into or from which vaporization and condensation may occur. This information can be obtained from the solubility tables included in chemical handbooks.

Applications of the Phase Rule. There are nine common types of vapor mixtures encountered in heat transfer, and these are given in Table 13.1. To demonstrate the use of the phase rule for determining the isothermal or nonisothermal nature of the condensation of a vapor mixture in a condenser three of the types will be illustrated. It is again emphasized that, in ordinary process design, condensers operate at sub-

TABLE 13.1. COMMON CONDENSATION REQUIREMENTS

Case	Type of components	Example	Degrees of freedom	Temp† during condensation
1	Pure vapor	Water	1	Isothermal
2	Two miscible	Butane-pentane	2	Decreasing
3	n-Miscible	Butane-pentane-hexane	n*	Decreasing
4	Vapor and noncondensable	Steam-air	2	Decreasing
5	n-Miscible + noncondensable	Butane-pentane-air	$n+1$	Decreasing
6	Two immiscible	Pentane-steam	1	Isothermal
7	n-Miscible + one immiscible	Butane-pentane-steam	n	Decreasing
8	n-Immiscible + noncondensable	Pentane-steam-air	2	Decreasing
9	n-Miscible + one immiscible + noncondensable	Butane-pentane-steam-air	$n+1$	Decreasing

* Where there are more than three degrees of freedom, the additional degrees represent concentrations which must be fixed. Thus a mixture of butane, pentane, and hexane requires fixing not only the butane in relation to hexane but also the amount of pentane in relation to hexane.
† For constant-pressure processes.

stantially constant pressure and that one degree of freedom is usually fixed by the operating pressure of the process.

For the analyses below reference should be made to Fig. 13.2 which shows the inlet end of a condenser. Directly under the inlet nozzle at 1-1' the inlet temperature is T_1, and at another section such as 2-2', after partial condensation has occurred, the temperature is designated as T_2, which may or may not be identical with T_1. When a vapor is to be condensed, it is convenient to consider condensation as the transfer of material between a gas phase and one or more liquid phases. Since condensation occurs because the heat-transfer surface is below the dew point of the vapor, it is proper to assume that the heat-transfer surface directly under the inlet nozzle is wet and supplies the liquid phases immediately after the vapor enters the condenser.

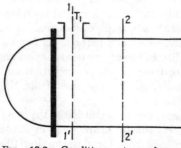

FIG. 13.2. Conditions at condenser inlet.

Case 2. *Condensation of a Mixture of Two Miscible Components.* Example: Butane-pentane.

$$F = C - P + 2$$
$$= 2 - 2 + 2 = 2 \qquad \text{(2 deg of freedom)}$$

At the inlet to the condenser the *pressure* p_1 and the *composition* C_1 of the vapor are fixed by the process operation preceding the condenser such as distillation. The system is completely defined by the process which has already fixed the two degrees of freedom p_1 and C_1, and T_1 at 1-1' is *fixed* therefore as a dependent variable as in the case of water boiling at 14.7 psia. Between 1-1' and 2-2' heat is removed and with it a quantity of the high-boiling compound and a lesser quantity of the low-boiling compound condense into the liquid phase. The composition C_2 of the vapor at 2-2' differs from C_1 at 1-1', the vapor being leaner in the high-boiling compound. Apply the phase rule again at 2-2'. There must still be two degrees of freedom, since it is desired to retain the same number of phases. The pressure is constant at p_1, but the composition has changed to C_2. According to the laws governing vapor-liquid equilibrium, the temperature T_2 at p_1, C_2 cannot be the same as T_1 at p_1, C_1, since the latter is a point on a vapor-liquid equilibrium line just as 212°F and 14.7 psia is a point on the equilibrium line for water. The temperature of condensation must consequently vary from 1-1' to 2-2', and the condensation is not isothermal.

Case 4. *Condensation of a Vapor from a Noncondensable Gas.* Example: Steam-air-water.

$$F = 2 - 2 + 2 = 2 \qquad \text{(2 deg of freedom)}$$

The operating pressure is fixed, and C_1 is fixed at the inlet to the condenser by the ratio of steam to air in the inlet gas. Since two degrees of freedom p_1 and C_1 are fixed, the temperature T_1 is again a dependent variable at 1-1'. Moving along to 2-2', where the pressure on the system is still p_1, some of the steam is condensed while the gas is not condensed and a new composition C_2 results. The temperature T_2 for p_1, C_2 must differ from T_1 at p_1, C_1 for the same reason as before. The temperature at 2-2' or any other point is actually the dew point for the mixture remaining in the gas phase and varies from the inlet to outlet as the composition of the vapor phase changes. Condensation is not isothermal.

Case 6. *Condensation of a Mixture of Two Immiscible Components* Example: Pentane-steam-water.

$$F = 2 - 3 + 2 = 1 \qquad \text{(1 deg of freedom)}$$

The condensate consists of two liquid phases. With but one degree of freedom the system is completely fixed at 1-1' by selecting the operating pressure p_1. The temperature T_1 then corresponds to p_1. At 2-2' the pressure is the same, p_1, and since this fixes the system, T_2 and T_1 must be identical, both being dependent upon p_1. Condensation is isothermal.

1. CONDENSATION OF A BINARY MIXTURE

Vapor-liquid Relationships. In the derivation of the Nusselt equation for a single component it was assumed that the rate at which the vapor enters the condensate film is nearly infinite. A vapor entering a condenser at its saturation pressure condenses because the surface is at a temperature below its dew point or saturation temperature. The rate at which the vapor passes from the vapor phase to the liquid phase, however, is dependent upon the mechanism of diffusion. This means that, if the temperature of the condensate film is lower than the saturation temperature of the bulk of the vapor, the pressure of the component at the condensate film is less than that of the vapor and a pressure differential is established. The direction of the differential promotes the flow out of the vapor phase.

In the condensation of a binary mixture the problem of diffusion is slightly more involved, although it does not produce a significant resistance in series with the resistance of the condensate film. In a binary mixture unless it is a constant boiling mixture such as 95 per cent ethanol-water, the higher boiling component condenses in greater proportion near the condenser inlet. The rate of condensation of both components is related to their individual pressure differentials between the bulk of the vapor and the condensate film. Colburn and Drew[1] pointed out that the *ratio* of the partial pressures exerted by the condensate film for a binary mixture is dependent upon the temperature of the film, which in turn is related to the temperature of the cold surface and the cooling water range. Thus for a given binary mixture entering a condenser not only the overall rate but also the chemical composition of the condensate are influenced by the temperature of the cooling water. This is not of particular concern in the total condensation of a vapor, but it can influence the composition of the product from a partial condenser. Colburn and Drew give equations for calculating the composition and temperature at the vapor-condensate interface for binary mixtures. It is customary, however, to assume the temperature of the condensate at the vapor-liquid interface to be the same as that of the vapor.

In the condensation of the overhead vapor coming from a binary dis-

[1] Colburn, A. P., and T. B. Drew, *Trans. AIChE*, **33**, 197–215 (1937).

tillation the vapor is almost entirely composed of the more volatile component and the presence of the second component establishes a condensing range as predicted by the phase rule. The second component, being less volatile, condenses more rapidly near the inlet than the outlet. It is conceivable, then, for a binary mixture with a condensing range of 20°F that the higher boiling component comes out in greater proportion during the first 10° than it does during the second 10°. Similarly, more of the heat may be removed from the vapor in the second 10° than in the first. The use of the LMTD is slightly in error on the unsafe side in the case of binary overhead vapors, although the size of the error does not usually justify its rejection. Where a number of compounds are contained in the overhead vapor as in multicomponent distillation, the temperature range and the distribution of the heat transfer differ greatly from the straight-line relationship of Q vs. t on which the LMTD is predicated. Methods of treating multicomponent mixtures are dealt with in the next section.

Film coefficients for the transfer of heat by condensation from binary mixtures may be treated in the same manner as heretofore for a single vapor using the weighted film properties of the mixture.

2. CONDENSATION OF A MULTICOMPONENT MIXTURE

Vapor-liquid Relationships in Mixtures. The phase rule has been used only qualitatively, but it is important for the identification of the various types of mixture problems. Except in binary mixtures or mixtures of several compounds whose boiling points in the pure condition do not differ greatly, the condensation of the vapor mixture occurs over a long temperature range. The fraction of the total heat load delivered during a fractional decrease in the vapor temperature need not be uniform over the entire condensing range, and this invalidates the use of the logarithmic mean alone or $F_T \times$ LMTD in the case of a 1-2 condenser. The solution of such problems requires the determination or calculation of the *condensing curve* for the mixture.

When a single vapor is in equilibrium with its liquid, the vapor and liquid have the same composition. For a mixture some of the components are more volatile than others (except in constant-boiling mixtures) and the vapor and liquid in equilibrium have different compositions, the percentage of the more volatile components being greater in the vapor. The following discussion applies particularly to mixtures which form *ideal solutions*, although suggestions for its application to nonideal solutions are included.

An ideal solution is one in which the presence of the several components has no effect upon the behavior of each and which is governed by Dalton's and Raoult's laws. For such a system Dalton's law states that the total pressure is the sum of the partial pressures in the phase above the liquid solution. Raoult's law states that the partial pressure of a component in the phase above a liquid solution is equal to the product of its pressure as a pure component and its mol fraction in the solution. The latter is not true of nonideal solutions in which the presence of the various components tends to reduce the partial pressures of the others so that the total pressure is not the sum of the products of the mol fractions and vapor pressures in the pure condition. Materials of an electrolytic or ionic nature deviate from the laws of ideal solutions to a great extent. For an ideal solution:

Dalton's law:

$$p_t = p_1 + p_2 + p_3 \qquad p_1 = p_t y_1 \tag{13.2}$$

and Raoult's law:

$$p_1 = p_{p1} x_1 \tag{13.3}$$

where p_t is the total pressure, p_1 the partial pressure of component 1, p_{p1} the vapor pressure of the pure component 1 at the temperature of the solution, y_1 the mol fraction of component 1 in the vapor phase, and x_1 the mol fraction of component 1 in the liquid. Subscripts 2, 3, etc., refer to other components. The mol fraction is the ratio of the number of mols of a single component to the total number of mols in the mixture and is sometimes abbreviated by *mf*. The mol per cent is the mol fraction multiplied by 100.

Solving Eqs. (13.2) and (13.3) for p_1,

$$p_1 = p_{p1} x_1 = p_t y_1 \tag{13.4}$$

Rearranging,

$$y_1 = \frac{p_{p1} x_1}{p_t} \tag{13.5}$$

Solutions which are ideal at moderate pressures appear to deviate from ideality at high pressures, each component tending in some degree to lower the pressure of the other. The total pressure is then no longer the summation of the partial pressures, and Eq. (13.5) is not valid. *Fugacities* or corrected *pressures*, designated by the letter *f*, are then introduced. The fugacities are the partial pressures of the compounds such that the various criteria of ideality may be retained, and their values originate with experimental pressure-volume-temperature studies on the actual chemical compounds.

$$f_{p1} x_1 = f_t y_1 \tag{13.6}$$

where f_t replaces the total pressure p_t

$$y_1 = \frac{f_{p1} x_1}{f_t} \tag{13.7}$$

or

$$y_1 = K_1 x_1 \tag{13.8}$$

where $K_1 = f_{p1}/f_t$

$$\Sigma y_1 = \Sigma(K_1 x_1) \tag{13.9}$$

K is called the *equilibrium constant*.

When a mixture is in vapor-liquid phase equilibrium, the vapor possesses a greater percentage of the more volatile components than the

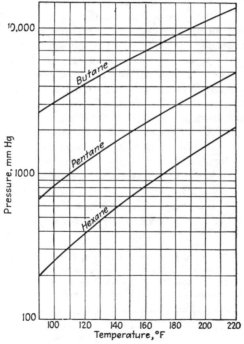

Fig. 13.3. Vapor pressure curves.

liquid. It is possible from Eqs. (13.4) to (13.9) at any given total pressure to calculate the mol fraction y_1 of the component in the vapor phase which may coexist with a mol fraction x_1 of the same component in the liquid phase. The vapor pressures of the pure components must be known at the temperature at which the vapor-liquid equilibrium is assumed to exist. Given butane, pentane, and hexane, as seen in Fig. 13.3, the butane possesses the highest vapor pressure at any given temperature. If several mixtures of the three compounds are formulated with equal ratios of pentane to hexane, the one with the largest percentage of butane would start to boil most nearly at the boiling point of pure butane. Other mixtures having a preponderance of hexane boil nearer the boiling point of hexane. Any mixture of the three compounds starts to boil at an initial temperature higher than the boiling point of pure butane and lower than the boiling point of pure hexane, since boiling starts only when the sum of the partial pressures, $p_1 + p_2 + p_3$, is infinitesimally greater than the total pressure on the system.

Boiling is presumed to start when the first bubble is forced out of the solution. This temperature is called the *bubble point* of the mixture, the term *boiling point* being reserved for a pure compound. If the mixture is boiled at *constant pressure* with a full expansion of the volume of vapor formed, the liquid phase does not disappear until the last droplet is vaporized. For the mixtures referred to above the last droplet consists mainly of hexane and it disappears at a higher temperature than that at which the initial bubble of the mixture formed. Upon complete vaporization the composition of the total vapor is the same as the composition of the liquid before vaporization began. Conversely, if the equilibrium were to start with all the mixture in the vapor phase, the liquid phase begins with the formation of the first droplet of condensate which is identical in composition to the last droplet vaporized. The temperature of formation of the first droplet by heat extraction from the vapor phase is again called the *dew point*. The difference between the bubble point and dew point is the *boiling range*, which must exist for any miscible mixture as predicted by the phase rule.

Example 13.1. Calculation of the Bubble Point, Dew Point, and Vapor Composition of a Mixture. The following mixture is to be heated and vaporized at atmospheric pressure. What is the temperature at which boiling will start (*i.e.*, the bubble point), and what is the composition of the first vapor formed?

Compound	Lb/hr	Mol. wt.	Mol/hr	mf
Butane, C_4H_{10}	1,625	58.12	28.0	0.077
Pentane, C_5H_{12}	16,200	72.15	224.0	0.613
Hexane, C_6H_{14}	9,750	86.17	113.0	0.310
	27,575		365.0	1.000

CONDENSATION OF MIXED VAPORS

Solution:

(a) Bubble point: At atmospheric pressure, $p_t = 14.7$ psi $= 760$ mm Hg.

$p_t = 760$ mm		Assume $T = 100°F$		Assume $T = 96°F$		Assume $T = 97°F$	
	x_1, mf	$p_{p1,100°F}$	$p_1 = p_{p1}x_1$	$p_{p1,96°F}$	$p_1 = p_{p1}x_1$	$p_{p1,97°F}$	$p_1 = p_{p1}x_1$
C_4*	0.077	3,170	244	2,990	230	3,040	234
C_5	0.613	790	484	725	444	740	454
C_6	0.310	250	77.5	229	71	234	72.4
	1.000		$p_t = \Sigma p_1 = 805.5$ Too high		745 Too low		760.4 Check

* The use of the letter C with a subscript is the usual abbreviation for a straight-chain organic molecule where the subscript indicates the number of carbon atoms. If not straight-chain as isobutane, the subscript designates the number of carbon atoms but the C is preceded by i-. Thus butane C_4, isobutane i-C_4, pentane C_5, isopentane i-C_5, etc.

The composition of the first bubble is found from Eq. (13.5), $y_1 = \dfrac{p_{p1}x_1}{p_t}$

	$\dfrac{p_{p1}x_1}{p_t} = y_1$
C_4	$234/760.4 = 0.308$
C_5	$454/760.4 = 0.597$
C_6	$72.4/760.4 = 0.095$
	$\overline{1.000}$

(b) Similarly at what temperature will the mixture start to boil if the system is under a pressure of 35 psia and what will the composition be? $p_t = 35$ psia $= 1810$ mm.

	x_1, mf	Assume $T = 150°F$		Assume $T = 149°F$		$\dfrac{p_{p1}x_1}{p_t} = y_1$
		$p_{p1,150°F}$	$p_1 = p_{p1}x_1$	$p_{p1,149°F}$	$p_1 = p_{p1}x_1$	
C_4	0.077	6100	469	6050	467	$467/1810 = 0.258$
C_5	0.613	1880	1153	1850	1135	$1135/1810 = 0.627$
C_6	0.310	680	211	670	208	$208/1810 = 0.115$
	1.000		$P_t = \Sigma p_1 = 1833$ Too high		1810 Check	$\overline{1.000}$

(c) Parts (a) and (b) may have been solved by Eqs. (13.6) to (13.9) from tabulated values of f_p/f_t which were computed from experimental data and are shown in Fig. 7 in the Appendix. In Fig. 7 values of the equilibrium constant K are plotted for a number of hydrocarbons against temperature and pressure. It is, in effect, the vapor pressure curve for the 18 compounds indicated. Thus if a focal point of $K = 1$ is chosen for any compound, a line connecting pressure and temperature through $K = 1$ gives the boiling temperature of the pure compound corresponding to any pressure. Since $K = 1$, from Eq. (13.8), $f_p/f_t = 1$, and the pressure of a pure compound f_p or p_p

must be identical with the total pressure f_t or p_t when there is no other fluid present. At 760 mm or 14.7 psia the bubble point computed from the fugacity data represented by Fig. 7 will be higher than in (a) owing to the mutual reduction of partial pressures resulting from nonideality.

BUBBLE POINT

	x_1, mf	Assume $T = 95°F$		Assume $T = 100°F$		Assume $T = 102°F$	
		$K_{95°F}$	$y_1 = K_1 x_1$	$K_{100°F}$	$y_1 = K_1 x_1$	$K_{102°F}$	$y_1 = K_1 x_1$
C_4	0.077	3.13	0.241	3.35	0.258	3.45	0.266
C_5	0.613	0.92	0.564	1.00	0.613	1.02	0.625
C_6	0.310	0.30	0.093	0.335	0.104	0.35	0.109
	1.000		$\Sigma y_1 = 0.898$		0.975		1.000
			Too low		Too low		Check

(d) The use of K values gives y_1 directly and permits use of the total mol fraction of $\Sigma y_1 = 1.00$ as the criterion for equilibrium. Similarly for 35 psia

	x_1	Assume $T = 150°F$		Assume $T = 153°F$	
		$K_{150°F}$	$y_1 = K_1 x_1$	$K_{153°F}$	$y_1 = K_1 x_1$
C_4	0.077	2.80	0.216	2.90	0.223
C_5	0.613	1.01	0.619	1.06	0.650
C_6	0.310	0.40	0.124	0.415	0.1285
	1.000		$\Sigma y_1 = 0.959$		1.0015
			Too low		Check

(e) The temperature at which the liquid phase disappears if boiling occurs in an enclosed vessel is the temperature at which only the last droplet is left. But this is the same as the physical picture of the dew point when the first droplet is formed. The initial compositions are then the vapor mol fractions or y's, and for the liquid formed $x_1 = y_1/K_1$.

Dew point: At $p_t = 14.7$ psia, 760 mm.

	y_1	Assume $T = 130°F$		Assume $T = 120°F$		Assume $T = 123°F$	
		$K_{130°F}$	$x_1 = \dfrac{y_1}{K_1}$	$K_{120°F}$	$x_1 = \dfrac{y_1}{K_1}$	$K_{123°F}$	$x_1 = \dfrac{y_1}{K_1}$
C_4	0.077	5.0	0.015	4.4	0.0175	4.60	0.0167
C_5	0.613	1.65	0.371	1.40	0.437	1.49	0.412
C_6	0.310	0.62	0.500	0.51	0.608	0.545	0.568
	1.000		$\Sigma x = 0.886$		1.062		0.9967
							Check

The dew point is 123°F.

At $p_t = 35$ psia

	y_1	Assume $T = 174°F$	
		$K_{174°F}$	$x_1 = \dfrac{y_1}{K_1}$
C_4	0.077	3.70	0.0208
C_5	0.613	1.38	0.444
C_6	0.310	0.58	0.533
	1.000		$\Sigma x = 0.998$

The dew point is 174°F.

The Operating Pressure of a Condenser. In Example 13.1 may be found the quantitative requirements for setting the operating pressure of a distilling column and condenser. If the condenser is operated at atmospheric pressure, the condensing range is from 123 to 102°F. The application of 1–2 condensers for these temperatures is not very satisfactory when cooling water is available at 85°F, since the water temperature range must be kept small to prevent a large temperature cross above the 102°F condensate outlet. It would be necessary for atmospheric pressure condensation to operate the condenser with a Δt of about 16.5°F with the use of a large amount of water. If the distilling column pressure were raised to 35 psia, the range of condensation would be from 174 to 153°F. The cooling-water range could be from 85 to 120°F, and the Δt would be about 57°F, requiring approximately one-quarter as much surface as at atmospheric pressure. It must be realized, however, that raising the pressure on the column increases the column first cost and also the temperature of the heating medium in the reboiler. The selection of the optimum process operating pressure is a matter of economic analysis. The total annual operating cost for several pressures, which includes utilities and fixed charges, is plotted against operating pressure, the optimum occurring when the total annual cost is a minimum.

Relative Volatilities. Another method of obtaining the composition during phase equilibrium is by *relative volatilities*. This method utilizes the principle that in a mixture of several components some are more and some are less volatile (have larger or smaller K's) than an intermediate compound. Although the K's may change greatly even over a small temperature range, the ratios of the equilibrium constants relative to each other remain nearly constant. By means of this premise it is possible to eliminate the successive trial-and-error calculation provided the first trial is reasonably close. For a three component system

$$y_1 = K_1 x_1 \qquad y_2 = K_2 x_2 \qquad y_3 = K_3 x_3$$
$$\frac{y_1}{y_2} = \frac{K_1 x_1}{K_2 x_2} \qquad \frac{y_3}{y_2} = \frac{K_3 x_3}{K_2 x_2} \tag{13.10}$$

$K_1/K_2 = \alpha_{1\text{-}2}$ is the relative volatility of compound 1 to compound 2 and $K_3/K_2 = \alpha_{3\text{-}2}$ is the relative volatility of compound 3 to compound 2.

For vaporization:
$$y_1 + y_2 + y_3 = 1.00 \tag{13.11}$$

Relative to y_2
$$\frac{y_1}{y_2} + 1 + \frac{y_3}{y_2} = \frac{1.00}{y_2} \tag{13.12}$$

Substituting,
$$\alpha_{1\text{-}2}\frac{x_1}{x_2} + 1 + \alpha_{3\text{-}2}\frac{x_3}{x_2} = \frac{1.00}{y_2}$$

Rearranging,
$$y_2 = \frac{x_2}{\alpha_{1\text{-}2}x_1 + x_2 + \alpha_{3\text{-}2}x_3} = \frac{x_2}{\Sigma \alpha x} \tag{13.13}$$

and since
$$\frac{x_2}{y_2} = \alpha_{1\text{-}2}\frac{x_1}{y_1}$$

$$y_1 = \frac{\alpha_{1\text{-}2}x_1}{\Sigma \alpha x} \qquad y_2 = y_2 \qquad y_3 = \frac{\alpha_{3\text{-}2}x_3}{\Sigma \alpha x} \tag{13.14}$$

For condensation:
$$x_1 + x_2 + x_3 = 1.00 \tag{13.15}$$

$$\frac{x_1}{x_2} + 1 + \frac{x_3}{x_2} = \frac{1}{x_2}$$

$$\frac{y_1}{y_2 \alpha_{1\text{-}2}} + 1 + \frac{y_3}{y_2 \alpha_{3\text{-}2}} = \frac{1}{x_2}$$

$$\frac{y_1}{y_2 \alpha_{1\text{-}2}} + 1 + \frac{y_3}{y_2 \alpha_{3\text{-}2}} = \frac{1}{x_2}$$

$$x_2 = \frac{y_2}{y_1/\alpha_{1\text{-}2} + y_2 + y_3/\alpha_{3\text{-}2}}$$

$$x_1 = \frac{y_1/\alpha_{1\text{-}2}}{\Sigma y/\alpha} \qquad x_2 = \frac{y_2}{\Sigma y/\alpha} \qquad x_3 = \frac{y_3/\alpha_{3\text{-}2}}{\Sigma y/\alpha} \tag{13.16}$$

Example 13.2. Calculation of the Bubble Point and Vapor Composition by Relative Volatilities. As before, make the assumption that the bubble point is 95°F, which is considerably off.

Bubble point: $p_t = 14.7$ psia.   Assume $T = 95°F$.

	x_1	$K_{95°F}$	$\alpha_{95°F}$	$\alpha_1 x_1$	$y_1 = \dfrac{\alpha_1 x_1}{\Sigma \alpha x}$
C_4	0.077	3.13	3.40	0.262	0.269
C_5	0.613	0.92	1.0	0.613	0.628
C_6	0.310	0.30	0.326	0.101	0.1035
	1.000		$\Sigma \alpha x = 0.976$		1.000

$$K_2 = \frac{y_2}{x_2} = \frac{(\alpha_2 x_2/\Sigma \alpha x)}{x_2} = \frac{0.628}{0.613} = 1.025$$

Look up the temperature of C_5 in Fig. 7 corresponding to a K_2 of 1.025 and $p_t = 14.7$ psia.

$K_2 = 1.025 \qquad T = 102°F \qquad$ (Checks Example 13.1 on first trial)

Dew point; $p_t = 14.7$ psia. Assume $T = 130°F$.

	y_1	$K_{130°F}$	$\alpha_{130°F}$	$\dfrac{y_1}{\alpha_1}$	$x_1 = \dfrac{y_1/\alpha_1}{\Sigma y/\alpha}$
C_4	0.077	5.0	3.03	0.0254	0.0175
C_5	0.613	1.65	1.00	0.613	0.419
C_6	0.310	0.62	0.376	0.824	0.5635
	1.000			$\Sigma y/\alpha = 1.462$	1.0000

$$K_2 = \frac{y_2}{x_2} = y_2 \bigg/ \frac{y/\alpha}{\Sigma y/\alpha} = \frac{0.613}{0.419} = 1.46$$

Look up the temperature of C_5 corresponding to a $K_2 = 1.46$ and $p_t = 14.7$ psia.

$$K_2 = 1.46 \qquad T = 122°F$$

The value computed from K values was 123°F, showing a small variation in the actual relationship of the volatilities.

Calculation of Compositions between the Dew Point and Bubble Point. The calculation of dew points and bubble points from mol fractions is usually an unnecessary step, and it is desirable to carry out multicomponent calculations on the total number of the mols directly.

Since
$$y_1 = K_1 x_1$$
it is the same as
$$\frac{V_1}{V} = K_1 \frac{L_1}{L}$$

where V_1 is the number of mols of a component in the vapor and L_1 the number of mols of the component in the liquid. V and L are the total number of mols of vapor and liquid, respectively.

At the dew point,
$$V_1 = K_1 L_1 \tag{13.17}$$

At the bubble point,
$$L_1 = \frac{V_1}{K_1} \tag{13.18}$$

In an enclosed vessel at equilibrium at *any temperature* between the dew point and bubble point if Y is the original number of mols of vapor consisting of Y_1, Y_2, and Y_3, etc., for each compound, the amount condensed is given by

$$Y = V + L \qquad Y_1 = V_1 + L_1$$

where V is the total mols of vapor remaining and L is the total mols of liquid formed. Then

$$V_1 = Y_1 - L_1$$

and

$$V_1 = K_1 L_1 \frac{V}{L}$$

$$L_1 = \frac{Y_1}{1 + K_1(V/L)} \qquad (13.19)$$

To determine the mols condensed at any given temperature between the dew point and bubble point, assume a ratio of V/L and calculate by Eq. (13.19) the number of mols of liquid formed for the assumed value of V/L. If the ratio of the mass of vapor remaining to the mols of liquid formed does not check the assumed value of V/L, a new assumption must be made.

Differential Condensation. The equilibrium in a condenser causes *differential condensation*. Consider a condenser as shown in Fig. 13.4 divided into a number of condensing intervals as 0-0 to 1-1, 1-1 to 2-2, etc. At 0-0 there is perhaps one droplet of condensate, but in the zone

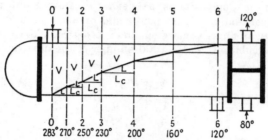

Fig. 13.4. Differential condensation in a condenser.

from 0-0 to 1-1 considerable condensate is formed. At 1-1 there is now a layer of condensate, and the total mols of vapor plus condensate is the same as at 0-0, but the mols of vapor is less and its composition differs from the original feed. The equilibrium at 1-1 differs from the dew point in so far as there is now a definite number of mols of liquid present instead of one droplet. In the interval from 1-1 to 2-2 where a new equilibrium is established, the vapor is in equilibrium not merely with the liquid which has been formed by condensation during the interval but also with a liquid whose composition consists of all the precondensed liquid from previous intervals. The similarity between this behavior and that which occurs in a constant-pressure batch vessel can be demonstrated. If L_c are the total mols of liquid condensed in the zone 0-0 to 1-1 before the zone from 1-1 to 2-2, L_{c_1} the mols of component 1 condensed from 0-0 to 1-1, and L' and L'_1 the total mols and mols of component 1 formed in the zone from 1-1 to 2-2 in Fig. 13.4, then at 2-2

$$\frac{V_1}{V} = K_1 \frac{L'_1 + L_{c_1}}{L' + L_c}$$

$$V_1 = Y_1 - L'_1$$

$$\frac{Y_1 - L'_1}{V} = K_1 \frac{L'_1 + L_{c_1}}{L' + L_c}$$

$$L'_1 = Y_1 - K_1 \frac{V(L'_1 + L_{c_1})}{L' + L_c}$$

$$L'_1 = \frac{Y - K_1(V/L' + L_c)L_{c_1}}{1 + K_1(V/L' + L_c)} \qquad (13.20)$$

At any point, however, $L' + L_c = L$, $L'_1 + L_{c_1} = L_1$, and Eq. (13.20) reduces to Eq. (13.19). Figure 13.4 is obviously idealized and does not take into account the hydrodynamics of such a condensate layer.

The Calculation of the Condensing Curve for a Multicomponent Mixture. The calculation of the weighted temperature difference is dependent upon the shape of the condensing curve which is a plot of vapor heat content vs. vapor temperature for the condensing range. The weighted temperature difference is then obtained by taking increments of the condensing range and computing the average temperature difference between the vapor and water temperatures. From a practical standpoint it is necessary to choose only a reasonable number of temperatures, since the solution of Eq. (13.19) is by trial and error. The best selection of the intervals should give equal increments of $dQ/\Delta t$ or UA, but this is rather difficult to accomplish by inspection. It is usually helpful to observe whether the vapor to be condensed has fronts or tails. Tails indicate that for a high dew point the main portion of the vapor does not condense until a considerably lower temperature is reached. Thus in example 13.1 if a small number of mols of C_9 or C_{10} was added to the mixture, the dew point would be considerably higher than before although the major heat load removed by condensation would start only at the temperature which previously was the dew point or near 130°F at 14.7 psia. With small amounts of C_9 and C_{10} the temperature of the vapor would fall rapidly upon the removal of a small amount of heat. Similarly in the case of fronts, the presence of the small amount of propane C_3 reduces the bubble point, although the heat removal between the old and new bubble points would not necessarily represent a significant percentage of the total heat load removed in the condenser.

The Weighted Temperature Difference. In the desuperheater-condenser and condenser-subcooler it was assumed that the counterflow temperature differences could be applied over the entire length of the shell to obtain the weighted Δt. In the case of a multicomponent mixture, because it requires an integration to obtain the weighted Δt, it is similarly very convenient to assume that the cooling medium is in counterflow with the multicomponent vapor even if a 1-2 condenser is employed. The percentage rise in the temperature of the cooling medium at any cross section of the shell is then taken as proportional to the percentage of the heat load removed from any condenser cross section to the exit. The weighted Δt is then the averaged temperature difference between the condensing curve (vapor heat content vs. T_v plot) and a straight line representing the cooling medium. When a 1-2 condenser is employed, if the value of F_T based on inlet and outlet temperatures is not very nearly 1.0, the assumption above may not be admissible. If q is the heat load for an interval on the condensing curve, then the weighted Δt is obtained from the total heat load Q divided by the summation of the values of $q/\Delta t_{av}$ where Δt_{av} is the average temperature difference for the interval.

Heat-transfer Coefficients for a Multicomponent Mixture. When a multicomponent mixture is condensed, the condensing range between the dew point and the bubble point may be greater than 100°. The liquid which is formed near the inlet differs greatly in composition from that formed near the outlet and must be cooled to the bubble point at the end of the shell before it drains from the condenser. The film coefficient in the condenser differs somewhat from the Nusselt assumptions in that the first liquid to condense is the higher boiler and higher boilers in any homologous chemical series are more viscous than the lower ones. For vertical or horizontal tubes this means that the Reynolds number for the inlet portion of the tubes might well be lower than might be calculated using the mean properties of the total mixture for the condensate film. Nevertheless, to avoid the tediousness of an integration to determine the changes in $\bar{h}$ due to changing liquid properties for differential changes in area dA, it is possible to use a method of averaging the inlet and outlet film coefficients. This can be done by using the mean properties of the mixture, or, if there is a great difference in the characteristics of the condensate between inlet and outlet, to calculate h at both terminals and take the mean. The compositions of the final condensate on the tubes of vertical and horizontal condensers condensing the same mixture are not identical, since the condensate is cumulative on vertical tubes. There appears to be little need, however, in view of the other assumptions which also apply, to take this into account except with viscous condensates.

In either vertical or horizontal condensers it is helpful to consider, as in the development of Eq. (13.20), that the phase equilibrium at a cross section exists between the residual vapor and all the liquid formed up to that point. In condensing a multicomponent mixture between its dew point and bubble point the vapor is cooled and the condensate formed at the inlet leaves at the temperature of the outlet, which may be colder by some 100°F or more. The vapor and condensate must be cooled sensibly as they travel the length of the shell, although this is not the same as desuperheating or subcooling, since it occurs concurrently with condensation rather than in distinct zones. The clean surface required for condensation is frequently calculated from the *entire* heat load, the weighted Δt, and a value of U_{Clean}, using h obtained from either of the two methods described above. The main problem of sensible-heat transfer appears to be in cooling the condensate rather than the vapor, since the vapor coefficient in the presence of condensation is quite high as discussed for steam in Chap. 12. One method of treating the sensible-heat transfer is to allow additional surface equivalent to the percentage which the sensible-heat load is to the total heat load. The sensible-heat-transfer surface is then an *additional* percentage of the condensing surface. This is equivalent to the use of a sensible-heat-transfer coefficient about one-half the condensing coefficient. Another method is to compute condensate cooling surface using the free-convection coefficient of about 50 but applied only to the average liquid cooling sensible-heat load. Actually both methods give about the same answer. The effectiveness of the surface required for sensible-heat transfer, and particularly for cooling the condensate, is assured by submerging all the additional surface allowance through the use of a loop seal or dam baffle as discussed previously. The overall clean coefficient is then the weighted coefficient based on the total clean surface. If the vapor enters above the dew point or condensate leaves below the bubble point, the desuperheating and subcooling zones are weighted along with the condensing zone as in Chap. 12.

Example 13.3. Condenser Calculations for a Multicomponent Mixture. The overhead vapor from a distilling column operating at 50 psia contains only saturated hydrocarbons such as propane, butane, and hexane and has the following analysis:

	Lb/hr	Mol/hr
C_3	7,505	170.5
C_4*	16,505	284.0
C_6	4,890	56.8
C_7	34,150	341.1
C_8	32,400	284.0
	95,450	1136.4

* A trace of C_5 has been combined half with C_4 and half with C_6 to simplify the calculation.

It is to be condensed in a 1-2 horizontal condenser using cooling water from 80 to 120°F. Pressure drops of 2.0 psi for the vapor and 10.0 psi for the water should be adequate. A minimum dirt factor of 0.004 should be provided. Available for the service is a 33 in. ID 1-2 condenser having 774 ¾ in. OD, 16 BWG tubes 16'0'' long and laid out on 1-in. triangular pitch. The bundle is arranged for four passes, and baffles are spaced, except at the inlet and outlet, 30 in. apart.

(a) Determine the condensing range.
(b) Compute the condensing curve.
(c) Compute the weighted Δt.
(d) Establish the suitability of the condenser.

Solution. (a) Condensing range: This is the temperature spread between the dew point [Eq. (13.17)] and the bubble point [Eq. (13.18)].

Dew point: Assume $T = 283°F$ *Bubble point:* Assume $T = 120°F$

	V_1	K_1 at 283°F	$\dfrac{V_1}{K_1}$	L_1	K_1 at 120°F	$K_1 L_1$
C_3	170.5	13.75	12.40	170.5	4.1	700
C_4	284.0	6.18	46.0	284.0	1.39	395
C_6	56.8	1.60	35.5	56.8	0.17	9.66
C_7	341.1	0.825	414.	341.1	0.06	20.44
C_8	284.0	0.452	628.	284.0	0.023	6.54
	1136.4		1147.0 Check	1136.4		1132.6 Check

Assume intervals at 270, 250, 230, 200, 160°, and the dew point, and solve for V/L by Eq. (13.19) to obtain the condensation in each interval and from it the heat load for the interval.

Range: 283 to 270°F
Trial: Assume $V/L = 4.00$.

	Y_1	$K_{270°}$	$\dfrac{K_1 V}{L}$	$1 + \dfrac{K_1 V}{L}$	$L_1 = \dfrac{Y_1}{1 + K_1(V/L)}$
C_3	170.5	12.75	51.0	52.0	3.28
C_4	284.0	5.61	22.4	23.4	12.13
C_6	56.8	1.40	5.60	6.60	8.60
C_7	341.1	0.705	2.82	3.82	89.3
C_8	284.0	0.375	1.50	2.50	113.7
	1136.4				$L = \Sigma L_1 = 227.0$

$V = 1136.4 - 227.0 = 909.4$
V/L calculated $= 4.00$
V/L assumed $= 4.00$ *Check*

CONDENSATION OF MIXED VAPORS 333

If the assumed and calculated values of V/L had not checked, a new value would have been assumed. A check in this type of calculation generally infers a variation of 0.01 or less as the ratio of V/L decreases. For the next range, 270 to 250°F, proceed as above and obtain the actual mols of condensation for the interval by subtracting the liquid in equilibrium at 270°F from that in equilibrium at 250°F. Then L_{c1} are the mols of the individual compounds formed before the interval and L_1' are the mols of the individual compounds formed in the interval. L_1' is obtained by subtracting L_{c1} from L_1, which is in turn obtained from a checked assumption of V/L where $\Sigma L_1 = L$. A summary of the point-to-point calculations is given in Table 13.2.

(b) Condensing curve: This step requires the calculation of the heat load between intervals. Except at the inlet and outlet of the condenser there is a change in both vapor and liquid quantities over the interval. The heat changes are determined from the change in enthalpies as given in Fig. 10. A representative interval is calculated following Table 13.2.

The LMTD would be 87.7°F, and the error resulting from its use would have been $(13/100.7) \times 100 = 12.9$ per cent on the safe side. In any system without serious fronts or tails the use of the LMTD is usually satisfactory, although there is no assurance that the error will always be on the safe side.

The condensing curve for true counterflow is shown as a straight line in Fig. 13.5. The actual condensing curve is shown as a curved line, and the area enclosed by the two represents the actual increase in temperature potential which is available.

TABLE 13.2. POINT-TO-POINT COMPOSITIONS
$L_1 = L_{c1} + L_1'$ $L = (L_1)$

T_{vapor} (DP)	283°		270°			250°			230°			200°
	Y_1	L_1	Y_1	L_{c1}	$L_{1'}$	Y_1	L_{c1}	$L_{1'}$	Y_1	L_{c1}	$L_{1'}$	
C_3	170.5	3.28	167.2	3.28	5.85	161.4	9.13	8.02	153.3	17.15	16.40	
C_4	284.0	12.13	271.9	12.13	20.81	251.1	32.94	26.04	225.0	58.98	46.6	
C_5	56.8	8.60	48.2	8.60	12.00	36.2	20.60	11.02	25.2	31.62	11.88	
C_7	341.1	89.3	251.8	89.3	94.5	157.3	183.8	63.6	93.7	247.4	51.6	
C_8	284.0	113.7	170.3	113.7	82.4	87.9	196.1	41.7	46.2	237.8	28.2	
	1136.4	227.0	909.4	227.0	215.6	693.9	442.5	150.3	543.4	593.0	154.7	
$V/L =$		4.00				1.567			0.916			0.520

			200°			160°			120°(BP)	
			Y_1	L_{c1}	$L_{1'}$	Y_1	L_{c1}	$L_{1'}$	Y_1	L_{c1}
C_3			136.9	33.55	40.2	96.7	73.8		170.5	
C_4			178.4	105.6	83.7	94.7	189.3		284.0	
C_5			13.3	43.5	9.30	4.0	52.8		56.8	
C_7			42.1	299.0	31.8	10.3	330.8		341.1	
C_8			18.0	266.0	14.30	3.7	280.3		284.0	
			388.7	747.7	179.3	209.4	927.0		1136.4	
$V/L =$			0.520			0.226				

HEAT LOAD FOR THE INTERVAL 270 TO 250°

	Mol. wt.	$H_{v,270°}$	Y_1	H_v	$H_{l,270°}$	L_{c1}	H_1
C_3	44	324	167.2	2,384,000	210	3.28	30,300
C_4	58	334	271.9	5,260,000	212	12.13	149,200
C_6	86	352	48.2	1,460,000	226	8.60	167,000
C_7	100	359	251.8	9,030,000	236	89.3	2,105,000
C_8	114	368	170.3	7,150,000	239	113.7	3,100,000
				25,284,000			5,551,500
				5,551,500			

$$H_{270°} = 30,835,500 \text{ Btu/hr}$$

	Mol. wt.	$H_{v,250°}$	Y_1	H_v	$H_{l,250°}$	L_{c1}	H_1
C_3	44	313	161.4	2,221,000	195	9.13	78,400
C_4	58	323	251.1	4,700,000	197.5	32.94	377,000
C_6	86	341	36.2	1,060,000	212.5	20.60	376,000
C_7	100	350	157.3	5,500,000	224	183.8	4,110,000
C_8	114	358	87.9	3,580,000	225	196.1	5,040,000
				17,061,000			9,981,400
				9,981,400			

$$H_{250°} = 27,042,400 \text{ Btu/hr}$$

$$Q_{270-250°} = 30,835,500 - 27,042,000 = 3,793,100 \text{ Btu/hr}$$

HEAT LOAD FOR THE ENTIRE RANGE

$T_{vapor,°F}$	H	$q = \Delta H$	Δt_W*	$t_W, °F$	Δt_{av}	$\dfrac{q}{\Delta t}$	$q_{cumulative}$
283	34,312,000	3,476,500	6.55	120			0
270	30,835,500	3,793,100	7.15	113.4	159.8	21,780	3,476,500
250	27,042,000	2,839,400	5.35	106.3	150.1	25,210	7,269,600
230	24,203,000	3,359,000	6.34	100.9	136.4	20,800	10,109,000
200	20,844,000	3,931,100	7.42	94.6	117.2	28,620	13,468,000
160	16,912,900	3,803,900	7.17	87.2	88.2†	44,550	17,399,100
120	13,109,000			80.0	54.9†	69,450	21,203,000
		21,203,000	39.98		$\Sigma UA =$	210,410	

* Water requirement: t_W is the water temperature.

$$\frac{21,203,000}{120 - 80} = 530,000 \text{ lb/hr of cooling water} \backsimeq 1060 \text{ gpm}$$

† LMTD

$$\text{Weighted } \Delta t = \frac{\Sigma q}{\Sigma UA} = \frac{21,203,000}{210,410} = 100.7°F$$

CONDENSATION OF MIXED VAPORS

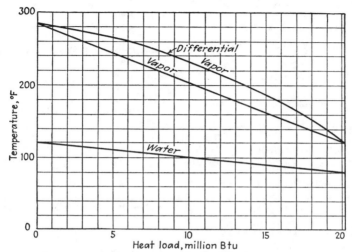

Fig. 13.5. Straight-line and differential condensing curves

The calculation of the exchanger for Example 13.3 follows:

Exchanger:

Shell side
ID = 33 in.
Baffle space = 30 in.
Passes = 1

Tube side
Number and length = 774, 16'0"
OD, BWG, pitch = ¾ in., 16 BWG, 1-in. tri.
Passes = 4

(1) Heat balance:
Q shell = 21,203,000 Btu/hr
Q water = 530,000 × 1(120 − 80) = 21,203,000 Btu/hr

Average molecular weight = $\dfrac{95,450}{1136.4}$ = 84

This corresponds very closely to hexane (mol. wt. = 86.2) whose properties will be used throughout.

Condensate sensible heat load = 95,450 × $\dfrac{0.6(283 - 120)}{2}$ = 4,670,000 Btu/hr

Submergence = 4,670,000 × $\dfrac{100}{21,203,000}$ = 22% (approx)

(2) Δt: Weighted Δt = 100.7°F
(3) T_c and t_c: The use of average temperatures will be satisfactory.

Hot fluid: shell side, vapor
(4') Unsubmerged tubes
= 774 × (1 − 0.22) = 604
$G'' = W/LN_t^{2/3}$ = 95,450/16 × 604$^{2/3}$
= 83.7   [Eq. (12.43)]

Cold fluid: tube side, water
(4) Flow area, $a_t' = 0.302$ in.^{2} [Table 10]
$a_t = N_t a_t'/144n$ [Eq. (7.48)]
= 774 × 0.302/144 × 4 = 0.406 ft^2

Hot fluid: shell side, vapor
Assume $\bar{h} = 200$
$T_v = (283 + 120)/2 = 201.5°F$
$t_w = t_a + \dfrac{h_o}{h_{io} + h_o}(T_v - t_a)$ [Eq. (5.31)]
$= 100 + \dfrac{200}{1120 + 200}(201.5 - 100)$
$= 115°F$
$t_f = \frac{1}{2}(T_v + t_w) = \frac{1}{2}(201.5 + 115)$
$= 158°F$
$s_f = 0.60$ [Fig. 6]
$\mu_f = 0.21$ cp [Fig. 14]
$k_f = 0.077$ Btu/(hr)(ft²)(°F/ft) [Table 4]
$\bar{h} = 206$ [Fig. 12.9]

Cold fluid: tube side water
(5) $G_t = w/a_t$
$= 530{,}000/0.406$
$= 1{,}300{,}000$ lb/(hr)(ft²)
$V = G_t/3600\rho = 1{,}300{,}000/3600 \times 62.5$
$= 5.79$ fps
(6) $h_i = 1355$ [Fig. 25]
$h_{io} = h_i \times \text{ID/OD}$
$= 1355 \times 0.62/0.75$
$= 1120$ Btu/(hr)(ft²)(°F) [Eq. (6.5)]

Clean overall coefficient U_c, condensation:

$$U_c = \frac{h_{io}h_o}{h_{io} + h_o} = \frac{1120 \times 206}{1120 + 206} = 174 \text{ Btu/(hr)(ft²)(°F)} \quad (6.38)$$

Clean surface required for condensation:

$$A_c = \frac{q_c}{U_c \, \Delta t} = \frac{21{,}203{,}000}{174 \times 100.7} = 1210 \text{ ft}^2$$

Clean surface required for subcooling:

$$A_s = 1210 \times 0.22 = 267 \text{ ft}^2$$

Total surface required:

$$A_C = 1210 + 267 = 1477 \text{ ft}^2$$

Weighted overall clean coefficient U_C:

$$U_C = \frac{Q}{A \, \Delta t} = \frac{21{,}203{,}000}{1477 \times 100.7} = 143$$

Design overall coefficient U_D:

$a'' = 0.1963$ ft²/lin ft (Table 10)
Total surface $= 774 \times 16'0'' \times 0.1963 = 2430$ ft²

$$U_D = \frac{Q}{A \, \Delta t} = \frac{21{,}203{,}000}{2430 \times 100.7} = 86.7 \text{ Btu/(hr)(ft²)(°F)}$$

Dirt factor R_d:

$$R_d = \frac{U_C - U_D}{U_C U_D} = \frac{143 - 86.7}{143 \times 86.7} = 0.00455 \text{ (hr)(ft²)(°F)/Btu} \quad (6.13)$$

Pressure Drop

(1') $a_s = \text{ID} \times C'B/144 P_T$ [Eq. (7.1)]
$= 33 \times 0.25 \times 30/144 \times 1.0$
$= 1.72$ ft²
Submergence may be neglected unless the calculated pressure drop is close.
$G_s = W/a_s$ [Eq. (7.2)]
$= 95{,}450/1.72 = 55{,}500$ lb/(hr)(ft²)

(1) At 100°F,
$\mu = 0.72 \times 2.42 = 1.74$ lb/(ft)(hr) [Fig. 14]
$D = 0.62/12 = 0.0517$ ft
$Re_t = DG_t/\mu$
$= 0.0517 \times 1{,}300{,}000/1.74 = 38{,}600$
$f = 0.00019$ ft²/in.² [Fig. 26]

CONDENSATION OF MIXED VAPORS

Pressure Drop

At $T_1 = 283°F$,
$\mu = 0.009 \times 2.42 = 0.0218$ lb/(ft)(hr)
$D_e = 0.73/12 = 0.0608$ ft [Fig. 15]
$Re_s = D_e G_s/\mu$
$\quad = 0.0608 \times 55{,}500/0.0218 = 155{,}000$
$f = 0.00125$ ft^2/in.^{2} [Fig. 29]
(**2'**) No. of crosses $(N + 1) = 12L/B$ [Eq. (7.43)]
$\quad = 12 \times 16/30 = 7$

$$\rho = \frac{84}{359 \times 743/492 \times 14.7/50}$$
$\quad\quad = 0.527$ lb/ft^3
$s = 0.527/62.5 = 0.00844$
$D_s = 33/12 = 2.75$ ft

(**3'**) $\Delta P_s = \dfrac{1}{2} \dfrac{fG_s^2 D_s(N+1)}{5.22 \times 10^{10} D_e s}$
  [Eq. (12.47)]
$\quad = \dfrac{1}{2} \dfrac{0.00125 \times 55{,}500^2 \times 2.75 \times 7}{5.22 \times 10^{10} \times 0.0608 \times 0.00844}$
$\quad\quad = 1.4$ psi

(2) $\Delta P_t = \dfrac{fG_t^2 L n}{5.22 \times 10^{10} D s \phi_t}$ [Eq. (7.45)]
$\quad = \dfrac{0.00019 \times 1{,}300{,}000^2 \times 16}{5.22 \times 10^{10} \times 0.0517 \times 1.0} \times \dfrac{\times 4}{\times 1.0}$
$\quad = 7.6$ psi

(3) $\Delta P_r = (4n/s)(V^2/2g')$ [Eq. (7.46)]
$\quad = 4 \times 4 \times 0.23 = 3.7$ psi [Fig. 27]

(4) $\Delta P_T = \Delta P_t + \Delta P_r$
$\quad = 7.6 + 3.7 = 11.3$ psi [Eq. (7.47)]

Summary

206*	h outside	1120
U_C	143	
U_D	86.7	
R_d Calculated 0.00455		
R_d Required 0.004		
1.4	Calculated ΔP	11.3
2.0	Allowable ΔP	10.0

* Condensation only.

The slightly excessive tube-side pressure drop for water should not be objectionable. The bundle should be submerged about 25 to 30 per cent.

3. CONDENSATION OF A MIXTURE OF MISCIBLES AND ONE IMMISCIBLE

This case arises in the steam distillation of organics which are miscible among themselves but not with water. The stripping of the volatile compounds from absorption oil is a typical example as shown in Fig 11.1.

The introduction of the steam permits part of the total operating pressure in the distilling column to be contributed by the steam, so that the mixture of volatiles and oil need not be raised to a high temperature to accomplish distillation. In this way and without resorting to vacuum a greater separation can be made between the volatiles and the oil, which boils over a somewhat higher range than that at which the distillation is carried out. If the overhead vapor is a mixture of a single compound and steam, the condensation occurs isothermally. If the mixture contains more than one compound miscible with the first but immiscible with water, there is a condensing range. Since the total pressure in the latter consists of water plus the miscibles, the equilibrium relationships for the miscibles corresponds to the sum of their partial pressures instead of the total pressure on the system. The partial pressure of the steam depends only upon the saturation pressure corresponding to its temperature in the mixture as given by the properties of saturated steam in Table 7. The total pressure is constant, but the relative partial pressures of the miscibles and steam change from point to point. The problems involved in the calculation of the condensing curve for this system are included in the more comprehensive problem demonstrated in Sec. 5 of this chapter. Ordinarily the calculation is so lengthy that the use of the LMTD or the use of $F_T \times$ LMTD is justifiable as the case may be.

Hazelton and Baker[1] carried out experimental work on a single vertical tube condensing benzene, toluene, and chlorbenzene with steam. It was found that the presence of the organic tended to favor the dropwise condensation of the steam. Film coefficients were found to be independent of the temperature drop across the condensate film and independent of the properties of the liquid condensed with water. Hazelton and Baker were able to correlate their work with the experimental results of others. For vertical tubes they obtained

$$\bar{h} = 79 \left[\frac{(\text{wt-}\%)_A \lambda_A + (\text{wt-}\%)_B \lambda_B}{(\text{wt-}\%)_A L} \right]^{\frac{1}{4}} \qquad (13.21a)$$

where A and B refer, respectively, to the organic and the water in the condensate film and L is the tube length in feet. They were able to correlate the work of several investigators on horizontal tubes by

$$\bar{h} = 61 \left[\frac{(\text{wt-}\%)_A \lambda_A + (\text{wt-}\%)_B \lambda_B}{(\text{wt-}\%)_A D_o} \right]^{\frac{1}{4}} \qquad (13.21b)$$

For tube bundles in horizontal exchangers it is unlikely that the coefficient differs greatly from the value predicted by Eq. (13.21b).

[1] Hazelton, R., and E. M. Baker, *Trans. AIChE*, **40**, 1–29 (1944).

4. CONDENSATION OF A VAPOR FROM A NONCONDENSABLE GAS

As in the case of a gas compressor intercooler (Chap. 9) if a mixture of a vapor and a gas is cooled in a constant-pressure operation, the temperature at which the first droplet of condensate appears is the *dew point*. The dew point is the saturation temperature of the vapor corresponding to its partial pressure in the mixture. The calculation of the dew point of a mixture of a vapor and a noncondensable gas was demonstrated in Example 9.3. The calculation of the compressor intercooler is covered in this section.

When a mixture of a vapor and a noncondensable gas is fed to a condenser and the temperature of the tubes is below the dew point, a film

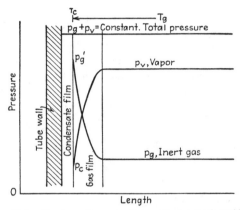

FIG. 13.6. Condensation potentials with noncondensables.

of condensate forms on the tubes. The relationship of the partial pressures is shown in Fig. 13.6. A film of noncondensable gas and vapor collects about the condensate film as suggested by the data of Othmer in Fig. 12.25. If an equilibrium is presumed to exist on the surface of the condensate film, the partial pressure of the vapor at the tube wall corresponds to the cold condensate p_c and the partial pressure of the vapor in the gas film lies between that at the condensate film p_c and that in the gas body p_v. In order for the vapor in the gas body to continue condensing into the condensate film, it must be driven across the gas film by the difference between the partial pressure of the vapor in the gas body and in the condensate. The passage of one component through another is called *diffusion* or *mass transfer*, and in a diffusion system the rate at which the steam condenses is no longer dependent entirely upon Nusselt's condensing mechanism but upon the laws governing diffusion. When vapor diffuses through a noncondensable gas film and liquefies at the tube wall, it also carries with it its latent heat of condensation. In addition to the partial-pressure difference promoting diffusion there is also a temperature difference between the gas body T_g and the condensate film T_c by which the gas is sensibly cooled. It will be shown that the rate of diffusion and the rate of heat transfer are not independent of each other.

Relationships between Heat and Mass Transfer. The theory of diffusion will be treated in greater detail in Chap. 17, which is devoted to diffusional heat transfer.

It was shown in Chap. 3 that there is an analogous behavior between heat transfer and fluid friction when a fluid flows in a tube. Another analogy exists for a system in which a transfer of matter by diffusion is accompanied by a transfer of heat. A number of authors have contributed to the refinement and extension of this analogy, but the derivations employed here are essentially those of Colburn,[1] Colburn and Hougen,[2] and Chilton and Colburn.[3] The paper of Colburn and Hougen forms the basis for design calculations.

It was shown in Chap. 9 that when dealing with gases it is advantageous to express the heat-transfer factor j_H for a fluid flowing inside a tube by

$$j_H = \frac{hD}{k}\left(\frac{c\mu}{k}\right)^{-1/3} = \frac{h}{cG}\left(\frac{c\mu}{k}\right)^{2/3}\left(\frac{DG}{\mu}\right) \tag{13.22}$$

A new factor j_h may then be defined as $j_h = j_H/(DG/\mu)$ or

$$j_h = \frac{h}{cG}\left(\frac{c\mu}{k}\right)^{2/3} \tag{13.23}$$

By employing j_h it is possible to eliminate the variation of k with temperature in making a sensible-heat-transfer calculation, since $(c\mu/k)$ is nearly constant over a wide temperature range. Since

$$Q = wc(t_2 - t_1) = h\pi\, DL\, \Delta t$$

and $G = 4w/\pi D^2$, on substituting for both h and G Eq. (13.23) becomes

$$j_h = \frac{h}{cG}\left(\frac{c\mu}{k}\right)^{2/3} = \frac{t_2 - t_1}{\Delta t}\frac{D}{4L}\left(\frac{c\mu}{k}\right)^{2/3} \tag{13.24}$$

Multiplying the last term by $\pi D/\pi D$,

$$j_h = \frac{h}{cG}\left(\frac{c\mu}{k}\right)^{2/3} = \frac{t_2 - t_1}{\Delta t}\left(\frac{a}{A}\right)\left(\frac{c\mu}{k}\right)^{2/3} \tag{13.25}$$

where a is the flow area $\pi D^2/4$, and A is the tube surface $\pi\, DL$.

[1] Colburn, A. P., *Trans. AIChE*, **29**, 174 (1933).
[2] Colburn, A. P., and O. A. Hougen, *Ind. Eng. Chem.*, **26**, 1178 (1934).
[3] Chilton, T. H., and A. P. Colburn, *Ind. Eng. Chem.*, **26**, 1183 (1934).

CONDENSATION OF MIXED VAPORS

When a vapor is absorbed from a gas, which is not saturated with molecules of the solvent, diffusion may occur in two directions: Molecules of the vapor may pass into the absorbent, and molecules of the absorbent may pass into the gas. In the passage of water vapor from the gas body into a condensate film consisting of liquid water alone the transfer of matter is in one direction and the mols of matter transferred from the gas to the liquid is given by

$$dN_d = d\left(\frac{Gap_v}{M_m p_t}\right) = K_G \, \Delta p \, dA \qquad (13.26a)$$

where A = diffusion surface, ft^2
a = flow area of gas and vapor, ft^2
G = mass velocity, lb/(hr)(ft^2)
K_G = mass diffusion coefficient, mol/(hr)(ft^2)(atm)
M_m = mean molecular weight of the vapor and noncondensable, lb/mol
N_d = material transferred, mol/hr
p_c = partial pressure in atmospheres of the vapor at condensate film, atm
p_v = partial pressure in atmospheres of the vapor in gas body, atm
p_t = total pressure on the system in atmospheres, atm
Δp = instantaneous driving potential in atmospheres, $p_v - p_c$, atm

The coefficient K_G is simply the dimensional rate constant which makes dN_d equal to the right side of the equation. K_G is determined experimentally, and therefore K_G and U are similar in function, K_G being to Δp in mass transfer what U is to Δt in heat transfer. If the initial concentration of the condensable vapor is small, it is convenient to use the simplification that the mass velocity of the mixture does not vary appreciably during the diffusion of the vapor out of the gas and that

$$\frac{G}{M_m} = \frac{G_i}{M_i}\left(\frac{p_t}{p_t - p_v}\right)$$

where the subscript i refers to the inert gas. Replacing $p_t - p_v$ by p_g, the pressure of the inert gas in the gas body, and keeping p_t constant in the tube with constant a, the differentiation of the second term in Eq. (13.26a) yields

$$dN_d = \frac{Ga \, dp_v}{M_m p_t} = K_G \, \Delta p \, dA \qquad (13.26b)$$

The second and third terms can be readily regrouped to give

$$\frac{K_G p_{gf}}{G/M_m} = \left(\frac{dp_v}{\Delta p}\right)\left(\frac{p_{gf}}{p_t}\right)\left(\frac{a}{dA}\right)$$

where p_{gf} is the log mean of p_g of the inert in the gas body and $p'_g = p_t - p_c$ the pressure of the inert at the condensate film. In integrated form it becomes

$$\frac{K_G p_{gf}}{G/M_m} = \left(\frac{p_1 - p_2}{\Delta p}\right)\left(\frac{p_{gf}}{p_g}\right)\left(\frac{a}{A}\right) \quad (13.27)$$

where p_1 and p_2 are the partial pressures of the diffusing component at A_1 and A_2.

When a fluid flows along a surface, the particles within the fluid exchange momentum with the stationary film at the surface, causing a pressure drop in the fluid in the direction of flow. This assumption led to Eq. (3.51) in the Reynolds analogy. It is entirely conceivable that a similar condition occurs when a vapor traveling along a surface condenses against the condensate film which it enters by moving at right angles to the direction of flow and giving up its momentum. In Eq. (3.51) and subsequent equations it became apparent that the ratio of the loss of momentum by skin friction to the total momentum of the stream depended upon how much stationary surface there was for the total amount of fluid flowing. For a given quantity of fluid flowing in a tube and a given total diffusion or heat-transfer surface, the amount of skin friction will be greater if the path consists of a long small-bore tube than for a short tube of large diameter. The index of these possibilities is the ratio A/a or, when used in a diffusion factor, its reciprocal a/A. Just as it was found in the refinement of the Reynolds analogy that the ratio of μ/k influences the heat transfer, so it can be inferred that the properties of the fluid also affect diffusion. The properties associated with skin friction are contained in the dimensionless Schmidt number $\mu/\rho k_d$, where k_d is the diffusion coefficient (diffusivity) in square feet per hour of one gas through another and μ and ρ are the viscosity and density of the mixture. If the influence of μ/k_d on diffusion is comparable to that of μ/k in heat transfer, then it is reasonable to multiply Eq. (13.27) by $(\mu/\rho k_d)$. Assuming $p_{gf}/p_g = 1.0$, designating the diffusion factor j_d, and arbitrarily using the two-thirds power

$$j_d = \left(\frac{p_1 - p_2}{\Delta p}\right)\left(\frac{a}{A}\right)\left(\frac{\mu}{\rho k_d}\right)^{2/3} \quad (13.28)$$

Attention is now directed to the similarity between Eqs. (13.25) and (13.28). From an extension of the Reynolds analogy to distillation, where the analogy between mass and heat transfer is very close, there is good reason to believe that j_d and j_h are the same function of the Reynolds number and equal. The relationship between diffusion and heat transfer is then obtained by equating Eqs. (13.25) and (13.28) and solving for K_G.

$$K_G = \frac{h(c\mu/k)^{2/3}}{cp_{gf}M_m(\mu/\rho k_d)^{2/3}} \quad (13.29)$$

The principal deduction from Eq. (13.29) is that the rates of diffusion and heat transfer do not occur independently. When the vapor concentration is high, as in many industrial applications, Eq. (13.28) must be calculated for incremental changes in the surface, since p_{gf}/p_g will no longer be unity.

The overall coefficient of heat transfer varies greatly during the condensation of a vapor from a noncondensable gas which is initially at its dew point because the potential for diffusion varies greatly as the vapor is removed from the gas body leaving a higher percentage of inert. At the inlet the composition of a mixture of vapor and noncondensable may be almost all vapor and the film coefficient may be very nearly that of the pure condensing coefficient for the vapor alone. But after much of the vapor has condensed, the outlet may consist of substantially pure noncondensable gas with a low accompanying film coefficient. It is often possible to have a variation of U_C for the condensation of steam from air from 1500 Btu/(hr)(ft²)(°F) at the inlet to a value of 15 at the outlet.

Inasmuch as the film coefficient varies from inlet to outlet, the heat distribution may also vary owing to a differential rate of change in the enthalpy of the vapor mixture as the temperature falls. In other words, although the temperature of the gas declines 50 per cent of the total gas-temperature range, it is probably untrue that 50 per cent of the total heat load will have been delivered. This is not merely the case of finding the true temperature difference but the heat-transfer coefficient also varies as $dQ = q$ varies. The surface is then defined by the fundamental equation

$$A = \int \frac{dQ}{(U \, \Delta t)} \quad (13.30)$$

Equation (13.30) cannot be integrated unless U and Δt are expressed as functions of Q. A much simpler method is to integrate numerically $dQ = q$ for small but finite intervals.

Diffusivities. For an excellent account of diffusion and the experimental methods of determining the diffusivity of one gas through another, the reader is referred to Sherwood's excellent book.[1] Gilliland[2] has established an empirical equation for the determination of the diffusivity of one gas through another which is given by

$$k_d = 0.0166 \frac{T^{3/2}}{p_t(v_A^{1/3} + v_B^{1/3})^2} \left(\frac{1}{M_A} + \frac{1}{M_B}\right)^{1/2} \tag{13.31}$$

where k_d = diffusivity, ft^2/hr
p_t = total pressure, atm
v_A, v_B = molecular volumes of diffusing and inert gases computed from the data on atomic volumes in Table 13.3
T = absolute temperature °K(°C abs)
M_A, M_B = molecular weights of the diffusing and inert gases, respectively

TABLE 13.3. ATOMIC VOLUMES

Bromine	27.0
Sulfur	25.6
Oxygen	7.4
In methyl esters	9.1
In higher esters and ethers	11.0
In acids	12.0
Carbon	14.8
Chlorine	24.6
Hydrogen	3.7
Nitrogen	15.6
In primary amines	10.5
In secondary amines	12.0
For benzene ring formation deduct	15
For naphthaline deduct	30
For the hydrogen molecule use	$v = 14.3$
For air use	$v = 29.9$

Example 13.4. Calculation of the Diffusivity of a Mixture. Calculate the diffusivity of a steam-CO_2 mixture at 267°F and 30 psig.

For steam, H_2O: $v_A = 2 \times 3.7 + 7.4 = 14.8$ $M_A = 18$
For CO_2: $v_B = 14.8 + 2 \times 7.4 = 29.6$ $M_B = 44$

$$T = 267°F \approx 273 + 130 = 403°K \quad 30 \text{ psig} \approx \frac{14.7 + 30}{14.7} = 3.04 \text{ atm}$$

$$k_d = 0.0166 \times \frac{403^{3/2}}{3.04(14.8^{1/3} + 29.6^{1/3})^2} \left(\frac{1}{18} + \frac{1}{44}\right)^{1/2} = 0.41 \text{ ft}^2/\text{hr}$$

Development of an Equation for Heat Transfer. Chilton and Colburn have shown that the results of their analogy, culminating in Eq. (13.29),

---

[1] Sherwood, T. K., "Absorption and Extraction," McGraw-Hill Book Company, Inc., New York, 1937.
[2] Gilliland, E. R., *Ind. Eng. Chem.*, **26**, 516 (1934).

hold on the safe side for flow inside tubes, flow across a single tube, and flow along plane surfaces. In each case the appropriate value of h is substituted in Eq. (13.29). It appears likely that these equations are valid in most cases of a vapor forming a nonviscous, noncontrolling condensate film. The use of h_o calculated for the gas from Fig. 28 and substituted in Eq. (13.29) for the condensation of steam from air and CO_2 in the shells of horizontal baffled condenser has been known to hold successfully in a number of applications.

To establish an equation which may be solved from point to point for U and Δt as in Eq. (13.30) it should be necessary only to sum up all the resistances in series at an average cross section in each increment of q. In the condensation of a vapor from a noncondensable gas the quantity of heat which leaves the gas film must equal the quantity picked up by the cooling water. The total heat flow across the gas film is the sum of the latent heat carried by vapor diffusion into the condensate film plus the sensible heat removed from the gas because of the temperature difference $T_g - T_c$. The heat load expressed in terms of the shell-side, tube-side, and overall potentials per square foot of surface when the mixture of gas and vapor flows in the shell is

$$\overset{(a)}{h_o(T_g - T_c)} + \overset{}{K_GM_v\lambda(p_v - p_c)} = \overset{(b)}{h_{io}(T_c - t_W)} = \overset{(c)}{U(T_g - t_W)} \quad (13.32)$$

where h_o = shell-side dry-gas coefficient, $Btu/(hr)(ft^2)(°F)$
h_{io} = tube-side water coefficient, $Btu/(hr)(ft^2)(°F)$
T_g = temperature of the gas, °F
T_c = temperature of the condensate, °F
t_W = water temperature, °F
p_v = partial pressure of vapor in the gas body, atm
p_c = partial pressure of the vapor at the condensate film, atm
M_v = molecular weight of vapor, dimensionless
λ = latent heat, Btu/lb

The possibility of subcooling the vapor at the tube wall has been omitted from the heat balance, since it is not usually significant compared with the larger latent effects. If subcooling is of consequence, it means that the amount of heat delivered at the tube wall is greater than that which entered the gas film.

In the application of Eqs. (13.29), (13.30), and (13.32) to the solution of an actual condenser it is assumed that there is a single value of T_g and T_c at any cross section and hence of p_v and p_c. In a condenser having several tube passes with water in the tubes this is obviously not possible. If there is a long condensing range, a small cooling-water range, and no temperature cross so that F_T for a 1-2 condenser would be substantially

1.0, then little harm results in the assumption of a true counterflow distribution of water temperatures. At whatever point the gas gives up half of the total heat load, the water can be assumed to have received half the total heat load. This may appear rather arbitrary, but in the final analysis it is a problem of maintaining $T_g - T_c$ and $p_v - p_c$ in their proper relation, and both require the presence of the same single value of T_c at any point. Otherwise the solution must be extended to cover each of the passes as a related condenser.

The Calculation of a Vapor-Noncondensable Mixture. The method of applying Eqs. (13.29), (13.30), and (13.32) are outlined below:

1. A complete exchanger must be assumed to fix the shell-side and tube-side flow areas. The surface is obtained by integration on the assumption of true counterflow.

2. From the process conditions compute h_o and h_{io} for the gas and cooling medium, respectively. The use of an average value for h_{io} is acceptable but not for h_o, since the mass velocity of the gas changes from point to point.

3. From the value of h_o obtain K_G/p_{gf} from Eq. (13.29).

4. Fix the first interval of calculation by fixing T_g, which also fixes the heat load q for the interval.

5. Assume values of T_c, the condensate temperature, so that Eqs. (13.32a) and (13.32b) balance. For each assumed value of T_c it is necessary to compute a new value of p_{gf}, since the pressure of the vapor at the condensate film is the saturation pressure corresponding to T_c.

6. When Eqs. (13.32a) and (13.32b) balance, the total heat load transferred per square foot of each is the same as the load which must have been transferred overall, $U(T_g - t_c)$.

7. From q obtained in 4 and $U(T_g - t_c)$ obtain dA for the interval.

8. Proceed with the next interval by assuming a lower value of T_g.

Example 13.5. Calculation of a Steam-Carbon Dioxide Condenser. Steam is to be condensed from carbon dioxide in the following exchanger:
21¼ in. ID shell with 12-in. baffle spacing.
246 tubes, ¾ in. OD, 16 BWG, 12'0" on 1-in. square pitch.
The tube bundle has four passes.
The hot stream is a mixture of 4500 lb of steam and 1544 lb of CO_2 at 30 psig entering at the dew point and leaving at 120°F. Cooling water will enter at 80°F and leave at 115°F.
The diffusivity of steam-CO_2 mixtures calculated by the Gilliland formula is 0.41 ft²/hr, and $(\mu/\rho k_d)^{2/3}$ may be taken for simplicity as constant at the average value of 0.62 between inlet and outlet.

(a) Determine the weighted Δt. (b) Determine the dirt factor for the condenser.

Solution. As a simplification it may be assumed that the heat-transfer coefficient at the average water temperature is constant throughout the exchanger. This does not alter the earlier assumption that the water temperature is considered in true counterflow with the condensate.

CONDENSATION OF MIXED VAPORS

Basis: One hour

Entering	Lb/hr	Mol/hr
CO_2	1544	35
H_2O	4500	250
Total	6044	285

Total pressure = 30 + 14.7 = 44.7 psia ≈ 3.05 atm where psi/14.7 = atm
Partial pressure of water = $250/285 \times 44.7$ = 39.2 psi ≈ 2.68 atm
Dew point = 267° from Table 7 ≈ 2.68 atm
Mean molecular weight, $M_m = 6044/285 = 21.2$

(a) Weighted temperature difference Δt:

Overall balances:

Inlet: Water vapor pressure, p_v = 2.68 atm
      Inert pressure, p_g = 3.05 − 2.68 = 0.37 atm
      Total pressure = 3.05 atm.
Exit: Partial pressure of water at 120°F = 0.1152 atm
      Water vapor pressure, p_v = 0.115 atm
      Inert pressure, p_g = 2.935 atm
      Total pressure = 3.05 atm
Pound mols steam inlet = 250

Pound mols steam exit = $35 \times \dfrac{0.115}{2.935} = 1.37$

Pound mols steam condensed = 250 − 1.37 = 248.63
Heat load: Assume points at 267, 262, 255, 225, 150, 120°F, and compute the heat load q for each interval.

For the interval from 267 to 262°F:

From Table 7, p_v at 262°F = 2.49 atm
              p_g = 3.05 − 2.49 = 0.56 atm

Mol steam remaining = $35 \times \dfrac{2.49}{0.56} = 156$

Mol steam condensed = 250 − 156 = 94
Heat of condensation = 94 × 18 × 937.3 + 0.46(267 − 262) × 94 × 18
                                                            = 1,590,000 Btu
Heat from uncondensed steam = 156 × 18 × 0.46(267 − 262) =   6,450
Heat from noncondensable = 1544 × 0.22 × 5.0   =   1,700
                                              Total for interval = 1,598,150

Heat balance:

Interval, °F	q
267–262	1,598,000
262–255	1,104,000
255–225	1,172,000
225–150	751,000
150–120	177,000
Total	4,802,000

Total water = $\dfrac{4{,}802{,}000}{115 - 80} = 137{,}000$ lb/hr

Water coefficient, h_i:

$$a_t = N_t \frac{a_t'}{144n} = 246 \times \frac{0.302}{144 \times 4} = 0.129 \text{ ft}^2 \quad (7.48)$$

$$G_t = \frac{w}{a_t} = \frac{137,000}{0.129} = 1,060,000 \text{ lb/(hr)(ft}^2)$$

$$V = \frac{G}{3600\rho} = \frac{1,060,000}{3600 \times 62.5} = 4.72 \text{ fps}$$

$$h_i = 1120 \quad \text{(Fig. 25)}$$

$$h_{io} = h_i \times \frac{\text{ID}}{\text{OD}} = 1120 \times \frac{0.62}{0.75} = 926 \quad (6.5)$$

Now proceed to determine $U \Delta t$ from point to point in the unit by assuming temperatures for the condensate film so that Eqs. (13.32a) and (13.32b) are equal. Shell-side coefficient for entrance gas mixture.

Mean properties for Point 1:

$$\text{Mean } c = \frac{1544 \times 0.22 + 4500 \times 0.46}{6044} = 0.407 \text{ Btu/(lb)(°F)}$$

$$\text{Mean } k = \frac{1544 \times 0.0128 + 4500 \times 0.015}{6044} = 0.0146 \text{ Btu/(hr)(ft}^2)(°F/ft)$$

$$\text{Mean } \mu = \frac{1544 \times 0.019 + 4500 \times 0.0136}{6044} = 0.015 \times 2.42 = 0.0363 \text{ lb/(ft)(hr)}$$

$$a_s = \text{ID} \times C' \frac{B}{144P_T} = 21.25 \times 0.25 \times \frac{12}{144 \times 1.0} = 0.442 \text{ ft}^2 \quad (7.1)$$

$$G_s = \frac{W}{a_s} = \frac{6044}{0.442} = 13,650 \text{ lb/(hr)(ft}^2) \quad (7.2)$$

$$Re_s = \frac{D_e G_s}{\mu} \quad D_e = \frac{0.95}{12} = 0.0792 \text{ ft} \quad \text{(Fig. 28)}$$

$$= 0.0792 \times \frac{13,650}{0.0363} = 29,800 \quad (7.3)$$

$$j_H = 102 \quad \text{(Fig. 28)}$$

$$\left(\frac{c\mu}{k}\right)^{1/3} = \left(\frac{0.407 \times 0.0363}{0.0146}\right)^{1/3} = 1.0$$

$$h_o = j_H \frac{k}{D_e}\left(\frac{c\mu}{k}\right)^{1/3} = 102 \times 0.0146 \times \frac{1.0}{0.0792} = 18.9 \quad (6.15b)$$

$$(\mu/\rho k_d)^{2/3} = 0.62 \quad \left(\frac{c\mu}{k}\right)^{2/3} = \left(\frac{0.407 \times 0.0363}{0.0146}\right)^{2/3} = 1.01$$

$$K_G = \frac{h_o(c\mu/k)^{2/3}}{cp_{gf}M_m(\mu/\rho k_d)^{2/3}} = \frac{18.9 \times 1.01}{0.407 \times p_{gf} \times 21.2 \times 0.62} = \frac{3.56}{p_{gf}}$$

Point 1:

$$T_g = 267°\text{F (entrance)} \quad p_v = 2.68 \text{ atm} \quad p_g = 3.05 - 2.68 = 0.37 \text{ atm}$$

$$t_W = 115°\text{F} \quad \Delta t = T_g - t_W = 267 - 115 = 152°\text{F}$$

Try

$$T_c = 244°\text{F} \quad p_c = 1.83 \text{ atm} \quad p_g' = 3.05 - 1.83 = 1.22 \text{ atm}$$

$$p_{gf} = \frac{p_g' - p_g}{2.3 \log p_g'/p_g} = \frac{1.22 - 0.37}{2.3 \log 1.22/0.37} = 0.715 \text{ atm}$$

$$h_o(T_g - T_c) + K_G M_v \lambda(p_v - p_c) = h_{io}(T_c - t_W) \quad (13.32)$$

$$18.9(267 - 244) + \frac{3.56}{0.715} \times 18 \times 933.8(2.68 - 1.83) = 926(244 - 115)$$

$$71,400 \neq 129,000 \quad \textit{No check}$$

CONDENSATION OF MIXED VAPORS 349

Try

$T_c = 220°F \qquad p_c = 1.17 \text{ atm} \qquad p'_g = 1.88 \text{ atm} \qquad p_{gf} = 0.93 \text{ atm}$

$18.9(267 - 220) + \dfrac{3.56}{0.93} \times 18 \times 933.8(2.68 - 1.17) = 926(220 - 115)$

$$98{,}400 = 97{,}500 \quad Check$$

$U \Delta t = \dfrac{98{,}400 + 97{,}500}{2} = 97{,}950$

$U = \dfrac{97{,}950}{267 - 115} = 644$

Having determined the conditions at the inlet, proceed from point to point in the exchanger. Since the partial pressure of water changes rapidly at high temperatures, it follows that most of the condensation will occur near the inlet. (Refer to the saturation curve for water.) To obtain a reasonable distribution of heat loads over the condenser, a second point very near the inlet has been chosen.

Point 2:

$T_g = 262°F$ and saturated $\qquad p_v = 2.49 \text{ atm} \qquad p_g = 3.05 - 2.49 = 0.56$

Mol steam remaining $= 35 \times \dfrac{2.49}{0.56} = 156$

New gas rate $= 1544 + 156 \times 18 = 4352 \text{ lb/hr}$
Mean properties: $M_m = 22.8 \qquad c = 0.382 \qquad k = 0.0143$
$\mu = 0.0154 \times 2.42 = 0.0373$

$G_s = \dfrac{W}{a_s} = \dfrac{4352}{0.442} = 9850 \text{ lb/(hr)(ft}^2)$

$Re = \dfrac{D_e G_s}{\mu} = 0.0792 \times \dfrac{9850}{0.0373} = 20{,}900$

$j_H = 83.5 \qquad \left(\dfrac{c\mu}{k}\right)^{1/3} = 1.0 \qquad \left(\dfrac{c\mu}{k}\right)^{2/3} = 1.0$

$h_o = 15.0$

$K_G = \dfrac{15.0 \times 1.0}{0.382 \times p_{gf} \times 22.8 \times 0.62} = \dfrac{2.80}{p_{gf}}$

Water rise $= 1{,}598{,}000/137{,}000 = 11.7°F$
$t_W = 115 - 11.7 = 103.3°F$

Try

$T_c = 182°F \qquad p_c = 0.534 \text{ atm} \qquad p'_g = 2.51 \text{ atm} \qquad p_{gf} = 1.305 \text{ atm}$

$15.0(262 - 182) + \dfrac{2.80}{1.305} \times 18 \times 937.3(2.49 - 0.534) = 926(182 - 103.3)$

$$71{,}900 = 72{,}800 \quad Check$$

$U \Delta t = \dfrac{71{,}900 + 72{,}800}{2} = 72{,}350$

$U = \dfrac{72{,}350}{262 - 103.3} = 456$

Points 3, 4, and 5 are calculated in the same manner.

Point 6:

$T_g = 120°F$ and saturated $\qquad p_v = 0.115 \text{ atm} \qquad p_g = 3.05 - 0.115 = 2.935 \text{ atm}$

Mol steam remaining $= 35 \times \dfrac{0.115}{2.035} = 1.37$

New gas rate = $1544 + 1.37 \times 18 = 1568.7$
Mean properties:
$M_m = 43.1$, $c = 0.214$, $k = 0.0102$,
$\mu = 0.016 \times 2.42 = 0.0387$
$G_s = \dfrac{w}{a_t} = \dfrac{1568.7}{0.442} = 3570$ lb/(hr)(ft²)
$Re = 0.0792 \times \dfrac{3570}{0.0387} = 7300$
$j_H = 47.5$ $\left(\dfrac{c\mu}{k}\right)^{1/3} = 0.935$
$\left(\dfrac{c\mu}{k}\right)^{2/3} = 0.872$
$h_o = 5.7$
$K_G = \dfrac{0.87}{p_{gf}}$
Water rise = $\dfrac{177{,}000}{137{,}000} = 1.3°F$
$t_W = 81.3 - 1.3 = 80°F$

Try

$T_c = 80.7°F$ $p_c = 0.0352$ atm $p'_g = 3.05 - 0.0352 = 3.015$ $p_{gf} = 2.97$
$5.7(120 - 80.7) + \dfrac{0.87}{2.97} \times 18 \times 1025.8(0.115 - 0.0352) = 926(80.7 - 80)$

$$654 = 648$$

$U\,\Delta t = \dfrac{654 + 648}{2} = 651$

$U = \dfrac{651}{(120 - 80)} = 16.2$

Having determined both q and $U\,\Delta t$, it need only be recalled that $A = \Sigma q/\Sigma U\,\Delta t$ and proceed to evaluate A. This may be done in any of several ways. More accurately, Σq should be plotted against $1/U\,\Delta t$, the area of the graph corresponding to A. A simpler and generally acceptable method involves tabulation of the results obtained so far and a numerical summation. This last will be demonstrated.

Point	T_g	T_c, °F	$U\,\Delta t$	$(U\,\Delta t)_{av}$	q	$A = \dfrac{q}{(U\,\Delta t)_{av}}$	Δt	Δt_{av}	$\dfrac{q}{\Delta t_{av}}$
1	267	220	97,950	. . . . . .	. . . .	. . . .	152		
2	262	182	72,350	85,150	1,598,000	18.8	158.7	155.5	10,300
3	255	145	45,900	59,075	1,104,000	18.7	159.7	159.2	6,930
4	225	101	12,900	26,000*	1,172,000	45.2	138.2	149.0	7,870
5	150	84	1,710	5,560*	751,000	135.1	68.7	99.3*	7,570
6	120	80.7	651	1,098*	177,000	161.0	40.0	53.3*	3,320
					4,802,000	$A_c = 378.8$ ft²		$\sum \dfrac{q}{\Delta t_{av}} =$	35,990

*LMTD.

CONDENSATION OF MIXED VAPORS

The LMTD based on a shell range from 267 to 120°F and tube range from 80 to 115°F is 84.0°F, but when weighted by the summation of $\Sigma UA = Q/\Delta t$,

$$\text{Weighted } \Delta t = \frac{Q}{\Sigma q/\Delta t_{av}} = \frac{4{,}802{,}000}{35{,}990} = 133°F$$

and

$$U_{\text{Clean}} = \frac{Q}{A\,\Delta t} = \frac{4{,}802{,}000}{378.8 \times 133} = 95.3$$

External surface/ft = 0.1963 (Table 10)
Total surface available = $246 \times 12'0'' \times 0.1963 = 580$ ft²

$$U_D = \frac{Q}{A\,\Delta t} = \frac{4{,}802{,}000}{580 \times 133} = 62.3$$

$$R_d = \frac{U_C - U_D}{U_C U_D} = \frac{95.3 - 62.3}{95.3 \times 62.3} = 0.0055$$

Remarks. The pressure drop for the shell side can be computed from the average based on the inlet and outlet total gas conditions. The 0.0055 dirt factor is greater

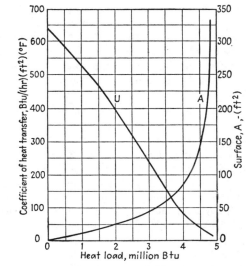

Fig. 13.7. Condensation of CO_2-steam mixture.

than is generally necessary, but large dirt factors are recommended because of the sensitivity of the calculation in the later stages. A plot of U and A vs. Q is shown in Fig. 13.7. Note that U_C varies from 644 to 16.2 from inlet to outlet.

It is not usually necessary to account for heat for subcooling all the condensate to the outlet temperature when the vapor is steam because of its high thermal conductivity when condensing on the tube wall. When the condensing vapor is organic, it will be necessary to allow extra surface as discussed in the preceding section.

5. CONDENSATION FROM A NONCONDENSABLE OF A MIXTURE IMMISCIBLE WITH WATER

The Influence of Components on the Condensing Curve. This section corresponds to Case 9 in Table 13.1. The methods used here refer particularly to steam distillation in the petroleum industry, although variations of the method can be applied to other problems as well.

When an organic vapor mixture immiscible with water enters a condenser and contains a noncondensable gas as well as steam, the calculation can be carried out using Eq. (13.20) with modifications. The total pressure is the sum of three distinct pressures: the pressure of the miscibles, pressure of the gas, and the pressure of the immiscible. The equilibrium among the miscibles is unaffected by the presence of the other components except that the equilibrium exists only at the sum of the partial pressures of the miscibles and not at the operating pressure of the condenser. Since the steam-gas and miscibles-gas mixtures attain their separate equilibrium relationships, the system may have two dew points, one for the steam and the other for the miscibles, although both need not be passed in the condenser. If there are a large number of miscible compounds, the calculation of the condensing curve will be considerably lengthier than the direct calculation of a multicomponent mixture.

A vapor mixture of miscibles is affected in definite ways by the presence of a noncondensable and steam. The noncondensable reduces the curvature of the condensation curve for the miscibles. This is due to the fact that the mols of noncondensable remain constant while the miscibles leave the vapor. The noncondensables, however, reduce the condensing coefficient for both the vapor of the miscibles and the steam. The dew point of the miscibles can be obtained from the sum of the pressures of the miscibles and the noncondensable by the summation of the K_1L_1 values with the mols of noncondensable included as a constant. Since the steam is immiscible with the miscibles, its dew point is a function only of the partial pressure and temperature at which the noncondensable becomes saturated with steam as found in tables of steam properties such as Table 7. Which of the two dew points occurs at the higher temperature depends upon the composition of the initial vapor. It is entirely possible to have either the steam or the miscibles condense first, or both may have dew points close together. Since the latent heat of the steam is about six to eight times as great as that of an oil, it is extremely important to determine just where the steam starts to condense, and the condensing curve may exhibit a greater divergence from a straight line on a heat capacity vs. T_v plot than is usually encountered with multicomponent mixtures of miscibles alone. The calculation of this type of problem can be facilitated by empirical methods. A problem of this type will be solved presently using an empirical method developed for petroleum fractions.

Development of an Empirical Solution. In the petroleum industry it is customary to take liquid samples of complex vapors and actually distill them in the laboratory at atmospheric pressure by either of two methods. The first is the ASTM distillation, which is carried out in a standardized manner with prescribed equipment as outlined in the ASTM Standards, 1930, Part II. A 100-cm^3 sample is distilled batchwise in a distillation flask with a thermometer submerged in the neck. The temperature is recorded at the *initial boiling point* (IBP) and for successive 10-cm^3 portions distilled over. The percentages distilled over are plotted against the temperature. This is called the ASTM distillation curve. Another method which is more accurate but much more elaborate is the *true boiling point* (TBP) distillation. This consists of introducing the sample of material into a round-bottom flask which is then connected to a Podbielniak distilling column containing a large number of theoretical distilling plates. The material is distilled over with a reflux ratio of 10:1 to 30:1. The TBP curve is plotted in a similar manner to the ASTM distillation except that a nearly complete separation takes place and the percentages distilled can be identified by their boiling points as pure compounds. The first material to come overhead in a TBP distillation has a lower initial boiling point than the ASTM, since the most volatile fraction will have been separated almost individually from the next less volatile fraction. The presence in an ASTM distillation of the unseparated fractions shortens the range of temperatures between the initial boiling point and the end point of the distillation by comparison with the TBP.

A number of correlations have been made by Piroomov and Beiswenger[1] and Packie,[2] of which the latter is used here. These studies undertake the problem of correlating a batch distillation such as the ASTM or TBP with the performance to be expected in a distilling column where the vaporization occurs by a continuous process. Thus suppose a still were continuously operated at its boiling point by the introduction of liquid feed which separates continuously into a vapor rising upward and a liquid which is withdrawn at the bottom. Consider such an equilibrium system. It is assumed that boiling produces a liquid of uniform composition and that the vapor is in equilibrium with the still liquid. A similar condition exists in distilling columns, reboilers, and condensers in which liquid continuously in equilibrium with the vapor above it is formed from

---

[1] Piroomov and Beiswenger, *Am. Petroleum Inst. Bull.* 10, p. 52, Jan. 3, 1920.
[2] Packie, J. W., *Trans. AIChE*, **37**, 51–78 (1941).

point to point. A curve which indicates the temperature of the liquid as a function of composition and for a given pressure when it is maintained in equilibrium with its vapor over its entire boiling or condensing range is called an *equilibrium flash* curve (EF). An EF curve is necessary for the calculation of the condensing curve, and it naturally differs from an ASTM curve in which the material previously vaporized is removed from the system. The method given below permits the conversion of an ASTM or TBP curve to an EF curve empirically as the result of a correlation of many flash distillations which have been compared with their respective ASTM and TBP distillations. The procedure for obtaining the condensing curve and the true temperature difference from the ASTM or TBP curve is as follows:

The true temperature difference:

1. Draw the ASTM distillation curve with per cent vaporized as the abscissa and temperature as the ordinate.
2. Draw the *distillation reference line* which is a straight line through the 10 and 70 per cent points of the ASTM or TBP curves.
3. Determine the slope of the distillation reference line by the following formula, where percentages represent points on the respective curves.

$$\frac{\text{Temp °F at } 70\% - \text{temp °F at } 10\%}{60} = °F/\%$$

4. Compute an average 50 per cent point for the distillation curve by averaging the 20, 50, and 80 per cent points, or if the 80 per cent point is lacking, use the 50 per cent point of the distillation reference line.
5. Enter the appropriate curves in Fig. 13.8 or 13.9, which is the key to the correlation. It relates the 50 per cent point of the ASTM distillation reference line to the 50 per cent point on the equilibrium flash reference line.
6. Using the slope calculated in 3 enter Fig. 13.10, which relates the slope of the ASTM reference line to the slope of the equilibrium flash reference line.
7. The 50 per cent point and slope determine the flash reference line, which is assumed to be straight through the 10 and 70 per cent points of the flash vaporization curve.
8. At the various percentages read temperatures from the ASTM or TBP curve and its reference line, and subtract the temperature for the reference line from that for the distillation curve noting the sign of the difference.

9. For each percentage distilled read on the appropriate curve in Fig 13.11 or 13.12 the corresponding value of the ratio

$$\frac{(\text{Temp on flash vaporization curve}) - (\text{temp on flash reference line})}{(\text{Temp on distillation curve}) - (\text{temp on distillation reference line})}$$

10. Multiply the ratio obtained in 9 by the difference obtained in 8 to obtain the difference between the temperatures on the flash vaporization curve and the flash reference line for a given percentage distilled. This enables a translation for the curvature of the ASTM to the EF curve.

11. For the same percentage distilled and used to obtain the product in 10, read a value of the temperature on the flash reference line and add

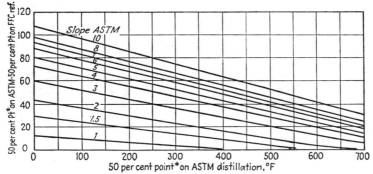

FIG. 13.8. *An average of the 20, 50, and 80 per cent values. Relation between 50 per cent points of ASTM and EF curves. (*Packie, Trans. AIChE.*)

to or subtract from it, according to the sign, the product obtained in 10. Proceed point by point to determine the entire flash vaporization curve.

12. The curve obtained thus is the EF or *flash vaporization curve* for 1 atm pressure, since it refers to the ASTM curve at atmospheric pressure. To obtain a curve at other pressures when the distilling column is to be operated at sub- or superatmospheric pressure, refer to Fig. 8 in the Appendix.[1]

[1] Figure 8 is a plot of the vapor pressure of hydrocarbons. It is not plotted by individual compounds but rather by the boiling points pure compounds have or may have at atmospheric pressure. Thus, isohexane boils at 140°F at atmospheric pressure. The oblique curve at 140°F may then be considered the vapor pressure curve for isohexane. Normal hexane, however, boils at 156°F, and its vapor pressure curve will be a curve in Fig. 8 proportionately between the 140 and 160° curves. Traveling upward or downward along the 140° curve one obtains a value for the pressure of the compund as the ordinate and the saturation temperature corresponding to that pressure. Thus at 14.7 psia isohexane has a boiling point of 140° read as the abscissa. If the pressure at 100°F is desired, move downward on the 140° curve to a temperature

Observe the temperature at the intersection of the flash reference and the distillation reference line. On the vapor pressure chart (Fig. 8) estimate the number of degrees of displacement in boiling point that the change in pressure would cause for a pure hydrocarbon with an atmospheric boiling point the same as the temperature at the intersection of the two reference curves. Displace each point of the flash vaporization curve by the same number of degrees by which a pure compound with atmospheric boiling point at the temperature of the intersection would be

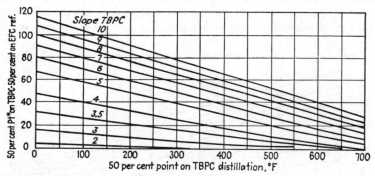

FIG. 13.9. Relation between 50 per cent points of TBP and EF curves. (*Packie, Transactions of American Institute of Chemical Engineers.*)

displaced by the change from atmospheric pressure. For operating pressures greater than 25 psi greater accuracy may be expected by employing the method of Katz and Brown.[1]

The method of calculating the heat-transfer coefficients for such systems is deferred following an illustrative example on the calculation of the weighted true temperature differences, so that the problems originating in the calculation of the condensing curve can be clarified first.

Example 13.6a. Calculation of a Hydrocarbon–Noncondensable–steam Mixture. The overhead of a distilling column operating at 5 psig contains 13,330 lb/hr of oil vapor, 90 lb/hr of noncondensable gas of 50 mol. wt., and 370 lb/hr of steam. On condensation the oil vapor has an average gravity of 50°API. The overhead is at a

---

of 100° read on the abscissa, giving a value at the left of 6.5 psia. In a mixture of organic compounds, if its boiling temperature is known at 14.7, Fig. 8 permits with equal facility the determination of the pressure and corresponding temperature for any new condition. This type of graph has certain inherent advantages over the more conventional Cox chart, particularly when it is preferable to divide arbitrarily a mixture of hydrocarbons into a number of fictitious pure compounds so as to approximate the total composition of the mixture.

[1] Katz, D. L., and G. G. Brown, *Ind. Eng. Chem.*, **25**, 1373–1384 (1933).

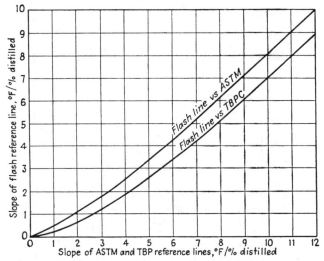

Fig. 13.10. Slope of ASTM and TBP reference lines to EFC reference lines. (*Packie, Transactions of American Institute of Chemical Engineers.*)

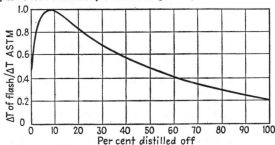

Fig. 13.11. Departure ΔT of the ASTM curve from its reference line. (*Packie, Transactions of American Institute of Chemical Engineers.*)

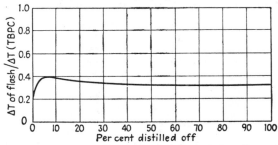

Fig. 13.12. Departure ΔT of the TBP curve from its reference line. (*Packie, Transactions of American Institute of Chemical Engineers.*)

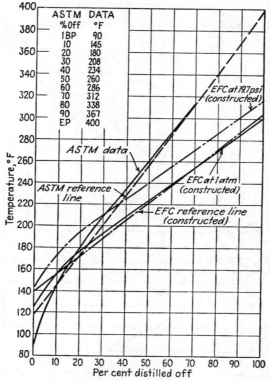

Fig. 13.13. Plot of ASTM curve and construction of the EFC.

temperature of 305°F and is to be completely condensed (as far as practicable) with water from 85 to 120°F.

Determine the weighted true temperature difference.

The ASTM distillation on the oil alone in the overhead is as follows:

% distilled	Temp, °F
IBP	90
10	145
20	180
30	208
40	234
50	260
60	286
70	312
80	338
90	367
End point	400

CONDENSATION OF MIXED VAPORS

Solution: Basis 1 hour.
(1) and (2) Plot the ASTM curve and reference line as shown in Fig. 13.13. Applying the method outlined, determine the dew point of (a) oil vapor and (b) steam, (c) the condensing curve, (d) the heat load, (e) the weighted true temperature difference, and lastly (f) the condenser.

(3) Slope of the ASTM $= \dfrac{70\% - 10\%}{60\%} = \dfrac{312 - 145}{60} = 2.79°F/\%$

(4) Average 50% point $= \dfrac{20 + 50 + 80\%}{3} = \dfrac{180 + 260 + 338}{3} = 259.3°F$

(5) For the 50 per cent point on the equilibrium flash curve (EFC) from Fig. 13.8 where the slope on the ASTM = 2.79°F/% and 259.3°F
50% point ASTM = 50% point flash curve = 38°F
50% on EFC = 259 − 38° = 221°F fixing first point on EFC

(6) From Fig. 13.10, upper curve
Slope of flash reference line = 1.65°F/%
10% on EFC = 50% − 40% = 221 − 40 × 1.65 = 155°F
70% on EFC = 50% + 20% = 221 + 20 × 1.65 = 254°F

(7) Draw this line as a reference through the 50 per cent point. Calculate the flash curve for different percentages off.

(8) ΔT ASTM = °F ASTM − °F ASTM reference
(9) ΔT ASTM × (factor Fig. 13.11) = ΔT EFC
(10) °F EFC reference + ΔT EFC = °F EFC
(11) 0% off:
ΔT ASTM = 90 − 117 = −27°F.
ΔT EFC = −27 × 0.50 = 13.5°F.
°F EFC = 139 − 13.5 = 125.5.
10% off:
ΔT ASTM = 145 − 145 = 0.
20% off:
ΔT ASTM = 180 − 173 = +7.0°.
ΔT EFC = +7.0 × 0.82 = 5.7°.
°F EFC = 172 + 5.7 = 177.7.
30% off:
ΔT ASTM = 208 − 201 = +7.0°.
ΔT EFC = +7.0 × 0.67 = 4.7°.
°F EFC = 188 + 4.7 = 192.7.
40% off:
ΔT ASTM = 234 − 229 = +5.0°.
ΔT EFC = 5.0 × 0.57 = 2.9°.
°F EFC = 204 + 2.9 = 206.9.
70% off:
ΔT ASTM = 312 − 312 = 0.

(12) Correct for pressure: Intersection ASTM reference and EFC reference at 170°F. From Fig. 8 in the Appendix follow oblique at 170° to intersection with 19.7 psi at 187°F. Add 187° − 170° = 17° to each point.

Calculation of the 80 per cent point:

The following vapor enters, 13,330 lb/hr oil 50°API, 370 lb/hr steam, 90 lb/hr gas 50 mol. wt.

360 PROCESS HEAT TRANSFER

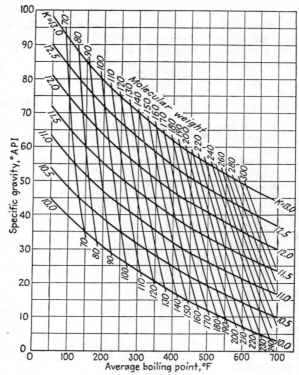

Fig. 13.14. Molecular weights of petroleum fractions. (*Watson, Nelson, and Murphy, Industrial and Engineering Chemistry.*)

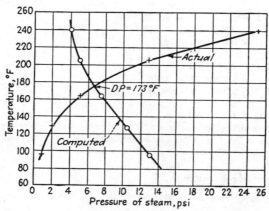

Fig. 13.15. Computed vs. actual vapor pressure of steam.

CONDENSATION OF MIXED VAPORS 361

For 80 per cent, $13{,}330 \times 0.80 = 10{,}664$ lb/hr oil uncondensed.
Average boiling point from the EFC at 1 atm is 269°F from Fig. 13.13.
Average boiling point from the EFC at 19.7 psia add 17°F = 269 + 17 = 286°F.
The molecular weight of the vapors from point to point is determined from Fig. 13.14.
For 50°API oil and 269°F, mol. wt. = 113.
The mols of oil still to be condensed = 10,664/113 =   94.3.
The following are also present: Mols gas = $99/50$ =   1.80
 Mols steam = $3790/_{18}$ =   20.6
 Mols total = 116.7

The total pressure is 19.7 psia, and the partial pressure of each component is proportional to its mol fraction.
Partial pressure of oil = $(94.3/116.7)19.7 = 15.9$ psia.
Partial pressure of NC gas = $(1.8/116.7)19.7 = 0.304$ psia.

The temperature at which the oil will condense in the presence of noncondensable corresponds to its partial pressure.

Refer to Fig. 8 in the Appendix, which relates the temperature at which a fraction will boil at a higher or lower pressure if the temperature at a given pressure is known. At 19.7 psia the boiling point is 286°F. At the intersection of this ordinate and abscissa follow the oblique to an ordinate of 15.9 psia and read the new abscissa, 277°F. These values are the actual condensing temperatures of the oil vapor in the presence of noncondensable gas and steam.

To determine the heat load it is also necessary to determine the temperature at which the steam starts to condense, since the ultimate condensate, water, is immiscible with the condensed oil. The steam may condense before the dew point of the oil is reached, during condensation of the oil, or after nearly all the oil has condensed. The dew point of the steam is a function only of its partial presure in the vapor and the saturation temperature of pure steam at the partial pressure.

At 19.7 psia, if only pure steam were present, it would condense at 227°F. The dew point in this case must be considerably lower.

CALCULATION OF DEW POINT OF THE STEAM

T, °F	$[p_t - (p_{\text{oil}} + p_{\text{NC}})] = p_{\text{steam}}$			p_{sat} (steam tables)
95	19.7	6.73	12.97	0.815
127	19.7	9.40	10.30	2.050
163	19.7	12.25	7.45	5.09
205	19.7	14.64	5.06	12.77
240	19.7	15.65	4.05	24.97

These are plotted in Fig. 13.15, the point of intersection, 173°F, being the dew point of the steam.

To establish the respective heat loads in the oil and steam zones the condensation of oil from inlet to the dew point of the steam is determined from Table 13.4. Where steam and oil condense together, it will be necessary to assume a temperature below the dew point of steam *and* the amount of oil condensed at that point. If at the assumed temperature the sum of the partial pressures adds up to the total pressure, 19.7 psia, the amounts of oil, steam, and noncondensable will be fixed.

At 173°F, the dew point of steam, $p_{sat} = 6.417$ psia
$p_{oil} + p_{NC} = 19.7 - 6.417 = 13.29$ psia

$$\begin{array}{r} & Mol/hr \\ \text{Oil} = & X \\ \text{NC gas} = {}^{90}\!/_{50} = & 1.8 \\ \text{Steam} = {}^{370}\!/_{18} = & 20.55 \\ \hline & 22.35 + X \end{array}$$

But the 20.55 mol steam ≈ 6.417 psia.

$$19.7 \times 20.55/(22.35 + X) = 6.417$$
$$X = 40.75 \text{ mols oil}$$

The partial pressures of the components, being in the mol ratios, are

$$\begin{array}{lr} & Psi \\ \text{Oil} \ldots\ldots\ldots\ldots\ldots & 12.74 \\ \text{NC gas} \ldots\ldots\ldots\ldots & 0.56 \\ \text{Steam} \ldots\ldots\ldots\ldots\ldots & 6.42 \\ \hline & 19.72 \end{array}$$

173°F and 12.74 psia are equivalent to 180°F at 14.7. From Fig. 13.14 the molecular weight of the vapors is 85.

$$\text{Lb/hr vapor} = 40.75 \times 85 = 3470 \text{ lb}$$
$$\% \text{ Condensed} = \frac{(13{,}330 - 3470)100}{13{,}330} = 74\%$$

It is now necessary to determine the amounts of steam and oil condensed at temperatures below the steam dew point. In constructing Table 13.4 the partial pressure of

Table 13.4. Oil Condensing Curve

%	Condensables, lb/hr	Av BP on EFC		50° API mol. wt.	Mol oil	Mol NC gas	Mol steam	Mol total	Total pressure, psia	Partial pressure oil, psia	Partial pressure NC gas, psia	Cond temp, °F
		14.7 psia, °F	19.7 psia, °F									
100	13,330	300	317	124	107.5	1.8	20.6	129.9	19.7	16.3	0.273	305
80	10,664	269	286	113	94.3	1.8	20.6	116.7	19.7	15.9	0.304	277
60	7,998	239	256	103	77.7	1.8	20.6	100.1	19.7	15.3	0.354	240
40	5,332	207	224	93	57.4	1.8	20.6	79.8	19.7	14.2	0.444	205
20	2,666	178	195	84	31.8	1.8	20.6	54.2	19.7	11.6	0.654	163
10	1,333	155	172	78	17.1	1.8	20.6	39.5	19.7	8.5	0.897	127
5	667	141	158	75	8.9	1.8	20.6	31.3	19.7	5.6	1.13	95

steam and the mols in the gas phase were considered constant. Actually, however, when the steam is condensed, it is the result of a continuously decreasing partial pressure. This correction is made below

CONDENSATION OF MIXED VAPORS

The reduction in the partial pressure of steam in the mixture increases the partial pressure of oil, thereby increasing the percentage condensation for a given temperature. Thus for 80 per cent condensation if both the steam and noncondensable gas were removed, the temperature would correspond to 195°F on the 19.7 psia EFC instead of 163°F when the steam and noncondensable gas are present.

Trial 1:

At 163°F, assume 90 per cent of oil vapor is condensed.

	Mol/hr	mf
Oil vapor $13,300 \times 0.10/78$..	17.1	$17.1/(18.9 + X)$
NC gas..................	1.8	$1.8/(18.9 + X)$
Steam...................	X	$X/(18.9 + X)$
Total..................	$18.9 + X$	

Then $19.7X/(18.9 + X)$ is the partial pressure of the steam. Since water is insoluble in the hydrocarbons, the partial pressure is also fixed by the saturation pressure given in the steam tables.

For 163°F $p_{steam} = 5.09$ psia

$$19.7X/(18.9 + X) = 5.09$$
$$X = 6.58 \text{ mols steam}$$

	Mol/hr	mf	$mf \times p_t = p_{partial}$
Oil vapor............	17.1	0.672	13.23
NC gas.............	1.8	0.070	1.38
Steam..............	6.58	0.258	5.09
Total.............	25.48	1.000	19.70

A temperature of 163°F at 13.23 psia is equivalent to a temperature of 188°F at 19.7 psia. From the EFC at 19.7 psia in Fig. 13.13 it will be seen that 16.5 per cent will be vaporized at 188°F or 83.5 per cent condensed. It was assumed that 90 per cent was condensed; hence the amount of oil assumed to be condensed was in error, requiring a new trial.

Trial 2:

At 163°F assume 85 per cent of vapor condensed.

	Mol/hr	mf	$mf \times p_t$
Oil vapor, $13,330 \times .15/81$..........	24.7	0.692	13.62
NC gas, $9\%_0$....................	1.8	0.050	0.99
Steam, X	$(X = 9.23)$	0.258	5.09
		1.00	19.70

364 *PROCESS HEAT TRANSFER*

163°F at 13.62 psia ≈ 185°F on the 19.7 psia EFC. This point corresponds to 15 per cent liquid or 85 per cent condensed. The assumption checks the curve.

Initial steam = $370/_{18}$ = 20.55 mols
Uncondensed steam at 163°F = 9.23
Mols condensed, 173 − 163°F = 11.32

Similarly

T, °F	Oil cond, %	Oil cond, lb	Steam cond, lb
173	74	9,863	0
163	85	11,350	204
127	97.5	13,000	357
95	100	13,330	370

Condensing Curve. The following enthalpy data have been taken from Fig. 11 in the Appendix and the steam tables. The limitations of this type of petroleum enthalpy data may be found in the literature. The inconsistency of the enthalpy datum temperatures for steam and the oil vapor does not affect this solution, since only differences are involved.

	Oil		Steam	
T_c, °F	H_v, vapor	H_l, liquid	H_g or H_v, gas or vapor	H_l, liquid
305	368	242	1197.0	Superheated
277	359	225	1184.1	Superheated
240	337	204	1167.0	Superheated
205	322	185	1150.6	Superheated
173	310	168	1135.4	140.9
163	306	163	1131.4	130.9
127	293	144	1116.6	94.9
95	283	128	1103.1	63.0

Heat load

305°F: H q
  Oil vapor = 13,330 × 368 = 4,920,000
  Steam = 370 × 1197.0 = 443,000
  NC gas = 90 × 0.46(273) = 11,300
                                         5,374,300   0

277°F:
  Oil vapor = 10,664 × 354 = 3,770,000
  Oil liquid = 2,666 × 225 = 595,000
  Steam = 370 × 1184.1 = 438,000
  NC gas = 90 × 0.46(245) = 9,300
                                         4,812,300   562,000

CONDENSATION OF MIXED VAPORS

240°F: H q
   Oil vapor = 7,998 × 337 = 2,695,000
   Oil liquid = 5,332 × 204 = 1,088,000
   Steam   =   370 × 1167 =   432,000
   NC gas   =   90 × 0.46(208) =   8,600
                                             4,223,600   588,700

205°F:
   Oil vapor = 5,332 × 322 = 1,715,000
   Oil liquid = 7,998 × 185 = 1,480,000
   Steam   =   370 × 1150.6 =   426,000
   NC gas   =   90 × 0.46(173) =   7,200
                                             3,628,200   595,400

173°F (dew point of steam):
   Oil vapor = 2,666 × 310   =   827,000
   Oil liquid = 10,664 × 168   = 1,790,000
   Steam   =   370 × 1135.4 =   420,000
   NC gas   =   90 × 0.46(141) =   5,800
                                             3,042,800   585,400

163°F:
   Oil vapor = 1,980 × 306   =   606,000
   Oil liquid = 11,350 × 163   = 1,850,000
   Steam   =   166 × 1131.4 =   188,500
   Water   =   204 × 130.9   =   26,700
   NC gas   =   90 × 0.46(131) =   5,400
                                             2,676,600   366,200

127°F:
   Oil vapor =   330 × 293   =   96,600
   Oil liquid = 13,000 × 144   = 1,870,000
   Steam   =   13 × 1116.6 =   14,400
   Water   =   357 × 94.9   =   33,900
   NC gas   =   90 × 0.46(95) =   3,900
                                             2,018,800   657,800

95°F:
   Oil liquid = 13,300 × 128   = 1,710,000
   Water   =   370 × 63   =   23,300
   NC gas   =   90 × 0.46(63) =   2,600
                                            = 1,735,900   282,900

These calculations are plotted in Fig. 13.16

Summary:
       Inlet to steam dew point:
         5,374,300 − 3,042,800 = 2,331,500 Btu/hr
       Steam dew point to outlet:
         3,042,800 − 1,735,900 = 1,306,900
       Total   3,638,400 Btu/hr

$$\text{Total water} = \frac{3{,}638{,}400}{120 - 85} = 104{,}000 \text{ lb/hr}$$

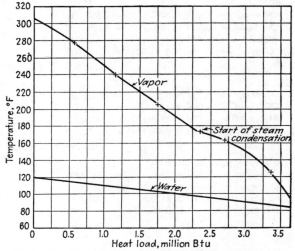

Fig. 13.16. Condensation of mixed hydrocarbons with gas and steam.

In this particular problem the condensing curve from the inlet to the dew point of the steam is nearly straight. When both the shell and tube steam temperature changes are proportional to the heat load, the LMTD may be applied.

Water temperature at dew point of steam,

$$t_W = 85 + \frac{1,306,900}{3,638,400} \times 35 = 97.5°F.$$

Weighted true temperature difference, Δt:
Inlet to dew point of steam:

$$\begin{array}{cccc} T_1 & 305 & 120.0 & t_2 \\ T_x & 173 & 97.5 & t_y \end{array} \quad \text{LMTD} = 122.2°F$$

$$UA = \frac{\Delta q}{\Delta t} = \frac{2,331,500}{122.2} = 19,050$$

Dew point of steam to outlet (from Fig. 13.16):

q	Δq	T_c	t_W	Δt_{av}	$\frac{\Delta q}{\Delta t_{av}} = UA$
2,331,500	....	173	97.5		
2,500,000	169,000	169	96	74	2,285
2,750,000	250,000	161	93	70.5	3,550
3,000,000	250,000	149	91	63	3,970
3,250,000	250,000	134	89	51.5	4,850
3,500,000	250,000	112	86	35.5	7,040
3,638,000	138,000	95	85	18.5	7,460
				$UA = \sum \frac{\Delta q}{\Delta t}$	29,155

CONDENSATION OF MIXED VAPORS

$$\text{Weighted } \Delta t = \frac{Q}{\Sigma \, \Delta q / \Delta t} = \frac{1{,}306{,}900}{29{,}155} = 44.8°\text{F}$$

Overall weighted temperature difference:

$$\text{Weighted } \Delta t = \frac{3{,}638{,}500}{29{,}155 + 19{,}050} = 75.5°\text{F}$$

The uncorrected LMTD is 60.1°F.

The Overall Coefficient of Heat Transfer. The condensing curve of Example 13.6a is typical of the condensation which may occur for the phase system indicated. It is repeated here that the steam dew point may be higher than the oil dew point and that the steam may start to condense first. When the latter occurs, the surface from the steam dew point to the oil dew point can be treated as the ordinary condensation of a vapor from a noncondensable with the superheated oil vapor combining with the gas to form the total noncondensables. If the vapor enters at a temperature above either dew point, it is cooled to the first dew point as a dry gas, although this does not happen after a distilling column, since the top plate vapor is always at a dew point.

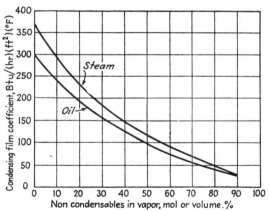

FIG. 13.17. Condensing coefficients for vapors and noncondensables.

Where steam is above its dew point, it is regarded as a noncondensable gas as a matter of safety even though the tube-wall temperature may be below the dew point of the steam and therefore wet. When oil vapor condenses in the presence of noncondensables, the problem is again one of diffusion except that the equations employed heretofore for diffusion are not very reliable when calculated from the properties of vapors of high molecular weight. A curve for the condensing and diffusion coefficient for oil vapors in the presence of noncondensable gases is plotted in Fig. 13.17 as a function of the percentage of noncondensables. Its origin is discussed presently.

When steam diffuses through a noncondensable gas and the vapor of miscibles, the rate of heat transfer is quite different from that of pure diffusion. An unpublished correlation has been made of the overall clean coefficients U_C obtained by calculating the pure diffusion of steam, using the method under Sec. 4 of this chapter, through gases with molecular weights between 16 and 50. The calculated data define a common line when U_C is plotted against the percentage of noncondensables even though the operating pressures on the systems varied. The curve obtained from this data is taken as the base curve, and to it is added a film resistance corresponding to an oil layer of 0.002 in., which gives an average resistance of 0.0020. The steam curve shown in Fig. 13.17 was arrived at in this manner. The oil diffusion curve was established by taking the pure condensing coefficient for a vapor and reducing it in the same percentage as the steam curve is reduced by the increase in the percentage of the noncondensables. The base value for the oil curve in Fig. 13.17 was 300, but any other condensing coefficient can be used and reduced accordingly. This will usually not be a necessary refinement. The use of the curves at low vacuums or very high pressures is questionable, but a large number of condensers have been designed at rates corresponding to those obtained by this method and have operated successfully.

While a method based on using fictitious coefficients as suggested above may not be so desirable as a method employing individual film calculations, it has the advantage of being rapid. The cooling of the condensate to the outlet temperature in horizontal condensers may be treated similarly to that in Sec. 2, by using the free convection rate U of about 50 and applied to the average liquid-cooling sensible heat load. The water is considered to be in the coldest passes, and the temperature rise of the water is figured accordingly.

Example 13.6b. Calculation of a Condenser for the Condensation from Noncondensable of a Mixture Immiscible with Water. The vapor of Example 13.6a is to be condensed with a fouling factor of at least 0.0030 and pressure drops not exceeding 2.0 psi for the vapor and 10.0 psi for the water.

Available for the purpose is a 27 in. ID exchanger containing 286 1 in. OD, 14 BWG tubes 12'0" long laid out on 1¼-in. triangular pitch. The bundle is arranged for eight tube passes and baffles are spaced 16 in. apart.

Solution:

Exchanger:

Shell side
ID = 27 in.
Baffle space = 16 in.
Passes = 1

Tube side
Number and length = 286, 12'0"
OD, BWG, pitch = 1 in., 14 BWG, 1¼ in. tri.
Passes = 8

CONDENSATION OF MIXED VAPORS 369

Clean surface requirements:

Heat load inlet to dew point of steam = 2,331,500 Btu/hr
$\Delta t = 122.2°$. h_{io} for water is 700 Btu/(hr)(ft^2)(°F).
From Table 13.4 at inlet

$$\text{NC gas + steam} = 1.8 + 20.6 = 22.4 \text{ mol/hr}$$
$$\text{Total} = 129.9 \text{ mol/hr}$$
$$\% \text{ NC gas} = \frac{22.4}{129.9} = 0.1735$$

From Fig. 13.17, $h_o = 205$ Btu/(hr)(ft^2)(°F)
At dew point of steam

	Mol/hr
NC gas + steam	22.4
Oil	40.75
Total	63.2

$$\% \text{ NC} = \frac{22.4}{63.2} = 0.354$$

From Fig. 13.17, $h_o = 140$
Log mean overall coefficient = 136.5 Btu/(hr)(ft^2)(°F)
$\Delta t = 122.2°\text{F}$

$$A_{C1} = \frac{Q}{U \Delta t} = \frac{2,331,500}{136.5 \times 122.2} = 139.5 \text{ ft}^2$$

At the dew point of the steam the water vapor is no longer considered a noncondensable gas. And since the remainder of the oil vapor is condensing, it is not to be considered a noncondensable in the presence of water vapor although the presence of each influences the rate of diffusion of the other.
At dew point of steam to outlet:

	Mol/hr
NC gas	1.8
Steam	20.6
Total	22.4

$$\% \text{ NC gas} = \frac{1.8}{22.4} = 0.080$$

From Fig. 13.17, $U_C = 212$ (weighted for oil and steam)
At outlet

$$\text{NC gas} = 1.8 \text{ mols}$$
$$\text{Steam} = \text{neg}$$

From Fig. 13.17, $U_C = 15$
Log mean overall coefficient = 74.5 Btu/(hr)(ft^2)(°F)
$\Delta t = 44.8°\text{F}$

$$A_{C2} = \frac{Q}{U \Delta t} = \frac{1,306,900}{74.5 \times 44.8} = 391 \text{ ft}^2$$

Subcooling rate: Consider all the condensate cooled over half the entire condensing range.

Heat of liquid (50°API) = $13,330 \times 0.55(305 - 95)/2 = 770,000$ Btu/hr
Assume the coldest water passes are in contact with the condensate.

Water rise = $\frac{770,000}{3,638,400} \times 35 = 7.4°\text{F}$

Condensate, 305 to 95°F

Water, 85 to 92.4°F
LMTD = 66.3°F
Use a value of $U = 50$ for free convection.

$$A_s = \frac{Q}{U \, \Delta t} = \frac{770{,}000}{50 \times 66.3} = 232.5 \text{ ft}^2$$

Total clean surface, $A_C = 139.5 + 391 + 232.5 = 763 \text{ ft}^2$

Clean overall coefficient U_C:

$$U_C = \frac{Q}{A_C \, \Delta t} = \frac{3{,}638{,}400}{763 \times 75.5} = 63.2 \text{ Btu/(hr)(ft}^2)(°F)$$

Design overall coefficient U_D:

External surface/ft of length = 0.2618 ft   (Table 10)
$A = 286 \times 12'0'' \times 0.2618 = 897 \text{ ft}^2$

$$U_D = \frac{3{,}638{,}400}{897 \times 75.5} = 53.7$$

Dirt factor:

$$R_d = \frac{U_C - U_D}{U_C U_D} = \frac{63.2 - 53.7}{63.2 \times 53.7} = 0.0028 \text{ (hr)(ft}^2)(°F)/\text{Btu}$$

The pressure drop is computed in the usual manner.
Submerge $(232.5/763)897 = 274 \text{ ft}^2$ of surface

Summary

*	h outside	700
U_C	63.2	
U_D	53.7	
R_d Calculated 0.0028		
R_d Required 0.003		
0.8	Calculated ΔP	5.6
2.0	Allowable ΔP	10.0

* Composite. Low dirt factor.

PROBLEMS

13.1. The following vapor enters a condenser at 40 psia:

	Mols/hr
C_4	6.4
C_5	219.7
C_6	2.3
	228.4

Assume a pressure drop of 5 psi in the condenser. Available for the service is a 25 in. ID 1-2 horizontal condenser containing 222 1 in. OD, 14 BWG, 12'0'' long tubes arranged for eight passes on 1¼-in. square pitch. Baffles are spaced 18 in. apart. Cooling is by water from 85 to 120°F.

(a) What is the true temperature difference?
(b) What are the dirt factor and pressure drops?

13.2. The following overhead vapor enters a condenser at 20 psia.

	Mols/hr
C_4	10.5
C_5	150.0
C_6	39.6
C_7	63.7
C_8	191.5

Cooling is effected by water from 85 to 120°F.
Available for the service is a 23¼ in. ID 1-2 horizontal condenser containing 308 ¾ in. OD, 14 BWG, 12'0" long tubes arranged for four passes on 1-in. square pitch. Baffles are spaced 24 in. apart.
(a) What is the true temperature difference?
(b) What are the dirt factor and the pressure drop?

13.3. The following overhead vapor enters a condenser at 50 psia.

	Mols/hr
C_4	31
C_5	245
C_6	124
H_2O	40
	440

Cooling is effected by water from 85 to 120°F.
(a) What is the true temperature difference?
(b) Calculate the size of 1-2 horizontal condenser required for the service providing a minimum dirt factor of 0.004 and a pressure drop of 2.0 for the vapor and 10.0 for water. Use 1-in. tubes on 1¼-in. square pitch and a minimum water velocity of 3.0 fps.

13.4. 2800 lb/hr of air is saturated with water vapor at 203°F at 5 psi, and the mixture enters a condenser where it is cooled to 100°F by water from 70 to 100°F.
Available for the service is a 27 in. ID 1-2 condenser containing 462 ¾ in. OD, 16 BWG, 12'0" tubes arranged for four passes on 1-in. triangular pitch. Baffles are spaced 24 in. apart.
(a) What is the true temperature difference?
(b) What are the dirt factor and the pressure drops?

13.5. An oil has the following ASTM distillation curve:

% distilled	°F
IBP	310
10	328
20	338
30	346
40	354
50	360
60	367
70	372
80	377
90	382
100	387

28,000 lb/hr of oil (mol. wt. = 154) and 4400 lb/hr of steam enter a condenser at 20 psia. Cooling will be effected by water from 85 to 100°F. Available for the service is a 31 in. ID 1-2 vertical condenser 16'0" tall containing 728 ¾ in. OD, 16 BWG, 16'0" long tubes arranged for two passes on 1-in. triangular pitch.

(a) What is the weighted Δt?
(b) What are the dirt factor and pressure drops?

13.6. A 48°API oil has the following ASTM distillation curve:

% distilled	°F
IBP	100
10	153
20	190
30	224
40	257
50	284
60	311
70	329
80	361
90	397
End point	423

15,300 lb/hr of oil vapor, 3620 lb/hr of steam, and 26,200 lb/hr of noncondensable gas (mol. wt. = 46) at 2.0 psi are to be condensed using cooling water with an inlet temperature of 85°F.

Available for the service are two 33 in. ID 1-2 horizontal condensers connected in series and containing 680 ¾ in. OD, 14 BWG tubes 16'0" long laid out for four passes on 1-in. square pitch. Baffles are at full pitch.

(a) Considering the condenser to be at atmospheric pressure, what is the weighted Δt?
(b) What are the dirt factors and the pressure drops?

NOMENCLATURE FOR CHAPTER 13

A	Heat-transfer surface, ft²
A_C	Clean heat-transfer surface, ft²
A_c, A_s	Heat-transfer surface for condensation and subcooling respectively, ft²
A_q	Heat-transfer surface corresponding to the heat load interval q, ft²
a	Flow area, ft²
B	Baffle spacing, in.
C	Specific heat of hot fluid in derivations, Btu/(lb)(°F); number of components in the phase rule, dimensionless
C'	Clearance between tubes, in.
c	Specific heat of cold fluid, Btu/(lb)(°F)
D	Inside diameter of tubes, ft
D_o	Outside diameter of tubes, ft
d	Inside diameter of tubes, in.
D_e	Equivalent diameter for heat transfer and pressure drop, ft
d_e	Equivalent diameter for heat transfer and pressure drop, in.
F	Degrees of freedom in the phase rule
f	Friction factor
f_p, f_t	Fugacity of the pure compound, fugacity at the total pressure, atm

CONDENSATION OF MIXED VAPORS 373

G	Mass velocity, lb/(hr)(ft²)
G_m	Mol mass velocity, mol/(hr)(ft²)
G'	Condenser loading for vertical condensers, lb/(hr)(ft)
G''	Condenser loading for horizontal condensers, lb/(hr)(ft)
g'	Acceleration of gravity, ft/sec²
H_v, H_i	Enthalpy of the vapor, enthalpy of the liquid, Btu/lb
h, h_i, h_o	Heat-transfer coefficient in general, for inside fluid, and for outside fluid, respectively, Btu/(hr)(ft²)(°F)
h_{io}	Value of h_i when referred to the tube OD, Btu/(hr)(ft²)(°F)
$\bar{h}$	The average condensing film coefficient, Btu/(hr)(ft²)(°F)
j_H	Factor for heat transfer, $(hD/k)(c\mu/k)^{-1/3}$, dimensionless
j_h	Factor for heat transfer, $(h/cG)(c\mu/k)^{2/3}$, dimensionless
j_d	Factor for diffusion, dimensionless
K	Equilibrium constant, dimensionless
K_1	Equilibrium constant for any single compound or the first compound, dimensionless
K_G	Diffusion coefficient, (lb-mol)/(hr)(ft²)(atm)
k	Thermal conductivity, Btu/(hr)(ft²)(°F/ft)
k_d	Diffusivity, ft²/hr
L	Tube length, ft or total liquid, mol or mol/hr
L_1	Liquid of a single compound or the first compound, mol or mol/hr
L_1'	Liquid of a single compound or the first compound condensed in an interval, mol or mol/hr
L_C	Length of tube on which condensation is effective, ft
L_c	Liquid of a single compound condensed prior to an interval, mol or mol/hr
M_A, M_B	Molecular weight of diffusing and inert gases, respectively
M_m	Mean molecular weight of vapor mixture, lb/mol
M_v	Molecular weight of diffusing component, lb/mol
mf	Mol fraction, dimensionless
N	Number of baffles, dimensionless
N_d	Number of mols transferred by diffusion, dimensionless
N_t	Number of tubes effective for condensation in a partially submerged bundle, dimensionless
n	Number of tube passes; number of compounds in a mixture
P	Number of phases in the phase rule
P_T	Tube pitch, in.
ΔP	Pressure drop in general, psi
$\Delta P_T, \Delta P_t, \Delta P_r$	Total, tube, and return pressure drop, respectively, psi
p	Partial pressure in general, atm
p_1, p_2	Partial pressures of the components; partial pressures for one compound at two different points, atm
p_t, p_p	Total pressure and pressure of the pure compound, atm
p_g, p_g'	Partial pressure of the inert gas in the gas body and at the condensate film, atm
p_{gf}	Log mean pressure difference of the inert gas between p_g and p_g', atm
Δp	Pressure difference of diffusing fluid between p_v and p_c, atm
Q	Heat flow, Btu/hr
q	Heat flow for interval, Btu/hr
q_c	Heat flow for condensation, Btu/hr

R_d	Combined dirt factor, $(hr)(ft^2)(°F)/Btu$
Re	Reynolds number, dimensionless
s	Specific gravity, dimensionless
T	Temperature of a gas mixture, °K
T_1, T_2	Inlet and outlet temperatures of hot fluid, °F
T_c	Temperature of condensate film; caloric temperature of hot fluid, °F
T_s	Steam temperature, °F
t, t_1, t_2	Cold-fluid temperature in general, inlet, and outlet, respectively, °F
t_c	Caloric temperature of cold fluid
t_f	Temperature of film, °F
t_W, t_w	Temperature of water, tube-wall temperature, °F
Δt	True temperature difference, °F
U, U_C, U_D	Overall coefficient of heat transfer, clean coefficient, and design coefficient, respectively, $Btu/(hr)(ft^2)(°F)$
U_c, U_s	Clean overall coefficient for condensation and subcooling, respectively, $Btu/(hr)(ft^2)(°F)$
V	Velocity, fps or total vapor, mol or mol/hr
V_1	Vapor of a single compound or the first compound, mol or mol/hr
v_A, v_B	Molar volumes of diffusing and inert gas, dimensionless
W	Weight flow of hot fluid, lb/hr
w	Weight flow of cold fluid, lb/hr
x, x_1, x_2	Mol fraction in liquid
Y, Y_1, Y_2	Total vapor feed or each component, mol or mol/hr
y, y_1, y_2	Mol fraction in vapor
α	Relative volatility
λ	Latent heat of vaporization, Btu/lb
μ	Viscosity, centipoises $\times 2.42 = lb/(ft)(hr)$
μ_w	Viscosity at the tube-wall temperature, centipoises $\times 2.42 = lb/(ft)(hr)$
ρ	Density, lb/ft^3
ϕ	Viscosity ratio, $(\mu/\mu_w)^{0.14}$

Subscripts (except as noted above)

a	Average
c	Condensate or condensing
f	Film
g	Gas (inerts)
i	Inert
l	Liquid
s	Shell side
t	Tube side
v	Vapor

CHAPTER 14
EVAPORATION

The Mechanism of Vaporization. Much of our present knowledge of the boiling phenomenon is obtained from the work of Jakob and Fritz[1] and later researches of Jakob.[2] When steam flows through a tube which is submerged in a pool of liquid, minute bubbles of vapor form at random points on the surface of the tube. The heat passing through the tube surface where no bubbles form enters the surrounding liquid by convection. Some of the heat in the liquid then flows toward the bubble, causing evaporation from its inner surface into itself. When sufficient buoyancy has been developed between the bubble and the liquid, the bubble breaks loose from the forces holding it to the tube and rises to the surface of the liquid pool. Kelvin postulated that, in order for this behavior to prevail, the liquid must be hotter than the saturation temperature in the incipient bubble. This is possible, since the spherical nature of the bubble establishes liquid surface forces on it so that the saturation pressure inside the bubble is less than that of the surrounding liquid. The saturation temperature of the bubble being lower than that of the liquid surrounding it, heat flows into the bubble. The number of points at which bubbles originate is dependent upon the texture of the tube surface, the number increasing with the roughness. Jakob and Fritz have detected the presence of superheated liquid close to the heating surface and have found the difference between the temperature of the superheated liquid and the saturation temperature of the vapor to be less for rough surfaces than for smooth surfaces.

Heat transfer by vaporization without mechanical agitation is obviously a combination of ordinary liquid free convection and the additional convection produced by the rising stream of bubbles. Under very small temperature differences between tube wall and boiling liquid the formation of bubbles proceeds slowly and the rate of heat transfer is essentially that of free convection as given in Eqs. (10.13) and (10.14). The surface tension as an influence on bubble formation and growth is another variable as shown in Fig. 14.1. The surface tension of water against air

[1] Jakob, M., and W. Fritz, *Forschr. Gebiete Ingenieurw.*, **2**, 434 (1931).
[2] Jakob, M., *Mech. Eng.*, **58**, 643 (1936); "Heat Transfer," Vol. 1, John Wiley & Sons, Inc., New York, 1949.

is approximately 75 dynes/cm at room temperature, whereas the majority of organics have surface tensions ranging from 20 to 30 dynes/cm at room temperature. The surface tensions of most liquids at their respective boiling points, however, are probably not so far apart as at room temperature. The surface tensions of the liquids against metals may also differ from their surface tensions against air, since the rate of vaporization of water is actually much greater than that of organics under identical conditions. If the surface tension of the liquid is low it tends to wet surfaces, so that the bubble in Fig. 14.1a is readily occluded by the liquid and rises. For liquids with intermediate surface tensions, as in Fig. 14.1b, a momentary balance may exist between the bubble and the tube wall, so that it is necessary to form larger bubbles before the buoyant force can free it from the surface. The bubble in Fig. 14.1c indicates the influence of high surface tension.

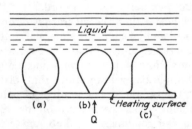
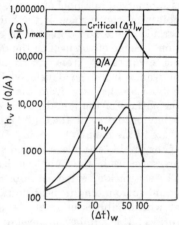

Fig. 14.1. Effect of interfacial tension on bubble formation. (*After Jakob and Fritz.*)

Fig. 14.2. Boiling curve of water from pools. (*After McAdams.*)

Consider the typical boiling coefficient curve of McAdams[1] based on the data on several investigators for water as shown in Fig. 14.2. From a $(\Delta t)_w$ above 5°F there is a relatively straight logarithmic relationship between the coefficient of vaporization and the temperature difference, where $(\Delta t)_w$ is the difference between the tube wall and the vapor temperatures. This relationship changes, however, at the *critical temperature difference*, which occurs at about 45°F for water evaporating from pools. At that temperature difference the hot surface and liquid approach the condition shown in Fig. 14.3. There is a predominance of vapor at the tube wall because of the high rate of heat throughput, and very little liquid actually contacts the hot tube wall. This condition is called

[1] McAdams, W. H., "Heat Transmission," p. 296, McGraw-Hill Book Company, Inc., New York, 1942.

blanketing or *vapor binding*, wherein the large amount of vapor formed at the tube wall actually serves as a gas resistance to the passage of heat into the liquid and reduces the film coefficient for vaporization as the temperature difference increases. Drew and Mueller[1] have reported the critical temperature differences for a number of organic compounds under varying conditions of surface and the organics exhibit a critical temperature difference of roughly 60 to 120°F. When vaporization takes place directly at the heating surface, it is called *nuclear boiling*, and when it takes place through a blanketing film of gas, it is *film boiling*.

A useful criterion of performance during vaporization is that of the maximum attainable heat flux defined as $(Q/A)_{max}$ or $(U \Delta t)_{max}$. The heat flux vs. $(\Delta t)_w$ is plotted for boiling water as the upper curve of Fig. 14.2. It represents the number of Btu per hour being transferred per square foot of surface with the maximum attainable flux corresponding to the heat flow at the critical temperature difference. For water at atmospheric pressure this occurs, according to McAdams,[2] when $h_v = 8800$ and $(Q/A)_{max} = 400,000$ and at higher pressures both values increase. More than $(Q/A)_{max}$ Btu/(hr)(ft²) cannot be forced through the heating surface because of the appearance of a gas film.

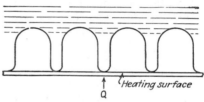

Fig. 14.3. Incipient blanketing. (*After Jakob and Fritz.*)

When a higher $(\Delta t)_w$ than the critical is employed, a smaller vaporization coefficient results and the flux similarly decreases. It follows then that the large temperature differences so favorable to conduction and convection may actually be a hindrance to vaporization.

The following factors are found to affect the rate of heat transfer by vaporization from pools and have defeated, to a great extent, the possibility of obtaining one or two simple correlations applicable to the majority of liquids: (1) nature of the surface and distribution of bubbles; (2) properties of the fluid such as surface tension, coefficient of expansion, and viscosity; and (3) the influence of the temperature difference upon the evolution and vigor of the bubbles.

Effect of Pressure and Properties on the Vaporizing Coefficient. Much of the present method of calculation of vaporization from pools consists of nothing more than the use of observed coefficients for individual liquids reported at atmospheric pressure. If the coefficient for vaporization from a pool has been reported for a fluid at atmospheric pressure, it can

---

[1] Drew, T. B., and A. C. Mueller, *Trans. AIChE*, **33**, 449–471 (1939).
[2] McAdams, *op. cit.*, p. 297.

be converted to subatmospheric pressures by the equation of Jakob:[1]

$$h_v = h_{v,14.7\text{psia}} \left(\frac{p}{p_{14.7\text{psia}}}\right)^{1/3} \tag{14.1}$$

For superatmospheric pressures up to 226 psi the coefficient is given by

$$h_v = h_{v,14.7\text{psia}} \left(\frac{p}{p_{14.7\text{psia}}}\right)^{1/6} \tag{14.2}$$

where h_v and p refer to the new conditions.

Since blanketing is caused by the accumulation of vapor bubbles, the pressure on the system is important in defining the size of individual bubbles. The influence of the viscosities and surface tensions of liquids on their respective atmospheric coefficients as functions of the absolute values of the properties has not been correlated. From experiments on single liquids outside single tubes with variable pressure Cryder and Finalborgo,[2] working at low fluxes, obtained a family of curves of nearly uniform slope when plotting h_v vs. $(\Delta t)_w$. Their mean equation is

$$\log \frac{h_{v,14.7\text{psia}}}{h_v} = 0.015(t_{14.7\text{psi}} - t) \tag{14.3}$$

where h_v and t refer to the new conditions.

Classification of Vaporizing Equipment. There are two principal types of tubular vaporizing equipment used in industry: *boilers* and *vaporizing exchangers*. Boilers are directly fired tubular apparatus which primarily convert fuel energy into latent heat of vaporization. Vaporizing exchangers are unfired and convert the latent or sensible heat of one fluid into the latent heat of vaporization of another. If a vaporizing exchanger is used for the evaporation of water or an aqueous solution, it is now fairly conventional to call it an *evaporator*. If used to supply the heat requirements at the bottom of a distilling column, whether the vapor formed be steam or not, it is a *reboiler*. When not used for the formation of steam and not a part of a distillation process, a vaporizing exchanger is simply called a *vaporizer*. When an evaporator is used in connection with a power-generating system for the production of pure water or for any of the evaporative processes associated with power generation, it is a *power-plant evaporator*. When an evaporator is used to concentrate a chemical solution by the evaporation of solvent water, it is a *chemical evaporator*. Both classes differ in design. Unlike evaporators it is the object of reboilers to supply part of the heat required for distillation and not a change in concentration, although a change generally cannot

---
[1] Jakob, M., *Tech. Bull. Armour Inst. Tech.*, **2**, No. 1 (1939).
[2] Cryder, D. S., and A. C. Finalborgo, *Trans. AIChE*, **33**, 346–361 (1937).

EVAPORATION 379

be avoided. Very often the term *evaporator* is also applied to a combination of several pieces of equipment each of which can also be defined as an evaporator.

Unfortunately certain classes of evaporators are still designed as part of an art rather than the rational summation of the individual resistances to heat flow as practiced heretofore. This is due to the high transfer coefficients with which certain classes of evaporators operate and the difficulty of identifying each of the small individual resistances which make up the overall resistance. As in the case of surface condensers in Chap. 12, numerous classes of evaporators are designed on the basis of *accepted* overall coefficients, and it is these classes and their processes which are treated in this chapter. Generally they involve vaporization from pools as compared with vaporization in the shell or tubes of a 1-2 exchanger. Evaporators which may be or are usually designed from individual coefficients will be treated in Chap. 15.

POWER-PLANT EVAPORATORS[1]

Introduction. One of the main purposes of power-plant evaporators is to provide relatively pure water for boiler feed. The principal features

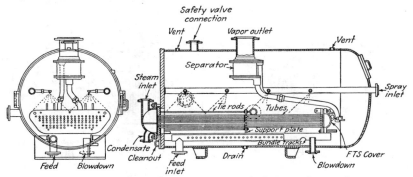

FIG. 14.4. Power-plant evaporator. (*The Lummus Company.*)

incorporated in power-plant evaporators are a tubular heating element, a space in which the liquid droplets which are carried up by the bursting bubbles may be *disengaged*, and a means of removing the scale from the outsides of the tubes. Three typical examples, each with merit, are shown in Figs. 14.4 to 14.6. In Fig. 14.4 the tube bundle is laid out flat and the feed is introduced at the bottom. In Fig. 14.5 the bundle is cylindrical and the feed is introduced just below the liquid level. All

[1] The author is indebted to Gerald D. Dodd of the Foster Wheeler Corporation for his generous assistance in the preparation of this section. Readers interested only in chemical evaporation may omit pages 379 to 393 without loss of context.

operate half full of water, the upper half being the disengaging space, and all are equipped with separators which return the entrained liquid below the liquid level in the evaporator.

The Treatment of Feed Water. When a pound of steam is evaporated from a pool of boiling water, most of the impurities originally in the water

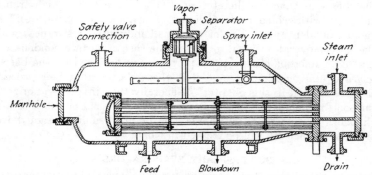

FIG. 14.5. Power-plant evaporator. (*Alco Products.*)

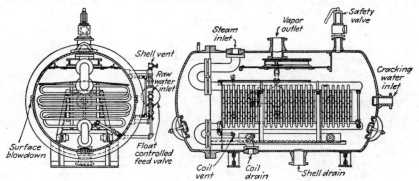

FIG. 14.6. Power-plant evaporator. (*Foster Wheeler Corporation.*)

remain behind and in time tend to form scale on the submerged heating surface. All natural water contains mineral salts, of which those of calcium and magnesium in particular form scale. The scale forms more rapidly on hot surfaces and is an additional resistance to the flow of heat. This is detrimental to the performance of evaporators which are designed for high rates of heat transfer. Water is classified as hard or soft principally from its behavior in household use. Soap reacts with the calcium and magnesium salts, forming insoluble compounds, but water which is considered soft in the home may still be unsatisfactory as feed to a continuous evaporator. In order to reduce the accumulation of the scale-forming materials, it is customary to vaporize only 90 per cent of the water feed continuously, the remaining 10 per cent being drained continuously as *blowdown*.

There are three principal chemical means by which waters are softened. These are (1) the cold lime-soda process, (2) the hot lime-soda process, and (3) the zeolite process. The last is by far the commonest on new installations. In the sodium zeolite process a complex sodium silicate reacts continuously with the scale-forming compounds of the feed water, replacing the positive ions of the calcium and magnesium compounds with sodium ions and at the same time retaining the calcium and magnesium compounds as calcium and magnesium zeolites. It is therefore necessary to have two zeolite beds in operation so that the calcium and magnesium can be removed from one bed while the other accumulates calcium and magnesium as zeolites. A detailed discussion of water conditioning is beyond the scope of this chapter, but since evaporation is itself a purification process, the question arises as to the need for preliminary chemical purification. The reasons are as follows: When bubbles of vapor are disengaged from the liquid at the surface of a pool, the vapor *entrains*, or carries off, some of the liquid water with it which has not been distilled and consequently contains the impurities of the concentrated *blowdown*. When these small quantities of carry-over are continuously fed to a boiler, they cause scale deposition on the boiler tubes. In addition they increase the quantity of blowdown which must be removed from the boiler where the temperature is greater than in the evaporator and which represents an added loss in sensible heat from the power generating system.

The softening of water prior to entering the evaporator is in no sense a guarantee of the purity of the final boiler feed since a number of factors influence the amount of entrainment occurring at the surface of the liquid. Significant is the ratio of the total solids to suspended solids or the ratio of total solids to sodium alkaline solids, particularly when accompanied by a large quantity of suspended solids or a minute quantity of inorganic solids. When the softening reduces the surface tension of the water appreciably, excessive priming or foaming invariably results. The control of foaming is consequently one of the most important considerations in evaporator design. The standard required purity of a typical evaporator product in the United States contains not more than two to four parts of mineral solids per U.S. gallon. Notwithstanding these precautions the formation of scale in an evaporator is such that it is considered normal to descale a power-plant evaporator every 24 hr.

There are three principal types of scale: (1) soft scale which can be washed from the tube, (2) hard scale which can be removed only manually, and (3) hard scale which can be *shocked* from the tube. All three evaporators in Figs. 14.4 to 14.6 are equipped for the removal of the last type of scale. The principle of scale removal by shocking is as follows: In Figs. 14.4 and 14.5 the distance between the tube sheets is fixed by external bars connecting them. In order to descale, all the hot water is removed from the shell and steam is circulated through the tubes, causing expansion of the tubes alone so that they bow if originally straight or straighten if originally installed with a bend at their centers. The hot

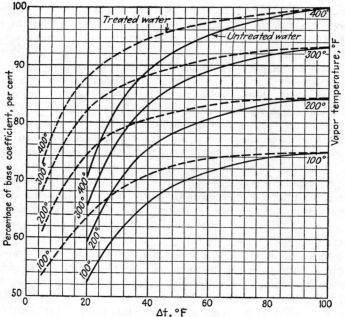

FIG. 14.7. Commercial evaporation coefficients for water.

scale is stressed owing to the expansion, and cold water is sprayed on the tubes, resulting in a sudden differential contraction which cracks the scale and causes it to be shed. If the evaporator is small, 100 to 600 ft^2, it is customary to flood the shell with cold water instead of spraying. For large installations the time required to fill and drain the shell would cause too long a shutdown period, and spraying is employed. Another variation of the power-plant evaporator is shown in Fig. 14.6 using serpentine tubes. The bundle is stressed between the vertical support pieces so that the scale is effectively removed from the curved portions as well as the flat portions of the tube.

EVAPORATION

Heat-transfer Coefficients in Power-plant Evaporators. The rates of heat transfer in power-plant evaporators, as mentioned before, are not treated on the basis of individual film coefficients. Because of the rapidity of scale formation and the nature of the scale resistance, the overall rates are based on the normal operating cycle of about one descaling every 24 hr. In Fig. 14.7 the percentages of a base average overall coefficient U_D are plotted against Δt which in an evaporator is the difference between the saturation temperature of the heating steam and the saturation temperature on the shell side. This temperature difference is always employed when evaporators are designed from overall coefficients. In the power-plant evaporator it is called the *temperature head*, and in chemical evaporators it is the *apparent temperature difference*, designated $(\Delta t)_a$. The values of U_D and Δt so defined permit the direct substitution of U_D in $Q = U_D A \Delta t$. The overall coefficient is greatly influenced by the pressure on the system, since the vapor volume of the bubbles is less at high pressure than at low pressures so that larger coefficients may be expected for the former. This is reflected by the curves which have been plotted in Fig. 14.7 as functions of the vapor temperature and which represent indirectly the operating pressure of the shell side. The curves level off abruptly at high-temperature heads owing to incipient vapor blanketing. The base value of the overall coefficient is varied in industry owing both to special problems suggested by a chemical analysis of the water and to changing competition among manufacturers.

A value of 700 Btu/(hr)(ft^2)(°F) is a fair average for the base value of U_D, although higher base coefficients have frequently been used.

Example 14.1. Calculation of Evaporator Surface. 10,000 lb/hr of distilled water is required from untreated water. Steam is available at 300°F, and the condenser will vent to the atmosphere. How much surface is required?

Assume a pressure drop through the condenser and lines of about 5 psi. The saturation temperature in the evaporator shell will be 19.7 psia or 226°F.

Heat balance:

$Q_{\text{evap}} = 10,000 \times 961 = 9,610,000$ Btu/hr
$Q_{300°F} = 10,550 \times 910 = 9,610,000$ Btu/hr

Temperature head:

$\Delta t = 300 - 226 = 74°F$

Overall coefficient:

From Fig. 14.7 at a heat head of 74°F and a vapor temperature of 226°F, the coefficient is 86.5 per cent of base. Using a base of 700 Btu/(hr)(ft^2)°F

$U_D = 700 \times 0.865 = 605$

$A = \dfrac{Q}{U_D \Delta t} = \dfrac{9,610,000}{605 \times 74} = 2150$ ft^2

Temperature Differences at Less than Maximum Flux. The preceding calculation is of value only to estimate the surface requirement. The surface will fill half of the shell or less, and the method of spacing the surface varies greatly with different manufacturers. Tubes, as a rule, are spaced farther apart in evaporators than in exchangers.

According to Fig. 14.7 it would be possible to obtain increasing overall coefficients up to a temperature head of 100°F although the vapor temperature is 212°F corresponding to atmospheric pressure. Previously it was stated that the critical temperature difference at atmospheric pressure was 45°F. A temperature head of Δt of 100°F corresponds to a critical temperature difference $(\Delta t)_w$ of about 75°F but the limitation of *the critical temperature difference holds only when operating at maximum flux.* In Fig. 14.7 the flux for 212°F vapor and $\Delta t = 100°F$ is

$$700 \times 0.85 \times 100 = 60{,}000 \text{ Btu}/(\text{hr})(\text{ft}^2)$$

If the flux has a value less than maximum, the temperature difference $(\Delta t)_w$ may be greater than the critical temperature difference. The flux is the principal index of vapor binding, and for this reason it is always restricted to a fraction of the maximum attainable when designing commercial evaporators. Restricting the flux to a *design* maximum amounts to an increase in the total surface for the purpose of reducing the amount of vaporization per unit of heat-transfer surface.

Multiple-effect Evaporation. In the production of distilled water the vapor formed in the evaporator is useful steam as well as relatively pure water. If a pound of steam is supplied to an evaporator as shown in Fig. 14.8a, it can be used to produce about 0.9 lb of steam from a pound of water. The remaining 0.1 lb of water contains the bulk of the impurities, and it is removed from the evaporator as evaporator *blowdown*. The 0.9 lb of vapor from the evaporator can be condensed by partially preheating the evaporator feed, or in the power plant, it can be mixed directly with cold return condensate before being fed to the boiler.

If, however, the original pound of steam was supplied to a process as shown in Fig. 14.8b and the vapor produced in the first evaporator was then used as a heat source in a second evaporator operating at a lower pressure than the first, an additional utilization could be made of most of the heat. If both evaporators in Fig. 14.8b were fed in *parallel* with raw water, about 0.85 lb of pure water would be formed in the first *effect* and about 0.75 lb would be formed in the second effect. For each pound of steam supplied about 1.6 lb of pure water would be produced. The original pound of steam can also be considered a pound of pure water. When the vapor formed in the first effect is reused as the heating medium in a second effect, it is a *double-effect* evaporator. When applied to three

EVAPORATION

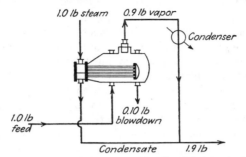

(a) SINGLE EFFECT-1 LB STEAM PRODUCES 1+0.9 LB DISTILLED WATER

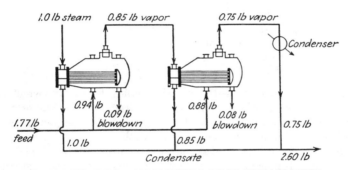

(b) DOUBLE EFFECT-1 LB STEAM PRODUCES 1+1.60 LB DISTILLED WATER

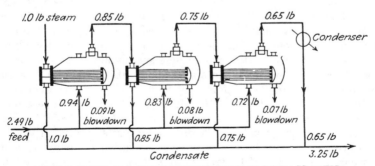

(c) TRIPLE EFFECT-1 LB STEAM PRODUCES 1+2.25 LB DISTILLED WATER

FIG. 14.8. Multiple-effect evaporation with parallel feed.

effects as shown in Fig. 14.8c, it is a *triple-effect* evaporator and the original pound of steam produces about 2.25 lb of pure water.

In order to maintain temperature differences for heat transfer between the vapor from one effect and the boiling liquid of the next effect, the pressure on each succeeding evaporator must be lower than its predecessor. The quantity of blowdown, which is arrived at from experience with water of different impurities, carries with it considerable sensible heat and thereby reduces the amount of vaporization which can be realized in succeeding effects. Due in part to this escape of heat from the system there is a limit to the number of effects which are justifiable. The fixed charges for additional effects ultimately dissipate the savings in energy which result from the use of a large number of effects.

Power-plant Evaporation Processes. Power-plant evaporation processes fall into four classes:

1. *Make-up* evaporators for boiler feed
2. *Process* evaporators for the production of purified water
3. *Heat-transformer* evaporators
4. *Salt-water distillers*

These processes are discussed below with typical flow sheets which contain all the necessary information computed from simple heat balances. Since only enthalpy differences are involved, it has not been felt necessary to include the individual heat balances for each case.

1. *Make-up Evaporators.* Make-up evaporators supply boiler feed water to replace leakage and losses from the system as process steam in plants or as discarded condensate. This is by far the largest class of evaporation process and is usually accomplished in a single-effect evaporator, although occasionally a double-effect evaporator may be used depending upon the characteristics of the condensate cycle in the power plant and the amount of make-up required. There is hardly a power plant built today which does not include such equipment. The evaporators themselves are small, containing about 100 to 1000 ft^2 of surface. Typical process examples of the use of single-effect make-up evaporators are shown in Figs. 14.9 and 14.10. In Fig. 14.9 a turbine operating on 150,000 lb/hr of steam at 400 psig and superheated to 800°F is bled in three stages to supply sufficient high-temperature superheated steam so that the make-up can be heated to the saturation temperature of the boiler corresponding to 400 psig. Naturally saturated steam at 400 psig cannot be used to heat feed up to its saturation temperature. It can be shown by an economic balance that it would be wasteful to use steam directly from the superheater at 800°F for feed-water heating in preference to the use of partially spent superheated steam from the turbine.

Bleed steam from the eighth stage of the turbine is run directly to the evaporator where it vaporizes 9000 lb/hr of make-up. The vaporized make-up and evaporator steam are both condensed in an evaporator-condenser by the bulk of boiler feed with a 5°F temperature approach. The remainder of the diagram is obtained by trial-and-error calculation for the optimum efficiency of the cycle.

A variation of this flow arrangement is shown in Fig. 14.10 in which some of the steam formed in the evaporator is combined with the make-up

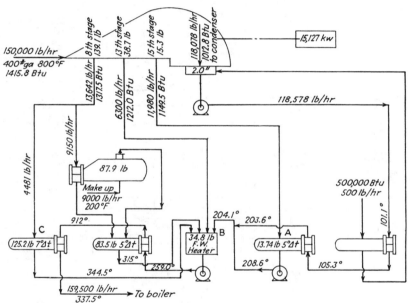

FIG. 14.9. Single-effect make-up evaporator with evaporator-condenser between two bleed heaters.

in a preheater. The remainder of the steam formed in the evaporator goes to an evaporator-condenser, where it is condensed by the boiler feed. The condensate from the evaporator tubes and the evaporator-condenser is then combined in the flash tank. In the flash tank the condensate from the evaporator comes out as superheated (339.4°F) liquid compared with the condensate from the evaporator-condenser (313.3°F), and a portion flashes back to the evaporator-condenser, thereby providing all the pressure differences necessary for the operation of the equipment. In Fig. 14.11 a triple-effect evaporator is shown for cases where an excessively large amount of make-up water is required.

388　　　　　　　　　*PROCESS HEAT TRANSFER*

The location of the evaporator-condenser is important. In the modern power plant the evaporator is heated by turbine bleed steam and the vapor produced discharges either into a stage heater on the next bleed point below or into an evaporator-condenser located between the two stage heaters A and C (see Fig. 14.9) where B is a feed-water heater.

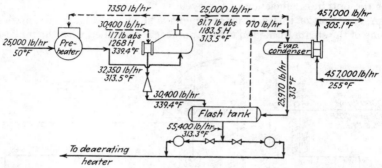

Fig. 14.10. Single-effect make-up evaporator with preheater, evaporator-condenser, and flash tank.

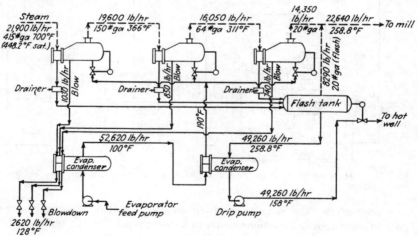

Fig. 14.11. Triple-effect make-up evaporator.

The latter arrangement is generally the most economical particularly if the vapor condensed is on the suction side of the boiler feed pump, *i.e.*, the low-pressure end of the bleed cycle. If there is sufficient temperature difference between the two stages in question, A and C, the evaporator will receive steam from the same bleed point from which the higher stage heater C receives steam. The evaporator-condenser is then located between that particular stage heater and the next lower one B. This arrangement permits taking all the heat for heating feed water in the evaporator-condenser and for the higher of the two feed heaters C, from the high bleed point without displacing the bleed steam from the next lower bleed point. It can readily be seen that, if the evaporator discharges into the next lower bleed heater, the heat absorbed in the next lower heater will come from the next higher bleed point and will have by-passed several stages of the turbine in so doing, resulting in a loss of kilowatts.

2. *Process Evaporators.* There are a number of industries which continuously require large quantities of distilled water. This type of plant employs double, triple, or quadruple effects and receives heat either

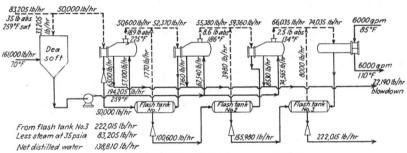

Fig. 14.12. Triple-effect process evaporator.

from a bleed point or from the boiler directly. The selection of the number of effects is closely allied to the relationship among the fixed charges and the steam operating cost. Multiple-effect evaporators with parallel feed need not have all effects simultaneously in operation and can therefore be adjusted if the demand for distilled water varies. Evaporators for this type of service are generally of medium size, about 500 to 2000 ft^2 per shell. A typical triple-effect process evaporator is shown in Fig. 14.12, where 83,205 lb/hr of steam at 35 psia (20 lb gage) and saturated is split to mix directly with the cold feed at 70°F and for vaporization in the evaporator. This process employs several of the elements discussed under make-up evaporators. The final product is 222,015 lb/hr of distilled treated water. In order to obtain a maximum quantity of vaporization from a given quantity of steam initially near atmospheric pressure, temperature differences are maintained from the first to the last effect by operating with the last effect under vacuum. This is accomplished by using a steam jet vacuum pump on the last effect. Thus the shell side of the first effect in Fig. 14.12 operates at 18.9 psia or 225°F, the second at 8.6 psia or 186°F, and the third at

2.5 psia or 134°F. This establishes differences in the effects of 34, 39, and 52°F. The operating pressures for each of the effects is determined by trial and error, so that all three effects will have the same surface as calculated by $A = Q/U_D \Delta t$. This procedure will be demonstrated in detail in the treatment of a chemical evaporator. If the vacuum were not applied, the maximum available temperature difference over the three effects would be from 259 to 212°F or 47 instead of 125°F as shown in the flow sheet.

The principle of vacuum evaporation is extensively used in chemical evaporation. Since the vapor from the last effect is at a low temperature, it has little value in the preheating of the feed, its temperature in this case being 134°F. The evaporator-condenser, therefore, operates with cooling-tower water instead of feed water. It will be noted that, although

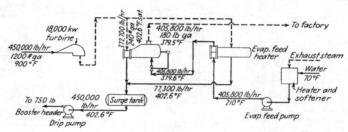

Fig. 14.13. Heat-transformer evaporator.

the heat from the last-effect vapor is rejected, the blowdown heat losses of such a system are reduced considerably by being at a low heat level. The steam required for the vacuum pump must also be taken into account in computing the efficiencies of vacuum processes.

3. *Heat-transformer Evaporators.* The heat-transformer evaporator is a single-effect system of one or more shells *in parallel* receiving steam from the exhaust of a high-pressure turbine or high-pressure engine. Flow sheets are shown in Figs. 14.13 and 14.14. The purpose of this type of evaporator is to condense steam from a high-pressure boiler which has passed through a high-pressure turbine and into the evaporator. The condensate is then returned directly to the high-pressure boiler by a pressure booster, thereby keeping the high-pressure circuit closed and continuously supplied with high-pressure boiler water and steam. Obviously high-pressure boiler and turbine installations are favorably affected by this circuit. By the condensation of the exhaust steam from the high-pressure turbine or engine the heat transfer in the evaporator is used to produce large quantities of process steam, all or a large part of which is never returned to the evaporator system. If the condensate is not

EVAPORATION 391

returned, it is because it may be difficult to collect or the vapor may be consumed in a chemical or heating process or it may be continuously contaminated.

This type of evaporator is relatively large, having been built in a single unit in sizes up to 11,000 ft^2 of surface and capable of producing 150,000 to 200,000 lb/hr of steam. Single evaporators of this size are 10 to 11 ft

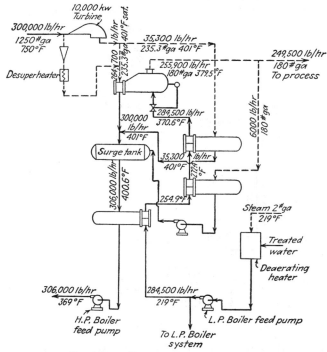

FIG. 14.14. Heat-transformer evaporator.

in diameter and 40 to 50 ft in length. There are not many of the larger installations, but where large quantities of process or heating steam are required, it is readily understood that a high-pressure boiler operating at 1400 psi throttle pressure and exhausting at 200 psi produces large quantities of "by-product" power. This power is charged only with the heat extracted on the passage of the steam through the turbine, the remaining charges being debited to the cost of producing high-pressure steam, the primary object of the steam cycle. When the energy from a high-pressure line is continuously used to produce low-pressure steam for purposes outside the original installation, it is called a *heat transformer* from its similarity to a stepdown electrical transformer and also a *reducing-valve* transformer. It is actually the only way in which high-pressure saturated steam can be converted to low-pressure saturated steam without superheat. The reason for the reducing-valve evaporator, however, is primarily to preserve the pressure on the high-pressure side.

Sometimes the heat going to process must be transported a considerable distance. To prevent condensation the vapor should be somewhat superheated before leaving the generating system, although the heat-transformer evaporator produces only saturated steam. For this purpose

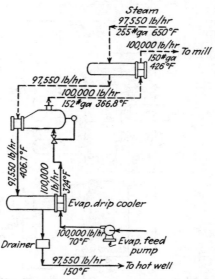

FIG. 14.15. Heat-transformer evaporator supplying superheat.

the arrangement shown in Fig. 14.15 is generally used to provide a heat exchanger or *reheater* which passes the steam to the heat-transformer evaporator through the tubes and the vapor from the evaporator through the shell, thereby providing some superheat.

4. *Salt-water Distillers.* A pound of fuel can normally yield about 10 lb of steam, and in a double-effect evaporator for use with salt water 10 lb of steam will yield a total of 18.5 lb of pure water. It is little wonder, then, that ships at sea commonly produce their own water requirements from sea water. Sea water contains about 3 per cent solids by weight corresponding to about 34,000 ppm compared with 340 ppm in fresh water. Instead of about 90 per cent vaporization it is customary to vaporize only about one-third of the feed. The remainder, containing

about 5 per cent solids or 51,000 ppm, is dumped overboard. Because of the largeness of the blowdown the use of a vacuum system with low evaporation temperatures is desirable, since low temperatures are also favorable to low rates of scaling. Unlike evaporators in stationary power

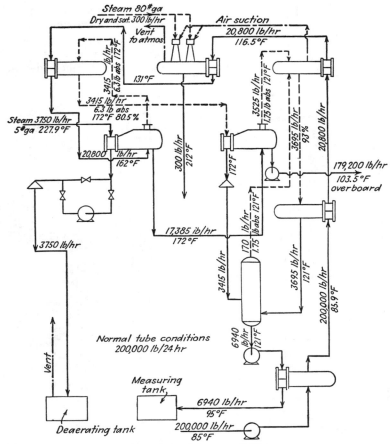

Fig. 14.16. Salt-water distiller.

plants salt-water evaporators or distillers operate for 600 to 700 hr without descaling. The flow sheet of a typical salt-water distiller is shown in Fig. 14.16.

Vacuum Operation of a Process. Evaporators frequently operate with the last effect under vacuum, and one of the important considerations is

394   PROCESS HEAT TRANSFER

the method of continuously maintaining the vacuum. The use of a mechanical compressor on the last stage is usually prohibitive because of the energy required for compression. The specific volume of water vapor at 2 in. Hg abs is 399.2 ft³/lb. Furthermore, there is little reason to operate a mechanical compressor when the required reduction in volume can be partially achieved by condensation. The latter is one of the principles embodied in the

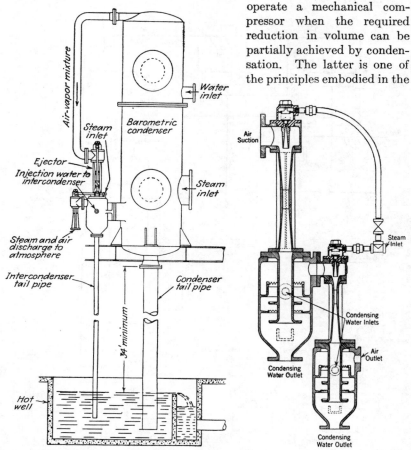

FIG. 14.17. Two-stage ejector with jet intercondenser, serving a barometric condenser. (*Foster Wheeler Corporation.*)

FIG. 14.18a. Two-stage ejector with jet inter- and aftercondensers. (*Foster Wheeler Corporation.*)

barometric condenser with air ejectors, an example of which is shown in Fig. 14.17. It is operated by the two steam jet ejectors.

A single ejector connected to a condenser is capable of maintaining a vacuum of about 26.5 in. Hg abs and can be made with several jets replacing the single-motive steam nozzle. This provides a more uniform distribution of steam in the mixing zone. When a vacuum of 26.5 to 29.3 in. is desired, it can be accomplished by means of a two-stage ejector as shown in Fig. 14.18a and b. For higher vacuums the use of a three-stage ejector is required. The two-stage detail in Fig. 14.18a is the same as that employed in Fig. 14.17. The condenser in Fig. 14.17 is a *barometric condenser* equipped with a water inlet and distribution plates, so that the water entering the inlet cascades or sprays over the incoming

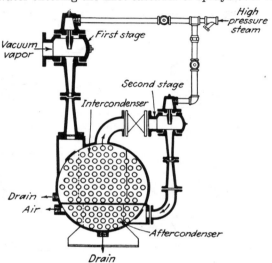

FIG. 14.18b. Two-stage ejector with surface inter- and aftercondensers. (*The Lummus Company.*)

stream from the evaporator or precondenser and removes a large part of the steam from the process as condensate. The remaining air with less steam flows overhead into the first stage ejector. After compression in the first stage the partial pressure of the steam will have been increased and most of the remaining vapor can be condensed by another direct contact with cooling water.

Again referring to Fig. 14.17, in order to remove the water and condensate from the assembly without losing vacuum it is necessary that a leg of liquid be maintained with a hydrostatic head $z\rho$ equal to the difference between the vacuum and atmospheric pressure, where z is the height and ρ the density. In this manner the upper surface of the liquid in the tail pipe is at a pressure corresponding to the vacuum and the liquid at the bottom of the tail pipe is at atmospheric pressure due to the weight of the hydrostatic head. Thus liquid under vacuum continuously enters the tail pipe, and liquid at atmospheric pressure continually leaves by way of the hot well at the bottom of the tail pipe. Atmospheric pressure corresponds to a hydrostatic head of 34 ft of water, and complete vacuum corresponds to zero hydrostatic head. To maintain a process at substantially complete vacuum requires that a leg of 34 ft of water be maintained between the barometric condenser and the hot well. If a vacuum of less than 29.92 in. Hg is to be maintained by the ejectors but a leg of 34 ft equivalent to 29.92 in. Hg has been provided, it means merely that the height of liquid in the tail pipe will automatically drop

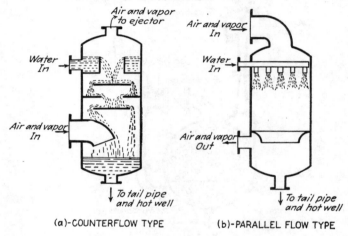

(a)-COUNTERFLOW TYPE   (b)-PARALLEL FLOW TYPE

FIG. 14.19. Barometric condensers.

to provide only the necessary hydrostatic difference between the operating vacuum of the ejector and the atmospheric pressure.

Barometric condensers are of two types, counterflow and parallel flow as shown in Fig. 14.19a and b. In counterflow types the temperature of the water at the liquid level may approach the vapor temperature more closely than in parallel-flow types. Counterflow types are preferable where water is at a premium or it is difficult to have the vapor enter from the top. If a pump is used to remove the tail liquid instead of a total barometric height, whatever head is supplied by the pump can be deducted from the total barometric height and the assembly is known as a *low-level* condenser.

The quantity of water required in the barometric condenser can be computed from

$$\text{Gpm} = \frac{Q}{500(T_s - t_w - t_a)} \tag{14.4}$$

where T_s = saturation temperature of the vapor, °F
 t_W = temperature of the water, °F
 t_a = degrees of approach, to T_s, °F

In counterflow barometric condensers t_a is taken as 5°F.

The two-stage ejector assembly in Fig. 14.18a will produce the same results as that of Fig. 14.18b. It differs only in that the condensation after each stage is accomplished by means of tubular surface instead of the direct contact of the cooling water with the steam or vapor mixture. The tubular intercondenser and aftercondenser are combined in a single shell, since the total surface required is usually quite small. Surface condensation is mandatory where the vacuum exhaust cannot be mixed with the cooling water for reasons of corrosion or chemical reaction.

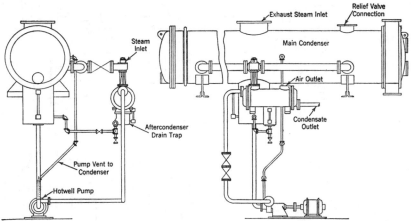

FIG. 14.20. Single-stage ejector with surface aftercondenser serving a surface condenser. (*Foster Wheeler Corporation.*)

In the operation of a steam turbine or engine the surface condenser discussed in Chap. 12 actually serves as the precondenser for the maintenance of vacuum across the turbine. The surface condenser yields not only condensate but also a mixture of air saturated with water vapor, which must be continuously removed. Failure to remove the air causes an increase in the pressure and temperature of the condenser, as well as a noncondensable air blanket which reduces the overall heat-transfer coefficient. The air removal can be accomplished as shown in Fig. 14.20 with a surface intercondenser and a removal pump. The hot-well pump is provided for low-level installation. A counterpart using a low-level barometric condenser can also be used. For the design and selection of ejectors Jackson[1] has presented a comprehensive discussion.

[1] Jackson, D. H., *Chem. Eng. Progress*, **44**, 347–352 (1948).

CHEMICAL EVAPORATION

Comparison between Power-plant and Chemical Evaporation. The purpose of the majority of power-plant evaporators is the separation of pure water from raw or treated water. The impurities are continuously withdrawn from the system as blowdown. In chemical industry the manufacture of heavy chemicals such as caustic soda, table salt, and sugar starts with dilute aqueous solutions from which large quantities of water must be removed before final crystallization can take place in suitable equipment. In the power-plant evaporator the unevaporated

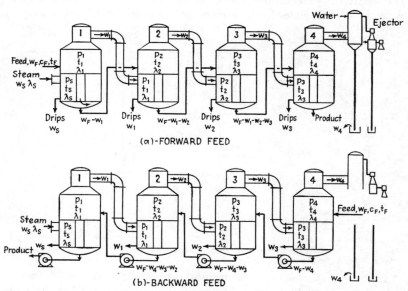

Fig. 14.21. Quadruple-effect chemical evaporator arranged for forward and backward feed.

portion of the feed is the residue, whereas in the chemical evaporator it is the product. This leads to the first of several differences between power-plant and chemical evaporation. These are as follows:

Absence of Blowdown. Chemical evaporators do not operate with blowdown, and instead of liquid being fed in parallel to each body it is usually fed to multiple-effect systems in series. Common methods of feeding are shown in Fig. 14.21a and b. The feed to the first effect is partially evaporated in it and partially in each of the succeeding effects. When the liquid feed flows in the same direction as the vapor, it is *forward feed*, and when fed in the reverse direction, it is *backward feed*. From the standpoint of effectively using the temperature potentials forward feed is preferable. If the liquid is very viscous, there is an advantage to the use of backward feed, since the temperature of the first effect is always the greatest and the corresponding viscosity will be less. The advantages and disadvantages of both will be discussed later. The absence of blowdown enables greater heat recovery in a chemical evaporator.

Boiling Point Rise (BPR). Although chemical evaporators are capable of high heat efficiencies, they are incapable under certain conditions of a high utilization of temperature potentials and consequently require greater surfaces. This is due to the fact that a concentrated aqueous solution undergoes a *boiling point rise* above the saturation temperature corresponding to pure water at the same pressure. Suppose steam enters the tubes or chest of a chemical evaporator at 45 psia and is to evaporate water from a caustic soda solution. The steam temperature is 274°F. If pure water is evaporated at 18 in. Hg, the temperature of the vapor formed would be 169°F. But because of the dissolved salt the liquor boils at 246°F at 18 in. Hg instead of at 169°F. The temperature difference across the heat-transfer surface is only $274 - 246 = 28°F$, and the difference of $246 - 169 = 77°F$ represents lost potential which cannot be attained owing to the presence of the dissolved material. The difference between the temperature of heating vapor and the saturation temperature corresponding to the pressure of the evaporating vapor is the *apparent temperature drop* $(\Delta t)_a$, or $274 - 169 = 105°F$ in the example above. Heat-transfer coefficients, which are reported on a basis of

$$U_D = \frac{Q}{A(\Delta t)_a},$$

are *apparent overall coefficients.* If coefficients are based on the temperature difference across the heating surface between the heating vapor and the evaporating liquid, as in most cases, $U_D = Q/A \ \Delta t$, where $\Delta t = 28°F$ in the example above. If solutions have boiling point rises above about 5°F, the latent heat of vaporization of steam from the solution differs from that obtained from the steam tables (Table 7) at the saturation pressure of the vapor. The latent heat of vaporization for steam from a solution can be computed either from Duhring's relationship or from the equation of Othmer.[1] According to Duhring's rule,

$$\frac{\lambda_w}{\lambda_s} = \frac{\Delta t'_w}{\Delta t'_s} \left(\frac{t'_w}{t'_s}\right)^2_{\text{abs.}} \tag{14.5}$$

[1] Othmer, D. F., *Ind. Eng. Chem.*, **32**, 841–856 (1940).

where λ_s = latent heat of 1 lb of pure water from solution at the temperature t and pressure p_s
λ_W = latent heat of 1 lb of pure water at the temperature t'_W but at p_s, the same pressure as t
t'_s, t'_W = boiling points of the solution and water at the same pressure, p_s, °R
$\Delta t'_s/\Delta t'_W$ = rate of change of the two boiling-point curves over the same pressure range

According to Othmer's method and based on the Clausius-Clapeyron equation,

$$\frac{\lambda_s}{\lambda_W} = \frac{d \log p_s}{d \log p_W} \qquad (14.6)$$

where p_s and p_W are the respective absolute vapor pressures of the solution and pure water over an identical range of temperatures.

The BPR can be computed but only for dilute solutions which are relatively ideal. For real solutions the data on boiling-point elevation must be obtained experimentally by measuring the vapor-pressure curve for a given concentration at two different temperatures. Additional determinations can be made for other concentrations if more than a single effect is used.

Fluid Properties. In the power-plant evaporator the water-softening process is modified in different localities so that the evaporator feed composition causes a minimum of foaming and other operating difficulties. In the chemical evaporator the residue, a concentrated solution, is the desired product, and usually no adjustment can be made to the solution to prevent foaming or to eliminate the deposition of scale. This must be taken into consideration in the design of the equipment. Furthermore, concentrated solutions, as discussed in Chap. 7, produce liquors of high viscosity. Particularly since boiling is a combination of vaporization and free convection the overall coefficient of heat transfer is a function of both the concentration and the temperature at which evaporation occurs. The influence of viscosity may be so great that a negligible Grashof group, $D^3\rho^2g\beta\,\Delta t/\mu^2$, results for evaporators operating with natural circulation. Under these circumstances the liquor cannot be relied upon to circulate very rapidly about the heating element and it is necessary to use *forced circulation* instead of *natural circulation* as presumed heretofore.

CHEMICAL EVAPORATORS

Chemical evaporators fall into two classes: natural circulation and forced circulation. Natural-circulation evaporators are used singly or in multiple effect for the simpler evaporation requirements. Forced-circulation evaporators are used for viscous, salting, and scale-forming solutions. Natural-circulation evaporators fall into four main classes:

1. Horizontal tube
2. Calandria vertical tube
3. Basket vertical tube
4. Long tube vertical

The discussion of evaporator design in this chapter deals only with those which are designed on a basis of flux and accepted overall coefficients. Those employing film coefficients are treated in the next chapter.

Horizontal-tube Evaporators. Horizontal-tube evaporators as shown in Fig. 14.22 are the oldest type of chemical evaporator. Although they once enjoyed widespread use, they have given way to other types. They consist of a round or square shell and a horizontal tube bundle which is usually square. They do not take very good advantage of the thermal currents induced by heating and therefore are not so acceptable as the types which have replaced them. The horizontal evaporator is the only distinct type of chemical evaporator employing steam in the tubes. The principal advantages of horizontal evaporators lie in the relatively small headroom they require and the ability to arrange the bundle so that air brought in with the steam does not collect and blanket useful surface.

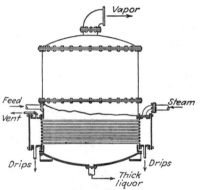

Fig. 14.22. Horizontal-tube evaporator. (*Swenson Evaporator Company.*)

The horizontal evaporator is least satisfactory for fluids which form scale or deposit salt, the deposit being on the outside of the tube, and it is therefore used only for relatively simple problems of concentration rather than for the preparation of a liquid for ultimate crystallization. It is well suited to processes in which the final product is a liquor instead of a solid, such as industrial sugar sirups, where the large volume of liquid stored in the evaporator can permit a close adjustment of the final density by changing the holdup in the evaporator. The tube length is determined by the size of the evaporator body itself. Because evaporation occurs on the outside of the tubes, eliminating a scale problem inside the tubes, the horizontal-tube evaporator uses smaller tubes than any other type, from $\frac{3}{4}$ to $1\frac{1}{4}$ in. OD.

Calandria-type Evaporators. The calandria evaporator is shown in Fig. 14.23. It consists of a short vertical-tube bundle, usually not more than 6'0" high, set between two fixed tube sheets which are bolted to the shell flanges. The steam flows outside the tubes in the steam chest, and there is a large circular space for downtake in the center of the bundle whereby the cooler liquid circulates back to the bottom of the tubes. The flow area of the downtake is from one-half the flow area of the tubes to an area equal to it. Tubes are large, up to 3 in. OD, to reduce the pressure drop and permit rapid circulation and are installed in tube sheets with ferrule-type packing. The layout of a typical calandria is shown in Fig. 14.24. One of the problems is to baffle the steam chest so that there is a relatively uniform tube coverage.

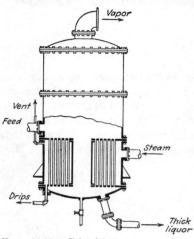

Fig. 14.23. Calandria-type evaporator. (*Swenson Evaporator Company.*)

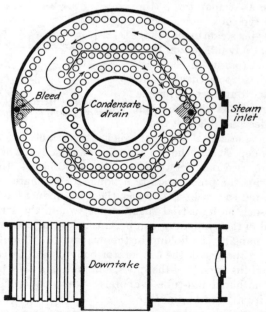

Fig. 14.24. Typical calandria baffling. Arrows indicate direction of steam flow. Shaded areas indicate location of noncondensable bleed points.

Another is to provide adequate bleed points so that no pockets of noncondensable gas develop. The condensate is removed at any convenient point. The space above the liquid level on the tube sheet serves primarily to disengage the liquid, which is carried along by the vapor. A common accessory of evaporators is a catchall which is installed on the vapor line for the purpose of removing entrained liquid and returning it to the bulk of the fluid. Two typical catchalls are shown in Fig. 14.25a and b. They operate on the principle of centrifugally removing the liquid droplets.

Calandria evaporators are so common they are often referred to as *standard* evaporators. Since scaling occurs inside the tubes, it is possible to use the standard evaporator for more rigorous services than the hori-

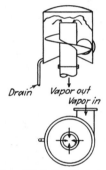   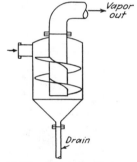

FIG. 14.25a. Catchall with bottom outlet.   FIG. 14.25b. Catchall with top outlet.

zontal tube evaporator and, in addition, a propeller can be installed in the dished or conical bottom to increase the rate of circulation.

Basket-type Evaporators. A basket-type evaporator is shown in Fig. 14.26. It is similar to a calandria evaporator except that it has a removable bundle which can be cleaned quite readily. The bundle is supported on internal brackets, and the downtake occurs between the bundle and the shell instead of in a central downtake. Because the tube sheets hang freely, the problem of differential expansion between the tubes and the steam chest shell is not important. This type frequently is designed with a conical bottom and may also have a propeller installed to increase circulation. As a result of these mechanical advantages the basket evaporator can be used for liquors which have a tendency to scale, although they are not recommended for liquids with high viscosities or great rates of scaling. The selection of a basket or calandria evaporator usually follows the established policies of the different industries in which they are used after years of experience, or modifications suggested by manufacturers. Some manufacturers have preferences for the one type in a certain application, whereas another will prefer the second type for the same service.

Long-tube Vertical Evaporators. A long-tube vertical evaporator is shown in Fig. 14.27. It consists of a long tubular heating element designed for the passage of the liquor through the tubes but once by natural circulation. The steam enters through a vapor belt as dis-

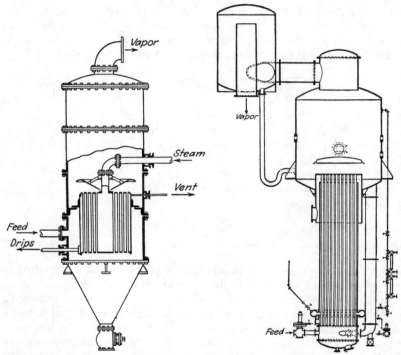

Fig. 14.26. Basket-type evaporator. (*Swenson Evaporator Company.*) Fig. 14.27. Long-tube vertical evaporator. (*General American Transportation Company.*)

cussed in Chap. 12, and the bundle is baffled so that there is a free movement of steam, condensate, and noncondensable downward. The upper tube sheet is free, and just above it there is a vapor deflector to reduce the entrainment. This type of evaporator is not especially adapted to scaling or salting liquors, but it is excellent for the handling of foamy, frothy liquors. The velocity of the vapor issuing from the tubes is greater than in the short-tube vertical types. Tubes are usually $1\frac{1}{4}$ to 2 in. OD and from 12 to 24 ft in length. When arranged for recirculation the apparatus will be as shown in Fig. 14.28. In this type, disengagement occurs outside the evaporator body. The calculation of this type of evaporator for aqueous solutions with known physical and thermal properties will be discussed in Chap. 15.

Forced-circulation Evaporators. Forced-circulation evaporators are made in a variety of arrangements as shown in Figs. 14.29 through 14.31. Forced-circulation evaporators may not be so economical to operate as natural-circulation evaporators, but they are necessary where the concen-

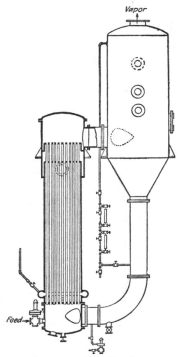
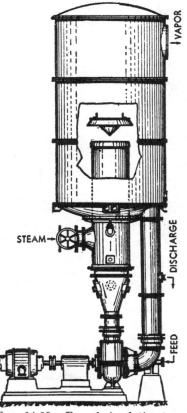

Fig. 14.28. Long-tube recirculation evaporator. (*General American Transportation Company.*)

Fig. 14.29. Forced-circulation-type evaporator with inside vertical heating element. (*Swenson Evaporator Company.*)

tration problem involves a solution with poor flow, scale, and thermal characteristics. Since the Grashof group varies inversely with the square of the viscosity, there is a limit to the viscosities of solutions which will naturally recirculate. With very viscous materials there is no alternative but to use this type of evaporator. Also, where there is a tendency to

406 PROCESS HEAT TRANSFER

form scale or deposit salts, the high velocities obtainable by the use of circulating pumps are the only means of preventing the formation of excessive deposits. Forced-circulation evaporators are well adapted to a close control of flow, particularly when a long time of contact may be injurious to the chemical in solution. Tubes for forced-circulation evaporators are smaller than in natural-circulation types, usually not exceeding 2 in. OD.

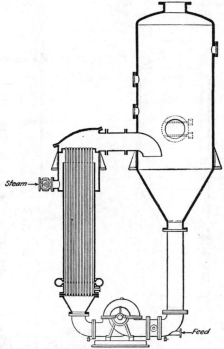

FIG. 14.30. Forced-circulation evaporator with vertical external element. (*General American Transportation Corporation.*)

In Fig. 14.27 the steam enters the bundle outside the evaporator body and contacts the tubes at the top of the bundle by means of the annular space provided. A deflector plate is installed above the upper tube sheet, and the circulating pump is installed at ground level. In Fig. 14.30 the same effect is produced by means of a vertical external bundle, which simplifies construction to a degree but which is not so compact. Figure 14.31 is a variation with a horizontal bundle which is particularly adaptable where the headroom is low.

EVAPORATION

Effect of Hydrostatic Head. Consider a pure fluid with a boiling surface somewhat above the top of a bundle of horizontal tubes. The boiling point is regarded as being set by the pressure at the liquid-vapor interface. If there is a great layer of liquid above the tube bundle, it will exert a hydrostatic pressure upon the liquid in contact with the tube surface. The added pressure upon the liquid raises the boiling temperature

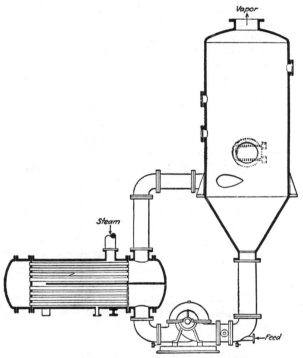

FIG. 14.31. Forced-circulation evaporator with horizontal external element. (*General American Transportation Corporation.*)

at the heat-transfer surface above that necessary to produce vapors of the saturation temperature corresponding to the pressure at the liquid-vapor interface. The effect of hydrostatic head, as in the case of BPR, reduces the useful temperature difference effective upon the heat-transfer surface. Since evaporators operate on fixed apparent temperature differences, the size of the heat-transfer surface must be increased accordingly for the presence of a hydrostatic head. The influence of temperature and pressure on the overall coefficient is essentially that shown in Fig. 14.7.

At high temperatures the liquid is less viscous and is more favorably suited to evaporation.

The effect of hydrostatic head may be estimated from

$$\Delta t_h = 0.03 \frac{T_R v}{\lambda_s} \Delta p \qquad (14.7)$$

where Δt_h = hydrostatic elevation of the boiling point, °F
T_R = solution boiling temperature, °R
v = specific volume of water vapor at T_R ft³/lb
λ_s = latent heat of vaporization corresponding to the saturation pressure, Btu/lb
Δp = hydrostatic head, ft

Usually Δp may be taken as corresponding to one-half the indicated liquid level. It is apparent that the influence of hydrostatic pressure will be greater as the vacuum upon the system is increased, since v varies considerably with the pressure whereas λ_s varies but little.

For all evaporators operating with natural circulation a loss in available capacity due to hydrostatic head cannot be avoided but the loss can be reduced by maintaining the lowest liquid levels consistent with the efficient operation of the equipment. If the froth is regulated to 10 in. above the upper tube sheet of vertical heating elements, good operating control can usually be effected. The achievement of good operating control is facilitated by the choice of the tube diameter and length and in general by designing for a high fluid velocity into the disengaging space. Problems of arrangement for natural circulation will be treated in Chap. 15.

Multiple-effect Chemical Evaporation. In the study of parallel-feed multiple-effect power-plant evaporators it was shown that in a triple-effect evaporator 1 lb of steam evaporated approximately 2.25 lb of water. The use of parallel feed is by no means the most economical and is used in chemical evaporation only when the feed solution is nearly saturated to begin with and evaporation is intended for the purpose of supersaturation. In chemical evaporation it is customary to employ forward feed, backward feed, or a modification of both, known as *mixed* feed. Returning to Fig. 14.21 there are certain opposing advantages and disadvantages resulting from the use of either forward or backward feed.

In forward feed if the feed liquor is at a higher temperature than the saturation temperature of the first effect, some evaporation will occur automatically as *vapor flashing*. Since a vacuum is usually maintained on the last effect, the liquor flows by itself from effect to effect and a liquor removal pump is required only on the last effect. Similarly, since the saturation temperature of the boiling solution in each effect is lower than the temperature of the effect preceding it, there is flashing or "free" evaporation in each succeeding effect, which reduces the overall steam requirement. In an evaporator the boiling film is the controlling resistance and the numerical value of the overall coefficient decreases with concentration because the viscosity is increasing. In forward feed the concentrated liquor is in the last effect, and obviously that effect has the lowest overall coefficient, since the liquor is most concentrated there and at the same time coldest.

When backward feed is employed, it overcomes the objection of having the most concentrated liquor in the coldest effect. Here the dilute liquor enters the last and coldest effect and leaves concentrated in the first effect, which is at the highest temperature. In this feed arrangement liquor must be heated in each effect as compared with solution flashing in each effect in forward feed. Furthermore, the feed must be pumped from effect to effect, which means that the number of places for air leakage, such as the pump and flanges, increases the maintenance and power cost. The temperature relations in backward feed usually offset these disadvantages in part, since the system is in counterflow and the economy of steam is greatest under these conditions.

If the feed liquor to a backward-feed evaporator is initially hot, its introduction into the last effect is wasteful, since the vapors which flash off in the last effect are lost to the condenser. In forward feed not only would these vapors flash but in each succeeding effect they would reevaporate additional water. The problem of the direction of feed to be employed is, as in most alternate problems of heat transfer, an economic one. Backward feed may or may not lead to smaller surface requirements, depending upon the extent of concentration and the viscosity of the desired final solution. The steam cost will be less for backward feed if the feed is cold and less for forward feed if the feed liquor is at approximately the operating temperature of the first effect or higher. The computation of problems by both methods will readily establish the most favorable operating relationship.

THE CALCULATION OF CHEMICAL EVAPORATORS

BY JOSEPH MEISLER[1]

Referring to Fig. 14.21a the surface and steam requirements for multiple-effect chemical evaporation can be computed by imposing a heat balance across each effect individually and a material balance

[1] The Air Reduction Company, Inc., and the Polytechnic Institute of Brooklyn.

over the whole system. The following nomenclature will be employed for a quadruple effect:

c_F = specific heat of feed, Btu/(lb)(°F)
t_F = temperature of feed, °F
w_F = feed, lb/hr
T_S = saturation temperature of steam to first effect, °F
W_S = steam to first effect, lb/hr
w_{1-4} = total water removed by evaporation, lb/hr
c_1, c_2, c_3, c_4 = specific heat of liquor in effects 1 to 4, Btu/(lb)(°F)
$t_1, t_2, t_3, t_4,$ = boiling points of liquor in effects 1 to 4, °F
w_1, w_2, w_3, w_4 = water removed in effects 1 to 4, lb/hr

Assume that there are no chemical heat effects as a result of the concentration (*i.e.*, negative heats of solution) and that there is no BPR.

Forward feed

Heat balance on first effect:

$$W_S \lambda_S + w_F c_F (t_F - t_1) = w_1 \lambda_1 \qquad (14.8)$$

Heat balance on second effect:

$$w_1 \lambda_1 + (w_F - w_1) c_1 (t_1 - t_2) = w_2 \lambda_2 \qquad (14.9)$$

Heat balance on third effect:

$$w_2 \lambda_2 + (w_F - w_1 - w_2) c_2 (t_2 - t_3) = w_3 \lambda_3 \qquad (14.10)$$

Heat balance on fourth effect:

$$w_3 \lambda_3 + (w_F - w_1 - w_2 - w_3) c_3 (t_3 - t_4) = w_4 \lambda_4 \qquad (14.11)$$

Material balance:

$$w_{1-4} = w_1 + w_2 + w_3 + w_4 \qquad (14.12)$$

The surface requirements will be

$$A_1 = \frac{Q}{U_D \Delta t} = \frac{W_S \lambda_S}{U_1 (T_S - t_1)} \qquad (14.13)$$

$$A_2 = \frac{w_1 \lambda_1}{U_2 (t_1 - t_2)}$$

$$A_3 = \frac{w_2 \lambda_2}{U_3 (t_2 - t_3)}$$

$$A_4 = \frac{w_3 \lambda_3}{U_4 (t_3 - t_4)}$$

EVAPORATION

Let
$$A_1 = A_2 = A_3 = A_4 \tag{14.14}$$

where U_1, U_2, U_3, and U_4 are the design overall coefficients in the respective effects. From the material balance and the heat balances there are five equations and five unknowns: W_S, w_1, w_2, w_3, and w_4. These may be solved for simultaneously.

Backward feed. Referring to Figure 14.21*b*:
Heat balance on fourth effect:

$$w_3\lambda_3 + w_F c_F(t_F - t_4) = w_4\lambda_4 \tag{14.15}$$

Heat balance on third effect:

$$w_2\lambda_2 + (w_F - w_4)c_4(t_3 - t_4) = w_3\lambda_3 \tag{14.16}$$

Heat balance on second effect:

$$w_1\lambda_1 + (w_F - w_4 - w_3)c_3(t_2 - t_3) = w_2\lambda_2 \tag{14.17}$$

Heat balance on first effect:

$$W_S\lambda_S + (w_F - w_4 - w_3 - w_2)c_2(t_1 - t_2) = w_1\lambda_1 \tag{14.18}$$

Material balance:

$$w_{1-4} = w_1 + w_2 + w_3 + w_4 \tag{14.19}$$

The surface relations will be the same as before, since it is practical to impose the restriction that the surface in each of the bodies be identical. It has also been shown by experience that under this condition the pressure differences between effects will be approximately equal. If steam enters the first effect of a quadruple-effect evaporator at atmospheric pressure and the last effect is at 26 in. Hg vacuum corresponding to 1.95 psia, the pressure difference between the steam and the first effect and from effect to effect will be $(14.7 - 1.95)/5$. This will enable selection of the saturation pressures in the individual effects. Since the heat-transfer coefficients will be different in the individual effects, it may be found that the surfaces defined by Eqs. (14.13) and (14.14) are unequal. This means that owing to the inequality in the overall coefficient in the different effects the Δt across each effect does not correspond to the assumption of an equal division of the total pressure differential. This will be particularly true when the overall coefficients in the different effects differ greatly or when there is considerable flashing in the first effect. To equalize the surface in each body the temperature differences in the individual effects can be adjusted so that a larger temperature difference will be employed in the effect having the lowest heat transfer coefficient, the heat loads in all effects remaining nearly equal.

Multiple-effect evaporators may be designed for minimum surface or minimum initial cost. These cases have been treated by Bonilla.[1] The design of multiple-effect evaporators for optimum conditions, however, is more the exception in industry than the rule, the trend being in the direction of standardization.

Example 14.2. Calculation of a Triple-effect Forward-feed Evaporator. It is desired to concentrate 50,000 lb/hr of a chemical solution at 100°F and 10.0 per cent solids to a product which contains 50 per cent solids. Steam is available at 12 psig, and the last effect of a triple-effect evaporator with equal heat-transfer surfaces in each effect will be assumed to operate at a vacuum of 26.0 in. Hg referred to a 30-in. barometer. Water is available at 85°F for use in a barometric condenser.

Assume a negligible BPR, an average specific heat of 1.0 in all effects, the condensate from each effect leaves at its saturation temperature and that there are negligible radiation losses. Calculate: (a) Steam consumption, (b) heating surface required for each body, (c) condenser water requirement. The accepted overall coefficients of heat transfer for the different effects will be $U_1 = 600$, $U_2 = 250$, and $U_3 = 125$ Btu/(hr)(ft^2)(°F),

Solution:

Total feed $w_F = 50,000$ lb/hr
Total solids in feed $= 0.10 \times 50,000 = 5000$ lb/hr

Total product $= \dfrac{5000}{0.50} = 10,000$ lb/hr

Total evaporation, $w_{1-3} = 50,000 - 10,000 = 40,000$ lb/hr
$c_F = 1.0$.

The balances applying to this problem are

First effect: $W_S \lambda_S + w_F(t_F - t_1) = w_1 \lambda_1$
Second effect: $w_1 \lambda_1 + (w_F - w_1)(t_1 - t_2) = w_2 \lambda_2$
Third effect: $w_2 \lambda_2 + (w_F - w_1 - w_2)(t_2 - t_3) = w_3 \lambda_3$
Material: $w_1 + w_2 + w_3 = w_{1-3}$
$t_F = 100°F$
T_S at 12 psig $= 244°F$
T_3 at 26 in. Hg (1.95 psia) $= 125°F$
Total temperature difference $= 119°F$

When a forward-feed multiple-effect evaporator employs equal surfaces in each effect, as noted above, experience indicates that the differences in the pressures between effects will be nearly equal. This will rarely be entirely true, but it forms an excellent starting point for the calculation of the pressures in the effects. Any discrepancies can be adjusted later.

Average pressure difference $= \dfrac{26.70 - 1.95}{3} = 8.25$ psi/effect

[1] Bonilla, C. F., *Trans.AIChE*, **41**, 529–537 (1945).

EVAPORATION

BREAKUP OF THE TOTAL PRESSURE DIFFERENCE

	Pressure, psia	ΔP, psi	Steam or vapor, °F	λ, Btu/lb
Steam chest, 1st effect	26.70	....	$T_S = 244$	$\lambda_S = 949$
Steam chest, 2d effect	18.45	8.25	$t_1 = 224$	$\lambda_1 = 961$
Steam chest, 3d effect	10.20 (20.7 in. Hg)	8.25	$t_2 = 194$	$\lambda_2 = 981$
Vapor to condenser	1.95 (26 in. Hg)	8.25	$t_3 = 125$	$\lambda_3 = 1022$

$$949 W_S + 50{,}000(100 - 224) = 961 w_1$$
$$961 w_1 + (50{,}000 - w_1)(224 - 194) = 981 w_2$$
$$981 w_2 + (50{,}000 - w_1 - w_2)(194 - 125) = 1022 w_3$$
$$w_1 + w_2 + w_3 = 40{,}000$$

Solving simultaneously,

$$w_1 = 12{,}400$$
$$w_2 = 13{,}300$$
$$w_3 = 14{,}300$$
$$w_{1-3} = w_1 + w_2 + w_3 = \overline{40{,}000}$$
$$W_S = 19{,}100$$

$$A_1 = \frac{W_S \lambda_S}{U_1(T_S - t_1)} = \frac{19{,}100 \times 949}{600 \times 20} = 1510 \text{ ft}^2$$

$$A_2 = \frac{w_1 \lambda_1}{U_2(t_1 - t_2)} = \frac{12{,}400 \times 961}{250 \times 30} = 1590$$

$$A_3 = \frac{w_2 \lambda_2}{U_3(t_2 - t_3)} = \frac{13{,}300 \times 981}{125 \times 69} = 1510 \quad \text{(Use 1600 ft}^2\text{/effect)}$$

Heat to condenser $= w_3 \lambda_3 = 14{,}300 \times 1022 = 14{,}710{,}000$ Btu/hr
Water requirement $= 14{,}710{,}000/(120 - 85) = 420{,}000$ lb/hr or $420{,}000/500$
$$= 840 \text{ gpm}$$

Economy, lb evaporation/lb steam $= 40{,}000/19{,}100 = 2.09$ lb/lb

Remarks. During operation the equal pressure-drop distribution may not maintain itself. This will occur if there is undue scaling in one of the effects, if a body is gas bound, or if liquor levels are not properly maintained. Another factor may be the withdrawal of a large quantity of steam from one of the effects as a source of low-pressure heating steam. Any deviation from an equal pressure-drop distribution does not mean that the entire multiple-effect assembly will fail to operate but instead that the unit will assume a new pressure distribution and operate with a reduced capacity and steam economy.

Nonalgebraic Solution of Evaporators. It will presently be shown that the use of a simultaneous algebraic solution such as that above can scarcely be applied advantageously to more complicated arrangements or to a great number of effects. In industrial problems it has been found preferable and timesaving to perform the heat and material balances on the evaporator by directly assuming the value of W_S and solving each effect by a direct balance on itself instead of through the use of simultaneous equations. If the total evaporation based on the assumed value of W_S does not equal the required quantity, a new value of W_S is assumed and the calculation repeated. This method is demonstrated in Example 14.2a, in which the hint that $W_S = 19,100$ from Example 14.2 is used.

Example 14.2a. Solution of Example 14.2 by Assuming W_S

Heat balance

1. Assume steam to 1st effect $W_S = 19,100$ lb/hr, $w_F = 50,000$ lb/hr
   19,100 lb at 12 psig = 19,100 × 949 = 18,100,000 Btu/hr
   Deduct for heating feed = 50,000(224 − 100) = 6,200,000
   Available for evaporation = 11,900,000 Btu/hr
   λ_1 at 224°F = 961 Btu/lb, $w_1 = 11,900,000/961 = 12,400$ lb/hr
   Transfer to 2d effect = 50,000 − 12,400 = 37,600 lb/hr
2. Vapors from 1st effect = 11,900,000 Btu/hr
   Add flash, 37,600(224 − 194) = 1,130,000
   Available for evaporation = 13,030,000 Btu/hr
   λ_2 at 194°F = 981 Btu/lb, $w_2 = 13,030,000/981 = 13,300$ lb/hr
   Transfer to 3d effect = 37,600 − 13,300 = 24,300 lb/hr
3. Vapors from 2d effect = 13,030,000 Btu/hr
   Add flash, 24,300(194 − 125) = 1,680,000
   Available for evaporation = 14,710,000 Btu/hr
   λ_3 at 125°F = 1022 Btu/lb $w_3 = 14,710,000/1022 = 14,300$
   Product = 24,300 − 14,300 = 10,000 lb/hr
4. Heat to condenser = 14,710,000 Btu/hr

If the quantities failed to check a new value for W_S could be assumed.

As a first trial the steam could have been estimated in the absence of the hint obtained from Exercise 14.2 by the relationship

$$W_S = \frac{w_e}{0.75 \times \text{number of effects}} \qquad (14.20)$$

where w_e is the total pounds of evaporation. Equation (14.20) is based on feed entering at its boiling point. If the feed enters below its boiling point, the factor 0.75 must be reduced somewhat. With the feed at only 100°F and the boiling point at 125°F the value must be reduced to 0.70.

Example 14.3. Backward-feed Multiple-effect Evaporator. Conditions are the same as Example 14.2 except using backward feed with overall coefficients of $U_1 = 400$, $U_2 = 250$, and $U_3 = 175$ Btu/(hr)(ft²)(°F).

Solution. As before,

$$w_1 = 50,000 \text{ lb/hr} \qquad w_{1-3} = 40,000 \text{ lb/hr}$$

and

$$c_F = 1.0$$

The balances applying to this problem are
Third effect: $w_2\lambda_2 + w_F(t_F - t_3) = w_3\lambda_3$
Second effect: $w_1\lambda_1 + (w_F - w_3)(t_2 - t_3) = w_2\lambda_2$
First effect: $W_S\lambda_S + (w_F - w_3 - w_2)(t_1 - t_2) = w_1\lambda_1$

EVAPORATION

Material:
$$w_1 + w_2 + w_3 = w_{1-3}$$
$$981w_2 + 50{,}000(100 - 125) = 1022w_3$$
$$961w_1 + (50{,}000 - w_3)(125 - 194) = 981w_2$$
$$949W_S + (50{,}000 - w_3 - w_2)(194 - 224) = 961w_1$$
$$w_1 + w_2 + w_3 = 40{,}000$$
$$w_1 = 15{,}950$$
$$w_2 = 12{,}900$$
$$w_3 = 11{,}150$$
$$w_{1-3} = w_1 + w_2 + w_3 = \overline{40{,}000}$$
$$W_S = 16{,}950$$

$$A_1 = \frac{16{,}950 \times 949}{400 \times 20} = 2010 \text{ ft}^2$$

$$A_2 = \frac{15{,}950 \times 961}{250 \times 30} = 2040 \text{ ft}^2$$

$$A_3 = \frac{12{,}900 \times 981}{175 \times 69} = 1050 \text{ ft}^2$$

Since the heating surfaces for the three effects are far from equal, the temperature differences employed in the effects must therefore be modified to meet the conditions of the problem. For the first trial the average surface was

$$\frac{(2010 + 2040 + 1050)}{3} = 1700 \text{ ft}^2$$

With a better distribution of temperatures and pressures, however, less than

$$3 \times 1700 = 5100 \text{ ft}^2$$

of surface may be expected, since the Δt in *both* of the first two effects will be improved at the expense of only the last effect.

Recalculation:

Assume an average surface of 1500 ft^2/effect, and find the temperature differences which will provide these surfaces. Assume

First effect: $T_S - t_1 = 28°F$, $A_1 = {}^{20}\!/_{28} \times 2010 = 1450$ ft^2
Second effect: $t_1 - t_2 = 41°F$, $A_2 = {}^{30}\!/_{41} \times 2040 = 1490$ ft^2
Third effect: $t_2 - t_3 = 50°F$, $A_3 = {}^{69}\!/_{50} \times 1050 = 1440$ ft^2
$T_S - t_3 = 119°F.$

The new temperature-pressure distribution is

	Pressure, psia	Steam or vapor, °F	λ, Btu/lb
Steam chest, 1st effect........	26.7	$T_S = 244$	949
Steam chest, 2d effect........	16.0	$t_1 = 216$	968
Steam chest, 3d effect........	16.4 in. Hg	$t_2 = 175$	992
Vapor to condenser........	26.0 in. Hg	$t_3 = 125$	1022

Solving again for w_1, w_2, w_3, and W_S,

$$w_1 = 15{,}450 \qquad A_1 = 1450$$
$$w_2 = 13{,}200 \qquad A_2 = 1470$$
$$w_3 = 11{,}350 \qquad A_3 = 1490 \qquad \text{(Use 1500 ft}^2\text{/effect)}$$
$$W_S = 16{,}850$$

Heat to condenser = $11{,}350 \times 1022 = 11{,}600{,}000$ Btu/hr
Water requirement = $11{,}600{,}000/(120 - 85) = 332{,}000$ lb/hr = $332{,}000/500$
$\hspace{10cm} = 664$ gpm
Economy, lb evaporation/lb steam = $40{,}000/16{,}850 = 2.37$ lb/lb

COMPARISON OF FORWARD AND BACKWARD FEED

	Forward	Backward
Total steam, lb/hr...............	19,100	16,850
Cooling water, gpm.............	840	664
Total surface, ft²...............	4,800	4,500

The operating conditions for both forward and backward feed are shown in Fig. 14.32a and b. Substantiating the simple reasoning, backward feed is more effective thermally than forward feed. Omitted, however, is the maintenance and investment on backward-feed pumps between each effect and the problems of air leakage and flow control. All of these are considerably greater in backward feed.

Optimum Number of Effects. The greater the number of effects the larger the amount of evaporation per pound of steam admitted to the first effect. The operating costs will be less the larger the number of effects. This is offset, however, by the increased first cost of the apparatus and increased maintenance charges for cleaning and replacement, both of which enter as fixed charges. Supervisory labor will be the same for the operation of any number of effects. The cost of condenser water must also be included, and it, too, will decrease the greater the number of effects employed. The optimum number of effects may be obtained by computing the process requirements with two, three, four, or up to six or eight effects and determining the fixed charges and operating cost resulting from each arrangement. When the total cost is plotted against the number of effects, a minimum will occur corresponding to the optimum number of effects. Actually, however, the number of effects in the various heavy chemical industries are fairly standardized. For example, table salt is concentrated in four effects in which parallel feed of liquor is used, caustic soda with two or three effects and backward feed, and sugar with five or six effects and forward feed. Except when introducing an entirely new chemical process it will rarely be necessary to carry out a complete economic analysis.

EVAPORATION 417

Bleed Steam. In certain industries and particularly in the production of sugar, there is a great need for low-pressure steam, say 10 to 15 psia, for the large amount of liquor preheating required throughout the plant. For the preheating it is found advantageous to use some of the steam from the first or subsequent effects for various preheating services such as in the decolorization of sugar sirup. Since any vapor formed in the first and later effects will have already been used one or more times for

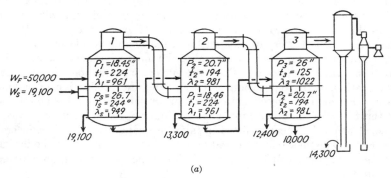

(a)

FIG. 14.32a. Example 14. Forward feed.

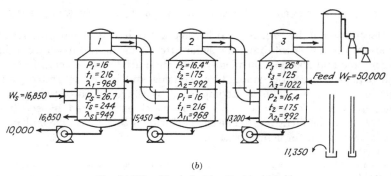

(b)

FIG. 14.32b. Example 14. Backward feed.

evaporation, the cost per Btu of the vapor is cheaper than that of fresh steam. Steam can be bled economically from one or more effects for these additional services while reducing the overall heating cost of the plant. To account for this in the heat-balance equations [Eqs. (14.8) through 14.18)] one need only include the term W_B for the pounds per hour of bleed steam in the effect in which it occurs. The quantity of bleed steam must also be introduced into the material balance of Eqs. (14.11) and (14.18). To prevent the introduction of one or more

418   PROCESS HEAT TRANSFER

unknowns than there are simultaneous equations, W_B should not be entered as another unknown but as a definite number of pounds of bleed steam or as a percentage of the total evaporation if the calculation is for an existing installation.

The Solution of Industrial Problems.[1] The preceding examples have served to introduce the elementary methods of calculation and the principles of multiple-effect evaporation. Actually, industrial problems are rarely so simple. Instead, the evaporator system must be integrated with the operation of the entire manufacturing process, and this complicates the calculation greatly. The element of experience is essential to the completion of a calculation within a reasonable period of time. In the remainder of this chapter several of the most widely met commercial problems will be analyzed. These are sugar concentration, waste-liquor evaporation in the paper pulp industry, and the production of caustic soda. The analyses involved in their solution should be readily adaptable to the majority of other problems. A fourth process, distillery waste concentration, has been omitted in favor of evaporation by thermocompression. The methods indicated may be employed either for the design of a new evaporator or for the calculation of performance in an existing evaporator.

THE CONCENTRATION OF CANE-SUGAR LIQUORS—FORWARD FEED

Description of the Process. It is the practice in cane-sugar production to strain the juice containing the sugar after it has been pressed from the sugar cane and clarified chemically. In the initial stage of clarification, called defecation, large amounts of colloids and inorganic and organic salts are removed. This is accomplished by the addition of lime to the juice at 200°F, forming a heavy slime which is removed by settling, decantation, and filtration. The clear solution is then fed to the multiple-effect evaporator for concentration.

Example 14.4. An evaporator installation is to have a capacity for concentrating 229,000 lb/hr of 13°Brix (degrees Brix is the per cent by weight of sugar in the solution) to 60°Brix, at which concentration it will be decolorized. This quantity of sugar solution results from milling 2300 tons of sugar cane per 20-hr day. The raw juices will be heated from 82 to 212°F by the use of exhaust vapors bled from the first and second effects. Steam will be available to the first effect at 30 psig.

Solution. The arrangement of equipment is shown in Fig. 14.33 and is justified by the fact that bled vapor is a cheaper preheating medium than fresh steam. In this analysis it is arbitrarily assumed that 37,500 lb/hr of 15 psig vapor is bled from the first effect for use in the vacuum pans in the crystallizing section of the plant. The

[1] Readers who are not directly interested in chemical evaporation problems may omit the remainder of this chapter without affecting the context of later chapters.

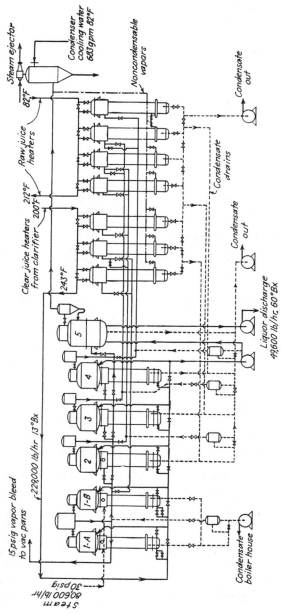

Fig. 14.33. Six-body sugar evaporator with preheaters.

BPR and specific heats of sugar solutions are given in Fig. 14.34. A flow diagram such as that in Fig. 14.33 is usually arrived at from experience or from preliminary studies of total surface requirements as in Table 14.1.

TABLE 14.1. AVERAGE EVAPORATION PER SQUARE FOOT HEATING SURFACE FOR SUGAR EVAPORATORS

Effects	Water evaporated, lb/(hr)(ft²)
1	14–16
2	6–8
3	5–6
4	4–5
5	3–4

In the matter of choosing the number of effects from experience, it is the practice to use a straight quadruple effect if little or no vapor bleeding is required and to add a "preevaporator" as a first effect if excessive bleeding of low pressure vapor are required for preheating feed, etc., or if relatively low-pressure live steam is bled for other plant processes, *i.e.*, vacuum pans.[1]

The forward-feed arrangement is common in the sugar industry, since stronger juices or sirups are sensitive to high temperatures. This procedure places a severe handicap on the last effect of a multiple-effect evaporator where the most viscous juice

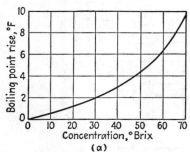
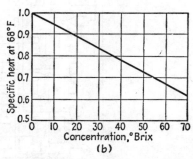

FIG. 14.34a. BPR of sugar solutions.   FIG. 14.34b. Specific heats of sugar solutions.

boils at the lowest temperature. However, because of the "caramelizing" tendency of high-concentration solutions this procedure is of paramount importance in the preparation of a product of high purity.

Table 14.2 is a convenient reference in estimating the temperature-pressure distribution and for presenting a summary of all the important process and design conditions. Items 2 to 9 inclusive are estimated first, and from these the heat balance can be made.

It is shown in item 2 of Table 14.2 that 30 psig steam is fed in parallel to effects 1A and 1B, the temperature 274°F in 3 being the saturation temperature corresponding to the steam pressure. The last-effect vapor body is designed to operate under a

[1] Futher discussion of the flow diagram will be given at the conclusion of the heat balance.

vacuum of 23 in. Hg (147°F). This is shown in 7 and 8. Item 13 shows that the sirup leaving the last effect is 60°Brix. From Fig. 14.34a this concentration corresponds to a boiling temperature of 147 or 154°F as shown in 5. For 6 of Table 14.2 it can be assumed tentatively that the °Brix concentration is divided evenly among all the effects; thus, the BPR for each effect can be read from Fig. 14.34a. The ultimate material distribution determined from the heat balance, however, will give a precise

TABLE 14.2. EVAPORATOR SUMMARY

Item	Effects					
	1A	1B	2	3	4	5
1. Steam flow, lb/hr	42,600	38,000				
2. Steam pressure, psi/in. Hg	30	30	15	5	4″	14½″
3. Steam temp, °F	274	274	250	227	205	181
4. Δt, °F	23	23	21	20	20	27
5. Liquor temp, °F	251	251	229	207	185	154
6. BPR, °F	1	1	2	2	4	7
7. Vapor temp, °F	250	250	227	205	181	147
8. Vapor pressure, psi/in. Hg	15	15	5	4	14½	23
9. λ, Btu/lb	946	946	960	975	990	1,010
10. Liquor in, lb/hr	229,000	190,200	154,000	117,100	87,800	64,000
11. Liquor out, lb/hr	190,200	154,000	117,100	87,800	64,000	49,600
12. Evaporation, lb/hr	38,800	36,200	36,900	29,300	23,800	14,400
13. °Brix (out)	15.7	19.4	25.5	34.3	46.5	60.0
14. A, ft²	3,500	3,500	5,000	5,000	5,000	3,500
15. U_D, Btu/(hr)(ft²)(°F)	478	425	310	264	219	138
16. $U_D \Delta t$, Btu/(hr)(ft²)	11,000	9,780	6,520	5,270	4,390	3,740

estimate of the Brix distribution in the different effects. Item 13 is a summary of the latter. In order to complete 2 to 9 inclusive, the distribution of temperature differences in each effect must be determined as in 4.

The total temperature difference in the evaporator system is 274° − 147 = 127°F. From the total must be subtracted the sum of all the BPR (*i.e.*, from effect 1B to the fifth effect inclusive), a sum of 16°F. This will give 127 − 16 = 111°F as the total "effective" temperature difference. The latter is distributed along the following lines: (a) The total is distributed in proportion to the values found in practice for a unit with the same number of effects. (b) The total is distributed on the basis of accepted average values of heat flux $U_D \Delta t = Q/A$ as found in industry and with values of U_D estimated from similar operations. Accepted values of the flux are given in (16) of Table 14.2. They may be redistributed for a different number of effects. (c) An equal distribution of the total pressure drop is assumed through the system. Either method will give a sufficiently close estimate of the temperature difference distribution for obtaining items 2 to 9 required for working the heat balance. It might be noted here that, if the estimated values of the temperature differences are found to be very much out of line with those found in practice, they must be revised. Having estimated the distribution of the temperature difference the saturated vapor pressure above the liquor or the saturated steam pressure in the following effect may be determined from the steam tables (Table 7). The latent heats corresponding to saturated vapor pressures will be obtained similarly.

Since it is planned to bleed vapors from the different effects for preheating both the raw and clarified sugar solutions, it is necessary to determine the quantities which will be bled from each effect. The raw sugar solutions will be preheated from 82 to 212°, a range of 130°F. Vapors from the fourth effect will be used to preheat the raw sugar solution from 82 to 144°F, vapors from the third effect from 144 to 184°, and vapors from the second effect from 184 to 212°. Heat quantities and surfaces for the juice heaters are calculated as in Table 14.3. Similarly juice from the clarifier is preheated by second-effect vapors from 200 to 220° and by first-effect evaporators from 220 to 243°F. In all instances the preheaters are designed to permit a duplication of surface, although in some instances overdesign obviously results. A spare heater is usually included in the system to allow continuity of operation during cleaning. In arriving at the temperature range for the individual preheaters a minimum terminal difference of 7°F has been used. Larger terminal differences may be employed if the spread of temperatures makes them permissible.

TABLE 14.3. SUGAR-JUICE HEATERS

Raw-juice heaters

1. $229{,}000(212 - 184)(0.91)$
   $= 5{,}840{,}000$ Btu/hr
   Vapor temp. $= 227°\text{F}$ $\Delta t = 26.6°\text{F}$
   $U_D = 231$
   Surface, $A = \dfrac{5{,}840{,}000}{26.6 \times 231} = 950 \text{ ft}^2$

2. $229{,}000(184 - 144)(0.90)$
   $= 8{,}250{,}000$ Btu/hr
   Vapor temp. $= 205°\text{F}$ $\Delta t = 37.6°\text{F}$
   Surface, $A = \dfrac{8{,}250{,}000}{37.6 \times 230} = 950 \text{ ft}^2$

3. $229{,}000(144 - 82)(.90) = 12{,}800{,}000$
   Vapor temp. $= 181°\text{F}$ $\Delta t = 62.2°\text{F}$
   Surface, $A = \dfrac{12{,}800{,}000}{62.2 \times 217} = 950 \text{ ft}^2$

Use 3 heaters at 1000 ft² each plus 1 heater as spare

Clear-juice heaters

1. $229{,}000(243 - 220)(0.91)$
   $= 4{,}800{,}000$ Btu/hr
   Vapor temp. $= 250°\text{F}$ $\Delta t = 15.8°\text{F}$
   $U_D = 234$
   Surface, $A = \dfrac{4{,}800{,}000}{15.8 \times 234} = 1300 \text{ ft}^2$

2. $229{,}000(220 - 200)(0.90)$
   $= 4{,}120{,}000$ Btu/hr
   Vapor temp. $= 227°\text{F}$ $\Delta t = 14.8°\text{F}$
   Surface, $A = \dfrac{4{,}120{,}000}{14.8 \times 214} = 1300 \text{ ft}^2$

Use 2 heaters at 1300 ft² each plus 1 heater at 1300 ft² as spare

Table 14.4 shows the heat and material balance for the evaporator. To effect a balance it is necessary to estimate correctly the steam flow to the first effect. In the simple case where no vapors are bled for process heating an assumption is made by using a quantity of steam which is about 20 per cent above theoretical, *i.e.*, the hourly evaporation divided by the number of effects. In the present case where excessive vapors are bled, the "*extrapolated* evaporation" is divided by the number of effects. The extrapolated evaporation is equivalent to the actual plus that which the vapors could effect if not bled for process work. Since only an estimate of the steam is

TABLE 14.4. HEAT BALANCE

Effect	Btu/hr	Evaporation, lb/hr
1A. Heat in steam = 42,600 × 929 × .97..............	38,400,000*	
Heating liquor = 229,000(251 − 243).91..........	1,670,000	
$\overline{}$ − 38,800	36,730,000/946	38,800
Liquor to 1B = 190,200		
1B. Heat in steam = 38,000 × 929 × .97..............	34,200,000*	
Heating liquor = 190,200(251 − 251)..........		
$\overline{}$ − 36,200	34,200,000/946	36,200
Liquor to 2d effect = 154,000		
2. Heat in 1A vapors...........................	36,730,000	
Heat in 1B vapors.............................	34,200,000	
Total heat available.........................	70,930,000	
First effect cond. flash = 80,600(274 − 250)..........	1,940,000	
	72,870,000	
Heat to vacuum pans = 37,500 × 946..............	−35,500,000	
	37,370,000	
Heat to clear-juice heaters.........................	− 4,800,000	
Heat to 2d effect...........................	32,570,000*	
Liquor flash = 154,000(251 − 229).85..............	+ 2,880,000	
$\overline{36,900}$	35,450,000/963	36,900
Liquor to 3d = 117,100		
3. Heat in 2d vapors...........................	35,450,000	
Cond. flash = 37,500(250 − 227)....................	865,000	
	36,315,000	
Heat to clear-juice heaters.........................	− 4,120,000	
	32,195,000	
Heat to raw-juice heaters.........................	− 5,840,000	
	26,355,000*	
Liquor flash = 117,100(229 − 207).83..............	+ 2,150,000	
$\overline{29,300}$	28,505,000/975	29,300
Liquor to 4th = 87,800		
4. Heat in 3d vapors...........................	28,505,000	
Cond. flash = 74,200(227 − 205)....................	1,630,000	
	30,135,000	
Heat to raw-juice heaters.........................	− 8,250,000	
	21,885,000*	
Liquor flash = 87,800(207 − 185).80..............	+ 1,540,000	
$\overline{23,800}$	23,425,000/989	23,800
Liquor to 5th = 64,000		
5. Heat in 4th vapors...........................	23,425,000	
Cond. flash = 101,400(205 − 181)....................	2,480,000	
	25,905,000	
Heat to raw-juice heaters.........................	−12,800,000	
	13,105,000*	
Liquor flash = 64,000(184 − 154).74..............	+ 1,470,000	
$\overline{14,400}$	14,575,000/1010	14,400
Prod. discharge = 49,600	Total hourly evaporation	179,400

* Heat supplied by heating element.

required, the following rule can be used to obtain the extrapolated evaporation: Add to the actual evaporation the equivalent evaporation of the bled vapors thus:

		Lb/hr
(a)	Actual evaporation...	179,400
(b)	Equivalent evaporation from vapors of 1st effect used for vacuum pans $(4 \times 35{,}500{,}000)/977$..	145,500
(c)	Equivalent evaporation from 1st effect vapors used for clarified-juice heaters, $(4 \times 4{,}800{,}000)/977$.......................................	19,700
(d)	Equivalent evaporation from 2d effect vapors used for clarified- and raw-juice heaters, $(3 \times 9{,}960{,}000)/977$.................................	30,600
(e)	Equivalent evaporation from 3d effect vapors used for raw-juice heaters, $(2 \times 8{,}250{,}000)/977$...	17,900
(f)	Equivalent evaporation from 4th effect vapors used for raw-juice heaters, $(1 \times 12{,}800{,}000)/977$..	13,100
	Extrapolated evaporation..	406,200

$$\text{Estimated steam quantity } \frac{406{,}200}{5} = 81{,}240 \text{ lb/hr}$$
$$\text{Actual steam required from final heat balance} = 80{,}600$$
$$\text{Error} = \phantom{00{,}0}640 \text{ lb/hr}$$

This method of estimating the steam quantity to the first effect gives quite consistent results and can be used as a "first" trial in Table 14.4. One or two trials will determine the exact steam flow for obtaining the required evaporation.

It will be noted from Table 14.4 that liquor and steam flow in series through all effects (in forward-feed arrangement) except for the first effect where the total steam is divided for parallel flow between effects 1A and 1B. The latter procedure is common where excessive evaporation in a single effect would require a very large vapor body. The division of the total steam in the heat balance between effects 1A and 1B is arbitrary, especially if equal heating surface will be supplied. However, it is the usual practice to divide the total steam in the ratio of 1.1/1.0 for effects 1A/1B to simplify calculations and to take into account the fact that the 1A effect will have a higher heat-transfer coefficient and thus use more steam for greater evaporation.

In effects 1A and 1B liquor must be preheated to its boiling point before evaporation commences. In all other effects the liquor flashes or vaporizes a portion of its water on entering the following body. To increase the steam economy still further, it is customary to flash the condensate to the vapor space of the following effect. Thus the total steam condensate is flashed to the vapor space in the first effect, and the vapors obtained from this flash are added to the total vapors leaving the 1A and 1B effects. Likewise the net condensate in the second effect $(38{,}800 + 36{,}200 - 37{,}500$, which was removed for vacuum pans) is flashed to the corresponding second-effect saturated vapor, etc. It will be noted that the total flash in each effect is cumulative, since the condensate flows in series to a common discharge header through a series of flash tanks. A well-designed condensate flash system may save as much as 10 per cent of the live-steam requirement.

The heat balance is completed when one obtains a product from the last effect equal in quantity and °Brix concentration to that for which the evaporator is being designed. Thus a material balance in the present case shows:

$$\text{Feed} = 229{,}000 \text{ lb/hr} \times 0.13 = 29{,}800 \text{ lb/hr sugar}$$
$$\text{Discharge} = \frac{29{,}800}{0.60} = 49{,}600 \text{ lb/hr of } 60°\text{Brix sugar}$$
$$\text{Evaporation} = 229{,}000 - 49{,}600 = 179{,}400 \text{ lb/hr water evaporated}$$

If an incorrect quantity of steam is assumed or a very poor temperature distribution is made, the required evaporation or product discharged will not be obtained. In the latter case, an apportionment of the difference between the required evaporation and that obtained in the first trial, based upon the assumed steam quantity and a recalculation of the heat balance, will give results close enough for practical purposes. In general, if the heat balance is off by less than 1 per cent in product or evaporation, it may be considered to check.

The vapors from the last effect can be used for process heating or carried to a condenser. Evaporator vapors are used to heat the juice before and after clarification by means of two sets of heaters. In Fig. 14.33 the heaters are grouped in sets of four and three. The first set is for preheating the raw juice with vapors of the second, third, and fourth effects. The second set is for preheating the clarified juice before it enters the first effect of the evaporator and uses vapors from the first and second effects. Each set of heaters operates with one less than that indicated, the omitted unit acting as a spare.

The first four effects of the evaporator system are designed to have bodies of the long-tube vertical-film type (Fig. 14.27), while the fifth effect is provided with a calandria-type body (Fig. 14.23). The first four effects have catchalls with tangential vapor inlets and bottom outlets, whereas the fifth effect is designed with a top vapor outlet. The entire evaporator unit is arranged to allow by-passing any unit from the system for cleaning or repair.

Vacuum is maintained by a barometric condenser and ejector. The gallons per minute of water required in the barometric condenser is calculated from Eq. (14.4). Where low vacuum and cold water are combined as in the present example, a value of 15°F may be used for the approach temperature difference t_a.

$$\text{Water} = \frac{14{,}575{,}000}{500(147 - 82 - 15)} = 583 \text{ gpm} \qquad (14.4)$$

Discussion. In Table 14.2 items 10 to 13 inclusive summarize the data obtained from the material and heat balance in Table 14.4. The surface design, *i.e.*, the heating element, vapor head, etc., is shown in 14 to 16 inclusive. Item 16 lists values of $U_D \Delta t$ employed commercially; dividing these values of $U_D \Delta t$ by the estimated Δt yields the values of U_D shown in 15. If the values of U_D are much out of line with correlated values for similar operation, then a redistribution of either Δt or $U_D \Delta t$ may be necessary. Both are affected simultaneously. Surfaces are found by dividing the starred heat loads in Table 14.4 by respective values of $U_D \Delta t$.

In designing heaters for both raw and clear juice in Table 14.3 the importance of equal size and duplication of units is emphasized. A careful division of the total heat load was made to allow large temperature differences when using the bled vapors of the different effects. Slightly larger transfer coefficients could have been used, but it is desirable to equalize the surfaces at the expense of arbitrarily varying the transfer coefficients.

A check of Eq. (14.7) will show that it is not imperative to account for the hydrostatic loss when setting up Table 14.2. Since the evaporator bodies are designed for high velocity and low liquor levels, it has been assumed that any capacity loss due to hydrostatic head is compensated for by a slightly conservative design of surface.

THE EVAPORATION OF PAPER PULP WASTE LIQUORS—BACKWARD FEED

Description of Paper-pulp Production Processes. There are three standard processes for producing paper pulp, namely, the soda, sulfate, and sulfite processes. All three produce wastes which are recovered either because of valuable residues or to avoid public nuisance.

1. In the soda process wood chips are cooked in digesters with caustic soda solution of about 12°Bé. After digestion, the pulp is washed clean with a hose in tanks having screen bottoms or else passed through a filter which is used for washing it. The liquor separated from the washed pulp, called *black liquor*, contains the resin and lignin of the wood and all the alkali used in the digester. The black liquor is evaporated, and solids are recovered for the value of the alkali.

The soda process is applied to both nonresinous and resinous woods. In the latter case the black liquors tend to foam and evaporators of the long-tube, vertical-film type are used almost exclusively. The solutions are not scale forming, undergo little elevation in boiling point (BPR), and the concentrated product is not excessively viscous. Liquors are usually concentrated to about a 40 per cent solid content by backward feed in a sextuple (or higher) effect evaporator, and the concentrate is subsequently fed to rotary incinerators. The latter produces *black ash*, which is leached, and the resulting solution is recausticized for use in the digesters.

2. The sulfate pulp process uses a mixture of sodium hydroxide, sodium carbonate, and sodium sulfide for digestion of wood chips. The sulfide is renewed by adding sodium sulfate and reducing it with carbon (thus, "sulfate" process). The process is used on resinous woods, and the content of inorganic material in the waste liquor is higher. It is usually concentrated to about 50 and often 55 per cent. The liquor is more viscous than the soda black liquor of same density, foams more, and has a higher elevation in boiling point owing to the larger content of inorganic solids. The usual practice is to use long-tube vertical-film-type evaporators with six effects or more and with backward feed. Because of the high viscosity in the first effect it is customary to divide the latter into two bodies with parallel steam flow. This procedure increases the liquor rate, which offsets the detrimental effect of the high viscosity on the heat-transfer rate.

3. The sulfite pulp process uses calcium or magnesium bisulfite as the active chemical in wood chip digestion. The active material is not

EVAPORATION

recovered.[1] When a mill is located in an isolated area the waste liquor is discarded; when a mill is located in nonisolated districts, evaporation of the liquors and burning of the concentrate are required to prevent surface or subsurface water polution. In recent years, however, much development work has been carried out on the fermentation of sulfite liquors for alcohol production.

Example 14.5. A unit concentrating soda pulp black liquor will be designed to evaporate 90,000 lb/hr of water from a feed of 144,000 lb/hr which enters at 170°F and contains 15.2 per cent solids. This is equivalent to a product discharge of 54,000 lb/hr containing 40.5 per cent solids. These quantities are based on the production of 150 tons of pulp per day, and the resulting black liquor contains 3500 lb/ton of total solids. Utilities available are dry saturated steam at 35 psig and condenser water at 60°F.

Solution. Economy studies in the industry indicate that a six-body sextuple-effect evaporator will be required. In this type of unit the liquor flow is arranged for divided feed to the fifth and sixth effects to eliminate the large vapor bodies which would otherwise be required were the total feed passed through in series. From the last two effects the liquor is pumped through the remaining bodies in series, backward flow. The discharge from the first effect, after passing through a liquid-level control valve to a liquor flash tank, is finally sent to a product storage tank. Figure 14.35 illustrates schematically the arrangement of the equipment for this process.

In this type of process all vapor, liquor, and condensate lines are designed with by-pass arrangements, permitting any one body to be cut out for cleaning or maintenance and still allowing the operation of the unit as a quintuple effect.

The vapors from the last effect are condensed in a multijet barometric condenser or are used for heating mill water to 110°F. If the latter scheme is followed, a surface condenser can be substituted for the multijet condenser.

The factors entering in making the heat balance are similar to those discussed under sugar evaporation. Referring to Table 14.5 it is important first to estimate correctly the temperature-pressure distribution, items 2 to 8.

The overall temperature difference is $280 - 125 = 155°F$ (corresponding to 35 psig and 26 in. Hg). For the estimated per cent solids distribution in each body, the BPR can be obtained from Fig. 14.36a. The estimated total BPR is 41°F, and the effective temperature difference (disregarding hydrostatic head factors in accordance with the discussion under sugar evaporators) is $155 - 41$, or 114°F. Specific-heat data are given in Fig. 14.36b. The distribution of temperature differences Δt in item 4 is in accordance with general practice, using higher temperature differentials in the first and last effects owing to the larger evaporation loads and the relatively even distribution in the other effects. Slight deviations in the estimated values of the temperature-difference distribution from what may be found in actual operation will not affect the heat balance or design of the evaporator too greatly. Poor temperature-difference estimates, however, will be reflected by poorly correlatable transfer coefficients. The accepted overall coefficients for multiple-effect soda–black-liquor processes are given

[1] Recent development by the Weyerhaeuser Timber Co. of a cyclic recovery system using the magnesium bisulfite base has increased interest in this method of pulp production. From unpublished paper on sulfite waste evaporation by D. Q. Kern and J. Meisler.

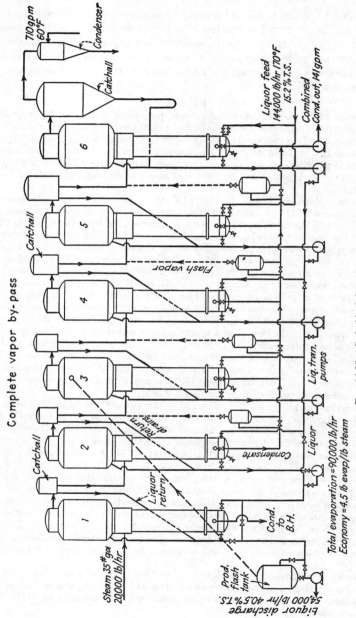

Fig. 14.35. Soda black-liquor evaporator.

TABLE 14.5. EVAPORATOR SUMMARY

All bodies will consist of 300 2 in. OD, 10 BWG tubes 24′0″ long

Item	Effects					
	1	2	3	4	5	6
1. Steam flow, lb/hr................	20,000					
2. Steam pressure, psi/in. Hg........	35	14.5	4	7″	16.5″	22″
3. Steam temp, °F...................	280	249	224	199	174	151
4. Δt, °F...........................	21	17	18	19	18	21
5. Liquor temp, °F..................	259	232	206	180	156	130
6. BPR, °F.........................	10	8	7	6	5	5
7. Vapor temp, °F...................	249	224	199	174	151	125
8. Vapor pressure, psi/in. Hg........	14.5	4	7″	16.5″	22″	26″
9. λ, Btu/lb.......................	946	962	978	994	1,008	1,022
10. Liquor in, lb/hr.................	73,400	88,300	101,100	113,000	72,000	72,000
11. Liquor out, lb/hr................	56,200	73,400	88,300	101,100	58,300	54,700
12. Evaporation, lb/hr...............	17,200	14,900	12,800	11,900	13,700	17,300
13. Total solids, %*.................	38.9	29.8	24.7	21.6	18.7	20.0
14. A, ft²........................	3,250	3,250	3,250	3,250	3,250	3,250
15. U_D, Btu/(hr)(ft²)(°F)..........	262	295	252	251	221	221
16. $U_D \Delta t$, Btu/(hr)(ft²).....	5,510	5,000	4,530	4,770	3,980	4,650

* 40.5 per cent from flash tank; 22,000 lb/hr vaporized in flash tank.

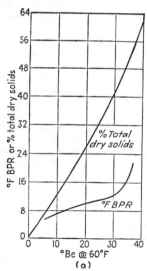

FIG. 14.36a. °Bé vs. °F BPR and total dry solids for soda black liquor.

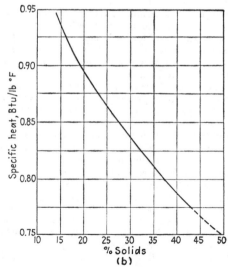

FIG. 14.36b. Specific heat vs. per cent solids for soda black liquor.

in Table 14.5. If the calculated coefficients differ appreciably from the accepted values, the Δt distribution will have to be altered accordingly.

Table 14.6 shows the procedure in determining the heat balance. It will be observed that this balance is one of trial and error. With some experience in backward-flow-evaporator calculations, however, not more than two trials are necessary to obtain the correct balance. A first trial is necessary for determining the approximate quantity of live steam to the first effect. Since the overall material balance (Table 14.5) shows that a total of 90,000 lb/hr of water is evaporated, the theoretical amount of steam for a six-effect evaporator would be 90,000/6, or 15,000 lb/hr. It is found, however, that the economy of multiple-effect black-liquor evaporators falls in the order of 75 per cent of theoretical or 6×0.75 which is 4.5. Thus, as an estimate

$$\frac{90{,}000 \text{ lb}}{4.5} = 20{,}000 \text{ lb/hr}$$

of steam will be used for a trial balance. This value is converted to its total evaporation potential by multiplying by its latent heat of vaporization and a factor of about 0.97 to take into account heat losses by radiation.

Since the feed is backward, the quantity of liquor being pumped from the second to the first effect is unknown. In fact, the only known liquor quantity is 54,000 lb/hr of product discharge leaving the flash tank. Ordinarily the black liquor in the first effect is not concentrated to the final desired solid content, but several of the last per cent of water are allowed to vaporize in a flash tank which is located ahead of the first effect and which is connected to one of the low-pressure evaporator bodies as shown in Fig. 14.35. In this instance the flash tank receiving the discharge from the first effect is piped to the vapor body of the third effect, which is at a saturation temperature of 199°F. Since the solution discharged from the first effect has a BPR of 10°F, its actual boiling temperature in the third effect is $199 + 10 = 209$°F and a flash corresponding to 259 to 209°F is effected.

If the quantity of liquor discharged from the first effect were known, the amount of additional water evaporated by flash could easily be calculated. An estimate is therefore made of this quantity as follows:

Divide the total evaporation by the number of effects, i.e., 90,000/6 = 15,000, and add to it about 15 per cent, giving a total of 17,200 lb/hr. The latter is an estimate of the amount of evaporation in the first effect. Since the flash is usually a few per cent of the product (here 2200/54,000 = 4.1 per cent), an estimated value of the flash is added to the evaporation in the first effect and the total added to the product discharge from the flash tank: thus $54{,}000 + (17{,}200 + 2200) = 73{,}400$ lb/hr. This quantity is the estimated discharge from the second effect and must be preheated from 232 to 259°F, the boiling point in the first effect. The latter is indicated in item 1b of the heat-balance calculations, Table 14.6. By subtracting the preheat from 1a and dividing the difference (16,270,000/946) by the latent heat of vaporization corresponding to the saturated vapor pressure in the first effect, the amount of actual evaporation in the first effect 1c is obtained. The latter value should correspond to that estimated. If it does not check, the value of the discharge liquor from the second effect 1b is revised to bring the estimated and the calculated evaporation rates in line. Moreover, since the preheat load 1b is usually in the order of 8 to 10 per cent of the steam load 1a only a small change is ordinarily required in the former to balance the estimated and calculated evaporation rates. The discharge to the flash tank is the difference between 1b and 1c. The amount of flashed vapor is equal to the heat load 1d divided by the latent heat of vaporization corresponding to

EVAPORATION

Table 14.6. Heat Balance

$$\text{Cooling water at } 60°F = \frac{17{,}750{,}000}{500(125 - 15 - 60)} = 710 \text{ gpm}$$

Effect	Btu/hr	Evaporation, lb/hr
1. *a.* Heat in steam = 20,000 × 924 × .97*............	17,900,000†	
b. Heating liquor = 73,400(259 − 232).82.........	1,630,000	
c. Evaporation = 17,200......................	16,270,000/946	17,200
d. To flash tank = 56,200(259 − 209).78........	2,200,000/978	2,200
e. Flashed vapor = 2,200		
f. Product = 54,000		
2. *a.* Heat in 1st vapors..........................	16,270,000†	
b. Heating liquor = 88,300(232 − 206).85.........	− 1,940,000	
c. Evaporation = 14,900......................	14,330,000/962	14,900
d. Liquor to 1*b* = 73,400		
3. *a.* Heat in 2d vapors...........................	14,330,000	
b. Condensate flash = 17,200(249 − 224).........	+ 430,000	
c. Total heat to 3d effect steam chest............	14,760,000†	
d. Heating liquor = 101,100(206 − 180).86........	2,250,000	
e. Evaporation = 12,800......................	12,510,000/978	12,800
f. Liquor to 2*b* = 88,300		
4. *a.* Heat in 3d vapors...........................	12,510,000	
b. Condensate flash = 32,100(224 − 199).........	+ 800,000	
c. Liquor flash from flash tank...................	+ 2,200,000	
d. Total heat to 4th effect.......................	15,510,000†	
e. Heating liquor = 113,000(180 − 143).88........	− 3,680,000	
f. Evaporation = 11,900......................	11,830,000/995	11,900
g. Liquor to 3*d* = 101,100		
5. *a.* Heat in 4th vapors...........................	11,830,000	
b. Condensate flash = 44,900(199 − 174).........	+ 1,120,000	
c. Total heat to 5th effect.......................	12,950,000†	
d. Liquor flash = 72,000(170 − 156).90.........	+ 900,000	
e. Evaporation = 13,700......................	13,850,000/1008	13,700
f. Liquor to 4*e* = 58,300		
6. *a.* Heat in 5th vapors...........................	13,850,000	
b. Condensate flash = 56,800(174 − 151).........	+ 1,310,000	
c. Total heat to 6th effect.......................	15,160,000†	
d. Liquor flash = 72,000(170 − 130).90.......	+ 2,590,000	
e. Evaporation = 17,300......................	17,750,000/1022	17,300
f. Liquor to 4*e* = 54,700		
g. Liquor to 4*e* from 5*f* = 58,300		
h. 4*e* = 113,000		

$$\text{Total hourly evaporation} = 90{,}000 \text{ lb}$$

$$\text{Economy} = \frac{90{,}000}{20{,}000} = 4.50 \text{ lb/lb}$$

* Radiation factor.
† For surface calculation.

saturated pressure in the third effect. Subtracting the flash 1e from 1d should give *the known product* in 1f. If a deviation from the product is obtained, slight changes are required in 1d, 1c, and 1b to complete the balance in the first effect.

In the second effect the value of 2a corresponds to the heat load of the vapors leaving the first effect and entering the second effect steam chest. The value of 2d is equal to 1b, which has already been determined. An estimate of 2b determines 2c. A check on the estimated value of 2b is obtained if $2b - 2c = 2d$. If they are unequal, a slight change in 2b will establish the equality.

The value of item 3a corresponds to the heat load of vapors leaving the second effect. Moreover, to increase steam economy the condensate from the second effect is discharged through a flash tank into a condensate header. The flash tank is piped to the steam chest of the third effect. Item 3b shows the equivalent heat load added to the steam chest of the third effect by flashing the condensate of the second effect into it.

Since item 3f equals 2b, an estimate of 3d determines the value of 3e. A check on the estimate 3d is determined by the equality of the values under $3d - 3e = 3f$.

Item 4a is the net heat load in the third effect vapors and 4b is the combined condensate flash from effects 2 and 3; *i.e.*, $17{,}200 + 14{,}900 = 32{,}100$ lb/hr. Item 4c corresponds to 1d, since it was stated above that the discharge from the first effect passed through a flash tank piped to the steam chest of the fourth effect (or to the vapor line of the third effect). The flashed vapors are therefore used as an additional source of heat for evaporation and are included in the total heat load in the vapors entering the fourth-effect steam chest 4d. Item 4e is estimated similarly to 3d, etc. It will be noted, however, that in using a temperature differential for preheating the liquor to the boiling temperature, 180°F, the liquor was preheated from 143 to 180°F. From Table 14.5 it will be observed that the boiling points in effects 5 and 6 are, respectively, 156 and 130°F. Since the feed is passed through the last two effects in parallel, the common temperature of the combined discharged liquor from these effects is taken as their arithmetic average; *i.e.*, $(156 + 130)/2 = 143°F$.

Item 5b in Table 14.6 is the combined condensate flash of effects 2, 3, and 4; *i.e.*, $17{,}200 + 14{,}900 + 12{,}800 = 44{,}900$ lb/hr. Since feed enters above its boiling-point temperature in both the fifth and sixth effects, a certain amount will flash on entering the two bodies. Items 5d and 6d show the flash calculations. The discharge from the fifth effect, 5f, combined with the discharge from the sixth effect, 6f, is pumped to the fourth effect.

Item 6b is the combined condensate flash of effects 2 to 5 inclusive. When added to 6a the sum gives the total heat load to the steam chest of the sixth effect.

If the heat balance is closed so that the total calculated hourly evaporation is found to be equal to that which is required in the material balance, then the sum of 5f and 6f will be equal to 4e and the estimated steam requirement will be satisfactory. If the heat balance does not close on the first trial, a revision of the steam estimate is made and 1 through 6 inclusive are repeated. To revise the steam requirements it is seen that the algebraic difference between the obtained and required evaporation divided by the estimated economy need only be added to the initial estimate of steam. Thus, if the calculated evaporation were 94,500 lb/hr, $(90{,}000 - 94{,}500)/4.5 = -1000$ or approximately $20{,}000 - 1000 = 19{,}000$ lb/hr of steam would be used for reworking the balance. Ordinarily not more than two trials need be made to obtain a closure in the balance.

Each body in Fig. 14.35 is of the long-tube vertical-film type Fig. 14.27.

The water rating of a multijet barometric condenser is quite similar to that for a countercurrent barometric type employed on sugar evaporators. The major difference is in the allowable approach to the vapor temperature. Whereas in the countercurrent barometric condenser it is permissible to use about 5°F approach for a good design, the multijet is designed conservatively for a 15°F approach. Thus, the gallons per minute required are

$$\frac{Q}{500(T_s - 15 - t_W)} = \frac{17,750,000}{500(125 - 15 - 60)} = 710 \text{ gpm}$$

In using a surface condenser it will not be desirable to approach the saturated vapor temperature closer than 15°F; thus, 710 gpm can be heated from 60 to 110°F for use as mill water.

CAUSTIC SODA CONCENTRATION—FORCED-CIRCULATION EVAPORATORS

Description of the Process. The concentration of caustic soda solutions obtained either from the reaction of soda ash and lime or from the electrolytic decomposition of brine present certain distinct problems: (1) Caustic solutions have a high elevation in boiling point which causes a large loss of available temperature difference in multiple-effect evaporators. (2) The concentrated solutions are highly viscous, reducing sharply the heat-transfer rate in natural-circulation evaporators. (3) Caustic solutions may have detrimental effects on steel, causing caustic embrittlement. (4) Caustic solutions may require the removal of large quantities of salt as the solution becomes concentrated.

In order to avoid caustic embrittlement of steel, solid nickel tubes are used along with nickel-clad plate-steel bodies. All auxiliary equipment which comes in contact with the caustic solution is also nickel clad. Due to the costliness of nickel as a material of construction, however, it is important to design this class of equipment for as small a heating surface as possible. Since the liquor-film heat-transfer coefficient is a function of the velocity of the caustic solution through the tubes, a high velocity must be established in order to obtain a large transfer coefficient in the tubes.

The process flow sheet is shown in Fig. 14.37. Cell liquor feed, along with filtrate and wash brine from the salt filter or centrifuge, is introduced into the filtrate feed tank a. From a the liquor is pumped into the second effect b of the evaporator where partial concentration and partial salt precipitation occur. The resulting semiconcentrated effluent from the second effect is pumped into the weak-liquor settler c along with the slurry underflow from the 50 per cent caustic settler d. The underflow from a contains all the precipitated salt from the system in the form of a slurry. This slurry is pumped to the filter or centrifuge to remove the precipitated salt and to wash the salt with brine. The filtrate entering the filtrate tank a is recirculated through the system with cell liquor feed

434 PROCESS HEAT TRANSFER

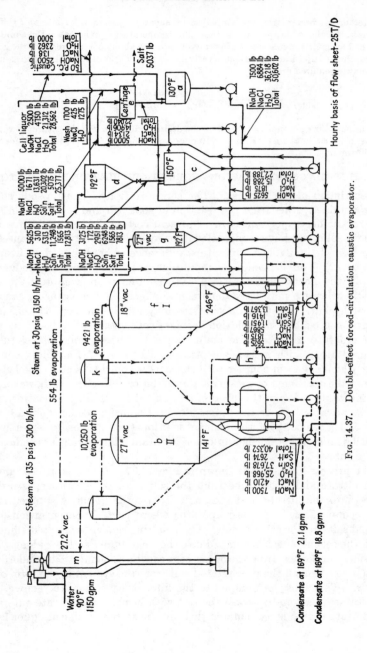

Fig. 14.37. Double-effect forced-circulation caustic evaporator.

as described above. The overflow from c is pumped into the first effect of the evaporator f for further concentration and further salt precipitation. The effluent from the first effect is at a temperature of 246°F and contains concentrated caustic solution and a suspension of salt. This

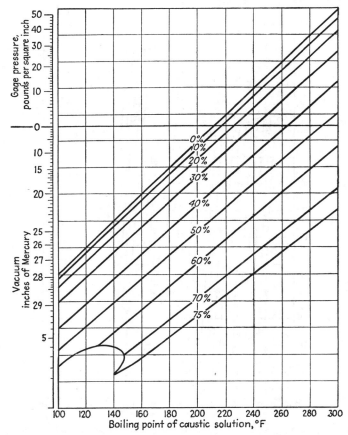

FIG. 14.38. Boiling-point–pressure relations of caustic soda solutions. (*Columbia Alkali Corporation.*)

slurry is pumped to the forced-circulation flash tank g where it is cooled to approximately 192°F and where further flash evaporation and further salt precipitation occurs. The flash-tank effluent is pumped to d. The underflow from this settler consists of a salt slurry which flows into c along with the second-effect effluent, and the salt is eventually removed by *a* as described above. The overflow from *d* is the product liquor containing 50% NaOH, 2.7% NaCl, and 47.3% water.

Steam, usually 20 psig, is introduced into the heating element of the first effect. Here the steam condenses and gives up its latent heat to the liquor circulated through the tubes, causing this liquor to boil when it

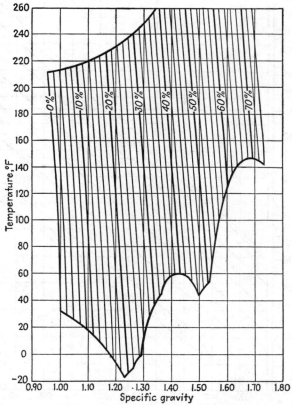

Fig. 14.39. Specific gravity of caustic soda solutions. (*Columbia Alkali Corporation.*)

reaches the vapor body of the first effect. The condensate from the first effect passes through a trap into a flash chamber *h*, where flashing occurs with the evolution of more vapors. The vapors from the first effect leave through a catchall *k*, returning the entrained liquor back to the bottom of the first-effect body. The stripped vapors leaving the catchall combine with the condensate flash vapors and are introduced into the second-effect heating element, where they are condensed and removed by means of a condensate pump. The condensation of these vapors produce further boiling in the second-effect vapor body. The vapors leaving the second body pass through another catchall l into a barometric condenser m, where they are totally condensed by contact with cold water. The

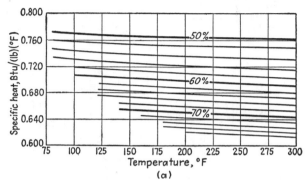

Fig. 14.40a. Specific heats of high-concentration caustic soda solutions.

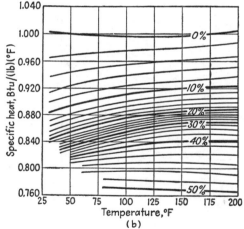

Fig. 14.40b. Specific heats of low-concentration caustic soda solutions. (*Columbia Alkali Corporation.*)

noncondensables from the system eventually reach the condenser through a system of vent piping, and there they are removed by a steam jet ejector n.

Example 14.6. Electrolytic cell liquor containing 8.75% NaOH, 16.60% NaCl, and 74.65% water at a temperature of 120°F is to be concentrated to a product containing 50% NaOH, 2.75% NaCl, and 47.25% water at a temperature of 192°F. The capacity of the unit is 25 tons per 20-hour day based on 100% caustic soda. Physical properties are given in Figs. 14.38 to 14.41.

Solution. Table 14.7 is a material balance. It will be observed that the calculations are conveniently based on 1 ton NaOH production per hour, and therefore the numerals in Fig. 14.37 are obtained by using a factor of 1.25 greater than those shown on Table 14.7.

From the material balance and from the temperature-pressure distribution in Table 14.8 the heat balance (Table 14.9) can be worked out rather easily for this evaporator.

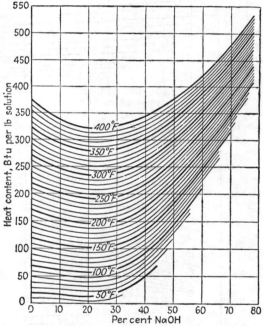

Fig. 14.41. Relative heat content of caustic soda solutions. (*Columbia Alkali Corporation.*)

In Table 14.8 the total temperature difference between the saturated live steam and 27 in. Hg vacuum is $274 - 115 = 159°F$. From Fig. 14.38 the estimated BPR of the solutions in the first and second effects is 77 and 26°F, respectively, corresponding to concentrations appearing in Table 14.7. Thus, the effective temperature difference is $159 - (77 + 26) = 56°F$. The latter is arbitrarily divided evenly between the two effects. In addition to the two effects of the evaporator, an additional amount of evaporation is obtained in the forced-circulation flash tank. The latter operates at 27 in. Hg and receives the concentrated slurry from the first effect and cools it from 246 to 192°F. The last column of Table 14.8 summarizes the thermal and material quantities for the flash tank.

Table 14.7. Caustic Evaporator Material Balance

Basis: 1 ton/hr NaOH

Cell liquor at 120°F	Wash at 80°F
8.75% NaOH = 2000	25% NaCl = 340
16.6% NaCl = 3800	75% H$_2$O = 1020
74.65% H$_2$O = 17050	Total wash = 1360
Total cell liquor = 22850	

	NaOH %	NaOH Lb	NaCl %	NaCl Lb	H$_2$O, Lb	Total, Lb
Overall operation:						
Cell liquor	8.75	2000	16.60	3800	17,050	22,850
Wash	....	....	25.00	340	1,020	1,360
Total in	....	2000	....	4140	18,070	24,210
Product	50.00	2000	2.75	110	1,890	4,000
Removed	....	....	....	4030	16,180	20,210
Weak-liquor settler operation:[a]						
From 2d effect	....	6000	....	5505	20,775	32,280
From 50% settler	50.00	2500	....	1390	2,360	6,500
Total in	....	8500	....	6895	23,135	38,780
Solids	24.62	8500	8.22	2835	23,135	34,470
Salt pptd	....	....	....	4060	....	....
Centrifuge operation:[b]						
From weak liquor settler	....	4000	....	5395	10,905	20,300
Wash	....	....	25.00	340	1,020	1,360
Total in	....	4000	....	5735	11,925	21,660
Solids	23.05	4000	9.84	1705	11,925	17,630
Salt pptd	....	....	....	4030	....	....
2d effect operation:[c]						
From centrifuge	23.05	4000	9.84	1705	11,925	17,630
Cell liquor	8.75	2000	16.60	3800	17,050	22,850
Evaporator feed	14.82	6000	13.60	5505	28,975	40,480
Evaporation	....	....	....	....	8,200	8,200
Product	....	6000	....	5505	20,775	32,280
Solids	19.95	6000	11.22	3370	20,775	30,045
Salt pptd	....	....	....	2135	....	....
1st effect operation:[d]						
From weak liquor settler	24.62	4500	8.22	1500	12,230	18,230
Evaporation	....	....	....	....	7,537	7,537
Product	....	4500	....	1500	4,693	10,693
Solids	47.05	4500	3.84	367	4,693	9,560
Salt pptd	....	....	....	1133	....	1,133
Flash-tank operation:[e]						
From 1st effect	....	4500	....	1500	4,693	10,693
Evaporation	....	....	....	....	443	443
Product	....	4500	....	1500	4,250	10,250
Solids	50.00	4500	2.75	248	4,250	8,998
Salt pptd	....	....	....	1252	....	1,252
From flash tank	....	4500	....	1500	4,250	10,250
50% out	50.00	2000	2.75	110	1,890	4,000
Underflow	....	2500	....	1390	2,360	6,250
Salt pptd	....	....	....	1252	....	1,252
Solids	50.00	2500	2.75	138	2,360	4,998

[a] Solution temperature (approximate) = 150°F.
[b] Solution temperature (approximate) = 145°F.
[c] Solution temperature (approximate) = 141°F.
[d] Solution temperature (approximate) = 246°F.
[e] Solution temperature (approximate) = 192°F.

It was noted that in a forced-circulation evaporator the surface is not a direct function of the temperature difference between the boiling solution and the saturated vapor temperature in the steam chest. The surface in a forced-circulation evaporator is determined by an economic balance between the size of the evaporator and the rate of circulation or, what amounts to the same thing, the power requirements of the pumps.

TABLE 14.8. CAUSTIC EVAPORATOR SUMMARY*

Item	Effect I	Effect II	Flash tank
1. Steam pressure, psi.................	30		
2. Steam temperature, °F.............	274	169	
3. Δt, °F............................	28	28	
4. Liquor temperature, °F............	246	141	192
5. BPR, °F........................	77	26	77
6. Vapor temperature, °F.............	169	115	115
7. λ, Btu/lb.........................	997	1,027	1,027
8. Feed, lb/hr*.....................	22,788	50,602	13,367
9. Product, lb/hr*...................	13,367	40,352	12,813
10. Evaporation, lb/hr*................	9,421	10,250	554
11. Heat flow, Btu/hr*................	11,890,000	11,020,000	
12. U_D, Btu/(hr)(ft²)(°F).............	700		
13. A, ft².........................	683	683	
14. Tubes, OD, in. and BWG..........	1, 16	1, 16	
15. Tube length, ft....................	7	7	
16. No. tubes........................	432	432	
17. Circulating pump, gpm............	3200 at 20 ft	3200 at 20 ft	167 at 45 ft
18. Apparent efficiency, %............	54	64	
19. BHP............................	38	35	8.2
20. Motor, hp.......................	40	40	10.0

* Correspond to actual flows in Fig. 14.37.

A method generally used to solve for surface requirements is indicated below. In designing for optimum surface it is customary to use about 8 fps velocity through the tubes of the heater. Furthermore, in this illustration tubes will be 1 in. OD, 16 BWG, solid nickel, 7'0" long arranged on a two-pass layout. The heaters must be designed for sufficient surface to transfer enough heat from the vapor or steam side to satisfy the heat quantities under 11 of Table 14.8.

For effect I. $0.8w(t_x - 246) = 11,840,000 = U_D A \Delta t$. That is, by circulating w lb/hr of liquid through the tubes and raising the temperature of the latter to t_x, a small temperature differential above the saturated-solution boiling temperature in the flash chamber, a gentle flash will occur with not too rapid a crystal growth. By assuming a given value t_x, both w and Δt are determined. Furthermore, by designing for a given velocity of 8 fps, the value of U_C, the clean overall coefficient, is determined from the thermal characteristics of the solution. This is illustrated below:

$$G = V(s \times 62.5 \times 3600) = 8 \times 1.5 \times 62.5 \times 3600 = 2,700,000, \text{lb/(hr)(ft}^2)$$

For this mass velocity a tube-film coefficient approximately equal to 1375 can be obtained from the methods employed in Chap. 7. Combining the latter with a steam film coefficient of approximately 1500 will give a value for U_C or U_D equal to 700. Setting up the following table:

EVAPORATION

t_x, °F	w, lb/hr	Δt	U_C	A, ft²	a_t, flow area per pass, ft²	G_{calc}	U_{calc}
251	2,970,000	25.4	700	670	0.87	3,420,000	
252	2,480,000	25.0	700	680	0.88	2,820,000	
252.5	2,290,000	24.7	700	686	0.89	2,570,000	700
253	2,120,000	24.5	700	695	0.90	2,520,000	

Thus the gain per minute for circulation is 3200, and from pressure-drop calculations in tubes, the static head, etc., a total dynamic head of 20 ft is found to be required.

TABLE 14.9. CAUSTIC EVAPORATOR HEAT BALANCE
Basis = 1 ton/hr NaOH

Effect	Btu/hr	Evaporation, lb/hr
1. a. Heat in steam = $10{,}500 \times 930 \times 0.974$	9,500,000	
b. Heating liquor = $18{,}230$ $\times (246 - 150) \times 0.83$	1,470,000	
c. Resultant heat	8,030,000	
d. Heat of concentrate	300,000	
e. Heat of vapors	7,730,000/997	7,750
2. a. Heat of vapors	7,730,000	
b. Condensate flash = $10{,}500(274 - 169)$	1,090,000	
c. Heat flow	8,820,000	
d. Heating liquor $= 40{,}480(141 - 130)0.92$	410,000	
e. Heat of vapors	8,410,000/1027	8,200
Flash tank:		
a. Liquor flash = $10{,}693(246 - 192)0.79$	460,000	
b. Heat of concentrate	10,000	
c. Heat of vapors	450,000/1027	443
Total evaporation	....	16,393
Condenser:		
Heat of flash tank vapors (above 110°F)	452,000	
Heat of vapors 2 (above 110°F)	8,450,000	
Total heat in (above 110°F)	8,902,000/20 = 445,000 lb	
Water $\Delta t = (110 - 90) = 20°$	= 890 gpm/(ton)(hr) of NaOH	

For effect II. $0.92w(t_x - 141) = 11{,}020{,}000 = U_D A \, \Delta t$. For a velocity of 8 fps $G = 8 \times 1.35 \times 62.5 \times 3600 = 2{,}430{,}000$ lb/(hr)(ft²). Using thermal characteristics for this solution gives an approximate $U_D = 700$ Btu/(hr)(ft²)(°F).

As for effect I:

t_x, °F	w, lb/hr	Δt	U_C	A, ft²	a_t, flow area per pass, ft²	G_{calc}	U_{calc}
146	2,400,000	25.4	700	620	0.80	2,790,000	700
146.5	2,160,000	25.2	700	683	0.89	2,430,000	

From the design coefficient above a smaller surface could have been used. However, since it is desirable to duplicate both the surface and pump of the first effect, the larger surface is chosen.

THERMOCOMPRESSION

The principle of thermocompression is continuously finding wider application in industry. Consider a single-effect evaporator. It is fed with steam and generates vapors which have nearly as much heat content as was originally present in the steam. These vapors are condensed with water as a convenient means of removal, but this is a severe waste of both Btu and water. Were it not for the fact that the temperature of the generated vapors is lower than the steam, it would be possible to circulate the vapor back into the steam chest and evaporate continuously without supplying additional steam. But apart from the heat balance, a temperature difference must exist between the steam and the generated vapors or no heat would have been transferred originally. If the vapors from the evaporator were compressed to the saturation pressure of the steam, however, the temperature of the vapors could be raised to that of the original steam. The cost of supplying the necessary amount of compression is usually small compared with the value of the latent heat in the vapors.

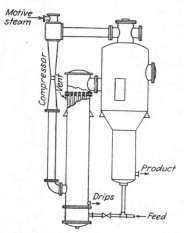

FIG. 14.42. Thermocompression evaporator. (*Buffalo Foundry & Machine Company.*)

The practice of recompressing a vapor to increase its temperature and permit its reuse is *thermocompression*. Where fuel is costly, the compression may be accomplished with a centrifugal compressor as in the Kleinschmidt still for the production of distilled water. Where steam is available at a pressure higher than that required in the evaporator, the

EVAPORATION

recompression can be effected in a steam jet booster. The latter operates on the principle of an ejector and is used to compress vapors instead of a noncondensable gas. A thermocompression evaporator of the type using a booster ejector is shown in Fig. 14.42.

The pressure at which steam is discharged from a thermocompressor depends upon the pressure and proportion at which the live steam and the vapors are supplied. It will be apparent from the analysis below that the higher the desired discharge pressure the greater the percentage of live steam required. Table 14.10 compares the proportions of live steam to entrained vapors for steam-chest pressures of 18 to 26 psig when live steam is available to the jet at 150 psig and the entrained vapors are taken off at 15 psig.

TABLE 14.10. THERMOCOMPRESSOR ECONOMY

Steam-chest pressure, psig	18	20	22	24	26
Entrainment ratio, lb entrained vapor/lb live steam	3.33	2	1.43	1.08	.87
Amount of live steam saved, %	66	67	59	52	46

Table 14.10 shows substantial savings of live steam for fixed quantities of low-pressure steam discharged from the compressor. Figure 14.43 represents the variation of the reciprocal of the entrainment ratio with steam-chest pressure for entrained vapors at different pressures.

Derivation of charts similar to Fig. 14.43 can be made by analyzing the thermodynamic conditions imposed on the jet thermocompressor. Referring to Fig. 14.44, high-pressure steam is caused to expand in the nozzle a from which it issues with a high velocity into the mixing space b where it transfers some of its momentum to the vapor sucked in. In the diffuser section d, which is the reverse of a nozzle, the mixed vapors are compressed to the steam-chest pressure p_3. The work of compression results from the conversion of kinetic energy of the high-velocity mixture into pressure head. Thus, vapor at low pressure p_2 is entrained and compressed to a higher pressure p_3 at the expense of energy in the motive steam. Using the treatment of Kalustian,[1]

let H_1 = enthalpy of steam at p_1, Btu/lb

H_2 = enthalpy of steam after isentropic expansion in nozzle to pressure p_2, Btu/lb

$H_{2'}$ = enthalpy of steam after actual expansion in nozzle to pressure p_2, Btu/lb

e_1 = efficiency of the nozzle

[1] Kalustian, P., *Refrig. Eng.*, **28**, 188–193 (1934).

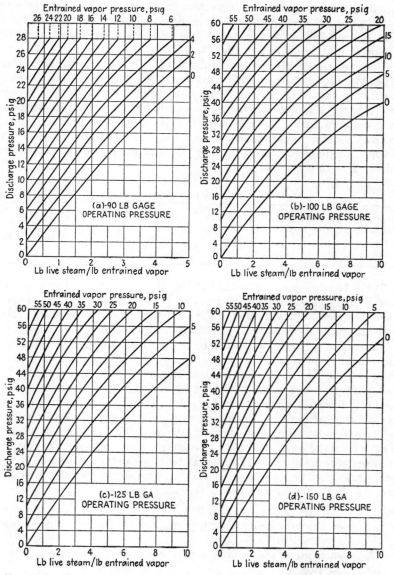

Fig. 14.43. Thermocompressor steam entrainment. (*Schutte & Koerting Company.*)

Then
$$e_1 = \frac{H_1 - H_{2'}}{H_1 - H_2} = \frac{\text{actual work of expansion}}{\text{theoretical work of expansion}} \quad (14.21)$$

Let H_3 = enthalpy of the mixture at start of compression in the diffuser section at p_2, Btu/lb

H_4 = enthalpy of the mixture after isentropic compression from p_2 to the discharge pressure p_3, Btu/lb

e_2 = efficiency of compression in the diffuser

Then
$$e_2 = \frac{H_4 - H_3}{\text{actual work of compression}}$$

Let M_2 = entrained vapor at pressure p_2 lb
M_1 = motive high-pressure steam at pressure p_1, lb

The actual work required for compressing the mixture of vapors in the diffuser is given by $(M_1 + M_2)(H_4 - H_3)/e_2$. The actual work obtain-

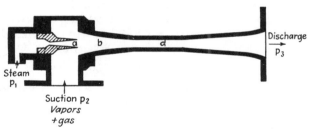

FIG. 14.44. Ejector.

able from the expansion of the high-pressure steam is less than the theoretical value $(H_1 - H_2)$ because of friction in the nozzle, which is taken into account by the nozzle efficiency e_1. There is a further loss in available energy in the transfer of momentum from the high-velocity jet to the relatively slow-moving entrained vapor. If e_3 is the efficiency of momentum transfer, the net available work from the jet is

$$M_1 e_1 e_3 (H_1 - H_2).$$

By equating the net available work obtained from the jet to the actual work required for compressing the mixture in the diffuser, one obtains

Entrainment ratio, $\dfrac{M_2}{M_1} = \left[\dfrac{(H_1 - H_2)}{(H_4 - H_3)} e_1 e_2 e_3 - 1 \right]$,

lb vapor entrained/lb steam (14.22)

Equation (14.22) must be solved by trial and error, since H_3 and H_4 are functions of x_3, the quality of the steam at p_3 which is in itself a function of M_2 and M_1. If one assumes no heat and no enthalpy change due to the mixing a balance on the mixing process gives

$$x_3(M_1 + M_2) = x_{2''}M_1 + x_4 M_2 \qquad (14.23)$$

$$x_3 = \frac{x_{2''}M_1 + x_4 M_2}{M_1 + M_2} \qquad (14.24)$$

where $x_{2''}$ is the quality of motive steam after expansion to p_2 and after it has lost its kinetic energy in the entraining process, x_4 is the quality of the entrained steam, and $x_{2''}$ is related to $x_{2'}$, the quality after expansion but before the entrainment step, by the equation

$$(1 - e_3)(H_1 - H_{2'}) = (x_{2''} - x_{2'})\lambda \qquad (14.25)$$

where λ is the latent heat of vaporization of the fluid at p_2. $x_{2'}$ is related to x_2, the quality after isentropic expansion, by

$$(1 - e_1)(H_1 - H_2) = (x_{2'} - x_2)\lambda$$

As an illustration, take the case where $p_1 = 150$ psig, $p_2 = 9$ psig, $H_1 = 1196$ Btu/lb (from Table 7), $H_2 = 1050$ Btu/lb (after isentropic expansion from p_1 to p_2), and $x_2 = 0.885$. It will be assumed that the values of e_1, e_2, e_3 are, respectively, 0.98, 0.95, and 0.85.

$$H_1 - H_{2'} = 0.98(1196 - 1050) = 143$$

and therefore $H_{2'} = 1053$. From Table 7 $\lambda = 954$ and from Eq. (14.22)

$$954(x_{2'} - x_2) = (1 - 0.98)(1196 - 1050)$$

and therefore $x_{2'} = 0.89$. From Eq. (14.25)

$$(1 - 0.85)(143) = 954(x_{2''} - 0.89)$$

and therefore $x_{2''} = 0.91$. Assuming $x_3 = 0.95$,

$$H_3 = 1159(0.95) = 1100.$$

Since the entrained steam at p_2 is assumed saturated, $x_4 = 1$ and therefore $H_4 = 1164$; hence from Eq. (14.22)

$$\frac{M_2}{M_1} = \frac{146}{64}(0.79) - 1 = 1.80 - 1 = 0.8$$

As a check on the assumed value of x_3, from Eq. (14.24) above

$$x_3 = \frac{x_{2''} + x_4(M_2/M_1)}{1 + M_2/M_1} = \frac{0.91 + 0.80}{1 + 0.80} = \frac{1.71}{1.8} = 0.95$$

(which checks the assumed value).

Since the entrainment ratio M_2/M_1 is a function of the thermal (expansion and compression) and mechanical (mixing) efficiencies of the compressor, specialized and good design of the different elements of the unit will increase the overall efficiency $e_1e_2e_3$. The closer this product approaches unity the higher the entrainment ratio or the greater the saving in live steam. In good design, overall efficiencies approach 0.75 to 0.8, with individual efficiencies in the order of $e_1 = 0.95 - 0.98$, $e_2 = 0.90 - 0.95$, and $e_3 = 0.80$ to 0.85.

To study the saving in steam, the reduction in water requirements and condenser size, a comparison will be made of a straight triple-effect evaporator and one involving the use of a thermocompressor in conjunction with a triple-effect evaporator.

THERMOCOMPRESSION SUGAR EVAPORATOR

Example 14.7. It is required to concentrate 100,000 lb/hr of a 10°Brix sugar solution to 30°Brix. Exhaust steam at 20 psig is available in limited quantity. Water is available at 85°F and also in limited quantity owing to cooling-tower capacity. The feed will enter at 230°F. The thermocompressor evaporator will recompress vapors from the second effect as shown in Fig. 14.45 using saturated steam at 150 psig.

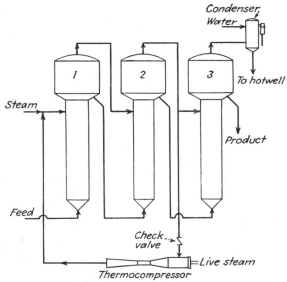

Fig. 14.45. Thermocompression sugar evaporator.

Solution. The analysis is shown in Tables 14.11 to 14.13. The live steam required for the thermocompressor is $M_1 = 32,200 - 14,300* = 17,900$.

Saving in live steam over straight triple effect = 20 per cent
Saving in water over straight triple effect = 20 per cent

* $M_1/M_2 = 1.25$ from Fig. 14.43. $M_1 + M_2 = 32,200$ from heat balance above, and therefore $M_2 = 32,200/(1.25 + 1) = 14,300$.

TABLE 14.11. EVAPORATOR SUMMARY

Effect	Straight triple effect			Thermocompression		
	1	2	3	1	2	3
Steam flow, lb/hr.................	22,400			17,900*		
Steam pressure, psi.	20	9	2	20	9	2
Steam temp, °F...................	258	237	217	258	237	217
Δt, °F.............................	20	18	22	20	18	22
Liquor temp, °F..................	238	219	195	238	219	195
BPR, °F.........................	1	2	3	1	2	3
Vapor temp, °F...................	237	217	192	237	215	192
Vapor pressure, psi/in. Hg........	9	2	10"	9	2	10"
λ, Btu/lb.......................	954	965	983	954	965	983
Liquor in, lb/hr.................	100,000	79,400	56,900	109,000	70,000	52,400
Liquor out, lb/hr................	79,400	56,900	33,300	70,000	52,400	33,300
Evaporation, lb/hr...............	20,600	22,500	23,500	30,000	17,600	19,100
°Brix (out).......................						30
Condenser water, gpm.............		455			365	

* Live steam to thermocompressor.

TABLE 14.12. HEAT BALANCE—STRAIGHT TRIPLE EFFECT

$$\text{Condenser water} = \frac{23,232,000}{500(192 - 5 - 85)} = 455 \text{ gpm}$$

Effect	Btu/hr	Evaporation, lb/hr
1. *a.* Heat in steam = (22,400)(940)(.97).........	20,400,000	
b. Heating liquor = 100,000(238 − 230).92.....	735,000	
c. Evaporation = 20,600..................	19,665,000/954	20,600
d. Liquor to 2d = 79,400		
2. *a.* Heat in 1st vapors......................	19,665,000	
b. Steam cond. flash = 22,400(258 − 237)......	470,000	
c. Liquor flash = 79,400(238 − 217).89........	1,485,000	
d. Heat for evaporation.....................	21,620,000/965	22,500
e. Evaporation = 22,500 lb/hr		
3. *a.* Heat in 2d vapors.......................	21,620,000	
b. Condensate flash = 20,600(237 − 217).......	412,000	
c. Liquor flash = 56,900(217 − 192).85........	1,200,000	
d. Heat for evaporation.....................	23,232,000/983	23,600
e. Product = 56,900 − 23,600 = 33,300 lb/hr		
f. Total evaporation........................		66,700

TABLE 14.13. HEAT BALANCE—THERMOCOMPRESSION EVAPORATOR

$$\text{Condenser water} = \frac{18{,}662{,}000}{500(192 - 5 - 85)} = 365 \text{ gpm}$$

Effect	Btu/hr	Evaporation lb/hr
1. a. Heat in steam = $(32{,}200)(940)(.97)$	29,300,000	
b. Heating liquor = $100{,}000(238 - 230).92$	735,000	
c. Heat for evaporation	28,565,000/954	30,000
d. Liquor to 2d = 70,000 lb/hr		
2. a. Heat in 1st vapors	28,565,000	
b. Heat removed entrained vapors by thermocompression = $\left(\frac{32{,}200}{1.25 + 1}\right) 954$	13,600,000	
c. Heat in 1st vapors	14,965,000	
d. Condensate flash = $32{,}200(258 - 237)$	677,000	
e. Liquor flash = $70{,}000(238 - 217).89$	1,310,000	
f. Heat for evaporation	16,952,000/965	17,600
g. Liquor to 3d = $70{,}000 - 17{,}600 = 52{,}400$ lb/hr		
3. a. Heat in 2d vapors	16,952,000	
b. Condensate flash = $30{,}000(237 - 217)$	600,000	
c. Liquor flash = $52{,}400(217 - 192).85$	1,110,000	
d. Heat for evaporation	18,662,000/983	19,100
e. Product = $52{,}400 - 19{,}100 = 33{,}300$ lb/hr		
f. Total evaporation		66,700

The saving in utilities illustrated in the above problem is of importance only if there is a shortage of them. In all circumstances, however, installation of a thermocompressor in conjunction with a multiple-effect evaporator involves balancing the initial outlay of the additional equipment against the reduction in utilities and in condenser size, change in evaporator body sizes, etc. If auxiliary motive power such as a turbine-driven pump is required in the plant, the available power from 150 psig live steam exhausting to 20 psig should be included in the economy.

Although the above treatment of thermocompression is of interest because of the large savings in utilities which it effects when used in conjunction with a multiple-effect evaporator, it is recommended also in applications where multiple-effect evaporation cannot be used. Thus the use of thermocompression evaporators is particularly advantageous in concentrating liquids which are sensitive to high temperatures. Low operating temperatures are assured because a high vacuum can be maintained. The steam used in a steam chest is usually below atmospheric pressure.

PROBLEMS

14.1. It is desired to concentrate a caustic soda solution from 12.5 to 40 per cent in a double-effect evaporator. 50,000 lb/hr of feed solution enters the evaporator at 120°F, and steam at 15 psig is available for effecting the concentration. Vacuum in the second effect is held at 24 in. Hg, and water from a cooling tower is available at 85°F. Estimate (a) the quantity of steam and water required for forward feed operation; (b) the quantity of steam and condenser water required for backward feed operation; (c) the surface required for operation in (a), assuming overall coefficients of heat transfer of 400 and 250 Btu/(hr)(ft²)(°F) for the first and second effects; (d) the surface in (b) for values of U_D equal to 450 and 350 in the two effects respectively; (e) if equal bodies were required for (c) and (d), what steam economy could be expected.

14.2. In Prob. 14.1 it is desired to concentrate the caustic soda solution in a forced-circulation double-effect evaporator unit. Only a limited amount of steam at 15 psig is available, and it is therefore decided to use turbine-driven circulation pumps taking 150 psig steam and exhausting at 15 psig. Assuming feed rate, vacuum, and water temperature as in Prob. 14.1 and forward-feed operation and equal surface for both effects, estimate (a) the water rate required for the turbine if its efficiency is taken at 70 per cent, (b) the pounds of 15 psig steam required for the desired concentration, (c) the make-up steam at 15 psig required if the turbine exhaust is used for the concentration and a solution circulation rate of 6 fps is assumed in the heaters of the two effects. (*Hint.* Assume $U_D = 400$; estimate surface, cross-sectional flow area, total circulation, and total head and power required.) (d) Estimate surface required for each effect.

14.3. A 20,000 lb/hr sugar solution at 180°F is to be concentrated from 12 to 30°Brix in a double-effect, forward-feed, calandria-type evaporator. Assuming that exhaust steam from a steam engine is available at 5 psig and the vacuum in the second effect is held at 23 in. Hg, estimate (a) steam economy; (b) surface of each body, assuming U_D of 500 and 200 for first and second effects, respectively, and equal surface for each effect; (c) water required for a barometric condenser, assuming that 90°F water is available and a 10°F approach to condensing temperature is permitted.

14.4. A triple-effect sugar evaporator is to be designed to concentrate 100,000 lb/hr of solution at 150°F from 14 to 50°Brix. Steam is available at 150 psig, and water at 75°F. Steam to the first effect will be 25 psig, and the vacuum in the last effect will be held at 24 in. Hg. Vapors will leave the first effect at 15 psig and will partly be compressed by a thermocompressor using 150 psig steam and partly enter the second effect as steam. Assuming equal surfaces for the three bodies, forward-feed arrangement, and a 5°F approach in the condenser, estimate (a) the total steam required to the first effect, (b) the total 150 psig steam used, (c) economy, (d) water required in the condenser, (e) surface required, assuming $U_D = 500, 300,$ and 150 for the three effects, respectively.

14.5. A paper mill producing 300 tons of pulp per 24 hr by the magnesium sulfite process concentrates a 12 per cent waste liquor to 55 per cent in a quintuple-effect long-tube evaporator. The solution entering the last two effects in parallel at 135° is evaporated by 45 psig saturated steam. Assume the following: (a) Feed contains 2800 lb total solids per ton pulp produced; (b) first effect is divided into two parallel bodies of 4500 and 7000 ft², respectively; (c) water is available at 75°F; (d) vacuum in the last-effect vapor body is 26 in. Hg; (e) 7°F water approach to condenser saturation temperature; (f) BPR is 18, 16, 13, 10, 5, 7°F, respectively, for all bodies; (g) product from first effect is flashed to fourth-effect steam chest; (h) vapor condensate flashes between effects.

EVAPORATION 451

Find (a) material and heat balance for the evaporator, (b) total water required (gallons per minute), (c) the total surface required, (d) steam economy of the evaporator.

14.6. Residual liquor from the electrolysis of brine in chlorine cells flows to a caustic concentration system at the rate of 1176 tons per 21-hr day. Concentration to 50% NaOH is effected in a triple-effect, forced-circulation evaporator. The latter boils off the water content in three successive stages, thereby crystallizing the salt content and increasing the percentage of caustic soda contained in the circulating solution.

The liquor flow is as follows:

Order of flow	Analysis of outgoing liquor flow			
	% crystallized salt	% dissolved salt	% dissolved NaOH	% water
1. Cell liquor at 160°F............	0	14.95	11.00	74.05
2. 2d effect evaporator............	3.56	15.47	13.97	67.00
3. 3d effect evaporator............	7.55	8.08	22.43	61.94
4. Weak-liquor settler.............	0*	8.00*	25.70*	66.30*
5. 1st effect evaporator...........	8.23	4.84	42.10	44.83
6. Flash tank.....................	11.35	2.40	44.30	41.95
7. 50% caustic settler.............	0*	2.70*	50.00*	47.30*

* Per cent changes owing to removal of crystallized salt at these stages to salt-recovery system.

Utilities:

Steam at 75 psig
Water to barometric condenser at 80°F
Condenser held at 27¼ in. Hg vacuum (third effect at 27 in. Hg vacuum)

Determine (a) the material balance in the system, (b) the heat balance in the system; estimate (c) the surface required ($Ans.$ three 1800 ft² external heaters), (d) total water required, (e) steam economy.

14.7. A 15°Brix sugar solution is to be concentrated to 60°Brix in a quadruple-effect calandria-type evaporator. Part of the vapors from the first, second, and third effects will heat the cold solution from 100 to 220°F in heat exchangers prior to injection into the first effect of the evaporator system. Dilute juice is concentrated at the rate of 500 gpm, using live steam at 25 psig. Estimate the following: (a) material balance through the evaporator system, (b) heat balance on the evaporator system assuming condensate flashing between effects and 26½ in. Hg vacuum in fourth effect, (c) steam economy, (d) water rate to a barometric condenser assuming 75°F water available, (e) surface required for the heaters, (f) surface required for the calandria bodies, assuming equal surface for each body.

NOMENCLATURE FOR CHAPTER 14

A	Heat-transfer surface, ft²
a_t	Tube-side flow area, ft²
BPR	Boiling point rise, °F
C	Specific heat of hot fluid in derived equations, Btu/(lb)(°F)
c	Specific heat, Btu/(lb)(°F)

D	Inside diameter of tubes, ft
e_1, e_2, e_3	Efficiency of nozzle, diffuser, and momentum transfer, respectively, dimensionless
G	Mass velocity, lb/(hr)(ft²)
g	Acceleration of gravity, ft/hr²
H	Enthalpy; H_1, motive steam, H_2 after isentropic expansion in nozzle, H_2' after actual expansion in nozzle; H_3 at start of compression in diffuser; H_4 after compression in diffuser, Btu/lb
M_1, M_2	High-pressure motive steam and entrained vapor, respectively, lb
p_1, p_2, p_3	Pressure of motive steam, suction, and ejector discharge, respectively, psi
p_s, p_W	Vapor pressure of solution and or pure water, respectively, psi
Δp	Hydrostatic pressure, ft
Q	Heat flow, Btu/hr
T	Temperature of the hot fluid, °F
T_R	Boiling temperature, °R
t	Temperature of fluid in general, °F
t_a, t_s, t_W	Approach temperature, temperature of solution and of pure water, °F
t_s', t_W'	Boiling point of solution and of pure water, °R
$\Delta t, (\Delta t)_a$	True temperature difference, apparent temperature difference, °F
$(\Delta t)_w$	Temperature difference between the wall and liquid, °F
Δt_h	Temperature rise due to hydrostatic heat, °F
$\Delta t_s, \Delta t_W$	Temperature change of boiling points of solution and pure water, °F
U	Overall coefficient of heat transfer, Btu/(hr)(ft²)(°F)
U_C, U_D	Clean and design overall coefficient of heat transfer, Btu/(hr)(ft²)(°F)
W	Weight flow of hot fluid, lb/hr
w	Weight flow of cold fluid, lb/hr
w_e	Evaporation, lb/hr
v	Specific volume, ft³/lb
x	Quality, subscripts correspond to $p_1, p_2,$ and p_3, per cent
z	Height, ft
β	Coefficient of thermal expansion, per cent/°F
λ	Latent heat of vaporization, Btu/lb
λ_s, λ_W	Latent heat of vaporization of solution and pure water, Btu/lb
μ	Viscosity, centipoises $\times 2.42 = $ lb/(ft)(hr)

Subscripts (except as noted above)

F	Feed
S	Steam
v	Vapor or vaporization
1, 2, 3, 4	Effect

CHAPTER 15

VAPORIZERS, EVAPORATORS, AND REBOILERS

In the preceding chapter calculations were included for only those types of evaporators which are usually designed on the basis of "accepted" overall heat-transfer coefficients. It is unfortunate that so important a class of apparatus is designed in this way, but the properties of concentrated aqueous solutions present problems which leave little alternative. When applied to solutions of liquids in water rather than solids in water, the evaporator may be calculated by methods similar to those of earlier chapters. Heat-transfer coefficients in the basket, calandria, and horizontal-tube evaporators are excluded from further discussion.

For the sake of clarity it is well to repeat the definitions adopted in Chap. 14 for dealing with *vaporizing exchangers*. Any unfired exchanger in which one fluid undergoes vaporization and which is not a part of an evaporation or distillation process is a *vaporizer*. If the vapor formed is steam, the exchanger is referred to as an *evaporator*. If a vaporizing exchanger is used to supply the heat requirements of a distillation process as vapors at the bottom of the distilling column, it is a *reboiler* whether or not the vapor produced is steam. The process requirements of evaporators have already been treated, and those of vaporizers and reboilers will be discussed presently. The principles and limitations which apply to vaporizers and vaporizer processes specifically apply also to evaporators and reboilers and should be regarded accordingly.

Vaporizing Processes. Vaporizers are called upon to fulfill the multitude of latent-heat services which are not a part of evaporative or distillation processes. The heat requirements are usually very simple to compute. Perhaps the commonest type of vaporizer is the ordinary horizontal 1-2 exchanger or one of its modifications, and vaporization may occur in the shell or in the tubes. If steam is the heating medium, the corrosive action of air in the hot condensate usually makes it advantageous to carry out the vaporization in the shell.

There are some fundamental differences between the operation and calculation of vaporization as treated in Chap. 14 and in the 1-2 horizontal or vertical vaporizer. In the power-plant evaporator, for example, the upper 50 or 60 per cent of the shell is used for the purpose of disengaging the liquid entrained by the bursting bubbles at the surface of the pool. The disengagement is further implemented by the use of a steam separator in the shell. The mechanical design and thickness of the evaporator shell, flanges, and tube sheets are based upon the product of the shell-side pressure and the diameter of the shell. In the majority of instances the pressure or vacuum is not great and the shell, flange, and tube-sheet thicknesses are not unreasonable. In the case of a vaporizer, however, operation is often at high pressure, and it is usually too expensive to provide disengagement space in the shell, since the inclusion of disengagement space at high pressures correspondingly increases the shell thickness. For this reason vaporizers are not usually designed for internal disengagement. Instead some external means, such as an inexpensive welded drum, is connected to the vaporizer wherein the entrained liquid is separated from the vapor.

When a vapor evaporates from the surface of a pool, as in the power-plant evaporator, it is possible to evaporate 100 per cent of the liquid fed to it without reducing the level of the pool provided the evaporator was originally filled to operating level with liquid. The reason less than 100 per cent of the feed is normally vaporized is because residue accumulates and it is necessary to provide a blowdown connection for its removal.

When a 1-2 exchanger is used as a vaporizer, it is filled with tubes and cannot be adapted for blowdown, since all the feed to a vaporizer is usually of value and a rejection as blowdown is prohibitive. If the feed were completely vaporized in the vaporizer, it would emerge as a vapor and any dirt which was originally present would be left behind on the tube surface over which total vaporization occurred, fouling it rapidly. If the 1-2 exchanger (vaporizer) were over-designed, that is, if it contained too much surface, disengagement would have to occur on the tubes and due to the excess surface the vapor would superheat above its saturation temperature. The latter is undesirable in most processes, since superheated vapors subsequently require surface elsewhere for desuperheating. In the case of a reboiler it will be shown later that superheat actually reduces the performance of the distilling column. These factors establish a rule which should always be employed for the calculation of vaporization processes: *The feed to a vaporizer should not be vaporized completely.* The value of this rule is apparent. If less than 100 per cent of the feed is vaporized in a 1-2 exchanger, the residual liquid can be counted on to prevent the accumulation of dirt directly on the surface of the heating element. A maximum of about 80 per cent vaporization appears to provide favorable operation in 1-2 exchangers, although higher percentages may be obtained in vessels having internal disengagement space.

Forced- and Natural-circulation Vaporizer. When liquid is fed to a vaporizer by means of a pump or gravity flow from storage, the vaporizer

VAPORIZERS, EVAPORATORS, AND REBOILERS

is fed by *forced circulation*. A typical example is shown in Fig. 15.1. The circuit consists of a 1-2 exchanger serving as the vaporizer and a disengaging drum from which the unvaporized liquid is withdrawn and recombined with fresh feed. The generated vapor is removed from the top of the drum.

Since it is desirable to vaporize only 80 per cent of the liquid entering the vaporizer, the total liquid entering will be 125 per cent of the quantity of vapor required. Even better cleanliness will result if less than 80 per cent of the liquid is vaporized, and this can be accomplished by *recirculation*. Suppose 8000 lb/hr of vapor is required. The liquid entering the vaporizer will be $1.25 \times 8000 = 10{,}000$ lb/hr. In this manner 80 per cent will be vaporized when 10,000 lb/hr of liquid passes into the vapor-

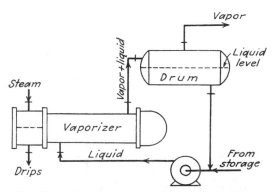

Fig. 15.1. Vaporizing process with forced circulation.

izer. The 2000 lb/hr of liquid which was not vaporized will be recombined with 8000 lb/hr of fresh liquid coming from storage or, if the storage is hot, returned directly to storage.

Now suppose that, for greater cleanliness and the attainment of a higher coefficient, the liquid is speeded through the vaporizer at a faster flow rate by means of a larger pump. It may be desired that liquid enter the vaporizer at a rate of 40,000 instead of 10,000 lb/hr suggested by the process. Since only 8000 lb/hr of vapor will be produced, only 8000 lb/hr of fresh liquid will continue to come from storage. The difference, 32,000 lb/hr, will have to be supplied by recirculation again and again of unvaporized liquid through the vaporizer. For 40,000 lb/hr entering the vaporizer, only 8000 lb/hr, or 20 per cent will actually be vaporized, and the higher velocity over the heat-transfer surface and the low percentage vaporized will permit longer operation without excessive fouling.

The advantage of recirculation can be computed economically. The high recirculating rate will increase the power cost but decrease the size of the equipment and the maintenance cost.

The vaporizer may also be connected with a disengaging drum without the use of a recirculating pump. This scheme is natural circulation and is shown in Fig. 15.2. It requires that the disengaging drum be elevated above the vaporizer. Recirculation is effected by the hydrostatic head difference between the column of liquid of height z_1 and the column of mixed vapor and liquid of height z_3. The head loss in the vaporizer itself due to frictional pressure drop corresponds to z_2. The hydrostatic head difference between z_3 and z_1 is available to cause liquid to circulate at

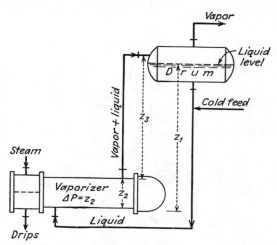

Fig. 15.2. Vaporizing process with natural circulation.

such a velocity that it produces a pressure drop z_2 in the vaporizer equal to the hydrostatic difference between z_3 and z_1. The cold feed is usually fed high on the return pipe so that the leg z_1 will have as great a density and hydrostatic pressure $z_1\rho$ as possible. If the pressure of the feed is greater than the operating pressure of the system, the feed will have to join the return pipe at a lower point unless a jet is used, so that there will be a sufficient liquid head above the junction to prevent feed from backing up directly into the drum. Piping the feed directly to the disengaging drum usually makes control of the operating variables more difficult. If there is a sufficiently great pressure difference between the feed and operating pressure, the feed can be used as the motive fluid in an ejector to increase the recirculation of liquid through the vaporizer. This is inefficient if the fresh liquid is compressed purposely to serve as the motive fluid, but if the fresh liquid is available under pressure for other reasons, the ejector permits recovery of some of the pressure head.

The advantages of forced circulation or natural circulation are in part economic and in part dictated by space. The forced-circulation arrangement requires the use of a pump with its continuous operating cost and fixed charges. As with forced-circulation evaporators, the rate of feed recirculation can be controlled very closely. If the installation is small, the use of a pump is preferable. If a natural-circulation arrangement is used, pump and stuffing-box problems are eliminated but considerably more headroom must be provided and recirculation rates cannot be controlled so readily. Natural-circulation steam generators are frequently laid out in accordance with Fig. 15.2 but employing a vertical

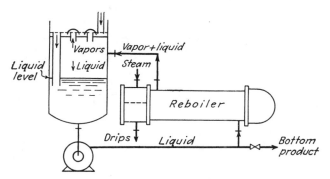

Fig. 15.3. Forced-circulation reboiler arrangement.

1-1 exchanger with vaporization in the tubes. Water is especially adaptable to natural-circulation arrangements, since the density differences between the liquid and vapor at a given temperature are very large.

Reboiler Arrangements. When reboilers are used, the space at the bottom of the column between the liquid level and the bottom plate is employed for disengagement and the bubble caps serve as separators. A typical arrangement for a forced-circulation reboiler is shown in Fig. 15.3. This type is called a *pump-through* reboiler. All the liquid on the bottom plate, frequently called *trapout* to distinguish it from bottom product, is carried by the downcomer below the liquid level of the column. The liquid can be recirculated through the reboiler as many times as is economically feasible, so that the percentage vaporized per circulation is kept low while the bottom product is withdrawn from a separate connection. In general, forced-circulation or pump-through reboilers are used only on small installations or those in which the bottoms liquid is so viscous and the pressure drop through the piping and reboiler so high that natural circulation is impeded.

By far the greater number of large reboiler installations employ natural circulation. This can be achieved in either of two simple ways as shown in Fig. 15.4a and b. In Fig. 15.4a all the liquid on the bottom plate is circulated directly to the reboiler, whence it is partially vaporized. The unvaporized portion, on being disengaged under the bottom plate, is withdrawn as bottom product. In Fig. 15.4b the liquid passes through the downcomer below the liquid level of the column as in forced circulation. The bottom liquid is free to recirculate through the reboiler as many times as the hydrostatic pressure difference between z_1 and z_3 will permit. Because there is no opportunity for recirculation in the arrange-

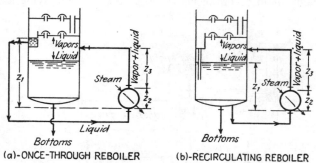

FIG. 15.4. Natural-circulation reboiler arrangements.

ment in Fig. 15.4a, it is called a *once-through* reboiler arrangement. Figure 15.4b is referred to as a *recirculating* reboiler.

Classification of Vaporizing Exchangers. There is a greater hazard in the design of vaporizing exchangers than any other type of heat exchanger. For this reason it is convenient to set up a classification based on the method of calculation as employed for each distinct type of service. Each of the common classes below is distinguishable by some difference in calculation.

A. Forced-circulation vaporizing exchangers
  1. Vaporization in the shell
    a. Vaporizer or pump-through reboiler with isothermal boiling
    b. Vaporizer or pump-through reboiler with boiling range
    c. Forced-circulation evaporator or aqueous-solution reboiler
  2. Vaporization in the tubes
    a. Vaporizer or pump-through reboiler with or without boiling range
    b. Forced-circulation evaporator or aqueous solution reboiler

B. Natural-circulation vaporizing exchangers
  1. Vaporization in the shell
     a. Kettle reboiler
     b. Chiller
     c. Bundle-in-column reboiler
     d. Horizontal thermosyphon reboiler
  2. Vaporization in the tubes
     a. Vertical thermosyphon reboiler
     b. Long-tube vertical evaporator

Heat-flux and Temperature-difference Limitations. It may be assumed that process conditions will always be established which provide for the vaporization of only part of the liquid fed to a vaporizer. When vaporizing liquids from pools, extremely high maximum fluxes have been obtained. For water a maximum flux is reported at 400,000 Btu/(hr)(ft^2) and for organics from 70,000 to 125,000 Btu/(hr)(ft^2), although these have been obtained only in laboratory-scale apparatuses with perfectly clean surfaces. It is again pointed out that the maximum flux occurs at the critical temperature difference and is a limitation of the maximum coefficient which may be attained. Beyond the critical temperature difference both the coefficient and the flux decrease, the decrease being due to the formation of a layer of gas on the tubes. It is the phenomenon of vapor blanketing which poses the principal difficulty in the design and operation of vaporizing exchangers.

Fluxes of magnitudes as high as those above are of little practical value in design. It will be recognized that vaporization in a 1-2 exchanger, occurring as it does without continuous disengagement, is very different from vaporization out of liquid pools. By restricting disengagement in the 1-2 exchanger, the possibility of vapor blanketing is increased greatly, so that it is necessary also to restrict the flux to an allowable value safely out of the range in which blanketing might occur. The flux is defined by Q/A or $U_D \Delta t$ but not by $h_v(\Delta t)_w$, where h_v is the vaporization coefficient and $(\Delta t)_w$ is the temperature difference between the tube wall and the boiling temperature. $h_v(\Delta t)_w$ is the flux based on the clean surface A_C, while Q/A is the flux based on the actual surface A, and A is greater than A_C in a vaporizer designed with a dirt factor. However, it is customary to restrict both Q/A and h_v to safe maxima, the two also serving to prevent the presence of too great a temperature difference $(\Delta t)_w$. The following restrictions will be observed throughout this chapter:

I. *Flux*
  a. The maximum allowable flux for forced circulation vaporizers and reboilers vaporizing organics is 20,000 Btu/(hr)(ft^2) and for natural circulation 12,000 Btu/(hr)(ft^2).

b. The maximum allowable flux for the vaporization of water or aqueous solutions of low concentration using forced or natural circulation is 30,000 Btu/(hr)(ft²).

II. *Film coefficient*
   a. The maximum allowable vaporizing film coefficient for the forced- or natural-circulation vaporization of organics is 300 Btu/(hr)(ft²)(°F).
   b. The maximum vaporizing film coefficient for the forced- or natural-circulation vaporization of water and aqueous solutions of low concentration is 1000 Btu/(hr)(ft²)(°F).

Relationship between Maximum Flux and Maximum Film Coefficient. The objects of the limitations above are the elimination of all vapor blanketing possibilities. Suppose it is desired partially to vaporize an organic compound boiling at 200°F in a forced-circulation vaporizer using steam at a temperature of 400°F so that $\Delta t = 200°F$ and the flow is such that a vaporizing coefficient of 300 Btu/(hr)(ft²)(°F) can be obtained. If the condensing steam coefficient is 1500, $U_C = 250$, and if $R_d = 0.003$, $U_D = 142$. The flux will be $142 \times 200 = 28{,}400$ Btu/(hr)(ft²), which exceeds limitation Ia. Since Q/A or $U_D \Delta t$ may not exceed 20,000, any change to permit compliance with Ia amounts to an increase in the total surface for vaporization. If the original steam and vapor temperatures are retained, the new overall coefficient U_D will be $20{,}000/200 = 100$ Btu/(hr)(ft²)(°F). The temperature difference $(\Delta t)_w$ may be greater than the critical temperature difference, since it does not occur at the maximum attainable flux and under these conditions the critical temperature difference can be exceeded within limits without the danger of vapor blanketing. There is no advantage to the use of very high temperature differences, however, since at the maximum allowable flux any increase in Δt must be offset by a decrease in the allowable value of U_D. Only when U_D is naturally small may the use of a high Δt be partially justified.

The test of whether or not a vaporizer exceeds the allowable flux is always determined by dividing the total vaporizing heat load by the total available surface for vaporization. By the same token the maximum value of U_D which may be anticipated is given by $U_D = (Q/A)(1/\Delta t)$ regardless of the dirt factor which results. When setting the temperature of the heating medium, it is seen that the use of a large Δt and corresponding $(\Delta t)_w$ also requires decreasing U_D which in turn gives a large value of $R_d = 1/U_D - 1/U_C$. The large dirt factor is not essential to the continued operation of the vaporizer from the standpoint of dirt but only as a preventive against vapor blanketing. Accordingly, when the temperature of the heating medium may be selected independently such as by setting the pressure in the case of steam, it need not be set at a value of Δt greater than that which gives a U_D corresponding to the **desired** dirt factor.

FORCED-CIRCULATION VAPORIZING EXCHANGERS

1. Vaporization in the Shell

a. **Vaporizer or Pump-through Reboiler with Isothermal Boiling.** The calculations employed in the solution of this type of vaporizer are common to the many simple vaporizing problems found in a plant whether or not connected with a distilling column. If a liquid is substantially pure or a constant-boiling mixture, it will boil isothermally. This usually applies to the bottom liquid of a distilling column separating a binary mixture into relatively pure compounds. For utility boiling operations, such as the vaporization of a cold liquid coming from storage, the liquid may not be at its boiling point and may require preheating to the boiling point. Since the shell of a forced-circulation vaporizer is essentially the same as any other 1-2 exchanger, the preheating can be done in the same shell as the vaporization. If the period of performance of a vaporizer is to be measured by a single overall dirt factor, it is necessary to divide the shell surface into two successive zones, one for preheating and one for vaporization, in much the same manner employed in condenser-subcoolers. The true temperature difference is the weighted temperature difference for the two zones, and the clean coefficient is the weighted clean coefficient as given by Eqs. (12.50) and (12.51).

If steam is the heating medium, only two tube passes are required and these need not be equally divided, since the return pass carries considerably less vapor than the first pass. If a hot stream such as gas oil is the heating medium, there is a problem in determining the true temperature difference in each zone. If the approach between the heating-medium outlet temperature and the vapor outlet temperature is not too small, the true temperature difference can be approximated by considering the temperature fall in each zone proportional to the heat removed from the heating medium. The method of using zones has also been discussed in Chap. 12 along with condenser-subcoolers.

Film Coefficients. Where there is a preheating zone, it can be computed by the use of Fig. 28 like any other heater with the cold fluid in the shell. The isothermal boiling film coefficient is also obtained by the use of Fig. 28 based on the premise that the heat must first be absorbed by the liquid by forced convection before passing into the vapor bubbles and that the liquid heating coefficient is the controlling coefficient in the sequence.

Pressure Drop. The shell-side pressure drop for the vaporizing zone is computed by introducing the average specific gravity into the denominator of Eq. (7.44). If the vaporizing liquid boils isothermally at t_s and receives heat from a medium having a range $T_1 - T_2$, two possibilities may be considered:

Case I: The vapor and heating medium may be in counterflow.
Case II: The vapor and heating medium may be in parallel flow.

CASE I: VAPORS AND HEATING MEDIUM IN COUNTERFLOW. Referring to Fig. 15.5, if W is the weight flow of the heating medium, C its specific heat, and T is the temperature of the heating medium at any tube length x,

$$WC \, dT = Ua'' \, dx(T - t_s) \quad (15.1)$$

where $a'' \, dx$ is the surface. Integrating T with respect to x,

$$\ln(T - t_s) = \frac{Ua''x}{WC} + C_1 \quad (15.2)$$

At $x = 0$,

$$C_1 = \ln(T_2 - t_s) \quad (15.3)$$

$$\ln \frac{T - t_s}{T_2 - t_s} = \frac{Ua''x}{WC} \quad (15.4)$$

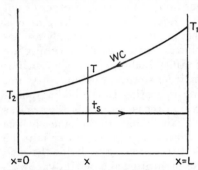

Fig. 15.5. Temperatures during vaporization of an isothermal fluid by a non-isothermal heating medium.

or

$$T = t_s + (T_2 - t_s)e^{Ua''x/WC} \quad (15.5)$$

When $x = L$, $T = T_1$,

$$Q = WC(T_1 - T_2) = WC(T_2 - t_s)(e^{Ua''L/WC} - 1) \quad (15.6)$$

The pounds evaporated w_e at any point x are given by

$$w_e = \frac{WC}{\lambda_s}(T_2 - t_s)(e^{Ua''x/WC} - 1) \quad (15.7)$$

If w is the weight flow of all the isothermal fluid and v_{av} its average specific volume, then for the entire mixture in passing from inlet to outlet

$$wv_{av} = \frac{WC}{\lambda_s}(T_2 - t_s)(e^{Ua''x/WC} - 1)v_v$$

$$+ \left[w - \frac{WC}{\lambda_s}(T_2 - t_s)(e^{Ua''x/WC} - 1)\right]v_l \quad (15.8)$$

where v_v is the specific volume of the vapor and v_l is the specific volume of the liquid.

For the pressure drop by assuming $1/\rho_{av} = v_{av}$

$$\int d\,\Delta P = \int \frac{fG^2}{2g\rho_{av}} dx = \int \frac{fG^2}{2g} v_{av} \, dx \quad (15.9)$$

$$\frac{2gw}{fG^2} \int d\,\Delta P = \int \left[\frac{WC}{\lambda_s}(T_2 - t_s)(e^{Ua''x/WC} - 1)v_v \, dx + wv_l \, dx \right.$$

$$\left. - \frac{WC}{\lambda_s}(e^{Ua''x/WC} - 1)v_l \, dx\right]$$

VAPORIZERS, EVAPORATORS, AND REBOILERS

Integrating and simplifying,

$$\frac{2g\,\Delta P}{fG^2 L} = v_{av} = \frac{WC}{\lambda_s w}[(v_v - v_l)(\text{LMTD}) - (T_2 - T_s)(v_v - v_l)] + v_l \quad (15.10)$$

Generally, however, it is simpler to apply the law of mixtures as given below, which closely resembles Eq. (15.10).

Case I: Vapors in counterflow to heating medium:

$$v_{av} = \frac{WC}{UA}(v_2 - v_l) - \frac{WC(T_2 - t_s)}{\lambda_s w}(v_v - v_l) + v_l \quad (15.11)$$

$$s_{av} = \frac{1}{v_{av}(62.5)} \quad (15.12)$$

Case II: Vapors in parallel flow to heating medium:

$$v_{av} = \frac{WC}{UA}(v_l - v_2) + \frac{WC}{\lambda_s w}(T_1 - t_s)(v_v - v_l) + v_l \quad (15.13)$$

$$s_{av} = \frac{1}{v_{av}(62.5)}$$

With the average gravity based on Eqs. (15.11) and (15.13), Eq. (7.44) is only part of the pressure drop, since no term has been included to account for the acceleration of the vapor through the vaporizer or the contraction loss at the outlet. Instead of using Eqs. (15.11) and (15.13), the pressure drop can be computed more rapidly by an arbitrary assumption of the value of the average specific gravity. At the start of the vaporizing zone the gravity is that of the liquid, whereas at the outlet it is considerably less, even if the percentage converted to vapor is not very high. Likewise with constant-flow area the velocity through the shell increases toward the outlet owing to the large volume of the exit fluid, although the mass velocity is presumably constant. On the other hand, the viscosity effective upon the pressure drop probably does not vary much over the vaporizing zone, being that of the liquid throughout. The pressure drop can be computed using the mass velocity as heretofore, the Reynolds number based on liquid properties at inlet conditions and the *mean* gravity between inlet and outlet. Similar reasoning can be applied to vaporization on the tube side. The comparison between the true average gravity defined by Eqs. (15.11) and (15.13) and the mean of the inlet and outlet gravities is shown below.

Example 15.1. Calculation of the Average Specific Volume. A vaporizer is to generate 10,000 lb/hr of 15 psig steam by removing heat from a bottom product. The bottom product is 150,000 lb/hr of approximately 42°API kerosene leaving a column at 400°F. The design coefficient in the vaporizer will be about 100.

Compare the true average specific volume with that obtained by the approximate method.

$t_s = 250°F \qquad \lambda_s = 945.3$ Btu/lb (Table 7)
Steam, $Q = 10{,}000 \times 945.3 = 9{,}450{,}000$ Btu/hr
Kerosene, $Q = 150{,}000 \times 0.63(400 - 300) = 9{,}450{,}000$ Btu/hr
$\Delta t = $ LMTD $= 91°F$ (5.14)

$$A = \frac{Q}{U_D \, \Delta t} = \frac{9{,}450{,}000}{100 \times 91} = 1040 \text{ ft}^2$$

$WC = 150{,}000 \times 0.63 = 94{,}500$ Btu/°F

From Table 7
$$v_l = 0.017 \qquad v_v = 13.75 \text{ ft}^3/\text{lb}$$

By the law of mixtures

Assume 80 per cent of the outlet fluid is vapor.

$v_2 = 0.80 \times 13.75 + 0.20 \times 0.017 = 11.0$ ft^3/lb
$$v_{\text{av}} = \frac{94{,}500(11.0 - 0.017)}{100 \times 1040} - \frac{94{,}500(300 - 250)}{945.3 \times 10{,}000}(13.75 - 0.017) + 0.017$$
$= 3.04$ ft^3/lb

By the approximate method

$v_l = 0.017 \qquad v_2 = 11.0$ ft^3/lb
$$v_{\text{av}} = \frac{0.017 + 11.0}{2} = 5.5 \text{ ft}^3/\text{lb}$$

Actual density $= \dfrac{1}{3.04} = 0.329$ lb/ft^3 $\qquad s = \dfrac{0.329}{62.5} = 0.0053$

Approximate density $= \dfrac{1}{5.5} = 0.182$ lb/ft^3 $\qquad s = 0.0029$

Since the pressure drop is inversely proportional to the gravity, the approximate method is safe. The acceleration loss is usually taken at about two velocities heads and can be omitted when using the approximate method.

While the condenser is usually the cleanest part of a distillation or vaporization-condensation assembly, the vaporizer is usually quite the opposite. Vaporizers tend to accumulate dirt, and for this reason high recirculation rates and large dirt factors will often be desirable. Preference should be given to the use of square pitch and a removable tube bundle. Although it may reduce the possibility of using a 1-2 vaporizing exchanger for other services, the baffle spacing can be increased or staggered from inlet to outlet to reduce the pressure drop of the fluid vaporizing in the shell.

Example 15.2. Vaporizer or Pump-through Reboiler with Isothermal Boiling. It is desired to provide 19,750 lb/hr of butane vapor at 285 psi using cold feed from storage at 75°F. The butane will boil isothermally at 235°F, and steam will be available at 100 psi.

VAPORIZERS, EVAPORATORS, AND REBOILERS

Available for the service is a 15¼ in. ID 1-2 exchanger with 76 1 in. OD, 16 BWG tubes 16'0" long laid out on 1¼-in. square pitch for two passes. The baffles are spaced 5 in. apart.
What will the dirt factor and pressure drops be?

Solution. To prevent total vaporization in the shell, the total liquid entering the vaporizer will be 19,750/0.80 = 24,700 lb/hr. The vapor should then be disengaged from the liquid in a drum, and the excess liquid returned to the pump suction for recombination with 19,750 lb/hr of new feed. Thus for every 19,750 lb/hr coming from storage at 75°F, 4950 lb/hr at 235°F will be mixed with it so that the inlet temperature will be 108°F.

Solution:

Exchanger:

Shell side	Tube side
ID = 15¼ in.	Number and length = 76, 16'0"
Baffle Space = 5 in.	OD, BWG, pitch = 1 in., 16 BWG, 1¼-in. square
Passes = 1	Passes = 2

(1) Heat balance:

Preheat:
Enthalpy of liquid at 108°F and 300 psia = 162 Btu/lb (Fig. 9)
Enthalpy of liquid at 235°F and 300 psia = 248 Btu/lb
$q_p = 24{,}700(248 - 162) = 2{,}120{,}000$ Btu/hr

Vaporization:
Enthalpy of vapor at 235°F = 358 Btu/lb
$q_v = 19{,}750(358 - 248) = 2{,}170{,}000$ Btu/lb
Butane, $Q = 2{,}120{,}000 + 2{,}170{,}000 = 4{,}290{,}000$ Btu/hr
Steam, $Q = 4880 \times 880.6 = 4{,}290{,}000$ Btu/hr (Table 7)

(2) Δt weighted: (Subscripts p and v indicate preheating and vaporizing.)
$(\Delta t)_p = $ LMTD $= 158.5$°F (5.14)
$(\Delta t)_v = $ LMTD $= 103.0$°F (5.14)

$\dfrac{q_p}{(\Delta t)_p} = 2{,}120{,}000/158.5 = 13{,}400$

$\dfrac{q_v}{(\Delta t)_v} = 2{,}170{,}000/103.0 = 21{,}100$

$\sum \dfrac{q}{\Delta t} = 34{,}500$

Weighted $\Delta t = \dfrac{Q}{\Sigma q/\Delta t} = \dfrac{4{,}290{,}000}{34{,}500} = 124.5$°F (12.51)

(3) T_c and t_c: Average values of temperatures will be satisfactory for preheat zone.

Hot fluid: tube side, steam	Cold fluid: shell side, butane
	Preheating:
(4) $a_t' = 0.594$ in.^{2} [Table 10]	**(4')** $a_s = $ ID $\times C'B/144P_T$ [Eq. (7.1)]
$a_t = N_t a_t'/144n$ [Eq. (7.48)]	$= 15.25 \times 0.25 \times 5/144 \times 1.25$
$= 76 \times 0.594/2 \times 144 = 0.157$ ft^2	$= 0.106$ ft^2
(5) $G_t = W/a_t$	**(5')** $G_s = w/a_s$ [Eq. (7.2)]
$= 4880/0.157$	$= 24{,}700/0.106$
$= 31{,}100$ lb/(hr)(ft^2)	$= 233{,}000$ lb/(hr)(ft^2)

Hot fluid: tube side, steam

(6) At $T_S = 338°F$,
$\mu = 0.015 \times 2.42 = 0.0363$ lb/(ft)(hr) [Fig. 15]
$D = 0.87/12 = 0.0725$ ft (Re_t is for pressure drop only)
$Re_t = DG_t/\mu$
$= 0.0725 \times 31,100/0.0363 = 62,000$

(9) h_{io} for condensing steam
$= 1500$ Btu/(hr)(ft^2)(°F)

Cold fluid: shell side, butane

(6') At $T_a = 172°F$ (average of 108 and 235°F)
$\mu = 0.115 \times 2.42 = 0.278$ lb/(ft)(hr) [Fig. 14]
$D_e = 0.99/12 = 0.0825$ ft [Fig. 28]
$Re_s = D_e G_s/\mu$
$= 0.0825 \times 233,000/0.278 = 69,200$
(7') $j_H = 159$ [Fig. 28]
(8') At 172°F(114°API)
$k(c\mu/k)^{1/3} = 0.12$ Btu/(hr)(ft^2)(°F/ft) [Fig. 16]
$\phi_s = 1.0$

(9') $h_o = j_H \dfrac{k}{D_e}\left(\dfrac{c\mu}{k}\right)^{1/3}$ [Eq. (6.15b)]
$= 159 \times 0.12/0.0825$
$= 231$ Btu/(hr)(ft^2)(°F)

Clean overall coefficient for preheating U_p:

$$U_p = \frac{h_{io}h_o}{h_{io} + h_o} = \frac{1500 \times 231}{1500 + 231} = 200 \text{ Btu/(hr)(ft}^2\text{)(°F)} \qquad (6.38)$$

Clean surface required for preheating A_p:

$$A_p = \frac{q_p}{U_p(\Delta t)_p} = \frac{13,400}{200} = 67.0 \text{ ft}^2$$

Vaporization:
(6') At 235°F,
$\mu = 0.10 \times 2.42 = 0.242$ lb/(ft)(hr) [Fig. 14]
$Re_s = 0.0825 \times 233,000/0.242 = 79,500$
(7') $j_H = 170$ [Fig. 28]
(8') At 235°F
$k(c\mu/k)^{1/3} = 0.115$ Btu/(hr)(ft^2)(°F/ft)
$\phi_s = 1.0$ (Fig. 16)

(9') $h_o = j_H \dfrac{k}{D_e}\left(\dfrac{c\mu}{k}\right)^{1/3}$ [Eq. (6.15)]
$= 170 \times 0.115/0.0825 = 237$

(9) h_{io} for condensing steam $= 1500$

Clean overall coefficient for vaporization U_v:

$$U_v = \frac{h_{io}h_o}{h_{io} + h_o} = \frac{1500 \times 237}{1500 + 237} = 205 \qquad (6.38)$$

Clean surface required for vaporization A_v:

$$A_v = \frac{q_v}{U_v(\Delta t)_v} = \frac{21,100}{205} = 103 \text{ ft}^2$$

Total clean surface A_C:

$$A_C = A_p + A_v = 67.0 + 103 = 170 \text{ ft}^2$$

VAPORIZERS, EVAPORATORS, AND REBOILERS

(13) Weighted clean overall coefficient U_C:

$$U_C = \frac{\Sigma UA}{A_C} = \frac{13{,}400 + 21{,}100}{170} = 203 \quad (12.50)$$

(14) Design overall coefficient:

Surface/lin ft of tube = 0.2618   (Table 10)
Total surface = $76 \times 16'0'' \times 0.2618 = 318$ ft^2

$$U_D = \frac{Q}{A\,\Delta t} = \frac{4{,}290{,}000}{318 \times 124.5} = 108.5$$

Check of maximum flux:

A total of 170 ft^2 are required of which 103 are to be used for vaporization. For the total surface required 318 ft^2 will be provided. It can be assumed, then, that the surface provided for vaporization is

$$^{103}\!/_{170} \times 318 = 193 \text{ ft}^2$$

The flux is $Q/A = 2{,}170{,}000/193 = 10{,}700$ Btu/(hr)(ft^2). (Satisfactory)

(15) Dirt factor:

$$R_d = \frac{U_C - U_D}{U_C U_D} = \frac{203 - 108.5}{203 \times 108.5} = 0.0043 \text{ (hr)(ft}^2\text{)(°F)/Btu} \quad (6.13)$$

Pressure Drop

(1) For $Re_t = 62{,}000$,
$f = 0.000165$ ft^2/in.^{2} [Fig. 26]
From Table 7, specific vol of steam at 115 psia = 3.88 ft^3/lb

$$s = \frac{1}{3.88 \times 62.5} = 0.00413$$

(2) $\Delta P_t = \frac{1}{2} \frac{fG^2 Ln}{5.22 \times 10^{10} D s \phi_t}$ [Eq. (7.45)]

$$= \frac{1}{2} \times \frac{0.000165 \times 31{,}100^2 \times 16 \times 2}{5.22 \times 10^{10} \times 0.0725 \times 0.00413 \times 1}$$

$$= 0.16 \text{ psi}$$

Preheat:
(1′) $Re_s = 69{,}200, f = 0.00145$ ft^2/in.^{2} [Fig. 29]
(2′) Length of preheat zone
$L_p = LA_p/A_C$
   $= 16 \times 67.0/170 = 6.3$ ft
(3′) No. of crosses, $N + 1 = 12L_p/B$ [Eq. (7.43)]
   $= 12 \times 6.3/5 = 15$
$s = 0.50$ [Fig. 6]
$D_s = 15.25/12 = 1.27$ ft

(4′) $\Delta P_s = \frac{fG_s^2 D_s (N+1)}{5.22 \times 10^{10} D_e s \phi_s}$ [Eq. (7.44)]

$$= \frac{0.00145 \times 233{,}000^2 \times 1.27 \times 15}{5.22 \times 10^{10} \times 0.0825 \times 0.50 \times 1.0}$$

$$= 0.70 \text{ psi}$$

Vaporization:
(1′) $Re_s = 79{,}500, f = 0.00142$ ft^2/in.^{2}
(2′) Length of vaporization zone
$L_v = 16 - 6.3 = 9.7$ ft
(3′) No. of crosses, $N + 1 = 12L/B$ [Eq. (7.43)]

$$= 9.7 \times {^{12}\!/_5} = 23$$

Mol. wt. = 58.1

Pressure Drop

$$\text{Density, } \rho = \frac{58.1}{359 \times {}^{695}\!/_{492} \times 14.7/300}$$
$$= 2.34 \text{ lb/ft}^3$$

$s_{\text{outlet liquid}} = 0.43$ [Fig. 6]
$\rho_{\text{outlet liquid}} = 0.43 \times 62.5 = 26.9 \text{ lb/ft}^3$
$s_{\text{outlet mix}} =$
$$\frac{24{,}700/62.5}{19{,}750/2.34 + 4950/26.9} = 0.046$$
$s_{\text{inlet}} = 0.50$
$s_{\text{mean}} = (0.50 + 0.046)/2 = 0.28$
$$\Delta P_s = \frac{0.00142 \times 233{,}000^2 \times 1.27 \times 23}{5.22 \times 10^{10} \times 0.0825 \times 0.28 \times 1.0}$$
$$= 1.9 \text{ psi}$$
ΔP_s (total) $= 0.7 + 1.9 = 2.6$ psi

Summary

1500	h outside	231/237
U_C	203	
U_D	108.5	
R_d Calculated 0.0043		
R_d Required		
0.16	Calculated ΔP	2.6
Neg	Allowable ΔP	5.0

b. Vaporizer or Pump-through Reboiler with Boiling Range. If a liquid undergoing vaporization is a mixture of a number of miscible compounds, it does not boil isothermally. Instead it has an initial boiling temperature (bubble point) and a final boiling temperature (dew point) at which the last bit of liquid is vaporized. When the mixture starts to boil at its bubble point, the more volatile compounds are driven out of the solution at a greater rate and as the volatile compounds enter the vapor phase the boiling temperature of the residual liquid rises. This means that throughout the vaporizer there is a temperature range over which boiling occurs, and the greater the percentage of the total liquid vaporized the more nearly the range extends from the bubble point to the dew point of the inlet liquid.

Because of the boiling range, sensible as well as latent heat must be absorbed simultaneously by the liquid as it proceeds through the vaporizer so that it will possess a range of boiling temperatures. Furthermore the sensible heat is absorbed over the same surface as the heat for vaporization in contrast to the preheater-isothermal vaporizer in which the two occur in separate zones. The calculation of the boiling coefficient in this case, however, is the same as for a preheater-vaporizer as calculated in Example 15.2. Here the heat from the tube wall is first absorbed by the liquid as sensible heat before it is transformed into vaporization. Since the rate of heat transfer from a hot liquid into an incipient vapor bubble is very great, it may be assumed that the sensible-heat-transfer coefficient as calculated from Fig. 28 for either direct vaporization or simultaneous sensible-heat transfer is the limiting resistance. The coefficient for the combined sensible-heat transfer and vaporization is calculated as if the entire vaporizing heat load were transferred as sensible heat to the liquid over its boiling range in the vaporizer.

The true temperature difference may be taken as the LMTD if the heating medium is isothermal. This assumes that the heat transferred is proportional to the change in temperature, namely, that one-half of the total load is delivered while the temperature rises one-half of the total temperature range for vaporization. If the bulk of a mixture consists of closely related compounds with some more or less volatile compounds, the assumption that the heat and temperature proportions are equal may cause considerable error. The true temperature difference can be obtained by graphical integration as treated in Example 13.3.

Film Coefficients. The sensible-heat-transfer coefficient should be regarded as the boiling coefficient when applying the restrictions of allowable flux and allowable coefficient even though it is computed from Fig. 28. When a liquid has a boiling range, the *average* flux, Q/A may be less than 20,000 but because of the variation in temperature difference, $U_D \Delta t_1$ at the greater terminal temperature difference may exceed 20,000. Discrepancies of this sort may cause bumping or erratic vaporization if the initial compounds to vaporize are very volatile compared with the bulk and tend to separate from the liquid too readily. A check of the inlet flux can prevent this difficulty. If the inlet flux does not exceed, say, 25,000 Btu/(hr)(ft²), there is no need to penalize the entire design by providing excess surface simply because of an excessive flux over the first few per cent of the vaporizing surface.

Pressure Drop. The pressure drop is computed in the same manner as for isothermal boiling, using the Reynolds number based on inlet conditions and a gravity which is the mean of inlet and outlet gravities. It is also quite possible that a fluid with a boiling range may enter a vaporizer below its bubble point. In such cases the surface is again divided into two consecutive zones, one for pure preheating and the other for vaporization of a mixture with a boiling range. The weighted coefficients and temperature difference may be obtained as before through the use of Eqs. (12.50) and (12.51).

c. **Forced-circulation Evaporator or Aqueous-solution Reboiler.** As seen in Chap. 14, the shells of 1-2 exchangers are not used in forced-circulation-evaporator processes, since the properties of water are excellent for natural-circulation equipment. However, a 1-2 exchanger can easily serve as a forced-circulation evaporator. In distillation processes such as the distillation of an acetone-water or alcohol-water solution the bottom product is nearly pure water. It may be advantageous in smaller operations of this nature to use a pump-through reboiler in preference to natural circulation, since the losses in the connecting piping may be unduly high and the use of larger connecting pipes does not assure smooth operation. The aqueous-solution reboiler can be calculated in the same manner as the pump-through reboiler, with or without a boiling range, except that the allowable flux and film coefficient is greater.

This type of equipment is usually designed with the dirt factor as the controlling resistance. The applicability of a method of computing water-vaporization rates is therefore of value only at low mass velocities. Since water vapor has a very low vapor density, low mass velocities must be employed whenever the allowable pressure drop is small. Film coefficients for boiling water and aqueous solutions can be obtained by the use of Fig. 28, although these will be about 25 per cent lower than those which have been obtained experimentally. Where data on the physical properties of aqueous solutions are lacking, they can be approximated by the methods of Chap. 7. If the mass velocity is very low, the value of the coefficient so obtained may be multiplied by 1.25 and the overall value of U_C should rarely exceed 600 Btu/(hr)(ft^2)(°F).

2. Vaporization in the Tubes

a. **Pump-through Vaporizer or Reboiler with or without Boiling Range.** The coefficients for vaporization with or without a boiling range can be obtained from Fig. 24 for organic liquids. The number of tube passes may be as great in horizontal exchangers as the pressure drop will allow. If the number of tubes in the final passes are greater than the number in the initial passes, it is possible to obtain a reduced pressure drop. The shell side, when employing steam, may be laid out on triangular pitch, since cleaning will be infrequent and the shell side can be cleaned by boiling out. The pressure drop can be computed using Eq. (7.45) with a Reynolds number based on the inlet properties and a specific gravity which is the mean between inlet and outlet. The tube fluid should flow upward.

b. **Forced-circulation Evaporator or Aqueous-solution Reboiler.** The tube-side data for the evaporation of water and aqueous solutions can also be obtained from Fig. 24. The boiling coefficients will be about 25 per cent greater than the computed values, and at low mass velocities the coefficient can be multiplied by a correction factor of 1.25.

This class also includes the forced-circulation evaporators similar to 1-1 exchangers. The calculation of the vertical long-tube evaporator will be treated as a natural-circulation vaporizer.

NATURAL-CIRCULATION VAPORIZING EXCHANGERS

1. Vaporization in the Shell

a. **Kettle Reboiler.** The kettle reboiler is shown in Fig. 15.6. It is a modification of the power-plant evaporator. The relation of the bundle to the shell is seen better from an end elevation. Another form of the kettle reboiler employing a tube sheet which covers the entire shell is

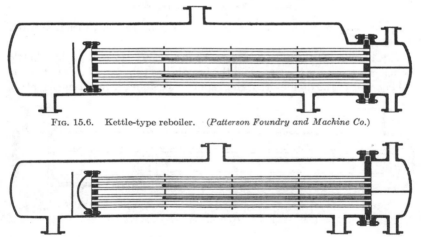

Fig. 15.6. Kettle-type reboiler. (*Patterson Foundry and Machine Co.*)

Fig. 15.7. Kettle-type reboiler with integral tube sheet. (*Patterson Foundry and Machine Co.*)

shown in Fig. 15.7. In this type the bundle is not circular but conforms to the shell as seen from an end elevation. The method of connecting this type of reboiler to the distilling column is shown in Fig. 15.8. Kettle reboilers are fitted with a weir to ensure that the liquid level in the reboiler is maintained and that the tube surface is not exposed. Since only about 80 per cent of the liquid bottoms entering at the inlet are vaporized, provision must be made for the removal of the bottoms product which is on the discharge side of the weir. There are a number of arbitrary rules on the volume required above the liquid level for disengagement and the maximum number of pounds per hour which may be vaporized from the liquid surface. If the top row of tubes is not higher than 60 per cent of the shell diameter, adequate disengaging space will

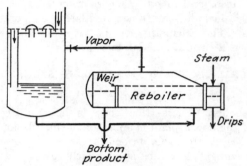

FIG. 15.8. Kettle reboiler arrangement.

be provided when a liquid level covering the top row of tubes is ensured by the weir.

b. Chiller. The chiller is shown in Fig. 15.9. It is a typical kettle reboiler, except for the weir, and the bundle has tubes to a height of about 60 per cent of the diameter. The vapor space above is for disengagement

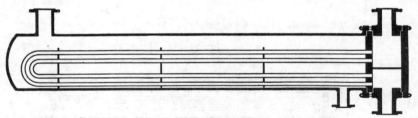

FIG. 15.9. Chiller. (*Patterson Foundry and Machine Co.*)

of the vapor from the liquid. Chillers are used in refrigeration processes of the vapor-compression type as shown in Fig. 15.10. The refrigeration cycle begins at the point a, where liquid refrigerant at a temperature higher than that of the condenser water and at high pressure passes to a constant-enthalpy throttle valve where its pressure is reduced. The pressure and temperature of the liquid on the downstream side of the valve are naturally less than on the high-pressure side. The expansion is adiabatic, and some of the liquid flashes into vapor cooling the remainder of the refrigerant on the low-pressure side at b. If the cold refrigerant is to be circulated directly to a refrigerator, the saturation temperature at b is often about 5 or 10°F lower than the temperature ultimately desired in the chamber being refrigerated. The partially vaporized refrigerant may enter the shell of a chiller, where the remainder is vaporized isothermally at low temperature by the fluid being chilled as it flows through the tubes. The vapor then passes to the compressor between c and d, where it is recompressed to the higher pressure (and temperature) such that it can be recondensed with the available cooling water.

Cold brine is often circulated to the chamber being refrigerated in preference to the refrigerant itself. Brines are usually sodium chloride

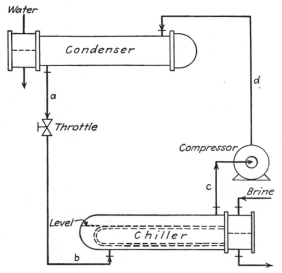

Fig. 15.10. Compression refrigeration system.

or calcium chloride solutions in concentrations up to 25 per cent by weight, depending upon the coldness of temperature to be maintained. They are cheap and have little susceptibility toward leakage. On accomplishing refrigeration by sensible heat absorption, brines remain in the liquid phase without developing high pressures and eliminate the need for costly vapor piping through the refrigerating system. In this way contaminants which might find their way into the refrigerant, particularly under vacuum, are kept out of it and from the compressor, condenser, and throttle. Brines, on the other hand, require that an additional temperature difference be maintained. When the refrigerant is circulated directly to the cold chamber, there is but one temperature difference between the chamber and the refrigerant. Using brines, however, there is one temperature difference between the refrigerant and brine and one between the brine and cold chamber, and this arrangement increases the cost of the refrigeration. For other aspects of refrigerating cycles, such as the selection of optimum conditions, reference may be made to standard thermodynamics texts. The chiller is frequently called the evaporator in the refrigeration process, although this use of the term is at variance with the nomenclature used herein. When used in large

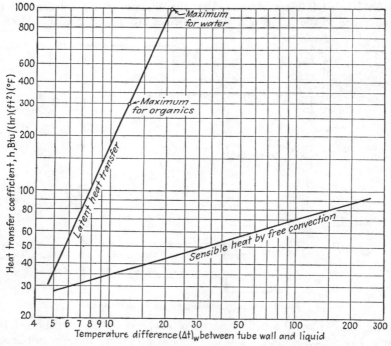

FIG. 15.11. Natural-circulation boiling and sensible film coefficients.

installations, the refrigerant may disengage into a separate drum instead of in the chiller shell.

Film Coefficients in Chillers and Kettle Reboilers. When a liquid is vaporized from a pool, the velocity of the liquid flowing over the heat-transfer surface is very low. At these low velocities the film coefficient for boiling is independent of the velocity and dependent instead upon the temperature difference between the tube wall and the saturation temperature of the boiling fluid. A curve showing this relationship is represented by the upper line in Fig. 15.11. The coefficients are again limited to 300 for organics and 1000 for water, except that the maximum allowable flux for organics is 12,000 Btu/(hr)(ft^2). The latter does not permit the use of large temperature differences for natural-circulation vaporizers and reboilers, which require more surface as a class than do forced-circulation types. The difference in cost of surface is usually more than offset by the saving in circulating pump power.

Chillers operate isothermally unless the vapor space of the chiller also serves as a flash chamber for the expansion, in which case the inlet temperature will be that at the throttle valve and the outlet temperature the saturation temperature of the refrigerant. The vaporization out of the liquid surface, however, is isothermal. Kettle reboilers also operate under nearly isothermal conditions, particularly when employed at the bottom of an aqueous-solution distilling column. When used with organic bottoms, however, there is usually a boiling range, and it is necessary to provide for the introduction of sensible heat. The sensible heat is regarded as entering the liquid by a modification of *free convection*,[1] and the curve for free convection to organics in natural circulation is given by the lower line of Fig. 15.11. It corresponds fairly closely to the results obtained from Fig. 10.4. In the case of kettle reboilers the total heat load is divided into sensible- and latent-heat fractions, and the surface required for each fraction is computed separately at its respective boiling or sensible coefficient. While there are several ways in which this procedure can be justified, it is nonetheless an empirical means of calculating coefficients comparable to those obtained in practice. This method is demonstrated in Example 15.4.

Pressure Drop in Kettle Reboilers and Chillers. The height of bottoms maintained in a distilling column naturally seeks its own level in the reboiler. If the reboiler is not mounted much below the liquid level in the column, there is a negligible hydrostatic head for fluid flow from the column to the reboiler, and hence the circulation rate is relatively small. This results in a low fluid velocity across the reboiler surface, and the pressure drop, as well as that of the connecting piping, may be considered negligible. The kettle reboiler, in fact, is the most suitable of all natural-circulation reboilers where the reboiler cannot be mounted sufficiently below the liquid level in the column to provide a large recirculation rate.

Example 15.3. Calculation of a Kettle Reboiler. 45,500 lb/hr of bottoms of 65°API gravity and small boiling range at 400°F enter a kettle reboiler from which 28,100 lb/hr of vapor is formed at an operating pressure of 200 psig. Heat is supplied by 28°API gas oil in the range from 575 to 475°F and 120 psig operating pressure. A pressure drop of 10 psi is permissible.

Available for the service is a 25 in. ID kettle reboiler containing a six-pass 15¼-in.

[1] The terms natural convection and natural circulation sometimes cause confusion. The former refers to heat transfer, and the latter to fluid flow.

circular bundle. The bundle contains 68 1 in. OD, 14 BWG tubes 12'0" long on $1\frac{1}{4}$-in. square pitch. The bundle is baffled only by quarter-circle support plates.
Will the reboiler be satisfactory?
What are the dirt factor and pressure drops?

Solution:

Kettle:

Shell side	Tube side
$15\frac{1}{4}$-in. circular bundle in 25 in. ID	Number and length = 68, 12'0"
$\frac{1}{4}$-circle support plates	OD, BWG, pitch = 1 in., 14 BWG, $1\frac{1}{4}$-in. square
	Passes = 6

(1) Heat balance:
Enthalpy of liquid at 400°F and 215 psia = 290 Btu/lb (Fig. 11)
Enthalpy of vapor at 400°F and 215 psia = 385 Btu/lb
Gasoline, $Q = 28,100 \times (385 - 290) = 2,670,000$ Btu/hr
Gas oil, $Q = 34,700 \times 0.77(575 - 475) = 2,670,000$

(2) Δt: Isothermal boiling
Δt = LMTD = 118°F (5.14)

(3) $T_c: \dfrac{\Delta t_c}{\Delta t_h} = \dfrac{475 - 400}{575 - 400} = 0.428$

$K_c = 0.37$ (Fig. 17)
$F_c = 0.42$
$T_c = 475 + 0.42(575 - 475) = 517°F$ (5.28)

Hot fluid: tube side, gas oil

(4) Flow area, $a_t' = 0.546$ in.^{2} [Table 10]
$a_t = N_t a_t'/144n$ [Eq. (7.48)]
$= 68 \times 0.546/144 \times 6 = 0.043$ ft^2

(5) $G_t = W/a_t$
$= 34,700/0.043$
$= 807,000$ lb/(hr)(ft^2)

(6) At $T_c = 517°F$,
$\mu = 0.27 \times 2.42 = 0.65$ lb/(ft)(hr) (extrapolated) [Fig. 14]
$D = 0.834/12 = 0.0694$ ft
$Re_t = DG_t/\mu$
$= 0.0694 \times 807,000/0.65 = 85,700$

(7) $j_H = 220$ [Fig. 24]

(8) At 517°F (28°API) [Fig. 16]
$k(c\mu/k)^{\frac{1}{3}} = 0.118$ Btu/(hr)(ft^2)(°F/ft)

(9) $h_i = j_H \dfrac{k}{D} \left(\dfrac{c\mu}{k}\right)^{\frac{1}{3}} \phi_t$ [Eq. (6.15d)]

$h_i/\phi_t = 220 \times 0.118/0.0694 = 374$

(10) $\dfrac{h_{io}}{\phi_t} = \dfrac{h_i}{\phi_t} \times \dfrac{\text{ID}}{\text{OD}}$ [Eq. (6.9)]
$= 374 \times 0.834/1.0 = 311$

The correction for $\left(\dfrac{\mu}{\mu_w}\right)^{0.14}$ is negligible.

Cold fluid: shell side, gasoline

(9') Assume $h_o = 300$ for trial.

(10') $t_w = t_c + \dfrac{h_{io}}{h_{io} + h_o}(T_c - t_c)$
[Eq. (5.31)]
$= 400 + \dfrac{311}{311 + 300}(517 - 400)$
$= 460°F$

$(\Delta t)_w = 460 - 400 = 60°F$
From Fig. 15.11, $h_v > 300$; hence use 300.

(13) Clean overall coefficient U_C:
$$U_C = \frac{h_{io}h_o}{h_{io} + h_o} = \frac{311 \times 300}{311 + 300} = 152 \text{ Btu/(hr)(ft}^2)(°\text{F)} \quad (6.38)$$

(14) Design overall coefficient U_D:
$$a'' = 0.2618 \text{ ft}^2/\text{lin ft} \quad \text{(Table 10)}$$
$$\text{Total surface} = 68 \times 12 \times 0.2618 = 214 \text{ ft}^2$$
$$U_D = \frac{Q}{A\,\Delta t} = \frac{2{,}670{,}000}{214 \times 118} = 105.5 \quad (5.3)$$

Check of maximum flux:
$$\frac{Q}{A} = \frac{2{,}670{,}000}{214} = 12{,}500 \text{ Btu/(hr)(ft}^2\text{)} \quad \text{(satisfactory)}$$

(15) Dirt factor R_d:
$$R_d = \frac{U_C - U_D}{U_C U_D} = \frac{152 - 105.5}{152 \times 105.5} = 0.0029 \quad (6.12)$$

Pressure Drop

(1) For $Re_t = 85{,}700$, $f = 0.00015 \text{ ft}^2/\text{in.}^2$ Negligible
                        [Fig. 26]
$s = 0.71$ [Fig. 6]

(2) $\Delta P_t = \dfrac{fG_t^2 Ln}{5.22 \times 10^{10} D s \phi_t}$

$= \dfrac{0.00015 \times 807{,}000^2 \times 12 \times 6}{5.22 \times 10^{10} \times 0.0694 \times 0.71 \times 1.0}$

$= 2.8$ psi

(3) $G_t = 807{,}000$, $V^2/2g' = 0.090$
                        [Fig. 27]
$\Delta P_r = \dfrac{4n}{s}\dfrac{V^2}{2g'}$ [Eq. (7.46)]

$= \dfrac{4 \times 6}{0.71} \times 0.09 = 3.1$ psi

(4) $\Delta P_T = \Delta P_t + \Delta P_r$ [Eq. (7.47)]
$= 2.8 + 3.1 = 5.9$ psi

Summary

311	h outside	300
U_C	152	
U_D	105.5	
R_d Calculated 0.0029		
R_d Required		
5.9	Calculated ΔP	Neg
10.0	Allowable ΔP	Neg

c. **Bundle-in-column Reboiler.** It will probably occur while examining the kettle reboiler (Fig. 15.6) that, if the bundle is to be submerged in the trapout, it can be inserted directly into the bottom of the column, as in Fig. 15.12. There is no objection from a heat-transfer standpoint. As seen in Example 15.3 only 214 ft² of surface was required for the transfer of 2,670,000 Btu/hr and this in a 15¼-in. circular bundle 16'0" long. The column required for 28,100 lb/hr of the vapor at 200 psig has a diameter of less than 3 ft. If the bundle is inserted in the bottom of such a column, many short tubes will be required and the height of the bottom of the column must be increased so as to maintain the same holdup space. Another obvious disadvantage lies in the size of the flanged connection which must be welded to the side of the column to accommodate the larger bundle. Internal supports are also required to keep the weight of the bundle from forming a cantilever with the column connection flange. These difficulties are usually surmountable when the column diameter is greater than 6 ft, but overall experience favors the use of external reboilers in preference to the small savings realized by eliminating the shell.

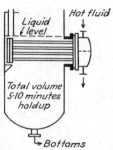

FIG. 15.12. Bundle-in-column reboiler.

The calculations for the bundle-in-column reboiler are identical with those for the kettle reboiler, using coefficients from Fig. 15.11.

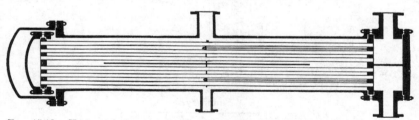

FIG. 15.13. Horizontal thermosyphon reboiler. (*Patterson Foundry and Machine Co.*)

d. **Horizontal Thermosyphon Reboiler.** This is perhaps the commonest type of reboiler. Figure 15.13 shows a horizontal thermosyphon reboiler. It consists of centrally located inlet and outlet nozzles, a vertical circular support plate between the nozzles, and a horizontal longitudinal baffle. Horizontal thermosyphons operate on the principle of divided flow as first outlined under condensers in Fig. 12.17, with halves of the entering fluid flowing away from each other below the longitudinal baffle and toward each other above it. Disengagement takes place in the column, and the reboiler may be connected by the arrangement of Fig. 15.4a or b. In Fig. 15.4a, as stated previously, all the liquid from the bottom plate is led directly to the reboiler. The rate of feed to the reboiler is the hourly trapout rate, which passes through the reboiler but once. In Fig. 15.4b the reboiler is connected to the bottoms, which are free to recirculate at a rate such that the frictional pressure drop in the reboiler and other resistances of the circuit just balance the hydrostatic head difference between the liquid and liquid-vapor legs. The hydrostatic head available in the latter arrangement, however, is less than in the once-through arrangement although a greater head is required for recirculation. The head is provided by raising the bottom liquid level in the column or by raising the column itself. Occasionally the reboiler may be set in a well, but this practice finds little favor in modern plants.

Film Coefficients in Horizontal Reboilers. The coefficients used for thermosyphons are substantially the same as those employed for kettle reboilers and are given by Fig. 15.11. When there is a boiling range, it is imperative that the overall clean coefficient be weighted for sensible- and latent-heat loads individually, although the procedure differs from the weighting of successive zones, since both sensible heating and boiling occur in the same temperature range. This problem was not encountered in forced-circulation reboilers and vaporizers because the rates of boiling and sensible-heat transfer are ordinarily very nearly identical. However, in a shell without forced convection the rate of sensible-heat transfer by free convection is usually less than one-sixth the natural-circulation boiling rate. In natural circulation, however, where both sensible-heat transfer and boiling occur over the same surface, the free convection coefficient is undoubtedly modified by the bubble movements which far exceed the agitation derived from ordinary free convection currents. To account for this modification, the sensible portion of the heat load is assumed to be transferred by ordinary free convection and the boiling portion is assumed to be transferred as natural-circulation vaporization.

Although the flow is not counterflow, it usually does not deviate greatly from it, because one or both of the fluids is substantially isothermal. If steam is the heating medium, the counterflow temperature difference applies directly. If a liquid is the heating medium rather than a vapor, the counterflow temperature difference applies only if the range on the material being vaporized is small and the approach between the heating medium and vaporizing medium inlet temperatures is appreciable. If F_T for a 1-2 exchanger exceeds 0.90 an insignificant error may be anticipated from the use of the 1-2 parallel flow–counterflow temperature difference.

Since the temperature differences for sensible heating and vaporization are the same, there is no weighted temperature difference. But the sensible heat q_s is transferred with a free convection coefficient of h_s, and the latent heat q_v is transferred with the considerably higher coefficient h_v. To permit obtaining a single dirt factor, as the index of performance or maintenance of the reboiler, the weighted coefficient may be obtained as follows: From $q = hA\,\Delta t$,

$$A_s(\Delta t)_s = \frac{q_s}{h_s}$$

$$A_v(\Delta t)_v = \frac{q_v}{h_v}$$

The weighted coefficient is then

$$h = \frac{Q}{q_s/h_s + q_v/h_v}$$

Since neither h_v nor h_s is influenced by the velocity through the reboiler, it will be of no consequence in the calculation whether the reboiler has been connected for once-through or recirculation operation.

Pressure Drop. There is an obvious need in the recirculation arrangement to keep the pressure drop through the thermosyphon as small as possible. When studying condensers it was observed that the greater the pressure drop through the condenser the higher the condenser must be elevated above the column to permit the gravity return of condensate. The effect of pressure drop on the elevation of the bottom tower liquid level above the reboiler is even more critical. The greater the pressure drop through the reboiler the higher the entire column and auxiliaries must be elevated above the ground level to produce sufficient hydrostatic head to overcome the pressure drop. Pressure drops of about 0.25 psi are generally allowed for the reboiler and attendant losses. If the column is small in diameter or height, a pressure drop as high as 0.50 psi may be allowed, but concessions of this nature to the design of the reboiler are rare. For a reboiler vaporizing a small fraction of the liquid entering it, the required elevation is greater, since the return leg to the column contains more liquid than vapor and the difference in density of the streams to and from the reboiler is small.

While half baffles may occasionally be used to increase the turbulence in the shell, the tubes are usually prevented from sagging by the vertical support plate between the inlet and outlet nozzles and additional quarter-circle support plates. The liquid entering the horizontal thermosyphon flows *one-half of the tube length* on the underside of the longitudinal baffle and one-half the tube length on the upper side, so that all the liquid travels the entire tube length but in each instance with a mass-velocity based on one-half the total flow. The length of path of each parallel stream equals

VAPORIZERS, EVAPORATORS, AND REBOILERS   481

the tube length, and it is sufficiently accurate to treat the pressure drop the same as for a shell without baffles and with axial flow as in Example 7.8. The diameter of a horizontal thermosyphon reboiler is greater than that which corresponds to the tube count of a conventional 1-2 exchanger because of the free space which must be provided in the upper half to allow the light mixture of vapor and liquid readily to reach the outlet nozzle. If the surface of a 25 in. ID layout sufficed for heat transfer, the tubes would be relocated in a 27 in. ID shell while retaining the tube pitch so as to leave a free vapor flow channel at the top of the shell and a lesser entrance channel at the bottom.

The equivalent diameter is computed directly by Eq. (7.3) from the wetted perimeter of the tubes, half the shell, and the width of the longi-

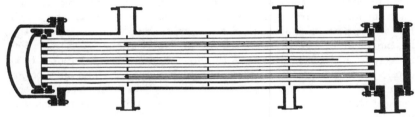

Fig. 15.14. Horizontal thermosyphon with double nozzles. (*Patterson Foundry and Machine Co.*)

tudinal baffle. The flow area is the difference between half a circle and the number of tubes in the upper or lower shell pass. In the absence of the actual tube layout these may be assumed to be equal. The Reynolds number is computed from the inlet liquid viscosity and the equivalent diameter. The pressure drop is based on the mean specific gravity between inlet and outlet, using a friction factor obtained from Fig. 26 for the *tube side*.

When there is a single inlet nozzle on the shell, it is customary not to use a tube length more than five times the shell diameter. Long, thin reboilers do not thermosyphon well. When a long, thin reboiler is indicated, it is usually equipped with two inlet nozzles as shown in Fig. 15.14 with a mass velocity based on one-fourth the flow rate in one-half the flow area. The following table will serve as a guide to well-proportioned horizontal thermosyphons:

Shell ID, in.	Tube length
12–17¼	8'0"
19¼–29	12'0"
31 and over	16'0"

When using a recirculating arrangement with a horizontal thermosyphon, recirculation can be computed approximately as the rate at which the pressure drop through the reboiler equals the hydrostatic difference of z_1 and z_3 in Fig. 15.4b although the rate does not affect the film coefficient. Instead of the recirculation rate it is preferable in reboilers to specify the *recirculation ratio*, which is defined as the pounds per hour of *liquid* leaving the reboiler compared with the pounds per hour of vapor alone. This should not be confused with the conventional definition of recirculation rates, which is defined as the ratio of the total hourly throughput to the hourly requirement.

In a recirculating reboiler the temperature range is not identical with that of a once-through arrangement. If the liquid recirculates, only a small amount of vapor is formed in each circulation and the vaporization occurs over a smaller temperature range, although the outlet temperatures in both arrangements are identical. In the recirculating arrangement the temperature difference is somewhat smaller. Usually the reduction in temperature difference is not significant unless the range of the heating medium is very close to that of the vaporization. A recirculation ratio four times or greater than the hourly once-through vapor rate is considered favorable from a cleanliness standpoint. The method of calculating recirculation ratios will be discussed in connection with the vertical thermosyphon reboiler, where a higher order of recirculation is usually obtained.

Example 15.4. Calculation of a Once-through Horizontal Thermosyphon Reboiler. 38,500 lb/hr of 60°API naphtha in a once-through arrangement is to enter a horizontal thermosyphon reboiler and produce 29,000 lb/hr of vapor in the temperature range from 315 to 335°F at an operating pressure of 5.0 psig. Heat will be supplied by 28°API gas oil with a range from 525 to 400°F.

Available for the service is a 21¼ in. ID reboiler containing 116 1 in. OD, 14 BWG tubes 12'0" long laid out on 1¼-in. square pitch. The bundle has a support plate above the single inlet nozzle and is arranged for eight passes.

What are the dirt factor and the pressure drops?

Solution:

Reboiler:

Shell side	Tube side
ID = 21¼ in.	Number and length = 116, 12'0"
Support plates = ¼ circles	OD, BWG, pitch = 1 in., 14 BWG, 1¼-in. square
Passes = divided	Passes = 8

(1) Heat balance:
  Enthalpy of liquid at 315°F and 19.7 psia = 238 Btu/lb   (Fig. 11)
  Enthalpy of liquid at 335°F and 19.7 psia = 252 Btu/lb
  Enthalpy of vapor at 335°F and 19.7 psia = 378 Btu/lb
  Naphtha $q_v = 29{,}000(378 - 252) = 3{,}650{,}000$
  $q_s = 38{,}500(252 - 238) = \underline{540{,}000}$
  $Q = \phantom{38{,}500(252 - 238) =\ } 4{,}190{,}000$ Btu/hr
  Gas oil, $Q = 51{,}000 \times 0.66(525 - 400) = 4{,}190{,}000$

VAPORIZERS, EVAPORATORS, AND REBOILERS 483

(2) Δt:

	Hot Fluid		Cold Fluid	
525	Higher Temp	335		190
400	Lower Temp	315		85
125	Differences	20		105

LMTD $= 131°\text{F}$

$R = \dfrac{125}{20} = 6.25 \qquad S = \dfrac{20}{525 - 315} = 0.095$

$F_T = 0.97$

$\Delta t = F_T \times \text{LMTD} = 0.97 \times 131 = 127°\text{F}$

(3) T_c:

$\dfrac{\Delta t_c}{\Delta t_h} = \dfrac{85}{190} = 0.447$ (Fig. 17)

$K_c = 0.42$
$F_c = 0.41$
$T_c = 400 + 0.41(525 - 400) = 451°\text{F}$ (5.28)
$t_c = 315 + 0.41(335 - 315) = 323°\text{F}$ (5.29)

Hot fluid: tube side, gas oil

(4) $a_t' = 0.546$ in.^{2} [Table 10]
$a_t = N_t a_t'/144 n$
$\quad = 116 \times 0.546/144 \times 8 = 0.055$ ft^2

(5) $G_t = W/a_t$
$\quad = 51{,}000/0.055$
$\quad = 928{,}000$ lb/(hr)(ft^2)

(6) At $T_c = 451°\text{F}$,
$\mu = 0.45 \times 2.42 = 1.09$ lb/(ft)(hr) [Fig. 14]

$D = 0.834/12 = 0.0695$ ft [Table 10]
$Re_t = DG_t/\mu$
$\quad = 0.0695 \times 928{,}000/1.09 = 59{,}200$

(7) $j_H = 168$ [Fig. 24]

(8) At $T_c = 451°\text{F}$ (28°API) [Fig. 16]
$k(c\mu/k)^{1/3} = 0.142$ Btu/(hr)(ft^2)(°F/ft)

(9) $h_i = (j_H k/D)(c\mu/k)^{1/3}\phi_t$ [Eq. (6.15a)]

$\dfrac{h_i}{\phi_t} = 168 \times \dfrac{0.142}{0.0695} = 343$

$\phi_t = 1.0$

(10) $h_{io} = h_i \times \text{ID/OD}$ [Eq. (6.6)]
$\quad = 343 \times 0.834/1.0$
$\quad = 286$ Btu/(hr)(ft^2)(°F)

Cold fluid: shell side, naphtha

Assume weighted $h_o = 200$
h_{io} from (10) $= 286$

$t_w = t_c + \dfrac{h_{io}}{h_{io} + h_o}(T_c - t_c)$ [Eq. (5.31)]

$\quad = 323 + \dfrac{286}{286 + 200}(451 - 323)$

$\quad = 382°\text{F}$

$(\Delta t)_w = 382 - 323 = 59°\text{F}$

From Fig. 15.11, $h_v = \; > 300$, use 300
$h_s = 60$
$q_v/h_v = 3{,}650{,}000/300 = 12{,}150$
$q_s/h_s = 540{,}000/60 \quad\;\, = \underline{9{,}000}$
$\phantom{q_s/h_s = 540{,}000/60 \quad\;\,\;} 21{,}150$

(10′) $h_o = 4{,}190{,}000/21{,}150$
$\quad = 198$ Btu/(hr)(ft^2)(°F)

Checks 200 assumed for h_o.

(13) Overall clean coefficient U_C:

$$U_C = \dfrac{h_{io} h_o}{h_{io} + h_o} = \dfrac{286 \times 198}{286 + 198} = 116 \text{ Btu/(hr)(ft}^2)(°\text{F}) \qquad (6.38)$$

(14) Design overall coefficient U_D:

$$\text{Surface per linear foot} = 0.2618 \quad \text{(Table 10)}$$
$$\text{Total surface} = 116 \times 12'0'' \times 0.2618 = 364 \text{ ft}^2$$
$$U_D = \frac{Q}{A \, \Delta t} = \frac{4{,}190{,}000}{364 \times 127} = 90.7$$

Check of maximum flux (based on total transfer through the surface):

$$\frac{Q}{A} = \frac{4{,}190{,}000}{364} = 11{,}500 \text{ vs. } 12{,}000 \text{ allowable}$$

(15) Dirt factor R_d:

$$R_d = \frac{U_C - U_D}{U_C U_D} = \frac{116 - 90.7}{116 \times 90.7} = 0.0024 \qquad (6.13)$$

Pressure Drop

(1) $Re_t = 59{,}200$, $f = 0.000168 \text{ ft}^2/\text{in.}^2$ [Fig. 26]
$s = 0.73$ [Fig. 6]

(2) $\Delta P_t = \dfrac{f G_t^2 L n}{5.22 \times 10^{10} D s \phi_t}$

$= \dfrac{0.000168 \times 928{,}000^2 \times 12 \times 8}{522 \times 10^{10} \times 0.0695 \times 0.73 \times 1}$

$= 5.3 \text{ psi}$

(3) $G_t = 928{,}000$ $V^2/2g' = 0.11$ [Fig. 27]
$\Delta P_r = 4 \dfrac{n}{s} \dfrac{V^2}{2g'} = \dfrac{4 \times 8}{0.73} \times 0.11 = 4.8 \text{ psi}$
[Eq. (7.46)]

(4) $\Delta P_T = \Delta P_t + \Delta P_r$ [Eq. (7.47)]
$= 5.3 + 4.8 = 10.1 \text{ psi}$

(1') $D'_e = 4 \times$ flow area/frictional wetted perimeter
Assume half of tubes above and half of tubes below longitudinal baffle.
Flow area $= \frac{1}{2}$ shell cross section $- \frac{1}{2}$ tube cross section

$= \dfrac{\pi}{8}(21.25^2 - 1.0 \times 116) = 132 \text{ in.}^2$

$a_s = {}^{132}\!/_{144} = 0.917 \text{ ft}^2$

Wetted perimeter $= \dfrac{\pi \times 21.25}{2} + \dfrac{\pi}{2} \times 1$
$\times 116 + 21.25 = 236.7 \text{ in.}$

$d'_e = 4 \times 132/236.7 = 2.23 \text{ in.}$
[Eq. (7.3)]

$D'_e = 2.23/12 = 0.186 \text{ ft}$
$G_s = (w/2)/a_s$
$= \frac{1}{2} 38{,}500 \times 0.917$
$= 21{,}000 \text{ lb/(hr)(ft}^2)$

For 60°API at 315 use data in Fig. 14 for 56°API gasoline as an approximation.
$\mu = 0.18 \times 2.42 = 0.435 \text{ lb/(ft)(hr)}$
$Re_s = D'_e G_s / \mu$
$= 0.186 \times 21{,}000/0.435 = 8950$
$f = 0.00028 \text{ ft}^2/\text{in.}^2$ [Fig. 26]
From Fig. 13.14, mol. wt. $= 142$

Density, $\rho = \dfrac{142}{359 \times {}^{795}\!/_{492} \times 14.7/19.7}$
$= 0.337 \text{ lb/ft}^3$

$s_{\text{outlet liquid}}$ at $335°F = 0.61$
$\rho_{\text{outlet liquid}} = 0.61 \times 62.5 = 38.1 \text{ lb/ft}^3$

$s_{\text{outlet mix}} = \dfrac{38{,}500/62.5}{29{,}000/0.337 + 9500/38.1}$
$= 0.071$

s_{inlet} at $315°F = 0.625$
$s_{\text{av}} = \frac{1}{2}(0.625 + 0.071) = 0.35$

Pressure Drop

$$\Delta P_s = \frac{fG_s^2(L_{\text{total}})}{5.22 \times 10^{10} D_e' s\phi_s} \quad [\text{Eq. (7.45)}]$$

$$= \frac{0.00028 \times 21{,}000^2 \times 12}{5.22 \times 10^{10} \times 0.186 \times 0.35 \times 1.0}$$

$$= 0.0004 \text{ psi}$$

Summary

286	h outside	30%60
U_C	116	
U_D	90.7	
R_d Calculated 0.0024		
R_d Required		
10.1	Calculated ΔP	Neg
10.0	Allowable ΔP	0.25

The dirt factor is somewhat low for continued service. The high pressure drop on the gas oil line is insignificant.

When a reboiler is overdesigned, it may operate by *breathing*. As liquid enters the reboiler, it may be completely vaporized very quickly because of the overdesign. New liquid replaces it and cools the surface down. The new liquid remains in the reboiler momentarily and is heated and completely vaporized also, so that intermittent bursts of vapor issue from the reboiler outlet instead of a smooth continuous flow of vapor and liquid mixture. This can be overcome by reducing the pressure on the steam if steam is the heating medium or by placing an orifice on the shell outlet flange so as to cause an increased pressure drop on the vapor.

Horizontal Thermosyphons with Baffles. Horizontal thermosyphons are occasionally designed with vertically cut baffles such as were discussed in connection with the 2-4 exchanger. The baffles do not affect the boiling film, but they do affect the sensible-heating coefficient, increasing it beyond the free-convection value. If the baffles are the usual 25 per cent, vertically cut, segmental baffles arranged for side-to-side flow, the sensible film coefficient can be computed from shell-side data as given in Fig. 28 and as discussed under the 2-4 exchanger. If the baffles are cut 50 per cent corresponding to quarter-circle support plates, the coefficient is treated on the basis of axial flow as above, using the equivalent diameter as calculated in Example 7.8 and the tube-side data of Fig. 24.

2. Vaporization in the Tubes

The members of this class are vertical units operating with a relatively large hydrostatic head and a low pressure drop. For this reason the vaporization usually occurs in the tubes of a one-tube-pass exchanger, which permits a greater recirculation rate than is common to horizontal units with vaporization in the shell. The three main classes of equipment employing this arrangement are the vertical long-tube evaporator, the vertical thermosyphon reboiler, and the unfired steam generator. The

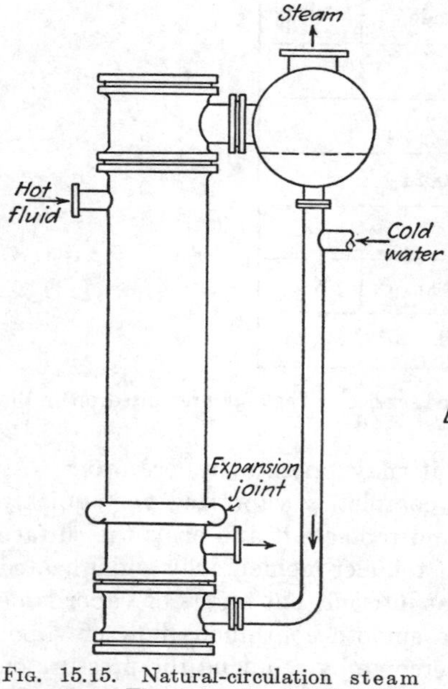

Fig. 15.15. Natural-circulation steam generator. This is similar to Fig. 14.27 but for foaming and scaling provisions.

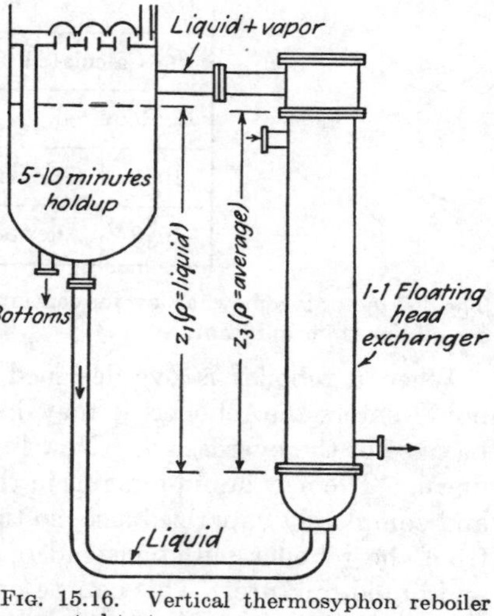

Fig. 15.16. Vertical thermosyphon reboiler connected to tower.

steam generator is shown in Fig. 15.15, and it is similar to the vertical long-tube evaporator in Fig. 14.27. The vertical thermosyphon reboiler is shown in Fig. 15.16, and it will presently be treated in detail to demonstrate the calculation common to natural-circulation units.

Recirculation Ratios. The recirculation ratio is attained when the sum of the resistances in the vaporization circuit is equal to the hydrostatic driving force on the vaporizing fluid. Referring to the vertical thermosyphon in Fig. 15.16, there are five principal resistances:

1. Frictional pressure drop through the inlet piping
2. Frictional pressure drop through the reboiler (z_2)
3. Expansion or acceleration loss due to vaporization in the reboiler

4. Static pressure of a column of mixed liquid and vapor (z_3) in the reboiler

5. Frictional pressure drop through the outlet piping

Expansion Loss Due to Vaporization. This is taken as two velocity heads based on the mean of the inlet and outlet specific gravities.

$$\Delta P_1 = \frac{G^2}{144 g \rho_{av}} \quad \text{psi} \quad (15.14)$$

Particularly where the recirculation ratio and the operating pressure are great, the difference in the densities between the inlet and outlet are not very large and the expansion loss is negligible.

Weight of a Column of Mixed Liquid and Vapor. This is difficult to evaluate if precision is required, since the expansion of the vapor is a function of the recirculation ratio, average specific volume of the vapor, coefficient of expansion of the liquid, etc. For nearly all practical cases it may be assumed that the variation of the specific gravity is linear between the inlet and the outlet. If v is the specific volume at any height x in the vertical tube of Fig. 15.17 whose total length is L and whose inlet and outlet specific volumes are v_i and v_o,

$$v = v_i + \frac{(v_o - v_i)x}{L} \quad (15.15)$$

If the weight of the column of mixture is m, the change in weight with height is dm, and if a is the cross-section flow area,

$$dm = \frac{a}{v} dx \quad (15.16)$$

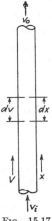

Fig. 15.17. Volume change in a single tube

If the static pressure of the column of liquid and vapor is designated by $z_3 \rho_{av}$ and the cross-section area a is unity,

$$z_3 \rho_{av} = \int_0^L \frac{dx}{v} = \int \frac{dx}{v_i + (v_o - v_i)x/L} \quad \text{psi} \quad (15.17)$$

Integrating and dividing by 144 to obtain the head per square inch

$$\frac{z_3 \rho_{av}}{144} = \frac{2.3L}{144(v_o - v_i)} \log \frac{v_o}{v_i} \quad \text{psi} \quad (15.18)$$

Rational solutions for the recirculation ratio can be established by taking all the heads in the circuit into account as functions of the mass velocity G, and upon solution for G the recirculation rate can be obtained directly. Because the gravity of the reboiler outlet mixture also varies with the recirculation ratio, the expression becomes complex and it is simpler to solve by trial and error. If the height of an existing reboiler is given, the recirculation ratio can be computed. If the recirculation ratio is given, the required head $z_1 \rho_l$ may be computed.

a. **Vertical Thermosyphon Reboiler.** The vertical thermosyphon is usually a conventional 1-1 exchanger with the channel end up. The upper tube sheet is placed close to the liquid level of the bottoms in the distilling column. Since the reboiler can be set close to the column, the frictional loss in the inlet and outlet piping is usually negligible. The recirculation ratio is determined from the difference between the hydrostatic head in the distilling column corresponding to the tube length of the reboiler and the weight of the vapor-liquid mixture. Recirculation ratios exceeding 4:1 are usually employed.

Example 15.5. Calculation of a Vertical Thermosyphon Reboiler. A vertical thermosyphon reboiler is to provide 40,800 lb/hr of vapor which is almost pure butane. In an arrangement identical with Fig. 15.16 the column operates at a pressure of 275 psig corresponding to a nearly isothermal boiling point of 228°F. Heat will be supplied by steam at 125 psig.

A recirculation ratio of 4:1 or greater should be employed.

What is the optimum exchanger to fulfill this requirement? ¾ in. OD, 16 BWG tubes on 1-in. triangular pitch will be used.

Solution:

(1) Heat balance:
Enthalpy of liquid at 228°F and 290 psia = 241 Btu/lb (Fig. 9)
Enthalpy of vapor at 228°F and 290 psia = 338 Btu/lb
Butane, $Q = 40{,}800(338 - 241) = 3{,}960{,}000$ Btu/hr
Steam, $Q = 4570 \times 868 = 3{,}960{,}000$ Btu/hr (Table 7)

(2) Δt: Isothermal boiling
$\Delta t = \text{LMTD} = 125°F$ (5.14)

(3) T_c and t_c: Both streams are isothermal.

Trial 1 (see Chap. 11 for method of approach):

(*a*) When establishing reboiler surface the first trial should always be taken for the maximum allowable flux

$$A = \frac{Q}{Q/A} = \frac{3{,}960{,}000}{12{,}000} = 330 \text{ ft}^2$$

Assume 16'0" long tubes. These will reduce the shell diameter and provide the cheapest reboiler. However, it will also require the greatest elevation of the column.

Number of tubes = $330/16'0'' \times 0.1963 = 105$ (Table 10)

(*b*) Since this will be a 1-1 exchanger, only one tube pass
From the tube counts: 105 tubes, 1 pass, ¾ in. OD, 1-in. triangular pitch
Nearest count: 109 tubes in a 13¼ in. ID shell

(*c*) Corrected coefficient U_D:
$A = 109 \times 16'0'' \times 0.1963 = 342 \text{ ft}^2$
$U_D = \dfrac{3{,}960{,}000}{342 \times 125} = 92.5$

VAPORIZERS, EVAPORATORS, AND REBOILERS

Recirculation ratio: Assume 4:1 recirculation ratio.

Static pressure of reboiler leg, $\dfrac{z_3\,\rho_{av}}{144} = \dfrac{2.3L}{144(v_o - v_i)} \log \dfrac{v_o}{v_i}$ psi (15.18)

Vapor density, $\rho_v = \dfrac{58}{359 \times {688}/{492} \times 14.7/290} = 2.27$ lb/ft^3

$v_{vapor} = \dfrac{1}{2.27} = 0.44$ ft^3/lb

$v_{liquid} = v_i = \dfrac{1}{62.5 \times 0.43} = 0.0372$ ft^3/lb (Fig. 6)

Weight flow of recirculated liquid $= 4 \times 40{,}800 = 163{,}200$ lb/hr
Total volume out of reboiler:

$$
\begin{aligned}
\text{Liquid, } 163{,}200 \times 0.0372 &= 6{,}100 \text{ ft}^3 \\
\text{Vapor, } 40{,}800 \times 0.44 &= 17{,}950 \text{ ft}^3 \\
\text{Total} &= 24{,}050 \text{ ft}^3
\end{aligned}
$$

$v_o = \dfrac{24{,}050}{(163{,}200 + 40{,}800)} = 0.1175$ ft^3/lb

Pressure of leg, $\dfrac{z_3\,\rho_{av}}{144} = \dfrac{2.3 \times 16}{144(0.1175 - 0.0372)} \times \log \dfrac{0.1175}{0.0372} = 1.60$ psi

Frictional resistance:

Flow area:

$a_t = N_t \dfrac{a_t'}{144 n} = 109 \times \dfrac{0.302}{144} = 0.229$ ft^2 (Table 10)

$G_t = \dfrac{w}{a_t} = \dfrac{163{,}200 + 40{,}800}{0.229} = 891{,}000$ lb/(hr)(ft^2)

At 228°F, $\mu = 0.10 \times 2.42 = 0.242$ lb/(ft)(hr) (Fig. 14)

$D = \dfrac{0.62}{12} = 0.0517$ ft

$Re_t = \dfrac{DG}{\mu} = \dfrac{0.0517 \times 891{,}000}{0.242} = 190{,}000$

$f = 0.000127$ ft^2/in.^{2} (Fig. 26)

$s_{mean} = \dfrac{(0.43 + 1/0.1175 \times 62.5)}{2} = 0.285$

$\Delta P_t = \dfrac{fG^2Ln}{5.22 \times 10^{10} Ds\phi_t} = \dfrac{0.000127 \times 891{,}000^2 \times 16}{5.22 \times 10^{10} \times 0.0517 \times 0.285 \times 1.0} = 2.09$ psi (7.45)

Total resistance $= 1.60 + 2.09 = 3.69$ psi
Driving force, $\dfrac{z_1 \rho_l}{144} = 16 \times 0.43 \times 62.5/144 = 2.98$ psi no check

The resistances are greater than the hydrostatic head can provide; hence the recirculation ratio will be less than 4:1. Of the resistances the frictional pressure drop may be reduced by the square of the mass velocity if the tubes are made shorter. The other alternative is to raise the liquid level in the column above the upper tube sheet.

Trial 2: Assume 12'0'' tubes and 4:1 recirculation ratio:

(a) Number of tubes $\simeq 330$ ft^2 = $330/12'0'' \times 0.1963 = 140$
(b) from the tube counts: 140 tubes, one pass, ¾ in. OD, 1-in. triangular pitch
Nearest count: 151 tubes in a 15¼ in. ID shell

(c) Corrected coefficient U_D:
$A = 151 \times 12'0'' \times 0.1963 = 356$ ft^2
$U_D = 3{,}960{,}000/356 \times 125 = 89.0$

Recirculation ratio: Assume 4:1 recirculation ratio
$v_i = 0.0372$ as before
$v_o = 0.1175$

Static pressure of leg, $\dfrac{z_3 \rho_{av}}{144} = \dfrac{2.3 \times 12}{144(0.1175 - 0.0372)} \log \dfrac{0.1175}{0.0372} = 1.20$ psi

Frictional resistance:

$a_t = 151 \times \dfrac{0.302}{144} = 0.316$ ft^2

$G_t = \dfrac{204{,}000}{0.316} = 645{,}000$ lb/(hr)(ft^2)

$Re_t = 0.0517 \times \dfrac{645{,}000}{0.242} = 138{,}000$

$f = 0.000135$ ft^2/in.^{2}

$\Delta P_t = \dfrac{0.000135 \times 645{,}000^2 \times 12}{5.22 \times 10^{10} \times 0.0517 \times 0.285 \times 1.0} = 0.88$ psi

Total resistance $= 1.20 + 0.88 = 2.08$ psi

Driving force, $\dfrac{z_1 \rho_t}{144} = 12 \times 0.43 \times \dfrac{62.5}{144} = 2.24$ psi

Since the driving force is slightly greater than the resistances, a recirculation ratio better than 4:1 is assured. With a mass velocity of 645,000 lb/(hr)(ft^2) equivalent to an inlet velocity $(V = G_t/3600\rho)$ of $645{,}000/3600 \times 62.5 \times 0.43 = 6.7$ fps the butane boiling coefficient may be computed as for forced circulation.

Hot fluid: shell side, steam

(9') Condensing steam
$h_o = 1500$ Btu/(hr)(ft^2)(°F)

Cold fluid: tube side, butane
(4), (5), (6) $Re_t = 138{,}000$
(7) $j_H = 330$ [Fig. 24]
(8) $k(c\mu/k)^{\frac{1}{3}} = 0.115$ Btu/(hr)(ft^2)(°F/ft)
(9) $h_i = (j_H k/D)(c\mu/k)^{\frac{1}{3}}$
    $= 330 \times 0.115/0.0517 = 735$
This exceeds the maximum. Use 300.
(10) $h_{io} = h_o \times \text{ID/OD} = 300$
    $\times 0.62/0.75 = 248$ Btu/(hr)(ft)(°F)

(13) Clean overall coefficient U_C:

$U_C = \dfrac{h_{io} h_o}{h_{io} + h_o} = \dfrac{1500 \times 248}{1500 + 248} = 213$ Btu/(hr)(ft^2)(°F) \hfill (6.38)

(14) Dirt factor R_d:
U_D has been obtained above.
$R_d = \dfrac{U_C - U_D}{U_C U_D} = \dfrac{213 - 89}{213 \times 89} = 0.0065$ (hr)(ft^2)(°F)/Btu \hfill (6.13)

Pressure Drop: The pressure drop through the reboiler has been computed, 0.88 psi. The head elevation z_1 will be 12 ft. The pressure drop on the shell using half-circle support plates is negligible.

VAPORIZERS, EVAPORATORS, AND REBOILERS

Summary

1500	h outside	248
U_C		213
U_D		89
R_d Calculated 0.0065		
R_d Required 0.004–0.006		
Neg	Calculated ΔP	0.88
Neg	Allowable ΔP	0.88

The large dirt factor must be retained because of the flux requirements. This is clearly an instance in which the high temperature of the steam yields no advantage. If the steam temperature were lower, a higher value of U_D could be used and the surface would remain the same.

The final reboiler will be

Shell side
    ID = 15¼ in.
    Baffle space = ½ circles
    Passes = 1

Tube side
    Number and length = 151, 12'0"
    OD, BWG, pitch = ¾ in., 16 BWG, 1-in. tri.
    Passes = 1

b. Long-tube Vertical Evaporator. The calculation in Example 15.5 can be applied directly to the long-tube vertical evaporator and the steam generator. The method of computing the recirculation ratio can also be applied directly to horizontal thermosyphons, although it is seen in practice that the recirculation ratio is usually quite low.

CALCULATIONS FOR DISTILLATION PROCESSES

The Reboiler Heat Balance. The heat requirements for a reboiler can be determined readily from heat balances on any continuous distillation column. A typical column is shown in Fig. 15.18 along with a condenser and reboiler. The function of reflux has already been discussed in Chap. 12. If R is the reflux ratio, i.e., the number of mols of condensate poured back into the column *per mol* of product withdrawn, the heat balance on the condenser is

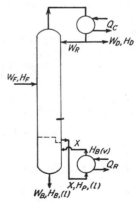

FIG. 15.18. Distilling column with condenser and reboiler.

$$(R + 1)W_D H_{D(v)} - (R + 1)W_D H_{D(l)} = Q_C \quad (15.19)$$

where W_D = distillate (overhead product), lb/hr
H_D = enthalpy of the distillate (overhead product), Btu/lb
Q_C = heat removed in the condenser, Btu/hr

Subscripts (l) and (v) refer to the liquid and vapor, respectively.

The heat balance on the entire column where the feed may be liquid or vapor is

$$\text{In} \qquad\qquad \text{Out}$$
$$W_F H_{F(l\text{ or }v)} + Q_R = Q_C + W_B H_{B(l)} + W_D H_{D(l)} \qquad (15.20)$$

where Q_R is the reboiler duty in Btu per hour and subscripts F and B refer to the feed and bottoms, respectively.

Rearranging,

$$Q_R = (R + 1)W_D H_{D(v)} - R W_D H_{D(l)} + W_B H_{B(l)} - W_F H_F \qquad (15.21)$$

Assuming that enthalpy data are available, the heat load for the reboiler can be determined if the distillate, feed and bottom quantities, and temperatures are known and if the reflux ratio is also given.

Example 15.6. Calculation of the Reboiler Duty. 20,000 lb/hr of a 50-50 mixture by weight of benzene and toluene is to be distilled at 5 psig total pressure to produce a distillate or overhead product containing 99.0 per cent by weight of benzene, the more volatile component, and a bottom product containing not more than 5 per cent of benzene. A reflux ratio of 2.54 mols of reflux per mol of distillate will be used. What heat load must the reboiler deliver?

First, how much distillate and product will be formed? Two balances may be applied to obtain this information: the overall material balance and a balance to determine how the total benzene in the feed is distributed between the distillate and the bottoms, thus:

Overall balance, $20{,}000 = W_D + W_B$
Benzene balance, $20{,}000 \times 0.50 = 0.99 W_D + 0.05 W_B$
Solving simultaneously, $W_D = 9570$ lb/hr
$\qquad\qquad\qquad\qquad\quad W_B = 10{,}430$ lb/hr

The enthalpies are obtained from Figs. 3 and 12 weighted for chemical composition at the respective temperatures

$$H_{B(l)} = 108.0 \text{ Btu/lb} \qquad \text{Latent heat} = 153.0 \text{ Btu/lb}$$
$$H_{D(l)} = 85.8 \qquad H_{D(v)} = 253.8$$
$$H_{F(l)} = 92.0$$

Substituting in Eq. (15.21),

$$Q_R = (2.54 + 1)9570 \times 253.8 - 2.54 \times 9570 \times 85.8 + 10{,}430 \times 108.0 - 20{,}000$$
$$\times 92.0 = 5{,}800{,}000 \text{ Btu/hr}$$

Vapor which must be generated in the reboiler = $5{,}800{,}000/153 = 37{,}900$ lb/hr

The liquid entering the reboiler in a once-through arrangement is the total quantity leaving the bottom plate or trapout. The trapout, in turn, is equal to the sum of the material vaporized in the reboiler and the bottoms product. The total quantity entering the reboiler is

$$37,900 + 10,430 = 48,330 \text{ lb/hr}$$

The percentage vaporized is $37,900 \times 100/48,330 = 78.5$ per cent and the vapor and liquid are disengaged under the bottom plate. The temperature during vaporization does not remain constant in the reboiler, although the boiling range for a binary mixture with a fairly pure bottom product is very small. The vapor-liquid equilibrium relationships have already been treated in Chap. 13. The general method of determining the boiling range will be demonstrated presently with a multicomponent mixture.

Temperature Potentials in Distillation. The operating pressure of a distillating column is usually determined by the approach of the bubble point of the overhead product to the average temperature of the cooling water available to the condenser. If condenser water is available from 80 to 100°F, it is impossible to remove the latent heat of condensation from a vapor at such a pressure that it condenses at 60°F. In such a case it is necessary to raise the operating pressure and obtain a higher saturation temperature at which the heat can flow into the cooling water. The higher the pressure the greater the temperature difference across the condenser and the smaller the condenser. This is offset by the increased cost of the reboiler, column, and condenser, all of which must be designed for a greater pressure. If steam is available in a plant at a definite maximum pressure, the greater the temperature difference in the condenser the smaller the temperature difference attainable in the reboiler.

The Quantitative Relationships in Fractional Distillation.[1] Only a brief discussion of the elements of fractional distillation are within the scope of this chapter, but they are of considerable value in analyzing the influence of the process variables on the size of the heat-transfer equipment and particularly the reboiler. Excellent treatments of distillation theory will be found in a number of standard references.[2] Con-

---

[1] Pages 493 to 502 are preliminary to the calculation of the reboiler duty in Examples 15.7 and 15.8.

[2] Badger, W. L., and W. L. McCabe, "Elements of Chemical Engineering," McGraw-Hill Book Company, Inc., New York, 1936. Perry, J. H., "Chemical Engineers' Handbook," 3d ed., McGraw-Hill Book Company, Inc., New York, 1950. Robinson and Gilliland, "Elements of Fractional Distillation," McGraw-Hill Book Company, Inc., New York, 1939. Walker, W. H., W. K. Lewis, W. H. McAdams, and E. R. Gilliland, "Principles of Chemical Engineering," McGraw-Hill Book Company, Inc., New York, 1937.

sider the distillation arrangement in Fig. 15.19 enclosed by the dash line 0-0'-3'-3. The upper horizontal dash line 1-1' cuts the distilling column between any two plates above the feed arbitrarily designated as the nth plate and the $n + 1$th plate. There is a quantity of vapor V mols/hr entering 1-1'-0'-0 with a vapor composition corresponding to that above the liquid of the composition of the nth plate and designated by V_n. The liquid L mols/hr leaving 1-1'-0'-0 are those coming from the $n + 1$th plate of the column and are designated by L_{n+1}. In addition some distillate W_D mols/hr is withdrawn. The material balance on 1-1'-0'-0 is the equation of the quantities in and out

$$V_n = L_{n+1} + W_D \quad (15.22)$$

The material balance applied to the mols of one individual component also holds.

$$V_n y_n = L_{n+1} x_{n+1} + W_D x_D \quad (15.23)$$

where y_n is the vapor mol fraction of one component in V_n, x_{n+1} is the liquid mol fraction of one component in L_{n+1}, and x_D is the mol fraction of the same component in the distillate. Solving for y_n, the vapor composition on any plate in the section above the feed plate, and eliminating V_n,

$$y_n = \left(\frac{L_{n+1}}{L_{n+1} + W_D} \right) x_{n+1} + \left(\frac{W_D}{L_{n+1} + W_D} \right) x_D \quad (15.24)$$

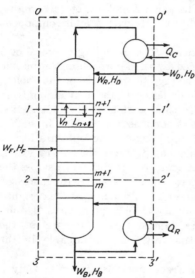

FIG. 15.19. Analysis of balances about a distilling column.

This equation defines the composition of the vapor throughout the upper section of the column. When a separation of components is to be accomplished and a distilling column is to be designed for it, certain data must be given, namely, the quantity and composition of the feed (W_F, x_F), the desired overhead distillate composition (x_D), and the bottoms-product composition (x_B). The quantities of distillate and bottoms (W_D and W_B) can be readily obtained by an overall material and component balance on the column as in Eqs. (15.22) and (15.23). In Eq. (15.24) there then remain three unknown variables, y_n, L_{n+1}, and x_{n+1}. If one of these variables can be determined at some point in the column, only two will remain unknown. Since the liquid composition of the feed and of the distillate (top plate) must be given at the start, only y_n and L_{n+1} are ordinarily unknown. Equation (15.24) is soluble when another relationship is introduced containing y_n and L_{n+1}. Referring to Fig. 15.19 a heat balance can be made about the upper portion of the column alone *without* including the condenser.

$$V_n H_{n(v)} + W_R H_{D(l)} = (W_R + W_D) H_{D(v)} + L_{n+1} H_{n+1(l)} \quad (15.25)$$

Similarly for the section below the feed, 2-2'-3'-3, the analogues of Eqs. (15.22) to (15.24), where any plate below the feed is taken as the mth plate and the plate above it as the $m + 1$th plate, the balances are

$$V_m + W_B = L_{m+1} \quad (15.26)$$
$$L_{m+1} x_{m+1} = V_m y_m + W_B x_B \quad (15.27)$$
$$y_m = \left(\frac{L_{m+1}}{L_{m+1} - W_B}\right) x_{m+1} - \left(\frac{W_B}{L_{m+1} - W_B}\right) x_B \quad (15.28)$$

From a heat balance at the bottom

$$Q_R + L_{m+1} H_{m+1(l)} = V_m H_{m(v)} + W_B H_{B(l)} \quad (15.29)$$

It has been observed that the ratio of the molar latent heats of vaporization to the absolute boiling points of related compounds is a constant This is Trouton's rule. If it is further assumed that the latent heats of all compounds are equal, the latent heats of any ideal solutions they may form are also equal per mol of mixture. By assuming that the sensible-heat change from plate to plate is negligible compared with the latent-heat changes, the calculation of distillation processes is greatly simplified. If the temperature difference between the distillate and bottoms is several hundred degrees, however, this assumption cannot be made. Following this assumption, if the molar latent heats of the vaporization and condensation throughout the tower are constant, the mols of material on each plate must also be constant. Then

$$H_{n(v)} = H_{B(v)} \quad (15.30)$$

and

$$V_n = W_R + W_D$$

And from Eq. (15.25) as a result of the equality in vapor heat content,

$$W_R H_{R(l)} = L_{n+1} H_{n+1(l)} \quad (15.31)$$

By assumption,

$$H_{R(l)} = H_{n+1(l)} \quad (15.32)$$

and

$$W_R = L_{n+1}$$

The mols of liquid and vapor flow have thus been proved constant above the feed. It is now necessary to determine whether it is also constant below the feed and, if constant, whether or not the value of V_m is also equal to V_n. Combining the heat balance for the section below the feed [Eq. (15.29)] with the overall heat balance [Eq. (15.21)],

$$V_m H_{m(v)} - L_{m+1} H_{m+1(l)} = (W_R + W_D) H_{D(v)} - W_F H_{F(l)} - W_D H_{D(l)} \quad (15.33)$$

and using the assumptions established previously,

$$V_m H_{m(v)} = (W_R + W_D) H_{D(v)} \quad (15.34)$$
$$H_{m(v)} = H_{D(v)} = H_{n(v)}$$
$$V_m = W_R + W_D = V_n \quad (15.35)$$

Equation (15.34) establishes the equality of vapor flow in both portions of the column.

Similarly

$$L_m H_{m(l)} = W_F H_{F(l)} + W_R H_{R(l)} \quad (15.36)$$
$$H_{m(l)} = H_{F(l)} = H_{R(l)}$$
$$L_m = W_F + W_R \quad (15.37)$$

Equation (15.24) may be generalized for the composition at any plate above the feed by

$$y = \left(\frac{L}{L + W_D}\right) x + \left(\frac{W_D}{L + W_D}\right) x_D = \frac{L}{V} x + \frac{W_D}{V} x_D \quad (15.38)$$

Equation (15.28) may be generalized for any plate below the feed by

$$y = \left(\frac{L}{L - W_B}\right) x - \left(\frac{W_R}{L - W_B}\right) x_B = \frac{L}{V} x - \frac{W_B}{V} x_B \quad (15.39)$$

The McCabe-Thiele[1] Diagram. If the composition of a binary mixture can be traced by the analysis of only one component throughout the column, the solutions of Eqs. (15.38) and (15.39) can be obtained rather simply, since these apply only to the more volatile component. The solution can be accomplished by plotting a vapor-composition curve at constant total pressure which relates the mol fraction of the more volatile component in the vapor arising from and in equilibrium with a given composition of liquid as calculated by Eq. (13.7). Such an equilibrium vapor-composition curve of ordinate y vs. the values of x as the abscissa for a benzene-toluene mixture, is plotted in Fig. 15.20. These are the equilibrium changes which would occur if an individual mixture of, say, 20 mol-per cent of benzene was maintained at its boiling point in a closed vessel and shows the difference in the vapor and liquid compositions

[1] McCabe, W. L., and E. W. Thiele, *Ind. Eng. Chem.*, **17**, 605 (1925).

required at equilibrium. The same is true at the other compositions, 40, 60 per cent, etc. These are also the changes which would occur to the liquid on the plates of a distilling column if no liquid entered and no vapor left the plates. Thus, a 40 mol-per cent benzene solution on a plate or $x_n = 0.40$ is in equilibrium with a vapor of 61.4 mol-per cent or

$$y_n = 0.614$$

If the vapor were completely condensed on the plate above $(n + 1)$ by a cooling coil and without the presence of liquid reflux, the concentration

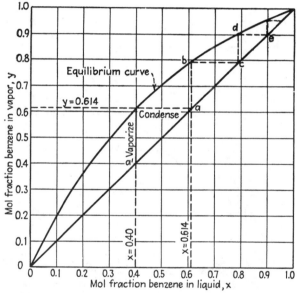

FIG. 15.20. Vapor-composition curve for mixtures of benzene and toluene at 5 psig.

of benzene in the condensed liquid would be $x_{n+1} = 0.614$. But the liquid reflux constantly enters and leaves the plates, and vapor constantly enters through the bubble cap and arises from it with a different composition. Although an equilibrium change occurs on each plate, there is never so broad an equilibrium change from plate to plate as indicated in Fig. 15.20 due to the continuous addition and removal of compounds from the plates. These have been accounted for in the balances from which Eqs. (15.38) and (15.39) were determined.

It is interesting to note that, if a diagonal is drawn across Fig. 15.20, it is the locus of the points at which a horizontal line joins a vapor and a liquid of the same composition. Thus if the 61.4 mol-per cent vapor were condensed by heat removal using a cooling coil rather than cold reflux, the new condensate at a would also be 61.4 mol-per cent. If the 61.4 mol-per cent condensate were vaporized again, it would describe the path ab and its condensation would be bc. The function of a plate in a distillating column is to provide a space for equilibrium to take place, and the number of abc paths required one after the other, as abc, cde, etc., to permit the separation of a volatile component of concentration x_F in the feed to a concentration x_D in the distillate is the number of theoretical

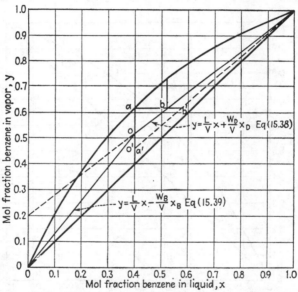

FIG. 15.21. Effect on the operating line of varying reflux ratio.

plates required for the separation from x_F to x_D. If the number of plates required to separate x_B from x_F is added to it, one obtains the total number of required theoretical plates to produce a bottoms and distillate of desired compositions from a feed of a given composition. Instead of following the diagonal line in Fig. 15.20 the composition between the liquid and the vapor passing throughout the column may be represented by the two lines in Fig. 15.21 with their identifying equations [Eqs. (15.38) and (15.39)] written directly thereon. Since these lines describe the compositions with which the column operates with reflux being poured back, they are the *operating lines*. The upper line represents the enrichment of **volatile** component above the feed plate and is called the *rectifying* sec-

VAPORIZERS, EVAPORATORS, AND REBOILERS

tion. The lower line represents the depletion of volatile component and is the *stripping* section.

The influence of the reflux can be seen more clearly with the aid of Fig. 15.21. If the amount of reflux is increased while x_D is fixed, the slope of the operating line for the rectification above the feed plate also increases, moving the point o downward to o'. When the vapor at a is condensed at b' instead of b, a greater composition change occurs per plate and a smaller number of plates are required for the separation. In terms of height a shorter column can be used. But again it must be pointed out that the greater the quantity of reflux and the shorter the column the greater the diameter of the column and the greater the heat load on the reboiler and condenser. For this reason the determination of the optimum reflux ratio is an economic consideration obtained by adding the operating costs and fixed charges of the equipment for different reflux ratios and determining the minimum operating cost as a function of the reflux ratio.

It is now apparent that the 45° diagonal line in Fig. 15.21 represents the smallest possible number of plates with which the separation of feed into distillate and bottoms can be effected. But the slope of the diagonal is 1.0, and in Eq. (15.38) if the slope is unity, $L/(L + W_D) = 1.0$ and $W_D = 0$, corresponding to a primed column in equilibrium from top to bottom but without feed and producing no distillate. On the other hand, suppose o were moved up to a. The number of equilibrium changes necessary to go from x_F to x_D with an operating line between x_D on the diagonal and a would be infinite, although the reflux would be a minimum. This last would require an infinitely high column, and if o were moved to a point even higher than a, it would be impossible to achieve the separation.

Minimum Reflux. The determination of the minimum reflux is a useful limitation for establishing a practical reflux ratio. The optimum reflux ratio, as mentioned before, is simply one of economic costs. If the operating line connects a and x_D, corresponding to minimum reflux, the value of the total mols of reflux at minimum reflux may be designated by W'_R. Then Eq. (15.38) may be written

$$y_F = \frac{W'_R}{V} x_F + \frac{W_D}{V} x_D \qquad (15.40)$$

but

$$W_D = V - W'_R \qquad (15.41)$$

or

$$\frac{W'_R}{V} = \frac{x_D - y_F}{x_D - x_F} \qquad (15.42)$$

where W'_R/V is the ratio of the mols of reflux at minimum reflux to the total vapor.

Thermal Condition of the Feed—Effect upon the Reboiler. In all the heat balances employed so far in this chapter it has been assumed that the feed was a liquid entering at its boiling point. This is not a requisite of a distillation process. The feed may enter the column directly from storage as a liquid colder than the boiling point corresponding to its composition, or it may come from the partial condenser of a preceding distilling column as a saturated vapor. It may even come from a preceding column as a liquid or vapor at a higher pressure than the operating pressure of the column to which it is fed. If a high-pressure liquid, it will flash on being released to the low-pressure column, giving a feed which is a mixture of liquid and vapor. If it is a high-pressure vapor and passes through a reducing valve before entering the low-pressure column, it may be superheated. How do these possibilities affect the column and the reboiler?

For the case of a liquid feed at its boiling point a liquid balance around the feed plate is

$$W_F = L_{m+1} - L_{n+1} \tag{15.43}$$

Define a new equation to correct for deviations from the balance in Eq. (15.43).

$$q = \frac{L_{m+1} - L_{n+1}}{W_F} \tag{15.44}$$

If $q = 1$, then as in Eq. (15.43) all the feed enters as liquid at its boiling point.

If $q < 1$, some of the feed is vapor, accounting for the discrepancy in the liquid balance.

If $q < 0$, qW_F must be negative. This means that there is less liquid downflow in the lower portion than in the upper section. This could occur only if some of the feed was superheated so that less fluid was required for the heat transfer.

If $q = 0$, $qW_F = 0$ and the feed is vapor at its boiling point.

If $q > 1$, the feed is cold liquid.

If these are the possibilities, how do they affect the intersection of the upper and lower operating lines of the column, since any point on the vertical line from x_F to a in Fig. 15.21 represents a liquid only? Employing Eqs. (15.23) and (15.27) and calling the coordinates of the intersection x', y',

$$V_n y' = L_{n+1} x' + W_D x_D$$
$$V_m y' = L_{m+1} x' + W_B x_B$$

and the component balance for the volatile component

$$W_F x'_F = W_D x'_D + W_B x'_B \tag{15.45}$$

And if $q = (L_{m+1} - L_{n+1})/W_F$, then $q - 1 = (V_m - V_n)/W_F$, from which

$$y' = \frac{q}{q-1} x' - \frac{x_F}{q-1} \tag{15.46}$$

If q is determined from Eq. (15.44), then Eq. (15.46) describes the point of intersection in terms of x and y. The various positions of the intersection in terms of q are shown in Fig. 15.22. Inspection of Fig. 15.22

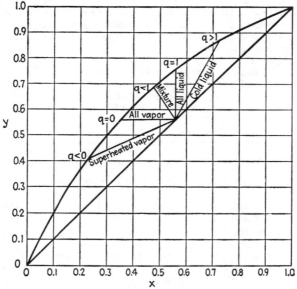

Fig. 15.22. Effect of thermal condition of feed.

indicates that for a given reflux ratio the colder the feed the fewer the number of plates required for the separation, since the reflux ratio in the lower section is increased by the condensation of the additional vapors at the feed plate. The increased reflux in the lower section must be offset by an increase in the size of the reboiler and the quantity of reboiler heat. This brings up the point of whether it is cheaper to supply the heat for cold feed in a feed preheater or in the reboiler. Very obviously it is cheaper to do so in nearly all cases in a separate preheater, since the temperature at which the feed absorbs the heat must be lower than the temperature at which the bottoms can absorb the additional heat in the reboiler. This might suggest an advantage to the use of vapor feed, but this is offset by the increased cost of transporting the vapor in larger lines and by an increased number of theoretical plates. Furthermore, vapor feed cannot contain sufficient heat to offset the need for a reboiler, since almost all the heat entering as vapor feed must be removed in the condenser to give product and the latent heat required for reflux must still be provided by the reboiler. If feed is available as a vapor, it is unwise to condense it first, since the reduced reboiler duty and operating cost of vapor feed more than justify the addition of extra theoretical plates.

Example 15.7. Distillation of a Binary Mixture. 20,000 lb/hr of a 50 weight per cent mixture of benzene and toluene is to be distilled at an operating pressure of 5.0 psig to produce a distillate containing 99.0 weight per cent benzene and bottoms containing 95.0 weight per cent toluene. How many plates are required?

Solution. First determine the pounds of distillate and bottoms obtained as in Example 15.6. Basis: One hour.

Material balance, $20{,}000 = W_D + W_B$
Benzene balance, $20{,}000 \times 0.50 = 0.99 W_D + 0.05 W_B$
Solving simultaneously:

$$W_D = 9570 \text{ lb/hr}$$
$$W_B = 10{,}430 \text{ lb/hr}$$

Compositions and Boiling Points

	Lb/hr	Mol. wt.	Mol/hr	x_1	p_{p1}, 214°F	$x_1 p_{p1}$	y_1
Feed:							
C_6H_6	10,000	78.1	128.0	0.543	1380	750	0.741
C_7H_8	10,000	93.1	107.5	0.457	575	262	0.259
	20,000	....	235.5	1.000	$\Sigma x p_p = 1012$		1.000
Distillate:					p_{p1} 195°F		
C_6H_6	9,474	78.1	121.5	0.992	1025	1015	0.996
C_7H_8	96	93.1	1.0	0.008	410	4	0.004
	9,570	....	122.5	1.000	$\Sigma x p_p = 1019$		1.000
Bottoms:					p_{p1} 246°F		
C_6H_6	520	78.1	6.7	0.059	2180	129	0.128
C_7H_8	9,910	93.1	106.3	0.941	940	882	0.872
	10,430	....	113.0	1.000	$\Sigma x p_p = 1011$		1.000

Minimum reflux: The equilibrium data plotted in Fig. 15.20 can be used.

$$\frac{W_R'}{V} = \frac{x_D - y_F}{x_D - x_F} = \frac{0.992 - 0.741}{0.992 - 0.543} = 0.558 \text{ mol/mol} \quad (15.42)$$

$$V = W_R' + W_D \quad \text{where} \quad W_D = 1$$
$$W_R' = 0.558 W_R' + 0.558 W_D$$
$$W_R' = 1.27 \text{ mol reflux/mol distillate}$$

VAPORIZERS, EVAPORATORS, AND REBOILERS 503

Assume 200 per cent of the theoretical minimum reflux as an economic reflux.

$$W_R = 1.27 \times 2 = 2.54 \text{ mol/mol distillate}$$

The intercept for the upper operating line is

$$\frac{W_D}{W_R + 1} x_D = \frac{1}{2.54 + 1} \times 0.992 = 0.280$$

Connecting the corresponding lines in Fig. 15.23:
  Thirteen plates are required.
  The feed plate is seventh from the top.
Total reflux = $122.5 \times 2.54 = 310.5$ mol

Heat Balances

Enthalpies are computed above 0°F using specific heats from Fig. 3 and latent heats from Fig. 12 of the Appendix.

	Mol/hr	Mol. wt.	Lb/hr	Temp, °F	Btu/lb	Btu/hr
Heat balance around condenser:						
Heat in:						
Top plate vapor..........	433	87.3	33,900	195	253.8	8,600,000
Heat out:						
Distillate...............	122.5	78.3	9,570	195	85.8	822,000
Reflux.................	310.5	78.3	24,330	195	85.8	2,090,000
Condenser duty, by difference.................	....	....	......	...	.....	5,688,000
						8,600,000
Overall heat balance:						
Heat in:						
Feed..................	235.5	84.8	20,000	214	92.0	1,840,000
Reboiler duty, by difference.................	....	....	......	...	.....	5,900,000
						7,640,000
Heat out:						
Distillate...............	122.5	78.3	9,570	195	85.8	822,000
Bottoms...............	113.0	92.8	10,430	246	108.0	1,130,000
Condenser duty.........	....	....	......	...	.....	5,688,000
						7,640,000

For simplicity, in the case of a binary distillation it may be assumed that the trapout and the bottoms liquid are at nearly the same temperature and the latent heat of vaporization per pound is that of the bottoms.

At 246°F, $\lambda = 153$ Btu/hr

Reboiler vapor = $5,800,000/153 = 37,900$ lb/hr

Trapout = $37,900 + 10,430 = 48,330$ lb/hr

Actually the trapout liquid on the lowest plate is at a lower temperature than the bottoms, and there is a boiling range in the reboiler. Since in this problem the aver-

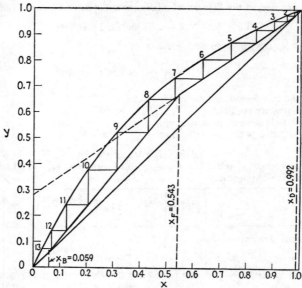

Fig. 15.23. Solution of Example 15.8.

age temperature change per plate is only $(246 - 195)/13 = 3.9°F$, the boiling range can be neglected, although in distillations with greater average temperature change per plate it must be taken into account. This point is treated in Example 15.8.

HEAT BALANCE AROUND REBOILER

	Mol/hr	Mol. wt.	Lb/hr	Temp, °F	Btu/lb	Btu/hr
Heat in:						
Trapout..................	522	92.8	48,330	246	108.0	5,230,000
Reboiler duty, by difference	...	....	......	...	.....	5,800,000
						11,030,000
Heat out:						
Reboiler vapor............	409	92.8	37,900	246	261.0	9,900,000
Bottoms.................	113	92.8	10,430	246	108.0	1,130,000
						11,030,000

The reboiler requirements are
  Total liquid to reboiler = 48,330 lb/hr
  Vaporization = 37,900
  Temperature (nearly isothermal) = 246°F
  Pressure = 5 psig
  Heat load = 5,800,000 Btu/hr

Reboiler Duty for a Multicomponent Mixture. The method of Example 15.7 may be extended to the calculation of the reboiler requirements for a multicomponent mixture (see Chap. 13). The feed quantity and composition as well as the distillate and bottoms compositions and the reflux ratio must be known. The steps in the calculation of the heat balance may be summarized:

1. Heat balance on condenser alone to determine Q_C

   Into condenser: Top tray vapor
   Out of condenser: Distillate and reflux

2. Overall heat balance on column to determine Q_R

   Into column: Feed and reboiler heat
   Out of column: Distillate, bottoms and condenser duty

3. Heat balance on reboiler alone: The quantity, composition, and temperature of the bottoms are known. The quantity, composition, and temperature of the trapout to the reboiler are *not* known. The heat required of the reboiler is known, but the quantity of the reboiler vapor is not known, although it has a composition which must be in equilibrium with the bottoms. The bottom-plate liquid will be at a lower temperature than the reboiler, and the liquid to the reboiler will be at this lower temperature. It is customary to assume this temperature difference and check the bottom-plate temperature afterward. For columns distilling a small range of components the difference will be as little as 5 to 10°F, and for wide ranges the liquid on the bottom plate may be 50°F or more below the reboiler temperature.

If X is the pounds of vapor formed and the subscript p refers to the bottom plate, the heat balance around the reboiler is

$$XH_{B(v)} + W_{B(l)}H_{B(l)} = XH_{P(l)} + W_{B(l)}H_{P(l)} + Q_R \qquad (15.47)$$

from which X can be obtained (see Fig. 15.18).

4. Since there is a material balance across the bottom plate, the bottom-plate liquid is the sum of the mols of vapor and mols of bottoms. Having determined the mols of vapor and the bottoms being known, the trapout is the sum.

Having the total mols in the bottom-plate liquid, the temperature on the plate can be checked by a bubble-point calculation. If the temperature assumed in (3) does not check the bubble point of the trapout, a new bottom-plate temperature must be assumed and (3) reworked.

Example 15.8. The Reboiler Duty for a Multicomponent Mixture. The feed, distillate, and bottoms quantities and compositions are given below. A reflux ratio of 2:1 will be used. What is the reboiler duty?

	Feed, mol/hr	Distillate, mol/hr	Bottoms, mol/hr
C_4	6.4	6.4	
C_5	223.8	219.7	4.1
C_6	51.6	2.3	49.3
C_7	71.9	.....	71.9
C_8	52.5	.....	52.5
C_9	54.7	.....	54.7
C_{10}	82.5	.....	82.5
C_{11}	76.6	.....	76.6
C_{12}	22.4	.....	22.4
	642.4	228.4	414.0

Solution:

DEW POINT OF OVERHEAD

	Mol/hr	$K_{40\text{ psia}}^{148°F}$	$\dfrac{V}{K}$
C_4	6.4	2.8	2.3
C_5	219.7	1.01	217.5
C_6	2.3	0.34	6.8
	228.4		228.4

BUBBLE POINT OF BOTTOMS

	Mol/hr	$K_{40\text{ psia}}^{330°F}$	KL	Lb/hr
C_5	4.1	5.8	23.8	1,700
C_6	49.3	3.0	148.0	13,900
C_7	71.9	1.68	120.8	13,030
C_8	52.5	.98	51.4	6,260
C_9	54.7	0.57	31.2	4,240
C_{10}	82.5	0.35	28.9	4,330
C_{11}	76.6	0.21	16.1	2,640
C_{12}	22.4	0.13	2.9	520
	414.0		423.1	46,620

Average mol. wt. = 46,620/423.1 = 110.3

The pressure at the top and bottom of the column actually differs because of the pressure drop of the vapor in the column itself. Usually the distillate is taken at 5 to 10 psi lower than the bottom.

VAPORIZERS, EVAPORATORS, AND REBOILERS

HEAT BALANCES

	Mol/hr	Mol. wt.	Lb/hr	Temp, °F	Btu/lb	Btu/hr
Heat Balance on Condenser						
Heat in:						
Top plate vapor........	685.2	71.3	48,894	148	286	13,980,000
Heat out:						
Distillate..............	228.4	71.3	16,298	124	129	2,100,000
Reflux (2-1)..........	456.8	71.3	32,596	124	129	4,200,000
Condenser duty, by difference.............	.....	.....	......	...	...	7,680.000
						13,980,000
Overall heat balance:						
Heat in:						
Feed.................	642.4	111.8	71,775	323	275	19,700,000
Reboiler duty, by difference...............	.....	.....	......	...	...	4,280,000
						23,980,000
Heat out:						
Distillate.............	228.4	71.3	16,248	124	129	2,100,000
Bottoms..............	414.0	134.0	55,477	330	290	14,200,000
Condenser duty.......	.....	.....	......	...	...	7,680,000
						23,980,000

Heat Balance on Reboiler
Assume 30° difference between reboiler and bottom plate giving a bottom-plate temperature of 300°F

Heat in:						
Trapout...............	619.7*	126.6	78,177	300	234	18,300,000
Reboiler duty...........	.....	.....	......	...	...	4,280,000
						22,580,000
Heat out:						
Reboiler vapor..........	205.7*	110.3	22,700	330	369	8,380,000
Bottoms...............	414.0	134.0	55,477	330	256	14,200,000
						22,580,000

$$* \quad X369 + 14{,}200{,}000 = X234 + 55{,}477 \times 234 + 4{,}280{,}000 \qquad \text{Eq. (15.47)}$$
$$135X = 3{,}080{,}000$$
$$X = 22{,}700 \text{ lb/hr of vapor}$$
$$= \frac{22{,}700}{110.3} = 205.7 \text{ mol/hr}$$

Trapout $= W_B + X = 414.0 + 205.7 = 619.7$ mol/hr

Average molecular weight $= \dfrac{78{,}177}{619.7} = 126.6$

Calculation of Bottom Plate Temperature

	y^*	Reboiler vapor $V = y205.7 +$ Bottoms $=$ Trapout			$K_{40\ psi}^{300°F}$	Mol $\times K$
C_5	0.056	11.6	4.1	15.7	4.5	70.6
C_6	0.350	71.9	49.3	121.2	2.25	273.0
C_7	0.285	58.7	71.9	130.6	1.20	156.8
C_8	0.122	25.0	52.5	77.5	0.66	51.1
C_9	0.074	15.2	54.7	69.9	0.38	26.6
C_{10}	0.068	14.1	82.5	96.6	0.22	21.2
C_{11}	0.038	7.8	76.6	84.4	0.13	10.9
C_{12}	0.007	1.4	22.4	23.8	0.07	1.8
	1.000	205.7	414.0	619.7		612.0

which checks 619.7

* Mol fraction of vapor in equilibrium with bottoms.

This checks the 30°F assumption between reboiler and bottom plate.
The reboiler requirements are

Vaporization.....................	22,700 lb/hr
Total liquor to reboiler............	78,177 lb/hr
Heat load......................	4,280,000 Btu/hr
Temperature range................	300–330°F
Operating pressure...............	40 psia

PROBLEMS

15.1. 20,000 lb/hr of hexane vapor is required from a liquid feed at 100°F and 85 psig. Heat is to be supplied by steam at 350°F.
Available for the service is a 17¼ in. ID 1-2 exchanger containing 160 ¾ in. OD, 16 BWG, 16'0" long tubes arranged for two tube passes on 1-in. square pitch. Shell baffles are spaced 16 in. apart.
What are the dirt factor and pressure drops?

15.2. 30,000 lb/hr of a mixture whose average properties correspond to hexane is vaporized from 40,000 lb/hr of feed at its bubble point at 105 psig. The boiling range is from 300 to 350°F, and heat is supplied by steam at 398°F.
Available for the service is a 19¼ in. ID 1-2 exchanger containing 132 1 in. OD, 14 BWG, 16'0" long tubes arranged for two tube passes on 1¼-in. square pitch. Shell baffles are spaced 12 in. apart.
What are the dirt factor and the pressure drops?

15.3. 45,000 lb/hr of 42°API kerosene enters a once-through horizontal thermosyphon reboiler. 32,000 lb/hr is vaporized between 285 and 330°F, using steam at 380°F. The required dirt factor is 0.0030, and the allowable pressure drop through the reboiler is 0.25 psi. Vaporization occurs at 5 psig, and the molecular weight of the vapor may be taken as 120. The latent heat of the vapor is 110 Btu/lb.
Available for the service is a 25 in. ID horizontal thermosyphon containing 324 ¾ in. OD, 16 BWG tubes, 12'0" long laid out on 1-in. square pitch for two passes.
Is the reboiler satisfactory?

15.4. 76,300 lb/hr of pentane bottoms (mol. wt. liquid = 77.2) is to be recirculated through a horizontal thermosyphon reboiler to provide 31,400 lb/hr of vapor (mol.

VAPORIZERS, EVAPORATORS, AND REBOILERS

wt. vapor = 74.7, latent heat = 153 Btu/lb). The tower and exchanger operate at 85 psig, and vaporization occurs from 225 to 245°F. Heat is supplied by 28°API gas oil from 470 to 370°F with a maximum pressure drop of 15 psi.

A 23¼ in. ID horizontal thermosyphon reboiler contains 240 ¾ in. OD, 13 BWG, 8'0" tubes with six passes on 1-in. square pitch. A pressure drop of 0.25 is permissible. What will the dirt factor and approximate recirculation ratio be?

15.5. A reboiler is to be designed for the vaporization service in Prob. 15.4 with a dirt factor of 0.004 except that the heat requirement will be supplied by a 50 psig steam line. Using ¾ in. OD, 13 BWG tubes, establish the layout of a reboiler which most reasonably fulfills the conditions.

15.6. A vertical thermosyphon reboiler must provide 120,000 lb/hr of butane vapor to a tower at 135 psig (boiling point = 178°F, latent heat = 122 Btu/lb) using steam at 12 psig in the shell.

Available for the service is a 35 in. ID 1-1 exchanger containing 900 ¾ in., 16 BWG, 16'0" long tubes on 1-in. triangular pitch. Shell baffles are half-circle supports on 24 in. centers.

What is the recirculation ratio?

15.7. For the process conditions in Prob. 15.6 determine the nearest vertical thermosyphon to yield approximately a 4:1 recirculation ratio and a dirt factor of 0.003 when the heat is supplied by steam at 250°F.

NOMENCLATURE FOR CHAPTER 15

A	Heat-transfer surface, ft²
A_p, A_s, A_v	Heat-transfer surface for preheating, sensible heating, and vaporization, ft²
a''	External surface per linear foot of pipe or tube, ft
a'_t	Flow area per tube, in.²
B	Baffle spacing, in.
C	Specific heat of hot fluid, Btu/(lb)(°F)
c	Specific heat of cold fluid, Btu/(lb)(°F)
D	Inside diameter of tube, ft
D_e, D'_e	Equivalent diameter for heat transfer and pressure drop, ft
D_s	Inside diameter of shell, ft
d_e	Equivalent diameter, in.
e	Napieran base, dimensionless
F_c	Caloric fraction, dimensionless
F_T	Temperature-difference factor, $\Delta t = F_T \times \text{LMTD}$, °F
f	Friction factor, ft²/in.²
G	Mass velocity, lb/(hr)(ft²)
g'	Acceleration of gravity, ft/sec²
H	Enthalpy, Btu/lb
h	Heat-transfer coefficient in general, Btu/(hr)(ft²)(°F)
h_i, h_o	Heat-transfer coefficient for inside fluid and for outside fluid, Btu (hr)(ft²)(°F)
h_{io}	Value of h_i when referred to the tube outside diameter, Btu/(hr) (ft²)(°F)
h_p, h_s, h_v	Film coefficient for preheating, sensible-heat transfer and vaporization, Btu/(hr)(ft²)(°F)
ID	Inside diameter of shell or tube, in.
j_H	Factor for heat transfer, dimensionless

K	Vapor-liquid equilibrium constant, dimensionless
K_c	Caloric constant, dimensionless
k	Thermal conductivity, Btu/(hr)(ft^2)(°F)/(ft)
L	Tube length, ft; liquid, mol/hr
LMTD	Logarithmic mean temperature difference, °F
m	Weight of a column of liquid and vapor, lb
N	Number of baffles
N_t	Number of tubes
n	Number of tube passes
OD	Outside diameter of tube, in.
P_T	Tube pitch, in.
ΔP	Pressure drop, psi
$\Delta P_T, \Delta P_t, \Delta P_r$	Total, tube, and return pressure drop, psi
p	Pressure, psi
p_p, p_t	Pressure of a pure component and total pressure on the system, atm or psi
q	Factor in distillation describing the condition of the feed, dimensionless
q_p, q_s, q_v	Heat flow for preheating, sensible heating, and vaporization, Btu/hr
Q	Heat flow, Btu/hr
Q_C, Q_R	Condenser and reboiler duties, Btu/hr
R	Reflux ratio; temperature group $(T_1 - T_2)/(t_2 - t_1)$, dimensionless
R_d	Dirt factor, (hr)(ft^2)(°F)/Btu
Re, Re'	Reynolds number for heat transfer and pressure drop, dimensionless
s, s_{av}	Specific gravity, average specific gravity, dimensionless
T_c	Caloric temperature of hot fluid °F
T_S	Steam temperature, °F
t_s	Isothermal boiling temperature, °F
T_1, T_2	Inlet and outlet temperatures of hot fluid, °F
t_c	Caloric temperature of cold fluid, °F
$\Delta t_c, \Delta t_h$	Cold and hot terminal differences, °F
Δt	True temperature difference, °F
$\Delta t_p, \Delta t_s, \Delta t_v$	True temperature difference for preheating, sensible-heat transfer, and vaporization, °F
$(\Delta t)_w$	Temperature difference between tube wall and boiling liquid, °F
U, U_C, U_D	Overall coefficient of heat transfer, clean coefficient, and design coefficient, Btu/(hr)(ft^2)(°F)
V	Velocity, fps; vapor, mol/hr
v, v_v, v_l	Specific volume, specific volume of vapor, specific volume of liquid, ft^3/lb
v_i, v_o, v_{av}	Inlet, outlet, and average specific volumes, ft^3/lb
W	Weight flow of hot fluid, lb/hr
W'_R	Minimum reflux, lb/hr
w	Weight flow of cold fluid, lb/hr
X	Vapor formed in reboiler, lb/hr
x	Distance, ft
x, y	Mol fraction of liquid and vapor
x', y'	Mol fraction at intersection of operating lines
z	Static height or frictional head, ft
λ	Latent heat of vaporization, Btu/lb
μ	Viscosity at the caloric temperature, lb/(ft)(hr)

μ_w	Viscosity at the tube wall temperature, lb/(ft)(hr)
ρ	Density, lb/ft³
ϕ'	Viscosity ratio, $(\mu/\mu_w)^{0.14}$

Subscripts (except as noted above)

B	Bottoms
D	Distillate
F	Feed
l	Liquid
m	Plate below the feed in a distilling column
$m+1$	Plate below the m plate
n	Plate above the feed
$n+1$	Plate below the n plate
p	Bottom plate
R	Reflux
s	Shell
t	Tube
v	Vapor

CHAPTER 16

EXTENDED SURFACES[1]

Introduction. When additional metal pieces are attached to ordinary heat-transfer surfaces such as pipes or tubes, they extend the surface available for heat transfer. While the extended surface increases the total transmission of heat, its influence as surface is treated differently from simple conduction and convection.

Consider a conventional double pipe exchanger whose cross section is shown in Fig. 16.1a. Assume that hot fluid flows in the annulus and cold

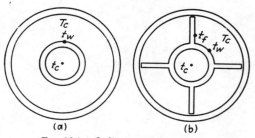

FIG. 16.1. Ordinary and finned tube.

fluid in the inner pipe, both in turbulent flow, and that the effective temperatures over the cross section are T_c and t_c, respectively. The heat transferred can be computed from the inner pipe surface, the annulus coefficient, and the temperature difference $T_c - t_w$, where t_w is the temperature of the outer surface of the inner pipe. Next assume that strips of metal are welded to the inner pipe as shown in Fig. 16.1b. Since the metal strips are attached to the cold tube wall, they serve to transfer additional heat from the hot fluid to the inner pipe. The total surface available for heat transfer no longer corresponds to the outer circumference of the inner pipe but is increased by the additional surface on the sides of the strips. If the metal strips do not reduce the conventional

---

[1] Comprehensive mathematical treatments of extended surfaces are given by Carslaw, H. S., and J. C. Jaeger, "Conduction of Heat in Solids," Clarendon Press, Oxford, 1947. Jakob, M., "Heat Transfer," John Wiley & Sons, Inc., New York, 1949.

EXTENDED SURFACES

annulus heat-transfer coefficient by appreciably changing the fluid-flow pattern, more heat will be transferred from the annulus fluid to the pipe fluid.

Pieces which are employed to extend the heat-transfer surfaces are known as *fins*. It will be shown in the case of pipes and tubes, however, that each square foot of extended surface is less effective than a square foot of unextended surface. Referring again to Fig. 16.1b, there is a temperature difference $T_c - t_f$ between the annulus fluid and the fin, and the heat which flows into the fin will be conducted by it to the inner

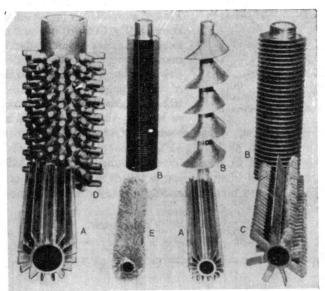

FIG. 16.2. Some commercial forms of extended surface. *a*. Longitudinal fins. (*Griscom-Russell Co.*) *b*. Transverse fins. (*Griscom-Russell Co.*) *c*. Discontinuous fins. (*Babcock and Wilcox Co.*) *d*. Pegs or studs. (*Babcock and Wilcox Co.*) *e*. Spines. (*Thermek Corporation.*) (*Gardner, Transactions of the ASME.*)

pipe. For heat to be conducted to the pipe, t_f must be greater than the pipe-wall temperature t_w. Then $T_c - t_f$ is less than $T_c - t_w$. Since the effective temperature difference between the fluid and fin is less than that between the fluid and pipe, it results in less heat transfer per square foot of fin than for the pipe. Furthermore, the temperature difference between the fluid and the fin changes continuously from the outer extremity to the base of the fin owing to the rate at which heat enters the fin by convection and the rate at which it is transferred to its base by conduction.

It will be found that two fundamental heat-transfer principles enter into the various relationships in fins: (1) to determine from the geometry and conductivity of the fin the nature of the temperature variation and (2) to determine the heat-transfer coefficient for the combination of fin and unextended surface. In the case of the double pipe exchanger, for example, the fins suppress the spiral eddies about the annulus, which in turn reduces the convection coefficient for the annulus below its conventional value as determined from Eq. (6.2).

Classification of Extended Surfaces. Fins of a number of industrial types are shown in Fig. 16.2. Pipes and tubes with longitudinal fins are marketed by several manufacturers and consist of long metal strips or channels attached to the outside of the pipe. The strips are attached either by grooving and peening the tube as in Fig. 16.3a or by welding continuously along the base. When channels are attached, they are integrally welded to the tube as in Fig. 16.3b. Longitudinal fins of this type are commonly used in double pipe exchangers or in unbaffled shell-and-tube exchangers when the flow proceeds along the axis of the tube. Longitudinal fins are most commonly employed in problems involving

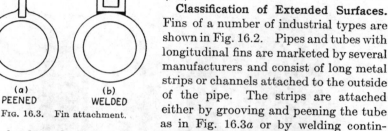

FIG. 16.3. Fin attachment.

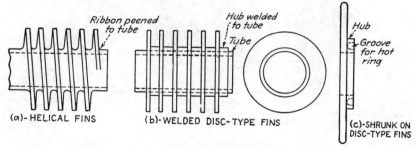

FIG. 16.4. Transverse fins.

gases and viscous liquids or when the smallness of one of a pair of heat-transfer streams causes steamline flow.

Transverse fins are made in a variety of types and are employed primarily for the cooling and heating of gases in *crossflow*. The helical fins in Fig. 16.4a are classified as transverse fins and are attached in a variety of ways such as by grooving and peening, expanding the tube metal itself to form the fin, or welding ribbon to the tube continuously. Disc-type fins are also transverse fins and are usually welded to the tube or shrunken to it as shown in Fig. 16.4b and c. In order to shrink a fin onto a tube a disc, with inside diameter slightly less than the outside diameter of the tube, is heated until its inside diameter exceeds the outside diameter of the tube. It is slipped onto the tube, and upon cooling, the disc shrinks to the tube and forms a bond with it. Another variation of the shrunk-on fin in Fig. 16.4c employs a hollow ring in its hub into which a hot metal ring is driven. Other types of transverse fins are known as *discontinuous fins*, and several shapes such as the *star* fin are shown in Fig. 16.5.

Spine- or peg-type fins employ cones, pyramids, or cylinders which extend from the pipe surface so that they are usable for either longitudinal flow or crossflow. Each type of finned tube has its own characteristics and effectiveness for the transfer of heat between the fin and the fluid inside the tube, and the remainder of this chapter deals with the derivation of the relationships and the applications of the commonest types. Perhaps the principal future uses will be in the field of atomic energy for the recovery of controlled heat of fission, in the reversing exchanger for commercial oxygen plants, in jet propulsion, and in gas-turbine cycles.

(a)- STAR-TYPE FIN

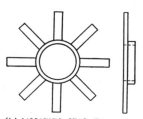

(b)-MODIFIED STAR FIN

FIG. 16.5. Discontinuous fins.

LONGITUDINAL FINS

Derivation of the Fin Efficiency. The simplest fin from the standpoint of manufacture as well as mathematics is the longitudinal fin of uniform thickness. For the derivation of its characteristics it is necessary to impose the limitations and assumptions given by Murray[1] and later by Gardner.[2]

1. The heat flow and temperature distribution throughout the fin is independent of time; *i.e.*, the heat flow is steady.
2. The fin material is homogeneous and isotropic.
3. There is no heat source in the fin itself.
4. The heat flow to or from the fin surface at any point is directly proportional to the temperature difference between the surface at that point and the surrounding fluid.
5. The thermal conductivity of the fin is constant.

[1] Murray, W. M., *J. Applied Mechanics*, **5**, A78–80 (1938).
[2] Gardner, K. A., *Trans. ASME*, **67**, 621–632 (1945).

516   PROCESS HEAT TRANSFER

6. The heat-transfer coefficient is the same over the entire fin surface.
7. The temperature of the fluid surrounding the fin is uniform.
8. The temperature of the base of the fin is uniform.
9. The fin thickness is so small compared with its height that temperature gradients across the width of the fin may be neglected.
10. The heat transferred through the outermost edge of the fin is negligible compared with that passing into the fin through its sides.
11. The joint between the fin and the tube is assumed to offer no bond resistance.

At any cross section as in Fig. 16.6 let T_c be the constant temperature of the hot fluid everywhere surrounding the fin and let t be the tempera-

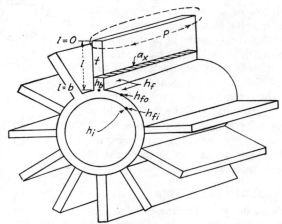

Fig. 16.6. Derivation of the longitudinal fin efficiency.

ture at any point in the fin and variable. Let Θ be the temperature difference driving heat from the fluid to the fin at any point in its cross section. Then

$$\Theta = T_c - t \tag{16.1}$$

If l is the height of the fin varying from 0 to b,

$$\frac{d\Theta}{dl} = -\frac{dt}{dl} \tag{16.2}$$

The heat within the fin which passes through its cross section by conduction is

$$Q = ka_x \frac{d\Theta}{dl} \tag{16.3}$$

EXTENDED SURFACES

where a_x is the cross-sectional area of the fin. This is equal to the heat which passed into the fin through its sides from $l = 0$ down to the shaded cross section. If P is the perimeter of the fin, the area of the sides is $P\,dl$ and the film coefficient from liquid to fin side, whether on fin surface or tube surface, is h_f.

$$dQ = h_f\Theta P\,dl \quad \text{or} \quad \frac{dQ}{dl} = h_f P\Theta \tag{16.4}$$

Differentiating Eq. (16.3) with respect to l

$$\frac{dQ}{dl} = ka_x \frac{d^2\Theta}{dl^2} \tag{16.5}$$

Equations (16.4) and (16.5) are equalities.

$$ka_x \frac{d^2\Theta}{dl^2} - h_f P\Theta = 0 \tag{16.6}$$

Rearranging,

$$\frac{d^2\Theta}{dl^2} - \frac{h_f P\Theta}{ka_x} = 0 \tag{16.7}$$

The direct solution of this equation is

$$\Theta = C_1 e^{\left(\frac{h_f P}{ka_x}\right)^{1/2} l} + C_2 e^{-\left(\frac{h_f P}{ka_x}\right)^{1/2} l} \tag{16.8}$$

Let

$$m = \left(\frac{h_f P}{ka_x}\right)^{1/2}$$

The general solution is

$$\Theta = C_1 e^{ml} + C_2 e^{-ml} \tag{16.9}$$

At $l = 0$

$$\Theta_e = C_1 + C_2 \tag{16.10}$$

where the subscript refers to the outer edge of the fin.

If no heat enters the fin at the extreme end, as qualified by assumption 10, $d\Theta/dl = 0$ when $\Theta = 0$ and $C_1 - C_2 = 0$.

$$C_1 = C_2 = \frac{\Theta_e}{2} \tag{16.11}$$

Equation (16.8) becomes

$$\frac{\Theta}{\Theta_e} = \frac{e^{ml} + e^{-ml}}{2} \tag{16.12}$$

or in general terms

$$\Theta = \Theta_e \cosh ml \tag{16.13}$$

At the base of the fin where $l = b$

$$\Theta_b = \Theta_e \cosh mb \tag{16.14}$$

where the subscript refers to the base of the fin.

Thus an expression has been obtained for the temperature difference between the constant fluid temperature and variable fin temperature in terms of the length of the fin. It is now necessary to obtain an expression for Q in terms of l. From Eq. (16.4) by differentiation with respect to the fin height l,

$$\frac{d^2Q}{dl^2} = h_f P \frac{d\Theta}{dl} \tag{16.15}$$

Substituting in Eq. (16.3),

$$Q = \frac{ka_x}{h_f P} \frac{d^2Q}{dl^2} \tag{16.16}$$

$$\frac{d^2Q}{dl^2} - \frac{h_f P}{ka_x} Q = 0 \tag{16.17}$$

As before, the solution is

$$Q = C_1' e^{ml} + C_2' e^{-ml} \tag{16.18}$$

At $l = 0$

$$C_1' + C_2' = 0 \qquad C_1' = -C_2'$$

and

$$\frac{dQ}{dl} = 0$$

$$\frac{dQ}{dl} = h_f P \Theta_e = mC_1' - mC_2' = 0 \tag{16.19}$$

$$C_1' = \frac{h_f P \Theta_e}{2m} \qquad C_2' = -\frac{h_f P \Theta_e}{2m}$$

$$Q = \frac{h_f P \Theta_e}{2m} e^{ml} - \frac{h_f P \Theta_e}{2m} e^{-ml} \tag{16.20}$$

In terms of hyperbolic functions

$$Q = \frac{h_f P \Theta_e}{m} \sinh ml \tag{16.21}$$

$$Q_b = \frac{h_f P \Theta_e}{m} \sinh mb \tag{16.22}$$

The ratio of heat load Q_b to the temperature difference Θ_b at the base is

$$\frac{Q_b}{\Theta_b} = \frac{h_f P \Theta_e \sinh mb}{m \Theta_e \cosh mb} \tag{16.23}$$

or

$$\frac{Q_b}{\Theta_b} = \frac{h_f P}{m} \tanh mb \tag{16.24}$$

Define h_b as the value of h_f to the fin surface alone when referred to the base area of the fin at $l = b$. Calling the ratio $h_b/h_f = \Omega$ the efficiency[1] *of the fin*, the average value of the heat-transfer coefficient at the base of the fin is given by Fourier's equation:

$$\frac{Q_b}{\Theta_b b P} = h_b \qquad (16.25)$$

The fin efficiency h_b/h_f may be defined by Eqs. (16.24) and (16.25).

$$\Omega = \frac{h_b}{h_f} = \frac{Q_b/\Theta_b b P}{\dfrac{mQ_b}{\Theta_b P} \cdot \dfrac{1}{\tanh mb}} = \frac{\tanh mb}{mb} \qquad (16.26)$$

Equation (16.26) applies only to the fin and not to the bare portion of the tube between the fins. To give the total heat removed by a finned tube, the heat flowing into the fin with a coefficient h_f must ultimately be combined with that flowing into the bare tube at the tube outside diameter. For this it is necessary to establish some reference surface at which the coefficient on the different parts can be reduced to the same heat flux. In an ordinary exchanger h_o is referred to the tube outside diameter. Because there is no simple reference surface on the outside of finned tubes it is convenient to use the inside diameter of the tube as the reference surface at which local coefficients are corrected to the same heat flux. By definition h_f is the coefficient to all the outside surface, whether it be the fin or bare tube. Beneath the bases of the fins there is naturally a greater heat flux than on the bare tube surface between fins, since the heat flowing through the bases of the fins is greater per unit of tube area. It may also be expected that part of the heat passing from the bases of the fins is conducted into the tube metal, so that the temperature difference between the annulus fluid and bare tube is not strictly constant. It is not usually necessary to correct for this effect, since the bare tube area so affected by the increased flux near the bases of the fins is ordinarily small compared with the total bare tube area. At the inside diameter of the tube, however, the heat from both the fins and the bare tube surface is assumed to have attained a uniform flux.

The total heat removed from the annulus liquid and arriving at the tube inside diameter is a composite of the heat transferred by the fins to the tube outside diameter and that transferred directly to the bare tube surface. These may be combined by means of the *weighted effi-*

---

[1] Another term which is used in the literature is the "fin effectiveness," which is the efficiency multiplied by the ratio of fin surface to fin base area.

ciency Ω'. If the heat transferred through the bare tube surface at the tube outside diameter is designated by Q_o, then

$$Q_o = h_f A_o \Theta_b \tag{16.27}$$

where A_o is the bare tube surface at the outside diameter exclusive of the area beneath the bases of the fins. If there are N_f fins on the tube, bPN_f is all of the fin surface. The total heat transfer at the outside diameter is given by

$$Q = Q_b + Q_o = h_b b P N_f \Theta_b + h_f A_o \Theta_b$$

$$= \left(\frac{h_b b P N_f A_o}{A_o} + \frac{h_f b P N_f A_o}{b P N_f} \right) \Theta_b$$

$$= \left(\frac{h_b}{A_o} + \frac{h_f}{b P N_f} \right) b P N_f A_o \Theta_b \tag{16.28}$$

Substituting Eq. (16.26) to eliminate h_b

$$Q = \left(b P N_f \frac{\tanh mb}{mb} + A_o \right) h_f \Theta_b \tag{16.29}$$

Calling h_{fo} the composite value of h_f to both the fin and bare tube surfaces when referred to the outside diameter of the tube, the weighted efficiency is by definition $\Omega' = h_{fo}/h_f$. Combining Eqs. (16.24) and (16.29)

$$\Omega' = \frac{h_{fo}}{h_f} = \left(\frac{b P N_f \dfrac{\tanh mb}{mb} + A_o}{b P N_f + A_o} \right) \tag{16.30}$$

But as in Eq. (6.2) the values of the coefficients vary inversely with the heat-flow area. If h_{fi} is the value of the composite coefficient h_{fo} referred to the tube inside diameter,

$$\frac{h_{fi}}{h_{fo}} = \frac{b P N_f + A_o}{A_i} \tag{16.31}$$

Substituting in Eq. (16.31)

$$h_{fi} = \left(b P N_f \frac{\tanh mb}{mb} + A_o \right) \frac{h_f}{A_i} \tag{16.32}$$

From Eq. (16.32)

$$h_{fi} = \left(\frac{\Omega A_f + A_o}{A_f + A_o} \right) \frac{(A_f + A_o)}{A_i} h_f \tag{16.33}$$

or simply

$$h_{fi} = (\Omega A_f + A_o) \frac{h_f}{A_i} \tag{16.34}$$

EXTENDED SURFACES

Thus an equation has been obtained which gives directly the heat-transfer coefficient on the inside of an extended tube which is the equivalent of a value of h_f on the outside surface of the tube. By substituting the physical and geometrical factors for a given tube and fin arrangement a *weighted efficiency curve* can be developed relating h_f to h_{fi} based on the inside tube surface.

Bonilla[1] has presented a graphical solution of Eq. (16.34) for the rectangular fin of uniform thickness. The method of derivation employed here is applicable to other types of longitudinal fins as well, although it is rather awkward for the longitudinal fin of triangular cross section. A table of the hyperbolic tangents is given in Table 16.1.

TABLE 16.1. HYPERBOLIC TANGENTS

mb	Tanh mb	$\dfrac{\text{Tanh } mb}{mb}$
0.0	0.0000	1.000
0.1	0.0997	0.997
0.2	0.1974	0.987
0.3	0.2913	0.971
0.4	0.380	0.950
0.5	0.462	0.924
0.6	0.537	0.895
0.7	0.604	0.863
0.8	0.664	0.830
0.9	0.716	0.795
1.0	0.762	0.762
1.1	0.801	0.728
1.2	0.834	0.695
1.3	0.862	0.663
1.4	0.885	0.632
1.5	0.905	0.603
2.0	0.964	0.482
3.0	0.995	0.333
4.0	0.999	0.250

Example 16.1. Calculation of the Fin Efficiency and a Weighted Efficiency Curve. To illustrate the use of Eq. (16.34) a weighted fin efficiency curve will be developed for one of the common types of finned tubes used in double pipe exchangers.

A double pipe exchanger employs a 1¼ in. OD, 13 BWG steel inner tube to which are attached 20 fins 20 BWG, ¾ in. high. Through suitable bushings the inner pipe is inserted into a 3-in. IPS outer pipe. Determine (*a*) the efficiency of the fins for various values of h_f, (*b*) the weighted efficiency curve.

[1] Bonilla, C. F., *Ind. Eng. Chem.*, **40**, 1098–1101 (1948).

Solution:

Total outside surface, A:
Fin surface, $A_f = (20 \times 0.75 \times 12 \times 2)/144 = 2.50$ ft²/lin ft
Bare tube surface, $A_o = (3.14 \times 1.25 - 20 \times 0.035)12/144 = 0.268$ ft²/lin ft
$A_f + A_o = 2.50 + 0.268 = 2.77$ ft²/lin ft
Total inside surface, $A_i = (3.14 \times 1.06 \times 12)/144 = 0.277$ ft²/lin ft

(a) Fin efficiencies:

Efficiency, $\Omega = \dfrac{\tanh mb}{mb}$ (16.26)

$$m = \left(\frac{h_f P}{k a_x}\right)^{\frac{1}{2}} = h_f^{\frac{1}{2}} \left(\frac{P}{k a_x}\right)^{\frac{1}{2}} = h_f^{\frac{1}{2}} \left(\frac{2 \times 20 \times 144}{25 \times 0.035 \times 12 \times 20}\right)^{\frac{1}{2}} = 5.24 h_f^{\frac{1}{2}}$$

h_f	$5.24 h_f^{\frac{1}{2}}$	mb	$\tanh mb$	$\Omega = \dfrac{\tanh mb}{mb}$
4	10.48	0.655	0.5717	0.872
16	20.96	1.31	0.864	0.660
36	31.44	1.97	0.962	0.498
100	52.4	3.28	0.995	0.303
400	104.8	6.55	1.00	0.152
625	131	8.19	1.00	0.122
900	157	9.82	1.00	0.102

The fin efficiencies are given in the last column. Note that, when the outside coefficient h_f is 4.0, both sides of the fin contribute heat-transfer surface which is 87.2 per cent as effective as bare tube surface. When $h_f = 100$, the surface is only 30.3 per cent as effective, and when $h_f = 900$ (water or steam), the extended surface is only 10.2 per cent as effective. This is because at the higher film coefficient the fin metal more nearly attains the temperature of the hot fluid and the temperature difference between the fluid and fin is correspondingly less for given values of T_c and t_c.

(b) The weighted efficiency curve directly gives the value of h_{fi} obtained at the tube inside diameter when the coefficient to the fin and the bare tube is h_f.

$$h_{fi} = (\Omega A_f + A_o) \frac{h_f}{A_i}$$ (16.34)

$h_f \times \quad (\Omega A_f \quad + \quad A_o) \; / \; A_i \; = \quad h_{fi}$
$\quad 4 \times (0.872 \times 2.50 + 0.268) \,/\, 0.277 = \quad 35.4$
$\quad 16 \times (0.660 \times 2.50 + 0.268) \,/\, 0.277 = \quad 110.8$
$\quad 36 \times (0.498 \times 2.50 + 0.268) \,/\, 0.277 = \quad 193.5$
$\quad 100 \times (0.303 \times 2.50 + 0.268) \,/\, 0.277 = \quad 370$
$\quad 400 \times (0.152 \times 2.50 + 0.268) \,/\, 0.277 = \quad 935$
$\quad 625 \times (0.122 \times 2.50 + 0.268) \,/\, 0.277 = 1295$
$\quad 900 \times (0.102 \times 2.50 + 0.268) \,/\, 0.277 = 1700$

These values of h_f and h_{fi} are plotted in Fig. 16.7. If the film coefficient to the fin is large, as with condensation or the cooling or heating of water in the annulus, there is little advantage to be gained from the use of fins particularly when a dirt factor must be included which may be the controlling resistance. Actually, a fin of a desired metal

EXTENDED SURFACES

can be designed to give a high efficiency for any reasonable value of h_f, but as h_f increases, the size and cost of the fin also increases. Greater efficiencies can also be obtained by using fins which are not uniformly thick such as are treated later on.

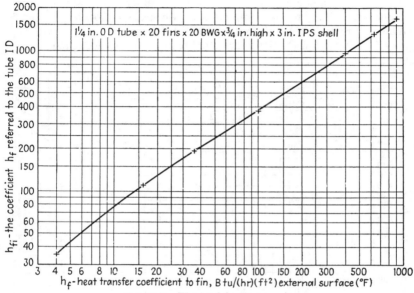

FIG. 16.7. Weighted fin-efficiency curve.

There are several manufacturers who supply completely assembled double pipe exchangers with fins on the inner pipe. Because the finned pipe or tube cannot be inserted through a packed gland as in Fig. 6.1,

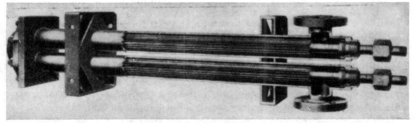

FIG. 16.8. Longitudinal fin double pipe exchanger. (*Griscom-Russell Co.*)

the method of confining the inner pipe is somewhat more complicated. A typical example of a standard preassembled double pipe finned exchanger is shown in Fig. 16.8, and as in Chap. 6, a single unit is called a *hairpin*. The cost of these units is extremely moderate, and they are preferable when one of the fluids is a gas, a viscous liquid, or a small stream. Operated in series or in parallel they are often superior to shell-and-tube exchangers, notwithstanding the increased number of leakable joints they introduce. When connected in any other manner but in series the true temperature difference is computed for series-parallel flow by Eq. (6.35). When operated with cooling water in the inner pipe a nonferrous pipe may be selected with either steel or nonferrous fins attached. Three

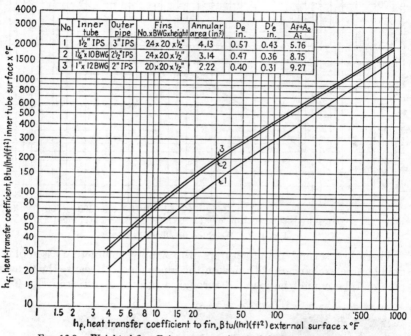

FIG. 16.9. Weighted fin-efficiency curves for steel double pipe exchangers.

weighted efficiency curves for three of the commonest sizes of double pipe finned exchangers are given in Fig. 16.9. These do not correspond to the offerings of any single manufacturer but are a composite of the three sizes ordinarily available from manufacturers' stocks.

Double-pipe-exchanger Coefficients and Pressure Drops. The addition of fins to the outsides of the inner pipes of double pipe exchangers generally reduces the value of the annulus film coefficient by smoothing the flow of fluid on the outside of the tube or pipe. Data obtained for bare tube surfaces cannot be used for the calculation of double pipe extended-surface exchangers. While a design curve or equation relating

h_f to the flow variables and heat-transfer properties can be determined by experiment in the usual manner (see Chap. 3), the correlation differs for extended surfaces. Starting with a double pipe finned exchanger, a fluid is passed through the annulus and heated by the dropwise condensation of steam, whose condensing coefficient is very high compared with the annulus coefficient. Any other heating medium can be substituted

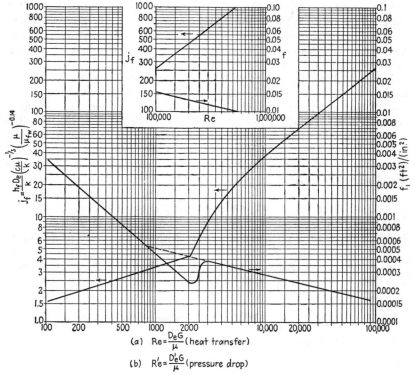

FIG. 16.10. Longitudinal fin heat transfer and pressure drop.

if its individual coefficient can be or has been determined accurately. For cooling experiments water is usually employed. Readings are taken of the oil-temperature change, the steam or water temperatures, and the fluid flow. The surface A_i is known from the inside diameter and length of the inner pipe, and the heat load, the LMTD, and U_i, the clean overall coefficient, can also be computed. The film coefficient for the annulus h_{fi} can then be determined from the following equation:

$$\frac{1}{U_i} = \frac{1}{h_i} + \frac{1}{h_{fi}} \tag{16.35}$$

The value of h_{fi} so obtained represents h_f to the fin and bare tube already combined and weighted. If four or more fins are attached to the inner pipe, the addition of a greater number does not appear to affect the heat-transfer coefficient h_f significantly. Naturally the weighted efficiency and the total heat transfer are directly influenced by the number of fins. To make the experimental data general for other fin and tube arrangements than those tested, the value of h_{fi} must be transferred from the inside diameter of the inner tube to the annulus. In the annulus it may then be resolved between the fin and bare tube surface by converting from h_{fi} to h_f and plotting a heat-transfer factor j_f, which includes the average value of h_f to both types of surface.

Figure 16.10a is an isothermal curve which was obtained in this manner from numerous heating and cooling experiments on different fin and tube double pipe arrangements and agrees with the published data of DeLorenzo and Anderson.[1] Friction factors are given in Fig. 16.10b. The equivalent diameter D'_e for pressure drop differs from D_e for heat transfer because of the inclusion of the perimeter of the outer pipe in the wetted perimeter. Both equivalent diameters are computed as in Chap. 6 for double pipe exchangers using four times the hydraulic radius.

To illustrate the method of treating experimental data on pipes or tubes with extended surface, a typical test point will be computed below. This is in the reverse order of the calculation of the surface requirements for the fulfillment of process conditions which will be demonstrated later.

Example 16.2. Calculation of a Heat-transfer Curve from Experimental Data. Experimental double pipe exchanger, steel hairpin.

Shell, 3 in. IPS
Finned inner tube: 1¼ in. OD, 13 BWG to which are attached 20 fins, 20 BWG, ¾ in. high
Length of finned inner tube, 10'0" per leg between return bends.

Data and observations:

Hot fluid, steam, $T_S = 302°F$ (in inner tube)
Cold fluid, 40.8°API kerosene, $t_1 = 151°F$
$\qquad\qquad t_2 = 185°F$
Weight flow, $w = 15,200$ lb/hr
The dropwise condensation of steam was promoted with oil.

Calculations:

ID of 3 in. IPS pipe = 3.068 in.

Annulus flow area, $a_a = \dfrac{\pi(3.068^2 - 1.25)^2}{4 \times 144} - \dfrac{20 \times 0.035 \times 0.75}{144} = 0.0395$ ft²

---

[1] DeLorenzo, B., and E. D. Anderson, *Trans. ASME*, **67**, 697–702 (1945).

Wetted perimeter $= \pi \times \frac{1.25}{12} - 20 \times \frac{0.035^*}{12} + 20 \times 0.75 \times \frac{2}{12} = 2.77$ ft

Equivalent diameter, $D_e = \frac{4 \times \text{flow area}}{\text{wetted perimeter}} = 4 \times \frac{0.0395}{2.77} = 0.057$ ft

Heat load, $Q = 15{,}200 \times 0.523(185 - 151) = 271{,}000$ Btu/hr

LMTD $= 133°F$

Internal surface/lin ft, $A_i = 0.277$ ft²/ft

$$U_i = \frac{271{,}000}{0.277 \times 20 \times 133} = 368 \text{ Btu}/(\text{hr})(\text{ft}^2)(°F)$$

$h_i = 3000$ (assumed value for dropwise condensation of steam)

$\frac{1}{U_i} = \frac{1}{h_i} + \frac{1}{h_{fi}} \qquad \frac{1}{368} = \frac{1}{3000} + \frac{1}{h_{fi}}$

$h_{fi} = 418$ Btu/(hr)(ft²)(°F)

From the efficiency curve (Fig. 16.7) for $h_{fi} = 418$, $h_f = 120$.

At the average temperature $\mu = 0.80$ cp $\times 2.42 = 1.94$ lb/(ft)(hr)

$\left(\frac{c\mu}{k}\right)^{1/3} = 2.34$

$$j_f = h_f \frac{D_e}{k}\left(\frac{c\mu}{k}\right)^{-1/3} = \frac{120 \times 0.057}{0.079 \times 2.34} = 37.0 \qquad (16.36)$$

$G_a = \frac{15{,}200}{0.0395} = 385{,}000$ lb/(hr)(ft²)

$Re_a = 0.057 \times \frac{385{,}000}{1.94} = 11{,}300$

This single point, $Re_a = 11{,}300$ vs. $j_f = 37$, does not fall exactly on the isothermal curve of Fig. 16.10a. It is off by more than is first apparent (37 vs. 41 or 9.8 per cent), since it represents a heating run and j_f should be divided by a viscosity correction to permit its representation on an isothermal curve which is applicable to both heating and cooling. The general implications of such a procedure have been discussed in Chaps. 5 and 6. In the correlations covered in Fig. 16.10a the viscosity correction was found to be $\phi_a = (\mu/\mu_{fw})^{0.14}$, where μ_{fw} is the viscosity taken at the temperature of the "wall" t_{fw}. Thus μ_{fw} replaces μ_w when the pipe wall has fins.

The Wall Temperature t_{fw} of a Finned Tube. The wall temperature of an extended-surface pipe or tube influences the values of the heat-transfer coefficients which are obtained when a fluid is heated and cooled over the same operating range. Since the fin and tube metals will not be at the same temperature, the use of any single temperature such as t_{fw} to replace the fin temperature t_f and the tube-wall temperature t_w is naturally fictitious. Kayan[1] has developed an ingenious experimental method of analyzing for the wall temperature which simulates heat transfer by means of an electrical analogy. Theoretical methods of obtaining the wall temperature are quite complex, and even the most typical applications of extended surfaces have not been covered by adequate derivations. The method employed below is semiempirical. Its principal

* This is the area of the bare tube beneath the bases of the fins. The extremities of the fins were not considered to receive heat in the derivation, and their lengths are not included in the perimeter.

[1] Kayan, C. F., Trans. ASME, **67,** 713–718 (1945); Ind. Eng. Chem., **40,** 1044–1049 (1948).

advantage lies in the brevity with which a solution can be obtained. It is presented here in the sequence which is most adaptable to design when an isothermal performance curve such as Fig. 16.10a is available. If an isothermal curve is to be developed from heating and cooling experiments, the sequence need merely be rearranged:

The true temperature difference Δt corresponds to the total resistance $1/U_{Di}$ in the design equation $(Q/A_i) = U_{Di} \Delta t$. The sum of the smaller temperature differences through each of the component resistances between the annulus and inner tube fluids must equal the true temperature difference. The dirt factors for the annulus and tube may not be combined in extended surface equipment, since they are applicable to such widely different actual surfaces that flux corrections must be employed. The component resistances are then the (1) annulus film, (2)

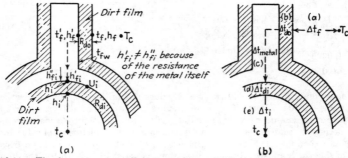

Fig. 16.11. The location of coefficients and temperature drops in longitudinal fins.

annulus dirt factor, (3) fin and tube metal, compositely, (4) tube-side dirt factor and (5) tube-side film. The component temperature differences are obtained by multiplying the component resistances by the flux Q/A_i when the latter is corrected to the appropriate surface at each point. The wall temperature t_{fw} occurs between temperature drops (2) and (3), and by obtaining this temperature μ_{fw} can be evaluated. While the clean coefficient U_i is obtained from individual coefficients calculated for properties at T_c and t_c, the summation of the component temperature differences in the design equation corresponds to Δt and not $T_c - t_c$.

The value of the wall t_{fw} is obtained by trial and error. First assume a value of t_{fw} which permits the calculation of ϕ_a, the annulus viscosity correction, and the annulus film coefficient corrected for heating or cooling. The reciprocal of the annulus film coefficient is the resistance in (1). The fouling factor is the resistance in (2). If the temperature after passing through these two resistances corresponds to the assumed value of t_{fw} **and** if the total temperature drop over the five resistances corresponds to Δt, the assumed value was correct. If no check is obtained, a new value of t_{fw} must be assumed and the procedure repeated. The location of the various temperatures and resistances involved in this method are shown in Fig. 16.11. The nomenclature differs from Eq. (16.35) in which there were no dirt factors. The steps in summing up the resistances and correcting for the different surfaces per linear foot of extended surface pipe or tube should be evident in the outline below.

1. Assume a temperature t_{fw} to establish a temperature difference between the annulus fluid and the wall.
2. Obtain $\phi_a = (\mu/\mu_{fw})^{0.14}$ where μ_{fw} is obtained at t_{fw}.
3. Obtain h_f/ϕ_a from a design curve and correct for the viscosity ratio ϕ_a and obtain h_f by $h_f = (h_f/\phi_a)\phi_a$. Then $R_f = 1/h_f$.
4. To obtain the value of h_f effective at the fin surface add to it the resistance through the annulus dirt factor R_{do}. For a dirt factor of R_{do},

$$\frac{1}{h'_f} = R_f + R_{do} \tag{16.37}$$

where h'_f is the effective value of h_f at the fin.

5. Obtain h'_{fi} for the value of h'_f in 4 from an appropriate weighted efficiency curve such as Fig. 16.9.
6. Multiply h'_f in 4 by the surface ratio $(A_f + A_o)/A_i$ giving h''_{fi}, which is h'_f corrected for flux but not for the resistance of the fin and wall metals.

$$h''_{fi} = \frac{h'_f(A_f + A_o)}{A_i} \tag{16.38}$$

The difference between h'_{fi} and h''_{fi} may be considered due to the resistance of the fin and tube wall metals.

$$R_{\text{metal}} = \frac{1}{h'_{fi}} - \frac{1}{h''_{fi}} \tag{16.39}$$

7. Obtain h_i for the tube fluid from any suitable tube-side data such as Fig. 24 or 25, where $\phi_t = (\mu/\mu_{fw})^{0.14}$ instead of $(\mu/\mu_w)^{0.14}$, and combine with the tube-side dirt factor, R_{di}. $R_i = 1/h_i$, and if h'_i is the value of h_i at the tube wall,

$$\frac{1}{h'_i} = R_i + R_{di} \tag{16.40}$$

8. From h'_i and h'_{fi} obtain U_{Di}, the corrected overall *design* coefficient based on the inside tube diameter.

$$\frac{1}{U_{Di}} = \frac{1}{h'_{fi}} + \frac{1}{h'_i} \tag{16.41}$$

9. Obtain the flux for the surface actually employed.

$$\frac{Q}{A_i} = U_{Di} \Delta t \tag{16.42}$$

10. Impose the flux from 9 on the individual resistances and determine whether the summation of the first two individual differences is the same as the assumed value

of $T_c - t_{fw}$ and if the sum of all differences equals the true temperature difference. In the same order as before, the temperature drops are

a. Annulus film:

$$\Delta t_f = \frac{Q/A_i}{h''_{fi}} \qquad (16.43a)$$

b. Annulus dirt, corrected to the inside diameter:

$$\Delta t_{do} = \left(\frac{Q}{A_i}\right) R_{do} \frac{A_i}{A_f + A_o} \qquad (16.43b)$$

c. Fin and tube metal:

$$\Delta t_{\text{metal}} = \left(\frac{Q}{A_i}\right) R_{\text{metal}} \qquad (16.43c)$$

d. Tube-side dirt:

$$\Delta t_{di} = \left(\frac{Q}{A_i}\right) R_{di} \qquad (16.43d)$$

e. Tube-side film:

$$\Delta t_i = \frac{Q/A_i}{h_i} \qquad (16.43e)$$

When the tube fluid is a gas, water, or a similar nonviscous fluid, the tube-side viscosity correction can be omitted and ϕ_a can be taken as 1.0. The metal of which the fin is fabricated greatly affects the efficiency, particularly since k for steel is 26 and for copper 220 Btu/(hr)(ft²)(°F/ft).

The Calculation of a Double Pipe Finned Exchanger. The method employed here closely follows that of the outline in Chap. 6. The true temperature difference for a double pipe exchanger is the LMTD calculated for true counterflow or a value computed from Eq. (6.35) for series-parallel flow. The effective tube length does not include the return bend between legs of a hairpin or any of the unfinned portion of the inner tube. The inside of the tube is used as the reference surface. Most manufacturers prefer to use the outside as the reference surface, since it has a larger numerical value. The solutions of problems can be converted to manufacturers' data by multiplying the inside surface by the ratio $(A_f + A_o)/A_i$ and dividing the overall coefficient U_{Di} by this ratio. Tube lengths of 12, 15, 20, and 24 ft are considered reasonable for extended-surface hairpins. Large tube lengths are permissible, since the fins on the inner pipe rest snugly on the outer pipe and there is no sagging. As explained previously, fouling factors cannot be combined as in ordinary double pipe exchangers because they are effective on widely different surfaces and each must be treated separately.

Example 16.3. Calculation of a Double Pipe Extended-surface Gas Oil Cooler. It is desired to cool 18,000 lb/hr of 28°API gas oil from 250 to 200°F in double pipe exchangers consisting of 3-in. IPS shells with 1½-in. IPS inner pipes on which are mounted 24 fins ½ in. high by 0.035 in. (20 BWG) wide. Water from 80 to 120°F will serve as the cooling medium. Pressure drops of 10.0 psi are allowable on both streams, and fouling factors of 0.002 for the gas oil and 0.003 for the water are required. How many 20-ft hairpins will be required?

EXTENDED SURFACES

Solution:

(1) Heat balance: Gas oil, $Q = 18{,}000 \times 0.53(250 - 200) = 477{,}000$ Btu/hr
(Fig. 4)
Water, $Q = 11{,}950 \times 1.0(120 - 80) = 477{,}000$ Btu/hr

(2) Δt: Assume true counterflow for the first trial. Unless the allowable pressure drop on either stream is exceeded, it will not be necessary to consider series-parallel arrangements calculated by Eq. (6.35).

$\Delta t = \text{LMTD} = 124°\text{F}$ (5.14)

(3) Caloric temperatures T_c and t_c:

$\dfrac{\Delta t_c}{\Delta t_h} = \dfrac{120}{130} = 0.92$ $K_c = 0.27$ $F_c = 0.47$ (Fig. 17)

$T_c = 200 + 0.47(250 - 200) = 224°\text{F}$ (5.28)
$t_c = 80 + 0.47(120 - 80) = 99°\text{F}$ (5.29)

Hot fluid: annulus, gas oil

(4′) 3 in. IPS, ID = 3.068 in. [Table 11]
1½ in. IPS, OD = 1.90 in. [Table 11]
Fin cross section, 20 BWG, ½ in. high
= $0.035 \times 0.5 = 0.0175$ in.²
[Table 10]

$a_a = \left(\dfrac{\pi}{4} \times 3.068^2 - \dfrac{\pi}{4} \times 1.90^2 - 24 \times 0.0175\right)$

$= 4.13$ in.²
$= 4.13/144 = 0.0287$ ft²
Wetted perimeter* = $(\pi \times 1.90 - 24 \times 0.035 + 24 \times 2 \times 0.5)$
$= 29.13$ in.
$d_e = 4 \times 4.13/29.13 = 0.57$ in.
$D_e = 0.57/12 = 0.0475$ ft

(5′) $G_a = W/a_a$
$= 18{,}000/0.0287$
$= 628{,}000$ lb/(hr)(ft²)

(6′) At $T_c = 224°\text{F}$,
$\mu = 2.50 \times 2.42 = 6.05$ lb/(ft)(hr)
[Fig. 14]
$Re_a = D_e G_a/\mu$
$= 0.0475 \times 628{,}000/6.05 = 4930$
(7′) $j_f = 18.4$ [Fig. 16.10]
(8′) At $T_c = 224°\text{F}$
$k(c\mu/k)^{\frac{1}{3}} = 0.25$ Btu/(hr)(ft²)(°F/ft)
[Fig. 16]

(9′) $h_f = j_f \dfrac{k}{D_e}\left(\dfrac{c\mu}{k}\right)^{\frac{1}{3}} \phi_a$ [Eq. (6.15)]

$\dfrac{h_f}{\phi_a} = 18.4 \times \dfrac{0.25}{0.0475} = 96.7$

Cold fluid: inner pipe, water

(4) $D = 1.61/12 = 0.134$ ft   [Table 11]
$a_t = \pi D^2/4 = \pi \times 0.134^2/4 = 0.0142$ ft²

(5) $G_t = w/a_t$
$= 11{,}950/0.0142$
$= 842{,}000$ lb/(hr)(ft²)
$V = G_t/3600\rho = 842{,}000/3600 \times 62.5 = 3.75$ fps

(6) At $t_c = 99°\text{F}$,
$\mu = 0.72 \times 2.42 = 1.74$ lb/(ft)(hr)
[Fig. 14]
$Re_t = DG_t/\mu$ (Re_t is for pressure drop only)
$= 0.134 \times 842{,}000/1.74 = 65{,}000$

(9) $h_i = 970 \times 0.82$
$= 795$ Btu/(hr)(ft²)(°F)   [Fig. 25]

---

* In the derivation, the outermost edge of the fins was assumed to have zero heat transfer.

532 PROCESS HEAT TRANSFER

Calculation of t_{fw} (numbers refer to the outline procedure):

(1) Assume $T_c - t_{fw} = 40°F$
$t_{fw} = 224 - 40 = 184°F$ (Fig. 14)

(2) At 184°F, $\mu_{fw} = 3.5$ cp
$$\phi_a = \left(\frac{\mu}{\mu_{fw}}\right)^{0.14} = \left(\frac{2.5}{3.5}\right)^{0.14} = 0.95$$

(3) $h_f = 96.7 \times 0.95 = 91.8$ Btu/(hr)(ft²)(°F)

(4) $R_{do} = 0.002$, $R_f = \dfrac{1}{91.8} = 0.0109$ (hr)(ft²)(°F)/Btu

$$\frac{1}{h'_f} = 0.002 + 0.0109 \tag{16.37}$$

$h'_f = 77.5$

(5) $h'_{fi} = 255$ (Fig. 16.9)

(6) $h''_{fi} = h'_f \dfrac{A_f + A_o}{A_i} = 77.5 \times 5.76 = 447$ (16.38) and (Fig. 16.9)

$$R_{\text{metal}} = \frac{1}{h'_{fi}} - \frac{1}{h''_{fi}} = \frac{1}{255} - \frac{1}{447} = 0.00169 \tag{16.39}$$

(7) Assume $\phi_t = 1.0$ for cooling water:
$R_{di} = 0.003 \quad R_i = \frac{1}{795} = 0.00126$

$$\frac{1}{h''_i} = 0.00126 + 0.003 \tag{16.40}$$

$h'_i = 235$

(8) $\dfrac{1}{U_{Di}} = \dfrac{1}{255} + \dfrac{1}{235}$ (16.41)

$U_{Di} = 122$

(9) To obtain the true flux the heat load must be divided by the *actual* heat-transfer surface.
For a 1½-in. IPS pipe there are 0.422 ft²/lin foot (Table 11)

Trial:

$A_i = \dfrac{Q}{U_{Di} \Delta t} = \dfrac{47{,}000}{122 \times 124} = 31.5$ ft²

Length of pipe required $= \dfrac{31.5}{0.422} = 74.8$ lin ft

Use two 20-ft hairpins = 80 lin ft
$A_i = 80 \times 0.422 = 33.8$ ft²
$\dfrac{Q}{A_i} = \dfrac{477{,}000}{33.8} = 14{,}100$ Btu/(hr)(ft²)

(a) Annulus film $\Delta t_f = \left(\dfrac{Q}{A_i}\right) / h''_{fi} = \dfrac{14{,}100}{447}$ $= 31.6°$

(b) Annulus dirt, $\Delta t_{do} = \left(\dfrac{Q}{A_i}\right) R_{do} \dfrac{A_i}{A_f + A_o} = 14{,}100 \times \dfrac{0.002}{5.76}$ $= 4.9$

$T_c - t_{fw} = \overline{36.5°}$

(c) Fin and tube metal, $\Delta t_{\text{metal}} = \left(\dfrac{Q}{A_i}\right) R_{\text{metal}} = 14{,}100 \times 0.00169 = 23.8$

(d) Tube-side dirt, $\Delta t_{di} = \left(\dfrac{Q}{A_i}\right) R_{di} = 14{,}100 \times 0.003$ $= 42.3$

(e) Tube-side film, $\Delta t_i = \left(\dfrac{Q}{A_i}\right) / h_i = \dfrac{14{,}100}{795}$ $= \underline{17.7}$

$\overline{120.3°}$

EXTENDED SURFACES 533

Assumed $T_c - t_{fw} = 40.0°$ Calculated $T_c - t_{fw} = 36.5°$ LMTD = 124°
The difference of $40 - 36.5 = 3.5°F$ will not materially change the value of $(\mu/\mu_{fw})^{0.14}$.

Note: The temperature drop of 120.3°F corresponds to a fouling factor based on 74.8 lin ft while the flux corresponds to 80 lin ft. The correction appears at the end of the solution.

Pressure Drop

(1') $d'_e = 4 \times 4.13/(29.13 + \pi \times 3.07)$
$= 0.43$ in.
$D'_e = 0.43/12 = 0.0359$ ft
$Re'_a = D'_e G_a/\mu$
$= 0.0359 \times 628{,}000/6.05 = 3730$
$f = 0.00036$ ft^2/in.^{2} [Fig. 16.10]
$s = 0.82$ [Fig. 6]

(2') $\Delta P_a = \dfrac{fG_a^2 Ln}{5.22 \times 10^{10} D'_e s \phi_a}$ [Eq. (7.45)]
$= \dfrac{0.00036 \times 628{,}000^2 \times 80}{5.22 \times 10^{10} \times 0.0359 \times 0.82 \times 0.95}$
$= 7.9$ psi

(1) $Re_t = 65{,}000$ (pipe)
$f = 0.000192$ ft^2/in.^{2} [Fig. 26]

(2) $\Delta P_t = \dfrac{fG_t^2 Ln}{5.22 \times 10^{10} D s \phi_t}$ [Eq. (7.45)]
$= \dfrac{0.000192 \times 842{,}000^2 \times 80}{5.22 \times 10^{10} \times 0.134 \times 1.0 \times 1.0}$
1.4 psi

Summary

0.002	Dirt Factor	0.003
255	h inside	235
U_{Di}	122	
7.9	Calculated ΔP	1.4
10.0	Allowable ΔP	10.0

Adjustment of the fouling factor. $U_{Di} = 122$ was based on 74.8 lin ft. Based on 80 lin ft $U_{Di} = 114$ and the difference is excess dirt factor equal to 0.00057. This can readily be added to the tube side to give a dirt factor of 0.00357, or it can be added to the annulus by taking $0.00057 \times 5.76 = 0.00328$. The total annulus dirt factor for the latter will then be $0.002 + 0.00328 = 0.00528$. For the new value of h'_f

$$\frac{1}{h'_f} = 0.00528 + \frac{1}{91.8} \tag{16.37}$$

$h'_f = 62$
$h'_{fi} = 220$ (Fig. 16.9)

$$U_{Di} = \frac{h'_{fi} h'_i}{h'_{fi} + h'_i} = \frac{220 \times 235}{220 + 235} = 114 \tag{16.41}$$

The corrected summary is

0.0053	Dirt Factor	0.003
220	h inside	235
U_{Di}	114	
7.9	Calculated ΔP	1.4
10.0	Allowable ΔP	10.0

Remarks: There is no need to readjust the calculation of t_{fw}, since only item (b) will be affected and this is usually insignificant in its contribution to the value of Δt. Thus the corrected value is

$$\Delta t_{do} = 14{,}100 \times \frac{0.0053}{5.76} = 13.0°$$

This gives a computed Δt of 128.4° vs. 124.0°, but this difference does not justify a new trial value of t_{fw}. The need for readjusting the fouling factor is of particular value when checking the performance of an existing hairpin or battery of hairpins for a new service.

As reported by a manufacturer the total surface would be $33.8 \times 5.76 = 194$ ft². The reported overall coefficient would be $114/5.76 = 19.8$ Btu/(hr)(ft²)(°F).

Extended-surface Shell-and-tube Exchangers. The use of extended surfaces in double pipe exchangers permits the transfer of a great deal of heat in a compact unit. The same advantages may be obtained from the use of longitudinally finned tubes in shell-and-tube arrangements equivalent to the 1-1, 1-2 or 2-4 exchanger. Because they are relatively uncleanable, extended-surface tubes are usually laid out on triangular pitch and are never spaced so closely that the fins of adjacent tubes intermesh. To prevent sagging and the possibility of tube vibration which might result from intermeshing, each tube in a bundle is supported individually. This cannot be done with conventional support plates because they introduce a certain amount of flow across the bundle which cannot be accomplished very well with longitudinal finned tubes. Support is accomplished, however, by welding or shrinking small circumferential rings about each tube which enclose the fins but at different points along the length of each tube. The rings prevent any tubes from intermeshing and at the same time afford a positive elimination of vibration damage. At several points along its length the entire bundle is then bound with circumferential bands which keep all of the finned tubes firmly pressed against the rings of adjacent tubes.

Longitudinal fin exchangers are relatively expensive and, since they are

EXTENDED SURFACES 535

not cleanable, can be used only for fluids which ordinarily have very low film coefficients and which are clean or form dirt that can be boiled out. This makes them ideal for gases at low pressure where the density is low and the allowable pressure drop is accordingly small.

The prototype of this exchanger is the 1-2 exchanger without baffles as calculated in Example 7.8. Longitudinal fin shell-and-tube exchangers are computed in the same manner as double pipe extended-surface exchangers using the same efficiency curves for identical tubes. Only the equivalent diameters for heat transfer and pressure drop differ. These are computed in the conventional manner for the entire shell using four times the hydraulic radius as discussed in Chap. 6 and demonstrated in Example 7.8.

Example 16.4. Calculation of a Longitudinal Fin Shell-and-tube Exchanger. 30,000 lb/hr of oxygen at 3 psig pressure and 250°F is to be cooled to 100°F using water from 80 to 120°F. The maximum allowable pressure drop for the gas is 2.0 psi and for the water 10.0 psi. Fouling factors of not less than 0.0030 should be provided for each stream.

Available for the service is a 19¼ in. ID 1-2 exchanger equipped with 70 16'0" tubes each with 20 fins ½ in. high of 20 BWG (0.035 in.) steel. The tubes are 1 in. OD, 12 BWG and are laid out on 2-in. triangular pitch for four passes.

Will the exchanger fulfill the service? What is the final gas side dirt factor?

Solution:

Exchanger:

   Shell side   *Tube side*
   ID = 19¼ in.   Number and length = 70, 16'0", 20 fins, 20 BWG, in.
Baffle Space = ring supports   OD, BWG, pitch = 1 in., 12 BWG, 2-in. tri.
   Passes = 1   Passes = 4

(1) Heat balance:
Oxygen at 17.7 psia, $Q = 30{,}000 \times 0.225(250 - 100) = 1{,}010{,}000$ Btu/hr
Water, $Q = 50{,}500 \times 1(100 - 80) = 1{,}010{,}000$ Btu/hr

(2) Δt:

	Hot Fluid		Cold Fluid	
250	Higher Temp	100	150
100	Lower Temp	80	20
150	Difference	20	130

LMTD = 64.6°F   (5.14)

$R = \dfrac{150}{20} = 7.5 \qquad S = \dfrac{20}{250 - 80} = 0.1175 \ (7.18) \qquad F_T = 0.87$ (Fig. 18)

$\Delta t = 0.87 \times 64.6 = 56.2°\text{F}$

(3) T_c and t_c: The average temperatures of 175 and 90°F will be adequate.

Hot fluid: shell side, oxygen

(4') $a_s = \frac{\pi}{4} \times 19.25^2$
$- 70 \left(\frac{\pi}{4} \times 1^2 + 20 \times 0.035 \times 0.5\right)$
$= 211.5$ in.^{2}
$= 211.5/144 = 1.47$ ft^2

Wetted perimeter $= 70(\pi \times 1 - 20 \times 0.035 + 20 \times 2 \times 0.5)$
$= 1570$ in.

$d_e = 4 \times 211.5/1570 = 0.54$ in. [Eq. (6.4)]
$D_e = 0.54/12 = 0.045$ ft

(5″) $G_s = W/a_s$
$= 30{,}000/1.47$
$= 20{,}400$ lb/(hr)(ft^2)

(6') At $T_a = 175$°F, $\mu = 0.0225 \times 2.42$
$= 0.0545$ lb/(ft)(hr) [Fig. 15]
$Re_s = D_e G_s/\mu$
$= 0.045 \times 20{,}400/0.0545 = 16{,}850$

(7') $j_H = 59.5$ [Fig. 16.10a]

(8') At 175°F, $c = 0.225$; $k = 0.0175$
$(c\mu/k)^{1/3} = (0.225 \times 0.0545/0.0175)^{1/3}$
$= 0.89$
$\phi_s = 1.0$ (for gases)

(9') $h_f = j_f \dfrac{k}{D_e} \left(\dfrac{c\mu}{k}\right)^{1/3} \phi_s$ [Eq. (6.15)]
$= 59.5 \times 0.0175 \times 0.89/0.045$
$= 20.5$
$R_{do} = 0.003$, $h_{do} = 1/0.003 = 333$
$h'_f = \dfrac{h_{do}h_f}{h_{do} + h_f} = \dfrac{333 \times 20.5}{333 + 20.5} = 19.3$
[Eq. (16.37)]
$h'_{fi} = 142$ Btu/(hr)(ft^2)(°F) [Fig. 16.9]

Cold fluid: tube side, water

(4) $a'_t = 0.479$ in.^{2} [Table 10]
$a_t = N_t a'_t/144n$ [Eq. (7.48)]
$= 70 \times 0.479/144 \times 4 = 0.0582$ ft^2
$D = 0.782/12 = 0.0652$ ft

(5) $G_t = w/a_t$
$= 50{,}500/0.0582$
$= 868{,}000$ lb/(hr)(ft^2)
$V = G_t/3600\rho = 868{,}000/3600 \times 62.5$
$= 3.86$ fps

(6) At $t_a = 90$°F,
$\mu = 0.80 \times 2.42 = 1.94$ lb/(ft)(hr) [Fig. 14]
$Re_t = DG_t/\mu$ (for pressure drop only)
$= 0.0652 \times 868{,}000/1.94 = 29{,}100$

(9) $h_i = 940 \times 0.96 = 903$ [Fig. 25]
$R_{di} = 0.003$, $h_{di} = 1/0.003 = 333$
$h'_i = \dfrac{h_{di}h_i}{h_{di} + h_i} = \dfrac{333 \times 903}{333 + 903}$
$= 243$ Btu/(hr)(ft^2)(°F) [Eq. (16.40)]

Overall design coefficient based on inside of tube U_{Di}:

$$U_{Di} = \frac{h'_{fi} h'_i}{h'_{fi} + h'_i} = \frac{142 \times 243}{142 + 243} = 89.6 \text{ Btu/(hr)(ft}^2\text{)(°F)} \quad (16.41)$$

Actual overall coefficient based on inside of tube:

Internal surface per lin ft $= 0.2048$ ft (Table 10)
$A_i = 70 \times 0.2048 \times 16'0'' = 230$ ft^2
$U_{Di} = \dfrac{Q}{A_i \Delta t} = \dfrac{1{,}010{,}000}{230 \times 56.2} = 78.2$

EXTENDED SURFACES

Adjustment of the fouling factor:

Excess fouling factor $= \left(\dfrac{1}{U_{Di\text{actual}}}\right) - \dfrac{1}{U_{Di}} = \dfrac{1}{78.2} - \dfrac{1}{89.6} = 0.00165$

Adding to the outside fouling factor: $\dfrac{A_f + A_o}{A_i} = 9.27$

$R_{do} = 0.003 + 9.27 \times 00165 = 0.0183$

$h'_{f,\text{actual}} = 0.0183 + \dfrac{1}{20.5} = 14.9$

$h'_{fi} = 113$

Check of actual overall coefficient:

$$U_{Di} = \dfrac{h'_{fi} h'_i}{h'_{fi} + h'_i} = \dfrac{113 \times 243}{113 + 243} = 77.3 \quad \text{Check vs. } 78.2$$

Pressure Drop

(1′) $d'_e = 4 \times 211.5/(1570 + \pi \times 19.25)$
$\qquad = 0.52 \text{ in.}$ [Eq. (6.5)]
$D'_e = 0.52/12 = 0.0433 \text{ ft}$
$Re'_s = D'_e G_s/\mu$
$\qquad = 0.0433 \times 20{,}400/0.0545 = 16{,}200$
$f = 0.00025 \text{ ft}^2/\text{in.}^2$ [Fig. 16.10b)]
Mol. wt. oxygen = 32
$\rho = \dfrac{32}{359 \times {}^{635}\!/_{492} \times 14.7/17.7}$
$\qquad = 0.083 \text{ lb/ft}^3$
$s = 0.083/62.5 = 0.00133$

(2′) $\Delta P_s = \dfrac{fG_e^2 Ln}{5.22 \times 10^{10} D'_e s \phi_s}$
[Eq. (7.45)]
$= \dfrac{0.00025 \times 20{,}400^2 \times 16}{5.22 \times 10^{10} \times 0.0433}$
$\qquad\qquad \times 0.00133 \times 1.0$
$= 0.6 \text{ psi}$

(1) For $Re_t = 29{,}100$
$f = 0.00021 \text{ ft}^2/\text{in.}^2$ [Fig. 26]

(2) $\Delta P_t = \dfrac{fG_tLn}{5.22 \times 10^{10} Ds\phi_t}$ [Eq. (7.45)]

$= \dfrac{0.00021 \times 868{,}000^2 \times 16 \times 4}{5.22 \times 10^{10} \times 0.0652 \times 1.0 \times 1.0}$
$= 3.0 \text{ psi}$

Summary

0.018	Dirt Factor	0.003
113	h inside	243
U_{Di}	78.2	
0.6	Calculated ΔP	3.0
2.0	Allowable ΔP	10.0

The large shell-side dirt factor might suggest that the exchanger is considerably oversized. However, the number of finned tubes which can be fitted on 2-in. triangular pitch into a 17¼-in. shell is only 54. This gives a required value of U_{Di} of 102 which is slightly greater than can be performed in a 17¼-in. exchanger.

TRANSVERSE FINS

Derivation of the Fin Efficiency. The different types of transverse finned pipes and tubes have been discussed at the beginning of the chapter. Expressions for their efficiencies are somewhat more difficult to derive than for longitudinal fins, since even the transverse fin of uniform cross section does not reduce to the simple differential equations of the longitudinal fin of uniform cross section. The derivations given here are those of Gardner,[1] which are ingenious because they develop a general expression that is applicable on modification to all the types of manufactured fins including the longitudinal fins. For the derivation of a general case, a transverse fin of varying cross section will be considered. The fluid surrounding the fin is again assumed to be hotter than the fin itself, and the flow of heat is from the outer fluid to the fin. The same assumptions apply as before. Referring to Fig. 16.12, let $\Theta = T_c - t$ where T_c is the constant hot-fluid temperature and t is the metal temperature at any point in the fin. The heat entering the two sides of the fin between 2-2' and 1-1' is dependent upon the surface between the two radii r_e and r_1. Thus the total surface between 2-2' and 0-0' is a function of r

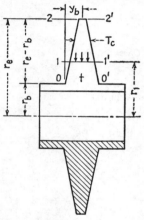

FIG. 16.12. Derivation of the transverse-fin efficiency.

$$dQ = h_f \Theta \, dA \qquad (16.44)$$

The heat which enters the fin between 2-2' and 1-1' flows toward its base through the cross section of the fin at 1-1'. Applying the Fourier equation

$$Q = -k a_x \frac{d\Theta}{dr} \qquad (16.45)$$

where k is the thermal conductivity and a_x is the cross-sectional area of the fin which in this case also varies with r

Differentiating Eq. (16.45),

$$-\frac{dQ}{dr} = \frac{d}{dr}\left(k a_x \frac{d\Theta}{dr}\right) = k a_x \frac{d^2\Theta}{dr^2} + \frac{k \, da_x}{dr} \frac{d\Theta}{dr} \qquad (16.46)$$

---
[1] Gardner, *op. cit.*

EXTENDED SURFACES

Equating Eqs. (16.44) and (16.46)

$$\frac{d^2\Theta}{dr^2} + \left(\frac{1}{a_x}\frac{da_x}{dr}\right)\frac{d\Theta}{dr} + \left(\frac{h_f}{ka_x}\frac{dA}{dr}\right)\Theta = 0 \qquad (16.47)$$

This second-order differential equation is somewhat more difficult to evaluate than Eq. (16.6), which had simple roots, and can be solved by *Bessel functions*.

A number of second-order differential equations arise in the solution of engineering problems which can be solved by a power series. The solution can be written as the sum of two arbitrary functions and two arbitrary constants as suggested by Eq. (16.9). In a general form the Bessel equation is

$$r^2\frac{d^2\Theta}{dr^2} + r\frac{d\Theta}{dr} + (r^2 - n^2)\Theta = 0 \qquad (16.48)$$

where n is a constant. Many types of functions which are independent solutions of the Bessel equation have been developed, and their properties are tabulated.[1] Douglass has provided the following solution for Eq. (16.48) into which Eq. (16.47) may be transformed by multiplying by r^2.

$$r^2\frac{d^2\Theta}{dr^2} + [(1 - 2m)r - 2\alpha_1 r]\frac{d\Theta}{dr}$$
$$+ [p^2 C_3^2 r^{2p} + \alpha_1^2 r^2 + \alpha_1(2m - 1)r + m^2 - p^2 n^2]\Theta = 0 \qquad (16.49)$$

where α_1, C_3, p, m, and n are constants, the last being the *order* of the Bessel function.

Equations (16.48) and (16.49) will have the same form when

$$a_x = C_4 r^{1-2pn} \qquad (16.50)$$

and

$$\frac{dA}{dr} = C_5 r^{2p(1-n)-1} \qquad (16.51)$$

where C_4 and C_5 are positive constants. If the cross section of the fin can be described by Eq. (16.50) and the surface from the outer edge to the cross section by Eq. (16.51), the general solution is found from the boundary conditions at the outer edge of the fin and at the base of the fin.

At the outer edge, $r = r_e$ and $\Theta = \Theta_e$
At the base, $r = r_b$ and $\Theta = \Theta_b$
For n equal to zero or an integer

$$\Theta = \Theta_b \left(\frac{u}{u_b}\right)^n \left[\frac{I_n(u) + \beta_1 K_n(u)}{I_n(u_b) + \beta_1 K_n(u_b)}\right] \qquad (16.52)$$

[1] Sherwood, T. K., and C. E. Reed, "Applied Mathematics in Chemical Engineering," p. 211, McGraw-Hill Book Company, Inc., New York, 1939.

where

$$\beta_1 = -\frac{I_{n-1}(u_e)}{K_{n-1}(u_e)} \tag{16.53}$$

and

$$u = -iC_3 r^p = r\sqrt{\frac{h_f}{ka_x}\frac{dA}{dr}} \tag{16.54}$$

where u_b and u_e are found by substituting r, a_x, and dA/dr for the edge and base, respectively, and $i = \sqrt{-1}$. Since Θ is a function of A as well as r, the heat transferred to the entire fin surface from the fluid is

$$Q = h_f \int_0^{A_f} \Theta \, dA \tag{16.55}$$

where A_f is the total fin surface. The same heat when transferred through the base of the fin to the tube is

$$Q = h_b \Theta_b A_f \tag{16.56}$$

Defining the fin efficiency as before,

$$\Omega = \frac{h_b}{h_f} = \frac{\int_0^{A_f} \Theta \, dA}{\Theta_b A_f} = \frac{2(1-n)}{u_b[1 - (u_e/u_b)^{2(1-n)}]}\left[\frac{I_{n-1}(u_b) - \beta_1 K_{n-1}(u_b)}{I_n(u_b) + \beta_1 K_n(u_b)}\right] \tag{16.57}$$

For a fin of given contour the exponent of r is known in Eq. (16.50). Equation (16.51) may be eliminated accordingly to permit the solution for n, the order of the Bessel function.

When $n = 0$, the introduction of Eqs. (16.52) and (16.53) and

$$dA = C_6 u^{1-2n} \, du$$

as obtained from Eqs. (16.51) and (16.54) into Eq. (16.57) gives the value of Ω for the annular fin of constant width.

Equation (16.57) reduces to

$$\Omega = \frac{2}{u_b[1 - (u_e/u_b)]^2}\left[\frac{I_1(u_b) - \beta_1 K_1(u_b)}{I_0(u_b) + \beta_1 K_0(u_b)}\right] \tag{16.58}$$

where

$$\beta_1 = \frac{I_1(u_e)}{K_1(u_e)} \tag{16.59}$$

and

$$u_b = \frac{(r_e - r_b)\sqrt{h_f/ky_b}}{\left(\dfrac{r_e}{r_b} - 1\right)} \tag{16.60}$$

$$u_e = u_b \left(\frac{r_e}{r_b}\right) \tag{16.61}$$

Plots of these equations are given in Fig. 16.13a.

For the case where n equals a fraction,

$$\Theta = \Theta_b \left(\frac{u}{u_b}\right)^n \left[\frac{I_n(u) + \beta_1 I_{-n}(u)}{I_n(u_b) + \beta_1 I_{-n}(u_b)}\right] \quad (16.62)$$

where

$$\beta_1 = \frac{I_{n-1}(u_e)}{I_{1-n}(u_e)} \quad (16.63)$$

If half the thickness of the fin, y, is given by

$$y = y_b \left(\frac{r}{r_b}\right)^{-\frac{2n}{1-n}} \quad (16.64)$$

$$u = (1-n)\left(\frac{r}{r_b}\right)^{\frac{1}{1-n}} \sqrt{\frac{h_f}{ky_b}}\, r_b \quad (16.05)$$

When n equals a fraction, Eq. (16.57) becomes

$$\Omega = \frac{2(1-n)}{u_b[1-(u_e/u_b)^{2(1-n)}]} \left[\frac{I_{n-1}(u_b) + \beta_1 I_{1-n}(u_b)}{I_n(u_b) + \beta_1 I_{-n}(u_b)}\right] \quad (16.66)$$

When $n = \frac{1}{3}$, corresponding to annular fins with constant heat flux at every cross section from r_b to r_e

$$\Omega = \frac{4}{3u_b[1-(u_e/u_b)^{4/3}]} \left[\frac{I_{-2/3}(u_b) + \beta_1 I_{2/3}(u_b)}{I_{1/3}(u_b) + \beta_1 I_{-1/3}(u_b)}\right] \quad (16.67)$$

$$\beta_1 = \frac{-I_{-2/3}(u_e)}{I_{2/3}(u_e)} \quad (16.68)$$

$$u_e = u_b \left(\frac{r_e}{r_b}\right)^{3/2} \quad (16.69)$$

$$y = y_b \left(\frac{r_b}{r}\right) \quad (16.70)$$

These values are plotted in Fig. 16.13b. For longitudinal fins and spines the efficiency curves and the fundamental equations are included in Fig. 16.13c and d. When a fouling factor is present replace h_f by h'_f.

In these graphs the efficiency Ω has been plotted as the ordinate and the principal shape element influencing the value of n in the Bessel equation has been plotted as the abscissa. The variations in the types of construction of transverse finned tubes are numerous and less standardized than for longitudinal fins, and no attempt will be made here to compute weighted efficiency curves of h'_f vs. h'_{fi}. A weighted efficiency curve can be computed, however, using Eq. (16.34) with the values of Ω being taken from Fig. 16.13a and b, in the same manner as before, when the frequent use of given fin and tube arrangements justifies their preparation.

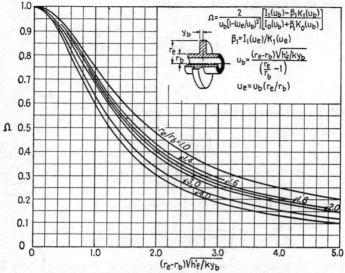

(a)-EFFICIENCY OF ANNULAR FINS OF CONSTANT THICKNESS

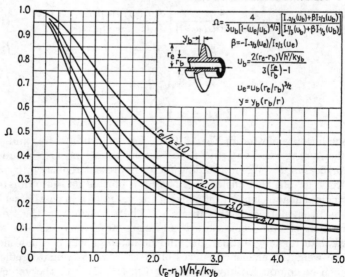

(b)-EFFICIENCY OF ANNULAR FINS WITH METAL AREA FOR CONSTANT HEAT FLUX

Fig. 16.13. Fin efficiencies. Where there is no foul-

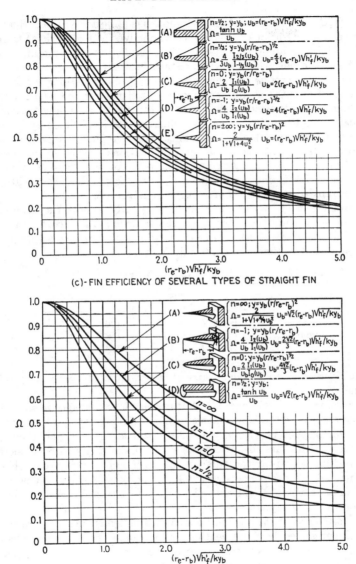

(c)- FIN EFFICIENCY OF SEVERAL TYPES OF STRAIGHT FIN

(d)- EFFICIENCY CURVES FOR FOUR TYPES OF SPINE

ing factor replace h_f' by h_f. (Gardner, *Transactions of the ASME.*)

The Optimum Thermal Fin. Schmidt[1] has undertaken the derivations for the shapes of various types of fins which give the highest heat transfer for the minimum amount of metal. As a result, the optimum fin may be considered the one which has a constant heat flux at any cross section between the outer edge and the base. For the common longitudinal and transverse fins these would correspond to tall narrow shapes with small base areas. Furthermore, the sides of the fins should have parabolic curvatures. Although the thermal efficiency for ideal fins may be high, the cost of their manufacture is usually excessive and they are not often structurally adaptable to industrial applications. The calculations of the optimum shapes have led, however, to the present type of manufacture using 20 BWG and lighter metal ribbons except where the conditions of heat transfer requires a more rugged construction.

The metal used in their manufacture greatly affects the economics of the fins. Schmidt has prepared the following table comparing the ratio of the quantity of metal required for fins with identical shapes where the weight and volume of a copper fin is taken as unity.

TABLE 16.2. OPTIMUM VOLUMES AND WEIGHTS OF METAL FINS

Metal	k	Gravity	Volume	Weight
Copper	222	8.9	1.0	1.0
Aluminum	121	2.7	1.83	0.556
Steel	30	7.8	7.33	0.43

Jakob[2] presents an excellent survey of optimum fins defined by other criteria.

Transverse Fin Exchangers. Transverse fin exchangers in crossflow are used only when the film coefficients of the fluids passing over them are low. This applies particularly to gases and air at low and moderate pressures. Tubes are also available which have many very small fins integrally shaped from the tube metal itself and which are usable in conventional 1-2 exchangers with baffled side-to-side flow. These can be calculated using a suitable shell-side heat-transfer curve and an appropriate efficiency curve.

Perhaps the most interesting applications of transverse fins are found in the larger gas cooling and heating services such as on furnaces (economizers), tempering coils for air conditioning, air-cooled steam condensers for turbine and engine work, and miscellaneous special services. An application which is growing in popularity is the air-cooled steam con-

---

[1] Schmidt, E., *Z. Ver. deut. Ing.* 885, **70** (1926).
[2] Jakob, *op. cit.*

denser as shown in Fig. 16.14 for localities with inadequate cooling-water supply. The steam enters the tubes, and an induced-draft fan circulates air over the transverse finned tubes. In this way it is possible to attain a closer approach to the atmospheric temperature than could be done with a reasonable surface composed entirely of bare tubes. In

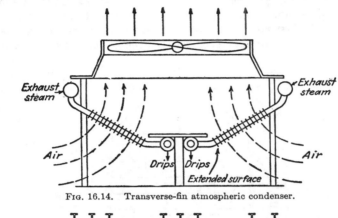

Fig. 16.14. Transverse-fin atmospheric condenser.

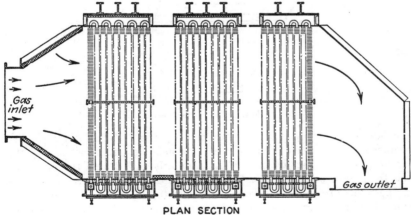

Fig. 16.15. Transverse-fin gas cooler. (*Foster Wheeler Corp.*)

Fig. 16.15 a plan is shown of a high-temperature gas cooler used in an aviation testing laboratory. With the exception noted before, all other transverse fin exchangers operate in crossflow. The true temperature differences for systems operating in crossflow naturally differ from other flow patterns except when one fluid is isothermal; then all systems are in true counterflow. Before proceeding to the calculation of a typical extended-surface gas exchanger, an investigation will be made of the influence of flow pattern on the true temperature difference in crossflow.

True Temperature Difference for Crossflow Arrangements. Consider a duct in which a hot gas (or other fluid) passes at right angles to a bundle carrying a cold tube liquid in a single pass as in Fig. 16.16. Assume that baffles indicated by the vertical lines are placed across the bundle in the direction of gas flow to prevent mixing over the length of the tubes. At point A in the first horizontal row of tubes and in the plane 1-1' there is an unique temperature difference between the gas-inlet temperature and

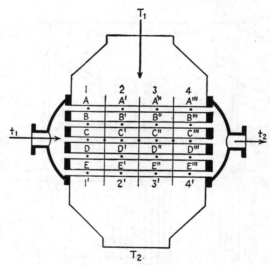

Fig. 16.16. Crossflow.

the liquid-inlet temperature. Moving to plane 2-2' there is a smaller temperature difference at A' between the gas-inlet temperature and the liquid at some higher temperature. Similarly at 3-3' and the point A'' the temperature difference is smaller than A', and further reductions of the temperature difference may be expected over the entire length of the tubes in the first tube row. Carrying the analysis to a lower row of tubes, the gas temperature off the first horizontal-tube row varies over the tube length, since the quantity of heat transferred in passing over the first row decreased from point to point with the decreasing temperature difference. Over the second horizontal row the temperature differences also vary but the temperatures of the gas leaving the row are different from those on leaving the preceding row of tubes. This picture is unlike that used in the derivation of the true temperature difference for a 1-2 exchanger as given in Chap. 7, in which the shell-side fluid was considered mixed and of homogeneous temperature at every cross section and where it was assumed that all the tube rows comprising each tube pass were at the same temperature.

Suppose, on the other hand, that the baffles in the exchanger in Fig. 16.16 are removed and that the tubes are not very long so that the gas passing over a row of tubes may be considered thoroughly mixed. This is much different from the preceding analysis, but it also differs from that of the 1-2 exchanger. In the 1-2 exchanger the uniform cross-section temperature varied along the length of the tube. In the present case the gas temperature at a cross section varies from row to row, and it can be seen that the true temperature difference is influenced considerably by whether one, both, or neither stream is mixed.

When a gas passes at right angles to a bundle composed of a single row of single-pass tubes, both fluids may be considered unmixed. However, it is not always possible to determine clearly if both streams are mixed, the definition being somewhat arbitrary in the case of tube bundles with long and short tube lengths. In the study below, four common theoretical mixing possibilities are treated to permit an estimation of the error resulting from the assumption of any particular one of them. Not all the possible crossflow temperature differences have been studied, but the principal derivations are available.

The derivations below are essentially those of Nusselt[1] and Smith,[2] and the graphs of the final equations are in the form developed by Nagle and Bowman, Mueller and Nagle.[3] These derivations involve the usual assumptions except as qualified for mixing. If T refers to the hot fluid, t to the cold fluid, and subscripts 1 and 2 to the inlet and outlet, respectively, it is convenient to define three parameters as follows:

$$\left. \begin{array}{l} K = \dfrac{T_1 - T_2}{T_1 - t_1} \\ S = \dfrac{t_2 - t_1}{T_1 - t_1} \end{array} \right\} \qquad R = \dfrac{K}{S} = \dfrac{T_1 - T_2}{t_2 - t_1}$$

$$(r) = \dfrac{\Delta t}{T_1 - t_1}$$

From Eq. (7.39) for counterflow define

$$(r)_{\text{counterflow}} = \dfrac{R - 1}{\ln (1 - S)/(1 - RS)} = \dfrac{(K - S)/S}{\ln (1 - S)/(1 - K)}$$

[1] Nusselt, W., *Tech. Mech. u. Thermodynam.*, **I**, 417–422 (1930).
[2] Smith, D. M., *Engineering*, **138**, 474–481, 606–607 (1934).
[3] See Chap. 7 for references.

Employing the factor F_T to correct the LMTD to the true temperature difference Δt in crossflow by the method introduced in Chap. 7,

$$F_T = \frac{(r)}{(r)_{\text{counterflow}}}$$

where $\Delta t = F_T \times \text{LMTD}$. Without a subscript (r) refers to the true value in crossflow.

Referring to Fig. 16.16 and applying the heat balances,

$$UA\Theta = \iint U(T - t)\, dx\, dy \tag{16.71}$$
$$WC(T_1 - T_2) = wc(t_2 - t_1) \tag{16.72}$$
$$\text{LMTD} = \frac{(T_1 - t_2) - (T_2 - t_1)}{\ln (T_1 - t_2)/(T_2 - t_1)} = \frac{\Delta t_h - \Delta t_c}{\ln \Delta t_h / \Delta t_c} = C_7 \Delta t_h \tag{16.73}$$

and where $C_7 = \left(1 - \dfrac{\Delta t_c}{\Delta t_h}\right) \Big/ \ln \dfrac{\Delta t_h}{\Delta t_c}$ is a function only of $\dfrac{\Delta t_c}{\Delta t_h}$. Let XY be the total tube area swept out as the shell fluid moves downward in the x direction and the tube fluid moves from left to right in the y direction.

Case A. Both Fluids Unmixed. Referring to Fig. 16.17a,

$$U(T - t)\, dx\, dy = -\frac{WC}{Y}\, dy\, \frac{\partial T}{\partial x}\, dx = \frac{wc}{X}\, dx\, \frac{\partial t}{\partial x}\, dy \tag{16.74}$$

If

$$x' = \frac{x}{X} \cdot \frac{UA}{WC} \quad \text{and} \quad y' = \frac{y}{Y} \frac{UA}{wc}$$

then

$$T - t = -\frac{\partial T}{\partial x'} = +\frac{\partial t}{\partial y'} \tag{16.75}$$

Solve when $T = T_1$ at $x' = 0$ and $t = t_1$ at $y' = 0$.
The solution is expressed in a doubly infinite series.

$$\frac{T - t_1}{T_1 - t_1} = 1 + \sum_{u=1}^{\infty} \sum_{v=1}^{\infty} \left[(-1)^{u+v} \frac{(u + v - 1)!v!}{(u!v!)^2} (x')^u (y')^v \right] \tag{16.76}$$

T_2 is obtained by integration as the mean of T when $x = X$ and y goes from 0 to Y. Then $WC(T_1 - T_2) = Q$, $x = X$, $x' = K/(r)$, $y = Y$, and $y' = S/(r)$.

$$(r) = \sum_{u=0}^{\infty} \sum_{v=0}^{\infty} \left\{ (-1)^{u+v} \frac{(u + v)!}{u!(u + 1)!v!(v + 1)!} \left[\frac{K}{(r)}\right]^u \left[\frac{S}{(r)}\right]^v \right\} \tag{16.77}$$

Values of F_T are plotted for $F_T = (r)/(r)_{\text{counterflow}}$ in terms of R and S in Fig. 16.17a.

EXTENDED SURFACES

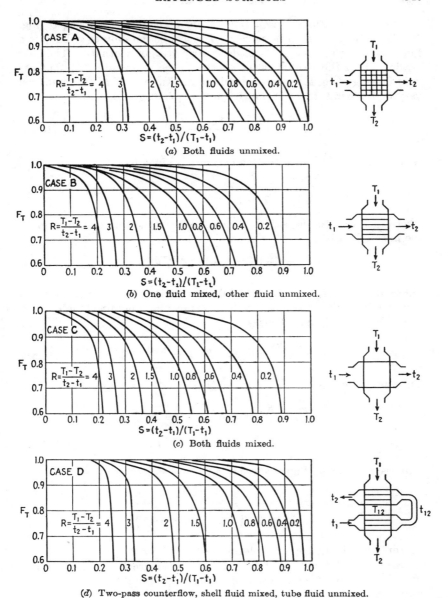

Fig. 16.17. Crossflow-temperature-difference correction factors. (*Bowman, Mueller, and Nagle, Transactions of the ASME.*)

Case B. One Fluid Mixed, Other Fluid Unmixed. Referring to Fig. 16.17b let the shell fluid be mixed. T is a function only of x, but t is a function of both x and y. Consider a strip at x_1 of width dx. The fluid flowing across the strip is at the constant shell temperature T_1, but the fluid flowing along the strip is unmixed and varies. If t is the unmixed varying temperature, at the inlet where $t = t_1$ at $y = 0$,

$$\frac{dt}{dy} = \frac{UX}{wc}(T - t) \tag{16.78}$$

At the outlet of the strip the temperature is given by

$$\frac{T - t_{x_1}Y}{T - t_1} = e^{-UXY/wc} \tag{16.79}$$

where t_{x_1} refers to the strip. The total heat transfer in the strip to the mixed fluid is

$$wc\frac{dx}{X}(t_1 - t_{x_2}Y) = WC\,dT \tag{16.80}$$

Substituting for $t_{x_2}Y$ in Eq. (16.80), integrating, and allowing for the boundary condition $t = t_1$ at $x = 0$, an expression is obtained for T which gives T_2 when $x = X$. Thus

$$\frac{T_1 - T_2}{T_1 - t_1} = 1 - e^{wc/WC}(1 - e^{-UXY/wc}) \tag{16.81}$$

The parameters have been chosen so that

$$\frac{wc}{WC} = \frac{K}{S} \quad \text{and} \quad \frac{UXY}{wc} = \frac{S}{(r)}$$

$$K = 1 - e^{-\frac{K}{S}\left(1 - e^{-\frac{S}{(r)}}\right)} \tag{16.82}$$

Expressing (r) as an implicit function

$$(r) = \frac{S}{\ln 1 \Big/ \left(1 - \frac{S}{K}\ln\frac{1}{1 - K}\right)} \tag{16.83}$$

Factors for this equation are plotted in Fig. 16.17b.

Case C. Both Fluids Mixed. Referring to Figure 16.17c, T is a function of x only and t is a function of y only. The total heat flow across a section dy is

$$U\,dy \int_0^x (T - t)\,dx$$

EXTENDED SURFACES 551

where T is variable and t is constant. The heat transferred is $wc\,dt$ and

$$\frac{wc\,dt}{UX\,dy} + t = \frac{1}{X}\int_0^X T\,dx \qquad (16.84)$$

Similarly,

$$\frac{wc\,dt}{UY\,dy} + t = \frac{1}{Y}\int_0^Y t\,dy \qquad (16.85)$$

Let $\dfrac{1}{X}\displaystyle\int_0^X T\,dx = \alpha_2$, $\dfrac{1}{Y}\displaystyle\int_0^Y t\,dy = \beta_2$ where α_2 and β_2 are constants. Substituting, integrating, and allowing for the boundary conditions $T = T_1$ at $x = 0$ and $t = t_1$ at $y = 0$, equations are obtained in α_2 and β_2.

$$\alpha_2 = \beta_2 - \frac{(r)}{K}(T_1 - \beta_2)(e^{-K/(r)} - 1) \qquad (16.86)$$

$$\beta_2 = \alpha_2 - \frac{(r)}{S}(t_1 - \alpha_2)(e^{-S/(r)} - 1) \qquad (16.87)$$

The integration also gives the temperature change for one fluid.

$$t_2 - t_1 = (\alpha_2 - t_1)(1 - e^{-S/(r)}) \qquad (16.88)$$

Solving simultaneously and using the temperature change times wc to obtain the heat transmitted,

$$(r)\left[\frac{K/(r)}{1 - e^{-K/(r)}} + \frac{S/(r)}{1 - e^{-S/(r)}}\right] = 1 \qquad (16.89)$$

Factors for this equation are plotted in Fig. 16.17c. Many of the cases for one crossflow and several series passes have not been solved so far. The case of greatest value is that of one crossflow and two series passes in parallel flow–counterflow when the shell fluid is unmixed and the series fluid is mixed between passes.

Case D. Shell Fluid Mixed, Two Pass, Counterflow, Fluid Unmixed Except between Passes. Let t_{12} be the mixed temperature between passes and T_{12} the corresponding shell temperature as shown in Fig. 16.17d. Call the upper pass I and the lower pass II, each with surface $A/2$. From Case B

$$1 - K_\mathrm{I} = e^{-K_\mathrm{I}/(S)\mathrm{I}}(1 - e^{-S_\mathrm{I}/(r)\mathrm{I}}) \qquad (16.90)$$

and

$$-K_\mathrm{II} = e^{-K_\mathrm{II}/S_\mathrm{II}}(1 - e^{-S_\mathrm{II}/(r)\mathrm{II}}) \qquad (16.91)$$

From the identity of flow in each path

$$\frac{K_\mathrm{I}}{S_\mathrm{I}} = \frac{K}{S} = \frac{K_\mathrm{II}}{S_\mathrm{II}} \qquad (16.92)$$

From the constancy of the heat-transfer coefficient and area in each pass

$$\frac{(r)_\mathrm{I}}{K_\mathrm{I}} = \frac{(r)_\mathrm{II}}{K_\mathrm{II}} \tag{16.93}$$

Let $M = (T_1 - t_{12})/(T_1 - t_1)$ and $N = (T_{12} - t_1)/(T_1 - t_1)$. Since the total heat transfer is the sum of both passes,

$$UA\,\Delta t = \frac{UA}{2}\Delta t_\mathrm{I} + \frac{UA}{2}\Delta t_\mathrm{II} \tag{16.94}$$

and

$$(r) = \tfrac{1}{2}M(r)_\mathrm{I} + \tfrac{1}{2}N(r)_\mathrm{II} \tag{16.95}$$

Also

$$T_1 - T_2 = (T_1 - T_{12}) + (T_{12} - T_2) \tag{16.96}$$
$$K = MK_\mathrm{I} + NK_\mathrm{II}$$
$$T_1 - t_1 = (T_1 - t_{12}) + (t_{12} - t_1)$$
$$1 = M + NS_\mathrm{II} \tag{16.97}$$
$$T_1 - t_1 = (T_1 - T_{12}) + (T_{12} - t_1)$$
$$1 = MK_\mathrm{I} + N \tag{16.98}$$

Solving all nine equations simultaneously and eliminating the eight unknowns K_I, S_I, $(r)_\mathrm{I}$, K_II, S_II, $(r)_\mathrm{II}$, M, and N produces a solution in terms of K, S, and (r). From Eqs. (16.92) and (16.93)

$$\frac{K_\mathrm{I}}{K_\mathrm{II}} = \frac{S_\mathrm{I}}{S_\mathrm{II}} = \frac{(r)_\mathrm{I}}{(r)_\mathrm{II}} \tag{16.99}$$

From Eqs. (16.90) and (16.91)

$$K_\mathrm{I} = K_\mathrm{II} \qquad S_\mathrm{I} = S_\mathrm{II} \qquad (r)_\mathrm{I} = (r)_\mathrm{II}$$

From Eqs. (16.94) and (16.96)

$$\frac{K_\mathrm{I}}{(r)_\mathrm{I}} = \frac{1}{2}\frac{K}{(r)} \qquad \frac{S_\mathrm{I}}{(r)_\mathrm{I}} = \frac{1}{2}\frac{S}{(r)}$$

Eliminating M and N between Eqs. (16.96) to (16.98) and applying Eq. (16.92),

$$K_\mathrm{I}^2\left(1 + \frac{S}{K} - S\right) - 2K_\mathrm{I} + K = 0 \tag{16.100}$$

$$K_\mathrm{I} = \frac{K}{1 \pm \sqrt{(1-K)(1-S)}} \tag{16.101}$$

Substituting in Eq. (16.90) using only the positive root,

$$\frac{\sqrt{(1-S)/(1-K)} - S/R}{1 - S/K} = e^{K/S}(1 - e^{\frac{1}{2}S/(r)}) \tag{16.102}$$

EXTENDED SURFACES

or

$$(r) = \frac{S}{2 \ln \dfrac{1}{1 - (S/K) \ln \sqrt{(1-S)/(1-K)} - (S/K)}{1 - S/K}} \quad (16.103)$$

Factors for this equation are plotted in Fig. 16.17d.

Dunn and Bonilla[1] have treated the case of fin transfer with no transverse mixing parallel to the extended surface and present graphs to permit a rapid solution for the system.

Comparison of Thermal Efficiency in Different Arrangements. Examination of the standard types of fluid-flow arrangements will indicate the relative thermal efficiency of each. There are three true fluid-flow patterns: counterflow, crossflow, and parallel flow. Consider a fluid being cooled from 400 to 200°F by a fluid being heated from 100 to 200°F so that the approach in terms of a 1-2 exchanger is zero or 200°F for a counterflow exchanger.

$$\begin{array}{cc} Hot\ fluid & Cold\ fluid \\ 400 & 200 = 200 = \Delta t_c \\ 200 & 100 = 100 = \Delta t_h \\ T_1 - T_2 = 200 & t_2 - t_1 = 100 \\ R = {}^{200}\!/_{100} = 2.0 & S = {}^{100}\!/_{300} = 0.333 \end{array}$$

Type of Flow	F_T
Counterflow, 1-1 exchanger	1.00
Parallel flow–counterflow, 2-4 exchanger	0.95
Crossflow, shell mixed and two unmixed tube passes in series	0.98
Crossflow, both fluids unmixed	0.90
Crossflow, shell fluid mixed and tube fluid unmixed	0.87
Parallel flow–counterflow, 1-2 exchanger	0.81
Crossflow, both fluids mixed	0.77
Crossflow, shell mixed and two unmixed parallel-flow tube passes	0
Parallel flow	0

The deductions which may be made from above are (1) that the presence of parallel flow greatly decreases the utilization of the thermal potential for heat transfer and (2) that mixing the fluid streams also decreases the utilization. The curves of F_T vs. R and S for crossflow with two parallel tube passes have not been included in Fig. 16.17, since it is impractical for even moderate temperature approaches and has little value as a means of heat recovery.

Film Coefficients to Transverse Fins. Almost all the data available on commercial scale runs have been made with air or flue gas. The heat-

---

[1] Dunn, W. E., and C. F. Bonilla, *Ind. Eng. Chem.*, **40**, 1101–1104 (1948).

transfer curve employed here has been transposed from the data of Jameson,[1] which agree within reasonable limits with the published results of Tate and Cartinhour[2] on economizers. It has been found that the heat-transfer coefficient is not influenced by the spacing of succeeding rows although nearly all the data were obtained for triangular pitch.

The actual heat-flow pattern in a transverse fin probably differs somewhat from the idealized picture used in deriving the fin efficiency. In triangular pitch the air or gas strikes the front and sides of the annular fins but not the back of the fin. As determined experimentally h_f is actually only an average value. The concentration of heat on the lead side of an annular fin probably introduces a potential for heat flow around the fin metal which was not considered in the derivation. Discontinuous fins such as stars or other types give generally higher coefficients than helical or disc fins, and this may be attributed in part to the greater ease with which the gas penetrates the space between adjacent discontinuous fins.

Because so many of the applications involve gases, it has become a common convention in extended-surface heat transfer to use Colburn's heat-transfer factor $j_h = (h/cG)(c\mu/k)^{2/3}$. The advantages of Colburn's form of the factor particularly with gases have been discussed in Chap. 9. For the sake of a consistent presentation the data of Jameson have been converted to the Sieder-Tate heat-transfer factor as shown in Fig. 16.18a with the value of the viscosity correction ϕ taken as 1.0 for gases. The equivalent diameter in Jameson's correlation has been defined by

$$D_e = \frac{2(A_f + A_o)}{\pi \text{ (projected perimeter)}} \qquad (16.104)$$

The projected perimeter is the sum of all the external distances in the plan view of a transverse finned tube. The mass velocity is computed from the free flow area in a single bank of tubes at right angles to the gas flow.

Pressure Drop for Transverse Fins. Unlike the heat transfer coefficient, the pressure drop is greatly influenced by the spacing of the succeeding rows of tubes, their layout and closeness. It is often possible in transverse-fin equipment that the vertical and transverse pitches will be unequal. Of a number of excellent correlations of pressure drop in crossflow that of Gunter and Shaw[3] is used here. It is equally satisfactory for crossflow calculations over bare tubes and the correlation is based on oils, water, and air. While there has been some objection to the broad-

---

[1] Jameson, S. L., *Trans. ASME*, **67**, 633–642 (1945).
[2] Tate, G. E., and J. Cartinhour, *Trans. ASME*, **67**, 687–692 (1945).
[3] Gunter, A. Y., and W. A. Shaw, *Trans. ASME*, **67**, 643–660 (1945).

ness of the correlation it gives relatively safe values for the pressure drop. The Reynolds numbers are computed on the basis of a *volumetric* equivalent diameter which reflects the proximity and arrangement of the succeeding rows, and the pressure-drop equation contains two dimensionless configuration factors. The volumetric equivalent diameter is defined by

$$D'_{ev} = \frac{4 \times \text{net free volume}}{A_f + A_o} \quad (16.105)$$

The net free volume is the volume between the center lines of two vertical banks of tubes less the volumes of the half-tubes and fins within the

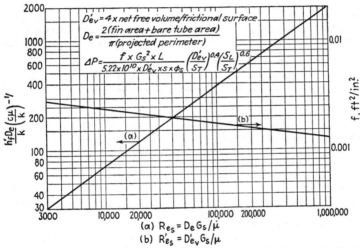

FIG. 16.18. Transverse-fin heat transfer and pressure drop. ((a) *Jameson* (b) *Gunter and Shaw, Transactions of the ASME.*)

center lines. A configuration factor is a dimensionless group which is ordinarily lost in the derivation but which may be included at the discretion of the experimenter. The factors used here are S_T and S_L where S_T is the pitch in a transverse bank and S_L is the center-to-center distance to the nearest tube in the next bank. The pressure-drop equation is thus

$$\Delta P = \frac{fG_s^2 L_p}{5.22 \times 10^{10} \times D'_{ev} s \phi_s} \left(\frac{D'_{ev}}{S_T}\right)^{0.4} \left(\frac{S_L}{S_T}\right)^{0.6} \quad (16.106)$$

where L_p is the length of the path. Friction factors are given in Fig. 16.18b.

Example 16.5. Calculation of a Transverse-fin Air Cooler. A duct 4'0" by 4'0" carries 100,000 lb/hr of air from a kiln drier at 250°F. The gas is to be used to preheat treated water from 150 to 190°F by revamping a section of the duct and installing 1 in. OD, 14 BWG tubes with ⅜ in. high annular brass fins of 20 BWG metal spaced ⅛ in. apart. Assume the spacing for the tubes is on 2¼-in. triangular pitch.

Use dirt factors of 0.0030 on both streams. The pressure head on the gas side is only a fraction of a pound per square inch, so that the overall pressure drop must be low.

How many tubes should be installed? How shall they be arranged?

Solution:

(1) Heat balance:
Air at 14.7 psia, $Q = 100{,}000 \times 0.25(250 - 200) = 1{,}250{,}000$ Btu/hr
Water, $Q = 31{,}200 \times 1.0(190 - 150) = 1{,}250{,}000$ Btu/hr

(2) Δt: Assume that in the final arrangement the water will flow in series from one vertical tube bank to the next, so that there will be as many passes as banks. If there are more than several banks in a 4- by 4-ft duct, the air may be considered mixed. Figure 16.17d will apply.

	Hot Fluid		Cold Fluid	
250	Higher Temp	190	60
200	Lower Temp	150	50
50	Difference	40	10

LMTD = 54.6°F (5.14)

$R = \dfrac{50}{40} = 1.25 \qquad S = \dfrac{40}{250 - 150} = 0.40 \qquad F_T = 0.985$ (Fig. 16.17d)

$\Delta t = 54.6 \times 0.985 = 53.7°F$

(3) T_c and t_c: The average temperatures of 225 and 170°F will be adequate.

Calculation of the duct equivalent diameter and flow area:

Equivalent diameter (refer to Fig. 16.19a):
$$d_e = \frac{2\,(\text{fin area} + \text{bare tube area})}{\pi\,(\text{projected perimeter})} \tag{16.104}$$

Fin area, $A_f = \dfrac{\pi}{4}(1.75^2 - 1^2) \times 2 \times 8 \times 12 \qquad = 310$ in.²/ft

Bare tube area, $A_o = \pi \times 1 \times 12 - \pi \times 1 \times 8 \times 0.035 \times 12 = \underline{27.2}$

Total $= \overline{337.2}$ in.²/ft, (2.34 ft²/ft)

Projected perimeter $= 2 \times \tfrac{3}{8} \times 2 \times 8 \times 12 + 2(12 - 8 \times 0.035 \times 12)$

$= 161.3$ in./ft

$d_e = \dfrac{2 \times 337.2}{\pi \times 161.3} = 1.33$ in.

$D_e = 1.33/12 = 0.111$ ft

EXTENDED SURFACES

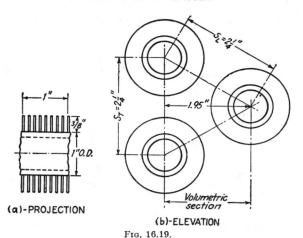

(a)-PROJECTION

(b)-ELEVATION

FIG. 16.19.

Flow area:
21 tubes may be fit in one vertical bank   (Fig. 16.19b)
20 tubes in alternating banks for triangular pitch
$a_s = (4 \times 12)(4 \times 12) - 21 \times 1 \times 48 - 21(2 \times 0.035 \times \frac{3}{8} \times 8 \times 48) = 1079$ in.^{2}
$= 1079/144 = 7.50$ ft^2

Hot fluid: duct, air
(4') $a_s = 7.50$ ft^2
$D_e = 0.111$ ft
(5') $G_s = W/a_s$
$= 100,000/7.50$
$= 13,300$ lb/(hr)(ft^2)
(6') At $T_a = 225°F$,
$\mu = 0.0215 \times 2.42 = 0.052$ lb/(ft)(hr)
[Fig. 15]
$Re_s = D_e G_s/\mu$
$= 0.111 \times 13,300/0.052 = 28,400$
(7') $j_f = 157$ [Fig. 16.18a]
(8') At $T_a = 225°F$,
$k = 0.0183$ Btu/(hr)(ft^2)(°F/ft)
$(c\mu/k)^{1/3} = (0.25 \times 0.052/0.0183)^{1/3} = 0.89$
$\phi_s = 1.0$ (for air)
(9') $h_f = j_f(k/D_e)(c\mu/k)^{1/3}$ [Eq. (6.15)]
$= 157 \times \dfrac{0.0183}{0.111} \times 0.89 = 23.1$

$R_{do} = 0.003;\ h_{do} = 1/0.003 = 333$
$h'_f = \dfrac{h_{do} h_f}{h_{do} + h_f} = \dfrac{333 \times 23.1}{333 + 23.1}$
$= 21.5$ Btu/(hr)(ft^2)(°F)
[Eq. (16.37)]

Cold fluid: tubes, treated water
(4) $a'_t = 0.546$ in.^{2} [Table 10]
Assume each bank carries all of the water.
$a_t = N_t(a'_t/144n)$ [Eq. (7.48)]
$= 21 \times 0.546/144 = 0.0795$ ft^2
$D = 0.834/12 = 0.0695$ ft
(5) $G_t = w/a_t$
$= 31,200/0.0795$
$= 392,000$ lb/(hr)(ft^2)
$V = G_t/3600\rho = 392,000/3600 \times 62.5$
$= 1.74$ fps
(6) At $t_a = 170°F$,
$\mu = 0.37 \times 2.42 = 0.895$ lb/(ft)(hr)
[Fig. 14]
$Re_t = DG_t/\mu$ (for pressure drop only)
$= 0.0695 \times 392,000/0.895 = 30,400$
(9) $h_i = 710 \times 0.94 = 667$ [Fig. 25]
$R_{di} = 0.003,\ h_{di} = 1/0.003 = 333$
$\phi_t = 1.0$
$h'_i = \dfrac{h_{di} h_i}{h_{di} + h_i} = \dfrac{333 \times 667}{333 + 667} = 222$
[Eq. (16.40)]

h'_{fi} (see Fig. 16.13a) k for brass = 60 Btu/(hr)(ft²)(°F/ft) (Table 3)

$$y_b = \frac{0.035}{2 \times 12} = 0.00146 \text{ ft}$$

$$(r_e - r_b)\sqrt{h'_f/ky_b} = \left(\frac{0.875 - 0.5}{12}\right)\sqrt{\frac{21.5}{60 \times 0.00146}} = 0.49$$

$$\frac{r_e}{r_b} = \frac{0.875}{0.50} = 1.75$$

$$\Omega = 0.91 \qquad \text{(Fig. 16.13a)}$$

$$h'_{fi} = (\Omega \times A_f + A_o)\frac{h'_f}{A_i} \qquad (16.34)$$

$$A_i = 0.218 \text{ ft}^2/\text{ft}$$

$$h'_{fi} = \left(0.91 \times \frac{310}{144} + \frac{27.2}{144}\right)\frac{21.5}{0.218} = 212$$

$$U_{D_i} = \frac{h'_{fi}h'_i}{h'_{fi} + h'_i} = \frac{212 \times 222}{212 + 222} = 108 \qquad (16.41)$$

Inside surface/bank = $21 \times 4 \times 0.218 = 18.3$ ft²

$$A_i = \frac{Q}{U_{D_i}\Delta t} = \frac{1{,}250{,}000}{108 \times 53.7} = 215 \text{ ft}^2$$

Use 215/18.3 = 11.8 banks, say 12 banks.[1]

Pressure Drop

(1) $D'_{e_v} = \dfrac{4 \times \text{net free volume}}{\text{frictional surface}}$

Refer to Fig. 16.19b.

Net free volume = $4 \times 4 \times \dfrac{1.95}{12} - \dfrac{1}{2}(21 + 20)\dfrac{\pi}{4} \times \dfrac{1^2 \times 4}{144} - \dfrac{1}{2}(21 + 20)\dfrac{\pi}{4}$

$$(1.75^2 - 1^2)\frac{0.035}{144} \times 8 \times 4$$

$$= 1.91 \text{ ft}^3$$

Frictional surface* = $\dfrac{1}{2}(21 + 20)2.34 \times 4 = 192$ ft²

$D'_{e_v} = 4 \times \dfrac{1.91}{192} = 0.040$ ft

$G_s = 13{,}300$ lb/(hr)(ft²)

$Re_s = 0.040 \times 13{,}300/0.052 = 10{,}200$

$f = 0.0024$ ft²/in.²   [Fig. 16.18b]

$\rho = \dfrac{29}{359 \times 685/492} = 0.058$ lb/ft³

$s = 0.058/62.5 = 0.000928$

$L_p = 12 \times 1.95/12 = 1.95$ ft

$\left(\dfrac{D'_{e_v}}{S_T}\right)^{0.4} = \left(\dfrac{0.040}{2.25/12}\right)^{0.4} = 0.538$

$\left(\dfrac{S_L}{S_T}\right)^{0.6} = 1.0$

(1) $Re_t = 30{,}400$
$f = 0.00020$ ft²/in.²   [Fig. 26]

---

[1] The effective path for pressure drop will be approximately 12×1.95 in. = 23.4 in.

* The walls of the duct may be neglected.

EXTENDED SURFACES

$$(2') \quad \Delta P_s = \frac{fG_s^2 L_p}{5.22 \times 10^{10} D'_{e_v} s \phi_s} \left(\frac{D'_{e_v}}{S_T}\right)^{0.4} \left(\frac{S_L}{S_T}\right)^{0.6}$$

$$= \frac{0.0024 \times 13{,}300^2 \times 1.95 \times 0.538}{5.22 \times 10^{10} \times 0.040 \times 0.000928 \times 1}$$

$$= 0.23 \text{ psi}$$

$$(2) \quad \Delta P_t = \frac{fG_t^2 L n}{5.22 \times 10^{10} D s \phi_t}$$

$$= \frac{0.00020 \times 392{,}000^2 \times 4 \times 12}{5.22 \times 10^{10} \times 0.0695 \times 1 \times 1}$$

$$= 0.41 \text{ psi}$$

Return losses will be negligible.

Double Extended Surface. Another type of heating element, which at first inspection appears to offer unlimited possibilities, is the double extended surface. Suppose two flowing fluids are separated by a metal wall and the surface is extended into both fluids by means of spines or pegs whose bases are superimposed. Per square foot of wall area it should be possible to add as much surface as desired as long as there are no restrictions along the axes of the spines. Such an arrangement will be called for only when the coefficients from both fluids to their respective fins is small. Under these conditions it is usually found that an unpractical length of fin is required for even a moderate total heat transfer. The fin efficiencies and total transfer can be determined by the methods already treated and by the use of Fig. 16.13c or d. The heat-transfer coefficients generally must be approximated from more conventional arrangements.

PROBLEMS

16.1. A double pipe exchanger consists of a 3-in. IPS outer pipe and a 1½-in. IPS inner pipe on which several arrangements of fins ½ in. high are under consideration. Compare the weighted efficiency curves with that in Fig. 16.9 when the following are used: (a) 18 steel fins, 20 BWG; (b) 18 copper fins, 20 BWG; (c) 18 steel fins, 16 BWG.

16.2. 4620 lb/hr of a 28°API gas oil will be used to preheat 5700 lb/hr of a 110°API butane reactor feed at elevated pressure from 260 to 400°F. The gas oil will enter at 575°F and leave at 350°F. A pressure drop of 10 psi is permissible on the gas oil, but on the reactor feed it should not exceed 2 to 3 psi. Dirt factors of 0.002 should be provided on each side. The gas oil is the controlling fluid and should flow in the annulus.

Available for the purpose are a number of 20-ft hairpin double pipe exchangers consisting of 2½-in. IPS outer pipes and 1¼ in. OD, 10 BWG inner tubes with 24 fins 20 BWG, ½ in. high.

How many sections should be used, and how shall they be arranged?

16.3. In a regenerative gas-absorption process 10,300 lb/hr of 15°Bé caustic soda ($s = 1.115$) leaves the regenerator at 240°F and is cooled to 170°F. The heat is absorbed by 10,300 lb/hr of 15°Bé caustic soda at 100°F being sent to the regenerator.

Available for the service are a number of 20-ft hairpins consisting of 3-in. IPS shells and 1½-in. IPS inner tubes with 24 fins 20 BWG, ½ in. high. Pressure drops of 10 psi are allowable. Dirt factors of 0.002 should be provided on each side.

How many sections are required, and how shall they be arranged?

16.4. 20,000 lb/hr of nitrogen at 0 gage pressure is to be heated from 100 to 175°F using exhaust steam at 212°F. Fouling factors of 0.002 should be provided for both. Available for the service is a 19¼ in. ID 1-2 exchanger containing 56 1¼ in. OD tubes, 12'0" long and having 24 fins 20 BWG, ½ in. high arranged for two passes.

Is the exchanger satisfactory? What are the pressure drops?

16.5. A textile impregnating room measures 50 by 100 by 12 ft. Because of the possibility of developing an explosive and toxic concentration it is necessary to change the air eight times an hour. To provide comfortable conditions the room should be kept at 75°F, although no provision will be made for humidity control. The lowest winter temperature to be anticipated is 30°F (Middle Atlantic). The air intake will be through an existing 4 by 4 ft duct in which a tempering coil is to be provided using exhaust steam at 212°F.

Available are 4'0" long helical fin tubes with male and female threads. These are ¾ in. OD, 16 BWG with ⅜ in. high brass fins 0.018 in. thick spaced six to the inch. Assume that they will be installed on 1¾-in. triangular pitch.

How many tubes are required? What static pressure must be available in the duct?

NOMENCLATURE FOR CHAPTER 16

A	Total outside heat-transfer surface, ft²
A_f	Surface of thin fins (both sides), ft²
A_i, A_o	Surface on inside of tubes and bare surface on the outside of a finned tube, respectively, ft²
a	Flow area, ft²
a'_t	Flow area/tube, in²
a_x	Cross-sectional area of fin at right angles to the heat flow, ft²
b	Length of fin from outer edge to the base, ft
C	Specific heat of hot fluid in derivations, Btu/(lb)(°F)
$C_1, C_2, C_3, C_4, C_5, C_6, C_7$	Constants, dimensionless
$C'_1, C'_2,$	Constants, dimensionless
c	Specific heat of cold fluid, Btu/(lb)(°F); a constant in the solution of the Bessel equation, dimensionless
D	Inside diameter of tubes or pipes, ft
d	Inside diameter of tubes, in.
D_e, D'_e	Equivalent diameter for heat transfer and pressure drop, ft
D'_{ev}	Volumetric equivalent diameter for crossflow, ft
e	Napieran base
F_T	Temperature-difference factor, $\Delta t = F_T \times$ LMTD, dimensionless
f	Friction factor, ft²/in²
G	Mass velocity, lb/(hr)(ft²)
g	Acceleration of gravity, 4.18×10^8 ft/hr²
h	Heat-transfer coefficient in general, Btu/(hr)(ft²)(°F)
h_b	Heat-transfer coefficient h'_f or h_f to fin delivered at the base of the fin, Btu/(hr)(ft²)(°F)
h_{di}, h_{do}	Dirt coefficient equivalent to the reciprocal of the dirt factor inside the tube R_{di} and outside the tube, R_{do}, respectively, Btu/(hr)(ft²)(°F)
h_f, h'_f	Heat-transfer coefficient on fin side of a pipe or tube, heat-transfer coefficient on fin side of a pipe or tube corrected for the dirt factor, Btu/(hr)(ft²)(°F)

EXTENDED SURFACES 561

h_{fi}, h'_{fi}	h_f and h'_f corrected to the inside surface of a pipe or tube, respectively, Btu/(hr)(ft^2)(°F)
h''_{fi}	Value of h'_{fi} if there were no resistance in the fin itself, Btu/(hr)(ft^2)(°F)
h_{fo}	The average film coefficient of the outside diameter of the tube weighting h_f to the fin surface and to the tube outside surface, Btu/(hr)(ft^2)(°F)
h_i, h'_i	Heat-transfer coefficient for the fluid inside the tube, h_i corrected for the dirt factor, Btu/(hr)(ft^2)(°F)
h_o	Heat-transfer coefficient for outside fluid, Btu/(hr)(ft^2)(°F)
$I_n(u)$	Modified Bessel function of the first kind and order n, dimensionless
i	$\sqrt{-1}$, dimensionless
j_f	Factor for heat transfer to finned pipes and tubes, dimensionless
j_H, j_h	The Sieder-Tate and Colburn heat-transfer factors, dimensionless
K	Temperature group for crossflow $(T_1 - T_2)/(T_2 - t_1)$, dimensionless
$K_n(u)$	Modified Bessel function of the second kind and order n, dimensionless
k	Thermal conductivity, Btu/(hr)(ft^2)(°F/ft)
L	Length of tube, ft
L_p	Length of path, ft
LMTD	Logarithmic mean temperature difference, °F
l	Height of fin at any point, ft
m	Either of two constants used in the derivation of longitudinal or annular fins, dimensionless
N_f	Number of fins (per tube)
N_t	Number of tubes in bundle
n	Number of tube passes; a constant denoting the order of the Bessel function, dimensionless
P	Perimeter, ft
ΔP	Pressure drop, psi
p	A constant in the Bessel equation, dimensionless; pressure, atm
Q	Total heat flow or total heat flow through bases of fins and bare tube surface, Btu/hr
Q_o	Heat flow, heat flow through bare tube surface, Btu/hr
R	Temperature group $(T_1 - T_2)/(t_2 - t_1)$, dimensionless
R_{do}, R_{di}	Dirt or fouling factor, outside tube and inside tube, (hr)(ft^2)(°F)/Btu
Re, Re'	Reynolds number for heat transfer and pressure drop, dimensionless
r	Radius, ft
(r)	Temperature group in crossflow, $\Delta t/(T_1 - t_1)$, dimensionless
$(r)_{\text{counterflow}}$	Temperature group in counterflow, LMTD/$(T_1 - t_1)$, dimensionless
S	Temperature group $(t_2 - t_1)/(T_1 - t_1)$, dimensionless

s	Specific gravity, dimensionless
T	Temperature of hot fluid, °F
T_s	Temperature of steam, °F
t	Temperature of fin or cold fluid in general, °F
T_c, t_c	Caloric temperature of hot fluid and cold fluid, respectively, from Eqs. (5.28) and (5.29), °F
t_w, t'_w, t_{fw}	Temperature of tube wall, temperature in fin metal, temperature of wall in finned tubes, °F
t_f, t'_f	Temperature of clean fin, temperature of fin beneath dirt film, °F
Δt	True temperature difference, °F
$\Delta t_c, \Delta t_h$	Hot and cold terminal differences, °F
$\Delta t_f, \Delta t_{do}, \Delta t_{di}, \Delta t_i$	Temperature drops across fin coefficient, outside dirt, inside dirt, and the inside fluid film corrected for viscosity and dirt, °F
U_{Di}	Overall design coefficient of heat transfer based on the inside tube surface, Btu/(hr)(ft²)(°F)
U_i	Overall clean coefficient of heat transfer based on inside tube surface, Btu/(hr)(ft²)(°F)
u	A function
V	Velocity of flow, fps
v	A function
W	Weight flow in general, weight flow of hot fluid, lb/hr
w	Weight flow of cold fluid, lb/hr
X, Y, x, y	Distances, ft
y	Half the width of a fin, ft
α	Constant in transverse fin and crossflow derivations
β	Constant in transverse fin and crossflow derivations
θ	Temperature difference between fluid and fin $T - t$, °F
μ	Viscosity at the caloric temperature, lb/(ft)(hr)
μ_w, μ_{fw}	Viscosity at the tube wall or fin temperatures, t_w or t_{fw}, lb/(ft)(hr)
ρ	Density, lb/ft³
ϕ	Viscosity ratio $(\mu/\mu_{fw})^{0.14}$, dimensionless
Ω	Fin efficiency h_b/h_f, dimensionless
Ω'	Weighted efficiency for fin and bare tube, dimensionless

Subscripts (except as noted above)

a	Annulus
b	Base (of fin)
e	Edge (of fin)
i	Interface of film or inside of pipe or tube
s	Shell
t	Tube
I	First of two passes
II	Second of two passes

CHAPTER 17

DIRECT-CONTACT TRANSFER: COOLING TOWERS

Introduction. In all preceding chapters the hot and cold fluids were separated by impervious surfaces. In tubular equipment, the tube limits the intimacy of contact between the hot and cold fluids and also serves as a surface upon which resistances accumulate as fouling and scale films. In order that a turbulent fluid in a tube may receive heat, particles in the eddying fluid body must contact a warm film at the tube wall, take in heat by conduction, and then mix with the eddying fluid body. A similar process occurs on the shell side, and the net heat exchange may occur through as many as seven individual resistances.

One of the principal reasons for employing tubes is to prevent the contamination of the hot fluid by the cold fluid. When one of the fluids is a gas and the other a liquid, an impervious surface is often unnecessary, since there may be no problem of mutual contamination, the gas and liquid being readily separable after mixing and exchanging heat. Fouling resistances are automatically eliminated by the absence of a surface on which they can collect and permit a direct-contact apparatus to operate indefinitely with uniform thermal performance. The greater intimacy of the direct contact generally permits the attainment of greater heat-transfer coefficients than are usual in tubular equipment.

Perhaps the outstanding application of an apparatus operating with a direct contact between a gas and a liquid is the cooling tower. It is usually a boxlike redwood structure with redwood internals. Cooling towers are employed to contact hot water coming from process cooling systems with air for the purpose of cooling the water and allowing its reuse in a process. The function of the wooden internals, or *fill*, is to increase the contact surface between the air and the water. A cooling tower ordinarily reduces the fresh cooling-water requirement by about 98 per cent, although there is some mutual contamination caused by the saturation of the air with water vapor.

The prospective use for direct-contact equipment in other services requiring rapid rates of heat transfer is perhaps greater than for any other type of heat-transfer apparatus. Although applied now almost exclusively to the humidification of air or the cooling of water, the principles of direct-contact heat transfer may be applied to the cooling or heating of other insoluble gases or liquids. This is especially true in the cooling of gases over long temperature ranges. The jet condenser referred to in Chap. 14 is also an example of direct contact as applied to condensation in which a large heat load may be condensed in an apparatus of small volume. A modification of the same principle might readily be applied to the condensation of organic vapors by a spray of water and particularly to the problem of condensing an oil vapor in the presence of a noncondensable gas. Future developments may well be anticipated for the recovery of atomic energy in commercial fission processes, in jet propulsion, and in gas-turbine cycles. This chapter treats the theory and develops the principal calculations applicable to direct-contact heat transfer.

Diffusion. If dry air at constant temperature is saturated by water at the same temperature in a direct-contact apparatus, the water vapor entering the air carries with it its latent heat of vaporization. The humidity of the air–water-vapor mixture increases during saturation because the vapor pressure of water out of the liquid is greater than it is in the unsaturated air and vaporization is the result. When the vapor pressure of water in the air equals that of the liquid, the air is saturated and vaporization ceases. The temperature of the water can be kept constant during air saturation if heat is supplied to it to replace that lost from it to the gas as latent heat of vaporization. Clearly, then, the heat transfer during the saturation of a gas with a liquid can be made to proceed without a temperature difference, although such a limitation is rarely encountered. It is seen, however, that there is a fundamental difference between this type of heat transfer and conduction, convection, or radiation.

When a movement of material is promoted between two phases by a vapor pressure (or concentration) difference, it is *diffusion* and is characterized by the fact that material is transferred from one phase to another or between both phases. This behavior is called *mass* or *material transfer* to set it apart from the ordinary concepts of heat transfer. While the phase-rule definitions apply to systems at equilibrium, if a phase is not homogeneous, *self-diffusion* may occur as the phase approaches homogeneity.

For the condensation of a vapor in the presence of a noncondensable gas it was expedient in Chap. 13 to introduce Colburn's analogy between heat transfer and mass transfer. This in turn was compared with Reynold's analogy between heat transfer and fluid friction as discussed in Chap. 3. The Reynold's analogy holds more closely for heat transfer and fluid friction than do the Colburn and other analogies for heat transfer and mass transfer. In the condensation of a vapor from a noncondensable gas an inert film near the tube wall retards the condensable vapor from reaching the condensate film at the tube wall. The rate at which the vapor passes through the inert film is a function of the pressure of the vapor in the gas body and in the condensate film adjoining the tube wall and follows the mechanism of diffusion.

Several of the illustrative examples solved in this chapter can be understood without a thorough background in the theory of diffusion, but the advantages of a good foundation are evident. The elements of simple diffusion are presented here in abbreviated form, drawing, wherever possible, upon the similarity of diffusion to convection heat transfer. Excellent treatments of the subject may be found in Sherwood[1] and Perry[2].

Diffusion Theory. Diffusion involves the passage of one fluid through another. Consider a gas, such as air, containing a small amount of acetone vapor which is soluble in water while the air may be considered insoluble in water. Suppose the air–acetone mixture is fed to a tower which has fresh water flowing continuously down its walls so that any acetone molecules which might bound into the water are removed by it from the gas body. How fast will acetone molecules be removed from the gas body?

An idealized picture of the problem is shown in Fig. 17.1. It may be assumed that a relatively stagnant air film forms at the liquid surface owing to the loss of momentum of the air molecules striking the liquid film and being dragged along by it. This is represented between 1-1 and 2-2′. The liquid film may also be considered relatively stagnant compared with the air body. This is the basis of the "two-film" theory. It may be conceded, because of the mutual solubility of acetone in water, that the rate at which acetone molecules can pass through the liquid film is exceedingly great. Thus the acetone molecules in the air film which arrive at the liquid film are depleted so rapidly by solution in the liquid film that the concentration of acetone molecules in the air film is less than it is in the gas body. This establishes a pressure or concentration

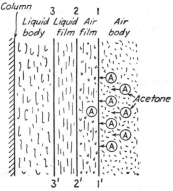

FIG. 17.1. Film theory showing principal resistances.

---

[1] Sherwood, T. K., "Absorption and Extraction," McGraw-Hill Book Company, Inc., New York, 1937.

[2] Perry, J. H., "Chemical Engineers' Handbook," 3d ed., Sec. 8, pp. 523–559, A. P. Colburn and R. L. Pigford, McGraw-Hill Book Company, Inc., New York. 1950.

gradient between the air body and the air film which continuously promotes the movement of acetone molecules in the direction of the liquid. This picture is analogous to that of the steady-state film theory of Chap. 3 in which a temperature gradient promoted heat transfer.

The air and liquid films are similar to thermal resistances in series. In the case of diffusion from 1-1' to 2-2' the concentration difference of the diffusing air-acetone mixture is the potential for the mass transfer of the acetone across the several resistances in series. Designating the diffusing gas by the subscript A and the inert or insoluble gas (air) by the subscript B the concentration of each gas may be expressed individually by its molar density, δ mols/ft^3. The rate at which the concentration of A in the air body decreases, $-d\delta_A$, depends upon four factors: (1) the number of mols of A, δ_A; (2) the number of mols of B, δ_B; (3) the relative difference of movement between the velocity of A, u_A and the velocity of B, u_B in the direction of diffusion; and (4) the length of the air film dl.

$$-d\delta_A = \alpha_{AB} \delta_A \delta_B (u_A - u_B)\, dl \tag{17.1}$$

where α_{AB} is the proportionality constant. If the net diffusion of the insoluble gas is zero, $u_B = 0$ and Eq. (17.1) reduces to

$$-d\delta_A = \alpha_{AB} \delta_A \delta_B u_A\, dl \tag{17.2}$$

If N_A is the number of pound-mols per hour transferred through A ft^2 of film surface,

$$\frac{N_A}{A} = u_A \delta_A \quad \text{lb-mol/(hr)(ft}^2) \tag{17.3}$$

$$-d\delta_A = \alpha_{AB} \delta_B \frac{N_A}{A}\, dl \tag{17.4}$$

and since

$$-d\delta_A = d\delta_B \tag{17.5}$$

$$N_A = \frac{A}{\alpha_{AB} \delta_B} \frac{d\delta_B}{dl} = \frac{A}{\alpha_{AB}} \frac{d\ln \delta_B}{dl} \tag{17.6}$$

and for equimolal diffusion of A into B.

$$\frac{N_A}{A} = -\frac{N_B}{A} = u_A \delta_A = -u_B \delta_B \tag{17.7}$$

From Eq. (17.1)

$$-d\delta_A = (\alpha_{AB} \delta_B u_A \delta_A - \alpha_{AB} \delta_A u_B \delta_B)\, dl = \alpha_{AB} u_A \delta_A (\delta_A + \delta_B)\, dl \tag{17.8}$$

$$u_A \delta_A = -\frac{1}{\alpha_{AB}(\delta_A + \delta_B)} \frac{d\delta_A}{dl} \tag{17.9}$$

DIRECT-CONTACT TRANSFER: COOLING TOWERS 567

From the continuity equation, input − output = accumulation. At steady state there will be no accumulation of A in the gas film. In pounds per cubic foot per unit time

$$-M_A \frac{\partial \delta_A}{\partial \theta} - M_A \frac{\partial (u_A \delta_A)}{\partial l} = 0 \tag{17.10}$$

where M_A is the molecular weight of gas A and θ is time.

$$\frac{\partial \delta_A}{\partial \theta} = \frac{\partial}{\partial l}\left[\frac{1}{\alpha_{AB}(\delta_A + \delta_B)} \frac{d\delta_A}{dl}\right] \tag{17.11}$$

Call $\delta_A + \delta_B = \delta$ the total number of mols per cubic feet.

$$\frac{\partial \delta_A}{\partial \theta} = \frac{\partial}{\partial l}\left(k_d \frac{\partial \delta_A}{\partial l}\right) \tag{17.12}$$

Defining the constant $k_d = 1/\alpha_{AB}\delta$ or

$$\alpha_{AB} = \frac{1}{k_d \delta} \tag{17.13}$$

k_d is the *diffusivity* or diffusion coefficient introduced in Chap. 13. It is basically defined, however, by Eqs. (17.12) and (17.13). Returning to Eq. (17.6),

$$N_A = k_d \, \delta A \, \frac{d \ln \delta_B}{dl} \tag{17.14}$$

And for a perfect gas, which the insoluble gas usually approximates,

$$\delta = \frac{n}{v} = \frac{p_t}{RT} \tag{17.15}$$

where n = total number of mols
p_t = total pressure
R = gas constant

$$N_A \, dl = \frac{k_d p_t A}{RT} d \ln \delta_B \tag{17.16}$$

Integrating over the length of the gas film from the gas body to the gas-film–liquid-film interface,

$$N_A = \frac{k_d p_t A}{RTl} \ln \frac{\delta_B}{\delta_{Bi}} = \frac{k_d p_t A}{RTl} \ln \frac{p_B}{p_{Bi}} \tag{17.17}$$

where the concentrations have also been expressed by the partial pressures and where the subscript B_i refers to the value at the interface 2-2′. For a driving force across the gas film consisting of the two pressures p_B and p_{Bi}, let p_{Bm} be the log mean driving pressure of the inert B. Then

$$p_{Bm} = \frac{p_B - p_{Bi}}{\ln p_B/p_{Bi}} \tag{17.18}$$

and

$$\ln \frac{p_B}{p_{Bi}} = \frac{p_B - p_{Bi}}{p_{Bm}} \tag{17.19}$$

$$N_A = \frac{k_d p_t A}{RTl p_{Bm}} (p_B - p_{Bi}) \tag{17.20}$$

This is similar to the ordinary transfer equation $Q = h_o A(T_c - t_w)$, using the nomenclature of Chap. 6, where $T_c - t_w$ is the temperature difference at a single cross section and the tube-wall temperature t_w corresponds to the interface concentration. Since the partial pressure of a component is proportional to its mol fraction in a mixture of perfect gases,

$$N_A = k_G A(p - p_i) = k_G A p_t (y - y_i) \tag{17.21}$$

where y and y_i are the mol fractions in the gas body and at the interface and

$$k_G = \frac{k_d p_t}{RTl p_{Bm}} \tag{17.22}$$

k_G may be compared with h for one of two fluids engaging in heat transfer.

The diffusivity can be determined experimentally by measuring the rate of evaporation of the diffusing gas A from a volumetric container into an inert which is passed over the vessel. Diffusivities can be computed by the Gilliland equation [Eq. (13.31)]. It will be noted that in Eq. (13.31) k_d is inversely proportional to the total pressure on the system p_t so that in Eq. (17.20) no correction need be made for pressure, since the product $k_d p_t$ in the numerator will be constant.

In an ordinary diffusion tower the potential difference $p - p_i$ or $y - y_i$ differs at every cross section or height of the tower as material is transferred out of the gas body. If the total transfer is the sum of a number of transfers through incremental surfaces with a differential potential on each increment, the differential equation for the entire height in which all the surface is contained becomes

$$dN_A = k_G(p - p_i) \, dA = k_G p_t (y - y_i) \, dA \tag{17.23}$$

Just as the total heat transfer Q can be calculated from a single individual film coefficient, the total surface, and the temperature difference between the fluid and the tube wall, so the material transfer can be determined from the change in the gas phase alone, using Eq. (17.21). The total material leaving the gas phase is obviously the same as that which enters the liquid phase. By a similar analysis to that above it can be shown that, for the liquid-film–liquid-body interface,

$$dN_A = k_L([c]_{Ai} - [c]_A) \, dA = k_L[c]_{av}(x_i - x) \, dA \qquad (17.24)$$

where $[c]$ is the concentration in the liquid with subscripts as above, x the mol fraction of the diffusing material in the liquid, and k_L is the liquid-side diffusion coefficient. Expressed as an equality,

$$dN_A = k_G p_t(y - y_i) \, dA = k_L[c]_{av}(x_i - x) \, dA \qquad (17.25)$$

k_G and k_L are the reciprocals of the two resistances in series comprising the gas and liquid films. In terms of overall coefficients and the overall potential for diffusion

$$\begin{aligned}dN_A &= K_G(p - p') \, dA = K_L([c]'_A - [c]_A) \, dA \\ &= K_G p_t(y - y') \, dA = K_L[c]_{av}(x' - x) \, dA\end{aligned} \qquad (17.26)$$

where p' = partial pressure of the diffusing vapor which would be in equilibrium with liquid of the concentration of the liquid body, atm

$[c]'_A$ = concentration of the diffusing vapor which would be in equilibrium with the partial pressure of the diffusing vapor in the gas body, lb-mols/ft^3

x' and y' = mol fractions corresponding to $[c]'_A$ and p', respectively, dimensionless.

The overall coefficients are expressed in either of the two ways

K_G = overall coefficient of mass transfer, lb-mols/(hr)(ft^2) (atmosphere of partial pressure difference)

K_L = overall coefficient of mass transfer, lb-mols/(hr)(ft^2) (concentration difference/ft^3)

K_G and K_L may be related to the individual films and to each other by means of Henry's law.

$$p' = C_H[c]'_A$$

where p' is the equilibrium partial pressure of A in the gas phase corresponding to a liquid concentration of $[c]'_A$ lb-mols/ft^3. C_H is the Henry's law proportionality constant. The law holds only for relatively dilute solutions.

Then

$$K_G = \frac{1}{(C_H/k_L) + (1/k_G)} \qquad (17.27)$$

$$K_L = \frac{1}{(1/k_L) + (1/C_H k_G)} \qquad (17.28)$$

K_G and K_L are really the same except for the dimensional differences in the respective equations in which they are employed. Equations (17.27) and (17.28) will be seen to resemble Eq. (6.7) for obtaining the overall coefficient of heat transfer from two individual film coefficients.

Colburn[1] has introduced the idea of a unit of mass transfer which is a measure of the number of interphase equilibrium changes required to effect a given amount of diffusion. It is identical with the concept of a theoretical plate in distillation under a particular condition. If G_m is the gas rate in mols per hour per square foot of tower cross section and dy is the change of concentration of the diffusing component, $dN_A = G_m\,dy$ and Eq. (17.26) can be written

$$G_m\,dy = K_G p_t(y - y')\,dA \tag{17.29}$$

If the surface per cubic foot of tower is a ft^2/ft^3, the total surface is $dA = a\,dV$, where V is the volume of the tower per square foot of tower cross section.

$$n_t = \int \frac{dy}{y - y'} = K_G a \frac{V}{G_m} \tag{17.30}$$

The integral of $dy/(y - y')$ over the entire height of tower gives the number of times the average potential may be divided into the total desired concentrate change. This is an index of the size of the absorption or desorption task which must be accomplished, and n_t is called the *number of transfer units*. When n_t is multiplied by G_m, it gives $K_G a V$, which is the total number of mols of material transferred. Different heights of various diffusion towers are required for the accomplishment of one transfer unit of diffusion depending upon how a particular tower is constructed and how much surface it contains per cubic foot of volume. From experiments on a particular type of tower of overall height Z it is possible to determine experimentally the number of transfer units of diffusion accomplished by it, and the height of a single transfer unit HTU will be

$$HTU = \frac{Z}{n_t} \tag{17.31}$$

The Wet-bulb Temperature. Humidification is a form of heat transfer as well as a form of diffusion. In air conditioning systems it is the increase in the moisture content of air and is usually accomplished by a spray washer. In a cooling tower air is also humidified but cold water and not moist air is its principal product. Particularly relative to the cooling tower, it is customary in the United States to define the moisture content of air by its wet-bulb temperature. This is a valuable concept,

---

[1] Colburn, A. P., *Trans. AIChE*, **35**, 211 (1934).

DIRECT-CONTACT TRANSFER: COOLING TOWERS

since it will be shown later that the wet-bulb temperature is also the lowest water temperature which can possibly be obtained by adiabatic humidification. The wet-bulb temperature is described here because it is a simple concept from which an excellent picture of simultaneous mass and heat transfer can be drawn. Refer to Fig. 17.2 which consists of a thermometer surrounded by a wick dipping into water at the same temperature as the surrounding air. The wick is always wet. A second thermometer is suspended in the surrounding air to indicate the *dry-bulb* temperature.

Assume that unsaturated air at any dry bulb is circulated over the wick. Because the wick is wet and the air is unsaturated, the partial pressure of the water vapor out of the wick is greater than that of the water vapor in the circulated air and water evaporates from the wick to the air. But the evaporation of water from the wick requires many Btu of latent heat, which must come from somewhere. When the air is initially circulated over the wick, the Btu could come from the wick itself by lowering the temperature of the wick below its initial temperature. If the initial wick temperature was the same as the dry bulb, any temperature drop in the wick would establish a temperature difference between the dry-bulb temperature and the lower temperature of the wick.

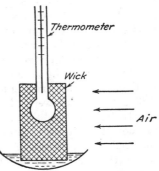

Fig. 17.2. The wet-bulb thermometer.

This causes sensible heat to flow from the air into the wick, thereby reducing the air temperature. An additional depression of the wick temperature continues as long as air is circulated, until a point is reached at which the temperature difference between the dry bulb and the wick causes just enough heat to flow into the wick to counterbalance the loss of heat from the wick by evaporation. Under this condition an equilibrium is established whereby the vaporization into a cubic foot or a pound of air is offset by the sensible heat removed from a cubic foot or a pound of air. Thus it should make no difference whether the air passes slowly or rapidly over the wick when the equilibrium is attained; the temperature of the wick should be depressed to the same extent, since the amount of water evaporating into each pound of dry air is offset by an equal sensible-heat removal from the same pound of air. The faster air is circulated over the wick, however, the greater will be the weight of water evaporated from the wick. This convenient temperature is called the *wet-bulb temperature*. The wet bulb naturally varies for air of a given dry bulb according to the extent to which it is saturated before contacting the wick, since the degree of saturation affects the amount of water vapor which can be evaporated into the air and consequently the amount of heat which can be removed from it.

The depression of the temperature of the wet-bulb thermometer, in the experiment described above, started when the water, presumably at the dry-bulb temperature, evaporated into the air at the same temperature. If the air had been saturated at the dry-bulb temperature, no water would have evaporated at all. If instead of humidifying air our primary interest was in a process for obtaining cold water by evaporating part of the water from the wick, the lowest water temperature attainable would be the wet-bulb temperature. Thus a cooling tower or any other humidifying apparatus which transferred no heat to or from its surroundings while the air and liquid were in contact could cool the water only to the wet-bulb temperature of the air, which is a function of the degree of saturation of the air or inert gas with the liquid being evaporated. In the northern part of the temperate zone, where a wet-bulb temperature of 75°F occurs in the summertime, it is impossible to cool water lower than 75°F except by refrigeration.

The process between the air and the wick can be described rather simply for the equilibrium condition when the wick has presumably reached the wet-bulb temperature. Let X be the absolute humidity of the original air, pounds of water per pound of dry air, which is related to the partial pressure of the water vapor in the original air by

$$X = \frac{p_{\text{water vapor}}}{p_t - p_{\text{water vapor}}} \times \frac{M_w}{M_a}$$

where M_w and M_a are the molecular weights of water and air. Let X' be the humidity (or vapor pressure) of the water out of the wick at the wet-bulb temperature. For simplicity, a diffusion overall mass-transfer coefficient K_X will be used in which the evaporative potential from the wick to the air is expressed in terms of the two humidities X' and X, whereas K_G employs the pressure in atmospheres to express the diffusion potential. Then if λ is the latent heat of vaporization, the heat leaving the wick per square foot of surface is given by

$$\frac{Q}{A} = K_X(X' - X)\lambda \qquad (17.32)$$

The heat of vaporization per square foot of wick is offset by the heat transferred from the air to the wick, which is given by

$$\frac{Q}{A} = h(t_{DB} - t_{WB}) \qquad (17.33)$$

where t_{DB} and t_{WB} are the dry-bulb and wet-bulb temperatures of the air, respectively, and h is the sensible-heat-transfer coefficient between the air and the wick.

Since Eqs. (17.32) and (17.33) are identities, and can be equated.

$$h(t_{DB} - t_{WB}) = K_X(X' - X)\lambda \qquad (17.34)$$

or

$$\frac{h}{K_X} = \frac{(X' - X)\lambda}{t_{DB} - t_{WB}} \qquad (17.35)$$

and the potential for K_X is pounds of water/pound of dry air. Using K_G for K_X the potential is atmospheres of water/atmosphere of air, which employs the *molar* ratio of water to air for a perfect gas. Weight and mol ratios are obviously not equal, although both are dimensionless. Thus a 1:1 mol ratio of water to air is equal to an $18/29$ or 0.62:1 weight ratio. The net effects of weight transfer must be the same as the product of molar transfer and the molecular weight of the diffusing vapor.

$$K_X(X' - X) = K_G M_A(p' - p) \qquad (17.36)$$

$$X' = \frac{p'}{p - p'}\frac{M_A}{M_B} = \frac{p'}{p_{Bm}}\frac{M_A}{M_B} = \frac{p'}{p_t}\frac{M_A}{M_B} \qquad (17.37)$$

The last is very nearly true for the temperature ranges in humidification

$$X = \frac{p}{p - p'}\frac{M_A}{M_B} = \frac{p}{p_{Bm}}\frac{M_A}{M_B} = \frac{p}{p_t}\frac{M_A}{M_B} \qquad (17.38)$$

$$K_X\left(\frac{p' - p}{p_{Bm}}\right)\frac{M_A}{M_B} = K_G M_A(p' - p) \qquad (17.39)$$

$$K_X = K_G p_{Bm} M_B \qquad (17.40)$$

Relation between the Wet-bulb and the Dew-point Temperatures. There is an interesting distinction between the wet-bulb and dew-point temperatures. Both are used in diffusion calculations. To obtain the wet bulb temperature the following is considered: As bone-dry air plus its accompanying vapor is circulated continuously over a wick at a given initial temperature, the air and vapor mixture are cooled by the passage of sensible heat from it into the wick. The air and vapor mixture get the heat back, however, in the form of vaporized water whose latent heat is equal to the heat which passed into the wick although the final temperature of the mixture is lower after circulation. The volume or weight of vapor per pound of *dry* air increases. Since no heat enters or leaves the system during the direct contact, it is an adiabatic process, and the wet bulb is obviously a temperature in such an adiabatic process.

The dew point is the temperature at which a gas of given vapor content deposits the first drop of condensate when cooled by a *constant-*

pressure process. In the temperature-entropy plot of Fig. 17.3 the saturation curve for water is described by *BEF*. The constant pressure cooling of water vapor occurs along the constant pressure path *AB*. Consider next the adiabatic contact between water and air. Being adiabatic, it is described by a fictitious path $A''C'''$, which is an adiabatic although not isentropic, since the mixture of air and water vapor is irreversible. The cooling of the air and evaporation of the water separately are not true adiabatic processes. Only the gross effects between the two are truly adiabatic. For the process of adiabatic saturation the path of the water vapor *alone* can be determined by experiment and is found to be along the line *AC*. The point *C* is the wet bulb, and since water vapor is added during the adiabatic saturation the partial pressure of the water vapor at the wet bulb *C*, is somewhat higher than *B*. When air is initially 70 per cent saturated or more, the wet bulb exceeds the dew point by less than 2 per cent of the difference between the dry bulb and the wet bulb. If gas at the dry-bulb temperature T_A possesses a higher degree of saturation, the point *A* moves to *A'* and the dew point is at the obviously higher temperature T'_B.

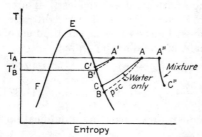

Fig. 17.3. Relation between wet-bulb and dew-point temperatures.

The Lewis Number. The relationship between the heat-transfer coefficient and the mass-transfer coefficient, h/K_x, enters into nearly every direct-contact problem. The analogy between heat transfer and fluid friction in Chap. 3 was arrived at by mathematical similarity. Heat transfer and mass transfer may also be related by comparing both with fluid friction. References to the subject will be found in Chap. 13. Only the most general application is developed here, although it appears that greater accuracies may be obtained from other analogies. For heat transfer Chilton and Colburn give

$$j_h = \frac{f}{2} = \frac{hD/k}{(DG/\mu)(c\mu/k)}\left(\frac{c\mu}{k}\right)^{\frac{2}{3}} = \frac{h}{cG}\left(\frac{c\mu}{k}\right)^{\frac{2}{3}}$$

where f is the dimensionless friction factor.

For mass transfer (diffusion)

$$j_d = \frac{f}{2} = \frac{K_x D/\rho k_d}{(DG/\mu)(\mu/\rho k_d)}\left(\frac{\mu}{\rho k_d}\right)^{\frac{2}{3}} = \frac{K_x}{G}\left(\frac{\mu}{\rho k_d}\right)^{\frac{2}{3}} \qquad (17.41)$$

Dividing,

$$\frac{j_d}{j_h} = \frac{K_xc(\mu/\rho k_d)^{2/3}}{h(c\mu/k)^{2/3}} = 1 \qquad (17.42)$$

or

$$\frac{h}{K_xc} = \frac{(\mu/\rho k_d)^{2/3}}{(c\mu/k)^{2/3}} = \left(\frac{k}{c\rho k_d}\right)^{2/3} = 1 \qquad (17.43)$$

and

$$\frac{h}{K_xc} = 1 \qquad (17.44)$$

Lewis[1] called attention to this relation, and the dimensionless group h/K_xc is known as the Lewis number Le. It conveys the extremely important information that the heat-transfer coefficient is to the mass-transfer coefficient as the value of the specific heat of the medium which serves for both heat transfer and mass transfer. For the case of the wick

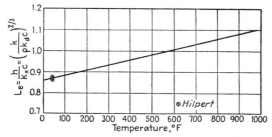

FIG. 17.4. Lewis numbers for the air-water system.

in a wet-bulb experiment if the mean specific heat of the moist air is 0.25 Btu/(lb)(°F), then the heat-transfer rate will be but one-quarter of the mass-transfer rate. It thus requires four times as great a potential for a given number of Btu to be transferred by sensible heat as for an equal number of Btu by mass transfer. The plot of Lewis number vs. temperature for the system air-water calculated from $Le = (k/\rho k_d c)^{2/3}$ is shown in Fig. 17.4. Two points obtained experimentally by Hilpert[2] are also included in the plot. Values of h/K_x for other systems are tabulated in Table 17.1.

The Effective Film. In the wet-bulb experiment it was seen that the same wet-bulb temperature would be attained regardless of the rate of gas flow over the wick. It has also been shown in Chap. 11 that the rate of heat transfer varies with the 0.8 power of the gas velocity. The same should be true for mass transfer. In the Eq. (17.44), if the ratio of the

[1] Lewis, W. K., *Trans. ASME*, **44**, 329 (1922).
[2] Hilpert, R., *Forschungsheft*, **3**, 355 (1932).

coefficients is constant, the film for both heat transfer and mass transfer must be affected to the same degree by a change in the gas rate. Sherwood[1] has presented a correlation for the vaporization of numerous liquids into air, including water, and obtained

$$\frac{D}{l} = 0.023 \left(\frac{DG}{\mu}\right)^{0.83} \left(\frac{\mu}{\rho k_d}\right)^{0.44} \qquad (17.45)$$

where D is the inside diameter of the tube and l is the thickness of the film. This may be compared with Eq. (6.2) of Sieder and Tate for heat transfer:

$$\frac{h_i D}{k} = 0.027 \left(\frac{DG}{\mu}\right)^{0.8} \left(\frac{c\mu}{k}\right)^{\frac{1}{3}} \left(\frac{\mu_w}{\mu}\right)^{0.14}$$

TABLE 17.1. AVERAGE VALUES OF h/K_X CALCULATED FROM WET-BULB DETERMINATIONS*
Diffusion into Air

Vapor	$\dfrac{h}{K_X}$	Calculated from Chilton-Colburn analogy, Eq. (17.43)
Benzene................	0.41	0.44
Carbon tetrachloride.....	0.44	0.49
Chlorobenzene..........	0.44	0.48
Ethyl acetate...........	0.42	0.46
Ethylene tetrachloride....	0.50	0.51
Toluene	0.44	0.47
Water.................	0.26	0.21

* Sherwood, T. K., *Trans. AIChE*, **28**, 107 (1932).

when $(\mu/\mu_w)^{0.14}$ is substantially 1.0 for gases. Other interesting extentions of the theory have been made by Arnold[2] and Chilton and Colburn.[3]

HUMIDIFICATION AND DEHUMIDIFICATION

Humidifiers: The Cooling Tower. The largest present-day uses of diffusional heat transfer are found in the cooling tower, the air-conditioning spray chamber, the spray drier, the spray tower, and the spray pond. The use of cooling towers has grown tremendously in the last twenty years owing to an increasing necessity. In many industrial localities cold fresh water is too scarce to permit its unlimited use as a cooling medium. The problem of supplying sufficient surface and subsurface cooling water has grown to such an extent that new plants are often

---

[1] Sherwood, *op. cit.*, p. 39.
[2] Arnold, J. H., *Physics*, **4**, 255 (1933).
[3] Chilton, T. H., and A. P. Colburn, *Ind. Eng. Chem.*, **26**, 1183 (1934).

required to develop the continual reuse of the limited water they may obtain from public or private sources. In some communities even river water, which may be present in abundance, requires precooling as discussed in Chap. 7. This is especially true of some of the rivers in the southern part of the United States having northern sources and which are heated to the dry-bulb temperature by the time they reach the South.

The available temperature of cooling water has been shown to be an important economic factor in the design of modern chemical and power plants. In the chemical plant it fixes the operating pressure on the condensers of distillation and evaporation processes and consequently on the equipment preceding them. In the power plant it fixes the turbine- or engine-discharge pressure and the ultimate recovery of heat. For these vital reasons the study of the cooling tower and the temperature of the water made available by it is of great importance in the planning of a process. The cooling tower is also the simplest member of a class of apparatus whose potentialities have scarcely been realized.

Classification of Cooling Towers. Modern cooling towers are classified according to the means by which air is supplied the tower. All employ stacked horizontal rows of fill to provide increased contact surface between the air and water. In *mechanical-draft* towers the air is supplied in either of the two ways shown in Fig. 17.5a and b. If the air is sucked into the tower by a fan at the top through louvers at the bottom, it is *induced draft*. If the air is forced in by a fan at the bottom and discharges through the top, it is *forced draft*. *Natural-circulation* towers are of two types, *atmospheric* and *natural draft* as in Fig. 17.5c and d.

Mechanical-draft Towers. Towers of this class are the commonest erected in the United States at present, and of these the vast majority are now induced-draft towers. The trend toward the induced-draft tower has been pronounced only in the last ten years, but it represents a logical transition, since there are advantages to its use which exceed all others except under very special conditions. In the forced-draft type as in Fig. 17.5b the air enters through a circular fan opening and a relatively large ineffective height and volume of tower must be provided as air-inlet space. The air distribution is relatively poor, since the air must make a 90° turn while at high velocity. In the induced-draft tower, on the other hand, the air can enter along one or more entire lengths of wall, and as a result the height of tower required for air entry is very small.

In the forced-draft tower the air is discharged at low velocity from a large opening at the top of the tower. Under these conditions the air possesses a small velocity head and tends to settle into the path of the fan intake stream. This means that the fresh intake air is contaminated by partially saturated air which has already passed through the tower before. When this occurs, it is known as *recirculation* and reduces the performance capacity of the cooling tower. In the induced-draft tower the air discharges through the fan at a high velocity so that it is driven up into the natural air currents which prevent it from settling at the air intake. In induced-draft towers, however, the pressure drop is on the intake side of the fan, which increases the total fan-power requirements. The higher velocity of discharge of the induced-draft tower also causes a

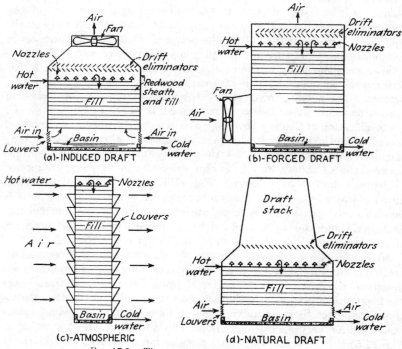

FIG. 17.5. The common types of cooling towers.

somewhat greater entrainment or *drift* loss of water droplets carried from the system by the air stream.

Natural-circulation Towers. In use in Europe and the Orient, the popularity of the types in Fig. 17.5c and d is declining in the United States. This is particularly true of the natural-draft tower.

The atmospheric tower avails itself of atmospheric wind currents. Air blows through the louvered sides in one direction at a time, shifting with the time of year and other atmospheric conditions. In exposed places having average wind velocities of 5 or 6 mph the atmospheric tower may prove to be the most economical type, and where power costs are high, it may even be preferable with wind velocities as low as 2½ to 3 mph. Since the atmospheric currents must penetrate the entire width of the tower, the towers are made very narrow by comparison with other types and must be very long to afford equal capacity. Towers of this type have been built which are over 2000 ft long. Drift losses occur over the entire side and are greater than for other types. These towers make less efficient use of the available potential, since they operate in crossflow whereas it was shown in Chap. 16 that the most effective use of potentials occurs in counterflow. When cooling water is desired at a temperature close to the wet bulb, this type is incapable of producing it. Atmospheric towers are consequently extremely large and have high initial costs, and when the air is becalmed, they may cease to operate. They have one great advantage, however, in that they eliminate the principal operating cost of mechanical draft towers, i.e., the cost of fan power. In areas with low average wind velocities the fixed charges and pumping costs offset the advantage. Representative average July wind velocities in the United States are shown in Fig. 17.6. An average velocity exceeding 5 or 6 mph is not a sufficient indication that an atmospheric tower is the best. With an average wind velocity of 5 mph the tower will operate at less than design capacity for part of the time. The placement of the tower in a 5-mph wind locality must be such that it is free of obstruction and can avail itself very fully of existing currents.

Natural-draft towers operate in the same way as a furnace chimney. Air is heated in the tower by the hot water it contacts, so that its density is lowered. The difference between the density of air in the tower and outside it causes a natural flow of cold air in at the bottom and the rejection of less dense, warm air at the top. Natural-draft towers must be tall for sufficient buoyancy and must have large cross sections because of the low rate at which the air circulates compared with mechanical draft. Natural-draft towers consume more pumping power. However, they eliminate the cost of fan power and may be more reliable in some localities than are atmospheric towers. In atmospheric towers emphasis must be placed on wind characteristics. In natural-draft towers primary consideration must be given to the temperature characteristics of the air. If it is customary for the air to rise to a high temperature during the day, at least relative to the hot-water temperature, the natural-draft tower will cease to operate during the hot portion of the day. The initial cost and fixed charges on these towers are rather great, and they seem to be passing out of use.

Closely allied to natural-circulation cooling towers is the spray pond consisting of a number of up-spray nozzles, which spray water into the

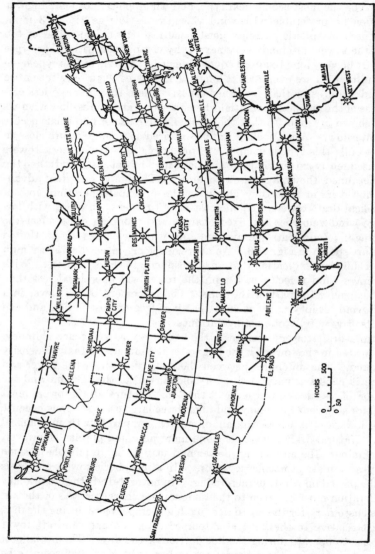

Fig. 17.6. Summer wind data. The length of each line indicates the number of hours in a normal July that the wind blows toward the center circles at each of 71 stations. Figures in circles are average July velocities in miles per hour. (*The Marley Company.*)

air without inducing air currents. These do not operate with an orderly attempt at air flow and consequently are not capable of producing water approaching the wet bulb as closely as cooling towers. Where the water must be cooled over a short range and without a close approach to the wet-bulb temperature, spray ponds may provide the most economical solution to a water-cooling problem. Drift losses in spray ponds are relatively high. Spray towers are also used widely. They are similar to atmospheric towers except that they employ little or no fill.

Cooling towers are occasionally equipped with bare-tube bundles, which are inserted just above the water basin at the bottom of the tower. These are referred to as *bare-tube* or *atmospheric coolers*. The primary cooling water flows inside the cooler while tower water is continuously circulated over it. The primary cooling water is thus contained in a totally closed system. The calculation of this type of apparatus will be treated in Chap. 20.

Cooling-tower Internals and the Role of Fill. If water passes through a nozzle capable of producing small droplets, a large surface becomes available for air-water contact. Since the water-air interface is also the heat-transfer surface, the use of a nozzle permits the attainment of considerable performance per cubic foot of contact apparatus. This is the principle of the spray pond and the spray tower. Consider a hypothetical spray tower as shown in Fig. 17.7a. Liquid fed to it falls through by gravity. If the tower is 16 ft high and no initial velocity is imparted to the droplet, it falls in approximate accord with the free-fall law, $z = \frac{1}{2}g\theta^2$, where z is the height, g the acceleration of gravity, and θ the time. A droplet of water will fall through the height of the tower in 1 sec. If liquid is fed at the rate of one droplet per second and there is no obstruction, one droplet will always be present in the tower and one droplet may be continuously removed from the tower per second. The effective surface in the tower of Fig. 17.7a is that of one droplet.

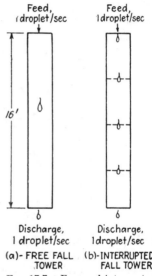

Fig. 17.7. Free and interrupted fall.

Now suppose that by introducing some geometrical forms on which a droplet may impinge or be deflected it is possible to make a droplet take 4 sec to fall through the height of the tower. Then as shown in Fig. 17.7b, *one* droplet is fed per second at the top and *one* droplet is continuously removed at the bottom but four droplets remain in the tower. The effective surface in the latter is that of four droplets or four times the free-fall surface. The function of fill is to increase the available surface in the tower either by spreading the liquid over a greater surface or by retarding the rate of fall of the droplet surface through the apparatus. In ordinary diffusion towers such as chemical absorbers the packing is introduced in the form of Raschig rings, Berl saddles, or other objects which are very compact and provide a surface on which the liquid spreads and exposes a large film. This is *film surface*. In the cooling tower, because of the requirement of a large air volume and small allowable pressure drop, it is customary to use spaced wooden slats of triangular or rectangular cross section, leaving the tower substantially unob-

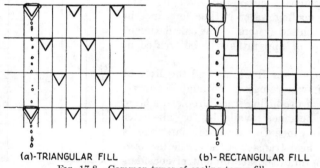

(a)-TRIANGULAR FILL   (b)- RECTANGULAR FILL
Fig. 17.8   Common types of cooling-tower fill.

structed. The packing, or *fill*, in a cooling tower is almost entirely fabricated in either of the forms of Fig. 17.8, and its purpose is to interrupt the descent of liquid. Although the free space between adjoining fill slats is relatively large, the horizontal projection of the fill does not permit the droplets of liquid to fall through the tower without impinging repeatedly on successively lower slats. Some of the liquid striking the top of a slat spatters, but a large part flows about it and breaks into turbulent flow at the bottom so as to form new droplets automatically and recreate new *droplet surface*. The surface on the sides of the fill is comparable to film surface in packed absorbers. There is, in addition, a large amount of *droplet* surface. More recently the trend has been toward the use of small rectangular slats. They are considerably cheaper to fabricate and install than those of triangular cross section and cause lower pressure drops. The mechanism of producing droplets below each horizontal row relies upon the draining liquid breaking into turbulent flow. Consequently, the method by which the droplets are formed at the top of the tower is of little consequence to the overall formation of surface as long as there is a uniform distribution of liquid over the entire cross section of the tower.

To demonstrate the effectiveness of these types of fill the free-fall analysis can be carried still further. According to the free-fall equation a droplet will fall 16 ft in 1 sec. In $\frac{1}{2}$ sec a droplet with zero velocity at the top falls approximately 4 ft, and its average velocity is 8 fps. In the second $\frac{1}{2}$ sec it falls the remaining 12 ft with an average velocity of 24 fps. The droplets traverse the lower three-quarters of the tower so quickly that the time of contact in the last three-quarters of the tower equals only that of the first quarter. The advantage of interrupted fall thus becomes apparent: Each time the fall is interrupted (say at each quarter of the tower in Fig. 17.7b) it is as if a droplet with zero velocity were starting to fall anew, and the interrupted tower is equivalent to the effective first quarters of four spray towers in series.

In most cooling towers the liquid is introduced by spraying the water upward at the top so that it travels up and then down before striking the first row of fill. This provides effective contact inexpensively, since the velocity of a droplet traveling upward must decrease to zero when its direction reverses. Another means of increasing the surface or time of contact in spray towers and cooling towers

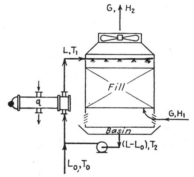

Fig. 17.9. Cooling-tower water and air flow.

is by atomizing the water instead of forming droplets. The same can be achieved with a nozzle instead of an atomizer by using a water-soluble wetting agent. This is not very practical, however, since very fine droplets cannot be caught in the drift eliminators without the expenditure of a large pressure drop. In the construction of cooling towers it is customary to employ droplets of such a size that drift losses can be guaranteed not to exceed 0.25 per cent of the total water circulated over the tower. With the expenditure of a greater fan power this loss can be reduced to 0.10 per cent.

The Heat Balance. Consider the flow diagram in Fig. 17.9. It consists of a cooling tower operating with a heat source in a closed circuit. Water from the basin of the tower is pumped through a battery of surface condensers, where its temperature is raised. The heated water passes back onto the tower along with make-up water, which compensates for the evaporation from the system owing to the saturation of the air passing through the tower.

For convenience, cooling towers are analyzed on the basis of 1 ft² of tower internal ground area. The air rate or *air loading* is taken as the *bone-dry* gas rate per unit of ground area G lb/(hr)(ft²). The *water loading* at the top of the tower is the water rate per unit of ground area L lb/(hr)(ft²), and the make-up rate is L_0 lb/(hr)(ft²). The heat load per hour per square foot q is accordingly the total hourly rate Q divided by the internal ground dimensions of the tower. Using the subscript 1 for the inlet and 2 for the outlet, the balance around the tower in terms of the gas for a 0°F datum is

$$q + L_0 C T_0 = G(H_2 - H_1) \tag{17.46}$$

where C is the specific heat of water and H is the enthalpy per pound of bone-dry air and including the heat of the vapor associated with the pound of bone-dry air. In terms of the water loading the heat balance is

$$q = LC(T_1 - T_2) + L_0 C(T_2 - T_0) \tag{17.47}$$

Combining both equations,

$$G(H_2 - H_1) = LC(T_1 - T_2) + L_0 C T_2 \tag{17.48}$$

The quantity of make-up water required to compensate for evaporation is

$$L_0 = G(X_2 - X_1) \tag{17.49}$$

Dividing Eq. (17.48) by Eq. (17.49)

$$\frac{G(H_2 - H_1)}{G(X_2 - X_1)} = \frac{LC(T_1 - T_2) + L_0 C T_2}{L_0} \tag{17.50}$$

$$L_0 \frac{H_2 - H_1}{(X_2 - X_1)} = LC(T_1 - T_2) + L_0 C T_2 \tag{17.51}$$

Combining Eq. (17.51) with Eq. (17.47)

$$LC(T_1 - T_2) + L_0 C T_2 = q + L_0 C T_0 \tag{17.52}$$

$$L_0 = \frac{q}{(H_2 - H_1)/(X_2 - X_1) - C T_0} \tag{17.53}$$

Equation (17.53) gives the quantity of make-up corresponding to any fixed terminal conditions. The enthalpy for saturated air appears on ordinary psychrometric charts. Caution is suggested in the use of these charts, since they invariably have differing datums. Some employ a 0°F air datum and a 32°F water vapor datum to permit the use of the Keenan and Keyes steam tables (Table 7). When a single chart is used, the enthalpy differences are all of sufficient accuracy for heat-transfer calculations. When part of the data is taken from one chart and part from another, serious error may result. For the solution of all air-water mixtures in this chapter the enthalpy of saturated air above 0°F can be computed by

$$H' = X't + X'\lambda + 0.24t \qquad (17.54)$$

where 0.24 is the specific heat of air.

For unsaturated air

$$H = Xt_{DP} + X\lambda_{DP} + 0.45X(t - t_{DP}) + 0.24t \qquad (17.55)$$

where 0.45 is the specific heat of water vapor from 0 to 120°F and the subscript DP refers to the dew point.

Example 17.1. Calculation of the Enthalpy of Saturated Air. What is the saturation enthalpy of air at 75°F?

At 75°F the saturation partial pressure of water is 0.4298 psia (Table 7).

Humidity, $X' = \dfrac{p_w}{p_t - p_w} \dfrac{M_w}{M_a} = \dfrac{0.4298}{14.696 - 0.4298} \times \dfrac{18}{29} = 0.0187$ lb water/lb air

Enthalpy above 0°F, $H' = 0.0187 \times 75 + 0.0187 \times 1051.5 + 0.24 \times 75$
$= 39.1$ Btu/lb dry air $\qquad (17.54)$

The enthalpies of Tables 17.2 and 17.3 have been computed in this manner.

TABLE 17.2. ENTHALPIES AND HUMIDITIES OF AIR-WATER MIXTURES AT 14.7 PSIA

Temp, °F	Vapor pressure, psia	Humidity, lb H₂O/lb air	Enthalpy, Btu/lb air	v air, ft³/lb	v air + H₂O, ft³/lb
40	0.1217	0.005	15.15	12.59	12.70
45	0.1475	0.0063	17.8	12.72	12.85
50	0.1781	0.0076	20.5	12.84	13.00
55	0.2141	0.0098	23.8	12.97	13.16
60	0.2563	0.0110	26.7	13.10	13.33
65	0.3056	0.0130	30.4	13.23	13.51
70	0.3631	0.0160	34.5	13.35	13.69
75	0.4298	0.0189	39.1	13.48	13.88
80	0.5069	0.0222	44.1	13.60	14.09
85	0.5959	0.0262	50.0	13.73	14.31
90	0.6982	0.0310	56.7	13.86	14.55
95	0.8153	0.0365	64.2	13.99	14.81
100	0.9492	0.0430	72.7	14.11	15.08
105	1.1016	0.0503	82.5	14.24	15.39
110	1.2748	0.0590	93.8	14.36	15.73
115	1.4709	0.0691	106.7	14.49	16.10
120	1.6924	0.0810	121.5	14.62	16.52
125	1.9420	0.0948	138.8	14.75	16.99
130	2.2225	0.1108	158.5	14.88	17.53
135	2.5370	0.1300	181.9	15.00	18.13
140	2.8886	0.1520	208.6	15.13	18.84
145	3.2810	0.1810	243.8	15.26	19.64
150	3.7180	0.2160	286.0	15.39	20.60

TABLE 17.3. ENTHALPIES AND HUMIDITIES AT VARIOUS ELEVATIONS
Elevations are feet above sea level
Humidity, X', lb water/lb dry air; enthalpies, H', Btu/lb dry air

Temp., °F	2000		4000		6000		8000	
	X'	H'	X'	H'	X'	H'	X'	H'
40	0.0056	15.8	0.0061	16.4	0.0065	16.8	0.0070	17.3
45	0.00680	18.4	0.0073	18.9	0.0079	19.6	0.0085	20.3
50	0.0082	21.1	0.0088	21.8	0.0095	22.6	0.0103	23.4
55	0.0099	24.3	0.0106	25.0	0.0115	26.0	0.0124	27.0
60	0.0119	27.7	0.0128	28.7	0.0138	29.8	0.0149	31.0
65	0.01420	31.5	0.0153	32.8	0.0165	34.1	0.0178	35.6
70	0.0170	35.9	0.0183	37.3	0.0197	38.9	0.0212	40.6
75	0.0202	40.7	0.0217	42.2	0.0234	44.4	0.0253	46.5
80	0.0239	46.1	0.0253	47.7	0.0278	50.5	0.0300	53.0
85	0.0284	52.5	0.0306	55.0	0.0330	57.7	0.0358	60.9
90	0.0334	59.4	0.0361	62.4	0.0390	65.7	0.0425	69.6
95	0.0394	67.5	0.0425	71.0	0.0460	75.0	0.0499	80.4
100	0.0465	76.8	0.0501	81.0	0.0544	85.7	0.0590	91.0
105	0.0545	87.2	0.0590	92.4	0.0640	98.1	0.0695	104.3
110	0.0645	99.9	0.0692	105.4	0.0751	111.9	0.0819	119.8
115	0.0750	113.4	0.0812	120.6	0.0885	128.8	0.0964	137.8
120	0.0880	129.8	0.0955	130.3	0.1040	147.0	0.1137	158.8
125	0.1029	148.0	0.1120	158.5	0.1220	170.0	0.1340	183.5
130	0.1208	169.8	0.1317	182.4	0.1440	196.7	0.1580	212.7
135	0.1412	195.1	0.1548	210.6	0.1698	227.9	0.1870	247.6
140	0.1665	225.7	0.1830	244.8	0.2010	265.5	0.2220	289.7
145	0.1965	261.8	0.2190	287.8	0.2420	314.8	0.2690	345.8
150	0.2320	304.5	0.2560	332.3	0.2842	365.0	0.3180	404.0

Heat Transfer by Simultaneous Diffusion and Convection. In the wet-bulb experiment the air and water were initially at the same temperature, although this was not a necessary requirement. By simply referring to Fig. 17.3 it can be seen that the equilibrium represented by the wet-bulb temperature is influenced by the absolute vapor content of the air.

In a cooling tower hot water is cooled by cold air. When water passes through the tower, the water temperature may fall below the dry bulb of the *inlet* air but not lower than the wet bulb of the inlet air. Consider the tower to be divided into two portions. In the upper portion hot water contacts outlet air which is colder than the water. Unlike the wet-bulb experiment, in this case the partial pressure of the water out of the liquid is greater than that in the exit air while the temperature of the water is *also* greater than that of the exit air. *Both* potentials work to decrease the water temperature by evaporation and sensible transfer to the air, thereby increasing the air enthalpy. In this manner, depending upon the amount of air and the amount of evaporation it is possible for the temperature of the water to fall to or *below* the dry bulb of the inlet air before reaching the bottom of the tower where the air enters. It is the fact that both potentials may operate adiabatically in the same direction while saturating air which makes the cooling tower so effective for cooling water. In the lower portion of the tower the water may possess a temperature equal to or lower than the dry bulb of the air it contacts, and sensible heat and mass transfer are in opposite directions, identical with the wet-bulb experiment. The limit to which the outlet water temperature may then fall in a cooling tower is that which is adiabatically in equilibrium with the inlet air, namely, the wet bulb.

The derivation of the performance of a cooling tower given below is essentially that of Merkel.[1] Since the total heat transfer in a cooling tower is the passage of heat by diffusion and convection from the water to the air,

$$q = q_d + q_c \qquad (17.56)$$

where q_d Btu/(hr)(ft^2) is the portion transferred by diffusion and q_c Btu/(hr)(ft^2) is the portion transferred by convection. In the definition of q it should be remembered that the area implied by its dimensions is the ground area of the tower and not the heat-transfer surface.

If λ is the average latent heat of vaporization of all the water vaporized in the tower,

$$q_d = L_0 \lambda \qquad \text{(nearly)} \qquad (17.57)$$

Combining with Eq. (17.47)

$$q_c = LC(T_1 - T_2) + LC(T_2 - T_0) - L_0 \lambda \qquad (17.58)$$

and

$$\frac{q_c}{q_d} = \frac{LC(T_1-T_2)+L_0C(T_2-T_0)-L_0\lambda}{L_0\lambda} = \frac{LC(T_1-T_2)+L_0C(T_2-T_0)}{L_0\lambda} - 1$$

$$(17.59)$$

But

$$L_0 = G(X_2 - X_1) \qquad (17.60)$$

and

$$G(H_2 - H_1) = LC(T_1 - T_2) + L_0 C(T_2 - T_0) \qquad (17.48)$$

from which

$$\frac{q_c}{q_d} = \frac{1}{\lambda}\left(\frac{H_2 - H_1}{X_2 - X_1}\right) - 1 \qquad (17.61)$$

This is an interesting relationship, since it states that the ratio of the heat transferred by convection to that transferred by diffusion, both in the

[1] Merkel, F., *Forschungsarb.*, **275**, 1-48 (1925).

same direction, is fixed by the inlet and outlet air conditions, which are either known or can be obtained by calculation. While Eq. (17.61) establishes the quantities of heat transfer by convection and diffusion, the ratio of the rates of heat and mass transfer has been set by the Lewis number.

Based on the overall rather than the individual films, the sensible-heat transfer from the water at temperature T to the air at temperature t is given by

$$dq_c = h(T - t)a\, dV \tag{17.62}$$

where a is the surface of the water per cubic foot of tower as both droplet and film surface and dV is the differential tower volume in which the surface exists. From this $a\, dV = dA$, where A is the heat-transfer surface. If c is the humid heat of the air defined by $c = 0.24 + 0.45X$,

$$dq_c = Gc\, dt \tag{17.63}$$
$$dq_d = \lambda\, dL \tag{17.64}$$

Since dL is the rate at which material diffuses, the differential form of Eq. (17.26) for weight flow is

$$\tfrac{1}{18} dL = K_G(p' - p)a\, dV \tag{17.65}$$

where p' is the partial pressure of water corresponding to the water temperature T and p is the vapor pressure in the air. For all practical purposes the humidity can be considered proportional to the partial pressure, at least in the range encountered in cooling-tower applications. Equation (17.65) becomes

$$dL = K_X(X' - X)a\, dV \tag{17.66}$$

where X' is the humidity at the water temperature T and X the humidity of the air.

Substituting in Eq. (17.64),

$$dq_d = K_X\lambda(X' - X)a\, dV \tag{17.67}$$

The water evaporated dL increases the humidity of the air stream above its value at the inlet by

$$dL = G\, dX \tag{17.68}$$

The combined heat transfer dq is then the sum of both modes of transfer

$$dq = dq_c + dq_d = h(T - t)a\, dV + K_X\lambda(X' - X)a\, dV \tag{17.69}$$

and

$$dq = G\, dH \tag{17.70}$$

Equation (17.70) is useful if it can be combined with Eq. (17.69), since it expresses the total heat transfer in the system in heat units alone. In order to avoid the appearance of both X and H in the same equation, the

X values can be factored out. Using an average value for the humid heat c and the latent heat λ and neglecting superheat, all of which are permissible in the relatively close range in which the cooling tower operates, for an air-water mixture consisting of 1 lb of air and X lb of water vapor,

$$H = 1ct + \lambda X \tag{17.71}*$$

$$G\,dH = G(c\,dt + \lambda\,dX) \tag{17.72}$$

Regrouping Eq. (17.69),

$$dq = K_x a\,dV\left[\left(\frac{hT}{K_x} + \lambda X'\right) - \left(\frac{ht}{K_x} + \lambda X\right)\right] \tag{17.73}$$

Add and subtract $c(T - t)$.

$$dq = K_x a\,dV\left[(cT + \lambda X') - (ct + \lambda X) + c(T - t)\left(\frac{h}{K_x c} - 1\right)\right] \tag{17.74}$$

Substitute Eq. (17.71) into Eq. (17.74).

$$dq = K_x a\,dV\left[(H' - H) + c(T - t)\left(\frac{h}{K_x c} - 1\right)\right] \tag{17.75}$$

dq can be expressed in terms of the enthalpy decrease of the total water quantity or the enthalpy increase of the total air mixture, both of which are equal.

$$dq = d(LCT) = G\,dH \tag{17.76}$$

The gas loading G remains constant throughout the tower because it is based on the bone-dry gas only. The liquid loading is not quite constant, however, owing to the evaporation of water into the bone-dry air. The saturation loss from water to air amounts to less than 2 per cent of the water circulated over the tower and may be considered constant without introducing a serious error.

Hence

$$d(LCT) = LC\,dT \tag{17.77}$$

and

$$LC\,dT = G\,dH \tag{17.78}$$

From Table 17.1, for water diffusing into air the Lewis number is approximately $h/K_x c = 1$, and the last term of Eq. (17.75) drops out, thus

$$LC\,dT = G\,dH = K_x(H' - H)a\,dV \tag{17.79}$$

By introducing the mass-transfer-rate equivalent $K_x = h/c$ both modes of heat transfer can be combined by the use of either coefficient, whichever is the easier to obtain. Actually $h/K_x c$ for the system air-water is not 1.0 as predicted by the Lewis number. The data of several investi-

* This is an approximation for Eqs. (17.54) and (17.55).

gators indicate that the value of the Lewis number is nearer 0.9. In practice Eq. (17.79) is always evaluated from diffusion potentials, which means that only the convection heat-transfer coefficient is in error if the theoretical value of the Lewis number is used. In most cooling-tower services the convection-heat transfer alone represents less than 20 per cent of the total heat load. The development of equations without the convenient simplification that the Lewis number equals 1.0 will be treated later.

The Analysis of Cooling-tower Requirements. Equation (17.79) is the key equation for the calculation of the design and the analysis of performance in cooling towers. K_x is the overall transfer-rate term analogous to U_C in exchangers, and it should be remembered that there is no dirt factor in direct-contact heat transfer. However, in tubular exchangers the heat-transfer surface is usually known or can be readily computed. In the cooling tower the value of a cannot be determined directly, since it is composed of a random disposition of droplet and film surface. The film surface is nearly independent of the thickness of the film, while the droplet surface is dependent upon both the portion of the liquid loading forming droplets and the average size of the droplets formed. In a tower having interrupted fall other factors will obviously enter. The inability to calculate a is overcome by experimentally determining the product $K_x a$ as a whole for a particular *type* of tower fill and at specific rates of flow for the diffusing fluids comprising the system.

In the development of diffusion theory it was shown that the number of transfer units n_t obtained from Eq. (17.30) provided a useful means of sizing the task to be accomplished in fulfilling a required amount of mass transfer by diffusion. For a particular type of fill if the height of one transfer unit (HTU) is known, the total height of tower required for the task per square foot of ground area is obtained as the product of $n_t(HTU)$. The units of mols and atmospheres are convenient for absorption calculations but the pound is more convenient in diffusion-heat transfer. Equation (17.79) is seen to be the analogue of Eq. (17.29) except that the portion of the total Btu transferred by sensible-heat transfer has first been converted by the Lewis number into an equivalent quantity of Btu mass transfer and then combined with the actual mass transfer since it is a fixed proportion of it. Consequently, $G\,dH$ is the total heat transfer in the diffusion tower. Rearranging Eq. (17.79),

$$\int \frac{dH}{H' - H} = K_x a \frac{V}{G} \qquad (17.80)$$

$$n_d = \int \frac{dT}{H' - H} = K_x a \frac{V}{L} \qquad (17.81)$$

While Eq. (17.80) resembles Eq. (17.30) except for its dimensions, it is not very convenient for use in cooling-tower calculations where the principal interest lies in the temperature of the water produced. Equation (17.80) can be transformed to Eq. (17.81) when multiplied by the ratio G/L and recalling that $C = 1.0$ for water. It is more convenient to use Eq. (17.81), whose value will be called the *number* of *diffusion units* n_d to avoid confusion with the number of transfer units n_t. If the height of one diffusion unit HDU is known for a given type of fill, the total height of tower required for a given service can be computed.

Determination of the Number of Diffusion Units. The number of diffusion units calculated by Eq. (17.81) is equal to $\int dT/(H' - H)$ and is determined only by the process conditions imposed upon a tower and not the

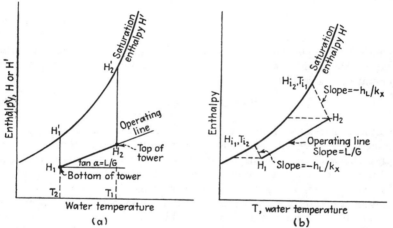

Fig. 17.10. (a) Graphical integration of $dT/(H' - H)$. (b) Correction for the liquid film resistance.

performance of the tower itself. Only the HDU is obtained experimentally. If a quantity of water of a given inlet temperature and a quantity of air of a given wet bulb are to be contacted, it will require a certain number of diffusion units as determined by the integration of Eq. (17.81) to reduce the water to any desired temperature. The number of diffusion units so obtained will be required in any type of tower, whether it be filled, packed, or empty. The height of the tower which is capable of providing the predetermined number of diffusion units varies for each type of fill and with the liquid and gas loadings.

Since the temperature of the water T is not a simple function of H' and H, it is more convenient to carry out the integration of Eq. (17.81) graphically or numerically. Referring to Fig. 17.10a, the tabulated saturation enthalpies of air H' are plotted against the water temperature T throughout the water range in the tower. The saturation values of H' form a curve which are the values of air saturated at the water temperature and which may be regarded to exist at the air film on the surface of the water. Since the saturation enthalpies also include the saturation humidities, this line is equivalent to the vapor pressure of water out of the water.

The next requirement is to determine the actual enthalpy at any point in the tower. Equation (17.78) states that $LC\,dT = G\,dH$, where $C = 1.0$ for water. It relates the change in enthalpy in the gas phase dH to the accompanying change in the water temperature dT in contact with the gas. This change can be represented by rearranging Eq. (17.78) to give

$$\frac{L}{G} = \frac{dH}{dT} \qquad (17.82)$$

which is the equation of a straight line in Fig. 17.10a whose slope is the ratio of the liquid to air loading L/G. The value of H at any point on the operating line is given by

$$H_2 = H_1 + \frac{L}{G}(T_2 - T_1) \qquad (17.83)$$

since the enthalpy of the inlet air H_1 is known or can be readily determined. As a visual aid it should be understood that the area on the graph between the saturation curve and the operating line is an indication of the size of the potential promoting the total heat transfer. A change in process conditions such that the operating line is moved downward to include a greater area between itself and the saturation line means that fewer diffusion units and a lesser height of any type of tower are ultimately required. At any temperature T in the column between T_1 and T_2 the potential driving heat out of the saturated film at the water surface and into the unsaturated air is the difference between the value of H' and H at that point. By taking small increments of T and dividing by the average difference of $H' - H$ for the increment, the number of diffusion units required to change the temperature of the water is obtained. When the increments are summed up, the total change in the water temperature gives the total value of n_d. As mentioned above, the starting point for the operating line H_1 is obtained from the condition of the atmospheric air entering the tower at the bottom.

When the number of diffusion units is obtained from Eq. (17.80) instead of Eq. (17.81), $(H' - H)$ can be obtained as the log mean of the enthalpy difference at the bottom and top of the tower. In Fig. 17.10a this is equivalent to taking the area between the operating line and a straight line drawn between H_1' and H_2'. The potential obtained in this manner is greater than the actual potential and causes errors which may be on the unsafe side. To obtain the number of diffusion units from the $\int dH/(H' - H)$, the expression must be multiplied by G/L.

Correction for the Liquid-film Resistance. From Eqs. (17.46) to (17.82) the derivation was based upon the assumption that the overall coefficient and the gas-side coefficient were identical. This, of course, implied that the liquid film does not offer any significant resistance to diffusion and that the gas side is controlling. This may be in error, particularly when the humidifying liquid is an aqueous solution. McAdams[1] has devised an excellent graphic method of allowing for the resistance of the liquid film. In writing Eq. (17.62) it was assumed that h was identical with h_G, which is the heat-transfer coefficient from the liquid-film–air-film interface to the air. When there is an appreciable liquid-film resistance, h in Eq. (17.62) should more correctly be written as an overall coefficient U, since it is the resultant of h_G and h_L, the latter being the convection coefficient from the liquid film to the interface. h without a subscript will be retained for cooling-tower applications in place of U because it is consistent with much of the literature. For the actual relationship when there is a significant liquid-film resistance, rewrite Eq. (17.62).

$$LC\, dT = h_L(T - T_i)a\, dV \qquad (17.62a)$$

which is the rate of transfer from the liquid body through the liquid film to the liquid-air interface. The rate of sensible transfer from the interface through the gas film to the gas body is given by

$$Gc\, dt = h_G(T_i - t)a\, dV$$

The analogue of Eq. (17.79) may be obtained in terms of the interface.

$$G\, dX = k_X(X_i - X)$$

Applying the Lewis number $h_G/k_X c = 1$, where k_X is the gas-side diffusion coefficient,

$$G\, dH = k_X(H_i - H)a\, dV \qquad (17.79a)$$

Equating (17.62a) and (17.79a) gives

$$\frac{h_L}{k_X} = \frac{H_i - H}{T_i - T}$$

Referring to Fig. 17.10b, a line of negative slope $-h_L/h_X$ is drawn from each of the terminal points of the operating line and the potential is the area included between the saturation line, the operating line, and the

[1] McAdams, W. H., "Heat Transmission," 2d ed., p. 290, McGraw-Hill Book Company, Inc., New York, 1942.

two lines of negative slope. The interface values of the enthalpies and temperatures are marked accordingly.

Cooling-tower Process Conditions. From the standpoint of tube corrosion 120°F is the maximum temperature at which cooling water ordinarily emerges from tubular equipment. If some of the liquid products in a plant are cooled below 120°F, the outlet-water temperature is usually lower than 120°F in order to prevent an appreciable temperature cross inside the tubular equipment. The temperature of the water to a cooling tower rarely exceeds 120°F and is usually less. When the temperature of the water from a process is over 120°F, the increased evaporation may justify the use of an atmospheric cooler to prevent the direct contact between hot water and air.

The lowest possible temperature to which water can be cooled in a cooling tower corresponds to the wet-bulb temperature of the air. This is *not* a practical limit, however, since the vapor pressure out of the water and in the air will be the same when the water attains the wet-bulb temperature, resulting in a zero diffusion potential for which an infinite tower is required. The difference between the water-outlet temperature from the tower T_2 and the wet-bulb temperature is called the *approach* in a cooling tower. Since most cooling towers operate over nearly the same water range, the approach is the principal index of how difficult the task will be as well as an indication of the number of diffusion units which will be required.

In Fig. 17.11 is shown a map of the United States on which are recorded the wet-bulb temperatures which are not exceeded for more than 5 per cent of the total hours for the four summer months from June through September. In the northeastern portion of the country the 5 per cent wet-bulb temperature is about 75°F, and it is customary in this region to cool the water in a tower to a 10° approach, or 85°F. On the Gulf Coast the 5 per cent wet-bulb temperature is about 80°F, and in these regions it is customary to invest more in cooling towers by cooling to a 5° approach, or again 85°F. Countless installations have justified the economics of these approaches. It must be realized, however, that, when a tower is designed for a definite approach to the 5 per cent wet bulb, there will be some hours during which the water issuing from the tower will be warmer than originally specified in the process conditions.

When a process involves volatile materials or vacuum operation, the 5 per cent wet bulb is discarded for another basis of selection. The approach may be selected as a compromise between the values in Fig. 17.11 and a compilation of reasonable maximum temperatures. Such a compilation is contained in Table 17.4. These are not the all-time maximums but temperatures which are rarely exceeded. As an example, dur-

Fig. 17.11. Summer wet-bulb temperature data. The wet-bulb temperatures shown will be exceeded not more than 5 per cent of the total hours during June to September, inclusive, of a normal summer. (*The Marley Company.*)

TABLE 17.4. REASONABLE MAXIMUM TEMPERATURES AND ABSOLUTE MAXIMUM WIND VELOCITIES IN THE UNITED STATES*

State	City	Reasonable maximum, Temp, °F			Max recorded wind velocity, mph
		Dry bulb	Wet bulb	Dew point	
Alabama	Mobile	95	82	80	87
Arizona	Phoenix	113	78	74	40
Arkansas	Little Rock	103	83	78	49
California	Fresno	110	75	66	41
	Laguna Beach	82	70	68	
	Oakland	94	68	67	50
	San Diego	88	74	72	43
	Williams	110	80	71	
Colorado	Denver	99	68	64	53
	Grand Junction	102	68	63	
Connecticut	Hartford	...	82	..	58
District of Columbia	Washington	99	84	80	55
Florida	Jacksonville	99	82	79	58
	Miami	92	81	79	87
Georgia	Atlanta	101	82	79	51
Idaho	Boise	109	71	65	43
Illinois	Chicago	104	80	77	65
	Moline	103	83	80	
Indiana	Evansville	102	82	79	60
Kansas	Wichita	110	79	74	68
Louisiana	New Orleans	95	83	81	66
	Shreveport	102	83	79	50
Massachusetts	Boston	96	78	76	60
Michigan	Detroit	101	79	76	67
Minnesota	St. Paul	103	79	75	78
Mississippi	Jackson	103	83	80	49
Missouri	Kansas City	109	79	76	57
	Kirksville	108	82	79	
	St. Louis	108	81	76	91
	Springfield	98	79	76	52
Montana	Helena	97	70	60	54
Nebraska	North Platte	104	76	71	73
	Omaha	108	80	75	53
Nevada	Elko	101	64	58	
	Reno	102	66	56	46
New Jersey	Camden	102	82	78	68
	Newark	99	81	77	
New Mexico	Albuquerque	98	68	66	63
New York	Albany	97	78	75	59
	Buffalo	93	77	74	73
	New York	100	81	78	73

DIRECT-CONTACT TRANSFER: COOLING TOWERS 597

TABLE 17.4. REASONABLE MAXIMUM TEMPERATURES AND ABSOLUTE MAXIMUM WIND VELOCITIES IN THE UNITED STATES.*—(*Continued*)

State	City	Reasonable maximum, Temp, °F			Max recorded wind velocity, mph
		Dry bulb	Wet bulb	Dew point	
North Carolina.............	Raleigh	98	82	80	45
	Wilmington	94	81	79	53
North Dakota...............	Fargo	105	..	..	58
Ohio.......................	Cincinnati	106	81	78	54
	Cleveland	101	79	76	60
Oklahoma...................	Tulsa	106	79	77	
Oregon.....................	Portland	99	70	68	43
Pennsylvania...............	Bellefonte	96	78	75	
	Pittsburgh	98	79	74	56
South Carolina.............	Charleston	98	82	80	81
South Dakota...............	Huron	106	76	74	56
	Rapid City	103	71	66	
Tennessee..................	Knoxville	100	79	76	59
	Memphis	103	83	80	58
Texas......................	Amarillo	101	75	70	65
	Brownsville	96	80	79	80
	Dallas	105	80	76	63
	El Paso	101	72	69	60
	Houston	100	81	79	63
	San Antonio	102	83	82	56
Utah.......................	Modena	97	66	61	
	Salt Lake City	102	68	64	53
Washington.................	Seattle	86	70	67	59
	Spokane	106	68	58	41
Wisconsin..................	Green Bay	99	79	75	53
	La Crosse	100	83	81	69
Wyoming....................	Rock Springs	91	62	58	

* Courtesy the Marley Company.

ing a 14-year period the wet-bulb temperature in New York City only twice exceeded 80°F for 7 hr duration or more. On one occasion a momentary wet bulb of 88°F was recorded. The two occasions on which the wet bulb exceeded 80°F were so rare that it is simpler to justify reducing the plant capacity or shutting down altogether during these infrequent occurrences rather than to use 80+°F as the basis for design. Since the 5 per cent wet bulb for New York is 75°F, only critical requirements can justify the use of a temperature between 75°F and the reasonable maximum of 81°F. Cooling towers are usually designed to withstand a wind velocity of 100 mph, which is equivalent to 30 lb/ft^2.

In the study of cooling towers the impression sometimes arises that the cooling tower cannot operate when the inlet air is at its wet-bulb temperature. This, of course, is not so. When air at the wet-bulb temperature enters the tower, it receives sensible heat from the hot water, and its temperature is raised thereby so that it is no longer saturated. Water then evaporates continuously into the air as it travels upward in the tower.

One of the objectionable characteristics of cooling towers is known as *fogging*. When the hot and saturated exit air discharges into a cold atmosphere, condensation will occur. It may cause a dense fog to fall over a portion of the plant with attendant safety hazards. If provision is made during the initial design, condensation can be reduced by any means which reduces the outlet temperature of the air. If it is desired to maintain a fixed range for the cooling water in coolers and condensers, fogging can be reduced by the recirculation of part of the basin water back to the top of the tower where it combines with hot water from the coolers and condensers. This reduces the temperature of the water to the tower, while the heat load is unchanged. The principal expense of the operation, aside from initial investment, will be that of the pumping power for the recirculation water, which does not enter the coolers and condensers.

Humidification Coefficients. An apparatus in which air and water can be brought into intimate contact may serve as a cooling tower or air humidifier. Considerable information can be found in the literature on the performance of a variety of packing and fills. Simpson and Sherwood[1] have given an excellent review of some of the pertinent data in the literature as well as some original data on fills suitable for cooling towers. Because of the outgrowth of modern diffusion calculations from absorption practices in the chemical industries, much of the data in the literature is represented by a plot of $K_G a$ vs. G for systems in which the gas film is controlling. This method now appears to be giving way to plots of HTU vs. G. Colburn[2] has recalculated much of the early data on this basis. The relation between the HTU and $K_G a$ is

$$HTU = \frac{Z}{K_G a} \frac{V}{G}$$

where Z is the total height containing n_t transfer units. Both calculations are based on the use of the pound-mols of water transferred and a driving potential expressed in atmospheres. There seems little reason to employ these units in humidification calculations, since the pounds of

---

[1] Simpson, W. M., and T. K. Sherwood, *Refrig. Eng.*, 535 (1946).
[2] Colburn, A. P. *Trans. AIChE*, **29**, 174 (1939).

water transferred and a driving potential in humidity units are so convenient. No data have appeared in the literature so far in which HDU, the height of a diffusion unit, is plotted against G for the humidification of air, but data have appeared with $K_x a$ plotted against G. The relation between the HDU and $K_x a$ is

$$HDU = \frac{Z}{K_x a} \frac{V}{L} \qquad (17.84)$$

and the relation between the $K_x a$ and $K_G a$ can be obtained from Eq. (17.40). The HDU or $K_x a$ are the performance characteristics of the given fill, or packing, and n_d is the size of the job required for the fulfillment of process conditions. In the case of packed towers containing small objects it is possible to report the data over wide ranges by an equation of the form

$$K_x a = C_1 G^\gamma \qquad (17.85)$$

If the value of $K_x a$ is multiplied by the ratio V/L, the number of diffusion units in a given height can be obtained, since $V = 1Z$.

If a tower is in operation and it is desired to determine its performance characteristics such as the HDU or $K_x a$, the number of diffusion units being performed must be calculated first from the observed inlet and outlet temperatures, humidities, and flow rates. The total packed, or filled, height divided by the value of n_d calculated from observed data will give the HDU. Of the data available in the literature only those for Raschig rings and Berl saddles are given here, since other types of packings and fill are less standardized and in some instances difficult to reproduce. The data in Table 17.5 have been published in part by McAdams[1] from tests by Parekh[2] and have been made available through the courtesy of Dean T. K. Sherwood. Data on fills producing droplet surface cannot be represented quite so readily as film-type packings, since the total droplet surface changes greatly with the number of droplets formed. This in turn is influenced by the liquid loading.

Caution is emphasized in the application of the information in the literature to use in actual services. Many of the data have been obtained on small packed, or filled, heights or on laboratory-scale apparatus of small cross section. A problem in the design of large towers is the attainment of a uniform distribution of both air and water over the entire cross section and height of the tower. A tower of large cross section should be easier to control than a small one because of the lower ratio of wall perimeter to cross section, which, in a sense, is a rough index of the fraction of the liquid flowing down the tower walls. Since the water is

---

[1] McAdams, W. H., *op. cit.*, p. 289.
[2] Parekh, M., Sc. D. Report in *Chem. Eng.*, MIT, (1941).

TABLE 17.5. HUMIDIFICATION CHARACTERISTICS: RASCHIG RINGS AND BERL SADDLES*

$$K_X a = C_1 G^\gamma$$

Where gas loading, G = lb/(hr)(ft² ground area)
Liquid loading, L = lb/(hr)(ft² ground area)
$K_X a$ = lb/(hr)(ft³)(lb/lb) potential

Packing, in.†	Depth, in.†	L	G	γ	$K_X a$	C_1
1 Raschig.....	24	500	250	0.50	226	14.3
		1500	250	0.50	468	29.6
		3000	250	0.50	635	40.2
1½ Raschig...	20.6	500	250	0.43	208	19.4
		1500	250	0.55	370	17.9
		3000	250	0.60	445	16.4
2 Raschig.....	19.1	500	250	0.47	190	14.3
		1500	250	0.54	301	15.3
		3000	250	0.53	351	18.9
½ Berl.......	15.5	500	250	0.61	320	11.1
		1500	250	0.61	468	16.3
		3000	250	0.61	595	20.7
1 Berl........	20.3	500	250	0.42	245	24.2
		1500	250	0.50	464	29.4
		3000	250	0.69	569	12.7
1½ Berl......	22	500	250	0.52	200	11.4
		1500	250	0.52	305	17.4
		3000	250	0.52	383	21.8

* From M. D. Parekh Report, MIT, courtesy of Professors T. K. Sherwood and W. H. McAdams.
† ½-in. Raschig rings do not follow the equation.

usually distributed only at the top, the tendency for more and more liquid to reach and descend along the wall and by-pass the packing increases with the height of the tower. A tower packed to a height of 20 ft does not actually deliver twice as many diffusion units as a tower packed to a height of 10 ft with the same packing, or fill. Another difficulty in large packed towers is the increased opportunity for *channeling*. With a large cross section there is a tendency for a disproportionate quantity of the liquid to descend through half the cross section while a disproportionate quantity of air rises in the other half. Still another note of caution refers particularly to data obtained on spray towers. As explained earlier, most of the performance occurs in the upper few feet of the tower. If a tower is equipped with sprays only at its top, the height of the tower will not be an indication of the available surface.

Consideration must also be given to the fact that many of the available data have been reported on experiments in which gas and liquid loadings of from 200 to 5000 lb/(hr)(ft^2) have been used. The twenty-fivefold range of variables provides considerable insight into the influence of these variables on performance. By the same token it is easier to obtain good data on a small packed height in which the approach to equilibrium is not very close and in which the precise experimental determination of the humidities is not too significant. None of these should be considered representative of the ranges employed in modern cooling towers. The purpose of a cooling tower is solely to produce cooling water, which, next to air itself, is the cheapest utility. The most important operating cost is that of the fan power for circulating the air. Accordingly a small allowable pressure drop of less than 2 in. of water is the standard practice. In all but extraordinary services the liquid loading on droplet-forming fills is from 1 to 4 gpm/ft^2 or 500 to 2000 lb/(hr)(ft^2). Gas loadings are between 1300 and 1800 lb/(hr)(ft^2), corresponding to gas velocities of roughly 300 to 400 fpm.

Another factor to be considered is that of flooding the fill in which the orderly counterflow of air and water is disrupted. Lobo, Friend, et al.[1] have made available the results of a study on the influence of liquid and gas loadings on the flooding points of packed towers. In slat-fill towers forming droplets there are two flooding points. One exceeds about 15 gpm/(ft^2) above which the liquid loading is so great that the water falls in curtains thereby reducing the production of water droplets. Another, occurring at higher liquid loadings, is the true flooding point in which the distribution of air and water is impaired. The liquid loading at the flood points is not independent of the gas loading. At the other extreme of low liquid loading is incipient wetting, in which the liquid flow is so small that the film surface may not be entirely wetted. In the case of the slat-fill tower few droplets are produced under these conditions and the surface is principally film surface.

Only limited performance data are available in the literature on commercial cooling towers. This is natural, since such information is usually confidential with fabricators, and operators only occasionally make test data available. By comparison with experimental towers the values of $K_x a$ or HDU obtained in large towers are relatively small. At gas and liquid loadings of about 1000 lb/(hr)(ft^2) many laboratory-scale data have been reported with values of $K_x a$ ranging from 200 to 600 lb/(hr)(ft^3)(lb/lb). It is doubtful that any full-sized apparatus erected in the United States at present is capable of a $K_x a$ exceeding 100 within the usual limits of pressure drop permitted in industry.

[1] Lobo, W. E., L. Friend, F. Hashmall, and F. Zenz, *Trans. AIChE*, **41**, 693 (1945).

The Calculation of Cooling-tower Performances.

Usually operators buy cooling towers rather than erect them themselves. This is undoubtedly the wisest policy, since it makes available to the operator a great deal of "know-how" in a field in which it is of great value. The operator will specify the quantity of water and the temperature range he requires for his process. The fabricator will propose a tower which will fulfill the conditions furnished by the operator for the 5 per cent wet bulb in the locality of the plant and guarantee the fan power with which it will be accomplished. With the first cost, fan power, and the approximate height of the pumping head the operator can compute the cost of the cooling-tower water based on a period of depreciation of about twenty years.

Assume that a cooling tower has been erected and placed in operation on this basis. An acceptance run is made to determine whether or not the cooling tower is meeting the guarantee. This consists of a wet-bulb determination on the windward side of the tower and the determination of the air rate by velometer or other pitot arrangement. The air leaving the cooling tower is always assumed to be saturated at its outlet temperature. Tests have shown it to be anywhere from 95 to 99 per cent saturated.

In a heat exchanger the performance is satisfactory if the measured overall coefficient during initial operation at the process conditions equals or exceeds the stipulated clean coefficient. In cooling towers the basis for design is one which is very rarely present. A cooling tower designed for a given approach to the 5 per cent wet bulb and erected in the fall of the year will probably not encounter the design condition for another 8 or 9 months. On the basis of its performance in autumn will it operate at the design conditions when they eventually return? A series of calculations answering questions of this nature will be undertaken presently.

Each performance calculation may be divided into two parts: (1) How many diffusion units correspond to the process requirement, and (2) how many diffusion units is or can the tower actually perform? Obviously at all conditions (2) should exceed (1).

Example 17.2. Calculation of the Number of Diffusion Units. A plant is being laid out in a restricted-water locality. The total heat load to be removed from several processes by the cooling tower is 26,000,000 Btu/hr. The locality has a 5 per cent wet-bulb temperature of 75°F. The water will leave the tower with a 10° approach to the wet bulb, or 85°F. Being water of ordinary air and mineral content it will emerge from equipment at a maximum temperature of 120°F. The water equivalent to this range is 1500 gpm.

A tower 24 by 24 ft has been erected with a fan capacity of 187,000 cfm. How many diffusion units must the tower be capable of performing to fulfill the process requirements?

Determine (a) by numerical integration and (b) by using the log mean enthalpy difference.

Solution. (a) On coordinates of enthalpy vs. water temperature such as Fig. 17.12 plot the saturation line from the data of Table 17.2.

Next determine the enthalpy of the inlet air which is one terminal of the operating line. In this problem it corresponds to a 75°F wet bulb or air saturated at 75°F. From Table 17.2 $H_1 = 39.1$ Btu/lb dry air at an outlet water temperature of 85°F.

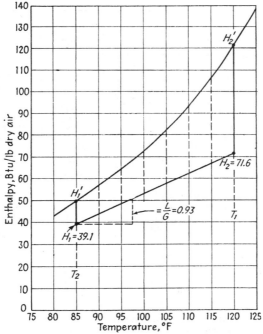

FIG. 17.12. Solution of Example 17.2.

The liquid and air loading will establish the slope of the operating line throughout the tower, starting at H_1. The liquid loading is simply the total hourly water rate divided by the ground area. The air loading was given in the proposal as 187,000 cfm at 75°F wet bulb. The density of the dry air in a cubic foot of mixture is $1/13.88 = 0.072$ lb/ft³ from Table 17.2.

Ground area $= 24 \times 24 = 576$ ft²

$L = 1500 \times 500/576 = 1302$ lb/(hr)(ft²)

$G = 187,000 \times 0.072 \times 60/576 = 1400$ lb/(hr)(ft²)

$\dfrac{L}{G} = \dfrac{1302}{1400} = 0.93$

In Fig. 17.12 from $H_1 = 39.1$ draw a line of positive slope equal to 0.93 or calculate H_2 from Eq. (17.83) and draw a line between H_1 and H_2.

$$H_2 = H_1 + \frac{L}{G}(T_2 - T_1) = 39.1 + 0.93(120 - 85) = 71.6 \text{ Btu}$$

The area between the saturation line and the operating line represents the potential for heat transfer. It can be determined either by counting squares, in which case a plot of $1/(H' - H)$ vs. T is more convenient, or by numerical integration. The latter is employed here:

T	H'	H	$H' - H$	$(H' - H)_{av}$	$\dfrac{dT(=5.0°F)}{(H' - H)_{av}}$
85	50.0	39.1	10.9		
90	56.7	43.7	13.0	11.45	0.418
95	64.2	48.4	15.8	14.4	0.347
100	72.7	53.1	19.7	17.7	0.282
105	82.5	57.7	24.8	22.2	0.225
110	93.8	62.4	31.4	28.1	0.178
115	106.7	67.0	39.7	35.55	0.1405
120	121.5	71.6	49.9	44.8	0.1115

$$n_d = K_X a \frac{V}{L} = \int \frac{dT}{H' - H} = 1.70$$

In order to fulfill process requirements the tower must be capable of performing 1.70 diffusion units. But to perform to that extent under process requirements it must also be capable of that much performance under any other conditions.

(b) Using the log mean enthalpy difference,
At the top of the tower $H'_2 - H_2 = 49.9$ Btu/lb
At the bottom of the tower $H'_1 - H_1 = 10.9$ Btu/lb

$$\text{Log mean }(H' - H) = \frac{49.9 - 10.9}{2.3 \log 49.9/10.9} = 25.8 \text{ Btu/lb}$$

$$n_d = \frac{K_X a V}{L} = \frac{dT}{H' - H} - \frac{120 - 85}{25.8} = 1.35 \text{ vs. } 1.70$$

The error is naturally larger the greater the range, and 35°F is nearly the extreme for a cooling tower. This method is acceptable only where the range is very small.

Example 17.3. Calculation of the Required Height of Fill. Suppose it is known that, for a particular fill at $L = 1302$ and $G = 1400$, $K_X a$ is 115 lb/(hr)(ft^3)(lb/lb). What height of fill should be supplied, and what is the HDU when a fill has a $K_X a$ of 115 lb/(hr)(ft^3)(lb/lb)?

Solution. Height of fill:
Since the loading is based on 1 ft^2 of ground area,

$$n_d = \frac{K_X a (1Z)}{L}$$

$$Z = \frac{n_d L}{K_X a} = 1.70 \times \frac{1302}{115} = 19.1 \text{ ft}$$

The height of a diffusion unit is Z/n_d.

$$HDU = \frac{19.1}{1.70} = 11.3 \text{ ft}$$

This last is contrasted to values of 3 ft obtained by London, Mason, and Boelter[1] on a streamline shape film tower of known surface or of 1½ ft obtained by Parekh (extrapolated) for Raschig rings.

Example 17.4. Determination of a Cooling-tower Guarantee. Under steady atmospheric conditions during an autumn test the 24- by 24-ft tower of Example 17.2 in actual operation showed a steady water temperature of 114.3°F on the tower and 79.3° off the tower. The wet bulb determined by psychrometer was 65°F. Air and water loadings were maintained at their design values.

Will this tower be capable of producing 1500 gpm of 120 to 85°F water when the wet bulb is 75°F?

Solution. This is a solution based on the observed data and is the same as before except that the operating line starting at H_1 corresponds to the 65°F wet bulb or 30.4 Btu/lb.

T	H'	H	$H' - H$	$(H' - H)_{av}$	$\dfrac{dT}{(H' - H)_{av}}$
79.3	43.4	30.4	13.0		
85	50.0	35.7	14.3	13.65	0.417
90	56.7	40.3	16.4	15.35	0.326
95	64.2	45.0	19.2	17.8	0.281
100	72.7	49.6	23.1	21.5	0.237
105	82.5	54.3	28.2	25.65	0.195
110	93.8	58.5	35.3	31.75	0.158
114.3	103.3	62.9	40.4	37.85	0.113
					$n_d = 1.72$

In order to fulfill the process requirements the cooling tower must be capable of providing 1.70 diffusion units when the wet bulb is 75°F. The same calculation applied to the *actual* test conditions shows it to be delivering 1.72 diffusion units, and it is therefore acceptable. The fan power can be checked independently from the electric circuit or from the steam rate for a turbine drive.

The Influence of Process Conditions upon Design. It is rewarding to study the effects of various process changes upon the height and cross section of the apparatus or the cost of its operation. Six of the considerations which affect the size of the tower are indicated in Fig. 17.13. They are analyzed best by means of the enthalpy-temperature diagram, since the area between the saturation and operating lines is a measure of the

[1] London, A. L., W. E. Mason, and L. M. K. Boelter, *Trans. ASME*, **62**, 41–50 (1940).

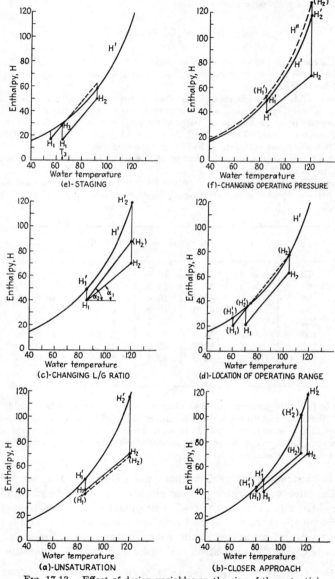

Fig. 17.13. Effect of design variables on the size of the potential.

total potential. The smaller the area the greater the height of tower required for fulfillment of the process conditions.

a. Unsaturation of the Inlet Air. Heretofore reference has been made only to the wet-bulb temperature of the inlet air and not to its dry bulb. In each case it has been assumed identical with the wet bulb, *i.e.*, adiabatically saturated. In Example 17.2 the enthalpy of the inlet air at the 75°F wet bulb was 39.1 Btu/lb. Suppose the air is at a dry bulb of 85°F when the wet bulb is 75°F. The air will be unsaturated, and its enthalpy will be 41.7 Btu/lb instead of 39.1. In Fig. 17.13a this will drop the operating line insignificantly from $H_1 - H_2$ to $(H_1) - (H_2)$, the crosshatch area representing the increase in potential. Failure to correct the enthalpy for the dry bulb gives results which are slightly on the safe side, and for this reason it is customary to specify only the wet bulb.

b. Close Approach. Both operating lines in Fig. 17.13b have the same L/G ratio (same slopes) and equal ranges of 35°F for the removal of the same process heat load. The operating line $(H_1) - (H_2)$ attempts to do the same cooling as $H_1 - H_2$ and with the same inlet air but between the temperatures of 115 and 80°F instead of 120 and 85°F. The area between the saturation curve and the operating line is greatly decreased by $(H_1) - (H_2)$. Similarly it might be desired to obtain water from 120 to 85°F with 80° wet-bulb instead of 75° wet-bulb air. This will raise the operating line $H_1 - H_2$ vertically and also decrease the potential.

c. Changing the L/G Ratio. If the ground area is too limited, such as when a cooling tower is erected atop a building, it may be necessary to employ a relatively high liquid loading without increasing the air quantity, since 400 fpm is about the maximum economical air velocity. This will decrease the cross section of the tower but increase the slope of the operating line $H_1 - H_2$ to $H_1 - (H_2)$ as in Fig. 17.13c, resulting in a decreased potential and a higher tower. This is the simple observation that the less air circulated per pound of water the less the extent of cooling.

d. Location of the Operating Range. The saturation line has a varying curvature. In Fig. 17.13d an operating line $H_1 - H_2$ is shown with a range from 105 to 70°F. Suppose it is desired to design a tower using the same inlet air but to cool water from 95 to 60°F. This would be impossible with the same L/G ratio, since the operating line $(H_1) - (H_2)$ would intersect the saturation line. Obviously heat transfer would stop at the intersection (H_2'), since the potential would be zero at that point. A considerably lower L/G ratio would be required, which in turn means more air must be circulated for the removal of the same number of Btu.

e. Staging. One of the means of overcoming the small L/G ratio in the preceding paragraph can be accomplished by the use of two towers. This is *staging*. The water at the top of the first tower is hot and contacts air of enthalpy H_2 along the operating line $H_1 - H_2$ as shown in Fig. 17.13*e*. The water is removed from the basin at temperature T_3 and is pumped over the second tower, which also uses atmospheric air of enthalpy H_1. The second tower operates between H_1 and H_3. In this way both operating lines may have large slopes without intersecting the saturation line. The fixed charges and operating cost of two towers increase the cost of water considerably, but water produced in this manner should be regarded as chilled water, and its cost and range compared with that of refrigerated water.

f. Elevation. Some plants are located at high elevations. Should this be mentioned in the process conditions? At reduced atmospheric pressure, as seen in Fig. 17.13*f*, the saturation line is higher, which in turn increases the potential and reduces the required size of tower if all other conditions are the same. This occurs because the partial pressure of the water is fixed whereas the total pressure has been decreased. The humidity of the saturated air at higher elevation is also greater.

The Influence of Operating Variables. Since a cooling tower uses the atmosphere as a cooling medium, it is also subject to the variations in the atmosphere. When operating at the design wet bulb, the tower should produce water of the range and at the temperature specified in the guarantee. When the wet bulb falls, however, it is the same as increasing the potential in the tower. If the heat load on the tower and the values of L and G circulated through the tower are all to be kept constant, the water still undergoes the same number of degrees of cooling in it but the inlet and outlet temperatures will be colder than guaranteed. The cooling tower is only able to remove the same heat load from the water by automatically reducing the potential difference, tne water temperatures declining accordingly with the wet bulb. From the standpoint of operating the coolers and condensers employed in a plant, this arrangement requires the simplest instrumentation. When a cooler receives water at a constant rate which is colder than that for which it was rated, the hot fluid is cooled below the desired outlet temperature. To prevent this from occurring in a cooler, the flow rate of the colder water through the cooler is reduced by by-passing, so that less water is used but with a constant design outlet temperature. The constant-temperature outlet water is then recombined with the by-pass water before returning to the cooling tower. In this way all coolers or condensers designed for water, say, from 85 to 120°F have outlet-water temperatures of 120°F

DIRECT-CONTACT TRANSFER: COOLING TOWERS

the year around. This was, in fact, the basis for the temperatures in Example 17.4.

If the water and air loadings on a tower are changed in any way, the number of diffusion units the tower is capable of providing are also altered. Ordinarily the loadings on a single cell cannot be changed very greatly. In fact a variation of 20 per cent from the mean of the design water loading is the maximum which may be anticipated, since the droplets are produced by nozzles which, in turn, have been sized for a given flow rate at a given head. The maximum discharge capacity will be about 120 per cent of design and when less than 80 per cent of the design water rate is used, the dispersion of the droplets is reduced along with the total quantity of water. The air loading can be regulated by varying the pitch of the fan blades, which can usually be rotated to a plus or minus 3° position from the mean. In summer the blades will be in a $+3°$ position and in winter at $-3°$. In winter the $-3°$ position of the fan delivers about 80 per cent of the $+3°$ air quantity, but the power saving is 40 per cent. If requested, the cooling-tower fabricator will specify the temperature range assumed by the cooling-tower water at 80 and 120 per cent of its design loading when the design air quantity is at the guarantee wet bulb. Sometimes, when the operating wet bulb is lower than the design wet bulb, it is desirable to use the increased potential to produce more cooling water of the original design temperature range. When the wet bulb varies, how much more water can be produced? This can be calculated from the 80, 100, and 120 per cent values, using the assumption that the number of diffusion units performed by a tower is dependent only upon the L/G ratio and not upon either L or G separately. This is a useful approximation within the range of cooling-tower operation and from 80 to 120 per cent of design capacity.

Example 17.5. The Recalculation of Cooling-tower Performance. In the guarantee of the 24- by 24-ft tower of Example 17.2 for cooling 1500 gpm from 120 to 85°F at a 75°F wet bulb the following data were given by the fabricator for overload and underload

Liquid loading, %	Temperature range, °F
120	122.2–87.2
100	120.0–85.0
80	117.5–82.5

When the wet bulb is 70°F, how much water from 120 to 85°F can the tower provide?

Solution. First reconstruct the portion of the actual performance curve from which the tower was designed. On the assumption that the performance curve is only a function of L/G and not L or G separately and, having already integrated the design condition, the same must be done for the overload and underload.

For 1500 gpm, $L/G = 0.93$
For 120 *per cent of design*, $L/G = 1.20 \times 0.93 = 1.115$
At 87.2°F, $H_2 = 39.1$ Btu
At 122.2°F, $H_2 = 39.1 + 1.115 \times 35 = 78.1$ Btu

T	H'	H	$H' - H$	$(H' - H)_{av}$	$\dfrac{dT}{(H' - H)_{av}}$
87.2	53.1	39.1	14.0		
90	56.7	42.0	14.7	14.35	0.195
95	64.2	47.6	16.6	15.65	0.320
100	72.7	53.3	19.4	18.0	0.278
105	82.5	58.8	23.7	21.55	0.232
110	93.8	64.3	29.5	26.6	0.188
115	106.7	70.0	36.7	33.25	0.150
120	121.5	75.6	45.9	41.3	0.121
122.2	128.2	78.1	50.1	48.0	0.0458
				$K_X a \dfrac{V}{L} =$	1.53

For 80 *per cent of design*, $L/G = 0.80 \times 0.93 = 0.74$
At 82.5°F, $H_1 = 39.1$
At 117.5°F, $H_2 = 39.1 + 0.74 \times 35 = 65.0$ Btu

T	H'	H	$H' - H$	$(H' - H)_{av}$	$\dfrac{dT}{(H' - H)_{av}}$
82.5	47.2	39.1	8.1		
85	50.0	40.8	9.2	8.7	0.286
90	56.7	44.6	12.1	10.7	0.465
95	64.2	48.3	15.9	14.0	0.357
100	72.7	52.0	20.7	18.3	0.273
105	82.5	55.6	26.9	23.8	0.210
110	93.8	59.3	34.5	30.7	0.163
115	106.7	63.0	43.7	39.1	0.128
117.5	121.5	65.0	56.5	50.1	0.050
				$K_X a \dfrac{V}{L} =$	1.92

The values of $K_X aV/L$ vs. L/G are plotted in Fig. 17.14.

How much water from 120 to 85°F can be circulated over the tower when the wet bulb falls to 70°F? In Fig. 17.14 both the ordinate $K_X aV/L$ and the abscissa L/G contain L, which in turn determines the desired water quantity. A value of L can be assumed, and if the value obtained for both $K_X aV/L$ and L/G falls on the curve of Fig. 17.14, the requirements of performance will be satisfied. The solution is consequently by trial and error.

DIRECT-CONTACT TRANSFER: COOLING TOWERS

Trial 1. Assume $L/G = 1.10$

$H_1 = 34.5$
$H_2 = 34.5 + 1.10 \times 35 = 73.0$

T	H'	H	$H' - H$	$(H' - H)_{av}$	$\dfrac{dT}{(H' - H)_{av}}$
85	50.0	34.5	15.5		
90	56.7	40.0	16.7	16.1	0.309
95	64.2	45.5	18.7	17.7	0.282
100	72.7	51.0	21.7	20.2	0.247
105	82.5	56.5	26.0	23.85	0.210
110	93.8	62.0	31.8	28.9	0.173
115	106.7	68.5	38.2	35.0	0.143
120	121.5	73.0	48.5	43.35	0.115
					1.48

From Fig. 17.14, for $K_X aV/L = 1.48$, $L/G = 1.19$
Assumed, $L/G = 1.10$ *No check*

Trial 2: Assume $L/G = 1.20$

$H_1 = 34.5$
$H_2 = 34.5 + 1.20 \times 35 = 76.5$

T	H'	H	$H' - H$	$(H' - H)_{av}$	$\dfrac{dT}{(H' - H)_{av}}$
85	50.0	34.5	15.5		
90	56.7	40.5	16.2	15.9	0.313
95	64.2	46.5	17.7	16.95	0.295
100	72.7	52.5	20.2	18.95	0.264
105	82.5	58.5	24.0	22.1	0.226
110	93.8	64.5	29.3	26.65	0.188
115	106.7	70.5	36.2	32.75	0.153
120	121.5	76.5	45.0	40.6	0.123
					1.56

From Fig. 17.14, for $K_X aV/L = 1.56$, $L/G = 1.08$
Assumed, $L/G = 1.20$ *No check*

By interpolation the conditions are satisfied when the value of $L/G = 1.14$. The total water which can be produced with an 85°F temperature will be $1500 \times \dfrac{1.14}{0.93} = 1840$ gpm.

Dehumidifiers. The dehumidification of hot humid air by washing with cold water is a common practice in air conditioning. Although it is generally carried out in spray washers, it can be effectively achieved through the use of a cooling tower or similar contacting device. Consider humid air saturated and at a higher temperature than the cold-water spray. The humidity or partial pressure of the water vapor in the hot air body is greater than the saturation humidity in the air film, which, in the absence of a liquid film resistance, is presumed to be saturated at the water temperature. The potential difference for mass transfer is then in the direction from the gas body to the water body, the reverse of the direction of humidification. The enthalpy at every point in the air body is greater than the corresponding value of the air film. This is shown graphically in Fig. 17.15. The operating line again has the slope of L/G, but it is above the saturation line.

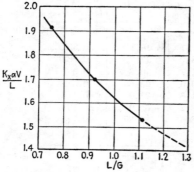

FIG. 17.14. Solution of Example 17.5.

The mass transfer of the water vapor from the air body into the liquid is a mechanism apart from condensation, although the heat of the water removed by mass transfer is equal to the latent heat of vaporization. In true diffusion the mass transfer from the gas body to the liquid is treated as a molecular phenomenon, in which case no dew is assumed to form.

Humidification and dehumidification may be considered identical except in the direction of the mass transfer. Studies indicate that the rate of dehumidification equals the rate of humidification, the same $K_x a$ data being applicable to either. Thus the number of diffusion units for dehumidification can likewise be determined by integrating the area between the upper operating line and the saturation line. The basic equation for the case of dehumidification where the Lewis number is unity becomes

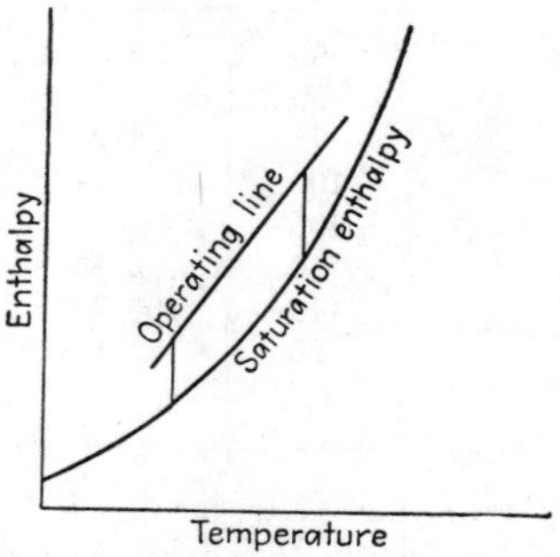

$$n_d = K_x a \frac{V}{L} = \int \frac{dT}{H - H'} \quad (17.86)$$

FIG. 17.15. Dehumidification potentials.

which differs from Eq. (17.81) only in the position of H and H'.

Heat Transfer from Gases. All the preceding illustrations have been concerned with cooling warm water. Suppose it is desired to cool a gas at a moderate pressure as in Chap. 9. If the gas is passed through a heat exchanger with a reasonable allowable pressure drop, a low heat-transfer coefficient results. This is particularly true where the static pressure on the gas is developed by a blower or fan. The same hot gas may be brought into contact with water in a packed or filled tower with droplet surface, and the heat transfer will be accomplished very cheaply. In many instances a cylindrical vessel containing a few feet of dumped packing material may replace a large tubular exchanger. The application of direct-contact heat transfer to the solution of gas-cooling problems has not yet received the widespread acceptance it appears to merit.

There is an unfounded prejudice in some quarters against the direct contact of water with high-temperature gases. It is sometimes felt that unpredictable results will be obtained such as frothing and boiling. Consider the typical cases shown in Fig. 17.16. In Fig. 17.16a a hot gas (air) enters a tower at 300°F with a dew point of 120°F and leaves at 200°F. The water has a range from 85 to 120°F. What will happen at

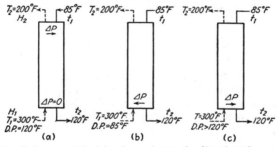

Fig. 17.16. Influence of the inlet dew point on the direction of mass transfer.

the bottom of the tower? The gas, by virtue of its dew point, has a vapor pressure of 1.69 psia. The water leaving the tower is at 120°F and has the same vapor pressure. There will be no potential for the mass transfer of water from the water body into the gas, so that there can be no vaporization, whether the temperature of the inlet gas be 200 or 2000°F. The mass transfer is established solely by the difference between the dew point of the inlet gas and the partial pressure corresponding to the outlet water temperature. At a differential height upward in the tower the relationship between the potentials changes: The water temperature is lower than the gas dew point, and water vapor starts to diffuse from the gas to the water. The diffusion process continues up to the gas outlet at the top.

In Fig. 17.16b the process conditions have been altered so that the dew point of the inlet gas is 85 instead of 120°F. Here the partial pressure of the outlet cold water is greater than that of the inlet gas, but the pressure exerted by the outlet water is still 1.69 psia, no matter how high the gas temperature may be. This case is somewhat different from that of a cooling tower, since the gas will be cooled and humidified whereas in the cooling tower it is heated and humidified. In Fig. 17.16c a gas enters with a higher dew point than that of the water and the movement of the vapor is from the gas to the water body throughout. Other interesting arrangements are possible. Analysis shows that sensible-heat transfer by direct contact must be accompanied by some mass transfer. When a gas is to be cooled over a great range, measured in hundreds of degrees, and the dew point corresponds to that of flue gases such as from 110 to 130°F, the amount of mass transfer may be very small.

Calculations without a Simplified Lewis Number. In the derivation of equations for the performance of cooling towers or other air-water humidifiers and dehumidifiers, a simple expression was obtained. This was owing very largely to the assumption that the Lewis number for air-water diffusion is nearly unity. It thus becomes possible to eliminate the last term of Eq. (17.75) from further evaluation in cooling-tower calculations. When $Le \neq 1.0$, Eq. (17.75), instead of reducing to Eq. (17.81), reduces to

$$n_d = K_X a \frac{V}{L} = \int \frac{dT}{(H' - H) + c(T - t)(Le - 1)} \quad (17.87)$$

Equation (17.87) cannot be evaluated in a simple, straightforward manner because of the last term of the denominator. In Fig. 17.4 the Lewis number for air-water has a value of 1.0 at 600°F. For the diffusion of water into other gases such as H_2 and CO_2 the Lewis number is about 2.0 and 0.6, respectively.

In the case of gas cooling if no simplifying assumption can be made for the Lewis number, it is necessary to return to the three basic equations for humidification and dehumidification. These are

$$dq_c = h_a \, dV(T - t) = Gc \, dT \quad (17.62)$$
$$dq_d = K_X a \, dV(X - X') = \lambda \, dL \quad (17.67)$$
$$L \, dt = G \, dH \quad (17.78)$$

where T now refers to the hot gas and t to the cold water. These equations are, respectively, the convection, diffusion, and overall heat balances. The first two can be related by the fact that $h = (Le)K_X c$. The overall heat balance given by Eq. (17.78) obviously is the sum of Eqs (17.62) and (17.67), respectively.

The problem is to determine the number of diffusion units corresponding to process requirements by the simultaneous integration of all three equations. This calls for a typical trial-and-error solution. Suppose a hot gas as in Fig. 17.16a contacts cold water in a counterflow tower. The process conditions leave two unknowns: the outlet air humidity and the quantity of water flowing over the tower. Although its temperature is known, the outlet-air enthalpy cannot be calculated until its humidity is known. Without the outlet humidity the total heat to be removed $G\,dH = L\,dt$ cannot be calculated, and without it L cannot be determined. The number of diffusion units is dependent upon the L/G ratio, which cannot be set without knowing L. The solution of a gas-cooling problem can be seen to hinge entirely upon the value of the outlet-air humidity. However, by assuming the outlet humidity for a required gas-outlet temperature the enthalpy can be determined, the heat balance closed, and the L/G ratio is found.

Neglecting mechanical problems of distribution, the required height of the tower is proportional to the number of diffusion units it must contain. Having obtained L by assuming the gas-outlet humidity, the quantity of water diffused in the tower and the quantity of heat transferred can be determined by $K_x aV/L$ or haV/L, both of which are related by the Lewis number. Starting at the bottom of the tower at the gas inlet, an increment $K_x a\,\Delta V/L$ can be assumed and the amount of mass and heat transfer occurring over the increment calculated, since L is now known and $K_x a\,\Delta V$ is the actual number of pounds of mass transfer for the interval. Working up the column by increments, Eqs. (17.62) and (17.67) can be integrated. Since Eq. (17.78) is the sum of Eqs. (17.62) and (17.67), a height should be reached in the tower at which all the heat transferred corresponds to Eq. (17.78). If the assumed enthalpy of the outlet-gas and the inlet-water temperature does not occur at the same height, the outlet enthalpy has been assumed incorrectly and a new assumption must be made.

Sherwood and Reed[1] have given the solution of the three differential equations by the method of W. E. Milne, which is straightforward but fairly lengthy. If an estimate of the outlet humidity can be made, the method outlined above may be simpler.

Example 17.6. Calculation of a Direct-contact Gas Cooler. A cooling operation consists of passing 50,000 lb/hr of bone-dry gas of 29 molecular weight (essentially nitrogen) over trays of hot material from whose surface oxygen must be excluded. The gas is heated in passing over the trays from 200 to 300°F and emerges with a dew point of 120°F. It then goes to a direct-contact cooler where it is cooled back to 200°F and dehumidified with water which is heated from 85 to 120°F. A pressure drop of about 2 in. of water is permissible.

(a) How many diffusion units are required to accomplish this process?

(b) Using some standard low-pressure-drop data in the literature calculate the dimensions of a direct-contact tower. In the solution of this problem data will be used for a simple fill described by Simpson and Sherwood.

Solution. Basis: 1 ft² of ground area.

---

[1] Sherwood, T. K., and C. E. Reed, "Applied Mathematics in Chemical Engineering," 134, McGraw-Hill Book Company, Inc., New York, 1939.

(a) **Tower fill of suspended fiber board will be used.** Such towers can easily operate with gas velocities of 450 fpm with reasonable pressure drops and drift elimination because they are film type towers. If the gas loading is assumed to be 1500 lb/(hr)(ft²), it will correspond to a gas velocity at the average temperature of the gas of 450 fpm. A higher gas rate cannot be justified, and a lower one might provide a tower of unnecessarily large cross section. This last can be checked only by a pressure-drop calculation once the height of the tower has been determined.

To close the heat balance and to determine the total heat transfer and liquid loading it is necessary to assume the outlet-gas humidity. Assumption: 20 per cent of the initial vapor content of the gas enters the water body.

At the gas inlet $X_1 = \dfrac{1.69}{14.7 - 1.69} \times \dfrac{18}{29} = 0.0807$ lb/lb

$G = 1500$ lb/hr

Total water in inlet gas $= 1500 \times 0.0807 = 121.05$ lb/hr

The inlet gas is at 300°F and a 120°F dew point. Use 0.25 Btu/(lb)(°F) for the specific heat of nitrogen

$H_1 = 0.0807 \times 120 + 0.0807 \times 1025.8 + 0.45 \times 0.0807(300 - 120) + 0.25$
$\times 300 = 174.0$ Btu/lb dry air (17.55)

20 per cent of the vapor has been assumed to diffuse into the water body.

Outlet gas humidity, $X_2 = \dfrac{121.05(1.0 - 0.20)}{1500} = 0.06456$ lb/lb

Dew point of outlet gas, $\dfrac{p_w}{14.7 - p_w} \times \dfrac{18}{29} = 0.06456$

$p_w = 1.388$ psia $\backsimeq$ 112.9°F dew point (Table 7)

The outlet gas has a temperature of 200°F and a 112.9°F dew point.

$H_2 = 0.06456 \times 112.9 + 0.06456 \times 1029.8 + 0.06456 \times 0.45(200 - 112.9)$
$+ 0.25 \times 200 = 126.4$ Btu/lb dry air (17.55)

Total heat load, $q = G(H_1 - H_2) = 1500(174.0 - 126.4) = 71{,}500$ Btu/hr

Water loading, $L = \dfrac{71{,}500}{120 - 85} = 2040$ lb/hr

This water loading corresponds to 4.0 gpm/(ft²), which is reasonable for this type of apparatus as explained earlier.

Interval 1:

$\dfrac{K_X a\, \Delta V}{L}$: 0 to 0.05

From Fig. 17.4 at 300°F, $Le = 0.93$

$C = 0.25 + 0.45 \times 0.0807 = 0.283$ Btu/(lb)(°F)

$haV = K_X a \dfrac{V}{L} L(Le)C = 0.05 \times 2040 \times 0.93 \times 0.283$
$= 26.9$ Btu/(hr)(°F)

$q_c = haV(T - t) = 26.9(300 - 120) = 4850$ Btu/hr

$\Delta T = \dfrac{4{,}850}{0.283 \times 1500} = 11.4$°F

$T_{0.05} = 300 - 11.4 = 288.6$°F

Since the dew point of the gas and the water outlet are the same in this problem, there is no diffusion in the first interval. In any other problem diffusion may occur in the first interval, and it can be treated the same as the second interval below:

$$\Delta t = \dfrac{4850}{2040} = 2.38\text{°F}$$
$$t_{0.05} = 120 - 2.38 = 117.6\text{°F}$$

DIRECT-CONTACT TRANSFER: COOLING TOWERS

The temperatures for the first interval are shown in Fig. 17.17.

Interval 2: $\dfrac{K_xa\,\Delta V}{L}$: 0.05 to 0.15

$h_aV = 0.10 \times 2040 \times 0.93 \times 0.283 = 53.8$ Btu/(hr)(°F)

For the interval:

$q_c = 53.8(288.6 - 117.6) = 9200$ Btu/hr

$\Delta T = \dfrac{9200}{0.283 \times 1500} = 21.7°\text{F}$

$T_{0.15} = 288.6 - 21.7 = 266.9°\text{F}$

$X'_{117.6°\text{F}} = 0.0748$ lb/lb

Lb water diffused during interval $= K_xaV(X - X')$

$K_xaV = K_xa\dfrac{V}{L} \times L = 0.10 \times 2040 = 204.0$ lb/(hr)(lb/lb)

$K_xaV(X - X') = 204.0(0.0807 - 0.0748) = 1.203$ lb/hr

Lb water remaining $= 121.05 - 1.203 = 119.85$ lb/hr

$\lambda_{117.6°\text{F}} = 1027$ Btu/lb

$q_d = 1.203 \times 1027 = 1235$ Btu/hr

$q = 9200 + 1235 = 10{,}435$ Btu/hr

$\Delta t = \dfrac{10{,}435}{2040} = 5.12°\text{F}$

$t_{0.15} = (117.6 - 5.1) = 112.5°\text{F}$

$X'_{112.5°\text{F}} = 0.0640$ lb/lb

$X_{112.5°\text{F}} = \dfrac{119.85}{1500} = 0.0798$ lb/lb

Interval 3: $\dfrac{K_xa\,\Delta V}{L}$: 0.15 to 0.25

$h_aV = 53.8$ Btu/(hr)(°F)

For the interval:

$q_c = 53.8(266.9 - 112.5) = 8300$ Btu/hr

$\Delta T = \dfrac{8300}{0.283 \times 1500} = 19.5°\text{F}$

$T_{0.25} = 266.9 - 19.5 = 247.4°\text{F}$

Lb water diffused during interval $= 204.0(0.0798 - 0.0640) = 3.22$ lb/hr

Lb water remaining $= 119.85 - 3.22 = 116.63$ lb/hr

$\lambda_{112.5°\text{F}} = 1030$ Btu/lb

$q_d = 3.22 \times 1030 = 3320$ Btu/hr

$q = 8300 + 3320 = 11{,}620$ Btu/hr

$\Delta t = \dfrac{11{,}620}{2040} = 5.70°\text{F}$

$t_{0.25} = 112.5 - 5.7 = 106.8°\text{F}$

$X'_{106.8°\text{F}} = 0.0533$ lb/lb

$X_{106.8°\text{F}} = \dfrac{116.63}{1500} = 0.0775$ lb/lb

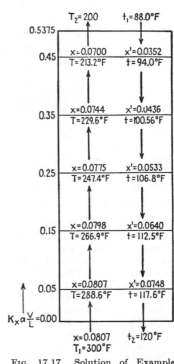

FIG. 17.17. Solution of Example 17.6.

618   PROCESS HEAT TRANSFER

The calculations of the remaining intervals until a gas temperature of 200°F is reached are shown in Fig. 17.17 from the summary table below.

Interval	$K_x a \dfrac{V}{L}$	T	t	H_2O diffused, lb	q_c	q_d
	0	300	120			
1	0.05	288.6	117.6	0	4,850	0
2	0.15	266.9	112.5	1.20	9,200	1,235
3	0.25	247.4	106.8	3.22	8,300	3,320
4	0.35	229.6	100.6	4.99	7.560	5,150
5	0.45	213.2	94.0	6.29	6,950	6,520
6	0.538	200.0	88.0	6.22	5,640	6,460
				21.92	42,500	22,685
	$n_d = 0.538$				$q = 65{,}185$ Btu/hr	

Assumed diffusion = 20 per cent
Calculated diffusion = $21.92 \times \dfrac{100}{121.05} = 18.0$ per cent

It is seen that the above summary is somewhat off, since the water temperature ended at 88°F whereas the inlet temperature was specified to be at 85°F. If an answer of greater accuracy is desired, the temperature and humidity are found to check when the diffusion is assumed to be 18½ per cent and 0.55 diffusion units are obtained. Plots can be made to show the variation of T and t with height.

(b) The following smoothed data are given for an experimental tower with vertical sheets of composition board:
Tower cross section: 41⅝ by 23⅞ in.
Filled height: 41⅜ in.
Sheet thickness: ⅛ in.
Horizontal spacing: ⅝-in. centers
Number of sprays: 18

L, lb water/(hr)(ft²)	G, lb air/(hr)(ft²)	$K_x a$	ΔP, in. H_2O
882	700	190	0.040
	1100	258	0.083
	1500	312	0.136
1178	700	200	0.049
	1100	290	0.095
	1500	373	0.150
1473	700	206	0.060
	1100	315	0.106
	1500	420	0.162

DIRECT-CONTACT TRANSFER: COOLING TOWERS 619

For $G = 1500$, extrapolate to $L = 2040$ on logarithmic coordinates. $K_Xa = 510$.

Tower height, $Z = \dfrac{n_d L}{K_X a} = 0.54 \times \dfrac{2040}{510} = 2.16$ ft

Cross section $= \dfrac{50{,}000}{1500} = 33.3$ ft^2

Note the small height. Even with an adequate factor for safety little height will be required to cool the gas.

The extrapolated pressure drop is 0.175 in. H$_2$O per 41⅜ in. of height.

SENSIBLE-HEAT TRANSFER

Heating and Cooling without Mass Transfer. When a hot gas contacts a nonvolatile cooling medium, the mass transfer is very small. In addition, with the exception of water, methanol, and ammonia, the latent heat of vaporization is also small, from 100 to 200 Btu/lb, so that the diffusion heat load is insignificant. In such case it is possible to neglect the mass transfer altogether and simplify the method of calculation for gas coolers.

Most performance data on packed towers have been obtained for mass transfer, but convection alone can be computed from mass-transfer data such as $K_X a$ or HDU by the simple relationship $haV/L = (Le)C(K_X aV/L)$. The enthalpy of the gas where there is no mass transfer is proportional to its temperature, and the potential for convection is the difference between the temperature of the gas and the temperature of the water at any cross section in the tower. Accordingly, $haV/L = \int dt/(T - t)$, and since only sensible-heat changes occur,

$$\frac{haV}{L} = \int \frac{dt}{\text{LMTD}} \qquad (17.88)$$

For a given system in a given tower if HDU is known, Eq. (17.88) can in turn be converted to the number of diffusion units by

$$n_d = (Le) \int \frac{dt}{\text{LMTD}} = \frac{haV}{L(Le)C} \qquad (17.89)$$

There is a dearth of information on the performance of different packings and fills during the saturation of gases with oils and other organics which are suitable as nonvolatile heat-transfer mediums for hot gases. This is also true for the Lewis numbers for such systems. Theoretical approximations can be obtained, but these may lead to considerable error.

In a cooling tower the convection-heat transfer represents only about 20 per cent of the total heat transfer. In the gas cooler of Example 17.6 convection accounted for 65 per cent of the total heat load. If the gas had had an inlet temperature of 1000°F and a dew point of 120°F, convection would have accounted for 95 per cent of the heat load. If a gas with a dew point of 120 to 130°F or less is to be cooled by water without a close approach, diffusion represents a small portion of the heat load and the entire problem can be treated as one of sensible-heat transfer. The dew points referred to above are within the range of flue gases. Some allowance must be made in the heat load and liquid loading, although the height of the tower will not be affected, since both diffusion and convection will occur in the same height. The use of this short method is demonstrated below.

Example 17.7. Approximate Calculation of a Gas Cooler. 50,000 lb/hr of the gas in Example 17.6 (dew point = 120°F) is to be cooled from an initial temperature of 500 to 200°F. How many diffusion units of any type of tower are required?
Solution. Assume that C for the mixture is 0.28 Btu/(lb)(°F).
Sensible-heat load = $50{,}000 \times 0.28(500 - 200)$ = 4,200,000
From Example 17.6 approximate diffusion = $\dfrac{50{,}000}{1{,}500} \times 22{,}685$ = 758,000
 4,958,000 Btu/hr

Actually, an allowance as high as 30 per cent of the sensible load can be made and the excess water compensated for by throttling when the tower is in operation.

$$\text{Total water quantity} = \frac{4{,}958{,}000}{120 - 85} = 142{,}000 \text{ lb/hr}$$

If the maximum liquid loading is taken as 2040 lb/(hr)(ft²), the required tower cross section will be

$$\frac{142{,}000}{2040} = 69.7 \text{ ft}^2$$

The new gas rate will be

$$\frac{50{,}000}{69.7} = 718 \text{ lb/(hr)(ft}^2)$$

Calculation of the number of diffusion units:

$$\frac{ha\,V}{L} = \int \frac{dt}{\text{LMTD}} \qquad (17.88)$$

The two terminal temperature differences are $(200 - 85)$ and $(500 - 120)$.

$$\text{LMTD} = \frac{(500 - 120) - (200 - 85)}{\ln(500 - 120)/(200 - 85)} = 222°\text{F}$$
$$dt = 120 - 85 = 35°\text{F}$$
$$\frac{haV}{L} = \frac{35}{222} = 0.16$$
$$n_d = \frac{haV}{L(Le)C}$$
$$= \frac{0.16}{0.93 \times 0.28} = 0.62 \text{ diffusion units}$$

For a tower employing fill as in Example 17.6:
By extrapolation for $G = 718$ and $L = 2040$, $K_X a = 215$

$$n_d = K_X a \frac{V}{L}$$

$$0.62 = 215 \times \frac{V}{2040} = 215 \times \frac{1Z}{2040}$$

$$Z = 6.8 \text{ ft high}$$

Ground dimensions $= (69.7)^{\frac{1}{2}} = 8\frac{1}{3} \times 8\frac{1}{3}$ ft

As before, the pressure drop will be negligible.

PROBLEMS

17.1. A cooling tower 30 by 30 ft (Example 15.2) has been designed to cool 1800 gpm of water from 110 to 85°F when the 5 per cent wet bulb is 75°F. The fans are capable of delivering 275,000 cfm of air. How many diffusion units are required? If the 5 per cent wet bulb was 80°F, how many diffusion units would be required?

17.2. A cooling tower 30 by 24 ft was designed to deliver 1200 gpm of water from 105 to 85°F when the 5 per cent wet bulb was 80°F. Fans are capable of delivering 230,000 cfm of air. (a) In an actual test at full loadings when the wet bulb was 70°F the water range was from 77.0 to 97.0°F. Was the tower fulfilling guarantee conditions? (b) If the range was 78.0 to 98.0°F, would the tower fulfill guarantee conditions?

17.3. In a Texas lube plant it is desired to install a water-cooled heat exchanger for a liquid with a waxy residue. A small portion of the cooling-tower water from a single cell is available, but there is some question that in winter the entering water will chill out wax and disrupt operation.

The tower operates in summer at an 80°F wet bulb with water on at 120°F and off at 85°F. Because of the main equipment on the water line the air and water quantities to the tower are fixed throughout the year at an $L/G = 0.86$. Similarly the load is also fixed.

As a winter extreme it is desired to know the temperature to the exchanger and on the tower when the wet bulb is 60°F.

17.4. A 6'0" ID tower 30 ft high packed with 10'0" of 3-in. coke was used to provide 40 gpm of water cooled from 120 to 80°F when contacted with 3600 cfm of air having a dry bulb of 85°F and a 75°F wet bulb.

To approximately what temperature could the same tower cool 6800 cfm of flue gas from an economizer if packed to a height of 20'0" with coke? In a previous test at a gas load of 1020 lb/(hr)(ft²) a diffusion coefficient of 555 lb/(hr)(ft³)(lb/lb) was obtained.

17.5. 100,000 lb/hr of nitrogen enters a direct-contact cooler at 350°F and a dew point of 130°F and is cooled to 200°F by water from 85 to 120°F. Assume that a maximum gas loading of 1400 lb/(hr)(ft²) and a maximum liquid loading of 2000 lb/(hr)(ft²) are permissible. What diameter of round tower is required? How many diffusion units are required?

NOMENCLATURE FOR CHAPTER 17

A	Heat-transfer surface, ft²
a	Surface of packing or fill, ft²/ft³
C_H	Henry's law constant, (atm)(ft³)/mol
C	Specific heat of hot liquid or humid heat of hot gas, Btu/(lb)(°F)
C_1	A constant

c	Specific heat of cold liquid or humid heat of cold gas, Btu/(lb)(°F)
$[c]$	Concentration in liquid, lb-mols/ft³
D	Inside diameter of tube, ft.
g	Acceleration of gravity, ft/hr²
G_m	Gas rate, lb-mols/(hr)(ft² of ground area)
G	Gas rate, lb/(hr)(ft² of ground area)
H, H'	Enthalpy of the gas, saturation enthalpy of the gas, Btu/lb dry air
HTU	Height of a transfer unit, ft
HDU	Height of a diffusion unit, ft
h	Overall coefficient of heat transfer h_G or h_L in a system in which one film actually controls, Btu/(hr)(ft²)(°F)
h_G, h_L	Coefficient of heat transfer from the gas to the gas-liquid interface, coefficient from the liquid to the gas-liquid interface, Btu/(hr)(ft²)(°F)
h_i, h_o	Coefficient of heat transfer referred to the inside diameter and the outside diameter of the tube, respectively, Btu/(hr)(ft²)(°F)
j_d, j_h	Factor for diffusion, factor for heat transfer, respectively, dimensionless
K_G	Overall coefficient of mass transfer, lb-mols/(hr)(ft²)(atm)
K_L	Overall coefficient of mass transfer, lb-mols/(hr)(ft²)(unit concentration)
K_X	Overall coefficient of mass transfer, lb/(hr)(ft²)(lb/lb)
k	Thermal conductivity, Btu/(hr)(ft²)(°F/ft)
k_d	Diffusivity, ft²/hr
k_G	Gas-film coefficient, lb-mol/(hr)(ft²)(atm)
k_L	Liquid-film coefficient, lb-mol/(hr)(ft²)(unit concentration)
k_X	Gas-film coefficient, lb/(hr)(ft²)(lb/lb)
L	Liquid loading, lb/(hr)(ft² of ground area)
$LMTD$	Logarithmic mean temperature difference, °F
L_0	Makeup water, lb/(hr)(ft² of ground area)
Le	Lewis number, h_G/k_Xc or h/K_Xc, dimensionless
l	Film thickness, ft
M	Molecular weight, dimensionless
M_w, M_a	Molecular weight of water and air, respectively, dimensionless
N_A	Gas rate of diffusion, lb-mol/hr
n	Number of mols
n_d	Number of diffusion units, dimensionless
n_t	Number of transfer units, dimensionless
p	Pressure of diffusing component in the gas body, atm
p'	Equilibrium pressure of the diffusing component, atm
p_{Bm}	Log mean pressure of inert gas, atm
p_t	Total pressure, atm or psi
p_w	Partial pressure of water, consistent with units of p_t
Q	Total heat transfer, Btu/hr
q	Heat transfer based on 1 ft² of ground area, Btu/(hr)(ft² ground area)
q_c, q_d	Heat transfer by convection and diffusion, respectively, based on 1 ft² of ground area, Btu/(hr)(ft² of ground area)
R	Gas constant, 1544 ft-lb/Btu
S	Entropy, Btu/°F
T	Temperature of the hot fluid, °F
T_0	Makeup water, temperature °F
t	Temperature of the cold fluid, °F

DIRECT-CONTACT TRANSFER: COOLING TOWERS

t_{DB}, t_{WB}, t_{DP}	Dry-bulb, wet-bulb, and dew-point temperatures, respectively, °F
U_C	Clean overall coefficient of heat transfer, Btu/(hr)(ft^2)(°F)
u	Velocity, ft/hr
V	Tower volume, ft^3
v	Specific volume, ft^3/lb
X, X'	Humidity of the gas, saturation humidity of the gas, lb/lb
x	Mol fraction of diffusing component in liquid phase
y	Mol fraction of diffusing component in gas phase
Z	Height of tower, ft
z	Height of free fall, ft
α_{AB}	Proportionality constant, hr/ft^2
γ	Exponent in Eq. (17.85)
δ	Molar density, lb-mols/ft^3
θ	Time, hr
λ	Average latent heat of vaporization, Btu/lb
λ_{DP}	Latent heat of vaporization at the dew point, Btu/lb
ρ	Density, lb/ft^3

Subscripts (except as noted above)

A	Diffusing component
B	Inert gas
i	Gas-liquid interface
1	Inlet
2	Outlet

CHAPTER 18

BATCH AND UNSTEADY-STATE PROCESSES

Introduction. The relationships of preceding chapters have applied only to the steady state in which the heat flow and source-temperature were constant with time. *Unsteady-state* processes are those in which the heat flow, the temperature, or both vary with time at a fixed point. Batch heat-transfer processes are typical unsteady-state processes in which discontinuous heat changes occur with specific quantities of material as when heating a given quantity of liquid in a tank or when a cold furnace is started up. Still other common problems involve the rates at which heat is conducted through a material while the temperature of the heat source varies. The daily periodic variations of the heat of the sun on various objects or the quenching of steel in an oil bath are examples of the latter. Other apparatuses based on the characteristics of the unsteady state are the regenerative furnaces used in the steel industry, the pebble heater, and equipment in processes employing fixed and moving-bed catalysts.

In batch processes for the heating of liquids the time requirement for heat transfer can usually be modified by increasing the circulation of the batch fluid, the heat transfer medium, or both. The reasons for using a batch rather than a continuous heat-transfer operation are dictated by numerous factors. Some of the common reasons are (1) the liquid being processed is not continuously available, (2) the heating or cooling medium is not continuously available, (3) the requirements of reaction time or treating time necessitates holdup, (4) the economics of intermittently processing a large batch justifies the accumulation of a small continuous stream, (5) cleaning or regeneration is a significant part of the total operating period, and (6) the simplified operation of most batch processes is advantageous.

In order to treat the commonest applications of batch and unsteady-state heat transfer systematically, it is preferable to divide processes between liquid (fluid) heating or cooling and solid heating or cooling. The commonest examples are outlined below.

1. Heating and cooling liquids
    a. Liquid batches
    b. Batch distillation

2. Heating and cooling solids
   a. Constant medium temperature
   b. Periodically varying temperature
   c. Regenerators
   d. Granular material in beds

HEATING AND COOLING LIQUIDS

1a. Liquid Batches

Introduction. Bowman, Mueller, and Nagle[1] have derived an expression for the time required to heat an agitated batch by the immersion of a heating coil when the temperature difference is the LMTD for counterflow. Fisher[2] has extended the batch calculation to include an external counterflow exchanger. Chaddock and Sanders[3] have treated agitated batches heated by external counterflow exchangers with the continuous addition of liquid to the tank and have also taken into account the heat of solution. Some of the derivations which follow apply to coils in tanks and jacketed vessels, although the method of obtaining overall coefficients of heat transfer for these elements is deferred until Chap. 20.

It is not always possible to distinguish between the presence or absence of agitation in a liquid batch, although the two premises lead to different requirements for the accomplishment of a batch temperature change in a given period of time. When a mechanical agitator is installed in a tank or vessel as in Fig. 18.1, there is no need to question that the tank fluid is agitated. When there is no mechanical agitator but the liquid is continuously recirculated, the conclusion that the batch is agitated is one of discretion. Where the heating element is an external exchanger, it is safer to assume agitation.

In the derivation of batch equations given below, T refers to the hot batch liquid *or* heating medium and t refers to the cold batch liquid or cooling medium. The following cases are treated here:

Heating and cooling agitated batches, counterflow

   Coil-in-tank or jacketed vessel, isothermal medium
   Coil-in-tank or jacketed vessel, nonisothermal medium

   External exchanger, isothermal medium
   External exchanger, nonisothermal medium

   External exchanger, liquid continuously added to tank, isothermal medium
   External exchanger, liquid continuously added to tank, nonisothermal medium

---

[1] Bowman, R. A., A. C. Mueller, and W. M. Nagle, *Trans. ASME*, **62**, 283–294 (1940).

[2] Fisher, R. C., *Ind. Eng. Chem.*, **36**, 939–942 (1944).

[3] Chaddock, R. E., and M. T. Sanders, *Trans. AIChE*, **40**, 203–210 (1944).

Heating and cooling agitated batches, parallel flow–counterflow

    External 1-2 exchanger
    External 1-2 exchanger, liquid continuously added to tank

    External 2-4 exchanger
    External 2-4 exchanger, liquid continuously added to tank

Heating and cooling batches without agitation

    External counterflow exchanger, isothermal medium
    External counterflow exchanger, nonisothermal medium
    External 1-2 exchanger
    External 2-4 exchanger

Heating and Cooling Agitated Batches. There are several ways of considering batch heat-transfer processes. If it is desired to accomplish a certain operation in a given time, the surface requirement is usually unknown. If the heat-transfer surface is known as in an existing installation, the time required to accomplish the operation is usually unknown. A third possibility arises when the time and surface are known but the temperature at the conclusion of the time is unknown. The following assumptions are involved in the derivations of Eqs. (18.1) through (18.23):

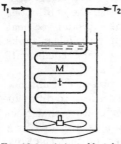

FIG. 18.1. Agitated batch.

1. U is constant for the process and over the entire surface.
2. Liqu d flow rates are constant.
3. Specific heats are constant for the process.
4. The heating or cooling medium has a constant inlet temperature.
5. Agitation produces a uniform batch fluid temperature.
6. No partial phase changes occur.
7. Heat losses are negligible.

Heating and Cooling Agitated Batches: Counterflow

COIL-IN-TANK OR JACKETED VESSEL, ISOTHERMAL HEATING MEDIUM. Consider the arrangement shown in Fig. 18.1 consisting of an agitated vessel containing M lb of liquid with specific heat c and initial temperature t_1 heated by a condensing medium of temperature T_1. The batch temperature t_2 at any time θ is given by the differential heat balance. If Q' is the total number of Btu transferred, then per unit of time

$$dQ = \frac{dQ'}{d\theta} = Mc\frac{dt}{d\theta} = UA\,\Delta t \qquad (18.1)$$

$$\Delta t = (T_1 - t) \qquad (18.2)$$

BATCH AND UNSTEADY-STATE PROCESSES

$$\frac{dt}{d\theta} = \frac{UA}{Mc} \Delta t \tag{18.3}$$

$$\int \frac{dt}{T_1 - t} = \frac{UA}{Mc} \int d\theta \tag{18.4}$$

Integrating from t_1 to t_2 while the time passes from 0 to θ,

$$\ln \frac{T_1 - t_1}{T_1 - t_2} = \frac{UA\theta}{Mc} \tag{18.5}$$

The use of an equation such as Eq. (18.5) requires the independent calculation of U for the coil or jacketed vessel as given in Chap. 20. With Q and A fixed by process conditions, the required heating time can be calculated.

COIL-IN-TANK OR JACKETED VESSEL, ISOTHERMAL COOLING MEDIUM. Problems of this type usually arise in low-temperature processes in which the cooling medium is a refrigerant which is fed to the cooling element at its isothermal boiling temperature. Conside the same arrangement as shown in Fig. 18.1 containing M lb of liquid with specific heat C and initial temperature T_1 cooled by a vaporizing medium of temperature t_1. If T is the batch temperature at any time θ,

$$\frac{dQ'}{d\theta} = -MC \frac{dT}{d\theta} = UA \Delta t \tag{18.6}$$

$$\Delta t = (T - t_1)$$

$$\ln \frac{T_1 - t_1}{T_2 - t_1} = \frac{UA\theta}{MC} \tag{18.7}$$

COIL-IN-TANK OR JACKETED VESSEL, NONISOTHERMAL HEATING MEDIUM. The nonisothermal heating medium has a constant flow rate W and inlet temperature T_1 but a variable outlet temperature

$$\begin{array}{ccc}(a) & (b) & (c)\end{array}$$

$$\frac{dQ'}{d\theta} = Mc \frac{dt}{d\theta} = WC(T_1 - T_2) = UA \Delta t \tag{18.8}$$

$$\Delta t = \text{LMTD} = \frac{T_1 - T_2}{\ln (T_1 - t)/(T_2 - t)}$$

$$T_2 = t + \frac{T_1 - t}{e^{UA/WC}}$$

Let $K_1 = e^{UA/WC}$ and equate (a) and (b) of Eq. (18.8).

$$Mc \frac{dt}{d\theta} = WC \left(\frac{K_1 - 1}{K_1}\right) (T_1 - t)$$

$$\ln \frac{T_1 - t_1}{T_1 - t_2} = \frac{WC}{Mc} \left(\frac{K_1 - 1}{K_1}\right) \theta \tag{18.9}$$

COIL-IN-TANK, NONISOTHERMAL COOLING MEDIUM:

$$\frac{dQ'}{d\theta} = -MC\frac{dT}{d\theta} = wc(t_2 - t_1) = UA\,\Delta t \qquad (18.10)$$

$$K_2 = e^{UA/wc}$$

$$\ln\frac{T_1 - t_1}{T_2 - t_1} = \frac{wc}{MC}\left(\frac{K_2 - 1}{K_2}\right)\theta \qquad (18.11)$$

EXTERNAL HEAT EXCHANGER, ISOTHERMAL HEATING MEDIUM. Consider the arrangement in Fig. 18.2 in which the fluid in the tank is heated by an external exchanger. Since the heating medium is isothermal, any type of exchanger with steam in the shell or tubes will be applicable. The advantages of forced circulation on both streams recommend this arrangement.

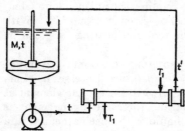

FIG. 18.2. Agitated batch with external 1-1 exchanger.

The variable temperature out of the exchanger t' will differ from the variable tank temperature t, and the differential heat balance is given by

$$\overset{(a)}{\frac{dQ'}{d\theta}} = \overset{(b)}{Mc\frac{dt}{d\theta}} = \overset{(c)}{wc(t' - t)} = UA\,\Delta t \qquad (18.12)$$

$$t' = T_1 - \frac{T_1 - t}{e^{UA/wc}}$$

Let

$$K_2 = e^{UA/wc}$$

$$\ln\frac{T_1 - t_1}{T_1 - t_2} = \frac{wc}{Mc}\left(\frac{K_2 - 1}{K_2}\right)\theta \qquad (18.13)$$

EXTERNAL EXCHANGER, ISOTHERMAL COOLING MEDIUM:

$$\ln\frac{T_1 - t_1}{T_2 - t_1} = \frac{WC}{MC}\left(\frac{K_1 - 1}{K_1}\right)\theta \qquad (18.14)$$

EXTERNAL EXCHANGER, NONISOTHERMAL HEATING MEDIUM. The differential heat balance is given by

$$\overset{(a)}{\frac{dQ'}{d\theta}} = \overset{(b)}{Mc\frac{dt}{d\theta}} = \overset{(c)}{wc(t' - t)} = \overset{(d)}{WC(T_1 - T_2)} = UA\,\Delta t \qquad (18.15)$$

BATCH AND UNSTEADY-STATE PROCESSES

There are two variable temperatures, t' and T_2, which appear in the LMTD and which must first be eliminated. Equating Eq. (18.15), (a) and (b),

$$t' = t + \frac{Mc}{wc}\frac{dt}{d\theta}$$

Equating Eq. (18.15), (a) and (c),

$$T_2 = T_1 - \frac{Mc}{WC}\frac{dt}{d\theta}$$

Let

$$K_3 = e^{UA\left(\frac{1}{wc} - \frac{1}{WC}\right)}$$

and

$$\ln\frac{T_1 - t_1}{T_1 - t_2} = \frac{K_3 - 1}{M}\frac{wWC}{(K_3 wc - WC)}\theta \qquad (18.16)$$

EXTERNAL EXCHANGER, NONISOTHERMAL COOLING MEDIUM:

$$K_4 = e^{UA\left(\frac{1}{WC} - \frac{1}{wc}\right)}$$

$$\ln\frac{T_1 - t_1}{T_2 - t_1} = \frac{(K_4 - 1)}{M}\frac{Wwc}{(K_3 WC - wc)}\theta \qquad (18.17)$$

EXTERNAL EXCHANGER, LIQUID CONTINUOUSLY ADDED TO TANK, ISOTHERMAL HEATING MEDIUM. The elements of the process are shown in Fig. 18.3. Liquid is continuously added to the tank at the rate of L_0 lb/hr and at the constant temperature t_0. It is assumed that no chemical heat effects attend the addition of the make-up liquid. Since M is the pounds of liquid originally in the batch and L_0 is in pounds per hour, the total liquid at any time is $M + L_0\theta$. The differential heat balance is given by

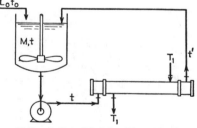

FIG. 18.3. Agitated batch with external 1-1 exchanger, liquid continuously added.

$$(M + L_0\theta)\frac{dt}{d\theta} + L_0 c(t - t_0) = wc(t' - t) \qquad (18.18)$$

and

$$wc(t' - t) = UA\,\Delta t$$

where

$$\Delta t = \text{LMTD} = \frac{t' - t}{\ln(T_1 - t)/(T_1 - t')}$$

Solving for t',

$$t' = \left(\frac{K_2 - 1}{K_2}\right) T_1 + \frac{t}{K_2}$$

Substituting in Eq. (18.18),

$$\int \frac{dt}{w\left(\frac{K_2 - 1}{K_2}\right)(T_1 - t) - L_0(t - t_0)} = \int \frac{d\theta}{M + L_0\theta}$$

$$\ln \frac{t_1 - t_0 - \frac{w}{L_0}\left(\frac{K_2 - 1}{K_2}\right)(T_1 - t_1)}{t_2 - t_0 - \frac{w}{L_0}\left(\frac{K_2 - 1}{K_2}\right)(T_1 - t_2)}$$

$$= \left[\frac{w}{L_0}\left(\frac{K_2 - 1}{K_2}\right) + 1\right] \ln \frac{M + L_0\theta}{M} \quad (18.19)$$

If the addition of liquid to the tank causes an average endothermic or exothermic heat of solution, $\pm q_s$ Btu/lb of make-up, it may be included by adding $\pm q_s/c_0$ to both the numerator and denominator of the left side. The subscript 0 refers to the make-up.

EXTERNAL EXCHANGER, LIQUID CONTINUOUSLY ADDED TO TANK, ISOTHERMAL COOLING MEDIUM:

$$(M + L_0\theta)c \frac{dT}{d\theta} + WC(T - T') = L_0C(T_0 - T) \quad (18.20)$$

$$\ln \frac{T_0 - T_1 - \frac{W}{L_0}\left(\frac{K_1 - 1}{K_1}\right)(T_1 - t_1)}{T_0 - T_2 - \frac{W}{L_0}\left(\frac{K_1 - 1}{K_1}\right)(T_2 - t_1)}$$

$$= \left[1 - \frac{W}{L_0}\left(\frac{K_1 - 1}{K_1}\right)\right] \ln \frac{M + L_0\theta}{M} \quad (18.21)$$

The heat-of-solution effects can be included by adding $\pm q_s/C_0$ to both the numerator and denominator of the left side.

EXTERNAL EXCHANGER, LIQUID CONTINUOUSLY ADDED TO TANK, NONISOTHERMAL HEATING MEDIUM. The heat balance is identical with Eq. (18.18) for heating except that Δt is written for the inlet and outlet temperatures of the heating medium.

$$\Delta t = \text{LMTD} = \frac{(T_2 - t) - (T_1 - t')}{\ln (T_2 - t)/(T_1 - t')} = \frac{(t' - t) - (T_1 - T_2)}{\ln (T_2 - t)/(T_1 - t')}$$

$$(T_1 - T_2)WC = (t' - t)wc = UA \, \Delta t$$

Let
$$K_5 = e^{\frac{UA}{wc}\left(1-\frac{wc}{WC}\right)}$$

$$\ln \frac{t_0 - t_1 + \dfrac{wWC(K_5 - 1)(T_1 - t_1)}{L_0(K_5WC - wc)}}{t_0 - t_2 + \dfrac{wWC(K_5 - 1)(T_1 - t_2)}{L_0(K_5WC - wc)}}$$

$$= \left[\frac{wWC(K_5 - 1)}{L_0(K_5Wc - wc)} + 1\right] \ln \frac{M + L_0\theta}{M} \quad (18.22)$$

The heat-of-solution effects can be included by adding $\pm q_s/c_0$ to both the numerator and denominator of the left side.

EXTERNAL EXCHANGER, LIQUID CONTINUOUSLY ADDED TO TANK, NON-ISOTHERMAL COOLING MEDIUM:

$$K_6 = e^{\frac{UA}{WC}\left(1-\frac{WC}{wc}\right)}$$

$$\ln \frac{T_0 - T_1 - \dfrac{Wwc(K_6 - 1)(T_1 - t_1)}{L_0(K_6wc - WC)}}{T_0 - T_2 - \dfrac{Wwc(K_6 - 1)(T_2 - t_1)}{L_0(K_6wc - WC)}}$$

$$= \left[\frac{Wwc(K_6 - 1)}{L_0(K_6wc - WC)} + 1\right] \ln \frac{M + L_0\theta}{M} \quad (18.23)$$

The heat-of-solution effects can be included by adding $\pm q_s/C_0$ to both the numerator and denominator of the left side.

Heating and Cooling Agitated Batches. 1-2 Parallel flow–Counterflow. The derivations for the preceding cases included assumption 7, which required that all external exchangers operate in counterflow. With non-isothermal cooling and heating mediums this will not always be advantageous, since it sacrifices the construction and performance advantages of multipass apparatus such as the 1-2 exchanger. The external 1-2 exchanger can be included by using the temperature difference as defined by Eq. (7.37).

$$\frac{UA}{wc} = \frac{1}{\sqrt{R^2 + 1}} \ln \frac{2 - S(R + 1 - \sqrt{R^2 + 1})}{2 - S(R + 1 + \sqrt{R^2 + 1})} \quad (7.37)$$

where

$$R = \frac{T_1 - T_2}{t' - t} = \frac{wc}{WC} \quad \text{and} \quad S = \frac{t' - t}{T_1 - t}$$

$$\frac{2 - S(R + 1 - \sqrt{R^2 + 1})}{2 - S(R + 1 + \sqrt{R^2 + 1})} = e^{(UA/wc)\sqrt{R^2+1}} = K_7$$

$$S = \frac{2(K_7 - 1)}{K_7(R + 1 + \sqrt{R^2 + 1}) - (R + 1 - \sqrt{R^2 + 1})} \quad (18.24)$$

and so S as well as R is a constant which is independent of the exchanger outlet temperatures.

EXTERNAL 1-2 EXCHANGER, HEATING. Using the same heat balance as defined by Eq. (18.15),

$$t' = t + \frac{M}{w}\frac{dt}{d\theta}$$

$$S = \frac{t' - t}{T_1 - t} = \frac{(M/w)(dt/d\theta)}{T_1 - t}$$

Rearranging,

$$\int \frac{dt}{T_1 - t} = \frac{Sw}{M} \int d\theta$$

$$\ln \frac{T_1 - t_1}{T_1 - t_2} = \frac{Sw}{M} \theta \tag{18.25}$$

where S is defined by Eq. (18.24).

EXTERNAL 1-2 EXCHANGER, COOLING:

$$\ln \frac{T_1 - t_1}{T_2 - t_1} = S \frac{wc}{MC} \theta \tag{18.26}$$

where S is defined by Eq. (18.24).

EXTERNAL 1-2 EXCHANGER, LIQUID CONTINUOUSLY ADDED TO TANK. HEATING:

$$(M + L_0\theta)c\frac{dt}{d\theta} = wc(t' - t) - L_0 c(t - t_0) \tag{18.27}$$

$$S = \frac{t' - t}{T_1 - t} = \frac{\left(\dfrac{M}{w} + \dfrac{L_0\theta}{w}\right)\dfrac{dt}{d\theta} + \dfrac{L_0}{w}(t - t_0)}{T_1 - t}$$

Simplifying

$$\frac{dt}{ST_1 + \dfrac{L_0}{w}t_0 - \left(S + \dfrac{L_0}{w}\right)t} = \frac{d\theta}{\dfrac{M}{w} + \dfrac{L_0\theta}{w}}$$

$$\ln \frac{ST_1 + \dfrac{L_0}{w}t_0 - \left(S + \dfrac{L_0}{w}\right)t_1}{ST_1 + \dfrac{L_0}{w}t_0 - \left(S - \dfrac{L_0}{w}\right)t_2} = \frac{Sw + L_0}{L_0} \ln \frac{(M + L_0\theta)}{M} \tag{18.28}$$

where S is defined by Eq. (18.24). The heat-of-solution effects can be included by adding $\pm q_s/c_0$ to the numerator and denominator of the left side.

EXTERNAL 1-2 EXCHANGER, LIQUID CONTINUOUSLY ADDED TO TANK, COOLING:

$$(M + L_0\theta)C \frac{dT}{d\theta} = L_0 C(T_0 - T) - WC(T - T') \quad (18.29)$$

$$\frac{1}{L_0 + WSR} \ln \frac{L_0 T_0 + WSRt_1 - (L_0 + WSR)T_1}{L_0 T_0 + WSRt_1 - (L_0 + WSR)T_2}$$

$$= \frac{1}{L_0} \ln \left(\frac{M + L_0\theta}{M} \right) \quad (18.30)$$

where S is defined by Eq. (18.24). The heat-of-solution effects can be included by adding $\pm q_s/C_0$ to the numerator and denominator of the left side.

HEATING AND COOLING AGITATED BATCHES, 2-4 PARALLEL FLOW–COUNTERFLOW. Equation (8.5) gives the true temperature relations for the 2-4 exchanger. This can be rearranged in terms of S to give

$$S = \frac{2(K_8 - 1)[1 + \sqrt{(1 - S)(1 - RS)}]}{(K_8 - 1)(R + 1) + (K_8 + 1)\sqrt{R^2 + 1}} \quad (18.31)$$

and

$$K_8 = e^{(UA/2wc)\sqrt{R^2+1}} \quad (18.32)$$

Since S cannot be expressed simply, Eq. (18.31) can be solved by trial and error, assuming different values of S until an equality is reached. The equations for heating and cooling are the same as those developed for the 1-2 exchanger except that the value of S from Eq. (18.31) replaces the value of S from Eq. (18.24). The heat-of-solution effects can be accounted for in the same manner as for the 1-2 exchanger.

Heating and Cooling without Agitation. It will be seen in Chap. 20 that agitation increases the film coefficient and thereby decreases the time requirement when heating or cooling liquids by means of a coil in tank. With external exchangers the presence of agitation, either intended or unavoidable, much to the contrary increases the time required to heat or cool a batch. This can be appreciated by simple analysis. Referring to Fig. 18.4, the batch with initial temperature t passes through an external exchanger and is returned to the tank where it builds up as a layer of temperature t_1. Such might be the case if the liquid was relatively viscous or the vessel tall and narrow. All the liquid enters the exchanger at the tank temperature t during the initial circulation and

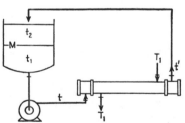

FIG. 18.4. Batch without agitation.

emerges with a temperature t_1 which is the feed temperature to the exchanger on the next recirculation. With agitation, however, the first liquid emerging from the exchanger mixes with the batch liquid and immediately raises its temperature above the initial temperature t. This in turn reduces the temperature difference in the exchanger and increases the time required for a given heat transfer.

Suppose the initial quantity of batch liquid is M lb and that it is circulated through the exchanger at the rate of w lb/hr. Since there is a discrete temperature change with each recirculation, the process is not described by a differential change. If the number of circulations required for the attainment of a final batch temperature is N, the time is given by $\theta = NM/w$.

EXTERNAL COUNTERFLOW EXCHANGER, ISOTHERMAL HEATING MEDIUM:

$$wc(t_1 - t) = UA \frac{(t_1 - t)}{\ln (T_1 - t)/(T_1 - t_1)}$$

For the initial circulation

$$t_1 = T_1 - \frac{1}{K_2}(T_1 - t)$$

For the first recirculation

$$t_2 = T_1 - \frac{1}{K_2}(T_1 - t_1)$$

In terms of T_1 and t

$$t_2 = T_1 - \frac{1}{K_2^2}(T_1 - t)$$

or

$$t_N = T_1 - \frac{1}{K_2^N}(T_1 - t) \tag{18.33}$$

When solved for N circulations,

$$\theta = \frac{NM}{w} \tag{18.34}$$

Provision can be made for the continuous addition of liquid by computing the mix temperature to the exchanger after each circulation. In such an event the size of the batch itself must be treated as increasing with each circulation, so that Eq. (18.34) does not apply unless M is increased by $L_0\theta$ pounds during the recirculation. The total time will be the sum of the individual calculations as above.

EXTERNAL COUNTERFLOW EXCHANGER, ISOTHERMAL COOLING MEDIUM:

$$T_N = t_1 - \frac{1}{K_1^N}(t_1 - T) \tag{18.35}$$

EXTERNAL COUNTERFLOW EXCHANGER, NONISOTHERMAL HEATING MEDIUM. The outlet temperature of the batch and the heating medium after each recirculation is not known. This case does not factor so simply as the preceding with an isothermal medium. Although the answer can be expressed in a series form, it is tedious to evaluate and greater speed can be attained by calculating the temperature changes after each recirculation. The temperature relations after each recirculation can be defined when

$$S = \frac{K_9 - 1}{K_9 R - 1} \qquad (18.36)$$
$$K_9 = e^{(UA/wc)(R-1)}$$

Initial circulation
$$t_1 = t + S(T_1 - t) \qquad (18.37)$$

Recirculation
$$t_2 = t_1 + S(T_1 - t_1)$$

Solve for each circulation by introducing the temperature from the preceding circulation.

EXTERNAL COUNTERFLOW EXCHANGER, NONISOTHERMAL COOLING MEDIUM. After each recirculation,

$$T_1 = T - S(T - t_1) \qquad (18.38)$$

EXTERNAL 1-2 EXCHANGER, HEATING AND COOLING. This case can be calculated in the same manner as the preceding but using S as defined by Eq. (18.24). An even greater simplification can be obtained by using the Ten Broeck chart in Fig. 7.25 and calculating each step separately. The continuous addition of liquid can be accounted for in each step along with its heat of solution.

EXTERNAL 2-4 EXCHANGER, HEATING AND COOLING. This is the same as the preceding except that S is defined by Eq. (18.31) or Fig. 8.7.

Example 18.1. Calculation of Batch Heating. 7500 gal of liquid benzene under pressure at 300°F is required for a batch extraction process. The storage temperature of the benzene is 100°F. Available as a heating medium is a 10,000 lb/hr 28°API oil stream at a temperature of 400°F. A pump connected to the tank is capable of circulating 40,000 lb/hr of benzene. Available for the service is 400 ft² of clean double pipe exchanger surface which, in counterflow streams, yields a U_C of 50 calculated for the flow rates above.

(a) How long will it take to heat the agitated batch using the double pipe counterflow assembly?

(b) How long will it take using a 1-2 exchanger with the same surface and coefficient?

(c) How long will it take using a 2-4 exchanger with the same surface and coefficient?

(d) How long will it take in (a) if the batch vessel is very tall and the batch is not considered to be agitated?

Solution:

(a) This corresponds to Eq. (18.16).
Specific gravity of benzene = 0.88
Specific heat of benzene = 0.48 Btu/(lb)(°F)
$M = 7500 \times 8.33 \times 0.88 = 55{,}000$ lb
$wc = 40{,}000 \times 0.48 = 19{,}200$ Btu/(hr)(°F)
$WC = 10{,}000 \times 0.60 = 6000$ Btu/(hr)(°F)

$$K_3 = e^{UA\left(\frac{1}{wc} - \frac{1}{WC}\right)}$$
$$= e^{50 \times 400 \left(\frac{1}{19{,}200} - \frac{1}{6000}\right)} = 0.101$$

Substitute in Eq. (18.16).

$$\ln \frac{400 - 100}{400 - 300} = \frac{(0.101 - 1)}{55{,}000} \times \frac{40{,}000 \times 6000}{(0.101 \times 6000 - 19{,}200)} \theta$$
$$\theta = 5.18 \text{ hr}$$

(b) This case corresponds to Eq. (18.25) in which S is defined by Eq. (18.24) and θ by Eq. (18.25).

$$R = \frac{wc}{WC} = \frac{19{,}200}{6000} = 3.20$$
$$K_7 = e^{UA/wc} \sqrt{R^2 + 1}$$
$$= e^{\frac{50 \times 400}{19{,}200} \sqrt{3.20^2 + 1}} = 33.1$$
$$S = \frac{2(33.1 - 1)}{33.1(1 + 3.20 + \sqrt{3.20^2 + 1}) - (1 + 3.20 - \sqrt{3.20^2 + 1})} = 0.266$$
$$\ln \frac{400 - 100}{400 - 300} = \frac{0.266 \times 40{,}000}{55{,}000} \theta$$
$$\theta = 5.63 \text{ hr}$$

(c) Use S from Eq. (18.31).

$$K_8 = e^{UA/2wc \sqrt{R^2+1}}$$
$$= e^{\frac{100 \times 200}{2 \times 19{,}200} \sqrt{3.20^2 + 1}} = 5.75$$

Solve Eq. (18.31) by trial and error:

$$S = \frac{2(5.75 - 1)[1 + \sqrt{(1 - S)(1 - 3.20S)}]}{(5.75 - 1)(3.2 + 1) + (5.75 + 1)\sqrt{3.20^2 + 1}}$$
$$= 0.282$$
$$\ln \frac{400 - 100}{400 - 300} = \frac{0.282 \times 40{,}000}{55{,}000} \times \theta$$
$$\theta = 5.33 \text{ hr}$$

(d) Use Eq. (18.37) and S from Eq. (18.36).

$$K_9 = e^{UA/wc(R-1)}$$
$$= e^{\frac{50 \times 400}{19{,}200}(3.20 - 1)} = 9.87$$
$$S = \frac{K_9 - 1}{K_9 R - 1} = \frac{9.87 - 1}{9.87 \times 3.20 - 1} = 0.290$$

$t_1 = t + S(T_1 - t) = 100 + 0.29(400 - 100) = 187°F$
$t_2 = t_1 + S(T_1 - t_1) = 187 + 0.29(400 - 187) = 249°F$
$t_3 = t_2 + S(T_1 - t_2) = 249 + 0.29(400 - 249) = 293°F$
$t_4 = t_3 + S(T_1 - t_3) = 293 + 0.29(400 - 293) = 324°F$

Actually, a fractional number of circulations are required. If the problem may be treated from the standpoint of total heat input to the batch, let
x = fractional circulation
$293(1 - x) + 324 = 300$
$x = 0.23$
Total circulations $= 3 + 0.23 = 3.23$

$$\theta = 3.23 \times \frac{55{,}000}{40{,}000} = 4.44 \text{ hr}$$

This value compares with 5.18 hr for the agitated batch.

1b. BATCH DISTILLATION

Introduction. Typical arrangements for batch distillation are shown in Figs. 18.5 and 18.6. The still pot is charged with a batch of liquid,

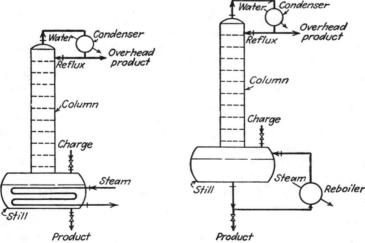

FIG. 18.5. Coil-heated batch still.

FIG. 18.6. Batch still with external reboiler.

and heat is supplied by a coil or a natural- or forced-circulation reboiler. In some high-temperature installations the still pot may be directly fired. Batch distillation is usually employed when there is an insufficiency of charge stock to warrant continuous operation and the assemblies are often relatively small. In batch distillation the composition and temperature of the residual liquid in the still constantly vary, and usually the same applies to the condensate except when the still has been charged with or forms a constant-boiling mixture. In batch distillation it is possible to obtain an initial overhead fraction which is purer than that obtainable with the same reflux by continuous distillation. This is particularly true when the overhead product is sold in different grades with a premium for purity. It is also possible by constantly varying the reflux ratio to obtain a nearly-uniform overhead composition although its quantity constantly decreases. The latter is usually too costly to be general.

The composition change during the batch distillation of a binary mixture is given by the Rayleigh equation.

$$\ln \frac{L_1}{L_2} = \int_{x_2}^{x_1} \frac{dx}{y - x} \qquad (18.39)$$

where L_1 = mols liquid charge to still
L_2 = mols residue after distillation
x_1 = mol fraction of volatile component in charge liquid
x_2 = mol fraction of volatile component in residue
y = mol fraction of vapor in equilibrium with x

The temperature must be obtained from a boiling-point curve if the mixture is not ideal and does not follow Raoult's and Henry's laws.

The Rayleigh equation does not contain any term with a unit of time. The time allotted to the distillation is therefore independent of any feed quantity. If a batch represents the intermittent accumulation of several hours' flow of charge stock, the rate of distillation must be such that the still pot will be vacated and ready to receive the next charge. If the distillation occurs infrequently, the rate of distillation can be determined economically from the optimum relationship between fixed and operating charges. In batch distillation the cost of labor is particularly large and favors a rapid distillation. The cost of the equipment, on the other hand, favors slower rates of distillation.

The Reboiler and Condenser. The design conditions for both the reboiler and the condenser are usually based upon the limiting conditions of operation. Very often batch distillation is controlled automatically by a program or elapsed-time controller (see Fig. 21.28) such that the heating medium is supplied at a rate producing a steady increase in the boiling temperature. If a heating medium such as steam were supplied to a batch reboiler at a fixed rate, much of it would not be condensed in the reboiler following the initial period of rapid vaporization whence the residue is heated sensibly. The charge stock is a mixture having volatile components which leave the residue at a decreasing rate as the distillation progresses. The boiling temperature of the residue consequently rises as the volatile components are depleted. As the sensible load in the still pot increases, the weighted heat transfer coefficient to the residue decreases. Suppose steam at 300°F is used to vaporize a charge having an initial boiling point of 200°F and the distillation is to be cut off at a residue composition corresponding to a boiling point of 250°F. A distillation curve can be prepared using the methods of Chaps. 13 and 15. Instantaneous coefficients can be computed for both the start and the finish from adjacent intervals of heat input. At the start when the film coefficient and heat load are high, the temperature difference is

$$300 - 200 = 100°F$$

but at the cutoff temperature, where the film coefficient and heat load are low, the temperature difference is only $300 - 250 = 50°F$. Both conditions must be tested for U and Δt to determine which one requires the greater surface.

The factors affecting the condenser are different. The water rate is usually kept constant. At the start of distillation the temperature of the overhead vapor may be near 200°F and the water, with a range from 85 to 120°F, provides an LMTD of 96.2°F. At the cutoff the overhead may be near 250°F, corresponding to a temperature difference of 150°F or greater, and the heat load and water range will be smaller. The condensing coefficient will change little over the entire run. The limiting condition for the condenser therefore is at the start of distillation, where the maximum heat load and minimum temperature difference usually occur together.

A common practice for obtaining the heat load for the reboiler and condenser without resorting to a distillation curve is to take the total heat load and divide it by the time allotted to the distillation. This will give a fictitious average hourly heat load which is greater than that at the cutoff but usually less than the initial heat load. The value of Q so obtained is combined with the value of U and Δt occurring at both the start and the cutoff temperatures, and the larger calculated surface plus some allowance for the error is used. If the heating element is also used to preheat the charge, it is entirely likely that the preheat rate may be limiting. The time for preheating can be obtained from one of the batch heating equations in the preceding section.

HEATING AND COOLING SOLIDS

2a. CONSTANT MEDIUM TEMPERATURE

Introduction. Since the appearance of Fourier's[1] early work the conduction of heat in solids has attracted the interest and attention of many

[1] Fourier, J. B., "Théorie analytique de la chaleur," 1822.

leading mathematicians and physicists. It is possible to present only some of the simplest and most representative cases here and to suggest the overall nature of the study. The reader is referred to the excellent books on the subject listed below.[1] They treat the subject in greater detail and provide the solutions for a number of specific problems as well as many with more complex geometry.

In the treatment of unsteady-state conduction the simplest types of problems are those in which the surface of the solid suddenly attains a new temperature which is maintained constant. This can happen only when the film coefficient from the surface to some isothermal heat-transfer medium is infinite, and although there are not many practical applications of this type, it is an important steppingstone to the solution of numerous problems. Ordinarily, heating or cooling involves a finite film coefficient or else a contact resistance develops between the medium and the surface so that the surface never attains the temperature of the medium. Moreover, the temperature of the surface changes continuously as the solid is heated even though the temperature of the medium remains constant. It is also possible that the temperature of the medium itself varies, but this class of problem will be treated separately in the next section. The cases treated in this section include those with finite film coefficients or contact resistances as well as those with infinite coefficients. The following are considered:

Sudden change of the surface temperature (infinite coefficient)

Wall of infinite thickness heated on one side
Wall of finite thickness heated on one side
Wall of finite thickness heated on both sides (slab)

Square bar, cube, cylinder of infinite length, cylinder with length equal to its diameter, sphere

[1] Boelter, L. M. K., V. H. Cherry, H. A. Johnson, and R. C. Martinelli, "Heat Transfer Notes," University of California Press, Berkeley, 1946. Carslaw, H. S., and J. C. Jaeger, "Conduction of Heat in Solids," Oxford University Press, New York, 1947. Grober, H., "Einfuhrung in die Lehre von der Warmeubertragung," Verlag Julius Springer, Berlin, 1926. Ingersoll, L. R., O. J. Zobel, and A. C. Ingersoll, "Heat Conduction with Engineering and Geological Applications," McGraw-Hill Book Company Inc., New York, 1948. McAdams, W. H., "Heat Transmission," 2d ed., McGraw-Hill Book Company, Inc., New York, 1942. Schack, A., "Der industrielle Warmeubergang," Verlag Atahleisen, Dusseldorf, 1929; English translation by H. Goldschmidt and E. P. Partridge, "Industrial Heat Transfer," John Wiley & Sons, Inc., New York, 1933. Sherwood, T. K., and C. E. Reed, "Applied Mathematics in Chemical Engineering," McGraw-Hill Book Company, Inc., New York, 1939. For a review of more recent methods see Dusinberre, G. M., "Numerical Analysis of Heat Flow," McGraw-Hill Book Company, Inc., New York, 1949, and Jakob, M., "Heat Transfer," Vol. 1, John Wiley & Sons, Inc., New York, 1949.

BATCH AND UNSTEADY-STATE PROCESSES 641

Changes due to media having contact resistances
  Wall of finite thickness
  Cylinder of infinite length, sphere, semiinfinite solid
  Newman's method for common and composite shapes
  Graphical determination of the time-temperature distribution

Wall of Infinite Thickness Heated on One Side. A wall of infinite thickness and at a uniform original temperature is subjected to surroundings with constant temperature T_s. It is assumed that there is no contact resistance between the medium and the surface it contacts, so that the face temperature of the wall is also T_s. This differs from ordinary quenching in which there is a very definite contact resistance. The general equation for conduction has been given by Eq. (2.13). For a wall of infinite thickness it reduces to unidirectional heat flow as given by Eq. (2.12). The group $k/c\rho$ is the thermal diffusivity consisting only of the properties of the conducting material. Calling this group α, the conduction may be represented by

$$\frac{\partial t}{\partial \theta} = \alpha \frac{\partial^2 t}{\partial x^2} \qquad (2.12)$$

Fourier has indicated that the time-temperature-distance relationship for a body of uniform temperature subjected to a sudden source of heat will be represented by the exponential $e^{-p\theta}e^{qx}$, where p and q are constants, θ is the time, and x the distance from the surface. With this as the starting point it is possible to set up a number of equations which describe the variation of the temperature with time and distance throughout a solid suddenly subjected to a heat source at one face. It is requisite, however, that the equation containing the exponential also fulfill all the boundary conditions imposed on the system. The most general equation of this type is given by

$$t = C_1 + C_2 x + C_3 e^{-p\theta} e^{qx} \qquad (18.40)$$

in which C_1, C_2, and C_3 are constants. A modification of Eq. (18.40) which describes the case in question and fulfills numerous boundary conditions is given by Schack as

$$t = C_1 + C_2 x + C_3 \frac{2}{\sqrt{\pi}} \int_{z=0}^{z=x/2\sqrt{\alpha\theta}} e^{-z^2} dz \qquad (18.41)$$

where $\frac{2}{\sqrt{\pi}} \int_0^z e^{-z^2} dz$ is readily recognized as the probability integral or Gauss's error integral with values from 0 to 1.0. The boundary conditions for an infinite wall heated on one face are that, when $x = x$ and $\theta = 0$, $t = t_0$ and, when $x = 0$ and $\theta = 0$, $t = T_s$, where t_0 is the initial uniform temperature of the solid. When $x = 0$ and $\theta = 0$, $t = T_s = C_1$ in which T_s is the temperature of the face of the slab directly upon being brought in contact with surroundings at T_s. When $x = x$ and $\theta = 0$, the temperature of the slab naturally possesses its original temperature t_0 or

$$t_{\theta=0} = C_1 + C_2 x + C_3 = t_0 \tag{18.42}$$

This can be valid, however, only if $C_2 = 0$, for otherwise t_0 would have to vary with x whereas it was assumed to be uniform.

Thus
$$t_0 = C_1 + C_3$$
or
$$C_3 = t_0 - T_s$$

Substituting for the constants in Eq. (18.41),

$$t = T_s + (t_0 - T_s) \frac{2}{\sqrt{\pi}} \int_0^{x/2\sqrt{\alpha\theta}} e^{-z^2} \, dz$$

or in abbreviated form

$$t = T_s + (t_0 - T_s) f_1 \left(\frac{x}{2\sqrt{\alpha\theta}} \right) \tag{18.43}$$

where $f_1(x/2\sqrt{\alpha\theta})$ denotes the value of the error integral in terms of the

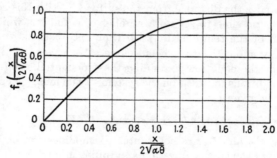

Fig. 18.7. Probability integral.

dimensionless group $x/2\sqrt{\alpha\theta}$. The values of the integral are plotted in Fig. 18.7. Equation (18.43) can be conveniently written

$$Y = \frac{T_s - t}{T_s - t_0} = f_1 \left(\frac{x}{2\sqrt{\alpha\theta}} \right) \tag{18.44}$$

Equation (18.44) is obviously the means by which the temperature t at any distance x and after time θ can be determined. A similar result can be obtained by dimensional analysis. For unit area the heat flow can be taken from Eq. (2.5).

$$\frac{dQ}{dA} = k\frac{dt}{dx} \tag{18.45}$$

where Q is the heat flow Btu/hr. To obtain dt/dx from the expression for t in Eq. (18.44) the derivative of the error integral becomes

$$\frac{\partial}{\partial x}\int_0^{x/2\sqrt{\alpha\theta}} e^{-z^2}\,dz = \frac{e^{-x^2/4\alpha\theta}}{2\sqrt{\alpha\theta}} \tag{18.46}$$

and Eq. (18.45) reduces to

$$\frac{Q}{A} = \frac{k(T_s - t_0)}{\sqrt{\pi\alpha\theta}}\,e^{-x^2/4\alpha\theta} \tag{18.47}$$

Values of the exponential are plotted in Fig. 18.8.

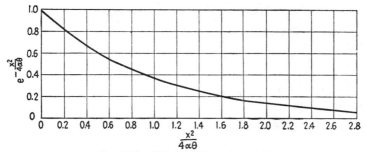

Fig. 18.8. Values of the exponential.

The heat flow through the surface is obtained when $x = 0$ or

$$\frac{Q}{A} = \frac{k(T_s - t_0)}{\sqrt{\pi\alpha\theta}} \tag{18.48}$$

and after θ hr the total heat which will have entered or left the wall will be

$$\frac{Q'}{A} = 2k(T_s - t_0)\sqrt{\frac{\theta}{\pi\alpha}} \tag{18.49}$$

where Q' is the number of Btu.

Example 18.2. Heat Flow through a Wall. It is desired to know the temperature of a thick steel wall 4 in. below its hot surface and 4 hr after its uniform temperature of 100°F was changed by suddenly applying a temperature of 1000°F to one face. How much heat will be passing into the wall at that time, and how much will already have passed into the wall?

Solution. Using Eq. (18.44):

Properties of steel: Assume

$t_{av} = 500°F$
$c = 0.12$ Btu/(lb)(°F)
$k = 24.0$ Btu/(hr)(ft^2)(°F/ft)
$\rho = 488$ lb/ft^3
$\alpha = \dfrac{k}{c\rho} = \dfrac{24.0}{0.12 \times 488} = 0.41$ ft^2/hr
$x = \dfrac{4}{12} = 0.333$ ft
$\theta = 4$

$$Y = \frac{T_s - t}{T_s - t_0} = f_1\left(\frac{x}{2\sqrt{\alpha\theta}}\right) = f_1\left(\frac{0.333}{2\sqrt{0.41 \times 4}}\right) = f_1(0.130) \quad (18.44)$$

From Fig. 18.7, for $x/2\sqrt{\alpha\theta} = 0.130$, $f_1(x/2\sqrt{\alpha\theta}) = 0.142$.

$$t = T_s + (t_0 - T_s)f_1\left(\frac{x}{2\sqrt{\alpha\theta}}\right) \quad (18.43)$$
$$t = 1000 + (100 - 1000)0.142 = 872°F$$

The heat flow crossing a plane 4 in. from the surface at 4 hr after applying the heat is given by Eq. (18.47).

$$\frac{Q}{A} = \frac{k(T_s - t_0)}{\sqrt{\pi\alpha\theta}} \quad (18.47)$$

$$\frac{Q}{A} = \frac{24(1000 - 100)}{\sqrt{\pi \times 0.41 \times 4}} = 9525 \text{ Btu/(hr)(ft}^2)$$

The total heat which flowed through a square foot of wall in the 4 hr is given by

$$\frac{Q'}{A} = 2k(T_s - t_0)\sqrt{\frac{\theta}{\pi\alpha}} \quad (18.49)$$

$$= 2 \times 24(1000 - 100)\sqrt{\frac{4}{3.14 \times 0.41}} = 76{,}000 \text{ Btu/ft}^2$$

Wall of Finite Thickness Heated on One Side. The equations developed in the preceding section for the infinitely thick wall can be applied also with limitations to walls of finite thickness. If the finite wall is relatively thick, the temperature-distance distribution for a short period after the heat has been applied will be nearly the same as for the infinite wall. As the period is lengthened, however, the penetration of heat through the wall to its cold face increases. Schack[1] has analyzed this problem by noting that a finite wall can be made to duplicate an infinite wall. This will occur if the same heat flow is removed from the remote face of the finite wall as would ordinarily flow through a plane

[1] Schack, *op. cit.*

in an infinite wall at the same distance from the hot surface. If t_0 is the temperature at the remote face of a finite wall l ft thick, the heat flow per square foot from the remote face will be

$$Q = h(t - t_0) \tag{18.50a}$$

And this is equal to the heat flow in an infinite wall at the distance $x = l$ from the hot face.

$$Q = k(T_s - t_0) \frac{e^{-x^2/4\alpha\theta}}{\sqrt{\pi\alpha\theta}} \tag{18.50b}$$

By equating the two the rate of heat removal required to reproduce the distribution of an infinite wall will be

$$h = \frac{k(T_s - t_0)e^{-x^2/4\alpha\theta}}{(t - t_0)\sqrt{\pi\alpha\theta}} \tag{18.51}$$

If the value of $l/2\sqrt{\alpha\theta}$ is nearly 1.0, the temperature increase in the remote surface will be very small. If the value of $l/2\sqrt{\alpha\theta}$ exceeds 0.6, it will be possible for most industrial applications to apply Eq. (18.44) to the finite wall directly.

Wall of Finite Thickness Heated on Both Sides. In a study of the time-temperature distribution during the annealing of optical glass, Williamson and Adams[1] developed equations for obtaining the center, center-line, or center-plane temperature of a number of shapes whose surfaces had suddenly been exposed to a heat source with infinite film coefficient. Included among these are the infinitely wide slab, the square bar, the cube, the cylinder with infinite length, the cylinder with length equal to diameter, and the sphere. Since only in the slab is the flow of heat along a single axis, it may be expected that equations for the other forms will be more complex. Where the flow of heat is symmetrical, it is considerably more convenient to use the center line or center plane as the distance reference. The surface conditions will then correspond to $\pm l/2$, and the center line or center plane to $l/2 = 0$. Williamson and Adams obtained equations in terms of a Fourier series. For the infinite slab the equation is

$$Y = \frac{T_s - t}{T_s - t_0} = \frac{4}{\pi}\left[e^{-\left(\frac{\pi}{2}\right)^{\frac{24\alpha\theta}{l^2}}} \sin\pi\frac{x}{l} + \frac{1}{3} e^{-9\left(\frac{\pi}{2}\right)^{\frac{24\alpha\theta}{l^2}}} \sin\frac{3\pi x}{l} \right.$$
$$\left. + \frac{1}{5} e^{-25\left(\frac{\pi}{2}\right)^{\frac{24\alpha\theta}{l^2}}} \sin\frac{5\pi x}{l} + \cdots \right]$$

As $4\alpha\theta/l^2$ increases, the series converges even more rapidly until at $4\alpha\theta/l^2 = 0.6$ only the first term is important. The solutions for all the

[1] Williamson, E. D., and L. H. Adams, *Phys. Rev.*, **14**, 99–114 (1919).

above shapes can be expressed in terms of a series. Williamson and Adams have presented their calculations in the simplified form

$$Y = \frac{T_s - t}{T_s - t_0} = f_2\left(\frac{4\alpha\theta}{l^2}\right) \quad (18.52)$$

or

$$t = T_s + (t_0 - T_s)f_2\left(\frac{4\alpha\theta}{l^2}\right) \quad (18.52a)$$

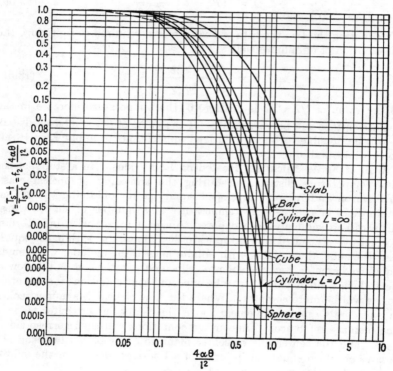

Fig. 18.9. Plot of $f_2(4\alpha\theta/l^2)$ for various shapes.

The series expression is presented by $f_2(4\alpha\theta/l^2)$, in which l is the principal depth or diameter and t the temperature at the center, center line, or center plane. For the various shapes and their respective final equations, the values of $f_2(4\alpha\theta/l^2)$ were tabulated by Williamson and Adams and have been plotted in Fig. 18.9.

Example 18.3. Center-line Temperature of a Shaft. Calculate the temperature at the center line of a long steel shaft 12 in. in diameter, initially at 100°F, 15 min after its surface temperature has been suddenly changed to 1000°F. As before, α may be taken as 0.41 ft²/hr.

$$\frac{4\alpha\theta}{l^2} = \frac{4 \times 0.41 \times 15\!/\!60}{(12\!/\!12)^2} = 0.41$$

From Fig. 18.9 when $L = \infty$

$$f_2\left(\frac{4\alpha\theta}{l^2}\right) = 0.155$$

From Eq. (18.52)

$$t = 1000 + (100 - 1000) \times 0.155 = 860°F$$

Wall of Finite Thickness Heated by a Fluid with Contact Resistance. This practical condition has been treated by Grober[1] along with the evaluation of the functions contained in the final equations. This corresponds to quenching in that it considers a contact resistance between the heating or cooling medium and the wall face or both faces in the case of an infinitely wide slab of finite thickness. The reciprocal of the contact resistance is the coefficient of heat transfer between the liquid and solid and, as mentioned earlier, causes the surface temperature to vary even though the temperature of the heating medium remains constant. The method of evaluating the film coefficient may be selected approximately from among the methods of earlier chapters or those in Chap. 20. In many cases it is difficult to obtain an analogous mechanism for computing heat transfer during a quenching operation. In a large vessel at nearly constant temperature, such as might be employed for the quenching of steel plates, the limiting coefficient between the oil and metal is that of free convection, and it varies continuously with time as the temperature difference between the metal and oil decreases. Consider a plate with an initial temperature t_0 suddenly plunged into a gas or liquid of constant surrounding temperature T_s. The temperature of the surface will be given by

$$t_{x=0} = T_s + (t_0 - T_s)f_3\left(\frac{4\alpha\theta}{l^2}, \frac{hl}{2k}\right) \tag{18.53}$$

$$Y = \frac{T_s - t}{T_s - t_0} = f_3\left(\frac{4\alpha\theta}{l^2}, \frac{hl}{2k}\right) \tag{18.53a}$$

The temperature of the center plane will be

$$t_{l/2} = T_s + (t_0 - T_s)f_4\left(\frac{4\alpha\theta}{l^2}, \frac{hl}{2k}\right) \tag{18.54}$$

or

$$Y = \frac{T_s - t_{l/2}}{T_s - t_0} = f_4\left(\frac{4\alpha\theta}{l^2}, \frac{hl}{2k}\right) \tag{18.54a}$$

[1] Grober, *op. cit.*

Grober has graphically evaluated the functions

$$f_3\left(\frac{4\alpha\theta}{l^2}, \frac{hl}{2k}\right) \quad \text{and} \quad f_4\left(\frac{4\alpha\theta}{l^2}, \frac{hl}{2k}\right)$$

which have been plotted by Schack as shown in Fig. 18.10 and 18.11 for the surface and centers of rectangular shapes. When presented in this form the plots are referred to as Schack charts.

Example 18.4. Quenching: The Schack Chart. Iron castings in the form of large slabs 10 in. thick are held at red heat (1100°F) before being hung vertically in 70°F air to cool.

It is desired to start further operations on them after 4 hr have elapsed.
(a) What will the surface temperature be after 4 hr?
(b) What will the center-plane temperature be after 4 hr?

Solution. In order to obtain the average coefficient from the slab to air by radiation and convection an assumption must be made of the surface temperature after 4 hr.

(a) Assume the temperature is 500°F after 4 hr. The coefficient from plate to air is the sum of the radiation and convection coefficients. From Eq. (4.32) and (4.33)

$$h_r = \frac{\epsilon\sigma(T_1^4 - T_2^4)}{T_1 - T_2}$$

where the temperatures are in °R.

For the initial coefficient at 1100°F,

$$h_r = \frac{0.70 \times 0.173 \times 10^{-8}(1560^4 - 530^4)}{1100 - 70} = 6.9 \text{ Btu/(hr)(ft}^2)(°F)$$

From Eq. (10.10)

$$h_c = 0.3(\Delta t)^{1/4}$$
$$= 0.3(1100 - 70)^{1/4} = 1.7 \text{ Btu/(hr)(ft}^2)(°F)$$

Total initial coefficient $= 6.9 + 1.7 = 8.6$ Btu/(hr)(ft²)(°F)

For the 4-hr coefficient at 500°F

$$h_r = 2.2$$
$$h_c = 1.35$$

Total coefficient after 4 hr $= 2.2 + 1.35 = 3.55$ Btu/(hr)(ft²)(°F)

The average coefficient is somewhat lower than the mean of the initial and final coefficients, since the radiation coefficient decreases rapidly as the temperature falls. The mean coefficient gives a surface temperature somewhat larger than the actual.

Mean coefficient, $h = (8.6 + 3.55)/2 = 6.1$ Btu/(hr)(ft²)(°F)

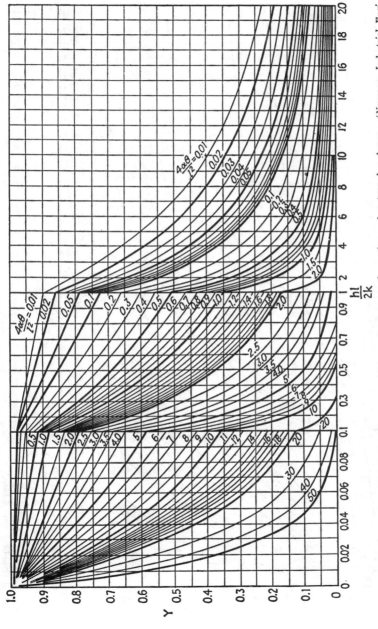

Fig. 18.10. Chart for determining the temperature history of points on the surfaces of rectangular shapes. (*Newman, Industrial Engineering Chemistry, after Schack.*)

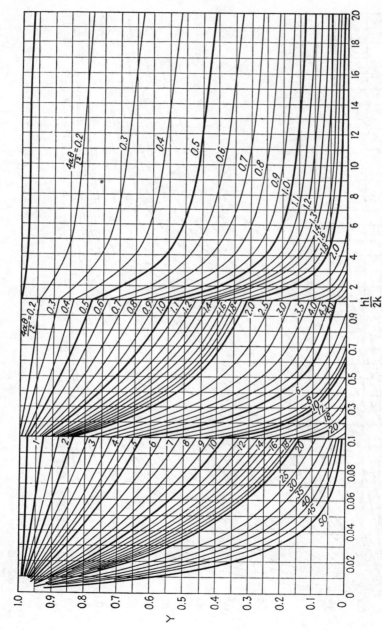

Fig. 18.11. Chart for determining the temperature history of points at the centers of rectangular shapes. (*Newman, Industrial Engineering Chemistry, after Schack.*)

From the Appendix, $k = 27$, $c = 0.14$, and $\rho = 490$ (approximately)

$$\alpha = \frac{k}{c\rho} = \frac{27}{0.14 \times 490} = 0.394$$

$$\frac{4\alpha\theta}{l^2} = \frac{4 \times 0.394 \times 4}{(10/12)^2} = 9.1$$

$$\frac{hl}{2k} = \frac{6.1 \times (10/12)}{2 \times 27} = 0.094$$

$$t_{x=0} = T_s + (t_0 - T_s)f_3\left(\frac{4\alpha\theta}{l^2}, \frac{hl}{2k}\right)$$
$$= 70 + (1100 - 70)f_3(9.1, 0.094) \tag{18.53}$$

From Fig. 18.10, $f_3(9.1, 0.094) = 0.42$

$$t_{x=0} = 70 + (1100 - 70)0.42 = 502°F$$

(b) The temperature of the center plane is given by

$$t_{l/2} = T_s + (t_0 - T_s)f_4\left(\frac{4\alpha\theta}{l^2}, \frac{hl}{2k}\right)$$
$$= 70 + (1100 - 70)f_4(9.1, 0.094) \tag{18.54}$$

From Fig. 18.11, $f_4(9.1, 0.094) = 0.43$

$$t_{x=0} = 70 + (1100 - 70)0.43 = 512°F$$

Finite and Semiinfinite Shapes Heated by a Fluid with a Contact Resistance. Referring to Eq. (18.52a) or (18.53a) Gurney and Lurie[1] noted that the relationships for heating various shapes with fluids having finite or infinite film coefficients could be represented by the four dimensionless groups $4\alpha\theta/l^2$, $(T_s - t)/(T_s - t_0)$, $2k/hl$, and $2x/l$. With these dimensionless parameters they made plots for the finite wall (slab), cylinder of infinite length, sphere, and semiinfinite solid. These are reproduced in Figs. 18.12 through 18.15. Gurney-Lurie charts are particularly useful, since they permit the calculation of the temperature not only at the surface and center of the object but at intermediate points as well. The ordinate Y is identical with the Schack chart for the same shape. Gurney-Lurie charts, however, are more difficult to interpolate than are Schack charts. The case of the semiinfinite solid is identical with the slab for the short initial period before the heat flow has reached the center plane of the slab. The use of these charts is demonstrated by the solution of an illustrative problem.

Example 18.5. The Gurney-Lurie Chart. Steel circular rods 8 in. in diameter and 12 ft long initially at 400°F are cooled in oil before being machined. The oil is maintained at a constant temperature of 200°F by means of a quench oil cooler. The average free convection coefficient from an 8-in. pipe to an oil of the properties of the quench oil is 50 Btu/(hr)(ft²)(°F).

What is the temperature 2 in. below the surface after 15 min?

[1] Gurney, H. P., and J. Lurie, *Ind. Eng. Chem.*, **15**, 1170 (1923).

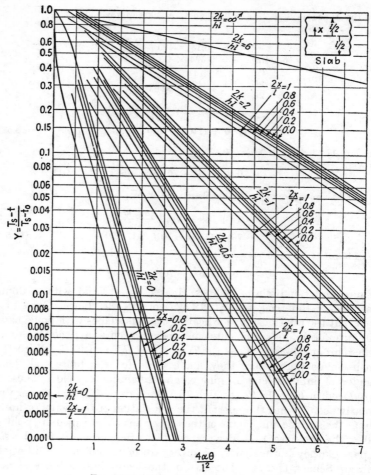

Fig. 18.12. Gurney-Lurie chart for slab.

Solution. This corresponds to an infinitely long cylinder, and Fig. 18.13 will be used. α will be evaluated at the mean temperature of 300°F

$$k = 25 \quad c = 0.12 \quad \rho = 490$$
$$\alpha = \frac{k}{c\rho} = \frac{25}{0.12 \times 490} = 0.425$$
$$\frac{4\alpha\theta}{l^2} = \frac{4 \times 0.45 \times {}^{15}\!/_{60}}{({}^{8}\!/_{12})^2} = 1.01$$
$$\frac{2x}{l} = \frac{4}{8} = 0.5$$
$$\frac{1}{Nu} = \frac{2k}{hl} = \frac{2 \times 25}{50 \times {}^{8}\!/_{12}} = 1.5$$

From Fig. 18.13

$$Y = 0.31 = \frac{T_s - t}{T_s - t_0}$$

$$t_{x=2\text{ in.}} = 400 + 0.31(200 - 400) = 338°F$$

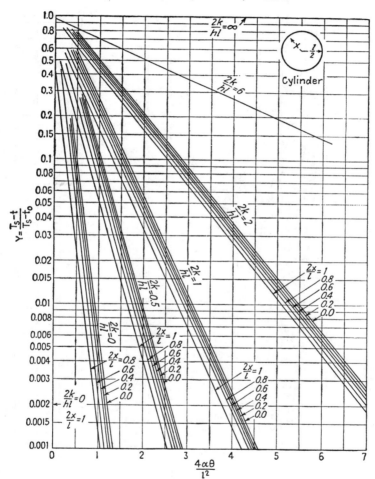

FIG. 18.13. Gurney-Lurie chart for cylinders.

Newman's Method for Common and Composite Shapes. Many of the objects which are regularly subjected to quenching do not conform with the simple shapes dealt with so far. Among such common shapes are the rectangular bar, the brick (rectangular parallelepiped), and the short cylinder of finite length. For heat flow along the x axis alone the temperature of the object is defined by Y or $Y_x = (T_s - t)/(T_s - t_0)$, where the subscript x indicates the direction of the heat flow. The center of the object is taken as the reference point. In a long rectangular bar

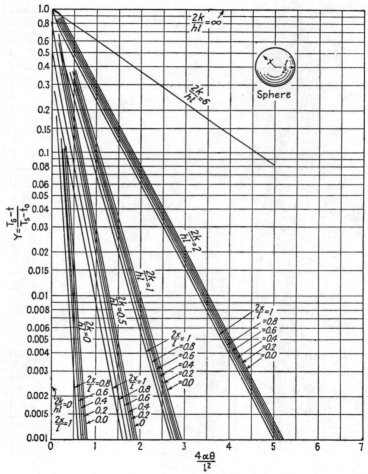

Fig. 18.14. Gurney-Lurie chart for spheres.

whose cross section is reckoned in the x and y directions, Newman[1] has proved that the temperature at any point in the cross section is given by $(T_s - t)/(T_s - t_0) = Y_x Y_y$, where both Y_x and Y_y have been calculated for the flow of heat through a finite slab. Similarly for a brick whose

[1] Newman, A. B., *Trans. AIChE*, **27**, 203 (1931).

cross section is reckoned by x and y the temperature at any point can be defined by $(T_s - t)/(T_s - t_0) = Y_x Y_y Y_z$, where the value of Y_z has been determined for the long dimension of the brick as a slab with heat flow

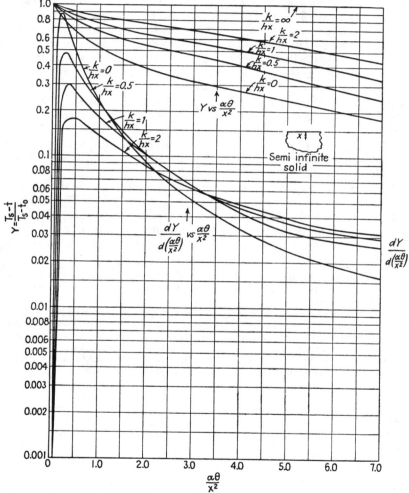

Fig. 18.15. Gurney-Lurie chart for semiinfinite solid.

in the z direction. For the cylinder of finite length if the length is l_x and the diameter l_z, the temperature can be defined by

$$\frac{T_s - t}{T_s - t_0} = Y_x Y_z.$$

In objects of finite length if one of the plane faces is insulated against the flow of heat, the calculation is treated as above except that the zero axis or reference point is moved to the insulated wall instead of the center and the distance perpendicular to the insulated face is doubled (l instead of $l/2$). If a pair of parallel faces is insulated so that there is no net heat flow in the x, y, or z direction, the entire direction is omitted. Similarly, it is possible that the film coefficient to the different faces of an object may not be identical on all sides, since the flow of fluid may be mostly parallel to some surfaces and perpendicular to others. This will not interfere with the solution of a problem for objects having parallel faces as long as the coefficient on any two parallel faces is the same and the heat-flow symmetry about the center is preserved. The values of Y for these shapes can be obtained from Schack or Gurney-Lurie charts for the finite slab or infinite cylinder and from which complex shapes can be built up accordingly. A Schack chart for the infinite cylinder has been prepared by Newman.[1]

To illustrate this method, consider a brick which has the dimensions l_x, l_y, and l_z in the x, y, and z directions, respectively. The extreme distances of the faces of the brick reckoned from its center are $l_x/2$, $l_y/2$, and $l_z/2$. At its center it is apparent that $l_x/2 = 0$, $l_y/2 = 0$, and $l_z/2 = 0$. For the cylinder, at the circumference the dimension is $l_x/2$ and at the parallel end faces $l_z/2$.

Example 18.6. The Application of Newman's Method to Heating a Brick. This problem uses the data of an example by Newman with numerical values obtained from Gurney-Lurie charts.

A firebrick 9 by 4.5 by 2.5 in. initially at 70°F is suspended in a flue through which furnace gases at 300°F are traveling at such a rate that the resulting film coefficient on all faces is 4.1 Btu/(hr)(ft²)(°F). Estimate the temperatures at the following points at the end of 1 hr: (a) the center of the brick, (b) any corner of the brick, (c) the center of the 9- by 4.5-in. faces, (d) the center of the 9- by 2.5-in. faces, (e) the center of the 4.5- by 2.5-in. faces, (f) the middle of the long edges.

The following data are supplied:

$k = 0.3$ Btu/(hr)(ft²)(°F/ft)
$\rho = 103$ lb/ft³
$c = 0.25$ Btu/(lb)(°F)

Solution:

$$\alpha = \frac{k}{c\rho} = \frac{0.3}{0.25} \times 10.3 = 0.01164 \text{ ft}^2/\text{hr}$$

$$\frac{4\alpha\theta}{l_x^2} = \frac{4 \times 0.01164 \times 1.0}{(9/12)^2} = 0.0828 \qquad \frac{2k}{hl_x} = \frac{2 \times 0.3}{4.1 \times 9/12} = 0.195$$

$$\frac{4\alpha\theta}{l_y^2} = \frac{4 \times 0.01164 \times 1.0}{(4.5/12)^2} = 0.3313 \qquad \frac{2k}{hl_y} = \frac{2 \times 0.3}{4.1 \times 4.5/12} = 0.390$$

$$\frac{4\alpha\theta}{l_z^2} = \frac{4 \times 0.01164 \times 1.0}{(2.5/12)^2} = 1.073 \qquad \frac{2k}{hl_z} = \frac{2 \times 0.3}{4.1 \times 2.5/12} = 0.702$$

[1] Newman, A. B., *Ind. Eng. Chem.*, **28**, 545-548 (1936).

From Fig. 18.12

	At the center	At the surface
$\dfrac{2x}{l} =$	0	1.0
	$Y_x = 0.98*$	$Y_x = 0.325*$
	$Y_y = 0.75$	$Y_y = 0.29$
	$Y_z = 0.43$	$Y_z = 0.245$

For all cases,
$$Y = \frac{T_s - t}{T_s - t_0} = \frac{300 - t}{300 - 70} = \frac{300 - t}{230}$$

(a) Center of brick ($l_x = 0;\ l_y = 0;\ l_z = 0$):
$$\frac{300 - t}{230} = 0.98 \times 0.75 \times 0.43 = 0.316$$
$t = 227.4°F$

(b) Corner of brick ($l_x/2,\ l_y/2,\ l_z/2$):
$$\frac{300 - t}{230} = 0.325 \times 0.29 \times 0.245 = 0.023$$
$t = 294.7°F$

(c) Center of 9- by 4.5-in. face ($l_x/2 = 0,\ l_y/2 = 0,\ l_z/2$):
$$\frac{300 - t}{230} = 0.98 \times 0.75 \times 0.245 = 0.18$$
$t = 258.5°F$

(d) Center of 9- by 2.5-in. face ($l_x/2 = 0,\ l_y/2,\ l_z/2 = 0$):
$$\frac{300 - t}{230} = 0.98 \times 0.29 \times 0.43 = 0.122$$
$t = 272°F$

(e) Center of 4.5- by 2.5-in. face ($l_x/2,\ l_y/2 = 0,\ l_z/2 = 0$):
$$\frac{300 - t}{230} = 0.325 \times 0.75 \times 0.43 = 0.105$$
$t = 275.8°F$

(f) Middle of long edge ($l_x/2 = 0,\ l_y/2,\ l_z/2$):
$$\frac{300 - t}{230} = 0.98 \times 0.29 \times 0.245 = 0.0695$$
$t = 284°F$

The Graphical Determination of the Time-temperature Distribution.
The time-temperature distribution for many practical problems has not appeared in the literature because their solutions are lengthy or their mathematics is extremely complex. A useful and concise method of treating such cases graphically has been developed by E. Schmidt.[1] Consider a slab of infinite width and finite thickness in which the heat flows only through the thickness. As before, the time-temperature relationship can be obtained by solution of the basic conduction equation

$$\frac{dt}{d\theta} = \alpha \frac{d^2 t}{dx^2} \tag{2.12}$$

* These values are read more easily from Schack charts.
[1] E. Schmidt, "Foppls Festschrift," pp. 179–189, Verlag Julius Springer, Berlin, 1924. See particularly, Sherwood and Reed, *op. cit.*, pp. 241–255.

The temperature in the slab at any point is a function of the time and distance. Divide the slab into a number of distance increments of Δx ft each, and consider an increment of time $\Delta \theta$. At a constant distance x from one face of the slab the incremental temperature change during a finite time change $\Delta \theta$ can be indicated by Δt_θ. For a constant value of θ the variation of the temperature with x can be indicated by Δt_x. Equation (2.12) can then be written

$$\frac{\Delta t_\theta}{\Delta \theta} = \alpha \frac{\Delta^2 t_x}{\Delta x^2} \qquad (18.55)$$

or rearranging,

$$\Delta t_\theta = \alpha \frac{\Delta \theta}{\Delta x^2} \Delta^2 t_x \qquad (18.56)$$

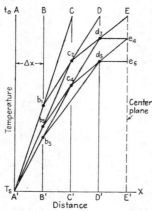

Fig. 18.16. Schmidt graphical method.

Referring to Fig. 18.16 in which the slab is divided into distance increments Δx, let $t_{n,m}$ represent the temperature at n distance increments from the origin $n\,\Delta x$ after m time increments, $m\,\Delta \theta$. When x is constant, the change of temperature with time in the layer $m\,\Delta x$ from the surface is

$$\Delta t_\theta = t_{n,m+1} - t_{n,m} \qquad (18.57)$$

When θ is a constant, the variation of the temperature with distance is

$$\Delta t_x = t_{n+1,m} - t_{n,m} \qquad (18.58)$$

and for the second change or difference between two differences

$$\Delta^2 t_x = \Delta(\Delta t_x) = (t_{n+1,m} - t_{n,m}) - (t_{n,m} - t_{n-1,m}) \qquad (18.59)$$

Substituting these in Eq. (18.56),

$$t_{n,m+1} - t_{n,m} = \alpha \frac{\Delta \theta}{\Delta x^2}[(t_{n+1,m} - t_{n,m}) - (t_{n,m} - t_{n-1,m})] \qquad (18.60)$$

If the increments of distance and time are chosen so that

$$\alpha \frac{\Delta \theta}{\Delta x^2} = \frac{1}{2} \qquad (18.61)$$

Equation (18.60) reduces to

$$t_{n,m+1} = \tfrac{1}{2}(t_{n+1,m} + t_{n-1,m}) \qquad (18.62)$$

Equation (18.62) is the basis of the graphic method. It implies that the temperature at any point at any time is the arithmetic mean of the two temperatures at $+$ and $-$ Δx during the preceding time increment $\Delta\theta$. A straight line drawn through the values of the temperatures at $(n-1)\Delta x$ and $(n+1)\Delta x$ intersects the vertical median at the arithmetic mean where the temperatures at $(n-1)\Delta x$ and $(n+1)\Delta x$ were the two previous temperatures.

The complete method is illustrated by Fig. 18.16. Consider a symmetrical object such as an infinite slab at initial temperature t_0 suddenly subjected at both faces to a cooling medium with zero contact resistance and temperature T_s. The heat flow is along the x axis. Since the temperature distribution about the center plane is symmetrical, only one-half of the slab need be considered and the half of the slab under consideration is broken up into the distance increments shown by the vertical lines. If the initial temperature upon contact is T_s at 0 and the temperature is t_0 at C, then at the time increment $\Delta\theta$ later the temperature in the plane B-B' will be the arithmetic mean of t_0 and T_s, b_1. Points C, D, and E remain unchanged during the first increment. During the second time increment the temperature at C in plane C-C' falls to c_2, D and E remaining unchanged. During the third time increment the temperature at b_1 falls to b_3 and the temperature at D falls to d_3. During this increment the temperature at the center plane does not change, since it is the mean of the values at $\pm\Delta x$ from the center plane, which are both still at t_0. During the fourth increment the temperature at C falls from c_2 to c_4, but the temperature at the center plane e_4 is the mean of the two identical values of d_3 at $\pm\Delta x$ from the center plane and therefore lies on a horizontal line. The procedure can be continued indefinitely with each horizontal line across the center plane representing two time increments.

Example 18.7. The Graphical Determination of the Time-Temperature Distribution. A steel slab 20 in. thick and at an initial temperature of 500°F is suddenly subjected to a temperature of 100°F on both faces. What is the temperature distribution after 20 min?

Solution. For simplicity take $\alpha = 0.40$ ft²/hr. Take distance increments Δx 2 in. apart.
$$\Delta x = \tfrac{2}{12} = 0.167 \text{ ft}$$

From the conditions of Eq. (18.61) take time increments such that $\alpha(\Delta\theta/\Delta x^2) = \tfrac{1}{2}$.

$$\Delta\theta = \frac{0.167^2}{2 \times 0.40} = 0.035 \text{ hr or } 2.1 \text{ min}$$

Number of steps required $= 20/2.1 = 9.5$

In Fig. 18.17 half of the slab is divided into distance increments of 2 in. and the time intervals are drawn accordingly. After 9 time increments corresponding to $9 \times 2.1 = 18.9$ min, the temperature at the center plane is 413°F. After 11 increments or 23.1 min it is 383°F. The value at 20 min can be obtained by plotting a time-temperature curve for the center plane alone from which a temperature of 406°F is obtained.

Time-Temperature Distribution with Contact Resistance. The instances in which the heat source is applied with negligible contact resistance are exceedingly rare. A graphical method was developed by Schmidt for a slab when the film coefficient to it was finite. Sherwood and Reed[1] have given an excellent presentation of the method. The presence of a finite film coefficient or contact resistance indicates that a temperature drop occurs between the bulk temperature of the surround-

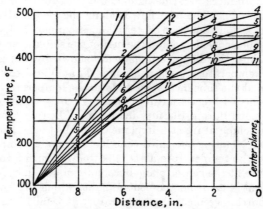

FIG. 18.17. Solution of Example 18.7 by the Schmidt method.

ing medium at T_s and the surface of the slab t_f. The heat balance across the surface is given by

$$k\left(\frac{dt}{dx}\right)_{x=0} = h(T_s - t_f) \qquad (18.63)$$

where T_s is again the temperature of the surroundings and t_f is the surface temperature. Accordingly, the temperature gradient across the surface is given by

$$\left(\frac{dt}{dx}\right)_{x=0} = \frac{T_s - t_f}{k/h} \qquad (18.64)$$

Any line cutting the surface of the slab on t vs. x coordinates such as in Fig. 18.16 must consequently have a numerical slope of $(T_s - t_f)/(k/h)$. Referring to Fig. 18.18, a slab is divided into distance increments Δx, except that they are marked off such that the surface of the slab corresponds to the middle of a distance increment. The reason will be apparent as the construction progresses. Next, the origin O is set off at a distance k/h ft to the left of the surface and at the temperature T_s.

[1] Sherwood and Reed, *op. cit.*

BATCH AND UNSTEADY-STATE PROCESSES   661

The slope of a line drawn from the origin O through the surface plane then has the slope $(T_s - t_f)/(k/h)$. By constructing the vertical at $\Delta x/2$ to the left of the surface, Schmidt's rule can be applied and each alternate time increment is represented by a line crossing the surface with the appropriate slope. The surface is thus the locus of values of t_f. The construction is carried out by the preceding Schmidt method until the center plane has been passed by a line drawn to $\Delta x/2$ at the right of the center plane. Since the distance increments are displaced by the distance $\Delta x/2$, the plane at $\Delta x/2$ to the right of the center plane is the

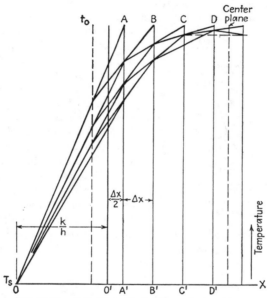

FIG. 18.18. Schmidt method with contact resistance.

mirror image of that at $\Delta x/2$ to the left with the temperatures drawn to corresponding points on the plane to the right. Thus for alternate increments of time the line slopes up and down.

Nonsymmetrical cases of heat flow can be studied by this approximate method when mathematical derivations lead to involved expressions. Such a condition arises when one face of a slab is exposed to a finite film coefficient while the other is exposed to a negligible resistance. Another common condition is that in which the faces are at two different temperatures or in which the temperatures undergo cyclical variations. For this last a number of repeated solutions are required, but they provide results which otherwise might be unobtainable. Schmidt has also provided

2b. PERIODICALLY VARYING TEMPERATURE

Periodic Variation of the Surface Temperature. There are a number of instances in which the temperature of the heating medium is not constant. In some, the medium and the surface temperatures vary harmonically. Typical examples of the latter are the temperature of the earth's surface or within an internal combustion cylinder, although there are not many practical applications. Figure 18.19 shows the variation at a distance x in a wall whose surface temperature varies as a sine function with time. Inasmuch as the heat must travel from the surface into the bulk of the material and the thermal diffusivity is finite, there is a time lag before the temperature-time variations at the surface are reproduced at the distance x. Naturally, the maximum amplitude at x is less than that at the surface, and it is still less at greater distances. It should be noted that Fig. 18.19 is a plot of time vs. temperature and not distance vs. temperature, so that the decline in the amplitude with distance is not indicated. If f is the number of complete periodic changes per hour, the time required for a point within the body to respond to a variation at the surface is evidently given by

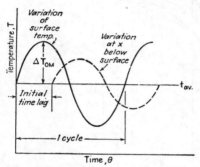

FIG. 18.19. Sine temperature variation at surface and at point x ft below.

$$\theta = \frac{x}{2}\sqrt{\frac{1}{\alpha \pi f}} \qquad (18.65)$$

where x is the distance in feet from the surface to the point. If the temperature variation is single harmonic, the surface temperature can be defined in terms of the maximum surface temperature by

$$\Delta t_{x=0} = \Delta t_{OM} \sin 2\pi f \theta \qquad (18.66)$$

where t_{OM} is the maximum temperature at the surface where $x = 0$. At the distance x the temperature has the same period but lags by $(x/2)\sqrt{1/\alpha \pi f}$ hr, the time initially required to travel from the surface to x. The maximum temperature at x is given by

$$\Delta t_{x=x,\max} = \Delta t_{OM} e^{-x\sqrt{\pi f/\alpha}} \qquad (18.67)$$

BATCH AND UNSTEADY-STATE PROCESSES   663

The equation for the variation of the temperature at any distance in typical problems requires the solution of a Fourier series which is beyond the limitations of space here. If the variation at the distance x at any time θ can be expressed as a sine function, it is given by

$$\Delta t_x = \Delta t_{0M} e^{-x\sqrt{\pi f/\alpha}} \sin\left(2\pi f\theta - x\sqrt{\frac{\pi f}{\alpha}}\right) \qquad (18.68)$$

If the variation can be expressed as a cosine function instead of a sine function, the temperature at any point x is given by

$$\Delta t_x = \Delta t_{0M} e^{-x\sqrt{\pi f/\alpha}} \cos\left(2\pi f\theta - x\sqrt{\frac{\pi f}{\alpha}}\right) \qquad (18.69)$$

The flow of heat through a flat wall for a half period can be determined from the basic conduction equation

$$dQ' = kA\left(\frac{dt}{dx}\right)_{x=0} d\theta$$

by differentiating Eq. (18.68) or (18.69) with respect to x, from which the following is obtained for Q for a half period:

$$Q = kA\, \Delta t_{0M} \sqrt{\frac{2}{\alpha\pi f}} \qquad (18.70)$$

Example 18.8. Calculations for a Wall with Periodic Temperature Variation. The surface of a thick brick wall is subjected to a daily temperature variation from 60 to 120°F. The cosine variation may be assumed. The cement is assumed to have the same properties of the bricks, which are $k = 0.3$ Btu/(hr)(ft²)(°F/ft), $\rho = 103$ lb/ft³, $c = 0.25$ Btu/(lb)(°F), and $\alpha = k/c\rho = 0.01164$ ft²/hr.
(a) What is the temperature lag 6 in. below the surface?
(b) What is the amplitude at this depth?
(c) What is the temperature deviation from the mean after 2 hr?
(d) What is the heat flow during the half period?

Solution:

(a) Temperature lag 6 in. below the surface:

$$\theta = \frac{x}{2}\sqrt{\frac{1}{\alpha\pi f}}$$

$$= \frac{6/12}{2}\sqrt{\frac{1}{0.01164\pi \cdot 1/24}} = 6.45 \text{ hr} \qquad (18.65)$$

(b) Amplitude:

$$\Delta t_{0M} = \frac{120 - 60}{2} = 30°\text{F}$$

$$\Delta t_{x=x} = \Delta t_{0M} e^{-x\sqrt{\pi f/\alpha}}$$

$$= 30 \cdot e^{-0.5\sqrt{\pi/24 \times 0.01164}} = 5.8°\text{F}$$

(c) Temperature deviation after 2 hr:

$$\Delta t_x = \Delta t_{0M} e^{-x\sqrt{\pi f/\alpha}} \cos\left(2\pi f\theta - x\sqrt{\frac{\pi f}{\alpha}}\right) \quad (18.69)$$
$$= 5.8 \cos (2 \times \pi \times \tfrac{1}{24} \times 2 - 1.635)$$
$$= 5.8 \cos 1.10 = 2.6°F$$

(d) Heat flow during the half period:

$$\frac{Q}{A} = k\,\Delta t_{0M}\sqrt{\frac{2}{\alpha\pi f}} \quad (18.70)$$
$$= 0.3 \times 30 \sqrt{\frac{2}{0.01164\pi\tfrac{1}{24}}} = 326 \text{ Btu/hr(ft}^2)$$

c. Regenerators

Introduction. A regenerator is an apparatus in which heat is alternately stored and removed. The broadest applications of regeneration have been in the blast and open-hearth furnaces of the steel industry. More recently regeneration applications have been extended to the "reversing exchanger" for large-scale air-separating plants in the Fischer-Tropsch process and other synthetic processes. A type of reversing exchanger which was popular in Germany employed Fränkl regenerator packing, which consists of metallic ribbons having corrugations at an angle of 45°. Two such ribbons are wound together in spirals which are then packed one above the other. Temperature histories and coefficients have been reported by Lund and Dodge.[1] Several of the proposed designs for the reversing exchanger also included the storage of heat by means of extended surfaces. Schack[2] presents an excellent survey of the literature on regenerators as well as his own work, which is recommended to the reader who is undertaking a comprehensive study of regenerator design. As treated here the regenerator is cited as another aspect of the unsteady state as well as to indicate the general methods of heat storage in processes. Actual regenerators require additional refinements in assumptions and practices.

Temperature Variations in Regenerators. The open-hearth furnace is charged with molten pig iron and scrap to supply the oxygen requirements for the oxidation of impurities. A mixture of hot fuel gas and air is burned above the pool of molten metal, thereby maintaining the temperature and removing the impurities arising from the pool. The products of combustion and oxidation are *waste gases* containing considerable heat which might profitably be recovered by preheating the fuel-gas–air mixture. When the waste gases leave the furnace, they are passed through

---

[1] Lund, G., and B. F. Dodge, *Ind. Eng. Chem.*, **40**, 1019–1032 (1948).
[2] Schack, *op. cit.*

a cold brick checkerwork to which they give up a large portion of their heat. In a matter of a few minutes the bricks become very hot, and at some optimum surface temperature the flow of waste gases is replaced by the passage of cold fuel-air mixture on its way to the furnace. This mixture is heated while cooling the brickwork back to a low temperature. In the meantime the flow of the waste gases continues over another brickwork only to be interrupted by the flow of more fuel-gas–air mixture. By employing a sufficient number of brick checkerworks it is possible to operate on a continuous basis with fuel-gas–air mixtures alternating with the waste gases in the transfer of heat.

It can be seen that the inlet temperature of either the waste gases or the fuel-gas–air mixture to the checkerwork is substantially constant but the outlet temperatures vary. This differs from preceding time-temperature relationships. The cyclic operation of a regenerator consists of two separate periods. During the heating period the brickwork absorbs heat, and during the cooling period the heat is removed, although the periods are usually of unequal duration. To simplify regenerator terminology the conventions of Schack and others will be followed. In these, gas refers to the hot medium (waste gas) and air to the cold medium. Heilgenstadt[1] has shown that the heat transfer in a generator can be represented by a modified form of the Fourier equation for 1 ft² of surface by

$$\frac{dq'}{dA} = H(T - t) \qquad \text{Btu/(period)(ft}^2) \qquad (18.71)$$

where q' is the heat transferred during the period from the gas to air whose mean temperatures are T and t, respectively. H is the overall coefficient of heat transfer for the *period*, Btu/(period)(ft²)(°F), and differs from U in the unit of time. Integrating, since $A = 1$, Eq. (18.71) becomes

$$q' = H(T - t) \qquad \text{Btu/(period)(ft}^2) \qquad (18.72)$$

Using θ_H and θ_C to indicate the duration of the gas and air half periods in hours, respectively, the heat transferred during the half periods is

$$q' = h_H(T - t_s)\theta_H = h_C(t_s - t)\theta_C \qquad (18.73)$$

where t_s is the mean temperature of the brickwork, which is assumed to be identical for both half-periods in an ideal regenerator and h_H and h_C are the gas and air film coefficients, Btu/(hr)(ft²)(°F). Solving for an ideal regenerator,

$$t_s = \frac{h_H T \theta_H + h_C t \theta_C}{h_H \theta_H + h_C \theta_C} \qquad (18.74)$$

[1] Heilgenstadt, W., *Mitt. Warmestelle Ver. deut. Eisenh.*, **73**, (1925).

From Eqs. (18.72) and (18.73)

$$H(T - t) = h_H T \theta_H - h_c t \theta_c \tag{18.75}$$

The value of H in Eq. (18.72) is thus the value for the ideal regenerator or H_{id}. Substituting in Eq. (18.73),

$$H_{id} = \frac{1}{1/h_H \theta_H + 1/h_c \theta_c} \tag{18.76}$$

In an actual regenerator the temperature of the gas decreases several times more rapidly than the air is heated. If T_s is the mean temperature of the surface during cooling only and t_s the mean during heating, the solution of Eq. (18.73) gives

$$H = H_{id}\left(1 - \frac{t_s - T_s}{T - t}\right) \tag{18.77}$$

Equation (18.71) can also be expressed in terms of the mean temperature of the brickwork.

$$dq' = h_H(T - t_s) \, dA \, d\theta \quad \text{Btu/(period)(ft}^2) \tag{18.78}$$

Considering the brick as a slab of finite thickness whose temperature varies periodically on both sides, Grober has defined a heat storage factor γ which is the ratio of the heat actually stored in the wall to that which could have been stored if the conductivity of the wall were infinite or

$$q' = q'_{\max} \gamma$$

where γ is $f_5(4\alpha\theta/l^2)$. Values of $f_5(4\alpha\theta/l^2)$ are plotted in Fig. 18.20. In terms of the brick material the weight of the brickwork for 1 ft² of heat-transfer surface is M/A, where the depth is one-half the thickness of the brick. If c_s is the specific heat of the solid, the heat absorbed by the brickwork is then

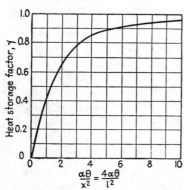

Fig. 18.20. Heat storage factor.

$$dq' = Mc_s\gamma \, dt_s \, dA \quad \text{Btu/(period)(ft}^2) \tag{18.79}$$

Equating (18.78) and (18.79),

$$Mc_s\gamma \, dt_s = h_H(T - t_s) \, d\theta \tag{18.80}$$

or rearranging,

$$\frac{dt_s}{d\theta} = \frac{h_H}{Mc_s\gamma}(T - t_s)$$

BATCH AND UNSTEADY-STATE PROCESSES 667

Differentiating with respect to θ,

$$\frac{d^2 t_s}{d\theta^2} = \frac{h_H}{Mc_s\gamma}\frac{dT}{d\theta} - \frac{h_H}{Mc_s\gamma}\frac{dt_s}{d\theta} \tag{18.81}$$

Solving for $dT/d\theta$,

$$\frac{dT}{d\theta} = \frac{dt_s}{d\theta} + \frac{Mc_s\gamma}{h_H}\left(\frac{d^2 t_s}{d\theta^2}\right)$$

from which it will be seen that the temperature of the gas is not only dependent upon the temperature of the wall but also upon its rate of change with time. From Eq. (18.73) and based on 1 ft^2 of surface,

$$q = h_H(T - t_s) \quad \text{Btu/hr}$$

in which q is the quantity of heat absorbed per hour under the potential difference $T - t_s$. Differentiating with respect to θ and substituting in Eq. (18.81),

$$\frac{dT}{d\theta} = \frac{h_H}{Mc_s\gamma}(T - t_s) + \frac{d(T - t_s)}{d\theta} \tag{18.82}$$

Integrating for a finite heating period θ,

$$\Delta T = \frac{q_{av}\theta}{Mc_s\gamma} + \frac{1}{h_H}(q_\theta - q_0) \tag{18.83}$$

where q_0 is the number of Btu transferred per hour at the start of the period and q_θ represents the transfer after time θ. If $T_{2,\theta=0}$ is the waste-gas exit temperature at the start of the period (the time for travel from inlet to outlet being negligible) and $T_{2,\theta=\theta}$ is the gas exit temperature at a finite time after the period has commenced, at the start

$$q_0 = \frac{WC}{A}(T_1 - T_{2,\theta=0}) \quad \text{Btu/(hr)(ft}^2) \tag{18.84}$$

and after θ hr

$$q_\theta = \frac{WC}{A}(T_1 - T_{2,\theta=\theta}) \tag{18.85}$$

where the inlet gas temperature T_1 is constant.

$$q_\theta - q_0 = \frac{WC}{A}(T_{2,\theta=0} - T_{2,\theta=\theta}) \quad \text{Btu/(hr)(ft}^2) \tag{18.86}$$

Substituting in Eq. (18.83),

$$\Delta T = \frac{q_{av}\theta}{Mc_s\gamma\left(1 - \dfrac{WC}{h_H A}\right)} \tag{18.87}$$

Although data are available on the heat-transfer coefficients to actual regenerators, they can be found roughly from Fig. 24 for smooth surfaces by using the equivalent diameter defined by Eq. (6.3) in place of D.

2d. Heat Transfer to Granular Materials in Beds

This case is of particular interest in processes employing fixed- and moving-bed catalysts. It can also be used to obtain coefficients for combination with regenerator equations when using granular materials. An excellent review of the literature on this subject has been given by Lof and Hawley.[1] Perhaps the most ambitious original undertaking in this field was by Schumann,[2] who formulated and solved the equations for the case of an incompressible fluid flowing through a bed of solids with infinite thermal conductivity. Schumann assumed that (1) any given particle could be considered to attain a uniform temperature throughout at any given instant, (2) the resistance to conduction in the solid itself was negligible, (3) the rate of heat transfer from fluid to solid at any point was proportional to the temperature difference between the two at the point, (4) the change in the volumes of fluid and solid with the temperature variation was negligible, and (5) thermal properties were independent of the temperature during a temperature-change cycle. Schumann calculated temperature-time curves in terms of the groups

$$\frac{T - t_0}{T_1 - t_0} \quad \frac{h'x}{C'v} = \frac{hx}{C'G} \quad \frac{h'\theta}{c'_s}(1 - f')$$

where
  T = gas temperature at any time, °F
  T_1 = inlet gas temperature, °F
  t = solid temperature at any time, °F
  t_0 = initial temperature of the solid bed, °F
  C' = specific heat of unit volume of gas at constant pressure, Btu/(ft^3)(°F)
  c'_s = specific heat of unit volume of solids, Btu/(ft^3)(°F)
  f' = fraction of voids in the solid bed
  G = mass velocity of fluid, lb/(hr)(ft^2)
  h' = volumetric heat-transfer coefficient, Btu/(hr)(ft^3)(°F)
  v = average volumetric fluid rate through bed, ft^3/(hr)(ft^2 of bed section)
  x = length of bed, ft

[1] Lof, G. O., and R. W. Hawley, *Ind. Eng. Chem.*, **40**, 1061–1070 (1948).
[2] Schumann, T. E. W., *J. Franklin Inst.*, **208**, 405 (1929).

Schumann's curves were later extended by Furnas,[1] and the data of both are given in Fig. 18.21. With a temperature-time curve of this form Furnas experimentally developed the equation for the volumetric heat-transfer coefficient.

$$h' = \frac{A'G^{0.7}T'^{4\,0.3}10^{0.68-3.56f'^2}}{D'^{0.9}_e} \qquad (18.88)$$

where T' = average of the inlet air and initial bed temperature, °F
A' = constant
D'_e = equivalent spherical diameter defined by Eq. (18.89), ft

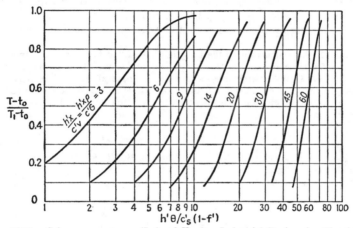

FIG. 18.21. Schumann curves. (*Lof and Hawley, Industrial Engineering Chemistry*)

$$D'_e = \left(\frac{6 \times \text{net volume of particles}}{\pi \times \text{number of particles}}\right)^{\frac{1}{3}} \qquad (18.89)$$

Lof and Hawley carried out experiments on granitic gravel ranging in size from 4 mesh to 1.5 in. and air rates from 12 to 66 standard cubic feet of air per minute per square foot of gross cross-sectional area. The thickness of the bed was 36 in. They obtained the following dimensional equation

$$h' = 0.79 \left(\frac{G}{D'_e}\right)^{0.7} \qquad (18.90)$$

Because the volumetric heat-transfer coefficient has not been introduced before, Table 18.1, computed by Lof and Hawley, is included for 0.394 in. particle diameters and an air rate of 50 standard cubic feet per minute per square foot, an inlet temperature of 200°F, and normal voids. The

[1] Furnas, C. C., *Trans. AIChE*, **24**, 142 (1930); *U.S. Bur. Mines Bull.* 361 (1932).

TABLE 18.1. SOLID BED COEFFICIENTS

Material	h', Btu/(hr)(ft³)(°F)
Iron ores	406–1200
Limestone	460
Coke	347
Blast-furnace charge	260
Coal	400
Firebrick	534
Iron balls	668
Gravel	399

authors suggest that Eq. (18.90) may be freely extrapolated to higher temperatures where the coefficient is increased by radiation.

The following problem has been adapted from Lof and Hawley.

Example 18.9. Calculation of the Length of a Bed. A bed of granitic-type gravel consisting of 1.0 in. particles has 45 per cent voids and is initially at a uniform temperature of 50°F. 60 lb/(hr)(ft²) cross-sectional area of hot air at 200°F is available for periods of 6 hr. It is considered wasteful from a heat-recovery standpoint to let the exit air leave the bed at a temperature above 90°F. How long a bed can be heated by the air stream?

Data:

Specific heat of gravel, 0.25 Btu/(lb)(°F) or 41.3 Btu/(ft³)(°F)
Actual density of gravel 165 lb/ft³
Specific heat of air 0.0191 Btu/(ft³)(°F)

Solution:

$$h' = 0.79 \left(\frac{G}{D'_e}\right)^{0.7} = 0.79 \left(\frac{60}{\frac{1}{12}}\right)^{0.7} = 79.4$$

$$\frac{h'\theta}{c'_s(1-f')} = \frac{79.4 \times 6}{41.3 \times 0.55} = 21.0$$

$$\frac{T - t_0}{T_1 - t_0} = \frac{90 - 50}{200 - 50} = 0.267$$

From Fig. 18.21,

$$\frac{hx\rho}{c'G} = 24.5 \qquad \rho = 0.0807 \text{ lb/ft}^3 \text{ air}$$

$$x = 24.5 \times \frac{0.0191 \times 60}{79.4 \times 0.0807} = 4.4 \text{ ft}$$

PROBLEMS

18.1. 6000 gal of liquid toluene is to be cooled from 300 to 150°F using water from 85 to 120°F. A pump in the tank is capable of circulating 30,000 lb/hr of toluene through an external exchanger. Available for the service is 400 ft² of heat-exchanger surface in various types, all affording approximately the same overall coefficient of 75 Btu/(hr)(ft²)(°F/ft).

(a) How long will it take to cool an agitated batch using, respectively, a double pipe exchanger, 1-2 exchanger, and 2-4 exchanger?

(b) How long will it take to cool a nonagitated batch using the same exchangers as in (a)?

(c) If the batch is cooled to 120°F, how long will it take in (a) and (b)?

(d) If the pumping rate is increased to 40,000 lb/hr and the coefficient increased to 90, how long will it take in (a) and (b)?

18.2. 6000 gal of liquid aniline is to be cooled from 300 to 150°F using water from 85 to 120°F. A pump in the tank is capable of circulating 30,000 lb/hr of aniline through an external exchanger. The overall clean coefficient is 40 Btu/(hr)(ft²)(°F).

(a) How much surface is required to accomplish the operation in 2 hr using double pipe, 1-2 and 2-4 exchangers with both agitated and nonagitated batches?

(b) How much surface is required for (a) if the time is reduced to 1 hr?

18.3. Cylindrical steel objects 4 in. long by 6 in. in diameter and at 1000°F are to be quenched by being dropped into a tank filled with recirculating water at a constant temperature of 100°F. The average overall coefficient on all faces will be about 75.

(a) How long will it take for the center of the cylinder to reach a temperature of 125°F?

(b) How long will it take for a depth 1 in. below the surface on the long axis to attain a temperature of 500°F?

18.4. A water main at 50°F passes through a mass of concrete at the same temperature, the whole being exposed to the atmosphere. The main is located 24 in. below the surface of the concrete. If the temperature were to fall to 0°F, how long would it take for the pipe to be in danger of freezing? Use $k = 0.65$ Btu/(hr)(ft²)(°F), $c = 0.20$, and $\rho = 140$ lb/ft³.

18.5. A steel slab 30 in. thick and at an initial temperature of 100°F is suddenly subjected to an air stream with a temperature of 300°F on both sides and a coefficient to steel of 5.0. What is the temperature distribution after ½ hr?

18.6. The local daily range of temperatures on a concrete roof is between 80 and 40°F. Assume that the roof is 12 in. thick and that the data in Example 18.4 apply.

(a) What variations occur 6 in. below the surface?

(b) What is the time lag?

(c) What is the temperature 4 hr after the maximum has been reached?

NOMENCLATURE FOR CHAPTER 18

A	Heat-transfer surface, ft²
C_0	Specific heat of hot fluid added to batch, Btu/(lb)(°F)
C	Specific heat of hot fluid, Btu/(lb)(°F)
C_1, C_2, C_3	Constants
c	Specific heat of cold fluid, Btu/(lb)(°F)
c_0	Specific heat of cold fluid added to batch, Btu/(lb)(°F)
c_s	Specific heat of solid, Btu/(lb)(°F)
C'	Specific heat of unit volume of gas, Btu/(ft³)(°F)
c'_s	Specific heat of unit volume of solids, Btu/(ft³)(°F)
f	Number of complete periodic temperature changes per hour, 1/hr
f'	Fraction of voids in solids, dimensionless
f_1, f_2, etc.	Abbreviations for various functions by which Y is expressed
H	Overall coefficient or heat transfer per period, Btu/(period)(ft²)(°F)
h	Heat-transfer coefficient, Btu/(hr)(ft²)(°F)
h'	Volumetric heat-transfer coefficient, Btu/(hr)(ft³)(°F)
h_H, h_C	Heat-transfer coefficient during heating and cooling half periods, respectively, Btu/(hr)(ft²)(°F)

h_c, h_r	Heat-transfer coefficient by convection and radiation, respectively, Btu/(hr)(ft^2)(°F)
K_1, K_2, etc.	Constants in batch heating and cooling equations, dimensionless
k	Thermal conductivity, Btu/(hr)(ft^2)(°F/ft)
L_0	Rate of liquid addition to a batch, lb/hr
L_1, L_2	Charge and residual quantities of liquid in batch distillation, mol
l	Length or width of object or diameter of cylinder, ft
M	Weight of batch liquid, lb; weight of solid, lb
m	Number of time increments, $\Delta\theta$
N	Number of circulations
n	Number of distance increments, Δx
Nu	Nusselt number, $hl/2k$
Q	Heat flow, Btu/hr
Q'	Heat, Btu
q	Heat flow during a heating or cooling period, Btu/(hr)(ft^2 of regenerator surface)
q'	Heat flow during a heating or cooling period, Btu/(period)(ft^2 of regenerator surface)
R	Temperature group $wc/WC = (T_1 - T_2)/(t_2 - t_1)$, dimensionless
S	Temperature group $(t_2 - t_1)/(T_1 - t_1)$, dimensionless
T	Temperature of hot fluid or batch at any temperature or time, °F
T_0	Temperature of hot fluid continuously added to batch, °F
T_s	Temperature of surrounding medium, °F
T_1, T_2	Initial and final temperatures of hot fluid, °F
t	Temperature of cold fluid, batch or solid, °F
t_f	Temperature of solid surface in the Schmidt method, °F
t_s	Temperature of brickwork in a regenerator, °F
t_n, t_m	Temperature of solid after n distance and m time increments, °F
t_0	Original temperature of solid, °F
t_{0M}	Maximum surface temperature, °F
Δt	True temperature difference, °F
$\Delta t_\theta, \Delta t_x$	Temperature difference after time θ and distance x, respectively, °F
$\Delta t_{n,m}$	Temperature difference after n distance and m time increments, °F
U	Overall coefficients of heat transfer, Btu/(hr)(ft^2)(°F)
v	Average volumetric fluid rate through bed, ft^3/(hr)(ft^2 of bed cross section)
W	Weight flow of hot fluid, lb/hr
w	Weight flow of cold fluid, lb/hr
x	Mol fraction of volatile component in liquid, dimensionless; distance, ft
x_1, x_2	Initial and final mol fractions of volatile component in liquid, dimensionless
Y	Temperature group $(T_s - t)/(T_s - t_0)$, dimensionless
Y_x, Y_y, Y_z	Temperature group Y in the directions x, y, and z, respectively, dimensionless
y	Mol fraction of volatile component in vapor, dimensionless; distance, ft
z	Function in the probability integral; distance, ft
α	Thermal diffusivity, $k/c\rho$, ft^2/hr
γ	Heat-storage factor, dimensionless
ϵ	Emissivity, dimensionless
θ	Time, hr

ρ	Density, lb/ft³
σ	Radiation constant, 0.173×10^{-8} Btu/(hr)(ft²)(°R⁴)

Superscripts and Subscripts (except as noted above)

0	Time or distance at start
p, q	Constants
x, y, z	Direction
θ	At time θ

CHAPTER 19

FURNACE CALCULATIONS

BY JOHN B. DWYER[1]

Introduction. The most important commercial applications of radiant-heat-transfer calculations are encountered in the design of steam-generating boilers and petroleum-refinery furnaces. Since the art of the construction of these units developed before the theory, empirical methods were evolved for the calculation of radiant-heat transfer in such furnaces. Various contributions[2] to the literature on general and specific radiant-heat transfer problems, especially those of H. C. Hottel, have made possible a more fundamental approach to furnace design. Several semitheoretical methods for furnace radiant-section heat-transfer calculations are now available. Often these methods can be adapted to the rapid solution of problems encountered in kilns, ovens, heat-treating and chemical furnaces, and miscellaneous equipment in which radiant-heat transfer is of importance.

It is the purpose of this chapter to present some of the empirical and semitheoretical methods of furnace radiant-section calculation, data for their use, and examples of their application. The limitations of these methods are pointed out, and their adaptability to miscellaneous heat-transfer problems indicated. Included is a brief description of several types of boilers and oil heaters currently in use. A discussion of the theoretical aspects of radiation from nonluminous gases is presented to illustrate the general scope of the problem, and the simplifications and assumptions used in reducing theories to practice are also pointed out.

While it is necessary to calculate the radiant-heat-transfer flux in order to design a furnace, many other factors often influence furnace arrangement, such as the permissible flux under various conditions and the extent and nature of boiler tube slagging on the efficiency of the surface. The precautions which must be taken to avoid the deposition

---

[1] The M. W. Kellogg Company. Acknowledgment is made of critical suggestions by W. E. Lobo of the same company.

[2] Perry, J. H., "Chemical Engineers' Handbook," 3d ed., pp. 483–498, McGraw-Hill Book Company, Inc., New York, 1950. McAdams, W. H., "Heat Transmission," Chap. 3 by H. C. Hottel, McGraw-Hill Book Company, Inc., New York, 1942. For additional references see McAdams, op. cit., p. 430.

of coke in oil-heating or vaporizing furnaces have a pronounced effect on the actual design of refinery furnaces. The art of furnace design, in fact, often far exceeds the importance of the calculations.

Steam-generating Boilers. There are two general types of steam-generating boilers, the *fire-tube* boiler and the *water-tube* boiler. The former consists of a cylindrical vessel having tubes passing through it which are rolled into the heads at each end of the vessel. The tube bundle is generally horizontal, and the upper section of the vessel is not tubed. Combustion gases pass through the tubes, and a water level is carried in the vessel to immerse the tubes completely but at the same time allowing disengaging space between the water level and top portion of the vessel. In occasional vertical-tube boilers of this type, the tubes must be immersed in water for that portion of their length required to reduce the temperature of the gases sufficiently to avoid overheating the uncooled upper portion of the tubes. Some of the water-cooled parts such as shell or tube sheets can be subjected to radiation from the combustion gases, since these parts may form a portion of the enclosure of the combustion chamber. The major mechanism of heat transfer from the gases to the tubes is convection. Fire-tube boilers seldom exceed 8 ft in diameter, and steam pressure is generally limited to 100 to 150 psig. Fire-tube boilers are used for low-capacity services up to 15,000 to 20,000 lb/hr of steam production for domestic, industrial, and process heating and for small-scale power generation as in locomotives, etc. Fuels employed may be coal, oil, or gas, and in some cases, local combustibles such as wood, sludge, etc., are used.

Water-tube boilers, as their name implies, have water within their tubes. Combustion of stoker or pulverized coal and coke or of gas or oil fuels provides radiation to the boiler tubes, and further heat transfer is accomplished by arranging the flow of hot gases over the tubes to provide convection-heat transfer. There are three important classifications of water-tube boilers: longitudinal drum, cross-drum straight tube, and cross-drum bent tube. The last is the most important of the three and will be discussed here briefly. Information on the other types can be found in Gaffert.[1]

Figure 19.1 shows a typical low-pressure boiler designed to generate 200,000 lb/hr of steam at 235 psig and 500°F. Since the saturation temperature at this pressure is only 401°F, 99°F of superheat is required. Only a small superheater is necessary, since the superheat duty is about 5 per cent of the total boiler duty. The radiant boiler tubes cover the entire wall and roof surface, forming a "water wall" by means of which

[1] Gaffert, G. A., "Steam Power Stations," 3d ed., McGraw-Hill Book Company, Inc., New York, 1946.

676   PROCESS HEAT TRANSFER

the temperature of the refractory walls is kept down, thus decreasing their maintenance. Often the water tubes are partially embedded in the walls. The radiant-section furnace walls sometimes are protected from overheating by circulating cooling air outside them. In the furnace shown, water is fed by gravity from the upper drums to headers at the bottom end of the water wall tubes on all four radiant walls. Circula-

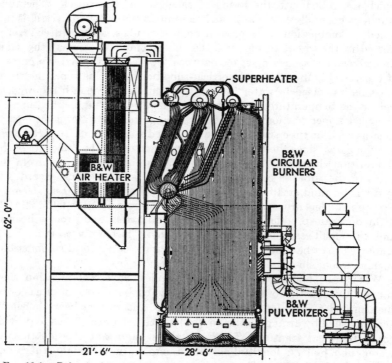

FIG. 19.1.   Pulverized-coal-fired low-pressure boiler.   (*Babcock & Wilcox Company.*)

tion is upward through these tubes, and the steam is disengaged from water in the upper drums, then passes through a steam separator before being superheated. In a low-pressure boiler, the convection tubes reduce the flue-gas temperature sufficiently that they proceed directly to the air preheater, obviating the need of an economizer (feed-water preheater). These convection tubes are the bent tubes running from the upper drums to the lower drum. Circulation in these tubes is, in general, downward in the left (cooler) bank and upward through the hotter bank.

A typical power-generating steam boiler is shown in Fig. 19.2. It has a capacity of 450,000 lb/hr of 900 psig steam delivered at 875°F. Since

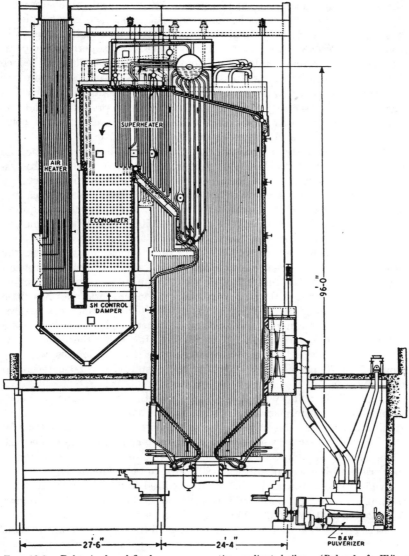

FIG. 19.2. Pulverized-coal-fired power-generation radiant boiler. (*Babcock & Wilcox Company.*)

the saturation temperature at 900 psig is 532°F, considerable superheat duty is required. Very little boiler convection surface can be placed between the radiant boiler and the superheater, since high-temperature combustion gases must be used to attain the required superheat temperature level with a reasonable amount of superheater tube surface. Since feed water must be brought essentially to the saturation temperature before it is admitted to the boiler drum, considerable heat is absorbed in the economizer section wherein the feed water is preheated, and the thermal efficiency of the unit is further increased by preheating the combustion air with the flue gases before they are sent to the stack.

Petroleum-refinery Furnaces. In atmospheric and vacuum crude distillation, thermal cracking, and modern high-temperature gas processing, the direct-fired tubular furnace is the primary factor in the refinery unit. Furnaces also are widely employed in various heating, treating, and vaporizing services. Refinery furnaces of various types are required for handling fluids at temperatures as high as 1500°F and at a combination temperature and pressure as severe as 1100°F and 1600 psig.

Oil or gas fuels are used exclusively in these furnaces, although in the near future the need may develop for firing with by-product petroleum coke. In general the thermal efficiency of refinery furnaces is considerably less than that of large boilers, since in many cases fuel has little worth in the refinery. With the trend toward utilization of a greater percentage of the crude oil produced, fuel is becoming scarcer and more valuable and refiners are recognizing the need for higher thermal efficiencies. It is expected that the range of thermal efficiencies will rise from 65 to 70 per cent employed in the past to 75 to 80 per cent in the future.

As in boilers, refinery furnaces usually contain both radiant- and convection-heat-transfer surface. Occasionally only radiant surface is employed for very low-capacity furnaces with duties up to 5,000,000 Btu/hr. Air preheaters have been used to a very limited extent because of the relative unimportance of fuel efficiency; however, even at moderate fuel prices their use generally can be shown to be economical.

In Fig. 19.3 is shown a box-type furnace fired from the end walls of the radiant section. Furnaces of this type might have capacities ranging from 25,000,000 to 100,000,000 Btu/hr heat input to the oil. Radiant tubes cover the side wall, roof, and bridge-wall (partition between radiant and convection section) surfaces. Oil is preheated in the bottom and top rows of the convection bank, then passes through the radiant tubes. After reaching an elevated temperature (900 to 1000°F) it is passed through a large number of convection-section tubes wherein it is maintained at a high temperature for sufficient time to accomplish the desired degree of cracking. These convection tubes are called the soaking section. The particular furnace shown employs flue-gas recirculation which serves to increase the convection-section duty and decrease the radiant-section duty. The amount of flue gas recirculated is controlled by two factors, namely, (1) limiting the radiant-section flux to prevent overheating the tubes and causing coke deposition within them and (2) controlling

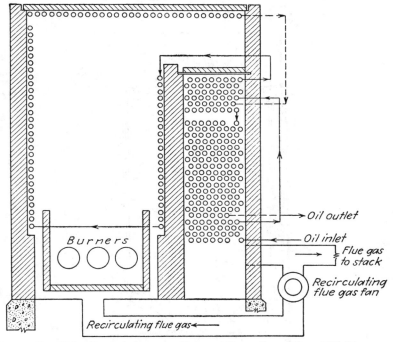

Fig. 19.3. Box-type furnace. (*Lobo and Evans, Transactions AIChE.*)

the temperature gradient through the convection soaker coil. The more nearly constant the oil temperature is held for a given outlet temperature the greater will be the "soaking factor" and extent of cracking. This assumes, of course, that the outlet temperature is the highest oil temperature. The endothermic reaction heat of cracking may result in a condition in which the oil temperature drops from soaker inlet to outlet. Such a temperature drop is inadvisable, particularly in vapor-phase cracking, since polymers formed in the vapor may condense on the tube walls and undergo further cracking to produce coke.

Figure 19.4 shows a De Florez type furnace, which is circular in cross section and employs vertical tubes. All the radiant tubes are equidistant from the burners, ensuring good circumferential heat distribution, but the flux may vary considerably from the bottom of the tubes to the top.

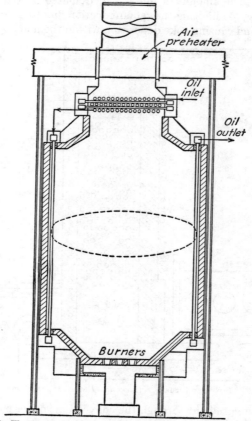

Fig. 19.4. De Florez circular furnace. (*Lobo and Evans, Transactions AIChE.*)

This furnace is fired from the bottom and has so little convection surface that an air preheater is employed to provide good thermal efficiency.

Figure 19.5 shows a double radiant-section box-type furnace. The convection-section tubes and those in one radiant section are employed for one service, while the other radiant section is independently controlled to perform another service. In Fig. 19.6 is an overhead convection-bank box-type furnace with the stack located on top of the con-

FURNACE CALCULATIONS

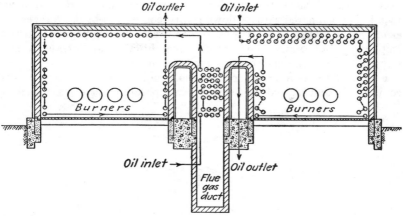

FIG. 19.5. Double-radiant-section box-type furnace. (*Lobo and Evans, Transactions AIChE.*)

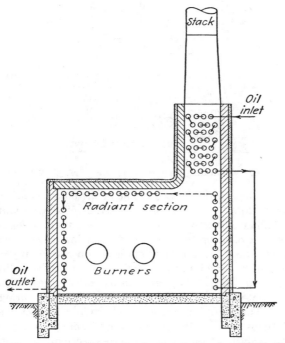

FIG. 19.6. Simple furnace with overhead convection bank. (*M. W. Kellogg Co.*)

682   PROCESS HEAT TRANSFER

vection bank. Such an arrangement saves on duct work and stack required as compared with a "down-draft" convection-bank arrangement as in Figs. 19.3 and 19.5. Figure 19.7 shows a somewhat similar furnace employing "A" frame type construction, utilizing an inherently stiff structural-steel arrangement to reduce steel costs.

Figure 19.8 presents a modern multiple radiant-section furnace. The convection bank is used for heating two separate oil streams. Each of

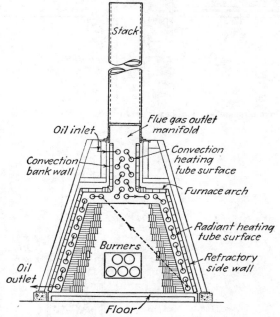

Fig. 19.7.   Oil-heating "A" frame furnace.   (*M. W. Kellogg Co.*)

these streams is identically heated in one of the outside radiant "heater" sections, and allowed to soak in one of the middle radiant "soaker" sections. Radiant soakers are preferred to convection soakers because they can be controlled better in so far as heat input is concerned. Furthermore, since the tubes can be seen during operation, any deformation of the tubes can be noted and tube failures with resultant fires avoided. Floor firing of the furnace permits the use of a large number of small burners distributed along the length of the tubes, ensuring even flux distribution. The small burners can be located close to the side-wall or bridge-wall tubes without danger of direct impingement of the burner flames on the tubes. As a result, the cross-sectional dimensions of the furnace can be reduced and tubes made longer than in a furnace fired from the end walls with large burners. A saving in the number of expensive return bends or "headers" employed can thus be realized.

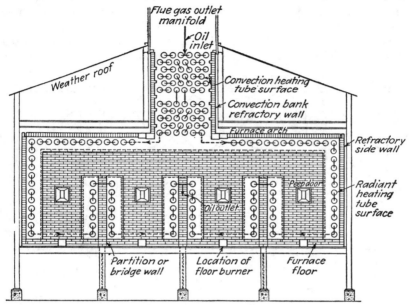

Fig. 19.8. Overhead convection bank oil-heating furnace provided with separately fired radiant sections. (*M. W. Kellogg Co.*)

Radiant-heat-transfer Factors. As pointed out at the conclusion of Chap. 4, the general equation for radiant-heat transfer can be represented by

$$Q = F_A F_\epsilon A \sigma (T_1^4 - T_2^4) \qquad (4.42)$$
$$= \mathfrak{F} \sigma A' (T_1^4 - T_2^4) \qquad (19.1)$$

where Q = heat flow by radiation alone to A', Btu/hr
T_1 = temperature of source, °R
T_2 = temperature of sink, °R
$\mathfrak{F}$ = factor to allow for both the geometry of the system and the nonblack emissivities of the hot and cold bodies, dimensionless
A' = effective heat-transfer surface of sink or cold body, ft²
σ = Stefan-Boltzmann constant, 0.173×10^{-8} Btu/(hr)(ft²)(°R⁴)

It is obvious that the application of such an equation to practical engineering problems must incorporate simplifications and assumptions.

684   PROCESS HEAT TRANSFER

It will be helpful to develop the bases for those simplifications and indicate the assumptions.

In general, the furnace consists of a heat receiver or *sink*, a heat *source*, and *enclosing surfaces* (the last being made up in some part by the sink and/or source). While there is complex interaction among these three essential parts, they can be evaluated best in the order given.

Heat Sink. The usual heat receiver for industrial furnaces is composed of a multiplicity of tubes disposed over the walls, roof, or floor of the

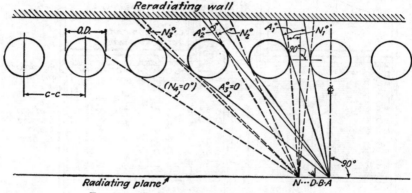

Fig. 19.9. Evaluation of the effectiveness factor α.

$$\alpha_{A,\text{ direct}} = \frac{90° - A_1° - A_2° - A_3°}{90°}$$

$$\alpha_{N,\text{ direct}} = \frac{90° - N_1° - N_2° - N_3° - N_4°}{90°}$$

$$\alpha_{\text{direct}} = \frac{\alpha_{A,\text{ direct}} + \alpha_{B,\text{ direct}} + \cdots \alpha_{N,\text{ direct}}}{N}$$

$$\alpha_{\text{total}} = \alpha_{\text{direct}} + \alpha_{\text{direct}}(1.00 - \alpha_{\text{direct}})$$

Note that $\alpha_{A,\text{ direct}}$, $\alpha_{B,\text{ direct}}$, etc., approach equality as distance between radiating plane and tube row increases.

furnace or located more centrally in the firebox. The most common case is the one in which bare tubes are arranged in a single row in front of a refractory wall. Although there are many rather arbitrary means of evaluating the *effective* heat-transfer surface of such arrangements, a rational development has been proposed by Hottel,[1] which now is used almost exclusively. It already has been stated that the elements of the furnace are best handled individually, and in evaluating the effective or "seen" surface of the tube rows, the assumption is made that the heat source is a radiating plane parallel to the tube row. End effects as described in Chap. 4 are eliminated by assuming both the plane of the

[1] Hottel, H. C., *Trans. ASME*, **53**, 265 (1931).

FURNACE CALCULATIONS

tubes and the radiating plane to be infinite in extent. All surfaces are assumed to be black.

In Fig. 19.9 is shown a method of evaluating the *effectiveness factor* α by which the surface of a plane replacing the tube row with assumed emissivity of 1.0 must be multiplied to obtain an equivalent (cold) plane surface. The plane replacing the tubes A_{cp} is equal to the number of tubes times length times center-to-center spacing. The first method shown is one suggested by Hottel,[1] which is simple and rapid. It is assumed that any heat loss through the refractory is equal to the heat transferred to the refractory by convection; hence all radiation impinging on the wall is reradiated. It can be seen that, as the tubes are spread farther apart, the fraction of the radiation originating at a point at the source which will be intercepted by the tubes will diminish, hence α will decrease. At the same time A_{cp} per tube is increased. The net effect is an increase in effective surface per tube but a decrease in effective surface per unit surface of furnace wall. The effectiveness of the tube is increased because a greater portion of its circumferential area is irradiated. The radiation not intercepted by the tubes reaches the refractory, from which it is reradiated. If it is assumed that the refractory is at a uniform temperature (not quite true), the radiation leaving the refractory will be intercepted to the same extent as the radiation from the source. Hence the total radiation absorbed by the tube row will be the fraction $[\alpha_{\text{direct}} + \alpha_{\text{direct}}(1 - \alpha_{\text{direct}})]$ of the radiation from the source. The illustration shows that several points along the radiating plane must be investigated (covering only one-half of the tube center-to-center distance because of the symmetry of the system) to obtain the average fraction of interception. Also by symmetry, only the angle between 90 and 180 need be investigated at each point.

A better understanding of the distribution of the radiant rate on the tube circumference is obtained from Fig. 19.10 in which the values are developed from the standpoint of the tube surface. The point on a tube which is located on the diameter perpendicular to the radiating plane and on the side of the tube facing the plane receives radiation through an angle of 180°, and hence the α value for this point is 1.00. Other points on the circumference can "see" through a smaller and smaller angle as one proceeds to the rear of the tube, until a point is reached at which no direct radiation from the plane is received. The effectiveness of each incremental area of the entire *circumferential* tube surface A is evaluated, and the sum of these effective areas must be divided by A_{cp} to obtain α. Again by symmetry, only half the tube circumference need be investi-

[1] Hottel, H. C., Personal communication.

gated, and it is rather apparent that the front of the tube absorbs considerably more heat than at other points on its circumference.

The interception of reradiation can be evaluated in the same manner, recalling that the intensity of the reradiation is $(1 - \alpha_{\text{direct}})$ times the intensity of the source. The total of direct radiation plus reradiation can be indicated in the polar coordinate plot to show the actual heat distribution. It is important to note the poor ratio between the average and the maximum value of α for the various points on the tube circumference. For a normal tube spacing of about 1.8 times outside diameter

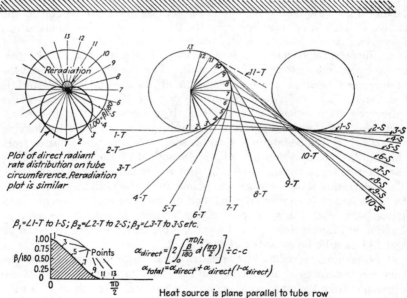

Fig. 19.10. Evaluation of α showing flux distribution on tube circumference.

in oil-refinery furnaces, the ratio is about 2.0, indicating that, at an average flux of 10,000 Btu/(hr)(ft²) on the total tube surface A, the maximum point flux (disregarding factors affecting heat distribution other than α) will be 20,000 Btu/(hr)(ft²) on the front face of the tube. The greater the ratio of center-to-center distance to outside diameter the lower the ratio of maximum rate to average rate, and since $\alpha_{\max}$ in all cases is 1.00, it can be seen readily from this development of α that the effective surface of the tubes increases. Except in special cases, however, the cheapest furnace results when tubes are spaced as close together as the physical and mechanical limitations of return bends will permit.

In the case of double rows of tubes, the back row receives about one-quarter of the total heat transfer. It is again also important to note that the ratio of maximum flux to the average flux (circumferential) for both rows of tubes becomes much worse than for a single row. The ratio can be obtained by dividing the total circumferential surface for the two rows by the total product of αA_{cp} for the two rows. Actually the tube emissivity is not 1.00 as assumed, and there will actually be some reflection from one tube to another. The net effect is to increase the effective emissivity of the row 2 or 3 per cent which is neglected in design practice.

Figure 19.11 presents values of α, direct and total, for single and double rows of tubes with refractory behind them. From the values of α_{direct} for the first and second row, it can be seen that a nest of tubes more than two rows deep can be assumed to have a value of α_{total} equal to 1.0. For convection banks whose tubes are radiated directly by the furnace A_{cp} is merely the width times the length of the opening.

In boiler furnaces tubes are sometimes half imbedded in the refractory, are sometimes finned, and occasionally are equipped with metal- or refractory-faced blocks. For a more detailed development of α values for such arrangements reference should be made to Hottel.[1] Still another serious complication arises in boiler furnaces from the slagging of the tubes.

A method of evaluating the effective radiant-heat-transfer surface for various boiler tubes has been presented by Mullikin.[2] For slagged tubes of any type, the effective surface is

$$(\alpha A_{cp})_s = A_{cp}\alpha F_c F_s F_\epsilon \quad (19.2)$$

where A_{cp} and α are as before with the subscript indicating a slagged condition and
where F_c = conductivity factor, dimensionless
F_s = slag factor, dimensionless
F_ϵ = emissivity factor, dimensionless

Note that the tube emissivity is introduced into the evaluation of the effective surface in this case while previously the emissivity was considered in the exchange factor $\mathfrak{F}$. Actually no confusion will result, since in the only method discussed here in which $(\alpha A_{cp})_s$ is employed F_ϵ is assumed to be unity. Practical values of F_c are 1.00 for bare and finned tubes, 0.70 for bare-faced metal blocks shrunk onto the tubes, and 0.33 for refractory-faced metal blocks. The slag factor will vary from a low of 0.6 to 0.9 or 1.0 for well-operated boilers. When the tubes are clean, F_s is 1.0, of course. From a review of published boiler tests the slag

---

[1] Hottel, *loc. cit.*
[2] Mullikin, H. F., *Trans. ASME*, **57**, 518 (1935).

factor appears to be ordinarily about 0.8 to 0.9. It is impossible to make generalizations about the slagging, which may occur with different coals (and some oil fuels) and with varying furnace temperatures and the slag

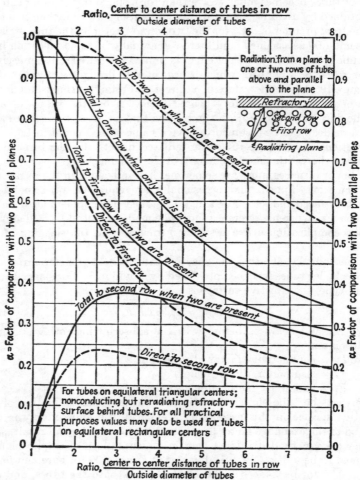

FIG. 19.11. Radiation between a plane and one or more tube rows parallel to the plane. (*Courtesy of Hottel.*)

factor in such cases is estimated from comparable information. A boiler in intermittent service tends to clean itself, while one in continuous service reaches an equilibrium state (steady or cyclic) at which the effects of the slag do not vary appreciably. Apparently there is considerable difference between the reduction in heat absorption due to dry ash on the tubes and that resulting when the slag is molten or running. A change in operating conditions may change the slag condition and cause variations in the accuracy of any equation which considered heat-transfer principles without the practical factors.

Heat Source. The heat to a furnace is provided primarily in the combustion reaction and in the sensible heat of the combustion air if it has been preheated. Gas fuels generally provide nonluminous flames. Oil fuels can be fired to provide flames of varying degrees of luminosity, depending upon burner design, extent of atomization, and percentage of excess air used. Pulverized coal burners produce a flame containing incandescent solid particles and of a greater degree of luminosity than the minimum obtainable with oil burners. Stoker firing provides an incandescent fuel bed.

The differences in the characteristics of the flames or heat patterns produced in the conventional firing of the various fuels have resulted in the development of methods of calculating radiant-heat transmission which apply on the one hand to oil- or gas-fired refinery furnaces and on the other hand to coal- (either stoker or pulverized) fired units. There is no simple, universally applicable method of calculating heat absorption in all types of furnaces. At the outset, then, the calculations must differentiate between furnaces fired with gas or oil and furnaces fired with solid fuels. It is justifiable to say that heat-absorption-calculation methods are further advanced in refinery-furnace design than in boiler work. Undoubtedly this is due at least in part to the greater complexity of the heat sources and sinks of boiler furnaces. While methods of calculation for coal-fired furnaces will be presented, the discussion of the heat source will be limited mainly to furnaces whose fireboxes are filled largely with nonluminous gases and whose bounding surfaces are not part of the heat source—as opposed to the case of a stoker-fired furnace. Luminous-flame calculations require information obtained either from experiment or from experience, and some data have been presented by Hottel.[1]

Consider for the moment a furnace in which gaseous fuel is fired by burners producing nonluminous flames. Further consider that the path of the combustion products through the furnace is very short compared with the dimensions of the plane perpendicular to the path. In such a furnace, equipped with many small burners to give good distribution of firing, one would expect little or no variation in the temperature of the gases from point to point in the furnace. No luminous flames are present to radiate to the tubes or refractory. The only primary sources of radiation are the combustion products themselves, and these are assumed to be

[1] Perry, J. H., "Chemical Engineers' Handbook," 3d ed., pp. 493–495, McGraw-Hill Book Company, Inc., New York, 1950.

uniform in temperature. There is a great difference in the emissivity of various gases at the same temperature. The diatomic gases such as O_2, N_2, and H_2 have very poor emissivities, so low that they may be considered zero in furnace-design work. On the other hand, H_2O, CO_2, and SO_2 have good emissivities (though much poorer than most solid materials), and CO has a fair emissivity. The sources of radiation may be referred to specifically as the radiating constituents in the combustion products. Ordinarily H_2O and CO_2 are the only radiating constituents which need be considered, since the small amount of sulfur in most fuels is negligible and furnaces are generally operated with sufficient excess air to eliminate CO.

The total radiation from a gas mass containing CO_2 and H_2O will depend upon the temperature of the gas and the number of radiating molecules present. The volume of the gas and the concentration of radiating molecules per unit of volume are therefore a measure of the radiation to be anticipated at a given temperature. In actuality the geometric shape of the gas mass must also be considered because of the angle factor involved in radiation.

The *mean beam length*, L ft, is the average depth of the blanket of flue gas in all directions for each of the points on the bounding surface of the furnace and is used instead of a cubical measure of the volume. The concentration of the radiating molecules is measured by their partial pressures. The emissivity of the gas mass in a furnace is a function of the product pL atm-ft, where p is the partial pressure of the radiating constituent. If more than one radiating constituent is present, the emissivities are additive, although a small correction must be made for the interference of one type of molecule with the radiation from the other. In calculating the emissivity of the gas mass, allowance must be made for the temperatures of both the source and the sink. For heat transfer to a black body, one would use the equation

$$Q_b = 0.173 F_{bA}\left[\epsilon_G\left(\frac{T_G}{100}\right)^4 - a_G\left(\frac{T_b}{100}\right)^4\right]A'_b \tag{19.3}$$

where Q_b = duty or heat transfer to the black body by radiation from gas, Btu/hr
A'_b = effective heat-transfer surface of black body, ft^2
a_G = absorptivity of gas at T_b, dimensionless
F_{bA} = factor to allow for the geometry of a system with a black-body sink, dimensionless
T_G = temperature of gas, °R
T_b = temperature of black body, °R
ϵ_G = emissivity of gas at T_G, dimensionless

Practically, a_G can be replaced by ϵ_G evaluated at T_b. When two radiating constituents are present, H_2O and CO_2, the equation can be rewritten (neglecting the correction factor for interference between the dissimilar molecules):

$$Q_b = 0.173 \mathfrak{F}_b \left[(\epsilon_C + \epsilon_W)_{T_G} \left(\frac{T_G}{100}\right)^4 - (\epsilon_C + \epsilon_W)_{T_b} \left(\frac{T_b}{100}\right)^4 \right] A_b' \quad (19.4)$$

where $(\epsilon_C + \epsilon_W)_{T_G}$ = emissivity of gas at T_G
  ϵ_C = emissivity of CO_2 at $p_{CO_2} \cdot L$ and T_G
  ϵ_W = emissivity of H_2O at $p_{H_2O} \cdot L$ and T_G
$(\epsilon_C + \epsilon_W)_{T_b}$ = emissivity (substituted for absorptivity)
  ϵ_C = emissivity of CO_2 at $p_{CO_2} \cdot L$ and T_b
  ϵ_W = emissivity of H_2O at $p_{H_2O} \cdot L$ and T_b

It should be noted that in addition to the correction factor which should be introduced to allow for interference, the emissivity of water vapor has been found by Egbert[1] to be a function of its partial pressure. Values of L for furnaces of various geometric shapes have been determined by Hottel, and Table 19.1 presents a useful digest of these values for furnace

TABLE 19.1. MEAN LENGTH OF RADIANT BEAMS IN VARIOUS GAS SHAPES

Dimensional ratio*	Mean length L, ft
Rectangular furnaces:	
1. 1-1-1 to 1-1-3 1-2-1 to 1-2-4	$\tfrac{2}{3} \sqrt[3]{\text{Furnace volume, ft}^3}$
2. 1-1-4 to 1-1-∞	1.0 × smallest dimension
3. 1-2-5 to 1-2-8	1.3 × smallest dimension
4. 1-3-3 to 1-∞-∞	1.8 × smallest dimension
Cylindrical Furnaces:	
5. $d \times d$	$\tfrac{2}{3}$ diameter
6. $d \times 2d$ to $d \times \infty d$	1 × diameter
Tube Banks:	
7. As in convection sections	L (ft) = $0.4 P_T - 0.567$ OD, in.

* Length, width, height in any order.

work. In industrial-furnace design the term *rate* is used synonymously with the term flux in preceding chapters and individual film coefficients are not considered. It is convenient to have charts which give the values of the radiant-heat-transfer flux q_c and q_w as functions of pL and T where

$$q_C = 0.173 \epsilon_C \left(\frac{T}{100}\right)^4 \text{ at } p_{CO_2} \cdot L \text{ and } T$$

$$q_W = 0.173 \epsilon_W \left(\frac{T}{100}\right)^4 \text{ at } p_{H_2O} \cdot L \text{ and } T$$

$$q_b = 0.173 \epsilon_b \left(\frac{T}{100}\right)^4 \text{ and } \epsilon_b = 1.00$$

[1] Egbert, R. B., Sc. D. Thesis in *Chem. Eng.*, MIT, 1941.

Such charts are presented in Figs. 19.12 and 19.13 from data by Hottel and Egbert. The correction for interference is also given in the insert graphs in per cent, and the corrected emissivity is equal to

$$\epsilon_G = \left[\frac{(q_c + q_w)_{T_G} - (q_c + q_w)_{T_S}}{(q_b)_{T_G} - (q_b)_{T_S}} \right] \frac{100 - \%}{100} \quad (19.5)$$

In many simple cases of radiant-heat transfer from gases to nonblack bodies no complications of angle factors need be introduced and the heat-transfer rate can be calculated simply by

$$q = \frac{Q}{A} = \epsilon_S [(q_c + q_w)_{T_G} - (q_c + q_w)_{T_S}] \frac{100 - \%}{100} \quad (19.6)$$

where the subscript S refers to the cold surface and the emissivity of the cold surface is ϵ_S. Such is the case in tube bundles as in furnace convection sections where radiant-heat transfer is often quite important. In this particular case the tubes are surrounded by the gas and no angle factor is required. The surface to be used is the circumferential tube area. (For banks of tubes there is additional radiant-heat transfer from the refractory side walls which add to the average rate or flux.)

Some conclusions may be drawn from the dependency of the gas emissivity upon pL. For furnaces of the same physical proportions but of differing sizes, one would expect the larger to have a higher rate of heat transfer with a given gas temperature (because of the greater value of L). The effect of increasing excess air is to decrease the value of p, hence the emissivity, and the radiant rate for a given gas temperature is thereby decreased. Though not strictly related to a discussion of the effects of excess air, it has been found by experience that refinery furnaces do not operate under optimum conditions because of the use of undue amounts of excess air. The combustion conditions of boilers are usually controlled more rigidly than refinery furnaces, because of the greater value of fuel in boiler plants.

Enclosing Surfaces. The part played by the refractory walls, arch, floor, etc., of the furnace in transferring heat from the gas to the cold surface is often difficult to visualize. The gas mass radiates in all directions. The emissivity of the gas evaluated from p and L, as discussed previously, is directional in that it denotes the radiation impinging on a point of the cold surface on the furnace enclosure. All of this radiation is directed from various sections of the gas mass *toward that particular point*. However, the various sections of the gas mass also radiate in other directions. Some of this radiation may be directed toward refractory surface (which is not cold), and the refractory in turn reradiates or

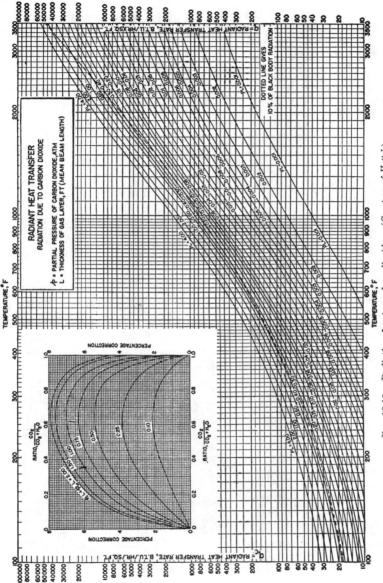

Fig. 19.12. Radiation due to carbon dioxide. (*Courtesy of Hottel.*)

694	PROCESS HEAT TRANSFER

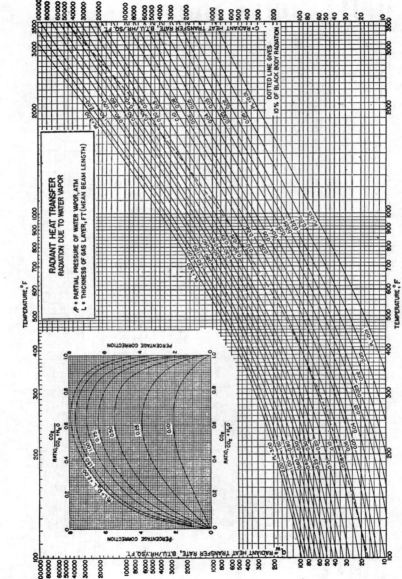

Fig. 19.13. Radiation due to water vapor. (*Courtesy of Hottel.*)

reflects the impinging radiation, some of it being redirected toward the point first considered. The gas mass is quite transparent to the reradiation (more so than to the reflection, since the spectral characteristics of the reradiation depend upon the characteristics of the surface of the refractory); hence the point on the cold surface receives more radiation than would be evaluated from the emissivity of the gas. The refractory is somewhat analogous to a reflector placed behind a light source.

An effective means of illustrating the effect of the refractory is to start with an enclosure containing no cold surface. It contains a gas of a certain emissivity at a given temperature. If a small aperture is made in the enclosure, the radiation streaming through the aperture will be equal to that coming from a black body at the gas temperature, no matter what the emissivity of the gas may be but provided the enclosure walls are well insulated and the system has reached thermal equilibrium. Now, if instead of the aperture one places a small section of cold surface within the enclosure, the radiation falling upon it will be equal to that which would come from black surroundings at the gas temperature. The effect of a very high ratio of refractory surface to cold surface is to produce a furnace emissivity of 1.0, even though the gas emissivity is low.

On the other hand, if the entire enclosing surface is cold and black, each point on the surface will receive only the radiation initially directed toward it, since radiation in other directions is completely absorbed and the furnace emissivity is equal to the gas emissivity. If the emissivity of the cold surface is less than 1, some of the radiation will be reflected, and though the net effect of a lower cold surface emissivity will be a decrease in heat transfer for a given temperature difference, the decrease will not be proportional to the decrease in emissivity. The reflections will be absorbed only partly by the gas, and the unabsorbed portion will be added to the primary radiation to some other portion of the cold surface. The lower the gas emissivity (absorptivity) the less will be the effect of a change in the emissivity of the cold surface. It might be mentioned that in a fire-tube boiler the radiant-heat transfer can be evaluated by application of these principles, and it is undoubtedly safe to assume that the emissivity of the tube surface is 1.0.

Qualitatively it has been demonstrated that the net radiant-heat transmission will be increased for a given gas emissivity, cold-surface emissivity, and gas and cold-surface temperatures by an increase in the ratio of refractory surface to cold surface. The random addition of refractory (in the form of a partition or bridge wall) does not provide a means of increasing the flux without increasing the furnace gas temperature, however. While the ratio of refractory to cold surface may thus be increased, it is increased at the expense of decreasing the mean beam length for the furnace. In so-called double radiant-section furnaces, each section is best handled as a separate furnace.

Quantitatively a rigorous evaluation of the effect of the refractory surface is very difficult and beyond the range of practicability for most engineering work. Hottel has developed an equation for the overall exchange factor which is included in Eq. (4.26).

$$F_\epsilon = \frac{1}{(1/\epsilon_F) + (1/\epsilon_S) - 1} \qquad (19.7)$$

in which ϵ_S is the emissivity of the cold surface, and ϵ_F, the effective emissivity of the furnace cavity (which will be greater than ϵ_G if any portion of the enclosure is not cold). Hottel further defines ϵ_F as a function of the gas emissivity, the ratio of refractory surface to cold surface, and an angle factor accounting for the geometrical relationships between various sections of cold and refractory surfaces.

In summary, the tube surface must be evaluated as equivalent plane surface. The emissivity of the gas mass is a function of its temperature, the temperature of the cold surface, the mean beam length of the furnace, and the partial pressures of the radiating constituents. The effective emissivity of the furnace is a function of the gas emissivity and the ratio (and relative arrangement) of refractory to cold surfaces. The overall exchange factor may be obtained from the furnace and cold surface emissivities for use in an equation of the Stefan-Boltzmann type for calculating the radiant heat transfer.

$$Q = 0.173 \mathcal{F} \left[\left(\frac{T_G}{100} \right)^4 - \left(\frac{T_S}{100} \right)^4 \right] \alpha A_{cp} \qquad (19.8)$$

Theoretically the proper average value of T_G must be used or the calculation of long furnaces must be carried out in a stepwise manner. Actually it is often satisfactory to consider the temperature of the gases leaving the furnace radiant section as the average temperature when the degree of turbulence of the gases is high. When highly luminous flames or stoker firing are employed additional data are required. It should be noted that methods for calculating the *average* heat transfer rates or fluxes in the radiant section give no measure of the uniformity of the rates on various tubes. Either experience or highly analytical calculations are required to estimate specific rates.

DESIGN METHODS

The common methods for calculating heat absorption in furnace radiant sections are surveyed below. Several are illustrated by calculations at the conclusion.

1. Method of Lobo and Evans.[1] This method makes use of the overall exchange factor ℱ and a Stefan-Boltzmann type equation. It has a good theoretical basis and is used extensively in refinery-furnace design work. It is also recommended for oil- or gas-fired boilers. The average deviation between the predicted and observed heat absorption for 85 tests on 19 different refinery furnaces varying widely in physical and operating characteristics was 5.3 per cent. The maximum deviation was 16 per cent. This method is illustrated by Example 19.1.

2. Method of Wilson, Lobo, and Hottel.[2] This is an empirical method which can be used for box-type furnaces fired with oil or refinery gas when fluxes lie between 5000 and 30,000 Btu/(hr)(ft^2) of circumferential tube surface. Other limitations are that the percentage of excess air be from 5 to 80 per cent and that tube-surface temperatures be at least 400°F lower than the radiant-section exit-gas temperature. The mean beam length should not be less than 15 ft. This method is widely used in industry and is recommended under the above limitations when the accuracy of the Lobo-Evans equation is not demanded. For most of the tests referred to under the Lobo-Evans method the average deviation was 6 per cent and the maximum deviation 33 per cent. This method is illustrated by Example 19.2.

3. The Orrok-Hudson[3] **Equation.** This is an early empirical equation for calculating heat absorption in the radiant section of a water-tube boiler. It has been replaced by more accurate expressions and is of limited value for design work. It can be used to estimate the effects of changes in firing rate or air-fuel ratio for an existing boiler fired with coal or oil if it is known that there will be no appreciable change in either the character or extent of the slagging of the furnace tubes. In such applications it may be necessary to adjust the constant in the equation to suit the known operating conditions. The use of this equation is illustrated in Example 19.3.

4. Wohlenberg[4] **Simplified Method.** This is an empirical method, though undoubtedly much sounder than the Orrok-Hudson equation for calculating radiant-heat absorption. It is useful only for coal firing. It is again repeated that a knowledge of the anticipated slagging is a prerequisite to the application of a heat-transfer equation to a boiler. Tests on seven large boilers indicated an average deviation of about 10 per cent when the slag factor was estimated from the furnace appearance. The maximum deviation was about 50 per cent for stoker firing, but better accuracy was obtained in furnaces fired with pulverized coal.

---

[1] Lobo, W. E., and J. E. Evans, *Trans. AIChE*, **35**, 743 (1939).
[2] Wilson, D. W., W. E. Lobo, and H. C. Hottel, *Ind. Eng. Chem.*, **24**, 486 (1932).
[3] Orrok, G. A., *Trans. ASME*, **47**, 1148 (1925).
[4] Wohlenberg, W. J., and H. F. Mullikin, *Trans. ASME*, **57**, 531 (1935).

APPLICATIONS

Method of Lobo and Evans. The equation given previously for radiant-heat transfer to the cold surface was

$$Q = 0.173 \mathfrak{F} \left[\left(\frac{T_G}{100}\right)^4 - \left(\frac{T_S}{100}\right)^4 \right] \alpha A_{cp} \qquad (19.8)$$

In addition some heat will be transferred by convection, and the total heat transfer to the cold surface is

$$\Sigma Q = 0.173 \mathfrak{F} \left[\left(\frac{T_G}{100}\right)^4 - \left(\frac{T_S}{100}\right)^4 \right] \alpha_{cp} A_{cp} + h_c A (T_G - T_S) \qquad (19.9)$$

where A = total tube surface, ft^2
 A_{cp} = equivalent cold plane surface, ft^2
 $\mathfrak{F}$ = overall exchange factor, dimensionless
 h_c = convection coefficient, Btu/(hr)(ft^2)(°F)
 ΣQ = total hourly heat transfer to the cold surface, Btu/hr
 T_G = temperature of flue gas leaving the radiant section, °R
 T_S = tube surface temperature, °R
 α = factor by which A_{cp} must be reduced to obtain effective cold surface, dimensionless

The convection term can be simplified by assuming that $h_c = 2.0$ and that for this term alone A is approximately $2.0 \alpha A_{cp}$. Since it is desired to divide all terms by $\mathfrak{F}$, a value of 0.57 will be used in its stead when the convection term is considered. Then

$$\frac{\Sigma Q}{\alpha A_{cp} \mathfrak{F}} = 0.173 \left[\left(\frac{T_G}{100}\right)^4 - \left(\frac{T_S}{100}\right)^4 \right] + 7(T_G - T_S) \qquad (19.10)$$

This relationship is shown graphically in Fig. 19.14. In addition to the above flux equation, a heat balance is necessary for the solution of the heat-absorption problem. The heat balance is as follows:

$$Q = Q_F + Q_A + Q_R + Q_S - Q_W - Q_G \qquad (19.11)$$

where Q = total radiant-section duty, Btu/hr
 Q_A = sensible heat above 60°F in combustion air, Btu/hr
 Q_F = heat liberated by fuel, Btu/hr (lower heating value)
 Q_G = heat leaving the furnace radiant sections in the flue gases, Btu/hr
 Q_R = sensible heat above 60°F in recirculated flue gases, Btu/hr
 Q_S = sensible heat above 60°F in steam used for oil atomization, Btu/hr

FURNACE CALCULATIONS

Q_W = heat loss through furnace walls, Btu/hr (1 to 10 per cent of Q_F, depending upon the size, temperature, and construction of the furnace. 2 per cent is a good design figure)

As a further simplification Q_S can be neglected, and the net heat liberation is

$$Q_F + Q_A + Q_R - Q_W = Q_{\text{net}} \qquad (19.12)$$

The heat leaving in the flue gases at the exit-gas temperature T_G is

$$Q_G = W(1 + G')C_{\text{av}}(T_G - 520)$$

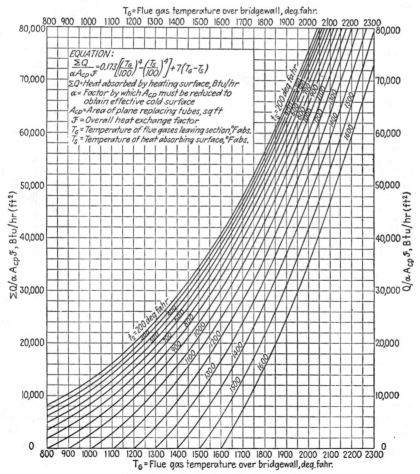

Fig. 19.14. Radiant-section heat flux.

where W = fuel rate, lb/hr
$(1 + G')$ = ratio of flue gas leaving the radiant section to fuel fired, lb/lb
G' = ratio of air to fuel, lb/lb
C_{av} = average specific heat of flue gases between $T_G°$R and 520°R, Btu/(lb)(°F)

In applying the equations, the equivalent cold plane surface is evaluated with the aid of Fig. 19.11. As mentioned earlier, A_{cp} is the surface of a

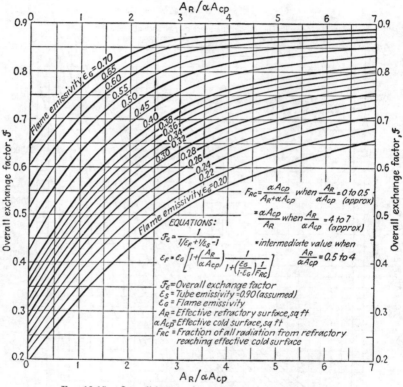

FIG. 19.15. Overall heat-exchange factor in radiant sections.

plane replacing the tube row and corresponds to the product of the number of tubes times their exposed length in feet times center-to-center spacing in feet. When the convection section is so located that it receives the benefit of direct radiation from the radiant section, it should be included in the equivalent cold plane surface. For a nest of tubes more than two rows deep, α may be taken as 1.0 and αA_{cp} is merely the product of the length times width of the opening of the convection bank.

FURNACE CALCULATIONS

When the convection bank is isolated from the radiant section, it is not included in the radiant-section calculations.

The gas emissivity is calculated from the mean beam length, partial pressures of the radiating constituents, tube temperature, and gas temperature (which usually has to be assumed in the first trial calculation). The overall exchange factor $\mathfrak{F}$ is shown in Fig. 19.15 as a function of gas ("flame") emissivity and the ratio of effective refractory surface A_R, where

$$A_R = A_T - \alpha A_{cp} \qquad (19.13)$$

where A_R = effective refractory surface, ft^2
A_T = total area of furnace surfaces, ft^2
αA_{cp} = equivalent cold plane surface, ft^2

The exit-gas temperature is obtained by a trial-and-error calculation until it satisfies both the heat-transfer and the heat-balance equation. To lessen the amount of calculation required in applying the method of Lobo and Evans, Fig. 19.14 gives the value of $\Sigma Q / \alpha A_{cp} \mathfrak{F}$ for various combinations of T_G and T_S.

It is recommended that, in furnaces where the path of the gas is more than 1.5 times the minimum dimension of the cross-sectional area for gas flow, a stepwise calculation be employed; typical of such a case would be a vertically fired, vertical cylindrical furnace whose height is twice its diameter. The top and bottom halves of the furnace would be calculated as separate furnaces except that the flue gases from the lower half would provide the heat liberation to the second.

In practice the total furnace duty is calculated first, including the sensible heat, heat of vaporization, and any heat of reaction. The efficiency of the furnace e is given by

$$e = \frac{\Sigma Q}{Q_F} \times 100 \quad \% \qquad (19.14)$$

It is determined from a balance between fuel costs and the first cost of the furnace (plus air preheater if one is used). The quantity of excess air to be used depends upon the type of fuel, type of burners, type of draft, and the temperature of the combustion air. In practice, however, 40 per cent excess air is used for designing natural- or induced-draft furnaces, 25 per cent excess air for forced-draft furnaces.

The use of air preheat is dictated by the temperature of the coolest material to be heated, the cost of fuel, and the cost of furnace surface and to some degree by conventional practices. Air preheaters are easily justified by the savings in fuel cost, although they complicate the operation of the furnace and require additional maintenance.

702   PROCESS HEAT TRANSFER

When the heat liberation has been determined, the design of a petroleum furnace is then established on the basis of the permissible *average* radiant-section rate as defined in Table 19.2 appearing at the conclusion of the chapter. The tube diameter is dependent upon consideration of the film coefficient, pressure drop, and radiant rate. The spacing of the tubes depends upon the characteristics of the headers or return bends. The closest possible spacing is used except when special requirements arise such as the need for an improved uniformity of flux on the tube circumference.

The most economical furnace design uses the maximum tube length which is compatible with the furnace cross section and which provides adequate clearance between the tubes and burners. In some refinery units tubes 60 ft long are used, although 40 ft is the more usual limit. In boilers tube lengths may even be longer.

Example 19.1. Calculation of a Furnace by the Method of Lobo and Evans. A furnace is to be designed for a total duty of 50,000,000 Btu/hr. The overall efficiency is to be 75 per cent (lower heating value basis). Oil fuel with a lower heating value of 17,130 Btu/lb is to be fired with 25 per cent excess air (corresponding to 17.44 lb air/lb fuel), and the air preheated to 400°F. Steam for atomizing the fuel is 0.3 lb/lb of oil. The furnace tubes are to be 5 in. OD on 8½-in. centers, in a single-row arrangement. The exposed tube length is to be 38'6''. The average tube temperature in the radiant section is estimated to be 800°F.

Design the radiant section of a furnace having a radiant-section average flux of 12,000 Btu/(hr) ft².

Solution. As in all trial-and-error solutions, a starting point must be assumed and checked. With experience, the first choice may come very close to meeting the desired conditions. For orientation purposes, one can make an estimate of the number of tubes required in the radiant section by assuming that

$$\frac{Q}{\alpha A_{cp}} = 2 \times \text{average flux} = 24{,}000 \text{ Btu/(hr)(ft}^2)$$

If the overall exchange factor is 0.57, $\Sigma Q/\alpha A_{cp}\mathfrak{F} = 24{,}000/0.57 = 42{,}000$; from Fig. 19.14 it can be seen that with a tube temperature of 800°F, an exit-gas temperature of 1730°F will be required to effect such a flux. The duty in cooling the furnace gases to 1730°F can be calculated, and from it the required number of tubes determined for the first approximation of the design.

Heat liberated by the fuel, $Q_F = \dfrac{50{,}000{,}000}{0.75} = 66{,}670{,}000$ Btu/hr

Fuel quantity $= \dfrac{66{,}670{,}000}{17{,}130} = 3890$ lb/hr

Air required $= 3890 \times 17.44 = 67{,}900$ lb/hr
Steam for atomizing $= 3890 \times 0.3 = 1170$ lb/hr
$Q_F = 66{,}670{,}000$ Btu/hr
$Q_A = 67{,}900 \times 82$ Btu/lb at 400°F $= 5{,}560{,}000$ Btu/hr (above 60°F)
$Q_S =$ negligible (1170 × 0.5 × 190°F) Btu/hr
$Q_F + Q_A \qquad\qquad = 72{,}230{,}000$ Btu/hr
$Q_W = 2\%$ of $Q_F \qquad = 1{,}330{,}000$ Btu/hr
$Q_{\text{net}} = Q_F + Q_A - Q_W = 70{,}900{,}000$ Btu/hr

FURNACE CALCULATIONS

Heat out in gases at 1730°F, 25 per cent excess air, 476 Btu/lb of flue gas,

$$Q_G = 476(3890 + 67{,}900 + 1170) = 34{,}500{,}000$$

$Q = Q_{net} - Q_G = 70{,}900{,}000 - 34{,}500{,}000 = 36{,}400{,}000$ Btu/hr (first estimate)
The surface/tube, $A = 38.5 \text{ ft} \times \pi \times 5/12 = 50.4 \text{ ft}^2$
The estimated number of tubes N_t is $N_t = \dfrac{36{,}400{,}000}{12{,}000 \times 50.4} = 60.1$
Try 60 tubes.
The layout of the cross section of the furnace may be as shown in Fig. 19.16.

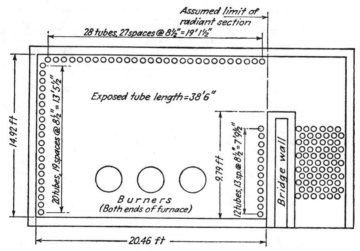

FIG. 19.16. Furnace of Examples 19.1 and 19.2.

Equivalent cold plane surface A_{cp}:

Center-to-center distance $= 8\frac{1}{2}$ in.

A_{cp} per tube $= \dfrac{8\frac{1}{2}}{12}$ in. $\times 38.5 = 25.7 \text{ ft}^2$

Total α to a single row, refractory backed, from Fig. 19.11:

Ratio of center-to-center/OD $= 8\frac{1}{2}/5 = 1.7 \qquad \alpha = 0.937 \qquad$ (Fig. 19.11)
αA_{cp}/tube $= 25.7 \times 0.937 = 25 \text{ ft}^2$
$\alpha A_{cp} = 60 \times 25 = 1500 \text{ ft}^2$

Refractory surface:

End walls	$= 2 \times 20.46 \times 14.92 =$	611 ft²
Side wall	$= 14.92 \times 38.5 =$	575 ft²
Bridge wall	$= 9.79 \times 38.5 =$	377 ft²
Floor and arch	$= 2 \times 20.46 \times 38.5 =$	1575 ft²
	$A_T =$	3138 ft²

$A_R = A_T - \alpha A_{cp} = 1638 \qquad \dfrac{A_R}{\alpha A_{cp}} = \dfrac{1638}{1500} = 1.09$

Mean beam length:

Dimension ratio = $38.5 \times 20.46 \times 14.92 = 3:2:1$ (approximately)
$L = \frac{2}{3} \sqrt[3]{\text{volume}} = \frac{2}{3} \sqrt[3]{38.5 \times 20.46 \times 14.92}$
$L = 15$ ft

Gas emissivity: From the analysis of the fuel, the steam quantity, and the assumption that the humidity of the air is 50 per cent of saturation at 60°F, the partial pressures of CO_2 and H_2O in the combustion gases with 25 per cent excess air are

$p_{CO_2} = 0.1084 \qquad p_{H_2O} = 0.1248 \qquad p_{CO_2}L = 0.1084 \times 15 = 1.63 \qquad p_{H_2O}L = 1.87$

From Fig. 19.12 and 19.13, the emissivity of the gas can be evaluated.

$$\epsilon_G = \left[\frac{(q_{CO_2} \text{ at } P_{CO_2}L + q_{H_2O} \text{ at } P_{H_2O}L)_{TG} - (q_{CO_2} \text{ at } P_{CO_2}L + q_{H_2O} \text{ at } p_{H_2O}L)_{TS}}{(q_b)_{TG} - (q_b)_{TS}} \right] \frac{100 - \%}{100} \quad (19.5)$$

% correction at $\dfrac{p_{CO_2}}{p_{CO_2} + p_{H_2O}} = \dfrac{0.1084}{0.2332} = 0.465$

$p_{CO_2}L + p_{H_2O}L = 3.56$

% correction = 8% (estimated)

$\epsilon_G = \left[\dfrac{(6500 + 14,500) - (650 + 1950)}{39,000 - 4400} \right] \left(\dfrac{100 - 8.0}{100} \right)$

$\epsilon_G = 0.489$

Overall exchange factor $\mathfrak{F}$:

$\mathfrak{F}$ at $\epsilon_G = 0.496$ and $\dfrac{A_R}{A_{cp}} = 1.09$

$\mathfrak{F} = 0.635$ from Fig. 19.15

Check of gas temperature required to effect assumed duty on assumed surface:

$\Sigma Q = 36,400,000$ Btu/hr assumed $\qquad \alpha A_{cp} = 1500$ ft² assumed

$\dfrac{\Sigma Q}{\alpha A_{cp} \mathfrak{F}} = \dfrac{36,400,000}{1500 \times 0.635} = 38,200$

T_G required (at $T_S = 800°$) = 1670°F compared with 1730°F assumed in heat balance)

The trial indicates that more duty than 36,400,000 Btu/hr will be performed, since this duty cools the flue gases to only 1730°F, while the flux corresponding to this duty could be effected by a gas temperature of 1670°F. Actually this is a fairly close check, and the number of tubes in the furnace need not be changed, since the final balance will be struck at about 1700°F for the assumed furnace. The duty would be 37,050,000 Btu/hr at this exit temperature, and assuming that $\mathfrak{F}$ does not change (it will go up slightly) $\Sigma Q / \alpha A_{cp} \mathfrak{F} = 39,000$, requiring a "driving" temperature of 1695°F, which is a close enough approximation. The circumferential flux will be $37,050,000/60 \times 50.4 = 12,280$ Btu/(hr)(ft²) as compared with the 12,000 flux specified. Such a difference is negligible.

In general, if the gas temperature required to effect the duty used in the heat balance falls below the temperature shown by the heat balance, the number of tubes assumed is too few or the actual flux will be higher than the assumed rate

Method of Wilson, Lobo, and Hottel. The limitations of this method have been pointed out. In the original publication several equations were presented, but the most useful is the following:

$$\frac{Q}{Q_F} = \frac{1}{1 + (G'/4200)\sqrt{Q_F/\alpha A_{cp}}} \qquad (19.15)$$

where again G' = lb air/lb fuel and the other nomenclature is similar to that given for the Lobo and Evans method. While the cold surface is properly evaluated so that the equation is applicable to furnaces having double as well as single rows of tubes, the effect of refractory surfaces is neglected. The effect of excess air on the radiant-section efficiency is measured only by G', the air-fuel ratio, and consequently the equation does not hold for fuels having heating values which are either very low or very high. The equation is recommended for rapid calculation (within its limitations) and for predicting the effects of changes in furnace operating conditions.

Example 19.2. Calculation of a Furnace by the Method of Wilson, Lobo, and Hottel. The furnace of Example 19.1 is to be fired with cracked gas fuel at 40 per cent excess air, using no air preheat. If the capacity of the burners limits the heat liberation to 50,000,000 Btu/hr, what will be the radiant-section duty? (The air-fuel ratio is 22.36 lb air/lb fuel.)

Solution:

$$Q = 50,000,000 \times \frac{1}{1 + (22.36/4200)\sqrt{50,000,000/1500}} \qquad (19.15)$$
$$= 25,300,000 \text{ Btu/hr}$$

The radiant-section average rate will be 8350 Btu/(hr)(ft²), and the exit-flue-gas temperature 1540°F by heat balance.

Wohlenberg Simplified Method. While this method is an empirical method, its derivation is of interest. Wohlenberg[1] developed a complex theoretical method for evaluating the heat absorption in boiler furnaces which took into account the many variables already discussed and in addition those factors unique to coal firing. The simplified method, however, relates the radiant-section absorption efficiency of any boiler furnace to a base or standard design by means of factors which correct for the differences in all of the characteristics between the two.

The base design is as follows:
Furnace volume = 8000 ft³
Release rate = 25,000 Btu/ft³ for pulverized coal firing
        = 40,000 Btu/ft³ for stoker firing

[1] Wohlenberg, W. J., and D. G. Morrow, *Trans. ASME,* **47,** 177 (1925). Wohlenberg, W. J., and E. L. Lindseth, *Trans. ASME,* **48,** 849 (1926).

Excess air = 20 per cent for pulverized coal firing
        = 40 per cent for stoker firing
Coal = Illinois bituminous
Fineness = 75 per cent through 200 mesh
Fraction cold = unity

The fraction cold is defined as the effective exposed radiant-heating surface in the furnace divided by the total exposed furnace surface (exclusive of the grate in stoker firing). The evaluation of the effective heating surface is by the method of Mullikin discussed previously. The absorption efficiency of the base furnace is 0.452 for pulverized fuel firing and 0.311 for stoker firing; the absorption efficiency e of any furnace is the ratio of heat absorbed to the heat liberated by the fuel (higher heating value) plus the heat in the combustion air.

Wohlenberg's equation is

$$e = F K_1 K_2 K_3 K_4 K_5 K_6 K_7 K_8 + C' \qquad (19.16)$$

where the various factors have the following significance: F is chosen according to type of firing, pulverized coal or stoker, and the influences of the other factors are

K_1 = furnace volume
K_2 = heat liberation rate, Btu/ft^3
K_3 = fraction cold
K_4 = excess air
K_5 = heating value of coal (HHV)
K_6 = pulverizing fineness
K_7 = furnace volume beyond K_1
K_8 = heat liberation rate beyond K_2
C' = air preheat

The equation does not apply to gas- or oil-fired boilers, although the methods proposed by Mullikin or Lobo and Evans may be used. The application of the simplified equation is straightforward and gives values which are in good agreement with the more complex method of Wohlenberg. It has been mentioned several times that the ash or slag on the boiler surfaces presents problems beyond those of ordinary heat-transfer calculations. Accordingly a basis of evaluating the influence of ash or slag under operating conditions must be established before these methods can be applied to rational design.

Orrok-Hudson Equation. This is very similar in form to the Wilson, Lobo, and Hottel equation, and as already pointed out, its greatest use is for comparing furnace performance under various operating conditions. The fractional heat absorption is

$$\frac{Q}{Q_F} = \frac{1}{1 + G' \sqrt{C_R/27}} \qquad (19.17)$$

FURNACE CALCULATIONS

where G', Q, and Q_F are defined as before and C_R is the pounds of fuel per hour per square foot of projected radiant-heating surface. For one tube, $C_R = $ (OD in./12) $\times$ exposed length.

Example 19.3. Calculation of Performance by the Orrok-Hudson Equation. What per cent increase in the radiant-section heat absorption may be expected in a boiler when the firing rate is increased 50 per cent?

The initial ratio of absorption to liberation is 0.38, and the excess air is expected to increase from 25 to 40 per cent as a result of the increased firing rate.

Solution:

$$\frac{Q_{F2}}{Q_{F1}} = 1.50 \qquad \frac{C_{R2}}{C_{R1}} = 1.5$$

$$\frac{G'_2}{G'_1} = \frac{140}{125}$$

Q_2/Q_1 is to be determined.

$$\frac{Q_1}{Q_{F1}} = \frac{1}{1 + G'\sqrt{C_R/27}} = 0.38$$

$$1 = 0.38 + 0.38 G'_1 \sqrt{C_{R1}/27}$$

$$G'_1 \sqrt{\frac{C_{R1}}{27}} = \frac{0.62}{0.38} = 1.63$$

$$G'_2 \sqrt{\frac{C_{R2}}{27}} = \frac{140}{125} G'_1 \sqrt{\frac{1.5 C_{R1}}{27}} = 1.12 \times 1.223 G'_1 \sqrt{\frac{C_{R1}}{27}}$$

$$G'_2 \sqrt{\frac{C_{R2}}{27}} = 1.37 G'_1 \sqrt{\frac{C_{R1}}{27}} = 1.37 \times 1.63 = 2.23$$

$$Q_2 = \frac{Q_{F2}}{1 + 2.23} = 0.31 Q_{F2}$$

$$\frac{Q_2}{Q_1} = \frac{0.31 Q_{F2}}{0.38 Q_{F1}} = \frac{0.31 \times 1.5 Q_{F1}}{0.38 Q_{F1}} = 1.22$$

Hence the radiant absorption will be increased only 22 per cent for an increase of 50 per cent in the heat liberated. In such a case the effects of the higher exit-flue-gas temperature on superheater tubes would have to be investigated.

Miscellaneous Applications. Radiant-heat transfer is of importance in sections of a furnace other than the radiant section proper. Radiant-heat transfer to the "shield" tubes (top two rows) of the convection bank of a furnace such as that shown in Fig. 19.16 may be evaluated by the method of Lobo and Evans. The procedure is the same as in the case of the calculations of the radiant section proper with the exception that the gas temperature to be used is that of the gases coming over the bridge wall. One would expect that the temperature to use would be the gas temperature *after* heat has been lost to the shield tubes by radiation. In this particular case, however, the mean radiating temperature is closer to that of the gases entering the cavity immediately above the shield tubes. Hence no trial and error is involved in calculating this heat transfer, since the radiant-section exit temperature is already known. It should be noted that, since the convection-heat transfer to the shield tubes will be evaluated independently, the value of $\Sigma Q/\alpha A_{cp}\mathfrak{F}$ corresponding to the gas temperature and the shield-tube-metal temperature should be reduced by the amount $7(T_G - T_S)$.

Once the gases have actually made their entrance into the rows of tubes, they will continue to lose heat by radiation, and in spite of the rather small mean beam length this radiation may account for 5 to 30 per cent of the total heat transfer in the entire convection section. This radiation can be evaluated by the equation

$$Q_{RC} = 0.173 \left[\left(\frac{T_G}{100}\right)^4 - \left(\frac{T_S}{100}\right)^4 \right] A \frac{1}{(1/\epsilon_G) + (1/\epsilon_S) - 1} \quad (19.18)$$

where Q_{RC} = radiant-heat flow at a point in the convection section in Btu per hour. Other terms were defined in the Lobo and Evans method. It should be noted that no correction for reradiation from the refractory side walls is made. (It is more convenient instead to use the method of Monrad[1] to apply a correction factor to the combined heat-transfer coefficient for convection and radiation.) With negligible loss in accuracy the last term can be rewritten

$$\frac{1}{(1/\epsilon_G) + (1/\epsilon_S) - 1} = \epsilon_G \epsilon_S \quad (19.19)$$

and the equation becomes, with further substitution,

$$\frac{Q_{RC}}{A} = \epsilon_S[(q_C + q_W)_{T_G} - (q_C + q_W)_{T_S}] \left(\frac{100 - \%}{100}\right) \quad (19.20)$$

The use of this equation is illustrated by Example 19.4. Another type of problem frequently encountered is one in which a kettle or tank is to be used for boiling a liquid and the vessel either is direct-fired or is heated by waste flue gases from another unit. It can be calculated by applying some of the principles already illustrated. Example 19.5 shows such an application.

Example 19.4. Calculation of the Equivalent Radiant Coefficient. In the convection section of a refinery furnace, tubes are 5 in. OD on 8½ in. centers, spaced on equilateral triangular pitch. The flue gases at the row of tubes under consideration are at 1500°F; the tube temperature is 650°F. The flue gases contain 10.84% CO_2 and 12.48% H_2O by volume. Calculate the radiant-heat transfer between the gas and tubes in terms of a coefficient which can be added to the convection-heat-transfer coefficient.

[1] Monrad, C. C., *Ind. Eng. Chem.*, **24**, 505 (1932).

Solution:

$$\epsilon_S = 0.90 \quad \text{(assumed)}$$
$$L = 0.4 \text{ (center to center)} - 0.567 \text{ (OD)} \quad \text{(from 7 in Table 19.1)}$$
$$= 0.4 \ (8.5) - 0.567 \ (5) = 3.400 - 2.835$$
$$= 0.565 \text{ ft}$$
$$p_{H_2O}L = 0.1248 \times 0.565 = 0.0704 \text{ atm-ft}$$
$$p_{CO_2}L = 0.1084 \times 0.565 = 0.0611 \text{ atm-ft}$$

$$q_{H_2O} \text{ at } T_G = 1050 \qquad q_{H_2O} \text{ at } T_S = 165$$
$$q_{CO_2} \text{ at } T_G = \underline{1700} \qquad q_{CO_2} \text{ at } T_S = \underline{160}$$
$$\phantom{q_{CO_2} \text{ at } T_G =} 2750 \phantom{q_{CO_2} \text{ at } T_S =} 325$$

$$\frac{Q_{RC}}{A} = 0.9(2750 - 325)$$

$$\frac{P_{CO_2}}{P_{CO_2} + P_{H_2O}} = \frac{0.1084}{0.1084 + 0.1248} = 0.465$$

$$P_{CO_2}L + P_{H_2O}L = 0.1315$$
$$\% \text{ correction} = 2\%$$
$$\frac{Q_{RC}}{A} = 0.9 \times 2425 \times 0.98 = 2140$$

The equivalent radiation coefficient is, then,

$$h_r = \frac{Q_{RC}}{A \ \Delta T} = \frac{2140}{1500 - 650} = 2.51 \text{ Btu/(hr)(ft}^2\text{)(°F)}$$

which is a very appreciable part of the total coefficient. To correct the total coefficient $h_c + h_r$ for radiation from the walls of the convection bank the method of Monrad mentioned previously is recommended.

Example 19.5. Calculation of a Heated Vessel. Design a simple vessel for continuously concentrating a solution whose boiling point is 480°F. The duty required is 500,000 Btu/hr. The source of heat available is 3050 lb/hr of flue gas at 1500°F. A quantity of 48 in. OD pipe formerly used in some discarded vessels is available.

Solution. The required unit might take the shape of that shown in Fig. 19.17. Flue gases flow parallel to the axis of the vessel in a tunnel beneath it. Although the cold surface is segregated from the refractory surface to a greater extent than in a furnace radiant section, the Lobo and Evans exchange factor can be used without affecting the accuracy of the solution nearly so much as the poor choice of a mean gas temperature.

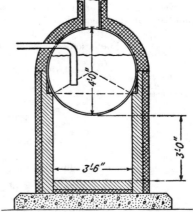

FIG. 19.17. Simple concentrator of Example 19.5.

Since the vessel will be very long compared with the cross-sectional area of the flue-gas path, a stepwise calculation is indicated unless the mean gas temperature can be evaluated properly. In a stepwise calculation, the heater could be broken up into sections of varying length, each having the same duty. Inasmuch as the heat trans fer is predominantly radiant and the cold surface is at a constant temperature. an adequate mean temperature is defined by

$$T_G = \sqrt[4]{T_{G,\,in}^2 T_{G,\,out}^2} \qquad (19.21)$$

The solution is worked on this basis. Assuming that overall heat losses equal 10 per cent of the duty, the exit-flue-gas temperature is found to be 900°F. The cold surface is assumed to have an emissivity of 0.9 and temperature of 500°F. The flue gases contain 7.1% CO_2 and 14.3% H_2O. The gas emissivity is found to be 0.259 at the hot end and 0.27 at the cold end; hence an average value of 0.265 can be used. Calculation of the gas emissivity was illustrated in Example 19.1. The equivalent surface per foot of length obviously lies between the values corresponding to the arc and the chord of the heated portion of the circumference; hence

$$\frac{\alpha A_{cp}}{\text{ft}} = \frac{3.5 + {}^{120}\!/_{360} \times \pi \times 4.0}{2} = 3.85 \text{ ft}^2/\text{ft}$$

$$\frac{A_R}{\text{ft}} = 3.0 + 3.6 + 3.0 = 9.6 \text{ ft}^2/\text{ft}$$

Arbitrarily neglecting end walls and also the side wall refractory over 3'0" above the floor,

$$\frac{A_R}{\alpha A_{cp}} = \frac{9.6}{3.85} = 2.49$$

$$\mathfrak{F} = 0.56 \text{ at } \epsilon_G = 0.265 \quad \text{and} \quad \frac{A_R}{\alpha A_{cp}} = 2.49$$

$$\frac{Q}{\alpha A_{cp} \mathfrak{F}} = 15,300 \text{ Btu/(hr)(ft}^2) \text{ at } T_G = 1174°F \quad \text{and} \quad T_S = 500°F$$

However, the convection coefficient is small, $1.0 \pm$ Btu/(hr)(ft²)(°F), and $A/\alpha A_{cp}$ is not 2.0 as in the assumptions for the Lobo and Evans equation.

$$\frac{Q}{\alpha A_{cp} \mathfrak{F}} \text{ (i.e., radiation only)} = 15,300 - 7(T_G - T_S) = 10,610 \text{ Btu/(hr)(ft}^2)$$

$$\frac{Q}{\alpha A_{cp}} = 10,610 \times 0.56 = 5940 \text{ Btu/(hr)(ft}^2) \text{ of } \alpha A_{cp}$$

Convection rate basis, αA_{cp}:

$$\frac{Q_c}{\alpha A_{cp}} = 1.0 \times (1174 - 500) \times \frac{4.2}{3.85} = 640$$

$$\frac{\Sigma Q}{\alpha A_{cp}} = 5940 + 640 = 6580 \text{ Btu/(hr)(ft}^2)$$

$$\alpha A_{cp} \text{ required} = \frac{500,000}{6580} = 76.0 \text{ ft}^2$$

$$\text{Length required} = \frac{76.0}{3.85} = 19.7 \text{ ft}$$

Inasmuch as the heat available is fixed in quantity and temperature level, the actual design should employ a liberal safety factor (say 25-ft length) and provision should be made to control the heat transfer either by means of a flue-gas by-pass or by admitting tempering air at the inlet to the unit.

Some Practical Aspects of Refinery Furnaces. While heat-transfer calculations are very important in the design of refinery furnaces, some of the other factors which enter into their design (and operation) require mention. The furnace must be designed around the burners. If the furnace is fired from the end wall so that the flames travel parallel to the tubes, the cross section of the furnace must be great enough to provide adequate clearance between the burners and the tubes. When such clearance is not provided, flames impinge on the tubes and may cause failures even when moderate pressures and temperatures are employed in the tubes.

When the burners are placed in the floor, side wall, or roof of a box-type furnace, a large number of small burners may be used and the clearances need not be so great. Since the space for locating the burners increases with the tube length, the maximum practical tube length may be used thereby decreasing the required number of return bends. Tube length is limited by the capacity of modern tube mills unless butt-welded tubes are used. The method of cleaning the tubes by turbining or the steam-air decoking method may also be a factor in determining the tube length. Ordinarily the design of the convection section requiring a reasonable flue-gas velocity for a good convection coefficient determines the tube length when the burner spacing is not controlling. The distribution of flue-gas flow in the convection section becomes more difficult when long tubes are used, and for furnaces with 50- to 60-ft tubes at least two flue-gas take-offs from the convection section should be provided. The care and operation of the burners and allied equipment have a pronounced effect on the cost of operating and maintaining the furnace. A furnace having a duty of 100,000,000 Btu/hr will burn in the neighborhood of a quarter of a million dollars worth of fuel per year, and such a furnace will use from 5,000,000 to 15,000,000 lb/year of steam for atomizing oil fuel depending upon whether or not the steam-oil ratio is carefully controlled. Oil burners should function properly when atomizing with 0.3 lb of steam per pound of oil.

Most burners require a certain amount of secondary air which is admitted through the burner register. It should be pointed out that, in general, the burner register should be used to control the air-fuel ratio (per cent excess air) while the stack damper should be used to control the draft at the top of the furnace. A slight draft (0.01 to 0.05 in. H_2O) should be maintained at that point at all times, so that leakage through the furnace roof or walls is inward. Outward leakage of flue gases damages the castings and steel supporting the brickwork, warps peep doors and explosion doors, and damages the furnace severely. The furnace should be provided with a draft-gage connection in the roof to guide the operator, but in the absence of such an instrument the appearance of the roof tile may be a guide. Dark lines at the joints between the bricks indicate an infiltration of air; bright lines indicate outward flow of flue gases.

The use of more excess air than necessary may often make it impossible to maintain the proper draft at the top of the furnace. Furthermore the higher the excess air the greater will be the fuel requirement for a given duty, and Orsat analyses of the flue gases should be made a part of the routine operation to ensure economical furnace performance. Excess oxygen in the flue gases increases the rate of oxidation of the tubes and tube supports. In some types of furnaces (depending upon the ratio of radiant- to convection-section surface) higher excess air results in a higher temperature of the flue gases entering the convection section and may cause premature failure of the tube sheets in the convection section.

The higher the radiant flux the lower the first cost of a furnace for a given duty. At average fluxes above 15,000 Btu/(hr)(ft^2), however, the savings diminish rapidly, and the problems of control and maintenance of the furnace increase disproportionately. The larger the furnace and the greater the ratio of refractory to cold surface, however, the lower will be the radiant-section flue-gas temperature for a given radiant flux. The average radiant flux permissible in a furnace depends upon the charge stock, extent of cracking, the heat-transfer coefficient within the tube, the ratio of $A/\alpha A_{cp}$ (which is the ratio of the rate on the face of the tube to the circumferential rate), and the distribution of fluxes along the length of the tubes and for tubes in various locations. All these details require consideration, and in Table 19.2 are presented some representative values of permissible average rates.

TABLE 19.2. PERMISSIBLE AVERAGE RADIANT RATES

Type of furnace	Allowable rate, Btu/(hr)(ft^2 of circumferential tube area)
Crude	10,000–16,000
Vacuum	5,000–10,000
Naphtha reforming	1,000–18,000
Gas oil cracking:	
Heating	10,000–15,000
Soaking	10,000
Visbreaking	10,000–12,000

It should be realized that average permissible rates are rule-of-thumb affairs and that, actually, the maximum point rate on a tube or tubes is the basic factor which should be prescribed for given conditions. Accordingly, the average permissible rate may be increased when tubes are spaced farther apart than normal, when the tubes are fired from both sides, or when any other specific steps are taken to improve the rate distribution. Tubes are generally spaced from one to one and one-half tube diameters from the face of a wall to the center line of the tubes.

FURNACE CALCULATIONS

One diameter is preferable from a practical standpoint, since it decreases the weight of the intermediate tube supports.

In the past, air preheaters have not been used so widely on refinery furnaces as on boiler furnaces. The use of air preheat is particularly desirable when the inlet temperature of the oil to be heated is quite high. When air preheat is employed, a greater proportion of the furnace duty is done in the radiant section than when the air is not preheated if a given radiant rate is to be maintained.

In furnace test work flue-gas temperatures should be determined with a high-velocity thermocouple, preferably of the multishielded type. Tube-metal temperatures can be measured by means of an optical pyrometer or by tube-skin thermocouples. It should be noted that the optical pyrometer may indicate temperatures higher than the true skin temperature; this is particularly true when the refractory surfaces are much hotter than the tubes, in which case some of the radiation received by the pyrometer consists of reflected radiation originating at the higher temperature refractory.

PROBLEMS

19.1. Using the Orrok-Hudson equation, calculate the efficiency of a boiler radiant furnace firing 1.0 ton/hr of coal at an air/fuel ratio of 16.0 lb air/lb fuel. The circumferential area of the boiler tubes is 7000 ft^2.

19.2. Calculate the maximum radiant rate on the circumference of a tube in the back (second) row of a group of double-row tubes in front of a furnace wall. The average circumferential rate on a tube in the front row is 10,000 Btu/(hr)(ft^2). All tubes are 4 in. OD on 7-in. centers, spaced on equilateral triangular pitch.

19.3. Using the Wilson, Lobo, and Hottel equation for a box-type heater, calculate the average circumferential heat-transfer rate in a furnace radiant section when (a) heat liberation is 142,000,000 Btu/hr, excess air = 30 per cent (20.75 lb air/lb fuel), $\alpha A_{cp} = 1970$ ft^2, $A = 4710$ ft^2, (b) as in (a) except 60 per cent excess air, (c) as in (a) except tubes (same outside dimension and C-C) are added to make $A = 6000$ ft^2, (d) as in (a) except heat liberation is 100,000,000 Btu/hr.

19.4. Using the Lobo-Evans method, calculate the fuel required (pounds per hour) in a furnace, having the following characteristics if an average radiant-section rate of 12,000 Btu/(hr)(ft^2) of circumferential tube area is to be developed:

Dimension of combustion chamber	15 by 30 by 40 ft
Tube OD	5 in.
Tube center-to-center spacing	10 in.
Number of tubes (arranged in single row)	90
Circumferential tube surface	4710 ft^2
Total wall area A_T	4300 ft^2
Fuel oil	Oil
Atomization medium	0.3 lb steam/lb oil
Heating value of fuel (LHV)	17,130 Btu/lb
Excess air	25%
Lb air/lb fuel	17.44
Estimated tube temperature	1000°F

Note. Assume the average specific heat [Btu/(lb)(°F)] of flue gases to be 0.28 between 60°F (H_2O as vapor) and the exit flue-gas temperature. For water-vapor and CO_2 concentrations in the flue gases, see Example 19.1.

19.5. Calculate the fuel requirement for a furnace in which 2000 lb/hr of aluminum chips are to be melted continuously. The chips are charged at 60°F. The furnace is 12 by 12 ft inside cross-sectional dimensions. The height between the level of the molten aluminum and the roof is 9 ft. The emissivity of the mass of aluminum may be taken as 0.30. Fuel oil is fired as in Prob. 19.4 with 17.44 lb air/lb fuel and 0.3 lb steam/lb fuel. Assume same average specific heat for flue gas as in Problem 19.4.

Note. Use the Lobo-Evans equation, correcting from a tube emissivity of 0.9 used in the derivation of the overall exchanger factor to an emissivity of 0.3. The value of F_e is found using the chart. ϵ_F is found by substituting 0.9 for ϵ_S in Eq. (19.7).

NOMENCLATURE FOR CHAPTER 19

A	Heat-transfer surface for convection, ft^2
A_{cp}	Area of a cold plane replacing a bank of tubes, ft^2
A_R	Area of effective refractory surfaces, ft^2
A_T	Total area of furnace surfaces, ft^2
A'	Effective or "seen" heat-transfer area of sink or cold body, ft^2
a_G	Absorptivity of gas, dimensionless
C	Specific heat, Btu/(lb)(°F)
C_R	Lb fuel/(hr)(ft^2 projected radiant-heating surface)
C'	Influence of air preheat, dimensionless
e	Furnace efficiency defined by Eq. (19.14), per cent
F	Factor for type of coal firing, dimensionless
$\mathfrak{F}$	The overall exchange factor to allow for the geometry and emissivities, dimensionless
$\mathfrak{F}_{bA}$	Factor to allow for the geometry of a system with a black-body sink, dimensionless
F_c	Conductivity factor, dimensionless
F_s	Slag factor, dimensionless
F_e	Emissivity factor, dimensionless
G'	Ratio of air to fuel, lb/lb; $(1 + G')$, ratio flue gas to fuel, lb/lb
h_c	Convection coefficient, Btu/(hr)(ft^2)(°F)
h_r	Equivalent radiation coefficient, Btu/(hr)(ft^2)(°F)
K_1, K_2	Constants
L	Mean length of beam, ft
N_t	Number of tubes
p	Pressure, atm
P_T	Tube pitch or center-to-center distance, in.
Q	Radiant-heat flow, Btu/hr
Q_{RC}	Radiant-heat flow at a point in the convection section, Btu/hr
Q_S	Sensible-heat flow, Btu/hr
Q_W	Heat loss through furnace wall, Btu/hr
ΣQ	Combined radiation and convection heat flow, Btu/hr
q	Heat flux, Btu/(hr)(ft^2)
T_S	Mean temperature of receiver, °F
T_1	Temperature of source, °R
T_2	Temperature of receiver, °R
W	Fuel rate, lb/hr

α	Effectiveness factor, dimensionless
β	Angle, deg
ϵ	Emissivity, dimensionless
ϵ_F	Effective emissivity of the furnace cavity, dimensionless
σ	Stefan-Boltzmann constant, 0.173×10^{-8} Btu/(hr)(ft^2)(°R^4)

Subscripts (except as noted above)

A	Air
b	Black body
C	Carbon dioxide
F	Fuel
G	Gas
R	Recirculated
S	Surface
W	Water

CHAPTER 20

ADDITIONAL APPLICATIONS

Introduction. There are a number of collateral uses for heat-transfer equipment which have not appeared in any of the preceding chapters. Some of these include the commonest and least expensive forms of heat-transfer surface such as coils, submerged pipes in boxes, and trombone coolers. For the most part the heat-transfer elements treated here are not very closely related to those discussed in earlier chapters, nor can their performances be calculated with equal accuracy. This is an important limitation when attempting to calculate the surface requirements for a close temperature approach. The following are treated here:

1. Jacketed vessels
2. Coils
3. Submerged-pipe coil
4. Trombone cooler
5. Atmospheric cooler
6. Evaporative condenser
7. Bayonet
8. Falling-film exchanger
9. Granular materials in tubes
10. Electric resistance heaters

1. JACKETED VESSELS

Vessels without Agitation. Few data are available in the literature for predicting coefficients within a jacket or between a jacket and a liquid inside a vertical cylindrical vessel in which no mechanical agitation occurs. During heating, the latter rely upon free convection which has not been correlated except as given in Chap. 10. Free convection-heating coefficients can be approximated for vessels of large diameter by Eqs. (10.8) through (10.11). No consistent data are available for cooling by free convection, although the coefficients are undoubtedly lower.

Colburn[1] has tabulated the results of a number of studies from which several broad generalizations can be drawn. For the transfer of heat from steam condensing in a jacket to water boiling within the vessel the clean overall coefficient is about 250 $Btu/(hr)(ft^2)(°F)$ for a copper vessel

---

[1] Perry, J. H., "Chemical Engineers' Handbook," 3d ed., pp. 481–482, McGraw-Hill Book Company, Inc., New York, 1950.

and 175 for a steel vessel. The difference is due to the conductivities and equivalent structural thicknesses of the two metals, respectively. The same coefficients might also be expected for boiling dilute aqueous solutions. For water-to-water heating or cooling, an overall coefficient of 100 Btu/(hr)(ft^2)(°F) appears to be in order provided neither stream is refrigerated. The coefficient is perhaps between 75 and 80 for aqueous solutions whose properties do not differ greatly from those of pure water. For heating or cooling nonviscous hydrocarbons the overall coefficient should be reduced to about 50. For fluids classified as medium organics in Table 8 a more probable range of coefficients is from 10 to 20. The jacket of a vessel may be baffled helically to assure positive circulation.

A heat-transfer coefficient selected from the groups above cannot be incorporated into the Fourier equation $Q = UA \, \Delta t$ except when the vessel operates under steady state. A jacketed vessel can be adapted for operation under steady-state conditions when there is a continuous influx and removal of the liquid entering it. Since a jacketed vessel is primarily a batch vessel, the temperature difference during batch heating or cooling is not constant. The coefficient must consequently be substituted into a suitable *unsteady*-state equation, such as Eqs. (18.5), (18.7), (18.9), or (18.11), which takes into account the time required to change the temperature of the batch while employing a temperature difference which varies with time.

Mechanically Agitated Vessels. Chilton, Drew, and Jebens[1] have published an excellent correlation on both jacketed vessels and coils under batch and steady-state conditions and employing the Sieder-Tate heat-transfer factor j with a Reynolds number modified for mechanical agitation. They employed a flat paddle. Although much of the work was carried out on a vessel 1 ft in diameter, checks were also obtained on vessels five times those of the experimental setup. The deviations on runs with water were the highest for the fluids tested, which included lube oils and glycerol, and were in some instances off by 17.5 per cent. Additional applications of this method have been discussed by Mack and Uhl[2] along with calculations. In jacketed vessels it was found that the correlation held up to the agitator rate at which air was beaten into the liquid. For water this corresponded to an agitator rate of 200 rpm, and for other liquids the rate was higher. In Fig. 20.1 a typical jacketed-vessel

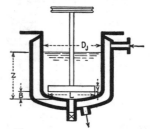

FIG. 20.1. Jacketed vessel.

[1] Chilton, T. H., T. B. Drew, and R. H. Jebens, *Ind. Eng. Chem.*, **36**, 510–516 (1944).
[2] Mack, D. E., and V. W. Uhl, *Chem. Eng.*, **54**, No. 9, 119–125; No. 10, 115–116 (1947).

assembly is shown. It consists of a vessel and jacket with suitable means for circulating the jacket fluid and a flat two-bladed paddle agitator.

The dimensions essential to the calculation are the height of the wetted portion of the vessel z, the diameter of the vessel D_j, the length of the paddle L, and the height of the bottom of the paddle above the bottom of the vessel B. Studies made by White and coworkers[1] indicate that power requirements can be determined as a function of the modified Reynolds number, $Re_j = L^2 N \rho / \mu$, where L is the length of the paddle in

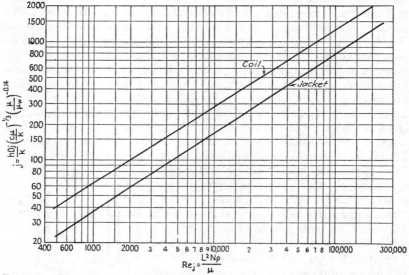

FIG. 20.2. Heat-transfer coefficients from jackets and coils. (*Chilton, Drew, and Jebens, Industrial and Engineering Chemistry.*)

feet, N the number of revolutions per hour, ρ the average density, and μ the viscosity of the liquid. Chilton, Drew, and Jebens have employed the same modified Reynolds number for heat transfer. Their results are given by the consistent equation

$$\frac{h_j D_j}{k} = 0.36 \left(\frac{L^2 N \rho}{\mu}\right)^{2/3} \left(\frac{c\mu}{k}\right)^{1/3} \left(\frac{\mu}{\mu_w}\right)^{0.14} \quad (20.1)$$

Equation (20.1) is plotted as the lower line in Fig. 20.2. It is seen that to scale up from pilot to plant size the change in the coefficient is given by

$$\frac{h_{j1}}{h_{j2}} = \left(\frac{L_1}{L_2}\right)^{1/3} \left(\frac{N_1}{N_2}\right) \quad (20.2)$$

[1] White, A. M., E. Brenner, G. A. Phillips, and M. S. Morrison, *Trans. AIChE*, **30**, 585 (1934).

ADDITIONAL APPLICATIONS

The power requirements are given by the dimensional equation of White and coworkers.

$$\text{hp} = 1.29 \times 10^{-4} D_j^{1.1} L^{2.72} N'^{2.86} y^{0.3} z^{0.6} \mu'^{0.14} \rho^{0.86} \qquad (20.3)$$

where y = width of the agitator, ft
N' = speed, rps
μ' = the viscosity, lb/ft $\times$ sec

These equations hold for paddle agitators centrally located with $L > 0.3 D_j$ and with a height $< L/6$. The coefficient in the jacket itself has not been determined, although it can be readily estimated for both steam and water. The use of this method for the solution of a typical problem is given below.

Example 20.1. Calculation of a Jacketed Vessel. A jacketed vessel has the following dimensions: A 12-in. IPS pipe is jacketed by a 14-in. IPS pipe both having dished bottoms. It is equipped with a paddle agitator 7.2 in. long and 1.2 in. high located at a distance of 1.8 in. from the bottom. The drive speed is 125 rpm. It is to be filled to a height of 10 in. with a batch of aqueous reacting liquor at 150°F which requires the addition of 32,600 Btu/hr to supply the endothermic heat of reaction and maintain temperature. A dirt factor of 0.005 should be provided. At what temperature must steam be available for the jacket?

Solution. This may be regarded as a steady-state problem, since Δt is constant.

$L = \dfrac{7.2}{12} = 0.6$ ft

$N = 125 \times 60 = 7500$ rev/hr

$\rho = 62.5$ lb/ft^3 (approximately)

At 150°F, $\mu = 0.44 \times 2.42 = 1.06$ lb/(ft)(hr) (Fig. 14)

$k = 0.38$ Btu/(hr)(ft^2)(°F/ft) (Table 4)

$c = 1.0$ Btu/(lb)(°F)

$Re_j = \dfrac{L^2 N \rho}{\mu} = \dfrac{0.6^2 \times 7500 \times 62.5}{1.06} = 160{,}000$

$j = 1100 = \dfrac{h_j D_j}{k}\left(\dfrac{c\mu}{k}\right)^{-1/3}\left(\dfrac{\mu}{\mu_w}\right)^{-0.14}$ (Fig. 20.2)

$\left(\dfrac{\mu}{\mu_w}\right)^{0.14}$ may be regarded as 1.0 for water.

$D_j = \dfrac{12.09}{12} = 1.01$ ft (Table 11)

$\left(\dfrac{c\mu}{k}\right)^{1/3} = \left(1 \times \dfrac{1.06}{0.38}\right)^{1/3} = 1.41$

$h_j = j \dfrac{k}{D_j}\left(\dfrac{c\mu}{k}\right)^{1/3}\left(\dfrac{\mu}{\mu_w}\right)^{0.14}$

$= 1100 \times \dfrac{0.38}{1.01} \times 1.41 \times 1.0 = 588$ Btu/(hr)(ft^2)(°F)

For the steam in the jacket referred to the inside diameter of the vessel,

$h_{oi} = 1500$ Btu/(hr)(ft^2)(°F)

Using the inside surface as reference,

$$U_C = \frac{h_j h_{oi}}{h_j + h_{oi}} = \frac{588 \times 1500}{588 + 1500} = 422 \text{ Btu/(hr)(ft}^2)(°F) \tag{6.38}$$

$$R_d = 0.005 \qquad h_d = \frac{1}{R_d} = \frac{1}{0.005} = 200$$

$$\frac{1}{U_D} = \frac{1}{U_C} + R_d$$

or

$$U_D = \frac{U_C h_d}{U_C + h_d} = \frac{422 \times 200}{422 + 200} = 136$$

For calculation of the heat-transfer surface consider the bottom to be a flat plate or use tables giving the surface as a function of diameter for elliptical heads.

$$A = \pi \times 1.01 \times 0.83 + \frac{\pi}{4} \times 1.01^2 = 3.43 \text{ ft}^2$$

The temperature difference is the LMTD, since both streams are isothermal.

$$Q = U_D A \, \Delta t$$

$$\Delta t = \frac{32{,}600}{136 \times 3.43} = 70°F$$

Since the reaction occurs at 150°F, the temperature of the steam must be

$$T_s = 150 + 70 = 220°F$$

Note: If a higher temperature of steam were required, the jacket wall temperature must be checked by Eq. (6.8) to assure that it is not higher than the boiling temperature of the vessel fluid. If vapor bubbles form at the inside surface, the film coefficient may be reduced by an unpredictable amount.

2. COILS

Introduction. Tube coils afford one of the cheapest means of obtaining heat-transfer surface. They are usually made by rolling lengths of copper, steel, or alloy tubing into helixes or double helical coils in which the inlet and exit are conveniently located side by side. Helical coils of either type are frequently installed in vertical cylindrical vessels with or without an agitator, although free space is provided between the coil and the vessel wall for circulation. When such coils are used with mechanical agitation, the vertical axis of the agitator usually corresponds to the vertical axis of the cylinder. Double helical coils may be installed in shells with the coil connections passing through the shell or shell cover. Such an apparatus is similar to a tubular exchanger, although limited to smaller surfaces. Another type of coil is the pancake coil, which is a spiral rolled in a single plane so as to lie horizontally near the bottom of a vessel to transfer heat by free convection. Examples of the single helix and pancake are shown in Fig. 20.3. The rolling of coils, particularly with diameters above 1 in., requires a special winding technique to prevent the tube from flattening into an elliptical cross section, since distortion reduces the flow area.

Tube-side Coefficients. Because of the increased turbulence it is to be expected that the tube-side coefficients for a coiled tube will be greater for a given weight flow than that for a straight tube. For a double pipe helical water-to-water exchanger Richter[1] obtained overall coefficients which are about 20 per cent greater than those computed for straight pipes using similar flow velocities. Jeschke[2] obtained data on the cooling of air in $1\frac{1}{4}$-in. steel coiled tubing. For ordinary use McAdams[3] suggests that straight tube data such as Eqs. (6.1) and (6.2) or Fig. 24 can be used when the value of h so obtained is multiplied by $1 + 3.5(D/D_H)$, where D is the inside diameter of the tube in feet and D_H is the diameter of the helix in feet. McAdams also suggests that in the absence of data on specific liquids the same correction can be applied to them. Precise corrections are not important, since in most instances it is customary to use either cold water or steam in the tube, neither of which is controlling. For water flowing in the tubes it is suggested that the uncorrected coefficients be obtained from Fig. 25.

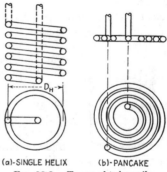

(a)-SINGLE HELIX (b)-PANCAKE

FIG. 20.3. Types of tube coils.

Outside Coefficients to Fluids without Mechanical Agitation. There is a dearth of data in the literature on the transfer of heat to helical coils by free convection. Colburn[4] has prepared a table of the available overall coefficients. The vertical helical coil is poorly adapted for free-convection heating, since the same liquid rises from the lowest to the highest turns successively, thereby reducing the effectiveness of the upper turns. Coefficients for pancake coils can be approximated from Eq. (10.7) or (10.11). To date, however, no standard method has appeared for calculating film coefficients for the outside of a single or double helical shell-and-coil type of exchanger. When employed for cooling fluids in vessels the effect of free convection is small.

[1] Richter, G. A., *Trans. AIChE*, **12**, Part II, 147–185 (1919).

[2] Jeschke, D., *Z. Ver. deut. Ing.*, **69**, 1526 (1925); *Z. Ver. deut. Ing., Erganzungheft*, **24**, 1 (1925).

[3] McAdams, W. H., "Heat Transmission," 2d ed., pp. 177, 184, McGraw-Hill Book Company, Inc., New York, 1942.

[4] Colburn, A. P., in Perry, *op. cit.*, p. 481.

Outside Coefficients to Fluids with Mechanical Agitation. Chilton, Drew, and Jebens[1] also obtained a correlation for heat transfer to fluids in vessels with mechanical agitation heated or cooled by submerged tube coils as shown in Fig. 20.4. Their equation for the coil is similar to that for the jacketed vessel with the same deviation and is given by

$$\frac{h_c D_j}{k} = 0.87 \left(\frac{L^2 N \rho}{\mu}\right)^{2/3} \left(\frac{c\mu}{k}\right)^{1/3} \left(\frac{\mu}{\mu_w}\right)^{0.14} \quad (20.4)$$

Equation (20.4) is represented by the upper line in Fig. 20.2. Its use is also similar. As with jacketed vessels, caution is again directed to the fact that for batch applications the value of h_c cannot be used with the Fourier equation. Instead, an appropriate equation must again be selected from Chap. 18. However, if the vessel is operated with continuous feed and continuous overflow, the value of h_c and U_D obtained from Eq. (20.4) can be substituted into the Fourier equation. If possible, the dirt factor should be made the limiting resistance.

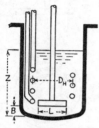

Fig. 20.4. Tube coil in a vessel.

Outside Coefficients Using Vertical Pipes. One of the disadvantages resulting from the use of a paddle-agitator and helical-coil arrangement is its low mixing efficiency. For good mixing and correspondingly higher transfer coefficients the agitator must impart vertical as well as horizontal flow lines. When using an agitator of the paddle or vertical-blade turbine types and radial tube banks with tubes arranged vertically in the vessel, the tubes act as baffles. Rushton, Lichtmann, and Mahony[2] investigated this type of arrangement employing a 4-ft tank and a liquid level 4 ft high. Four banks of four 1-in. IPS vertical pipes were arranged at right angles about both a 16-in. six-blade turbine and a 12-in. four-blade turbine. The maximum coefficients were obtained when the mixing turbine was located at a height 2 ft above the bottom. The heat-transfer coefficients have been reported for water as functions of the Reynolds number by the following dimensional equations:

For the 16-in. six-blade turbine:

$$\text{Heating: } h_c = 0.00285 \, \frac{L^2 N \rho}{\mu} \quad (20.5a)$$

$$\text{Cooling: } h_c = 0.00265 \, \frac{L^2 N \rho}{\mu} \quad (20.5b)$$

---

[1] Chilton, T. H., T. B. Drew, and R. H. Jebens, *loc. cit.*

[2] Rushton, J. H., R. S. Lichtmann, and L. H. Mahony, *Ind. Eng. Chem.*, **40**, 1082–1087 (1948).

ADDITIONAL APPLICATIONS

For the 12-in. four-blade turbine:

$$\text{Heating: } h_c = 0.00235 \left(\frac{L^2 N \rho}{\mu}\right)^{0.7} \quad (20.6a)$$

$$\text{Cooling: } h_c = 0.00220 \left(\frac{L^2 N \rho}{\mu}\right)^{0.7} \quad (20.6b)$$

where L is the diameter of the turbine.

Example 20.2. Calculation of a Tube Coil. The thermal conditions of Example 20.1 will be used. 32,600 Btu/hr will be introduced into an isothermal liquid at 150°F using steam at 220°F. The coils will consist of turns of ½ in. OD copper tubing with a mean coil diameter of 9.6 in. How many turns are required?

Solution:

$L = 0.6$ ft $\quad N = 7500$ rev/hr $\quad \rho = 62.5$ ft^3/lb $\quad \mu = 1.06$ lb/(ft)(hr)
$k = 0.38$ Btu/(hr)(ft^2)(°F/ft) $\quad c = 1.0$ Btu/(lb)(°F)

$Re_j = \dfrac{L^2 N \rho}{\mu} = 160{,}000$

$j = 1700$ (upper line, Fig. 20.2)

$D_i = 1.01$ ft $\quad \left(\dfrac{c\mu}{k}\right)^{1/3} = 1.41, \quad \left(\dfrac{\mu}{\mu_w}\right)^{0.14} = 1.0$ for water

$h_c = j \dfrac{k}{D} \left(\dfrac{c\mu}{k}\right)^{1/3} \left(\dfrac{\mu}{\mu_w}\right)^{-0.14}$

$ = 1700 \times \dfrac{0.38}{1.01} \times 1.41 = 900$ Btu/(hr)(ft^2)(°F)

For steam,

$h_{oi} = 1500$

$U_C = \dfrac{h_c h_{oi}}{h_c + h_{oi}} = \dfrac{900 \times 1500}{900 + 1500} = 562$ Btu/(hr)(ft^2)(°F)

$R_d = 0.005 \quad h_d = \dfrac{1}{0.005} = 200$

$U_D = \dfrac{U_C h_d}{U_C + h_d} = \dfrac{562 \times 200}{562 + 200} = 147.5$

$A = \dfrac{Q}{U_D \Delta t} = \dfrac{32{,}600}{147.5 \times (220 - 150)} = 3.16$ ft^2

External surface/lin ft = 0.1309 ft^2/ft (Table 10)
Per turn = $\pi \times 0.8 \times 0.1309 = 0.328$ ft^2

Turns = $\dfrac{3.16}{0.328} = 9.6$

3. SUBMERGED-PIPE COIL

Introduction. This is one of the simplest and cheapest methods of obtaining both cooling and condensing surface. A series of pipes are connected by standard fittings and are submerged in a concrete or wood trough with water circulating about the pipes as shown in Fig. 20.5. Coolers of this type are of considerable value when the hot fluid is corrosive or erosive such as when carrying suspended abrasive particles. The calculations of the trough side are naturally only approximate, but since water flows in the trough, it is not the limiting resistance except when the coil is employed to condense steam.

Temperature Difference in the Submerged-pipe Coil Cooler. Since the flow of water outside the coil is mostly along the axes of the tubes, the true temperature difference depends upon the arrangement of the tube passes. The trough is usually arranged for one pass. If the pipes are connected by a header at each end with a single pass so that the pipe fluid is in counterflow with the water, the true temperature difference is given by the LMTD. If the pipes are connected by return elbows in a multipass arrangement, the flow pattern may be treated as parallel flow–counterflow, with the 1-2 exchanger correction applying provided the trough liquid is reasonably mixed at every point along the length of the pipes. For crossflow arrangements a corresponding correction can

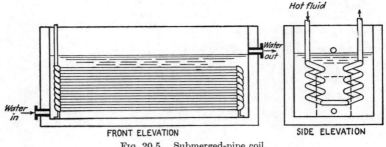

FIG. 20.5. Submerged-pipe coil.

be obtained from Fig. 16.17 or 20.7b in the event that any of these flow patterns apply. Coefficients for the pipe side can be obtained from Fig. 24.

Trough Heat-transfer Coefficients. Trough heat-transfer coefficients are usually difficult to evaluate. There are no conventional arrangements and few published data for this type of equipment. In installations with little or no baffling, a large portion of the cooling water by-passes the surface between the coil and the trough enclosure. Owing to the low water velocities usually encountered, pipe coils tend to foul at a rapid rate because of the growth of algae slime and other scales. The use of large fouling factors is requisite not only from the standpoint of dirt but as a means of providing additional design safety. Fouling factors of less than 0.01 should be avoided, in which case the maximum value of the overall design coefficient will be less than 100.

Ordinarily there is considerable free space in the cross section of the trough, so that the actual linear velocity of the water over the coil may be extremely small. At the extreme of low velocities the trough-side

ADDITIONAL APPLICATIONS   725

coefficient will approach the free convection coefficient from a pipe to water. To estimate the minimum possible coefficient Fig. 10.4 can be used. However, if there is any sort of distribution of the water, there need be no hesitancy in using some multiple of the values obtained from Fig. 10.4.

Slurries and Suspensions. Pipe coils are becoming increasingly common in modern catalytic processes. The catalyst is often a finely divided dust which forms a slurry or a suspension with the carrying liquid. Because of the possibility that the solid will settle, slurries flow with high velocities to maintain turbulence, and the possibility of solids being thrown out of the liquid at elbows can be lessened by using streamline fittings. Many slurries are extremely erosive, the solids having an abrasive action on the metal, and in this respect the pipe coil is ideally suited, since it is possible to use extra heavy or double extra heavy steel pipe in assembling the coil. If the continuous erosion causes serious contamination to the product or poisons the catalyst, it is possible to resort to harder alloy pipes fabricated to IPS dimensions.

Very often data are available separately on the carrying liquid and the solid and not on the combined slurry dispersion. With a slurry it is fairly common to consider the controlling film coefficient to be that which exists between the liquid and the tube wall. The transfer of heat from the liquid to the catalyst particles is not considered to offer an appreciable resistance. The film coefficient can consequently be calculated for the combined heat load in the conventional manner using Eq. (6.2) or Fig. 24 and based upon the properties of the liquid alone except for the viscosity. The presence of the solid changes the viscosity of the liquid to an unpredictable extent, since at low concentrations some fine dusts tend to adsorb large quantities of liquid and greatly increase the viscosity. Other solids appear to be only superficially wetted and do not change the viscosity significantly. Using clays and similar catalysts it may be assumed that a concentration of 2 or 3 lb of solid per gallon of mixture does not increase the viscosity of the liquid by more than 100 per cent, and this higher value is suggested in the absence of actual viscosity data on a slurry.

Example 20.3. Calculation of a Submerged-pipe Coil Slurry Cooler. A spent clay catalyst comes from a catalytic reaction vessel as a slurry dispersed in an oil whose properties correspond to 28°API gas oil. The clay is dispersed to the extent of 1 lb/gal. It enters a submerged pipe coil at 675°F and leaves at 200°F. The viscosity of the mixture at 400°F is 2.3 centipoises. The combined flow is 33,100 lb/hr, and cooling will be effected by water of low mineral content from 120 to 140°F. Pipe lengths will consist of 3-in. extra heavy steel pipe 24 ft long.

What will be the size of the pipe coil?

Solution:

$T_{av} = \frac{1}{2}(675 + 200) = 437.5°F \qquad c = 0.64$ Btu/lb
Oil, $Q = 33{,}100 \times 0.64(675 - 200) = 10{,}200{,}000$ Btu/hr*
Water, $Q = 510{,}000 \times 1.0(140 - 120) = 10{,}200{,}000$ Btu/hr

Δt: It may be assumed that one stream carries all of the pipe liquid. Whether in crossflow or parallel flow–counterflow the true temperature-difference relations have not been derived. A case closely allied to the pipe coil will be treated for the two- (or more) pass trombone cooler. In any event the true temperature difference will be very nearly the same as the LMTD, since the outlet temperatures of both streams do not approach each other closely and the average temperatures are far apart.

$$\text{LMTD} = 230°F \tag{5.14}$$

T_c and t_c: The use of average temperatures will be satisfactory, since the outside coefficient cannot be determined very accurately.

Hot fluid: tube side, oil
$a'_t = 6.61$ in.²/pipe [Table 11]
Try all pipes in series.
Flow area, $a_t = 6.61/144 = 0.0458$ ft²
Mass vel, $G_t = W/a_t$
$\quad = 33{,}100/0.0458 = 723{,}000$ lb/(hr)(ft²)
The viscosity is given at 400°F, which is near enough for this calculation.
$\mu = 2.3 \times 2.42 = 5.56$ lb/(ft)(hr)
$D = 2.9/12 = 0.242$ ft [Table 11]
$Re_t = DG_t/\mu$
$\quad = 0.242 \times 723{,}000/5.56 = 31{,}300$
$j_H = 100$
$k(c\mu/k)^{\frac{1}{3}} = 0.245$ Btu/(hr)(ft²)(°F/ft)
  [Fig. 16]
$h_i = j_H[k(c\mu/k)^{\frac{1}{3}}]/D$ [Eq. (6.15)]
$\quad = 100 \times 0.245/0.242 = 101$
OD = 3.5 in. [Table 11]
$h_{io} = h_i \times \text{ID/OD} = 101 \times 2.9/3.5$
$\quad = 83.7$ Btu/(hr)(ft²)(°F) [Eq. (6.5)]

Cold fluid: trough side, water
Assume poor circulation and a minimum coefficient.
For a trial assume $h_o = 150$ Btu/(hr)(ft²)(°F)

$t_w = t_{av} + \dfrac{h_{io}}{h_{io} + h_o}(T_{av} - t_{av})$
  [Eq. (6.8)]
$\quad = 130 + \dfrac{83.7}{83.7 + 150}(437.5 - 130)$
$\quad = 240°F$

$t_f = \frac{1}{2}(t_w + t_{av})$
$\quad = \frac{1}{2}(240 + 130) = 185°F$
Refer to Fig. 10.4:
$\Delta t = 240 - 130 = 110°F$
$d_o = 3.5$ in.
$(\Delta t/d_o) = 110/3.5 = 31.4$
$h_o = 150$ (trial checks)

$U_C = \dfrac{h_o h_{io}}{h_o + h_{io}} = \dfrac{150 \times 83.7}{150 + 83.7} = 53.8$ Btu/(hr)(ft)(°F)

$R_d = 0.010 \qquad h_d = \dfrac{1}{.010} = 100$

$U_D = \dfrac{U_C h_d}{U_C + h_d} = \dfrac{53.8 \times 100}{53.8 + 100} = 35.0$

$A = \dfrac{Q}{U\,\Delta t} = \dfrac{10{,}200{,}000}{35.0 \times 230} = 1265$ ft²

External surface/lin ft = 0.917 ft²/ft (Table 11)

Pipe lengths required $= \dfrac{1265}{0.917 \times 24} = 58$

---

* This is safe, since the specific heat of the solids is only about 0.2 Btu/(lb)(°F) and 33,100 lb/hr represents the combined flow.

These must be arranged in series either in a single helical stack 29 tubes high or, if headroom is not available, in several lower stacks. Actually some boiling occurs outside part of the pipe surface where $t_w > 212°F$ so as greatly to increase the trough coefficient.

4. TROMBONE COOLERS

Introduction. Trombone coolers are sometimes called *trickle coolers, cascade coolers, horizontal-film coolers, S-type coolers,* and *serpentine coolers.* A sketch is shown in Fig. 20.6. Trombone coolers consist of a bank of standard pipes one above the other in series and over which water trickles downward, partly evaporating as it travels. They have been used extensively in the heavy chemical, brewing, coke, petroleum, and ice-making industries. They are frequently made of ceramic materials for cooling corrosive gases at atmospheric pressure, such as HCl and NO_2. When calculating heat flow through ceramics the resistance of the pipe wall must be included. The trombone cooler presents two problems: (1) the evaluation of the outside film coefficient and (2) calculation of the crossflow true temperature difference.

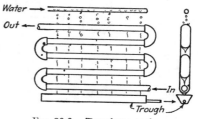

FIG. 20.6. Trombone cooler.

Temperature Difference in the Trombone Cooler. The flow arrangement of Fig. 20.6 in crossflow differs from any of the cases in Fig. 16.17 in that the fluid flowing outside the multipass pipes is not mixed over the length of the pipes whereas the multipass pipe fluid is mixed. Bowman, Mueller, and Nagle[1] have prepared correction factors F_T by which the true temperature difference Δt can be obtained as the product

$$F_T \times \text{LMTD}$$

for both the return-bend and helical types of trombone arrangement. These are given in Fig. 20.7a and b and are based, respectively, on the following:

Return bend arrangement:

$$\frac{1}{1-K} = e^{K_1 R}[\cosh K_1 R + (1 - K_1) \sinh K_1 R]$$

where $K_1 = 1 - e^{-S/2(r)}$.

Helical arrangement:

$$\frac{1}{1-K} = e^{K_1 R}(e^{K_1 R} + K_1^2 R)$$

[1] Bowman, Mueller, and Nagle, *Trans. ASME*, **62**, 291 (1940).

where again $K_1 = 1 - e^{-S/2(r)}$ and

$$K = \frac{T_1 - T_2}{T_1 - t_1}, \qquad R = \frac{T_1 - T_2}{t_2 - t_1}, \qquad (r) = \frac{LMTD}{T_1 - t_1}, \qquad S = \frac{t_2 - t_1}{T_1 - t_1}$$

While these apply specifically to units having but two pipe passes only, a small error arises when the corrections are applied to units having a greater number of pipe passes.

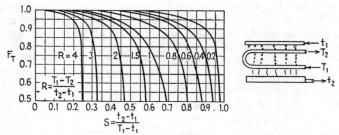

FIG. 20.7a. Mean-temperature-difference correction factor for two-pass trombone.

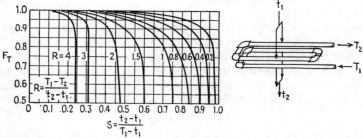

FIG. 20.7b. Mean-temperature-difference correction factor for two-pass helical trombone. (*Bowman, Mueller, and Nagle, Transactions ASME.*)

Outside Film Coefficients. The main published work on this type of apparatus was carried out by McAdams and coworkers at MIT and summarized by McAdams, Drew, and Bays.[1] (1) It is assumed that no evaporation occurs from the surface of the water although it is exposed to the atmosphere. (2) It is assumed that half of the liquid flows down each side of each pipe in streamline flow. The criterion of streamline flow is a Reynolds number, $4G'/\mu$, of less than 2100, where $G' = w/2L$,

[1] McAdams, W. H., T. B. Drew, and G. S. Bays, Jr., *Trans. ASME*, **62**, 627–631 (1940).

ADDITIONAL APPLICATIONS 729

w is the water rate in pounds per hour, and L is the length of each pipe in the bank in feet. The equation for the transfer coefficients within ±25 per cent is given by the dimensional equation

$$h = 65 \left(\frac{G'}{D_0}\right)^{\frac{1}{3}} \qquad (20.7)$$

where D_0 is the outside diameter of the pipe in feet. When the value of the Reynolds number exceeds 2100, it is to be expected that the rates will be somewhat higher. Any appreciable evaporation will also increase the film coefficient. Large fouling factors and low outlet-water temperatures are recommended, particularly when the water has a large mineral content.

Example 20.4. Calculation of a Trombone SO₂ Cooler. 3360 lb/hr of SO₂ gas issues from a sulfur burner at 450°F and is to be precooled to 150°F in a high-conductivity pipe-trombone cooler. 3-in. IPS pipes will be used. Owing to space limitations the straight length of the pipe may not exceed 8'0''. Cooling water is available at 85°F and should not be heated above 100°F because of scale and corrosion from the outside.

How many pipe lengths are required for the bank when the overall fouling factor is 0.010?

Solution:

$T_{av} = 300°F \qquad C = 0.165$ Btu/(lb)(°F) (Fig. 3)
SO₂, $Q = 3360 \times 0.165(450 - 150) = 166{,}500$ Btu/hr
Water, $Q = 11{,}100 \times 1.0(100 - 85) = 166{,}500$ Btu/hr

	Hot Fluid		Cold Fluid
450	Higher Temp	100	350
150	Lower Temp	85	65
300	Differences	15	285

LMTD = 169°F (5.14)

R will be $^{300}\!/_{15} = 20.0$, which is beyond the values plotted in Fig. 20.7b. The graph can be used by replacing R by $1/R$ and S by RS. This interchanges the two streams without affecting the temperature relationships.

$R = 20 \qquad \frac{1}{R} = 0.05$

$S = \frac{(t_2 - t_1)}{T_1 - t_1} = \frac{15}{365} = 0.0412 \qquad RS = 20 \times 0.0412 = 0.824$

$F_T = 0.98$
$\Delta t = 0.98 \times 169 = 166°F$

T_c and t_c: Average values of the temperatures will suffice since the gas-film coefficient does not vary greatly from inlet to outlet. $T_{av} = 300°F$ and $t_{av} = 92.5°F$.

Hot fluid: tube side, SO_2
$a'_t = 7.38$ in.²   [Table 11]
All pipes will be in series.
Flow area, $a_t = 7.38/144 = 0.0512$ ft²
$G_t = W/a_t$
$\quad = 3360/0.0512 = 65{,}600$
$D = 3.068/12 = 0.256$ ft   [Table 11]
At $T_a = 300°F$,
$\mu = 0.017 \times 2.42 = 0.041$ lb/(ft)(hr)
                                 [Fig. 15]
$Re_t = DG_t/\mu$
$\quad = 0.256 \times 65{,}600/0.041 = 410{,}000$
$j_H = 790$
$k = 0.0069$ (nearest value available in Table 5)
$(c\mu/k)^{1/3} = (0.165 \times 0.041/0.0069)^{1/3}$
$\quad\quad\quad\quad\quad = 0.99$

$h_i = j_H \dfrac{k}{D}\left(\dfrac{c\mu}{k}\right)^{1/3}$ [Eq. (6.15)]
$\quad = 790 \times \dfrac{0.0069}{0.256} \times 0.99 = 21.1$

$h_{io} = h_i \times \text{ID/OD}$
$\quad = 21.1 \times 3.068/3.50$
$\quad\quad\quad\quad = 18.5$ Btu/(hr)(ft²)(°F)

Cold fluid outside, water
$G = w/2L$
$\quad = 11{,}100/2 \times 8 = 694$ lb/(hr)(ft)
At $t_a = 92.5°F$,
$\quad \mu = 0.80 \times 2.42 = 1.94$ lb/(ft)(hr)
$Re = 4G'/\mu$
$\quad = 4 \times 694/1.94 = 1430$ (streamline)
$D_o = 3.5/12 = 0.292$ ft
$h_o = 65(G'/D_o)^{1/3}$
$\quad = 65(694/0.292)^{1/3}$
$\quad\quad = 868$ Btu/(hr)(ft²)(°F)

$U_C = \dfrac{h_o h_{io}}{h_o + h_{io}} = \dfrac{868 \times 18.5}{868 + 18.5} = 18.1$ Btu/(hr)(ft²)(°F)

$R_d = 0.010 \quad\quad h_d = \dfrac{1}{0.010} = 100$

$U_D = \dfrac{U_C h_d}{U_C + h_d} = \dfrac{18.1 \times 100}{18.1 + 100} = 15.3$

$A = \dfrac{Q}{U\,\Delta t} = \dfrac{166{,}500}{15.3 \times 169} = 65.6$ ft²

External surface/lin ft $= 0.917$

Number of pipe lengths $= \dfrac{65.6}{0.917 \times 8} = 8.95$ (use 9)

5. ATMOSPHERIC COOLERS

Introduction. The atmospheric cooler is an improvement over the trombone cooler particularly for large services. Atmospheric coolers are located in cooling towers beneath the fill (see Chap. 17). They are also known as *spray coolers, bare-tube coolers,* and *open* or *atmospheric coolers* and provide tubular heat-transfer surface for certain services at about one-third the cost of conventional shell-and-tube equipment. A typical arrangement in an induced-draft cooling tower is shown in Fig. 20.8. Perhaps the broadest use of atmospheric coolers lies in the cooling of the jacket water of internal-combustion engines. In a survey of 106 plants in the natural gasoline industry, Kallam[1] found the atmospheric cooler to be the most widely used type, accounting for 44.5 per cent of all installations. Other methods of jacket-water cooling are the direct circulation of the jacket water over the cooling tower, induced-draft air-cooled extended-surface units, and the use of a conventional exchanger between the cold recirculated cooling tower water and the jacket water. Calculations applicable to these methods have already been discussed.

When piped directly from the atmospheric cooler to the water jacket, the water travels in a closed circuit and the same water is continuously heated and cooled. Naturally this minimizes corrosion and the deposition of slime in both the jacket and the tube bundle. Since jacket water is usually recirculated over a short temperature range of 10 to 20°F and

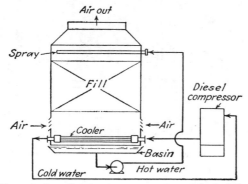

FIG. 20.8. Atmospheric cooler showing an arrangement for jacket water cooling.

at a temperature to the cooler above 120°F, the advantages of the closed circuit are important.

Other methods of cooling jacket water are less advantageous. The high temperature of the recirculated jacket water prohibits its direct circulation over the cooling-tower fill without an atmospheric cooler because the air-saturation loss would be very great. When jacket water is cooled by cooling-tower water in an external exchanger, the pumping cost of the cooling-tower water is increased by the added pressure drop through the exchanger. The equipment cost is also increased. When using an atmospheric cooler it is possible to circulate less cooling water then required for sensible-heat transfer, since some evaporation occurs on the outside of the tubes of the atmospheric cooler. When used for the condensation of exhaust steam from turbines or engines, the atmospheric

[1] Kallam, F. L., *Petroleum Refiner*, **27**, 371–378 (1948).

732 PROCESS HEAT TRANSFER

cooler is then called an *evaporative condenser*. This modification will be discussed in the next section.

An atmospheric cooler is a bank of tubes in a supported frame between two cast headers with removable bolted covers. A sketch including alternate nozzle connections is shown in Fig. 20.9. The headers are usually cast in one piece including the nozzles. They vary in length and width to conform with the internal structure dimensions of the cooling-tower supporting structure. Generally the tubes are from 8 to 20 ft long with cooler widths of 3 to 6 ft. The width is small, so that the major tower-supporting beams can be erected between adjoining cooler

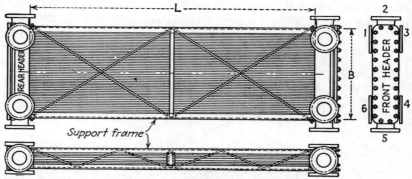

Fig. 20.9. Atmospheric cooler with alternate nozzle arrangements. (*Foster Wheeler Corporation.*)

sections. Thus a 24- by 24-ft cooling tower might be equipped with four sections having 20-ft tube lengths and 5-ft widths. The tubes are rolled directly into the headers and can be replaced when corroded by simply unbolting the header covers and reaming out the corroded tube. The outside structural framework of the cooler also supports the tubes. If the headers are of a weldable material, the framework is welded directly to them. If the material cannot be welded, the framework is bolted to the headers.

The tubes are laid out in horizontal rows with about three to seven superimposed horizontal rows per header. The vertical pitch between the horizontal rows is usually less than the horizontal pitch, since an adequate water-flow area must be provided to permit the free drainage of water into the basin. Where a larger number of rows is required, it is customary to use two sections one above the other. Tube passes may be orientated in any convenient manner.

The advantages of the atmospheric cooler are lower first cost and lower pumping cost. The great disadvantage lies in the rapid rate of scale formation on the outside of the tubes due to evaporation. The rate is such that succeeding horizontal rows are always superimposed rather than staggered for accessibility during cleaning and scale removal. While scale may be loosened by thermal shock, the shock does not necessarily cause it to drop off the tubes and waters of different mineral contents present separate scale problems. Scale removal by scraping is often the only solution, and the number of horizontal tube rows per section is kept small in anticipation of the scale. Notwithstanding these precautions, pieces of scale frequently fall from the upper tube rows and lie across the lower rows, interfering with the even distribution of water over the tubes.

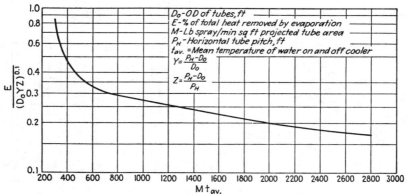

Fig. 20.10. Atmospheric coolers: Heat removed by evaporation. (*Kallam, Refiner.*)

Kallam recommends fouling factors for design which are identical with those of Table 12 although the service period is very much shorter.

The Calculation of Atmospheric Coolers. The method of calculation employed here is based on the correlation of Kallam.[1] While this method has been cited by Nelson,[2] the original article presents results without experimental data. Kallam states that 45 tests were run on three coolers embracing the following variations: cooling water rate, 3 to 38 lb/(min) (ft² of projected tube surface); water velocity in tubes, 2 to 14 fps; wind velocity at cooling tower, 10 to 1000 fpm; temperature of cooling water to atmospheric cooler (off the fill), 60 to 78°F; temperature of cooling water off cooler, 88 to 120°F (to basin); temperature of water to tubes, 109 to 153°F; temperature of water from tubes, 90 to 141°F; dry-bulb

[1] Kallam, F. L., *Petroleum Refiner*, **19**, 371–382 (1940).
[2] Nelson, W. L., "Petroleum Refinery Engineering," 3d ed., McGraw-Hill Book Company, Inc., New York, 1949.

temperature, 60 to 76°F; wet-bulb temperature, 54 to 66°F; heat removed by evaporation of cooling water, 0.4 to 50.0 per cent. No deviations have been given, and Kallam states that the relative humidity of the air above the atmospheric cooler did not influence the percentage of evaporation off the cooler. While the method is entirely empirical, the author knows of two instances in which coolers calculated by Kallam's method proved to be satisfactory in practice. In order to set the temperature range for the cooling-tower water over the cooler, an economic balance must be carried out between the cost of both the cooling tower and the atmospheric cooler for several temperatures to determine the economic optimum.

For the outside film coefficient Kallam employs two dimensionless configuration terms, $Y = (P_H - D_o)/D_o$ and $Z = (P_H - D_o)/P_H$, where P_H

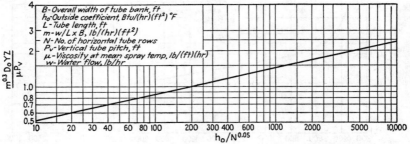

FIG. 20.11. Atmospheric coolers: Outside film coefficients. (*Kallam, Refiner.*)

is the horizontal pitch and D_o is the outside diameter of the tubes. Calculation curves are given in Figs. 20.10 and 20.11, although the actual groups employed are both arbitrary and dimensional. There are, in addition, two flow terms which are based on different cooler outside areas. These are the *projected tube area* of a horizontal row used for computing the water spray M in pounds per minute per square foot of projected tube area and the overall ground area of the section calculated as the product of the tube length L and the distance between the outer sides of the two extreme tubes in a horizontal row B. The latter is used to compute the spray in pounds per hour per square foot of *base* area between tube sheets m. Figure 20.10 gives the percentage of the total spray water evaporated and is plotted as a function of the spray quantity and its mean temperature over the atmospheric section. This permits adjustment of the spray quantity so that the combined sensible- and *latent*-heat effects will balance the tube-side heat load. This will be demonstrated in the following illustration:

Suppose that it is desired to remove 1,000,000 Btu/hr from a bare tube cooler consisting of two sections with 20 ft long tubes and a 4 ft wide bundle and that the design of the cooling tower is such that the water temperature will rise from 70 to 80°F. Assume 1 in. OD tubes with 2-in. horizontal pitch and 1¾-in. vertical pitch. If no evaporation occurred, it would be necessary to circulate 100,000 lb/hr or 200 gpm.

To determine the quantity of spray with evaporation:

Number of tubes per horizontal row $= \dfrac{4 \times 12}{2} = 24$

Projected area per tube $= \frac{1}{12} \times 20 = 1.67 \text{ ft}^2$
Total projected area $= 2 \times 24 \times 1.67 = 80 \text{ ft}^2$
$D_o = \frac{1}{12} = 0.0833 \text{ ft}$
$P_H = \frac{2}{12} = 0.167 \text{ ft}$
$Y = \dfrac{P_H - D_o}{D_o} = \dfrac{0.167 - 0.0833}{0.0833} = 1.0$
$Z = \dfrac{P_H - D_o}{P_H} = \dfrac{0.167 - 0.0833}{0.167} = 0.5$
$(D_o Y Z)^{0.1} = (0.0833 \times 1.0 \times 0.5) = 0.7278$

Since the average water temperature over the bundle is 75°F, $Mt_{av} = 75M$. It is now necessary to assume values of the percentage evaporated E as a function of the spray loading, both of which determine the heat load.

Trial 1:

Assume $E = 20$ per cent.

$$\dfrac{E}{(D_o Y Z)^{0.1}} = \dfrac{0.20}{0.7278} = 0.275$$

From Fig. 20.10

$$Mt_{av} = 1040$$
$$M = {}^{1040}\!/_{75} = 13.9 \text{ lb/(min)(ft}^2)$$

Heat removed from cooler sensibly $= 13.9 \times 80 \times 60 \times 10 = 667{,}000$ Btu

Evaporation $= \dfrac{1{,}000{,}000 - 667{,}000}{1{,}000{,}000} \times 100 = 33\%$

The computed evaporation of 33 per cent does not check the assumed value of 20 per cent, and a new trial must be made. An assumption of 18.5 per cent checks with a spray of 17.05 lb/(min)(ft²). The heat removed sensibly will be 818,000 Btu/hr, and the spray to be circulated will be

$$\dfrac{818{,}000}{8.33 \times 60} = 163.6 \text{ gpm}$$

compared with a total of 200 gpm when vaporization is not considered. The outside film coefficient can be obtained from Fig. 20.11 or the dimensional equation

$$h_o = 190 N^{0.05} \left(\frac{m^{0.3} D_o YZ}{\mu P_V} \right)^{4.4} \tag{20.8}$$

where P_V = vertical pitch, ft
$m = w/LB$,
w = total hourly flow over the section
μ = viscosity, lb/(ft)(hr)
N = number of horizontal rows

Temperature Difference in the Atmospheric Cooler. The correct evaluation of the true temperature difference is extremely important in this type of apparatus, since the approach of both outlet streams is usually quite close. As an example of crossflow it is necessary to decide whether the spray liquid is mixed or unmixed in accordance with the derivations covered by Eqs. (16.71) through (16.103). The spray fluid in commercial towers can usually be considered unmixed, and the true temperature difference can be obtained through the use of Fig. 16.17b.

Example 20.5. Calculation of an Atmospheric Jacket Water Cooler. 295 gpm of jacket water from a diesel compressor are to be cooled from 143° to 130°F. It is possible to convert a 12 × 16 ft cooling tower to permit the insertion of two atmospheric sections side-by-side consisting of 1 in. OD × 16 BWG tubes 12 0″ long on 1⅞-in. horizontal pitch and 1½-in. vertical pitch. The distance between the extreme horizontal tubes in each section will be 4′0″.

Individual dirt factors of 0.003 for the cooling tower water and 0.001 for the jacket water should be provided.

What will the spray requirement be and how many horizontal rows of tubes should be provided in the sections?

Solution:

Outside film coefficient:

Number of tubes per row/section = 4 × 12/1.875 = 25.6 (say 25)
Total projected area = 2(1²⁵⁄₁₂ × 1 × 25) = 50 ft²
Assume a spray of 28 lb/(min)(ft²)
Average temperature over section = ½(90 + 110) = 100°F
$Mt_{av} = 100 \times 28 = 2800$
From Fig. 20.10
$$\frac{E}{(D_o YZ)^{0.1}} = 0.171$$

$D_o = \dfrac{1}{12} = 0.0833$ ft $P_H = \dfrac{1.875}{12} = 0.1562$ ft

$Y = \dfrac{0.1562 - 0.0833}{0.0833} = 0.874$

$Z = \dfrac{0.1562 - 0.0833}{0.1562} = 0.466$

ADDITIONAL APPLICATIONS

$(D_oYZ)^{0.1} = (0.0339)^{0.1} = 0.712$
$E = 0.171 \times 0.712 = 0.12\%$ evaporation
Heat load $= 295 \times 500(143 - 130) = 1,920,000$ Btu/hr
Sensible heat $= 1,920,000(1 - 0.12) = 1,690,000$ Btu/hr
Final spray temperature: $1,690,000 = 28 \times 60 \times 50(t_2 - 90)$
$t_2 = 110°$F Checks for 28 lb/(min)(ft²)
Total spray $= 28 \times 60 \times 50 = 84,000$ lb/hr
$$m = \frac{w}{LB} = \frac{84,000}{2 \times 4 \times 12} = 875 \text{ lb/(hr)(ft}^2)$$
$m^{0.3} = 7.62$
$\mu_{\text{water}} = 0.76$ cp
$$\frac{m^{0.3}D_oYZ}{\mu P_V} = \frac{7.62 \times 0.0339}{2.42 \times 0.76 \times 0.125} = 1.12$$
From Figure 20.11,
$$\frac{h_o}{N^{0.05}} = 300$$

Assume three horizontal rows/section

$$N^{0.05} = 3^{0.05} = 1.057$$
$$h_o = 300 \times 1.057 = 317 \text{ Btu/(hr)(ft}^2)(°F)$$

Tube-side coefficient

Assume an even number of passes (using vertical partitions) arranged to give a tube-side velocity of about 8 fps. Try four tube passes per section or eight total passes.

Flow area/pass, $a_t = 3 \times 25 \times \dfrac{0.594}{4 \times 144} = 0.0775$ ft²

Mass velocity, $G_t = \dfrac{W}{a_t} = 295 \times \dfrac{500}{0.0775} = 1,900,000$ lb/(hr)(ft²)

Velocity $= \dfrac{G_t}{3600\rho} = \dfrac{1,900,000}{3600 \times 62.5} = 8.45$ fps

ID $= 0.87$ in.
$h_i = 2300 \times 0.93 = 2140$ Btu/(hr)(ft²)(°F)   (Fig. 25)
$h_{io} = h_i \times \dfrac{\text{ID}}{\text{OD}} = 2140 \times \dfrac{0.87}{1.0} = 1860$ (6.5)

Overall coefficients:

$$U_C = \frac{h_o h_{io}}{h_o + h_{io}} = \frac{317 \times 1860}{317 + 1860} = 271 \text{ Btu/(hr)(ft}^2)(°F)$$

External surface/lin ft $= 0.2618$ ft²/ft   (Table 11)
Total surface $= 2(3 \times 25 \times 12 \times 0.2618) = 471$ ft²

Δt:

Hot Fluid		Cold Fluid	
143	Higher Temp	110	33
130	Lower Temp	90	40
13	Differences	20	7

$$\text{LMTD} = 36.7°\text{F} \tag{5.14}$$
$$R = {}^{13}\!\!\!/_{20} = 0.65 \qquad S = 20/53 = 0.377$$
$$F_T = 0.97 \qquad\qquad \text{(Fig. 16.17}b\text{)}$$
$$\Delta t = F_T \times \text{LMTD} = 0.97 \times 36.7 = 35.6°\text{F}$$
$$U_D = \frac{Q}{A\,\Delta t} = \frac{1{,}920{,}000}{471 \times 35.6} = 114.5$$

Dirt factor:

$$R_d = \frac{U_C - U_D}{U_C U_D} = \frac{271 - 114.5}{271 \times 114.5} = 0.0051 \; (\text{hr})(\text{ft}^2)(°\text{F})/\text{Btu}$$

The assumption of three horizontal rows is satisfactory, since a dirt factor of 0.004 was required.

The pressure drop on the tube side can be computed from Eqs. (7.45) through (7.47). The characteristics which the cooling tower must possess to permit the heat removal from the cooling tower water can be determined by the methods of Chap. 17.

6. EVAPORATIVE CONDENSERS

Atmospheric coolers were formerly used for the condensation of steam from turbines and noncondensing steam engines. They are less frequently used for this purpose now. Although Kallam's method is probably satisfactory for the condensation of organic vapors where the condensing coefficient is small by comparison, it is unsatisfactory in the case of steam condensation. The condensation of steam requires the use of high liquid loadings in the cooling tower because of the large heat flux associated with the condensation of a pound of steam. In this range the outside coefficients predicted from Fig. 20.11 are probably unreliable. It is safer to employ a conventional clean value for the overall coefficient than to attempt its calculation. A value of U_C of from 350 to 450 Btu/(hr)(ft^2)(°F) should give a reasonable condenser.

7. BAYONET EXCHANGERS

Introduction. A bayonet consists of a pair of concentric tubes, the outer of which has one end sealed as shown in Fig. 20.12. Bayonets were formerly known as *field* tubes. Both the outer and inner tubes extend from separate stationary tube sheets and extend either into shells or directly into vessels. The surface of the outer tube is the principal heat-transfer surface. Bayonet exchangers are excellently adapted to the condensation of vapors under moderate and very low vacuum, outside the bayonets and in exchanger shells, since the tubes are larger and heavier, usually 1 in. OD inside 2 in. OD, and the individual bayonets are generally self-supporting without the use of support plates. These qualities combine to cause a negligible pressure drop to the flow of vapors under vacuum outside bayonets. An example of a vacuum condenser arrangement is shown in Fig. 20.13, while the exchanger shown in Fig. 20.14 is known as a *suction* heater or *suction tank* heater.

ADDITIONAL APPLICATIONS   739

Suction heaters are installed in tanks used for the storage of viscous liquids and semiplastics such as molasses, heavy lube oils, fuel oils, and asphalt. If a liquid is viscous, it would be necessary to maintain all the liquid in the tank at a temperature corresponding to a feasible pump-

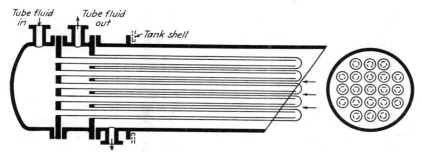

FIG. 20.12.   Section through a bayonet (suction-type) exchanger.

ing viscosity. This could be achieved by means of a steam coil, which would be required to supply sufficient heat to offset the tank losses. Since viscous fluids afford little free convection, the coil would probably be supplemented by continuous agitation and if the stored material should

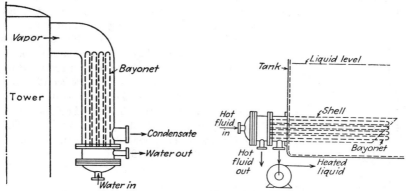

FIG. 20.13. Bayonet exchanger as vacuum condenser.

FIG. 20.14. Tank suction heater with bayonet tubes.

cool to too low a temperature, its ultimate removal might cause a serious problem. When a suction heater is installed as shown in Fig. 20.14, heat need be supplied only while the pumps are operating. As hot liquid is removed by the pump, it is replaced by the hydrostatic pressure of the liquid remaining in the tank. Bayonets are extremely effective for this type of service, but they are also rather expensive, and ordinarily the job can be accomplished more cheaply by substituting an unbaffled U-tube bundle for the bayonets.

Temperature Difference in the Bayonet Exchanger. Bayonet exchangers introduce some interesting temperature differences which have

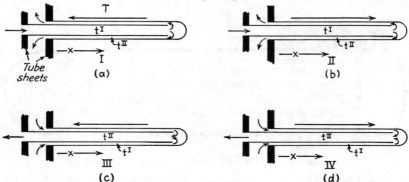

Fig. 20.15. Arrangements of parallel flow and counterflow in bayonet exchangers.

not been encountered in any of the flow patterns discussed thus far. Consider the single bayonet shown in Fig. 20.15a with hot fluid on the shell side traveling in the direction shown. The temperature diagram is shown in Fig. 20.16. Cold liquid flows into the bayonet by the inner tube and is rejected through the annulus in the opposite direction. There is a film coefficient on the shell side of the bayonet and a coefficient on the inside of the outer tube yielding the overall coefficient U_C. There is additional heat transfer between the cold medium entering the inner tube and the heated cold medium in counterflow with it in the annulus. The flow areas of the inner tube and annulus differ, although the weight flow is the same in each and the bayonet fluid has different film coefficients in the inner tube and in the annulus. These two coefficients produce a second overall coefficient u. The temperature distribution in Fig. 20.15a is but one of four possibilities. The remainder are shown in Fig. 20.15b, c, and d. The true temperature difference is seen to be related not only to the flow pattern but also to the ratio of the two overall coefficients U and u.

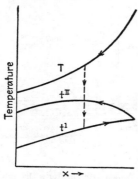

Fig. 20.16. Temperature diagram for Case I. (*Hurd, Industrial Engineering Chemistry.*)

The derivation employed here is that of Hurd[1] and agrees with an

[1] Hurd, N. L., *Ind. Eng. Chem.*, **38**, 1266–1271 (1946).

ADDITIONAL APPLICATIONS

unpublished derivation of K. A. Gardner. As reference surfaces the outside diameters of the outer and inner tubes are used to define the cylindrical areas. The area at the sealed end itself is neglected. Letting P be the perimeter of the outer tube and p the perimeter of the inner tube, the differential heat balance on each tube-side stream gives

$$U(T - t^{II})P\,dx - u(t^{II} - t^{I})p\,dx = -wc\,dt^{IV} \qquad (20.9)$$
$$up(t^{II} - t^{I})\,dx = wc\,dt^{I} \qquad (20.10)$$

Adding

and

$$UP(T - t^{II})\,dx = wc(dt^{I} - dt^{II}) \qquad (20.11)$$

$$wc(t^{II} - t^{I}) = WC(T_1 - T) \qquad (20.12)$$

Taking a heat balance on the right-hand end,

$$\frac{dT}{dx} = \frac{UP}{WC}(T - t^{II}) \qquad (20.13)$$

Differentiating

$$\frac{d^2T}{dx^2} = \frac{UP}{WC}\left(\frac{dT}{dx} - \frac{dt^{II}}{dx}\right) \qquad (20.14)$$

Differentiating Eq. (20.12) with respect to x and rearranging,

$$\frac{dt^{II}}{dx} = \frac{dt^{I}}{dx} - \frac{WC}{wc}\frac{dT}{dx} \qquad (20.15)$$

From Eqs. (20.10) and (20.12)

$$\frac{dt^{I}}{dx} = \frac{up}{wc}\frac{WC}{wc}(T_1 - T) \qquad (20.16)$$

Substituting Eqs. (20.15) and (20.16) in Eq. (20.14) and letting

$$(T_1 - T) = \tau,$$

$$-\frac{d^2\tau}{dx^2} + \left(\frac{UP}{WC} - \frac{UP}{wc}\right)\frac{d\tau}{dx} + \frac{UPup}{(wc)^2}\tau = 0 \qquad (20.17)$$

Substituting $d\tau/dx = X$ and factoring,

$$-X^2 + \left(\frac{UP}{WC} + \frac{UP}{wc}\right)X + \frac{UPup}{(wc)^2} = 0 \qquad (20.18)$$

Solving the quadratic

$$X = \frac{1}{2}\frac{UP}{WC}\left[1 + \frac{WC}{wc} \pm \sqrt{\left(1 + \frac{WC}{wc}\right)^2 + 4\frac{up}{UP}\left(\frac{WC}{wc}\right)^2}\right] \qquad (20.19)$$

Designate the radical by the letter ψ, and calling the plus and minus solutions X_1 and X_2, respectively,

$$X_1 = \frac{1}{2}\frac{UP}{WC}\left(1 + \frac{WC}{wc} + \psi\right) \tag{20.20}$$

$$X_2 = \frac{1}{2}\frac{UP}{WC}\left(1 + \frac{WC}{wc} - \psi\right) \tag{20.21}$$

Equation (20.17) is a second-order differential equation whose solution is

$$\tau = T_1 - T = C_1 e^{X_1 x} + C_2 e^{X_2 x} \tag{20.22}$$

From the terminal conditions

$$T = T_1 \quad \text{when} \quad x = L$$

and

$$T = T_2 \quad \text{when} \quad x = 0$$

from which

$$C_1 = -\frac{(T_1 - T_2)e^{X_2 L}}{e^{X_1 L} - e^{X_2 L}} \quad \text{and} \quad C_2 = \frac{(T_1 - T_2)e^{X_1 L}}{e^{X_1 L} - e^{X_2 L}} \tag{20.23}$$

Differentiating Eq. (20.22) with respect to x and substituting in Eq. (20.13),

$$C_1 X_1 e^{X_1 x} + C_2 X_2 e^{X_2 x} = -\frac{UP}{WC}(T - t^{\text{II}}) \tag{20.24}$$

Evaluated at $x = 0$ it becomes

$$C_1 X_1 + C_2 X_2 = -\frac{UP}{WC}(T_2 - t_2) \tag{20.25}$$

Substituting Eqs. (20.20), (20.21), and (20.23) in Eq. (20.25),

$$\frac{1}{2}(T_1 - T_2)\left[1 + \frac{WC}{wc} - \psi\left(\frac{e^{X_1 L} + e^{X_2 L}}{e^{X_1 L} - e^{X_2 L}}\right)\right] = t_2 - T_2 \tag{20.26}$$

Solving for $e^{X_1 L}/e^{X_2 L}$ and taking logarithms

$$(X_1 - X_2)L = \ln\left[\frac{(T_2 - t_2) + \frac{1}{2}(T_1 - T_2)\left(1 + \frac{WC}{wc}\right) + \frac{1}{2}\psi(T_1 - T_2)}{(T_2 - t_2) + \frac{1}{2}(T_1 - T_2)\left(1 + \frac{WC}{wc}\right) - \frac{1}{2}\psi(T_1 - T_2)}\right] \tag{20.27}$$

Eliminating X_1 and X_2 by Eq. (20.20) and eliminating L by introduction of the Fourier equation $WC(T_1 - T_2) = UPL\,\Delta t$, where Δt is the true temperature difference between the shell side and tube side,

$$\Delta t = \frac{(T_1 - T_2)\psi}{\ln\left[\dfrac{(T_2 - t_2) + \dfrac{1}{2}(T_1 - T_2)\left(1 + \dfrac{WC}{wc}\right) + \dfrac{1}{2}\psi(T_1 - T_2)}{(T_2 - t_2) + \dfrac{1}{2}(T_1 - T_2)\left(1 + \dfrac{WC}{wc}\right) - \dfrac{1}{2}\psi(T_1 - T_2)}\right]} \quad (20.28)$$

Substituting the value of ψ and eliminating wc/WC by introducing its equivalent $(T_1 - T_2)/(t_2 - t_1)$,

$$\Delta t = \frac{\sqrt{(T_1 - T_2 + t_2 - t_1)^2 + 4\dfrac{up}{UP}(t_2 - t_1)^2}}{\ln\left[\dfrac{T_2 - t_2 + T_1 - t_1 + \sqrt{(T_1 - T_2 + t_2 - t_1)^2 + 4\dfrac{up}{UP}(t_2 - t_1)^2}}{T_2 - t_2 + T_1 - t_1 - \sqrt{(T_1 - T_2 + t_2 - t_1)^2 + 4\dfrac{up}{UP}(t_2 - t_1)^2}}\right]}$$
(20.29)

which gives the true temperature difference in terms of the terminal differences. In order to establish a graphical representation, dimensionless parameters will again be used.

$$R = \frac{T_1 - T_2}{t_2 - t_1} = \frac{wc}{WC} \qquad V = \frac{1}{2}\left(\frac{T_1 + T_2 - t_1 - t_2}{t_2 - t_1}\right)$$

where V is the arithmetic mean temperature difference per unit temperature change in the tube fluid.

$$F = \frac{up}{UP}$$

where F is the ratio of the conductances per unit of tube length and the true temperature difference per unit change in tube side temperature is $\Delta t_D = \Delta t/(t_2 - t_1)$. Using these ratios Eq. (20.29) can be written

$$\Delta t_D = \frac{\sqrt{(R + 1)^2 + 4F}}{\ln\left[\dfrac{2V + \sqrt{(R + 1)^2 + 4F}}{2V - \sqrt{(R + 1)^2 + 4F}}\right]} \quad (20.30)$$

This equation also expresses Δt_D for Case IV in Fig. 20.15d. For Cases II and III the equivalents of Eqs. (20.29) and (20.30) are

$$\Delta t = \frac{\sqrt{(T_1 - T_2 - t_2 + t_1)^2 + 4\dfrac{up}{UP}(t_2 - t_1)^2}}{\ln\left[\dfrac{T_2 - t_2 + T_1 - t_1 + \sqrt{(T_1 - T_2 - t_2 + t_1)^2 + 4\dfrac{up}{UP}(t_2 - t_1)^2}}{T_2 - t_2 + T_1 - t_1 - \sqrt{(T_1 - T_2 - t_2 + t_1)^2 + 4\dfrac{up}{UP}(t_2 - t_1)^2}}\right]}$$
(20.31)

and

$$\Delta t_D = \frac{\sqrt{(R-1)^2 + 4F}}{\ln\left[\dfrac{2V + \sqrt{(R-1)^2 + 4F}}{2V - \sqrt{(R-1)^2 + 4F}}\right]} \quad (20.32)$$

Equations (20.30) and (20.32) can be rearranged to the form

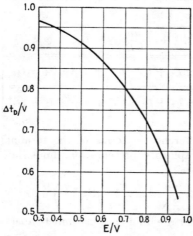

FIG. 20.17. Solution of Eq. (20.33). (*Hurd, Industrial Engineering Chemistry.*)

$$\frac{\Delta t_D}{V} = \frac{2(E/V)}{\ln\left(\dfrac{1 + \dfrac{E}{V}}{1 - \dfrac{E}{V}}\right)} \quad (20.33)$$

where for Cases I and IV

$$E = \tfrac{1}{2}\sqrt{(R+1)^2 + 4F} \quad (20.34)$$

and where for Cases II and III

$$E = \tfrac{1}{2}\sqrt{(R-1)^2 + 4F} \quad (20.35)$$

The solution of Eq. (20.33) is given in Fig. 20.17, and the solution for Eqs. (20.34) and (20.35) is given by Fig. 20.18a and b, respectively. Because this method of obtaining the true temperature difference differs so greatly from the methods of preceding chapters, the procedure is summarized.

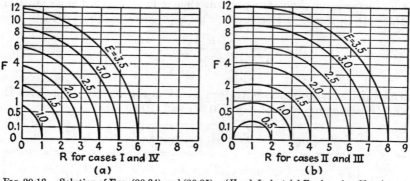

FIG. 20.18. Solution of Eqs. (20.34) and (20.35). (*Hurd, Industrial Engineering Chemistry.*)

1. Determine R and V, and after having determined U and u, determine F.

2. From R and F determine E in Fig. 20.18a or b for the arrangement identified in Fig. 20.15. If V is equal to or greater than $3E$, proceed directly to step 4, using V as identical with the Δt_D.

3. Calculate E/V, and read $\Delta t_D/V$ from Fig. 20.17. Multiply by V to obtain Δt_D.

4. Multiply Δt_D by $(t_2 - t_1)$ to obtain Δt for use in the Fourier equation.

Example 20.6. Calculation of the True Temperature Difference. 1000 lb/hr of fluid of 0.5 specific heat is to be heated from 50 to 100°F by means of 250 lb/hr of hot water in the shell which will be cooled from 200 to 100°F. The bayonet consists of 1 in. OD tubes within 2 in. OD tubes. U and u are 60 and 120 Btu/(hr)(ft²)(°F) as computed by methods described below. Assume the arrangement of Case III.

$$R = \frac{200 - 100}{100 - 50} = 2.0$$

$$V = \frac{1}{2}\left(\frac{200 + 100 - 50 - 100}{100 - 50}\right) = 1.50$$

$$F = \frac{120 \times 1}{60 \times 2} = 1.0$$

In Fig. 20.18b for $R = 2.0$ and $F = 1.0$, the abscissa and ordinate intersect at $E = 1.10$. Thus $E/V = 1.10/1.50 = 0.739$, and from Fig. 20.17 $\Delta t_D/V = 0.783$ and

$$\Delta t_D = 0.783 \times 1.50 = 1.175.$$

$$\Delta t_D = \frac{\Delta t}{t_2 - t_1} \qquad \Delta t = 1.175 \times 50 = 58.8°F$$

The LMTD is 72°F for counterflow.

Bayonet-exchanger Heat-transfer Coefficients. The bayonet exchanger presents no new problem in the calculation of overall heat-transfer coefficients. The overall coefficient u for the outside diameter of the inner tube is found in the same manner employed for the double pipe exchanger in Chap. 6. The equivalent diameter for the annulus is based on the perimeter at the outside diameter of the inner tube and considered effective at the outside diameter of the inner tube. To compute the annulus coefficient at the inside diameter of the outer tube, the equivalent diameter is based on the perimeter at the inside diameter of the outer tube and is considered effective at the inside diameter of the outer tube. The flux is then corrected to the outside diameter of the outer tube by the simple ratio of the inside diameter to outside diameter.

Shell-side coefficients are treated in the same manner as former prototypes. Condensation and vaporization are treated by the methods of Chaps. 12 through 15. The use of a bayonet exchanger for sensible-heat transfer in a shell without baffles usually simulates longitudinal flow along the axis of the bayonet and can be computed by the methods of Example 7.5. For segmentally baffled bayonet bundles use Fig. 28.

8. FALLING-FILM EXCHANGERS

The falling-film exchanger is usually a conventional 1-1 exchanger designed to operate vertically as shown in Fig. 20.19a. Liquid enters the channel at the top at such a rate that the tubes do not flow full of liquid, but instead, liquid descends by gravity along the inner walls of the

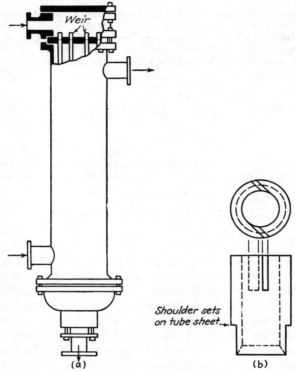

FIG. 20.19. (a) True counterflow falling-film exchanger with circular flow weirs. (b) Detail of a circular flow weir for insertion into top of tubes.

tubes as a thin film. Obviously this produces a much greater linear velocity for a given weight flow than could be obtained if the tube flowed full. The vertical tubes could be made to flow full if desired by feeding the liquid upward or through the use of a loop seal as discussed in Chap. 15. McAdams[1] quotes Bays[2] and coworkers as having determined the coefficients for heating water in film flow in 1.5 to 2.5 in. ID pipes 0.4 to

---

[1] McAdams, *op. cit.*, p. 203.
[2] McAdams, Drew, and Bays, *loc. cit.*

6.08 ft long. The coefficients are given within ± 18 per cent by the dimensional equation

$$h_i = 120 \left(\frac{w}{\pi D}\right)^{1/3} \tag{20.36}$$

For other liquids McAdams cites a dimensionless empirical equation attributed to T. B. Drew for turbulent flow

$$\frac{h_i}{(k^3\rho^2 g/\mu_f^2)^{1/3}} = 0.01 \left[\left(\frac{c\mu}{k}\right)\left(\frac{4G'}{\mu_f}\right)\right]^{1/3} \tag{20.37}$$

where $G' = w/\pi D$ and μ_f is evaluated at the average film temperature. For streamline flow where the Reynolds number $4G'/\mu_f$ is less than 2000, Bays and McAdams[1] give

$$h_i = 0.67 \left[\left(\frac{k^3\rho^2 g}{\mu_f^2}\right)\left(\frac{c\mu_f^{5/3}}{kL\rho^{2/3}g^{1/3}}\right)\right]^{1/3} \left(\frac{4G'}{\mu_f}\right)^{1/9} \tag{20.38}$$

Very often a swirl is imparted to the falling film by means of overflow pieces having nearly tangential notches which protrude above the upper tube sheet. A typical weir is shown in Fig. 20.19b.

The shell-side coefficient can be computed in the usual manner, using Fig. 28 for sensible coefficients or by the methods of Chaps. 12 or 15 for condensation or evaporation.

9. GRANULAR MATERIALS IN TUBES

The transfer of heat to or from fluidized solids has become increasingly important in modern catalytic and adsorptive processes. The transfer of heat to granular materials in beds was discussed as an unsteady-state problem in Chap. 18. When a dry granular material travels through a vertical or horizontal tube, the flow mechanism is akin to rodlike flow as discussed in Chap. 3. Based upon an earlier derivation of Graetz,[2] Drew[3] established the theoretical equation for a fluid in rodlike flow to be

$$\frac{t_2 - t_1}{T_s - t_1} = 1 - 0.692 e^{-5.78\frac{\pi k L}{wc}} - 0.131 e^{-30.4\frac{\pi k L}{wc}} - 0.0536 e^{-74.8\frac{\pi k L}{wc}} \tag{20.39}$$

where T_s = constant temperature of the wall
  w = granular flow rate, lb/hr

For values of $(t_2 - t_1)/(T_s - t_1)$ greater than 0.55 only the first term of the series need be considered. Equation (20.39) reduces to

$$\frac{t_2 - t_1}{T_s - t_1} = 1 - 0.692 e^{-5.78\pi(kL/wc)} \tag{20.40}$$

[1] Bays, G. S., and W. H. McAdams, *Ind. Eng. Chem.*, **29**, 1240–1246 (1937).
[2] Graetz, L., *Ann. Physik*, **25**, 337 (1885).
[3] Drew, T. B., *Trans. AIChE*, **26**, 26–80 (1931).

or
$$\frac{T_s - t_2}{T_s - t_1} = 0.692 e^{-5.78\pi(kL/wc)} \qquad (20.41)$$

In a study of heat transfer to sand, Brinn and coworkers[1] experimentally verified the rodlike nature of the flow of several granular materials in tubes. They also made the following additional observations:

1. The transfer of heat to a fluid in rodlike flow is identical with the *unsteady-state* heating of an infinitely long cylinder with negligible resistance between the heat source and the cylinder as defined by an equation of the form of Eq. (18.52). The series expression for the infinitely long cylinder and Eq. (18.52) can be shown to be equivalent when the values of the exponents in the infinite series of Eq. (20.39) are identical. This requires that the Graetz number wc/kL and the group $4\alpha\theta/D^2$* be equal. In these groups α is the thermal diffusivity, θ the time in hours, and D the inside diameter of the cylinder in feet. Since the linear velocity of the material in the tubes is given by

$$V = \frac{4w}{\rho \pi D^2} \quad \text{ft/hr}$$

and the volume of granular material by

$$v_s = \frac{\pi D^2}{4} L$$

The time is

$$\theta = \frac{4v_s}{V} \pi D^2$$

The transformation is given by

$$\frac{4\alpha\theta}{D^2} = \left(\frac{4k}{c\rho D^2}\right)\left(\frac{\pi D^2 L}{4w/\rho}\right) = \frac{\pi k L}{wc}$$

2. For a given weight flow through the tube the temperature change is independent of the diameter. This leads to the interesting point that a desired value can be obtained for the heat-transfer coefficient by changing the tube diameter.

3. If there are other resistances in series with the rodlike flow, it is not possible to obtain the overall resistance and hence U by the arithmetic addition of the resistances. The solutions for such cases are given graphically in Fig. 20.20.

[1] Brinn, M. S., S. J. Friedman, F. A. Guckert, and R. L. Pigford, *Ind. Eng. Chem.*, **40**, 1050–1061 (1948).

* The inside diameter of the tube D here replaces the general length l in Eq. (18.52), since only cylinders are under consideration.

4. It is not necessary to obtain the film coefficient h_i for the granular material flowing inside the tube in calculating the size or performance of the system. It may be obtained as a matter of interest, however, from Eq. (20.41) and the following, which defines the arithmetic mean film coefficient,

$$h_i = \frac{2k}{\pi D}\left[\frac{1 - (T_s - t_2)/(T_s - t_1)}{1 + (T_s - t_2)/(T_s - t_1)}\right]\frac{wc}{kL} \qquad (20.42)$$

While the preceding has applied to granular materials in rodlike flow in which the temperature of the wall is uniform, Brinn *et al.* have also solved the equations and given charts for the cases in which the temperature of the medium in the tube jacket is not uniform and also for the cases where the film coefficient from the jacket fluid to the tube is finite and infinite. To this extent they have employed the two temperature parameters R' and S' as an index of the temperature variation of both the rodlike and jacket flow and a parameter β which is the reciprocal of the Nusselt number, so that

$$R' = \frac{wc}{WC} \qquad S' = \frac{t_2 - t_1}{T_1 - t_1} \qquad \beta = \frac{2k}{h_{oi}D}$$

where h_{oi} is the jacket coefficient referred to the inside diameter of the tube. R is replaced by R' and S by S', since wc, t_2 and t_1 always refer to the granular material whether or not it is the colder of the two. Obviously when $R' = 0$, it is equivalent to a constant jacket temperature, and when $\beta = 0$, it corresponds to a negligible contact resistance between the tube and the jacket fluid or an infinite value of h_{oi}. Graphs for values of $\beta = 0$ and with $R' = 0$, 0.5, and 1.5 are shown in Fig. 20.20a, b, c, and d. The reciprocals of these graphs correspond to the Gurney-Lurie chart of Fig. 18.13.

Data are usually lacking on the thermal conductivity of the granular material. Its value differs from the thermal conductivity of the solid, since the conduction of heat is obviously affected by granulation. Brinn *et al.* have found an excellent agreement in the value of the thermal conductivity for a granular solid when determined by heating or cooling a stationary bed similar to that of Chap. 18 or heating or cooling a granular solid in a tube under rodlike flow. For Ottawa sand and crushed ilmenite ore they report values of 0.172 and 0.132 Btu/(hr)(ft^2)(°F/ft), respectively, for stationary beds and 0.155 and 0.141, respectively, for vertical rodlike flow. The densities of the granular materials were, respectively, 101 and 103 lb/ft^3 for the Ottawa sand and 149 and 168 for the ilmenite ore, although the density of the moving material is usually increased by movement. Specific heats for granular solids have been reported for

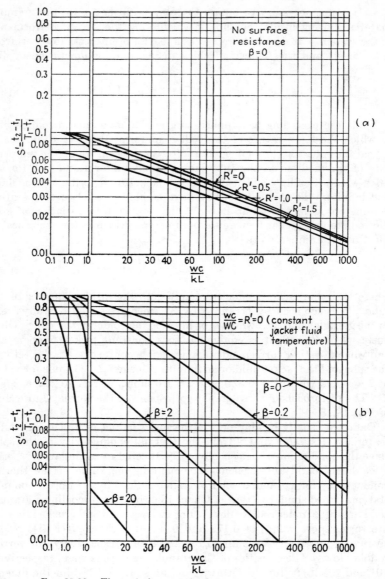

Fig. 20.20. Theoretical curves for heat transfer to fluids in rodlike flow in

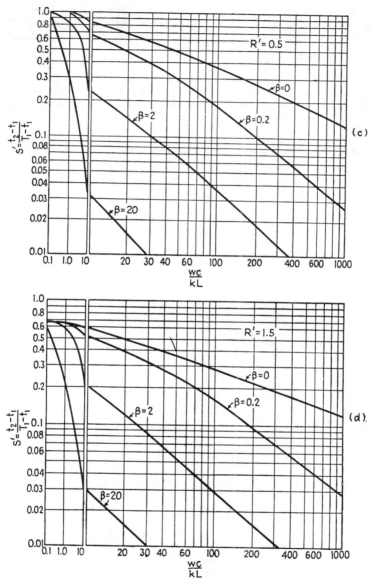

cylindrical pipes for various values of R' and β. (*Brinn, et al., Industrial and Engineering Chemistry.*)

several materials, and these are somewhat higher than those of the solid. For quartz, silica, and Ottawa sand they range from 0.175 to 0.20 Btu/(lb)(°F). Examples given below have been taken from the original reference.

Example 20.7. Calculation of Sand Cooling with Negligible Resistance. A sand has a specific heat 0.20 Btu/(lb)(°F) and a thermal conductivity of 0.15 Btu/(hr)(ft^2)(°F/ft). It is to be cooled from 284 to 104°F at the rate of 1000 lb/hr by water from 86 to 104°F. A bundle will consist of vertical tubes 10'0" long and with inside diameters of 2.0 in.
 (a) How many tubes are required?
 (b) How many tubes are required if air from 86 to 104°F is the cooling medium?
Solution. (a) Assume h_{oi} for water is 500 Btu/(hr)(ft^2)(°F).

$$\beta = \frac{2k}{h_{oi}D} = \frac{2 \times 1.05}{500 \times \frac{2}{12}} = 0.0036 \quad \text{(nearly zero)}$$

The water cooling rate is

$$\frac{1000 \times 0.2(284 - 104)}{104 - 86} = 2000 \text{ lb/hr}$$

$$R' = \frac{wc}{WC} = \frac{1000 \times 0.2}{2000 \times 1} = 0.1$$

$$S' = \frac{t_2 - t_1}{T_1 - t_1} = \frac{104 - 284}{86 - 284} = 0.91$$

From Fig. 20.20b

$$\beta = 0 \quad R' = 0.1 \quad S' = 0.91$$

$$\frac{wc}{kL} = 8.33$$

$$w = \frac{8.33 \times 0.15 \times 10}{0.20} = 62.5 \text{ lb/hr}$$

This is the rate per tube.

$$\text{Total number of tubes} = \frac{1000}{62.5} = 16$$

(b) For air, assume $h_{oi} = 9$.

$$\beta = 0.2$$

The air rate using 0.25 for the specific heat of air is

$$\frac{1000 \times 0.2(284 - 104)}{0.25(104 - 86)} = 8000 \text{ lb/hr}$$

Interpolating between Fig. 20.20b and c,

$$\beta = 0.2 \quad R' = 0.1 \quad S' = 0.91$$

$$\frac{wc}{kL} = 5.23$$

$$w = 39 \text{ lb/hr}$$

and 26 tubes are needed. The film resistance increases the size of the unit by 62 per cent.

10. ELECTRIC-RESISTANCE HEATING

Introduction. The transformation of fuel energy into work (and then into electricity) is an inefficient process, only a portion of the fuel energy

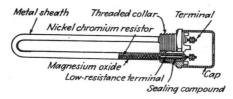

Cutaway view of a two-terminal (single heat) immersion heater
(a)-IMMERSION HEATER

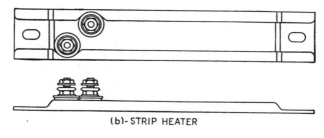

(b)-STRIP HEATER

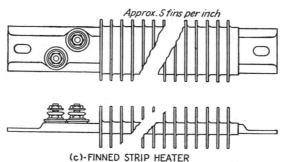

(c)-FINNED STRIP HEATER

FIG. 20.21. Common process resistance heating elements. (*Westinghouse Electric Corporation.*)

finally being available as work. The transformation of fuel energy into steam can be accomplished theoretically with 100 per cent efficiency. These are the practical deductions of the Carnot cycle, and electricity costs ten to fifteen times as much as steam in localities which are not served by hydroelectric power. For this reason electric resistance heating has not been considered a part of process heat transfer, although the comparison of the cost of electricity and coal need not necessarily preclude its use. There are innumerable applications in which electric-resistance heating can be made to operate more effectively than fluid heat transfer particularly in batch operations. There are still other advantages which may be derived from the compactness of the standard types of electric heating elements: the ease with which they provide high temperatures, the elimination of combustion hazards, and their ready application to automatic regulation and control. It is fairly safe to conclude, however, that most of the applications of electric-resistance heating are small compared with those of the majority of industrial processes.

The three commonest resistance heating elements are the *immersion heater*, the *strip heater*, and the *finned-strip heater*. These are shown in Fig. 20.21. Others are available such as the cartridge heater which are treated by the same general method. Electric heaters consist of resistance wires embedded in a refractory which is then encased in a metal sheath. Immersion heaters are used for direct introduction into liquids and are made with copper sheathing for water heating and steel sheathing for oil and organics. The temperature of the copper sheathing may not exceed 350°F, while the steel-sheathed heaters may operate at a maximum temperature of 750°F. The strip heater is used most for the conduction heating of solids, and it can also be used for heating air and gases. Strip heaters are available with two maximum operating temperatures: 750°F for steel-sheathed heaters and 1200°F for chrome-steel-sheathed heaters. By far the best suited for air and gas heating is the finned-strip heater combining electric heating and extended surface and available with maximum operating temperatures of 750 and 950°F.

The technique of handling electric-heating calculations has been developed by the electrical industry and is not identical with the methods employed in either the mechanical or chemical industries. This is doubtless due to the greater standardization of electric equipment.

The laws governing electric-resistance heating, neglecting the effect of the resistance on the circuit, are among the simplest in elementary physics. These are Ohm's law, $I = E/R^\omega$ and Joule's law, $P = EI$, where I is the current in amperes, E the voltage, R^ω the resistance in ohms, and P the power in watts. The rate at which electrical energy is converted into heat is given by $I^2 R^\omega$. The rate of energy delivery is the watt or kilowatt, and this is equivalent to the *Btu per hour*. Similarly the total quantity of energy is the *kilowatt-hour* or *watt-hour*, and this is equivalent to Btu without a time unit.

In electric-resistance heating the element is capable of attaining a very high temperature. The ultimate temperature attained by a resistance heater is that which causes the heat energy to be dissipated at the same

ADDITIONAL APPLICATIONS

rate as it is produced. To prevent the element from burning out, the cold fluid or solid must be capable of receiving the heat at a sufficiently rapid rate to keep the sheath from exceeding its maximum allowable

TABLE 20.1. IMMERSION HEATERS

Water heaters, copper sheath, max temp 350°F						Oil heaters, steel sheath, max temp 750°F					
Watts	Watts/ in.^{2} immersion	Immersed length, in.	Collar, in. IPS	No. heats	No. tubes	Watts	Watts/ in.^{2} immersion	Immersed length, in.	Collar, in. IPS	No. heats	No. tubes
1,000	35	8	1¼	1	1	1000	20	9	1¼	1	1
2,000	35	10	2	3	2	2000	20	17	2	3	2
3,000	35	15	2	3	2	3000	20	25	2	3	2
4,000	35	20	2	3	2	4000	20	34	2	3	2
5,000	35	24	2	3	2	5000	20	41	2	3	2
7,500	35	30	2	3	4	Forced Circulation					
10,000	35	35	2	3	4						
						5000*	35	25	2	3	2
						6000*	35	30	2	3	2
						8000*	30	33	2	3	4

* 230 volts only. All others 115 or 230 volts.

TABLE 20.2. STRIP HEATERS

Steel sheath, max surface temp 750°F				Chrome steel sheath, max surface temp 1200°F			
Watts	Watts/ in.	Total length, in.	Heated length, in.	Watts	Watts/ in.	Total length, in.	Heated length, in.
150	37	8	4	250	62	8	4
250	31	12	8	350	44	12	8
350	25	18	14	500	36	18	14
500	25	24	20	750	37	24	20
750	28	31	27	1000	38	31	27
1000	31	36	32	1500	38	43	39
1250	32	43	39				

temperature. In other words, the design of electric elements is dictated by the flux which must be dissipated to the cold material. The flux is the deliverable energy in Btu per hour per square foot of surface. In electrical units it is expressed in watts per square inch of element surface or watts per linear inch of element corresponding to watts per square inch for elements of uniform surface per inch of length.

Flux data for the various types of heater are given in Tables 20.1 through 20.4. The wattage indicates the output of the element but does

TABLE 20.3. FINNED-STRIP HEATERS

Steel sheath, max surface temp 750°F				Chrome steel sheath, max surface temp 950°F			
Watts	Watts/in.	Total length, in.	Heated length, in.	Watts	Watts/in.	Total length in.	Heated length, in.
250	31	12	8	250	62	8	4
350	25	18	14	350	44	12	8
500	25	24	30	500	36	18	14
750	25	31	27	750	37	24	30
1000	31	36	32	1000	38	31	27
1250	32	43	39	1500	47	36	32

TABLE 20.4. STRIP HEATERS
Clamp-on service watts per square inch of contact surface

Steel sheath, max surface temp 750°F		Chrome steel sheath, max surface temp 1200°F	
Total length, in.	Watts/in.²	Total length, in.	Watts/in.
12*	21	8*	42
18	16	12	29
24	16	18	24
31	19	24	25
36	21	31	25
43	22	43	26

* Passes.

not indicate its temperature in so doing. Included is the watts per square inch or watts per inch which the various elements may dissipate to the cold fluid or solid. For finned- and ordinary strip heaters the watts per inch removed by air as a function of the air velocity and initial temperature are given in Figs. 20.22 and 20.23. At low air velocities it is seen that there is little advantage to the use of finned-strip heaters in preference to bare-strip heaters. Since the outlet-air temperature also affects the sheath temperature, Fig. 20.24 relates the surface temperature of strip heaters as a function of the flux and the ambient-air temperature. When dealing with elements operating at such high temperatures, the

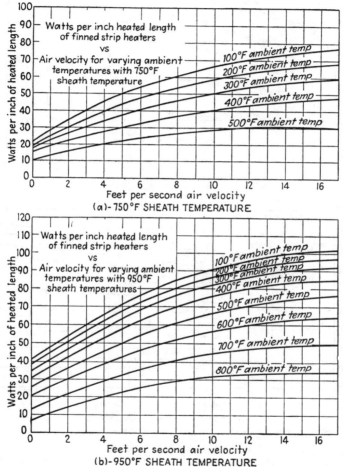

Fig. 20.22. Finned-strip heater air-heating curves. (*Westinghouse Electric Corporation.*)

effect of radiation is of greater importance than in ordinary steam-heated equipment. Figure 20.25 relates the losses obtained experimentally under varying conditions. The use of these graphs and tables is shown explicitly by solving typical problems to which they apply.

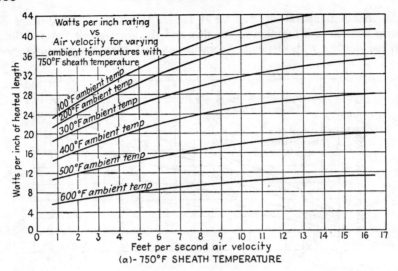

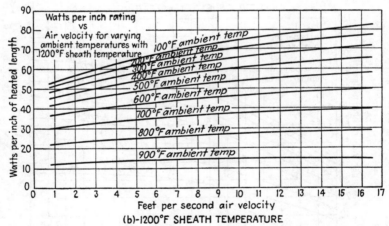

Fig. 20.23. Strip-heater air-heating curves. (*Westinghouse Electric Corporation.*)

Example 20.8a. Immersion Water Heater. A tank 2 by 3 by 2 ft high is filled with water to a height of 18 in. It is required to heat the water from a tap temperature of 50°F to a demand temperature of 150°F within 2 hr. Thereafter, the tank must supply 8 gal/hr of water at 150°F.

What immersion elements are required?

Solution:

Unsteady state.

Lb water = $2 \times 3 \times 18/12 \times 62.5 = 562$ lb

ADDITIONAL APPLICATIONS

$Q' = 562.5(150 - 50)1.0 = 56,250$ Btu

$Q = \dfrac{56,250}{2} = 28,125$ Btu/hr $= \dfrac{28,125}{3412} = 8.25$ kwhr

Losses:

From surface of water (Fig. 20.25c) $= 2 \times 3 \times 260$ $\quad\quad\quad\quad = 1.56$ kwhr
From sides of vessel (Fig. 20.25c) $= 5.5(2 \times 2 \times 2 + 2 \times 2 \times 3)/1000 = \underline{0.11}$
From bottom $=$ negligible

Total losses $= 1.67$ kwhr

Total requirement $= 8.25 + 1.67 = 9.92$ kwhr

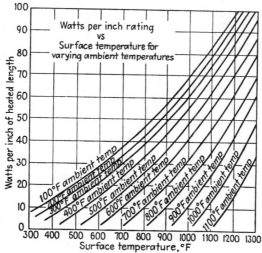

Fig. 20.24. Surface temperature vs. flux for strip heaters. (*Westinghouse Electric Corporation.*)

Steady state: This is the same as heating 8 gal/hr of water.

$Q = 8 \times 8.33(150 - 50)1.0 = 6670$ Btu/hr or 1.95 kwhr
Losses $= \quad\quad\quad\quad\quad\quad\quad\quad\quad\quad\quad\quad\quad\quad 1.67$
$\quad\quad\quad\quad\quad\quad$ Total requirement $= \overline{3.62}$ kwhr

The unsteady-state requirement controls the selection of the immersion heater requiring 10 kwhr. The nearest size in Table 20.1 is the 4000-watt or 4-kw immersion element with a 20-in. immersed length and wired for three heats. Three are required. The rate of absorption of heat by water is high, and there is little need ever to check the sheath temperature although the same rule is not universally true for organics.

Example 20.8b. Strip Heater for Air Heating. An oven is charged hourly with 100 lb of steel which is air dried at 370°F by means of strip heaters located inside it. Four air changes per hour are required for the removal of the moisture. The oven is 3 by 4 by 2 ft insulated with 2 in. of rock wool. The inlet air and steel are at 70°F. What arrangement of strip heaters is required?

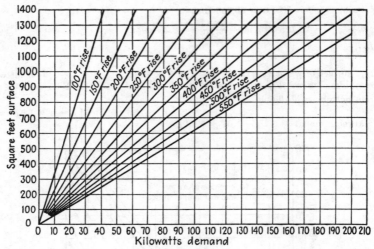

Fig. 20.25a. Heat losses through insulated oven walls. Curve based on an insulation 1 in. thick of standard high-grade material, such as 85 per cent magnesia, rock wool, Filinsul, etc. If insulation is 2 in. thick, divide curve values by 2; if 4 in. thick, divide by 4; etc.

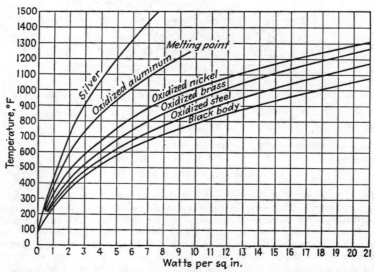

Fig. 20.25b. Heat losses from smooth solid surfaces. Heat losses from a horizontal surface laid flat on one side are top surface, 110 per cent of curve values; bottom surface, 55 per cent of curve values; averages of losses from top and bottom surfaces, 82½ per cent of curve values.

ADDITIONAL APPLICATIONS   761

Solution:

Heat to steel charge = $100 \times 0.12(370 - 70)$ = 3600 Btu/hr
Heat to air = $4 \times 2 \times 3 \times 4 \times 0.075 \times 0.25(370 - 70)$ = 540
                                                          Total = 4140 Btu/hr or 1.22 kw

Losses:

From Fig. 20.25a for 52 ft² of oven outside surface and a temperature rise of 300°F the loss is 5 kw for 1 in. thick insulations. For 2 in. thick insulation the loss is 5/2 = 2.50
                                                          Total   3.72 kw

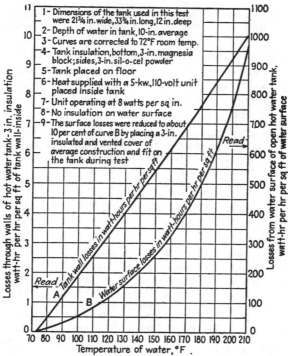

FIG. 20.25c. Heat losses from open hot-water tanks. (*Westinghouse Electric Corporation.*)

From Table 20.2 it is seen that the energy and space requirement can be fulfilled by four heaters of 1000 watts capacity each. The distribution will be poor, however, because so few elements are employed, and the use of eight elements of 500 watts is preferable. It has been assumed that steel-sheathed elements are used. The surface temperature must be checked to assure that this element will not exceed a surface temperature of 750°F, since air absorbs heat relatively slowly.

From Table 20.2 the required dissipation is 25 watts/in. From Fig. 20.24 for 370°F air the surface of the sheath will be 680°F and hence is satisfactory. If the temperature had exceeded 750°F, a chrome-steel element would have been necessary.

Example 20.8c. Finned-strip Heater. Determine the number of finned-strip heaters required to raise the temperature of 270 cfm of air flowing in an 18- by 18-in. duct from 70 to 120°F. Density of air at 70°F is 0.075 lb/ft^3.

$Q = 270 \times 0.075 \times 60 \times 0.25(120 - 70) = 15{,}200$ Btu/hr at 4.47 kw
Velocity $= 270/1.5 \times 1.5 \times 60 = 2.0$ fps

Refer to Fig. 20.22a. The air is capable of removing 33 watts/in which is the maximum dissipation which may be expected. Any group of heaters providing 5 kw which do not require a dissipation of more than 33 watts/in. and which will fit into the duct will be satisfactory. Thus in Table 20.3 elements of 350 watts with a total length each of 18 in. each are satisfactory.

Consider the same service handled by conventional strip heaters without fins. From Fig. 20.23a the air will remove 26 watt/in. of heated length. From Table 20.2 the 18 in. long element with 350 watts capacity will also be satisfactory and cheaper.

Clamp-on Applications. One of the best examples of the versatility of the strip heater is the ease with which it can be clamped-on or bound to various vessels. When used in clamp-on services, it has been found that the temperature drop across the surface-to-surface contact is 14 to 19°F per unit of flux.

Example 20.8d. Clamp-on Plastic Heating. It is desired to heat a cast-iron plate 26 by 12 by 1 in. at 70°F on which a plastic with specific heat of 0.22 Btu/(lb)(°F) is to be softened. It is necessary that one 70-lb batch of the plastic be heated on the plate after it has attained a steady state rate from 70 to 300°F in 1 hr. The plate will be heated on its underside by clamp-on strip heaters in 1 hr. How many are required?

Solution. The specific gravity of cast iron is 7.2, and its specific heat is 0.13 Btu/(lb)(°F).

Unsteady state:

Weight of plate $= 26 \times 12 \times 1 \times 62.5 \times 7.2/1728 = 81$ lb
$Q = 81(300 - 70)0.13 = 2400$ Btu/hr or 0.705 kw
Losses: From Figure 20.25b for a black body the radiation is 1.5 watts/in.^{2}. The radiation from the top is actually 110 per cent of this value, and from the bottom of the plate it is 55 per cent for an average of 82.5 per cent.
Total radiation loss $= 2 \times 26 \times 12 \times 1.5 \times 0.825/1000 = 0.80$ kw
Total requirement $= 0.70 + 0.80 = 1.50$ kw

Steady state: Input required to plate

$Q = 70 \times 0.22(300 - 70) = 3550$ Btu/hr or 1.04 kw
Losses $=$ 0.80
              Total requirement $= 1.84$ kw

The steady state is controlling. The requirements are satisfied by four 24-in. strip heaters, but the sheath temperature must now be checked. Since the temperature drop per unit flux density is 14 to 19°F, assume an average of 16.5°F. For clamp-on strips 24 in. long the watts per square inch deliverable are 16. The total temperature drop is the product of the two.

$$\Delta t = 16.5 \times 16 = 264°F$$

ADDIITONAL APPLICATIONS 763

The sheath temperature is then $300 + 264 = 564°F$, which is satisfactory for steel-sheathed elements with a 750°F maximum.

NOMENCLATURE FOR CHAPTER 20

A	Heat-transfer surface, ft^2
B	Height of agitator above vessel bottom, ft; width of an atmospheric cooler, ft
C	Specific heat of hot fluid, Btu/(lb)(°F)
C_1, C_2	Constants
D	Inside diameter of pipe or tube, ft
D_j	Diameter of inside of vessel, ft
D_o	Outside diameter of pipe or tube, ft
E	Temperature group defined by Eq. (20.34) or (20.35), dimensionless; voltage; evaporation, per cent
F	The ratio up/UP, dimensionless
G	Mass velocity, lb/(hr)(ft²)
G'	Liquid loading for trombone coolers $w/2L$, lb/(hr)(ft); for falling films $w/\pi D$, lb/(hr)(ft)
g	Acceleration of gravity, ft/hr²
h	Heat-transfer coefficient, Btu/(hr)(ft²)(°F)
h_i, h_o	Heat-transfer coefficient referred to the inside diameter and outside diameter of pipe or tube, Btu/(hr)(ft²)(°F)
h_{oi}, h_{io}	h_o referred to the inside diameter; h_i referred to the outside diameter, Btu/(hr)(ft²)(°F)
I	Current, amp
j	Factor for heat transfer, dimensionless
K	Temperature group in crossflow $(T_1 - T_2)/(T_2 - t_1)$, dimensionless
K_1, K_2	Constants
k	Thermal conductivity, Btu/(hr)(ft²)(°F/ft)
L	Tube length, ft; length of agitator paddle, ft
LMTD	Log mean temperature difference, °F
M	Spray, lb/(min)(ft² projected tube area)
m	Spray, lb/(hr)(ft² cooler cross-sectional area)
N	Agitator speed, rev/hr; number of horizontal tube rows
N'	Agitator speed, rpm
P	Perimeter of large pipe or tube, ft; power, watts
P_H, P_V	Horizontal tube pitch, ft; vertical tube pitch, ft
p	Perimeter of smaller pipe or tube, ft;
Q	Heat flow, Btu/hr
Q'	Heat, Btu
R	Temperature group $(T_1 - T_2)/(t_2 - t_1)$, dimensionless
R'	Temperature group in rodlike flow, dimensionless
R_d	Dirt factor, (hr)(ft²)(°F)/Btu
R^ω	Resistance, ohms
(r)	Temperature group LMTD/$(T_1 - t_1)$, dimensionless
S	Temperature group $(t_2 - t_1)/(T_1 - t_1)$, dimensionless
S'	Temperature group in rodlike flow, dimensionless
T	Temperature of the hot fluid in general, °F
T_1, T_2	Inlet and outlet temperatures of the hot fluid, °F
T_s	Constant hot-fluid temperature, °F
t	Temperature of the cold fluid in general, °F

t_f, t_w	Temperature of the film, temperature of the wall, °F
t_1, t_2	Inlet and outlet temperature of the cold fluid, °F
Δt	True temperature difference, °F
Δt_D	Temperature group $\Delta t/(t_2 - t_1)$, dimensionless
U	Overall coefficient of heat transfer, Btu/(hr)(ft²)(°F)
U_D, U_C	Overall design coefficient; overall clean coefficient, Btu/(hr)(ft²)(°F)
u	Overall coefficient of heat transfer, Btu/(hr)(ft²)(°F)
V	Velocity, fps; temperature group, dimensionless
v_s	Volume of granular material, ft³
W	Weight flow of hot fluid, lb/hr
w	Weight flow of cold fluid, lb/hr
X	A function
x	Distance, ft
Y	Space group $(P_H - D_o)/P_H$, dimensionless
y	Thickness of agitator, ft
Z	Space group $(P_H - D_o)/D_o$, dimensionless
z	Height of liquid in vessel, ft
α	Thermal diffusivity, ft²/hr
β	Dimensionless group, $2k/h_{oi}D$
θ	Time, hr
μ	Viscosity, lb/(ft)(hr)
ρ	Density, lb/ft³
τ	A function
ψ	A function

Subscripts (except as noted above)

c	Coil
j	Jacket

CHAPTER 21

THE CONTROL OF TEMPERATURE AND RELATED PROCESS VARIABLES

Introduction. The even operation of a process is dependent upon the control of the process variables. These are defined as conditions in the process materials or apparatuses which are subject to change. Because there may be several materials and several pertinent operating factors which may change in the simplest of processes, the maintenance of control over an entire process is an important aspect of process design. Many of the advances in processing technology during recent years have been due in part to the widespread use of automatic-control mechanisms. Naturally a comprehensive study of so broad a field of endeavor is beyond the scope of this text, and it is the intention here to introduce in a practical way only the most elementary principles of process control.

Process Variables. When the flow sheet is laid out for a process, the temperatures, pressures, and fluid-flow quantities are theoretically fixed in accordance with heat, pressure, and material balances. The translation of the flow sheet into an operable plant requires that special provision be made to assure the *relative* constancy of the various quantities and qualities. It is impossible to achieve absolute constancy in even the simplest of industrial operations and these do not include the multitude of complex operations which are normally encountered. Take the simple case of a storage tank to which one pump continuously supplies liquid and an identical pump continuously removes liquid. Because of differences in suction and discharge, both pumps, acting independently, pump at different rates and the liquid level in the storage tank cannot be expected to remain constant.

Similar factors influence nearly every other steady-state condition. Consider the utilities alone such as high- and low-pressure steam, cooling water, electricity, compressed air, and fuel supply. When any single unit process in a plant is either shut down or started up, it may affect the supply of the utilities to the other unit processes. Furthermore, when the furnaces in a powerhouse are periodically refired, the temperature, pressure, and quantity of steam throughout the plant may show some variation. Similarly a sudden change in the steam demand at some point in the plant for steaming out large vessels may cause a sufficient speed variation to affect the performance of turbine-driven pumps, compressors, and generators including their discharge rates and pressures. Or the temperature of cooling-tower water, varying with atmospheric conditions, might affect the total heat transfer at critical points in the process. Add to these the variations resulting from changes in the composition of the feed materials such as boiling points, specific heats, or viscosities, and additional fluctuations may be anticipated in the pressure, temperature, and fluid flow of the process streams.

Automatic control is employed to measure, suppress, correct, and modify changes of the four principal types of process variation:
1. Temperature control
2. Pressure control
3. Flow control
4. Level control

There are, in addition, other[1] controllable variables such as specific gravity, thermal conductivity, speed, and composition. Because of its importance to heat transfer particular attention is given in the following pages to temperature control and its relationship to the other principal process variables.

Self-acting and Pilot-acting Controllers. It is the object of all controllers to regulate the process variables, and to do so they must be capable of first measuring the variables. Some instruments are equipped to *indicate* the variable in a continuously readable form, and others, *recorders*, are equipped with pen and ink on a traveling chart calibrated for time. These are not essential parts of the control but additional conveniences.

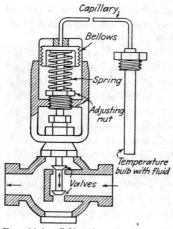

Fig. 21.1. Self-acting temperature controller.

Controllers or regulators are either *self-acting* or *pilot-acting*. A self-acting temperature controller is shown in Fig. 21.1. A bulb filled with a fluid having a favorable coefficient of thermal expansion is placed in a stream or vessel at the point where the temperature is to be controlled. The bulb is connected to the controller by means of a capillary tube. As an example of its operation suppose that the temperature of a stream

---

[1] See also Perry, J. H., "Chemical Engineers' Handbook," 3d ed., pp. 1263–1340 by R. W Porter and D. M. Considine, McGraw-Hill Book Company, Inc., New York, 1950.

leaving the shell of a heater is to be controlled by adjusting the flow of steam to the tubes of the heater. Any increase in the temperature of the stream causes an increase in the pressure of the bulb fluid, which is transmitted in turn to the bellows of the regulator assembly. The larger area of the bellows amplifies the force acting against the spring and causes the valve to move in a direction which decreases the flow of steam. If the same type of regulator were used to control the flow of cooling water instead of steam, the valve would be arranged to open when the bulb pressure increased. The control points of self-acting controllers can be varied by means of adjusting nuts, which change the tension on the spring.

In general the temperature-sensitive element consists of a bulb and capillary filled or partly filled with a volatile fluid, an inert gas, or an expanding liquid. Thermocouple and thermal-expansion elements also find wide application in temperature control. With the exception of the liquid-filled bulb the total amount of fluid contained in the capillary is very small compared with the amount confined in the bulb. If the capillary were accidentally to lie across a steam pipe, it would not materially affect the measurement of the instrument, since the amount of fluid present where the capillary contacts the pipe is exceedingly small. Under certain conditions compensation must be included for variations in the ambient temperature at the point where the variation is registered.

From a mechanical standpoint self-acting types of instruments are the simplest. Most of these fall into the regulator rather than the instrument class, because the regulator does not actually measure the variable but acts upon deviations from the set control point. Many involve no calibration in their manufacture whatever but are adjusted to produce the desired effect in the field. Typical of this class are weight-loaded reducing or back-pressure valves, spring-loaded diaphragm valves for reducing or back-pressure regulation, and the temperature regulator discussed above.

The second class of instruments are the pilot-acting controllers in which the detection of a small variation actuates a more powerful force by which the variable is corrected. This class is capable of a greater degree of sensitivity than the regulators because they eliminate some of the lags which would be inherent in self-acting mechanisms actuated by the force of a large volume of fluid. Most pilot-acting instruments use compressed air to effect the larger controls, while some employ electric relays and motor-driven controllers. The latter are used particularly for isolated services or where compressed air is not available.

The majority of pilot-acting instruments operate on the principle shown in Fig. 21.2 or a modification of it. The controller operates between two fluids, one being controlled and another which adds or removes heat to or from the fluid being controlled. In Fig. 21.2 compressed air at 15 to 20 psig enters the control head, part passing to a

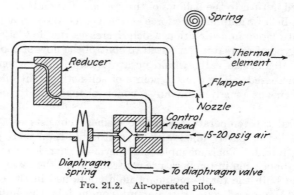

Fig. 21.2. Air-operated pilot.

diaphragm valve and part through the reducer. The discharge air from the reducer is at a low pressure and continuously escapes through the nozzle by impinging against a flat flapper. While air escapes through the nozzle, the diaphragm spring is contracted, since it is also affected by the escape of air through the nozzle. This holds the valve in the control head against the left end of the control head. When a change occurs in the temperature of the fluid being controlled a capillary spiral or capillary helix connected to a temperature-sensitive element such as a bulb is displaced in response to the temperature change. The movement is used to bring the flapper toward the nozzle, thereby obstructing the escape of air and expanding the diaphragm spring. This may in turn be used to relieve the air pressure on a diaphragm valve effecting the control, thereby permitting it to close. A movement of the flapper of 0.002 in. is sufficient to actuate the control. A typical diaphragm valve is shown in Fig. 21.3, although diaphragm valves are available which either open or close when subjected to the action described.

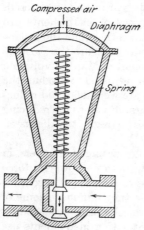

Fig. 21.3. Diaphragm valve.

Lags. Instrumentation would pose no problem if only a single variable were to be controlled. Usually, however, the regulation of a process

CONTROL OF TEMPERATURE

requires the control of pressures, temperatures, and flow rates and the maintenance of adequate liquid levels to assure flow continuity. The greatest impediment to precise control results from a series of time lags between the measurement of a variation and its correction. These are inherent in all automatic control systems. Haigler[1] has presented a visual picture of the time lags in a simple heat-transfer system from which Fig. 21.4 has been taken. Starting with the measurement of the variable at a, the lags may be traced alphabetically.

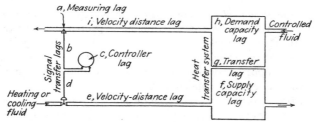

FIG. 21.4. Sources of lags in automatic control.

a. Measuring lag
b. Signal transfer from sensitive element to instrument
c. Controller operation lag
d. Signal transfer from controller to regulator
e. Velocity-distance lag until adjusted stream reaches apparatus
f. Capacity lag until the supply side fluid is adjusted
g. Transfer lag through tubing from supply to demand side
h. Capacity lag until the demand side is corrected
i. Velocity-distance lag until the measuring element detects the correction

All these lags are not of equal importance, but they do present difficulties in the accomplishment of close control such as is required by modern high-temperature, high-pressure, and catalytic processes. Numerous questions arise in automatic control such as the extent to which the regulating valve on the correcting fluid should be opened and whether or not the opening should vary with the size of the deviation measured. How fast should it open? How much of the time-temperature relationship of the response of the correction should an instrument be able to anticipate? These are particularly important in the larger processes in which all the process variables must be controlled simultaneously in less time than would accompany the operation of several related simple instruments.

[1] Haigler, E. D., *Trans. ASME,* **60,** 633–640 (1938).

Automatic-control Mechanisms. For a general background and mathematical development the literature on instrumentation should be consulted. Available to deal with the lags inherent in process control are four basic mechanisms. These are

1. *Two-position control* (also called *on-and-off* or *open-and-shut*), in which the overall flow of fluids, Btu, watts, etc., is limited to predetermined maximum and minimum values or completely turned on or off when the measured variable exceeds or falls below the control point setting.

2. *Proportional control* (also *throttling*), in which the *magnitude* of the corrective action taken by the instrument is proportional to the extent of the deviation of the measured variable from the control point setting.

3. *Floating control*, in which the *rate* of corrective action taken by an instrument is proportional to the extent of the deviation of the measured variable from the control-point setting.

4. *Proportional control* and *automatic reset*, in which the corrective action taken by an instrument is a combination of proportional control plus a corrective action which is proportional to the rate of the deviation and the duration of that deviation from the control-point setting. Another basic mechanism, the *derivative function*, is one which applies a correction which is proportional to the rate of the deviation and which is unaffected by the amount or duration of the deviation.

An example of two-position control may be found in the simple constant-temperature bath. When the temperature falls below a fixed minimum, heat is supplied until the bath reaches a fixed maximum temperature. The heat supply is then turned off until the bath again reaches the minimum whence heat is again supplied (on and off). For proportional control consider a vessel of small capacity containing a heating element. Cold water continuously enters and hot water leaves the vessel. The control-point temperature has been fixed. Suppose the temperature of the cold-water supply suddenly drops very much. For each degree that the water in the vessel deviates from the control point, a *greater* amount of heat will be supplied. In floating control, the *rate* at which the valve moves is proportional to the deviation of the temperature from the control setting. If the temperature is below the control point, the valve opens at a constant *rate* or a rate which increases with the extent of deviation. As the control point is reached, the valve closes but at a rate in proportion to the deviation. Floating control is not well suited to operations requiring rapid temperature control, since the correction in the temperature near the control point (narrow band) is slow and cyclical. Some floating-control mechanisms are modified to give a rapid action outside a fixed narrow band of deviations and a slow action inside the narrow band. A combination of floating and proportional control will make the action more rapid. *Proportional control and automatic reset* correct the action of proportional control when a major deviation may continue with additional minor deviations in the same variable. Suppose a major demand change occurs. In proportional control the valve position is fixed by the magnitude of the deviation. If the valve is correcting for a large deviation while an additional minor deviation occurs, the correction for the minor deviations would be made with a valve which is open primarily to the extent of the major deviation. The reset shifts the proportioning band so that the minor corrections can be made as if the major deviation were the control point and not in proportion to the total deviation from the control point.

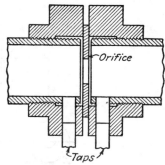

Fig. 21.5. Orifice taps.

Flow Control. The temperature control of a fluid can be accomplished by changing the flow of a heating or cooling medium. In many instances, however, the control of flow may be of greater importance than the control of temperature, and the methods of automatically maintaining flow warrant preliminary consideration. Unlike the expansion of a bulb, flow variations are detected by means of the differential pressures they cause

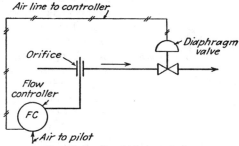

Fig. 21.6. Simple flow control.

across an orifice. Usually the flange and orifice assembly shown in Fig. 21.5 is tapped and fitted with tubes leading to a capillary helix or other device which is sensitive to the differential pressure. A displacement changes the position of the flapper, and the remainder of the control is similar to that shown in Fig. 21.2.

The simplest arrangement of flow control is shown in Fig. 21.6. The flow of compressed air to the pilot controller mechanism is usually assumed to be present without being indicated, and very often the controller may also record the variations, thus being a *flow-recorder controller*. In establishing the primary flow control for a process it is important to consider whether the pumps will be of the reciprocating or centrifugal

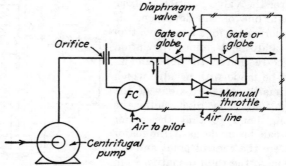

FIG. 21.7. Flow control with centrifugal pump.

types. If a centrifugal pump is employed, flow control can be effected by throttling against the pump pressure as shown in Fig. 21.7. In the pilot shown in Fig. 21.2 the compressed air leaving the control head kept the diaphragm valve open. About 70 per cent of the controllers are of this type, and about 30 per cent open if the air fails. The latter type is particularly useful when controlling the flow of reflux to distilling towers. The use of a by-pass around the controller is a standard practice which permits the continued operation of the process with manual control if the controller or its air supply should fail. When using reciprocating pumps, throttling against the displacement of the pump may cause destructive pressures to be built up.

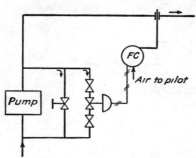

FIG. 21.8. Flow control with reciprocating pump.

The flow controller in these cases is installed in a by-pass, returning to the reciprocating pump suction as shown in Fig. 21.8. In this manner constant discharge is maintained by automatically altering the amount of by-passing. The second line having but a single valve allows manual by-passing when both the valves adjoining the diaphragm valve must be closed because of controller or air failure.

When using automatic controllers, the flow may be intentionally divided so that the bulk of the fluid flows through a continual by-pass and only a small amount through the controlling line. This reduces the cost of the controller by reducing its size. However, the use of a continual by-pass as an integral part of the flow-control system introduces several unfavorable factors in the maintenance of uniform control. Except where the flow lines are extremely large and the control valve is expensive, it is undesirable to use a continual by-pass. Like temperature control, pressure and level control are often intimately associated with the control of flow.

TEMPERATURE CONTROL

Instrumentation Symbols. Most of the common arrangements of process apparatus require several interrelated instruments to assure reasonable control. To simplify representation, the instrumentation flow-plan symbols employed here are those suggested in a survey questionnaire by a symbols subcommittee of the Instrument Society of America. The legend for a number of the common symbols is given in Fig. 21.9. The first letters which appear as instrument designations are

F = flow
L = level
P = pressure
T = temperature

The second or third letters are

C = controller
G = glass (as LG)
I = indicator
R = recorder
S = safety
V = valve
W = well

Often it is desirable to assemble all the instruments at a single point away from the locality at which the measurement and control actually occur. These are referred to as *board*-mounted instruments as differentiated from locally mounted instruments and are designated by a horizontal line through the instrument. Often the instruments on a single control board are at some distance from the point of measurement. In such cases the response is amplified by a pneumatic transmitter. This is particularly requisite for the control of low-pressure fluids at a centralized board.

In the typical instrumentation diagrams which follow, the air lines to the instruments and the by-passes around the controllers have been omitted. Except where necessary to manual control, the indicating or recording features have also been omitted, although they are essential to the instrumentation of the process as a whole.

Whenever a control valve is installed directly in the inlet or outlet line of an exchanger, cooler, or heater, the fluid passing through the

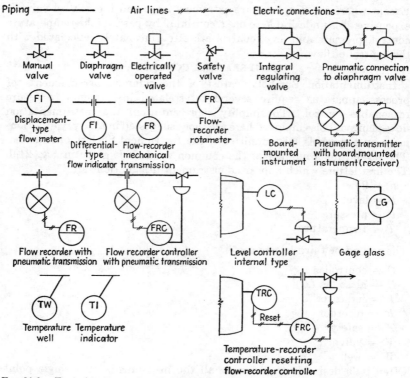

Fig. 21.9. Typical instrument symbols. Instrument lines are usually lighter than piping or equipment.

apparatus must still retain sufficient available pressure so that, when operating, the control valve is capable of increasing or decreasing the flow of the line. Thus an exchanger which at design flow utilizes nearly all of the available static pressure on a line has little pressure available for overcoming the effect of the control valve. An instrument is useless which operates under such conditions that the control valve is wide open and unable to open farther to effect control. Thus when selecting a pump for a process the total head must be not only sufficient to overcome

CONTROL OF TEMPERATURE 775

the sum of the static heads and pressure drops of the equipment but also enough to overcome the drop through the control valve.

Coolers. Figures 21.10 to 21.12 show three methods of providing control for coolers. Figure 21.10 is used where the maximum possible cooling is desired or where there is an abundance of cooling water. When an overhead volatile product from a distilling column passes through an aftercooler and is to be sent to storage, it is cooled as low as possible, since any decrease in the evaporation and vent loss may constitute a significant economy. In such cases only a manual throttling valve is provided on the outlet-water line. If water is abundant, the manual valve is left completely open. The only instrumentation suggested is an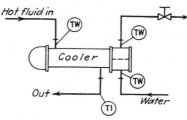

FIG. 21.10. Simple cooler.

industrial thermometer or temperature indicator on the hot-fluid outlet. This enables the operator to make routine checks to ascertain that the temperature of the liquid is well below the maximum allowable storage temperature. Thermometers are useful at the other nozzles but only when checking the performance of the cooler or other apparatus. Thermometer wells are usually provided for this purpose as plugged bosses

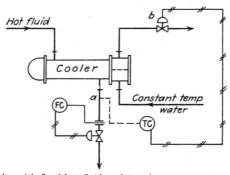

FIG. 21.11. Cooler with fixed hot fluid outlet and near-constant water temperature.

in the necks of nozzles. If the nozzle necks are not conveniently located for testing, the thermometers can be inserted in Tees as near the cooler as possible. However, when checking exchanger performance and guarantee, the temperatures should be taken at the nozzle necks if at all possible.

The arrangement in Fig. 21.11 is used when a fixed hot-fluid outlet temperature and flow quantity are desired. The flow of the hot fluid is controlled independently by the flow controller. The cooling water is assumed to have a relatively constant temperature, so that the hot-fluid outlet temperature is controlled by the quantity of cooling water. The temperature of the hot-fluid discharge is measured by a temperature-sensitive element at a and transferred to the temperature controller which actuates an air-operated valve on the cold-water discharge at b. If the temperature at a varies, the valve at b opens or closes accordingly. This same arrangement can be used for gases provided a sufficiently sensitive element is used.

In Fig. 21.12 an arrangement is shown for cooling gases below their dew points such as occurs in compressor aftercoolers. The gas is cooled

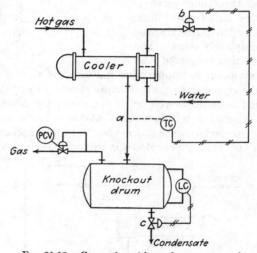

FIG. 21.12. Gas cooler with condensate removal.

and its temperature is measured at a and controlled at the water discharge b. The cooled gas and condensate pass into the knockout drum, where a liquid level is continuously maintained by the level controller. The simplest type of level controller consists of an external float whose vertical movement actuates the control. When the level rises above the control point or range of the level controller, it opens the valve at c and lets the liquid run out. The purpose of the level controller is to maintain a liquid seal which prevents the escape of gas from the condensate outlet and also prevents the condensate from backing up into the shell. The back-pressure regulator or pressure-control valve (PCV) releases gas from the knockout drum without upsetting the system flow, pressure, or liquid level.

Exchangers. Exchangers do not always require special temperature control. Since their purpose in a process is to provide the maximum recovery of heat, there is no reason to restrict their performance by the use of controls. It will ordinarily suffice to use manual valves which are kept wide open. The principal controls are usually provided on both the cooler and heater which are adjacent to the exchanger. For example, when a stream is to serve as the feed to a distilling column, it may enter the system through a bottoms-to-feed exchanger as in Fig. 11.1 and then through a feed preheater. The temperature control of the stream will be effected in the preheater by a flow adjustment of the steam entering the preheater. Similarly, when a fluid is cooled in an exchanger, it usually passes through a cooler and its temperature is controlled by a flow adjustment of the water. It is not possible to control both the flow quantities and outlet temperatures of both streams passing through an exchanger at the exchanger itself, since one adjustable quality must always be present. Thus, if the outlet temperatures of both streams are to be controlled and the flow or temperature of one stream may vary, the flow or outlet temperature of the other stream must also vary.

There are, however, a number of instances in which the outlet temperature of one of the streams must be controlled, and such cases are treated here. Most of the problems of exchanger instrumentation are encountered when the two streams are of unequal size, the one being very much larger than the other. By the same token the larger stream possesses a short temperature range and the smaller stream a very large temperature range. Figures 21.13 and 21.14 are typical of this application. In Fig. 21.13 the hot fluid is considered the large stream and the cold fluid the small stream. The hot fluid is to be flow- and temperature-controlled, and its inlet temperature is assumed to be subject to small variations. The cold fluid is flow-controlled, and its outlet temperature is assumed to be reasonably constant. The net Btu removal or injection in this case can best be effected on only a portion of the large stream, and to this end the by-pass is provided for temperature control. It will be seen, however, that the flow control of the hot fluid is carried out on the entire stream. This arrangement provides flexible temperature control, allowing any proportion to be by-passed. It also permits the elimination of any problems of excessive pressure drop which might result if the entire large stream were passed through the exchanger. It is also advantageous where an exchanger is repeatedly used for several different services. As an example, a particular distilling column and its auxiliaries may be used for a time with one feed stock and then changed to another stock for another period. The controls shown in Fig. 21.13 will enable the exchange of a fairly uniform heat load in the exchanger for the several stocks. The pressure-control valve permits a constant controlled flow of the fluid being temperature-controlled, and the only varying qualities are the inlet temperature of the temperature-controlled fluid and the outlet temperature of the secondary fluid.

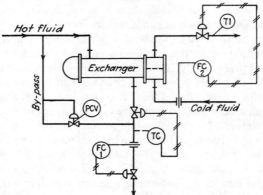

FIG. 21.13. Exchanger with temperature-controlled large stream.

Heaters. Few types of apparatus cause greater difficulty than heaters using low-pressure steam with temperature control on the cold-fluid outlet temperature. The difficulty can be eliminated, however, when proper means for removing the condensate and air are provided. All heaters

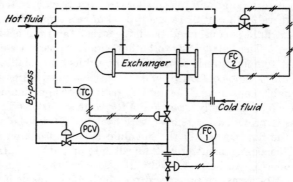

FIG. 21.14. Exchanger with temperature-controlled small stream.

using steam at superatmospheric pressures should be equipped with air vents at the highest accessible points. Heaters operating with a cold-fluid outlet temperature exceeding 212°F as in Fig. 21.15 generally offer little difficulty, because the condensate is not subcooled below the pressure necessary to force it through the outlet trap.

CONTROL OF TEMPERATURE

In practice, heaters are often designed for small temperature differences. Added to this is the fact that they are usually overdesigned, particularly when clean, so that the cold-fluid outlet temperature and the steam condensate temperature are nearly the same. The cold fluid and the condensate may both have outlet temperatures below 212°F (corresponding to 0 psig) indicating that more heat is being removed from the steam than is desired thereby subcooling the condensate to vacuum. Without temperature control a condensate level may build up in the heater to cover a portion of the tubes, reduce the available surface and the transferred heat load, and reduce the outlet temperature on the hot

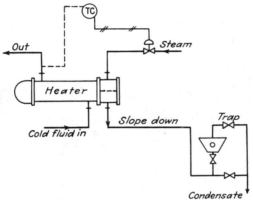

FIG. 21.15. Heater for high cold-fluid outlet temperature.

fluid. If the corrosive action resulting when a condensate level is maintained in the tubes is not excessive, the steam flow can be controlled manually or by a self-regulating flow controller with a simple trap for condensate removal provided the heat demand does not vary too frequently or rapidly for manual adjustment.

With temperature control a system for the removal of low-pressure steam embodying the principles shown in Fig. 21.16 must be employed. However, a pressure-control valve may be used to maintain the pressure in the heater. When the pressure-control valve is installed in the condensate outlet line, the temperature-control valve then varies the condensate level in the heater to compensate for varying heat loads. Very often an arrangement like that shown in Fig. 21.15 is unsuited for heaters providing a low cold-fluid outlet temperature. Without condensate covering some of the tubes, the total surface, containing some excess, is always exposed and the steam has to condense at subatmospheric pressure in order to reduce the LMTD and provide the desired cold-fluid outlet temperature. On attaining an abnormally high fluid outlet temperature, the temperature-control valve throttles the steam until a condensate level builds up inside the heater. By the time the cold-fluid outlet temperature has decreased to normal, condensate covers the tubes, causing the condensate in the lower portion to be subcooled and a resultant vacuum within the heater. When the temperature controller next detects a subnormal cold-fluid outlet temperature, it must admit enough steam to raise the pressure sufficiently to remove all the subcooled condensate and decrease the condensate level in the heater. In doing

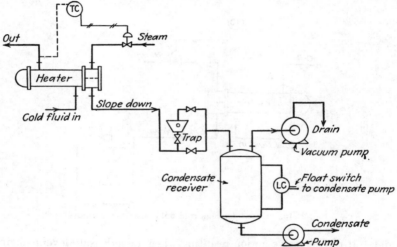

Fig. 21.16.   Heater using low-pressure steam.

this too many tubes may be uncovered by the steam pressure, causing the cold-fluid outlet temperature to rise above the control point. This procedure causes cycling and a virtually uncontrolled cold-fluid outlet temperature. The arrangement in Fig. 21.15 is suitable for heaters using high-pressure steam in which the condensing temperature and pressure, after passing through the control valve, is sufficiently high to blow the condensate out.

Total Condensers. Figure 21.17 shows the arrangement for a condenser with gravity flow of reflux. The principal disadvantage to the use of gravity flow lies in the fact that the condenser and accumulator must be elevated above the tower, requiring additional structural support. The overhead condensate drains into the accumulator, which is provided with a manual vent for continuous bleeding if the operation is at elevated

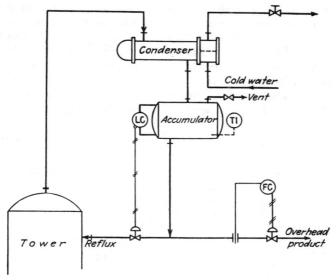

Fig. 21.17. Condenser with gravity flow of reflux.

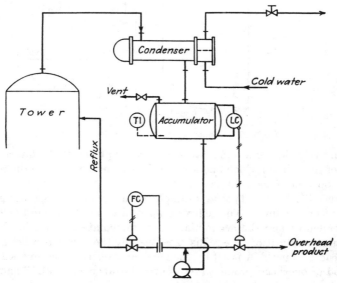

Fig. 21.18. Condenser with pumped reflux.

pressure or left wide open if operation is at atmospheric pressure. If there is any tendency for noncondensables to enter the system, it may be necessary to bleed them at the top of the condenser. The condensate builds up a level in the accumulator, and reflux is regulated by the level controller. The overhead product is removed at a fixed rate by the flow controller. It is seen that there is no positive temperature control other than manual in the event that the reflux temperature should fall appreciably below that of the top tray. This may be compensated for, how-

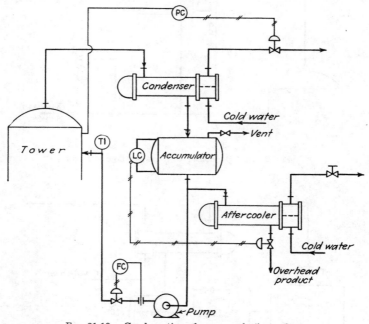

Fig. 21.19. Condensation of a pure volatile product.

ever, by introducing a temperature controller at one of the intermediate trays of the column which operates on the steam line going to the reboiler at the bottom of the column.

Figure 21.18 is a typical example of a condenser operating with pumped-back reflux. The reflux pump is often at the ground level, and the condenser and the accumulator are immediately above it. In this particular application, the reflux rate is set by the flow controller and the overhead product flow is set by the level controller. The reversal of the method of overhead product flow control between Figs. 21.17 and 21.18 is due to the smaller liquid head in the gravity-flow system. Figure 21.19

shows an arrangement for the condensation of a pure volatile product which must be subcooled in an aftercooler to prevent the flashing of the overhead product when its liquid pressure is relieved. A particular feature of this type of arrangement is the use of a pressure controller for the control of temperature by direct connection between the water outlet flow and the tower pressure. When the system contains noncondensables this arrangement cannot be employed.

Figure 21.20 shows an arrangement for a system having noncondensables. The instruments used here differ in the manner by which they

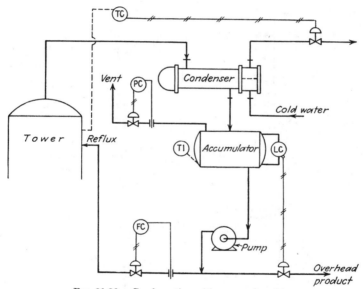

Fig. 21.20. Condensation with noncondensables.

effect control. First, the noncondensables are prevented from building up in the system by means of the pressure controller on the accumulator, which provides a continuous automatic vent to the atmosphere or to the next step in the process if the saturated noncondensables are subjected to further treatment. The tower overhead temperature is maintained by the controller adjusting the rate of water flow to the condenser and likewise adjusting the reflux temperature to maintain the overhead product purity.

Partial Condensers. A partial condenser arrangement is shown in Fig. 21.21. The partial condenser is used for volatile materials to enable reflux to be pumped back to the tower at substantially the top tray temperature while the remainder of the overhead is condensed and cooled so as to prevent reflashing when the pressure is released in the storage tank. The elements of the control can be identified by comparison with Figs. 21.19 and 21.20.

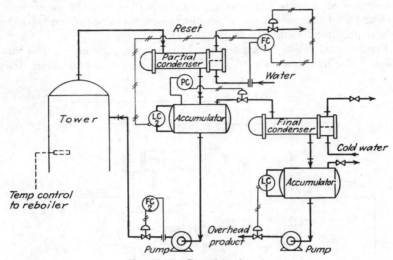

Fig. 21.21. Partial condenser.

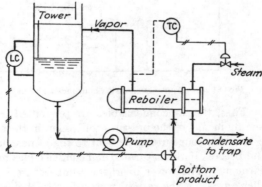

Fig. 21.22. Pump-through reboiler with small bottoms rate.

Pump-through Reboilers. A reboiler can be critical in a distillation process, and its control is extremely important. Furthermore, steam-heated reboilers are subject to the same operating difficulties encountered with heaters. Because pump-through reboilers are used only for small

CONTROL OF TEMPERATURE

services, the requirements of control are usually more sensitive than those for larger operations.

Figure 21.22 shows the arrangement for a pump-through reboiler where the amount of bottom product is small compared with the total flow through the reboiler. Figure 21.23 shows the arrangement for cases where the bottom product is an appreciable part of the total flow. In this case the bottom product rate is set separately by the level controller

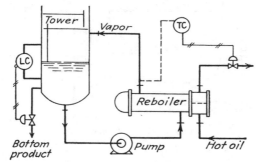

FIG. 21.23. Pump-through reboiler with high bottoms rate.

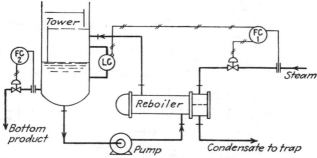

FIG. 21.24. Pump-through reboiler with pure bottoms or constant-boiling mixture.

so as not to interfere with the reboiler operation. In both Figs. 21.22 and 21.23 it has been assumed that the liquids have appreciable boiling ranges so as to make the use of a temperature controller effective. If the bottom product is substantially pure or a constant boiling mixture, as in Fig. 21.24, the use of temperature control is impractical, since the temperature will remain constant independently of the steam supply. Only the flow rate changes, and to overcome this limitation the flow of steam is adjusted by the level controller. The reset permits partial response of the flow controller so as to increase the level controller range.

Natural-circulation Reboilers and Evaporators. Because of their similarity the chiller- and kettle-type reboiler are included in this classification. The instrumentation for the chiller is shown in Fig. 21.25. The valve on the temperature controller should be sized for a small

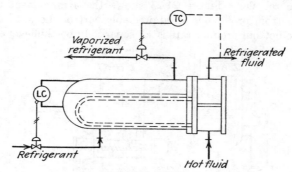

Fig. 21.25. Chiller.

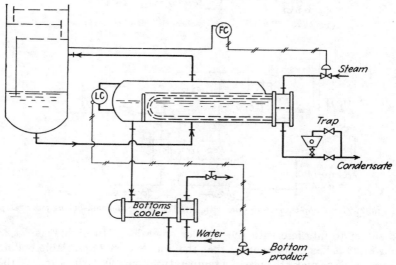

Fig. 21.26. Kettle-type reboiler.

pressure drop so as not to reduce the compressor suction or unnecessarily increase the power cost. Figure 21.26 shows the instrumentation for the kettle reboiler. The functions of the level control are evident. The horizontal thermosyphon is shown in Fig. 21.27 with once-through flow as compared with a recirculating arrangement. The arrangements of

Figs. 21.26 and 21.27 may also serve as the basis for the instrumentation of horizontal and vertical evaporators.

Batch Processes. Batch processes are usually very simple to control.[1] Most batch processes do not require automatic control of any sort.

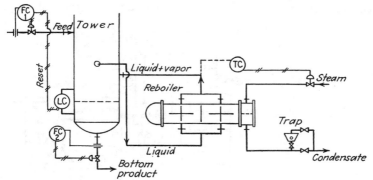

FIG. 21.27. Thermosyphon reboiler.

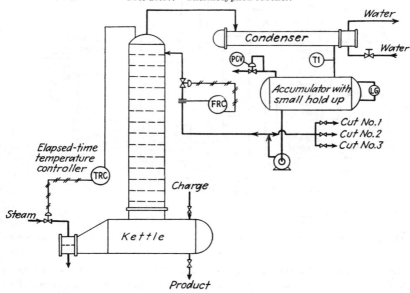

FIG. 21.28. Batch distillation with elapsed time-temperature controller.

Batch distillation presents several interesting problems which are shown in Fig. 21.28. During batch distillation the still and possibly the over-

[1] See particularly Perry, J. H., "Chemical Engineers' Handbook," 3d ed., McGraw-Hill Book Company, Inc., New York, 1950.

788 PROCESS HEAT TRANSFER

head temperatures change continuously as material is taken off overhead, and the related problem is then the control of temperature and pressure. Instead of using a temperature controller, an elapsed time-temperature controller or program controller is used. It sets a time period for the distillation temperatures and throttles the steam while the still tempera-

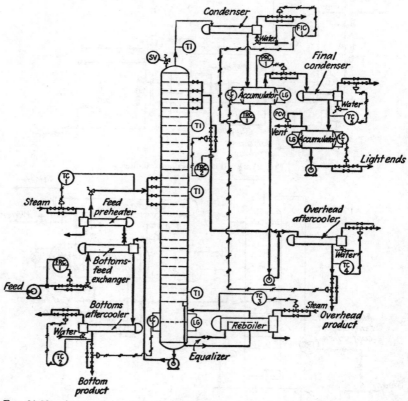

Fig. 21.29. Automatic control of a distillation process with feed containing some noncondensable gas.

ture rises through a timed cycle of temperature rises. The pressure control is maintained simply by a back-pressure regulator (pressure-control valve) on the assumption that there is some small amount of volatiles or noncondensables. The reflux is flow-controlled during the entire operation.

Continuous Distillation. Figure 21.29 shows the instrumentation for a continuous distillation charged with a feed having noncondensables.

With the exception of steam distillation which closely resembles it, other types of feed permit the instrumentation to be simplified. Included in Fig. 21.29 are the by-passes about the instrument controllers and the safety or relief valves on the exchangers. Safety valves are of the expansion-spring type. If it is possible in any way for the flow of the cold liquid to be halted because a valve has been accidentally turned off or the instrument control fails, a safety valve is placed on the inlet side of the cold fluid. It permits the expansion of the halted cold liquid to be relieved inasmuch as the hot fluid may continue to flow and prevents destructively large pressures from being built up in the liquid lines. After the small amount of liquid expansion has been relieved, the valve reseats itself. Attention should also be focused on the hook-up of the reboiler. Instead of the overflow going to the product outlet, as in the recirculating hook-up, the overflow is used to equalize the liquid level on a once-through arrangement. This is preferable in the event that the reboiler is overdesigned and more than the required amount o vaporization may occur. The temperature indicators shown in the column are multiple thermocouples connected usually to a single indicating instrument. The instruments which would make a convenient panel board for the control of the process are shown by a horizontal line for board mounting.

Conclusion. The elements of process control presented here have been among the simplest. The steppingstones by which an overall modern process can be controlled involve far more than the combination of several single effects. For these complex applications the many compound instruments of particular manufacturers are required to amplify the measurements or surpress the lags resulting from a large number of controls.

APPENDIX OF CALCULATION DATA

CONTENTS

Table 1.	Conversion factors	793

Thermal Conductivities[1]

Table 2.	Thermal conductivities of common materials	795
Table 3.	Thermal conductivities of metals	799
Table 4.	Thermal conductivities of liquids	800
Table 5.	Thermal conductivities of gases and vapors	801
Fig 1.	Thermal conductivities of hydrocarbon liquids	803

Specific Heats[1]

Table 3.	Specific heats of metals	799
Fig. 2.	Specific heats of liquids	804
Fig. 3.	Specific heats of gases at 1 atm	805
Fig. 4.	Specific heats of hydrocarbon liquids	806
Fig. 5.	Specific heats of hydrocarbon gases and vapors	807

Specific Gravities[1]

Table 3.	Specific gravities of metals	799
Table 6.	Specific gravities and molecular weights of liquids	808
Fig. 6.	Specific gravities of hydrocarbons	809

Equilibrium Data[2]

Fig. 7.	Equilibrium constants for hydrocarbons	810
Fig. 8.	Vapor pressures of hydrocarbons	811

Enthalpies and Latent Heats

Fig. 9.	Enthalpies of pure hydrocarbons	812
Fig. 10.	Enthalpies of light hydrocarbons	813
Fig. 11.	Enthalpies of petroleum fractions	814
Fig. 12.	Latent heats of vaporization	815
Table 7.	Thermodynamic properties of steam	816

Viscosities[1]

Fig. 13a.	Viscosity conversion chart	820
Fig. 13b.	Viscosity correction chart for gases	820
Fig. 14.	Viscosities of liquids	823
Fig. 15.	Viscosities of gases and vapors	825
Fig. 16.	Values of $k(c\mu/k)^{1/3}$ for hydrocarbons	826

[1] For aqueous and organic solutions see page 161 on which is given approximate formulas.

[2] See also Fig. 13.14.

Temperature Difference Corrections[1]

Fig. 17.	Caloric temperature factor.	827
Fig. 18.	LMTD correction factors for 1-2 exchangers	828
Fig. 19.	LMTD correction factors for 2-4 exchangers	829
Fig. 20.	LMTD correction factors for 3-6 exchangers	830
Fig. 21.	LMTD correction factors for 4-8 exchangers	831
Fig. 22.	LMTD correction factors for 5-10 exchangers.	832
Fig. 23.	LMTD correction factors for 6-12 exchangers.	833

Heat-transfer Data[2]

Fig. 24.	Tube-side heat-transfer curve.	834
Fig. 25.	Tube-side water-heat-transfer curve.	835
Fig. 26.	Tube-side fraction factors.	836
Fig. 27.	Tube-side return-pressure losses.	837
Fig. 28.	Shell-side heat-transfer curve.	838
Fig. 29.	Shell-side fraction factors.	839
Table 8.	Approximate overall heat-transfer coefficients.	840

Layout Data

Table 9.	Tube-sheet layouts (tube counts).	841
Table 10.	Heat-exchanger and condenser-tube data.	843
Table 11.	Steel-pipe dimensions (IPS).	844

Fouling Factors

Table 12.	Fouling factors.	845

[1] See Index for corrections.
[2] See Index for other data.

TABLE 1. CONVERSION FACTORS AND CONSTANTS

Energy and power:

Btu = 0.252 kg-cal
Btu = 0.293 watt-hr
Btu = 0.555 pcu (pound centigrade unit)
Btu = 778 ft-lb
Btu/min = 0.236 hp
Hp = 42.4 Btu/min
Hp = 33,000 ft-lb/min
Hp = 0.7457 kw
Hp-hr = 2543 Btu
Kw = 1.3415 hp
Watt-hr = 3.415 Btu

Fluid flow:

Bbl/hr = 0.0936 cfm
Bbl/hr = 0.700 gpm
Bbl/day = 0.0292 gpm
Bbl/day = 0.0039 cfm
Cfm = 10.686 bbl/hr
Gpm = 1.429 bbl/hr
Gpm = 34.3 bbl/day
Gpm $\times s$ (specific gravity) = $500 \times s$ lb/hr

Heat-transfer coefficients:

Btu/(hr)(ft^2)(°F) = 1.0 pcu/(hr)(ft^2)(°C)
Btu/(hr)(ft^2)(°F) = 4.88 kg-cal/(hr)(m^2)(°C)
Btu/(hr)(ft^2)(°F) = 0.00204 watts/(in.^{2})(°F)

Length, area, and volume:

Bbl = 42 gal
Bbl = 5.615 ft^3
Cm = 0.3937 in.
Ft^3 = 0.1781 bbl
Ft^3 = 7.48 gal
Ft^3 = 0.0283 m^3
M^3 = 6.290 bbl
M^3 = 35.314 ft^3
Ft = 30.48 cm
Ft = 0.3048 m
Gal = 0.02381 bbl
Gal = 0.1337 ft^3
Gal = 3.785 liter
Gal = 0.8327 gal (Imperial)
In. = 2.54 cm
Liter = 0.2642 gal
Liter = 1.0567 qt
M = 3.281 ft
Ft^2 = 0.0929 m^2
M^2 = 10.76 ft^2

Pressure:

Atm = 33.93 ft of water at 60°F
Atm = 29.92 in. Hg at 32°F
Atm = 760 mm Hg at 32°F
Atm = 14.696 psi
Atm = 2116.8 lb/ft^2
Atm = 1.033 kg/cm^2
Ft of water at 60°F = 0.4331 psi
In. of water at 60°F = 0.0361 psi
Kg/cm^2 = 14.223 psi
Psi = 2.309 Ft of water at 60°F

Temperature:

Temperature °C = $\frac{5}{9}$(°F − 32)
Temperature °F = $\frac{9}{5}$(°C + 32)
Temperature °F absolute (°R) = °F + 460
Temperature °C absolute (°K) = °C + 273

Thermal conductivity:

Btu/(hr)(ft^2)(°F/ft) = 12 Btu/(hr)(ft^2)(°F/in.)
Btu/(hr)(ft^2)(°F/ft) = 1.49 kg-cal/(hr)(m^2)(°C/m)
Btu/(hr)(ft^2)(°F/ft) = 0.0173 watts/(cm^2)(°C/cm)

Viscosity (additional factors are contained in Fig. 13):

Poise = 1 g/(cm)(sec)
Centipoise = 0.01 poise
Centipoise = 2.42 lb/(ft)(hr)

Weight:

Lb = 0.4536 kg
Lb = 7000 grains
Ton (short or net) = 2000 lb
Ton (long) = 2240 lb
Ton (metric) = 2205 lb
Ton (metric) = 1000 kg

Constants:

Acceleration of gravity = 32.2 ft/sec^2
Acceleration of gravity = 4.18 × 10^8 ft/hr^2
Density of a cubic foot of water = 62.5 lb/ft^3

APPENDIX OF CALCULATION DATA

TABLE 2. THERMAL CONDUCTIVITIES OF SOME BUILDING AND INSULATING
MATERIALS*
$k = \text{Btu}/(\text{hr})(\text{ft}^2)(°\text{F}/\text{ft})$

Material	Apparent density ρ, lb/ft^3 at room temperature	°F	k
Aerogel, silica, opacified...................	8.5	248	0.013
		554	0.026
Asbestos-cement boards....................	120	68	0.43
Asbestos sheets...........................	55.5	124	0.096
Asbestos slate............................	112	32	0.087
	112	140	0.114
Asbestos.................................	29.3	−328	0.043
	29.3	32	0.090
	36	32	0.087
	36	212	0.111
	36	392	0.120
	36	752	0.129
	43.5	−328	0.090
	43.5	32	0.135
Aluminum foil, 7 air spaces per 2.5 in........	0.2	100	0.025
		351	0.038
Ashes, wood..............................		32–212	0.041
Asphalt..................................	132	68	0.43
Boiler scale (ref. 364).....................			
Bricks			
Alumina (92–99% Al$_2$O$_3$ by weight) fused...		801	1.8
Alumina (64–65% Al$_2$O$_3$ by weight)........		2399	2.7
(See also Bricks, fire clay)................	115	1472	0.62
	115	2012	0.63
Building brickwork.......................		68	0.4
Chrome brick (32% Cr$_2$O$_3$ by weight)......	200	392	0.67
	200	1202	0.85
	200	2399	1.0
Diatomaceous earth, natural, across strata	27.7	399	0.051
	27.7	1600	0.077
Diatomaceous, natural, parallel to strata	27.7	399	0.081
	27.7	1600	0.106
Diatomaceous earth, molded and fired.....	38	399	0.14
	38	1600	0.18
Diatomaceous earth and clay, molded and fired................................	42.3	399	0.14
	42.3	1600	0.19
Diatomaceous earth, high burn, large pores	37	392	0.13
	37	1832	0.34

TABLE 2. THERMAL CONDUCTIVITIES OF SOME BUILDING AND INSULATING MATERIALS.*—(Continued)

Material	Apparent density ρ, lb/ft^3 at room temperature	°F	k
Bricks: (Continued)			
Fire clay, Missouri....................	......	392	0.58
		1112	0.85
		1832	0.95
		2552	1.02
Kaolin insulating brick	27	932	0.15
	27	2102	0.26
Kaolin insulating firebrick	19	392	0.050
	19	1400	0.113
Magnesite (86.8% MgO, 6.3% Fe$_2$O$_3$, 3% CaO, 2.6% SiO$_2$ by weight)............	158	399	2.2
	158	1202	1.6
	158	2192	1.1
Silicon carbide brick, recrystallized	129	1112	10.7
	129	1472	9.2
	129	1832	8.0
	129	2192	7.0
	129	2552	6.3
Calcium carbonate, natural................	162	86	1.3
White marble........................	......	......	1.7
Chalk.............................	96	......	0.4
Calcium sulphate (4H$_2$O), artificial..........	84.6	104	0.22
Plaster, artificial......................	132	167	0.43
Building............................	77.9	77	0.25
Cambric, varnished....................	......	100	0.09
Carbon, gas..........................	......	32–212	2.0
Cardboard, corrugated.................	......	......	0.037
Celluloid............................	87.3	86	0.12
Charcoal flakes.......................	11.9	176	0.043
	15	176	0.051
Clinker, granular.....................	......	32–1292	0.27
Coke, petroleum......................	......	212	3.4
		932	2.9
Coke, powdered.......................	......	32–212	0.11
Concrete, cinder.......................	......	......	0.20
1:4 dry............................	......	......	0.44
Stone.............................	......	......	0.54
Cotton wool.........................	5	86	0.024
Cork board..........................	10	86	0.025
Cork, ground........................	9.4	86	0.025
Regranulated......................	8.1	86	0.026

TABLE 2. THERMAL CONDUCTIVITIES OF SOME BUILDING AND INSULATING MATERIALS.*—(*Continued*)

Material	Apparent density ρ, lb/ft^3 at room temperature	°F	k
Diatomaceous earth powder, coarse	20.0	100	0.036
	20.0	1600	0.082
Fine	17.2	399	0.040
	17.2	1600	0.074
Molded pipe covering	26.0	399	0.051
	26.0	1600	0.088
4 vol. calcined earth and 1 vol. cement, poured and fired	61.8	399	0.16
	61.8	1600	0.23
Dolomite	167	122	1.0
Ebonite			0.10
Enamel, silicate	38		0.5–0.75
Felt, wool	20.6	86	0.03
Fiber insulating board	14.8	70	0.028
Fiber, red	80.5	68	0.27
With binder, baked		68–207	0.097
Gas carbon		32–212	2.0
Glass			0.2–0.73
Boro-silicate type	139	86–167	0.63
Soda glass			0.3–0.44
Window glass			0.3–0.61
Granite			1.0–2.3
Graphite, dense, commercial		32	86.7
Powdered, through 100 mesh	30	104	0.104
Gypsum, molded and dry	78	68	0.25
Hair, felt, perpendicular to fibers	17	86	0.021
Ice	57.5	32	1.3
Infusorial earth (see Diatomaceous earth)			
Kapok	0.88	68	0.020
Lampblack	10	104	0.038
Lava			0.49
Leather, sole	62.4		0.092
Limestone (15.3 vol % H$_2$O)	103	75	0.54
Linen		86	0.05
Magnesia, powdered	49.7	117	0.35
Magnesia, light carbonate	19	70	0.04
Magnesium oxide, compressed	49.9	68	0.32
Marble			1.2–1.7

TABLE 2. THERMAL CONDUCTIVITIES OF SOME BUILDING AND INSULATING MATERIALS.*—(Continued)

Material	Apparent density ρ, lb/ft³ at room temperature	°F	k
Mica, perpendicular to planes		122	0.25
Mill shavings			0.033–0.05
Mineral wool	9.4	86	0.0225
	19.7	86	0.024
Paper			0.075
Paraffin wax		32	0.14
Petroleum coke		212	3.4
		932	2.9
Porcelain		392	0.88
Portland cement (see Concrete)		194	0.17
Pumice stone		70–151	0.14
Pyroxylin plastics			0.075
Rubber, hard	74.8	32	0.087
Para		70	0.109
Soft		70	0.075–0.092
Sand, dry	94.6	68	0.19
Sandstone	140	104	1.06
Sawdust	12	70	0.03
Scale (ref. 364)			
Silk	6.3		0.026
Varnished		100	0.096
Slag, blast furnace		75–261	0.064
Slag wool	12	86	0.022
Slate		201	0.86
Snow	34.7	32	0.27
Sulphur, monoclinic		212	0.09–0.097
Rhombic		70	0.16
Wallboard, insulating type	14.8	70	0.028
Wallboard, stiff pasteboard	43	86	0.04
Wood shavings	8.8	86	0.034
Wood, across grain			
Balsa	7–8	86	0.025–0.03
Oak	51.5	59	0.12
Maple	44.7	122	0.11
Pine, white	34.0	59	0.087
Teak	40.0	59	0.10
White fir	28.1	140	0.062
Wood, parallel to grain			
Pine	34.4	70	0.20
Wool, animal	6.9	86	0.021

* From L. S. Marks, "Mechanical Engineers' Handbook," McGraw-Hill Book Company, Inc., New York. 1941.

TABLE 3. THERMAL CONDUCTIVITIES, SPECIFIC HEATS, SPECIFIC GRAVITIES OF METALS AND ALLOYS

$k = \text{Btu}/(\text{hr})(\text{ft}^2)(°\text{F}/\text{ft})$

Substance	Temp, °F	k^*	Specific heat,† Btu/(lb)(°F)	Specific gravity
Aluminum	32	117	0.183	2.55–7.8
Aluminum	212	119	0.1824	
Aluminum	932	155	0.1872	
Antimony	32	10.6	0.0493	
Antimony	212	9.7	0.0508	
Bismuth	64	4.7	0.0294	9.8
Bismuth	212	3.9	0.0304	
Brass (70-30)	32	56	0.1315‡	8.4–8.7
Brass	212	60	0.1488‡	
Brass	752	67	0.2015‡	
Copper	32	224	0.1487	8.8–8.95
Copper	212	218	0.1712	
Copper	932	207	0.2634	
Cadmium	64	53.7	0.0550	8.65
Cadmium	212	52.2	0.0567	
Gold	64	169.0	0.030	19.25–19.35
Gold	212	170.8	0.031	
Iron, cast	32	32	0.1064	7.03–7.13
Iron, cast	212	30	0.1178	
Iron, cast	752	25	0.1519	
Iron, wrought	64	34.6	See Iron	7.6–7.9
Iron, wrought	212	27.6	See Iron	
Lead	32	20	0.0306	11.34
Lead	212	19	0.0315	
Lead	572	18	0.0335	
Magnesium	32–212	92	0.255	1.74
Mercury	32	4.8	0.0329	13.6
Nickel	32	36	0.1050	8.9
Nickel	212	34	0.1170	
Nickel	572	32	0.1408	
Silver	32	242	0.0557	10.4–10.6
Silver	212	238	0.0571	
Steel	32	26	See Iron	7.83
Steel	212	26	See Iron	
Steel	1112	21	See Iron	
Tantalum	64	32	0.0342	16.6
Zinc	32	65	0.0917	6.9–7.2
Zinc	212	64	0.0958	
Zinc	752	54	0.1082	

* From L. S. Marks, "Mechanical Engineers' Handbook," McGraw-Hill Book Company, Inc., New York, 1941.

† From K. K. Kelley, *U.S. Bur. Mine Bull.* 371 (1939).

‡ Weighted value for copper and zinc.

Table 4. Thermal Conductivities of Liquids*

$k = \text{Btu}/(\text{hr})(\text{ft}^2)(°\text{F}/\text{ft})$

A linear variation with temperature may be assumed. The extreme values given constitute also the temperature limits over which the data are recommended.

Liquid	°F	k	Liquid	°F	k
Acetic acid 100%	68	0.099	Heptyl alcohol (n-)	86	0.094
50%	68	0.20		167	0.091
Acetone	86	0.102	Hexyl alcohol (n-)	86	0.093
	167	0.095		167	0.090
Allyl alcohol	77–86	0.104			
Ammonia	5–86	0.29	Kerosene	68	0.086
Ammonia, aqueous 26%	68	0.261		167	0.081
	140	0.29	Lauric acid	212	0.102
Amyl acetate	50	0.083			
Alcohol (n-)	86	0.094	Mercury	82	4.83
	212	0.089	Methyl alcohol 100%	68	0.124
	86	0.088	80%	68	0.154
	167	0.087	60%	68	0.190
Aniline	32–68	0.100	40%	68	0.234
			20%	68	0.284
Benzene	86	0.092	100%	122	0.114
	140	0.087	Chloride	5	0.111
Bromobenzene	86	0.074		86	0.089
	212	0.070			
Butyl acetate (n-)	77–86	0.085	Nitrobenzene	86	0.095
Alcohol (n-)	86	0.097		212	0.088
	167	0.095	Nitromethane	86	0.125
(iso-)	50	0.091		140	0.120
Calcium chloride brine 30%	86	0.32	Nonane (n-)	86	0.084
15%	86	0.34		140	0.082
Carbon disulphide	86	0.093	Octane (n-)	86	0.083
	167	0.088		140	0.081
Tetrachloride	32	0.107	Oils		
	154	0.094	Castor	68	0.104
Chlorobenzene	50	0.083		212	0.100
Chloroform	86	0.080	Olive	68	0.097
Cymene (para)	86	0.078		212	0.095
	140	0.079	Oleic acid	212	0.0925
Decane (n-)	86	0.085	Palmitic acid	212	0.0835
	140	0.083	Paraldehyde	86	0.084
Dichlorodifluoromethane	20	0.057		212	0.078
	60	0.053	Pentane (n-)	86	0.078
	100	0.048		167	0.074
	140	0.043	Perchloroethylene	122	0.092
	180	0.038	Petroleum ether	86	0.075
Dichloroethane	122	0.082		167	0.073
Dichloromethane	5	0.111	Propyl alcohol (n-)	86	0.099
	86	0.096		167	0.095
Ethyl acetate	68	0.101	Alcohol (iso-)	86	0.091
Alcohol 100%	68	0.105		140	0.090
80%	68	0.137			
60%	68	0.176	Sodium	212	49
40%	68	0.224		410	46
20%	68	0.281	Sodium chloride brine 25.0%	86	0.33
100%	122	0.087	12.5%	86	0.34
Benzene	86	0.086	Stearic acid	212	0.0786
	140	0.082	Sulfuric acid 90%	86	0.21
Bromide	68	0.070	60%	86	0.25
Ether	86	0.080	30%	86	0.30
	167	0.078	Sulfur dioxide	5	0.128
Iodide	104	0.064		86	0.111
	167	0.063	Toluene	86	0.086
Ethylene glycol	32	0.153		167	0.084
			β-trichloroethane	122	0.077
Gasoline	86	0.078	Trichloroethylene	122	0.080
Glycerol 100%	68	0.164	Turpentine	59	0.074
80%	68	0.189			
60%	68	0.220	Vaseline	59	0.106
40%	68	0.259			
20%	68	0.278	Water	32	0.330
100%	212	0.164		86	0.356
				140	0.381
Heptane (n-)	86	0.081		176	0.398
	140	0.079			
Hexane (n-)	86	0.080	Xylene (ortho-)	68	0.090
	140	0.078	(meta-)	68	0.090

*From Perry, J. H., "Chemical Engineers' Handbook," 3d ed., McGraw-Hill Book Company, Inc., New York, 1950.

APPENDIX OF CALCULATION DATA

TABLE 5. THERMAL CONDUCTIVITIES OF GASES AND VAPORS*
$$k = \text{Btu}/(\text{hr})(\text{ft}^2)(°F/\text{ft})$$

The extreme temperature values given constitute the experimental range. For extrapolation to other temperatures, it is suggested that the data given be plotted as $\log k$ vs. $\log T$ or that use be made of the assumption that the ratio c_μ/k is practically independent of temperature (or of pressure, within moderate limits).

Substance	°F	k	Substance	°F	k
Acetone	32	0.0057	Dichlorodifluoromethane	32	0.0048
	115	0.0074		122	0.0064
	212	0.0099		212	0.0080
	363	0.0147		302	0.0097
Acetylene	−103	0.0068	Ethane	−94	0.0066
	32	0.0108		−29	0.0086
	122	0.0140		32	0.0106
	212	0.0172		212	0.0175
Air	−148	0.0095	Ethyl acetate	115	0.0072
	32	0.0140		212	0.0096
	212	0.0183		363	0.0141
	392	0.0226	Alcohol	68	0.0089
	572	0.0265		212	0.0124
Ammonia	−76	0.0095	Chloride	32	0.0055
	32	0.0128		212	0.0095
	122	0.0157		363	0.0135
	212	0.0185		413	0.0152
			Ether	32	0.0077
Benzene	32	0.0052		115	0.0099
	115	0.0073		212	0.0131
	212	0.0103		363	0.0189
	363	0.0152		413	0.0209
	413	0.0176	Ethylene	−96	0.0064
Butane (n-)	32	0.0078		32	0.0101
	212	0.0135		122	0.0131
(iso-)	32	0.0080		212	0.0161
	212	0.0139			
Carbon dioxide	−58	0.0068	Heptane (n-)	392	0.0112
	32	0.0085		212	0.0103
	212	0.0133	Hexane (n-)	32	0.0072
	392	0.0181		68	0.0080
	572	0.0228	Hexene	32	0.0061
Disulphide	32	0.0040		212	0.0109
	45	0.0042	Hydrogen	−148	0.065
Monoxide	−312	0.0041		−58	0.083
	−294	0.0046		32	0.100
	32	0.0135		122	0.115
Tetrachloride	115	0.0041		212	0.129
	212	0.0052		572	0.178
	363	0.0065	Hydrogen and carbon dioxide	32	
Chlorine	32	0.0043	0% H₂	.....	0.0083
Chloroform	32	0.0038	20%	.....	0.0165
	115	0.0046	40%	.....	0.0270
	212	0.0058	60%	.....	0.0410
	363	0.0077	80%	.....	0.0620
Cyclohexane	216	0.0095	100%	.....	0.10

TABLE 5. THERMAL CONDUCTIVITIES OF GASES AND VAPORS.*—(Continued)

Substance	°F	k	Substance	°F	k
Hydrogen and nitrogen	32		Nitric oxide	−94	0.0103
0% H$_2$		0.0133		32	0.0138
20%		0.0212	Nitrogen	−148	0.0095
40%		0.0313		32	0.0140
60%		0.0438		122	0.0160
80%		0.0635		212	0.0180
Hydrogen and nitrous oxide	32		Nitrous oxide	−98	0.0067
0% H$_2$		0.0002		32	0.0087
20%		0.0170		212	0.0128
40%		0.0270			
60%		0.0410	Oxygen	−148	0.0095
80%		0.0650		−58	0.0119
Hydrogen sulphide	32	0.0076		32	0.0142
				122	0.0164
Mercury	392	0.0197		212	0.0185
Methane	−148	0.0100			
	−58	0.0145	Pentane (n-)	32	0.0074
	32	0.0175		68	0.0083
	122	0.0215	(iso-)	32	0.0072
Methyl alcohol	32	0.0083		212	0.0127
	212	0.0128	Propane	32	0.0087
Acetate	32	0.0059		212	0.0151
	68	0.0068			
Methyl chloride	32	0.0053	Sulphur dioxide	32	0.0050
	115	0.0072		212	0.0069
	212	0.0094			
	363	0.0130	Water vapor	115	0.0120
	413	0.0148		212	0.0137
Methylene chloride	32	0.0039		392	0.0187
	115	0.0049		572	0.0248
	212	0.0063		752	0.0315
	413	0.0095		932	0.0441

* From Perry, J. H., "Chemical Engineers' Handbook," 3d ed., McGraw-Hill Book Company, Inc., New York, 1950.

APPENDIX OF CALCULATION DATA

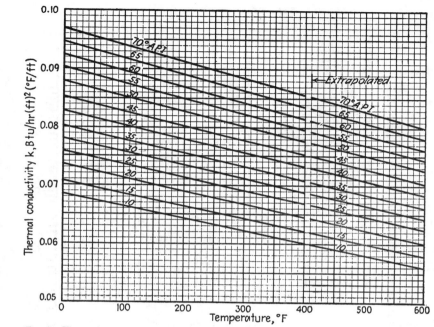

FIG. 1. Thermal conductivities of hydrocarbon liquids. (*Adapted from Natl. Bur. Standards Misc. Pub. 97.*)

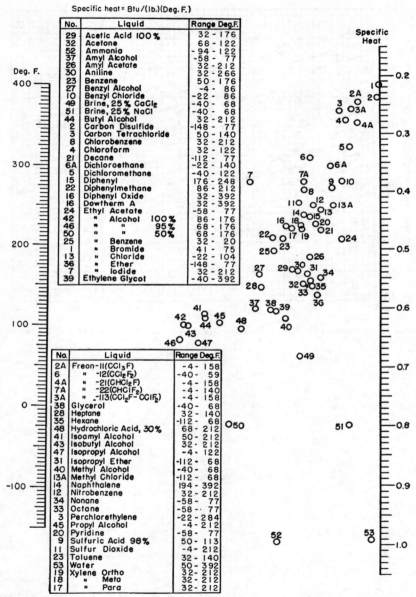

Fig. 2. Specific heats of liquids. (*Chilton, Colburn, and Vernon, based mainly on data from International Critical Tables. Perry, "Chemical Engineers' Handbook," 3d ed., McGraw-Hill Book Company, Inc., New York, 1950.*)

APPENDIX OF CALCULATION DATA

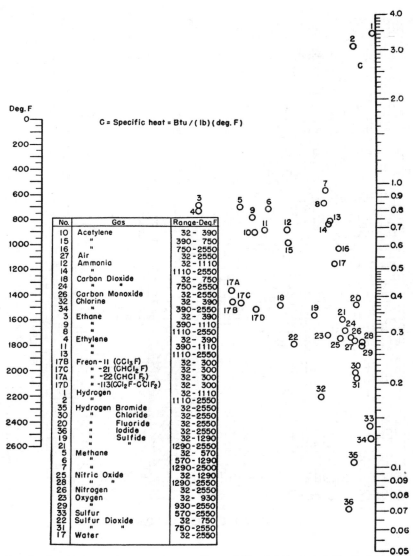

Fig. 3. Specific heats of gases at 1 atm. (Perry, "*Chemical Engineers' Handbook,*" 3d ed., McGraw-Hill Book Company, Inc., New York, 1950.)

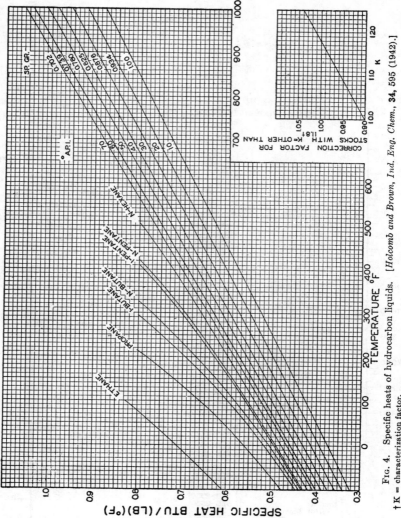

Fig. 4. Specific heats of hydrocarbon liquids. [*Holcomb and Brown, Ind. Eng. Chem.,* **34,** 595 (1942).]

† K = characterization factor.

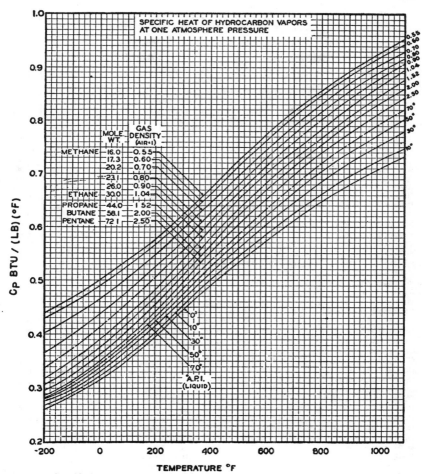

Fig. 5. Specific heats of hydrocarbon vapors at 1 atm. [*Holcomb and Brown, Ind. Eng. Chem.*, **34**, 595 (1942).]

Table 6. Specific Gravities and Molecular Weights of Liquids

Compound	Mol. wt.	s*	Compound	Mol. wt.	s*
Acetaldehyde	44.1	0.78	Ethyl iodide	155.9	1.93
Acetic acid, 100%	60.1	1.05	Ethyl glycol	88.1	1.04
Acetic acid, 70%		1.07	Formic acid	46.0	1.22
Acetic anhydride	102.1	1.08	Glycerol, 100%	92.1	1.26
Acetone	58.1	0.79	Glycerol, 50%		1.13
Allyl alcohol	58.1	0.86	n-Heptane	100.2	0.68
Ammonia, 100%	17.0	0.61	n-Hexane	86.1	0.66
Ammonia, 26%		0.91	Isopropyl alcohol	60.1	0.79
Amyl acetate	130.2	0.88	Mercury	200.6	13.55
Amyl alcohol	88.2	0.81	Methanol, 100%	32.5	0.79
Aniline	93.1	1.02	Methanol, 90%		0.82
Anisole	108.1	0.99	Methanol, 40%		0.94
Arsenic trichloride	181.3	2.16	Methyl acetate	74.9	0.93
Benzene	78.1	0.88	Methyl chloride	50.5	0.92
Brine, CaCl₂ 25%		1.23	Methyl ethyl ketone	72.1	0.81
Brine, NaCl 25%		1.19	Naphthalene	128.1	1.14
Bromotoluene, ortho	171.0	1.42	Nitric acid, 95%		1.50
Bromotoluene, meta	171.0	1.41	Nitric acid, 60%		1.38
Bromotoluene, para	171.0	1.39	Nitrobenzene	123.1	1.20
n-Butane	58.1	0.60	Nitrotoluene, ortho	137.1	1.16
i-Butane	58.1	0.60	Nitrotoluene, meta	137.1	1.16
Butyl acetate	116.2	0.88	Nitrotoluene, para	137.1	1.29
n-Butyl alcohol	74.1	0.81	n-Octane	114.2	0.70
i-Butyl alcohol	74.1	0.82	Octyl alcohol	130.23	0.82
n-Butyric acid	88.1	0.96	Pentachloroethane	202.3	1.67
i-Butyric acid	88.1	0.96	n-Pentane	72.1	0.63
Carbon dioxide	44.0	1.29	Phenol	94.1	1.07
Carbon disulfide	76.1	1.26	Phosphorus tribromide	270.8	2.85
Carbon tetrachloride	153.8	1.60	Phosphorus trichloride	137.4	1.57
Chlorobenzene	112.6	1.11	Propane	44.1	0.59
Chloroform	119.4	1.49	Propionic acid	74.1	0.99
Chlorosulfonic acid	116.5	1.77	n-Propyl alcohol	60.1	0.80
Chlorotoluene, ortho	126.6	1.08	n-Propyl bromide	123.0	1.35
Chlorotoluene, meta	126.6	1.07	n-Propyl chloride	78.5	0.89
Chlorotoluene, para	126.6	1.07	n-Propyl iodide	170.0	1.75
Cresol, meta	108.1	1.03	Sodium	23.0	0.97
Cyclohexanol	100.2	0.96	Sodium hydroxide, 50%		1.53
Dibromo methane	187.9	2.09	Stannic chloride	260.5	2.23
Dichloro ethane	99.0	1.17	Sulfur dioxide	64.1	1.38
Dichloro methane	88.9	1.34	Sulfuric acid, 100%	98.1	1.83
Diethyl oxalate	146.1	1.08	Sulfuric acid, 98%		1.84
Dimethyl oxalate	118.1	1.42	Sulfuric acid, 60%		1.50
Diphenyl	154.2	0.99	Sulfuryl chloride	135.0	1.67
Dipropyl oxalate	174.1	1.02	Tetra chloroethane	167.9	1.60
Ethyl acetate	88.1	0.90	Tetra chloroethylene	165.9	1.63
Ethyl alcohol, 100%	46.1	0.79	Titanium tetrachloride	189.7	1.73
Ethyl alcohol, 95%		0.81	Toluene	92.1	0.87
Ethyl alcohol, 40%		0.94	Trichloroethylene	131.4	1.46
Ethyl benzene	106.1	0.87	Vinyl acetate	86.1	0.93
Ethyl bromide	108.9	1.43	Water	18.0	1.0
Ethyl chloride	64.5	0.92	Xylene, ortho	106.1	0.87
Ethyl ether	74.1	0.71	Xylene, meta	106.1	0.86
Ethyl formate	74.1	0.92	Xylene, para	106.1	0.86

* At approximately 68°F. These values will be satisfactory, without extrapolation, for most engineering problems.

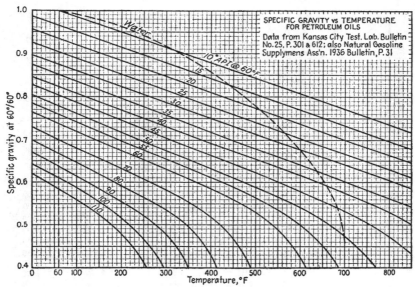

Fig. 6. Specific gravities of hydrocarbons.

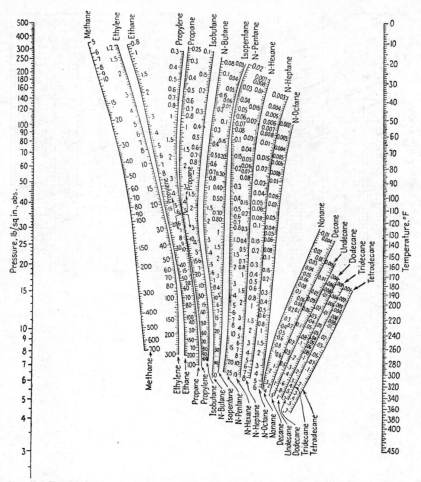

FIG. 7. Equilibrium constants for hydrocarbons. [*Scheibel and Jenny, Ind. Eng. Chem.*, **37**, 81 (1945).]

APPENDIX OF CALCULATION DATA

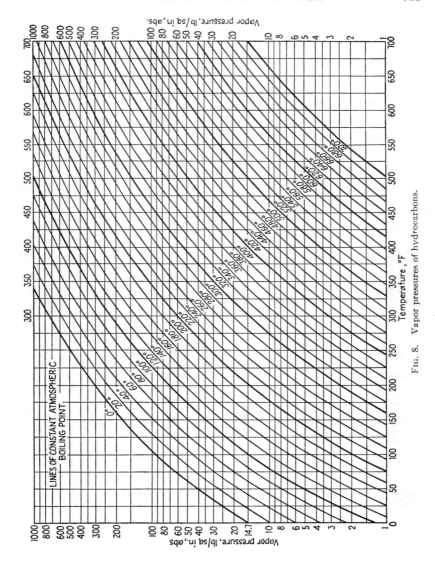

Fig. 8. Vapor pressures of hydrocarbons.

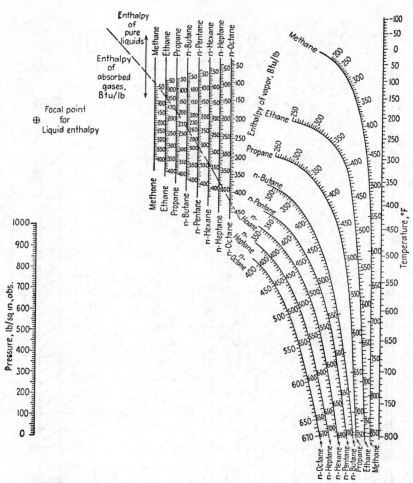

FIG. 9. Enthalpies of pure hydrocarbons. [*Scheibel and Jenny, Ind. Eng. Chem.*, **37**, 992 (1945).]

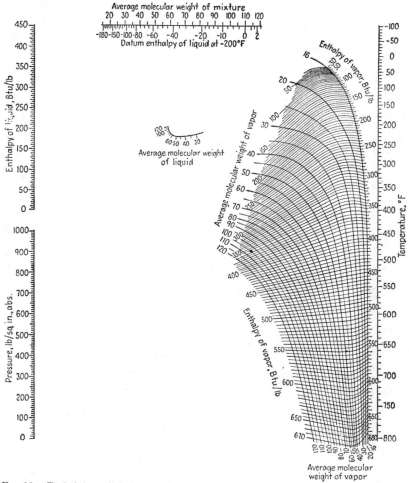

Fig. 10. Enthalpies of light hydrocarbons. [Scheibel and Jenny, Ind. Eng. Chem., **37**, 993 (1945).]

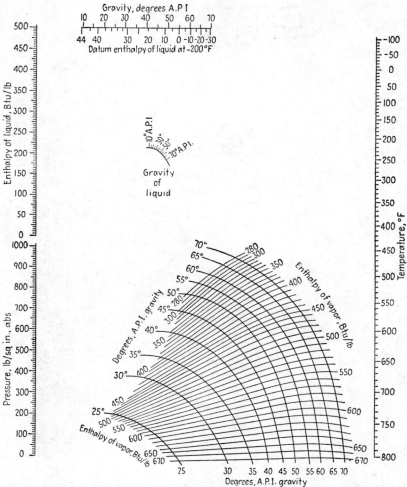

FIG. 11. Enthalpies of petroleum fractions. [*Scheibel and Jenny, Ind. Eng. Chem.*, **37**, 994 (1945).]

APPENDIX OF CALCULATION DATA

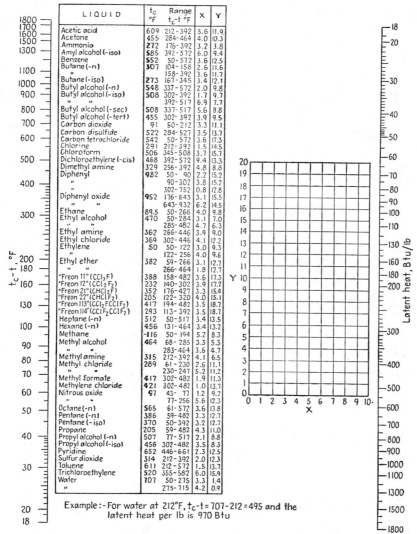

Fig. 12. Latent heats of vaporization. [*Reproduced by permission of Chilton, Colburn, and Vernon, personal communication (revised) 1947.*]

Example: — For water at 212°F, $t_c - t = 707 - 212 = 495$ and the latent heat per lb is 970 Btu

TABLE 7. THERMODYNAMIC PROPERTIES OF STEAM
Dry saturated steam: pressure table*

Abs press, psi	Temp, °F	Specific volume		Enthalpy			Entropy			Internal energy		Abs press., psi
		Sat. liquid	Sat. vapor	Sat. liquid	Evap	Sat. vapor	Sat. liquid	Evap	Sat. vapor	Sat. liquid	Sat. vapor	
p	t	v_f	v_g	h_f	h_{fg}	h_g	s_f	s_{fg}	s_g	u_f	u_g	p
1.0	101.74	0.01614	333.6	69.70	1036.3	1106.0	0.1326	1.8456	1.9782	69.70	1044.3	1.0
2.0	126.08	0.01623	173.73	93.99	1022.2	1116.2	0.1749	1.7451	1.9200	93.98	1051.9	2.0
3.0	141.48	0.01630	118.71	109.37	1013.2	1122.6	0.2008	1.6855	1.8863	109.36	1056.7	3.0
4.0	152.97	0.01636	90.63	120.86	1006.4	1127.3	0.2198	1.6427	1.8625	120.85	1060.2	4.0
5.0	162.24	0.01640	73.52	130.13	1001.0	1131.1	0.2347	1.6094	1.8441	130.12	1063.1	5.0
6.0	170.06	0.01645	61.98	137.96	996.2	1134.2	0.2472	1.5820	1.8292	137.94	1065.4	6.0
7.0	176.85	0.01649	53.64	144.76	992.1	1136.9	0.2581	1.5586	1.8167	144.74	1067.4	7.0
8.0	182.86	0.01653	47.34	150.79	988.5	1139.3	0.2674	1.5383	1.8057	150.77	1069.2	8.0
9.0	188.28	0.01656	42.40	156.22	985.2	1141.4	0.2759	1.5203	1.7962	156.19	1070.8	9.0
10	193.21	0.01659	38.42	161.17	982.1	1143.3	0.2835	1.5041	1.7876	161.14	1072.2	10
14.696	212.00	0.01672	26.80	180.07	970.3	1150.4	0.3120	1.4446	1.7566	180.02	1077.5	14.696
15	213.03	0.01672	26.29	181.11	969.7	1150.8	0.3135	1.4415	1.7549	181.06	1077.8	15
20	227.96	0.01683	20.089	196.16	960.1	1156.3	0.3356	1.3962	1.7319	196.10	1081.9	20
25	240.07	0.01692	16.303	208.42	952.1	1160.6	0.3533	1.3606	1.7139	208.34	1085.1	25
30	250.33	0.01701	13.746	218.82	945.3	1164.1	0.3680	1.3313	1.6993	218.73	1087.8	30
35	259.28	0.01708	11.898	227.91	939.2	1167.1	0.3807	1.3063	1.6870	227.80	1090.1	35
40	267.25	0.01715	10.498	236.03	933.7	1169.7	0.3919	1.2844	1.6763	235.90	1092.0	40
45	274.44	0.01721	9.401	243.36	928.6	1172.0	0.4019	1.2650	1.6669	243.22	1093.7	45
50	281.01	0.01727	8.515	250.09	924.0	1174.1	0.4110	1.2474	1.6585	249.93	1095.3	50
55	287.07	0.01732	7.787	256.30	919.6	1175.9	0.4193	1.2316	1.6509	256.12	1096.7	55
60	292.71	0.01738	7.175	262.09	915.5	1177.6	0.4270	1.2168	1.6438	261.90	1097.9	60
65	297.97	0.01743	6.655	267.50	911.6	1179.1	0.4342	1.2032	1.6374	267.29	1099.1	65
70	302.92	0.01748	6.206	272.61	907.9	1180.6	0.4409	1.1906	1.6315	272.38	1100.2	70
75	307.60	0.01753	5.816	277.43	904.5	1181.9	0.4472	1.1787	1.6259	277.19	1101.2	75
80	312.03	0.01757	5.472	282.02	901.1	1183.1	0.4531	1.1676	1.6207	281.76	1102.1	80
85	316.25	0.01761	5.168	286.39	897.8	1184.2	0.4587	1.1571	1.6158	286.11	1102.9	85
90	320.27	0.01766	4.896	290.56	894.7	1185.3	0.4641	1.1471	1.6112	290.27	1103.7	90
95	324.12	0.01770	4.652	294.56	891.7	1186.2	0.4692	1.1376	1.6068	294.25	1104.5	95
100	327.81	0.01774	4.432	298.40	888.8	1187.2	0.4740	1.1286	1.6026	298.08	1105.2	100
110	334.77	0.01782	4.049	305.66	883.2	1188.9	0.4832	1.1117	1.5948	305.30	1106.5	110
120	341.25	0.01789	3.728	312.44	877.9	1190.4	0.4916	1.0962	1.5878	312.05	1107.6	120
130	347.32	0.01796	3.455	318.81	872.9	1191.7	0.4995	1.0817	1.5812	318.38	1108.6	130
140	353.02	0.01802	3.220	324.82	868.2	1193.0	0.5069	1.0682	1.5751	324.35	1109.6	140
150	358.42	0.01809	3.015	330.51	863.6	1194.1	0.5138	1.0556	1.5694	330.01	1110.5	150
160	363.53	0.01815	2.834	335.93	859.2	1195.1	0.5204	1.0436	1.5640	335.39	1111.2	160
170	368.41	0.01822	2.675	341.09	854.9	1196.0	0.5266	1.0324	1.5590	340.52	1111.9	170
180	373.06	0.01827	2.532	346.03	850.8	1196.9	0.5325	1.0217	1.5542	345.42	1112.5	180
190	377.51	0.01833	2.404	350.79	846.8	1197.6	0.5381	1.0116	1.5497	350.15	1113.1	190
200	381.79	0.01839	2.288	355.36	843.0	1198.4	0.5435	1.0018	1.5453	354.68	1113.7	200
250	400.95	0.01865	1.8438	376.00	825.1	1201.1	0.5675	0.9588	1.5263	375.14	1115.8	250
300	417.33	0.01890	1.5433	393.84	809.0	1202.8	0.5879	0.9225	1.5104	392.79	1117.1	300
350	431.72	0.01913	1.3260	409.69	794.2	1203.9	0.6056	0.8910	1.4966	408.45	1118.0	350
400	444.59	0.0193	1.1613	424.0	780.5	1204.5	0.6214	0.8630	1.4844	422.6	1118.5	400
450	456.28	0.0195	1.0320	437.2	767.4	1204.6	0.6356	0.8378	1.4734	435.5	1118.7	450
500	467.01	0.0197	0.9278	449.4	755.0	1204.4	0.6487	0.8147	1.4634	447.6	1118.6	500
550	476.94	0.0199	0.8424	460.8	743.1	1203.9	0.6608	0.7934	1.4542	458.8	1118.2	550
600	486.21	0.0201	0.7698	471.6	731.6	1203.2	0.6720	0.7734	1.4454	469.4	1117.7	600
650	494.90	0.0203	0.7083	481.8	720.5	1202.3	0.6826	0.7548	1.4374	479.4	1117.1	650
700	503.10	0.0205	0.6554	491.5	709.7	1201.2	0.6925	0.7371	1.4296	488.8	1116.3	700
750	510.86	0.0207	0.6092	500.8	699.2	1200.0	0.7019	0.7204	1.4223	598.0	1115.4	750
800	518.23	0.0209	0.5687	509.7	688.9	1198.6	0.7108	0.7045	1.4153	506.6	1114.4	800
850	525.26	0.0210	0.5327	518.3	678.8	1197.1	0.7194	0.6891	1.4085	515.0	1113.3	850
900	531.98	0.0212	0.5006	526.6	668.8	1195.4	0.7275	0.6744	1.4020	523.1	1112.1	900
950	538.43	0.0214	0.4717	534.6	659.1	1193.7	0.7355	0.6602	1.3957	530.9	1110.8	950
1000	544.61	0.0216	0.4456	542.4	649.4	1191.8	0.7430	0.6467	1.3897	538.4	1109.4	1000
1100	556.31	0.0220	0.4001	557.4	630.4	1187.8	0.7575	0.6205	1.3780	552.9	1106.4	1100
1200	567.22	0.0223	0.3619	571.7	611.7	1183.4	0.7711	0.5956	1.3667	566.7	1103.0	1200
1300	577.46	0.0227	0.3293	585.4	593.2	1178.6	0.7840	0.5719	1.3559	580.0	1099.4	1300
1400	587.10	0.0231	0.3012	597.7	574.7	1173.4	0.7963	0.5491	1.3454	592.7	1095.4	1400
1500	596.23	0.0235	0.2765	611.6	556.3	1167.9	0.8082	0.5269	1.3351	605.1	1091.2	1500
2000	635.82	0.0257	0.1878	671.7	463.4	1135.1	0.8619	0.4230	1.2849	662.2	1065.6	2000
2500	668.13	0.0287	0.1307	730.6	360.5	1091.1	0.9126	0.3197	1.2322	717.3	1030.6	2500
3000	695.36	0.0346	0.0858	802.5	217.8	1020.3	0.9731	0.1885	1.1615	783.4	972.7	3000
3206.2	705.40	0.0503	0.0503	902.7	0	902.7	1.0580	0	1.0580	872.9	872.9	3206.2

* Abridged "Thermodynamic Properties of Steam" by Joseph H. Keenan and Frederick G. Keyes. John Wiley & Sons, Inc., New York, 1937.

APPENDIX OF CALCULATION DATA

TABLE 7. THERMODYNAMIC PROPERTIES OF STEAM.—(Continued)
Dry saturated steam: temperature table*

Temp., °F	Abs press., psi	Specific volume			Enthalpy			Entropy			Temp., °F
		Sat. liquid	Evap.	Sat. vapor	Sat. liquid	Evap.	Sat. vapor	Sat. liquid	Evap.	Sat. vapor	
t	p	v_f	v_{fg}	v_g	h_f	h_{fg}	h_g	s_f	s_{fg}	s_g	t
32	0.08854	0.01602	3306	3306	0.00	1075.8	1075.8	0.0000	2.1877	2.1877	32
35	0.09995	0.01602	2947	2947	3.02	1074.1	1077.1	0.0061	2.1709	2.1770	35
40	0.12170	0.01602	2444	2444	8.05	1071.3	1079.3	0.0162	2.1435	2.1597	40
45	0.14752	0.01602	2036.4	2036.4	13.06	1068.4	1081.5	0.0262	2.1167	2.1429	45
50	0.17811	0.01603	1703.2	1703.2	18.07	1065.6	1083.7	0.0361	2.0903	2.1264	50
60	0.2563	0.01604	1206.6	1206.7	28.06	1059.9	1088.0	0.0555	2.0393	2.0948	60
70	0.3631	0.01606	867.8	867.9	38.04	1054.3	1092.3	0.0745	1.9902	2.0647	70
80	0.5069	0.01608	633.1	633.1	48.02	1048.6	1096.6	0.0932	1.9428	2.0360	80
90	0.6982	0.01610	468.0	468.0	57.99	1042.9	1100.9	0.1115	1.8972	2.0087	90
100	0.9492	0.01613	350.3	350.4	67.97	1037.2	1105.2	0.1295	1.8531	1.9826	100
110	1.2748	0.01617	265.3	265.4	77.94	1031.6	1109.5	0.1471	1.8106	1.9577	110
120	1.6924	0.01620	203.25	203.27	87.92	1025.8	1113.7	0.1645	1.7694	1.9339	120
130	2.2225	0.01625	157.32	157.34	97.90	1020.0	1117.9	0.1816	1.7296	1.9112	130
140	2.8886	0.01629	122.99	123.01	107.89	1014.1	1122.0	0.1984	1.6910	1.8894	140
150	3.718	0.01634	98.06	97.07	117.89	1008.2	1126.1	0.2149	1.6537	1.8685	150
160	4.741	0.01639	77.27	77.29	127.89	1002.3	1130.2	0.2311	1.6174	1.8485	160
170	5.992	0.01645	62.04	62.06	137.90	996.3	1134.2	0.2472	1.5822	1.8293	170
180	7.510	0.01651	50.21	50.23	147.92	990.2	1138.1	0.2630	1.5480	1.8109	180
190	9.339	0.01657	40.94	40.96	157.95	984.1	1142.0	0.2785	1.5147	1.7932	190
200	11.526	0.01663	33.62	33.64	167.99	977.9	1145.9	0.2938	1.4824	1.7762	200
210	14.123	0.01670	27.80	27.82	178.05	971.6	1149.7	0.3090	1.4508	1.7598	210
212	14.696	0.01672	26.78	26.80	180.07	970.3	1150.4	0.3120	1.4446	1.7566	212
220	17.186	0.01677	23.13	23.15	188.13	965.2	1153.4	0.3239	1.4201	1.7440	220
230	20.780	0.01684	19.365	19.382	198.23	958.8	1157.0	0.3387	1.3901	1.7288	230
240	24.969	0.01692	16.306	16.323	208.34	952.2	1160.5	0.3531	1.3609	1.7140	240
250	29.825	0.01700	13.804	13.821	216.48	945.5	1164.0	0.3675	1.3323	1.6998	250
260	35.429	0.01709	11.746	11.763	228.64	938.7	1167.3	0.3817	1.3043	1.6860	260
270	41.858	0.01717	10.044	10.061	238.84	931.8	1170.6	0.3958	1.2769	1.6727	270
280	49.203	0.01726	8.628	8.645	249.06	924.7	1173.8	0.4096	1.2501	1.6597	280
290	57.556	0.01735	7.444	7.461	259.31	917.5	1176.8	0.4234	1.2238	1.6472	290
300	67.013	0.01745	6.449	6.466	269.59	910.1	1179.7	0.4369	1.1980	1.6350	300
310	77.68	0.01755	5.609	5.626	279.92	902.6	1182.5	0.4504	1.1727	1.6231	310
320	89.66	0.01765	4.896	4.914	290.28	894.9	1185.2	0.4637	1.1478	1.6115	320
330	103.06	0.01776	4.289	4.307	300.68	887.0	1187.7	0.4769	1.1233	1.6002	330
340	118.01	0.01787	3.770	3.788	311.13	879.0	1190.1	0.4900	1.0992	1.5891	340
350	134.63	0.01799	3.324	3.342	321.63	870.7	1192.3	0.5029	1.0754	1.5783	350
360	153.04	0.01811	2.939	2.957	332.18	862.2	1194.4	0.5158	1.0519	1.5677	360
370	173.37	0.01823	2.606	2.625	342.79	853.5	1196.3	0.5286	1.0287	1.5573	370
380	195.77	0.01836	2.317	2.335	353.45	844.6	1198.1	0.5413	1.0059	1.5471	380
390	220.37	0.01850	2.0651	2.0836	364.17	835.4	1199.6	0.5539	0.9832	1.5371	390
400	247.31	0.01864	1.8447	1.8633	374.97	826.0	1201.0	0.5664	0.9608	1.5272	400
410	276.75	0.01878	1.6512	1.6700	385.83	816.3	1202.1	0.5788	0.9386	1.5174	410
420	308.83	0.01894	1.4811	1.5000	396.77	806.3	1203.1	0.5912	0.9166	1.5078	420
430	343.72	0.01910	1.3308	1.3499	407.79	796.0	1203.8	0.6035	0.8947	1.4982	430
440	381.59	0.01926	1.1979	1.2171	418.90	785.4	1204.3	0.6158	0.8730	1.4887	440
450	422.6	0.0194	1.0799	1.0993	430.1	774.5	1204.6	0.6280	0.8513	1.4793	450
460	466.9	0.0196	0.9748	0.9944	441.4	763.2	1204.6	0.6402	0.8298	1.4700	460
470	514.7	0.0198	0.8811	0.9009	452.8	751.5	1204.3	0.6523	0.8083	1.4606	470
480	566.1	0.0200	0.7972	0.8172	464.4	739.4	1203.7	0.6645	0.7868	1.4513	480
490	621.4	0.0202	0.7221	0.7423	476.0	726.8	1202.8	0.6766	0.7653	1.4419	490
500	680.8	0.0204	0.6545	0.6749	487.8	713.9	1201.7	0.6887	0.7438	1.4325	500
520	812.4	0.0209	0.5385	0.5594	511.9	686.4	1198.2	0.7130	0.7006	1.4136	520
540	962.5	0.0215	0.4434	0.4649	536.6	656.6	1193.2	0.7374	0.6568	1.3942	540
560	1133.1	0.0221	0.3647	0.3868	562.2	624.2	1186.4	0.7621	0.6121	1.3742	560
580	1325.8	0.0228	0.2989	0.3217	588.9	588.4	1177.3	0.7872	0.5659	1.3532	580
600	1542.9	0.0236	0.2432	0.2668	617.0	548.5	1165.5	0.8131	0.5176	1.3307	600
620	1786.6	0.0247	0.1955	0.2201	646.7	503.6	1150.3	0.8398	0.4664	1.3062	620
640	2059.7	0.0260	0.1538	0.1798	678.6	452.0	1130.5	0.8679	0.4110	1.2789	640
660	2365.4	0.0278	0.1165	0.1442	714.2	390.2	1104.4	0.8987	0.3485	1.2472	660
680	2708.1	0.0305	0.0810	0.1115	757.3	309.9	1067.2	0.9351	0.2719	1.2071	680
700	3093.7	0.0369	0.0392	0.0761	823.3	172.1	995.4	0.9905	0.1484	1.1389	700
705.4	3206.2	0.0503	0	0.0503	902.7	0	902.7	1.0580	0	1.0580	705.4

* Abridged from "Thermodynamic Properties of Steam" by Joseph H. Keenan and Frederick G. Keyes, John Wiley & Sons, Inc., New York.

TABLE 7. THERMODYNAMIC PROPERTIES OF STEAM.—(Continued)
Properties of superheated steam*

Abs press., psi (sat temp)		200	300	400	500	600	700	800	900	1000	1100	1200	1400	1600
		\multicolumn{13}{c}{Temp, °F}
1 (101.74)	v	392.6	452.3	512.0	571.6	631.2	690.8	750.4	809.9	869.5	929.1	988.7	1107.8	1227.0
	h	1150.4	1195.8	1241.7	1288.3	1335.7	1383.8	1432.8	1482.7	1533.5	1585.2	1637.7	1745.7	1857.5
	s	2.0512	2.1153	2.1720	2.2233	2.2702	2.3137	2.3542	2.3923	2.4283	2.4625	2.4952	2.5566	2.6137
5 (162.24)	v	78.16	90.25	102.26	114.22	126.16	138.10	150.03	161.95	173.87	185.79	197.71	221.6	245.4
	h	1148.8	1195.0	1241.2	1288.0	1335.4	1383.6	1432.7	1482.6	1533.4	1585.1	1637.7	1745.7	1857.4
	s	1.8718	1.9370	1.9942	2.0456	2.0927	2.1361	2.1767	2.2148	2.2509	2.2851	2.3178	2.3792	2.4363
10 (193.21)	v	38.85	45.00	51.04	57.05	63.03	69.01	74.98	80.95	86.92	92.88	98.84	110.77	122.69
	h	1146.6	1193.9	1240.6	1287.5	1335.1	1383.4	1432.5	1482.4	1533.2	1585.0	1637.6	1745.6	1857.3
	s	1.7927	1.8595	1.9172	1.9689	2.0160	2.0596	2.1002	2.1383	2.1744	2.2086	2.2413	2.3028	2.3598
14.696 (212.00)	v		30.53	34.68	38.78	42.86	46.94	51.00	55.07	59.13	63.19	67.25	75.37	83.48
	h		1192.8	1239.9	1287.1	1334.8	1383.2	1432.3	1482.3	1533.1	1584.8	1637.5	1745.5	1857.3
	s		1.8160	1.8743	1.9261	1.9734	2.0170	2.0576	2.0958	2.1319	2.1662	2.1989	2.2603	2.3174
20 (227.96)	v		22.36	25.43	28.46	31.47	34.47	37.46	40.45	43.44	46.42	49.41	55.37	61.34
	h		1191.6	1239.2	1286.6	1334.4	1382.9	1432.1	1482.1	1533.0	1584.7	1637.4	1745.4	1857.2
	s		1.7808	1.8396	1.8918	1.9392	1.9829	2.0235	2.0618	2.0978	2.1321	2.1648	2.2263	2.2834
40 (267.25)	v		11.040	12.628	14.168	15.688	17.198	18.702	20.20	21.70	23.20	24.69	27.68	30.66
	h		1186.8	1236.5	1284.8	1333.1	1381.9	1431.3	1481.4	1532.4	1584.3	1637.0	1745.1	1857.0
	s		1.6994	1.7608	1.8140	1.8619	1.9058	1.9467	1.9850	2.0212	2.0555	2.0883	2.1498	2.2069
60 (292.71)	v		7.259	8.357	9.403	10.427	11.441	12.449	13.452	14.454	15.453	16.451	18.446	20.44
	h		1181.6	1233.6	1283.0	1331.8	1380.9	1430.5	1480.8	1531.9	1583.8	1636.6	1744.8	1856.7
	s		1.6492	1.7135	1.7678	1.8162	1.8605	1.9015	1.9400	1.9762	2.0106	2.0434	2.1049	2.1621
80 (312.03)	v			6.220	7.020	7.797	8.562	9.322	10.077	10.830	11.582	12.332	13.830	15.325
	h			1230.7	1281.1	1330.5	1379.9	1429.7	1480.1	1531.3	1583.4	1636.2	1744.5	1856.5
	s			1.6791	1.7346	1.7836	1.8281	1.8694	1.9079	1.9442	1.9787	2.0115	2.0731	2.1303
100 (327.81)	v			4.937	5.589	6.218	6.835	7.446	8.052	8.656	9.259	9.860	11.060	12.258
	h			1227.6	1279.1	1329.1	1378.9	1428.9	1479.5	1530.8	1582.9	1635.7	1744.2	1856.2
	s			1.6518	1.7085	1.7581	1.8029	1.8443	1.8829	1.9193	1.9538	1.9867	2.0484	2.1056
120 (341.25)	v			4.081	4.636	5.165	5.683	6.195	6.702	7.207	7.710	8.212	9.214	10.213
	h			1224.4	1277.2	1327.7	1377.8	1428.1	1478.8	1530.2	1582.4	1635.3	1743.9	1856.0
	s			1.6287	1.6869	1.7370	1.7822	1.8237	1.8625	1.8990	1.9335	1.9664	2.0281	2.0854
140 (353.02)	v			3.468	3.954	4.413	4.861	5.301	5.738	6.172	6.604	7.035	7.895	8.752
	h			1221.1	1275.2	1326.4	1376.8	1427.3	1478.2	1529.7	1581.9	1634.9	1743.5	1855.7
	s			1.6087	1.6683	1.7190	1.7645	1.8063	1.8451	1.8817	1.9163	1.9493	2.0110	2.0683
160 (363.53)	v			3.008	3.443	3.849	4.244	4.631	5.015	5.396	5.775	6.152	6.906	7.656
	h			1217.6	1273.1	1325.0	1375.7	1426.4	1477.5	1529.1	1581.4	1634.5	1743.2	1855.5
	s			1.5908	1.6519	1.7033	1.7491	1.7911	1.8301	1.8667	1.9014	1.9344	1.9962	2.0535
180 (373.06)	v			2.649	3.044	3.411	3.764	4.110	4.452	4.792	5.129	5.466	6.136	6.804
	h			1214.0	1271.0	1323.5	1374.7	1425.6	1476.8	1528.6	1581.0	1634.1	1742.9	1855.2
	s			1.5745	1.6373	1.6894	1.7355	1.7776	1.8167	1.8534	1.8882	1.9212	1.9831	2.0404
200 (381.79)	v			2.361	2.726	3.060	3.380	3.693	4.002	4.309	4.613	4.917	5.521	6.123
	h			1210.3	1268.9	1322.1	1373.6	1424.8	1476.2	1528.0	1580.5	1633.7	1742.6	1855.0
	s			1.5594	1.6240	1.6767	1.7232	1.7655	1.8048	1.8415	1.8763	1.9094	1.9713	2.0287
220 (389.86)	v			2.125	2.465	2.772	3.066	3.352	3.634	3.913	4.191	4.467	5.017	5.565
	h			1206.5	1266.7	1320.7	1372.0	1424.0	1475.5	1527.5	1580.0	1633.3	1742.3	1854.7
	s			1.5453	1.6117	1.6652	1.7120	1.7545	1.7939	1.8308	1.8656	1.8987	1.9607	2.0181
240 (397.37)	v			1.9276	2.247	2.533	2.804	3.068	3.327	3.584	3.839	4.093	4.597	5.100
	h			1202.5	1264.5	1319.2	1371.5	1423.2	1474.8	1526.9	1579.6	1632.9	1742.0	1854.5
	s			1.5219	1.6003	1.6546	1.7017	1.7444	1.7839	1.8209	1.8558	1.8889	1.9510	2.0084
260 (404.42)	v				2.063	2.330	2.582	2.827	3.067	3.305	3.541	3.776	4.242	4.707
	h				1262.3	1317.7	1370.4	1422.3	1474.2	1526.3	1579.1	1632.5	1741.7	1854.2
	s				1.5897	1.6447	1.6922	1.7352	1.7748	1.8118	1.8467	1.8799	1.9420	1.9995
280 (411.05)	v				1.9047	2.156	2.392	2.621	2.845	3.066	3.286	3.504	3.938	4.370
	h				1260.0	1316.2	1369.4	1421.5	1473.5	1525.8	1578.6	1632.1	1741.4	1854.0
	s				1.5796	1.6354	1.6834	1.7265	1.7662	1.8033	1.8383	1.8716	1.9337	1.9912
300 (417.33)	v				1.7675	2.005	2.227	2.442	2.652	2.859	3.065	3.269	3.674	4.078
	h				1257.6	1314.7	1368.3	1420.6	1472.8	1525.2	1578.1	1631.7	1741.0	1853.7
	s				1.5701	1.6268	1.6751	1.7184	1.7582	1.7954	1.8305	1.8638	1.9260	1.9835
350 (431.72)	v				1.4923	1.7036	1.8980	2.084	2.266	2.445	2.622	2.798	3.147	3.493
	h				1251.5	1310.9	1365.5	1418.5	1471.1	1523.8	1577.0	1630.7	1740.3	1853.1
	s				1.5481	1.6070	1.6563	1.7002	1.7403	1.7777	1.8130	1.8463	1.9086	1.9663
400 (444.59)	v				1.2851	1.4770	1.6508	1.8161	1.9767	2.134	2.290	2.445	2.751	3.055
	h				1245.1	1306.9	1362.7	1416.4	1469.4	1522.4	1575.8	1629.6	1739.5	1852.5
	s				1.5281	1.5894	1.6398	1.6842	1.7247	1.7623	1.7977	1.8311	1.8936	1.9513

* Abridged from "Thermodynamic Properties of Steam," by Joseph H. Keenan and Frederick G. Keyes, John Wiley & Sons, Inc., New York, 1937.

APPENDIX OF CALCULATION DATA 819

TABLE 7. THERMODYNAMIC PROPERTIES OF STEAM.—(*Continued*)
Properties of superheated steam*

Abs Press., psi (sat temp)		500	550	600	620	640	660	680	700	800	900	1000	1200	1400	1600
450 (456.28)	v	1.1231	1.2155	1.3005	1.3332	1.3652	1.3967	1.4278	1.4584	1.6074	1.7516	1.8928	2.170	2.443	2.714
	h	1238.4	1272.0	1302.8	1314.6	1326.2	1337.5	1348.8	1359.9	1414.3	1467.7	1521.0	1628.6	1738.7	1851.9
	s	1.5095	1.5437	1.5735	1.5845	1.5951	1.6054	1.6153	1.6250	1.6699	1.7108	1.7486	1.8177	1.8803	1.9381
500 (467.01)	v	0.9927	1.0800	1.1591	1.1893	1.2188	1.2478	1.2763	1.3044	1.4405	1.5715	1.6996	1.9504	2.197	2.442
	h	1231.3	1266.8	1298.6	1310.7	1322.6	1334.2	1345.7	1357.0	1412.1	1466.0	1519.6	1627.6	1737.9	1851.3
	s	1.4919	1.5280	1.5588	1.5701	1.5810	1.5915	1.6016	1.6115	1.6571	1.6982	1.7363	1.8056	1.8683	1.9262
550 (476.94)	v	0.8852	0.9686	1.0431	1.0714	1.0989	1.1259	1.1523	1.1783	1.3038	1.4241	1.5414	1.7706	1.9957	2.219
	h	1223.7	1261.2	1294.3	1306.8	1318.9	1330.8	1342.5	1354.0	1409.9	1464.3	1518.2	1626.6	1737.1	1850.6
	s	1.4751	1.5131	1.5451	1.5568	1.5680	1.5787	1.5890	1.5991	1.6452	1.6868	1.7250	1.7946	1.8575	1.9155
600 (486.21)	v	0.7947	0.8753	0.9463	0.9729	0.9988	1.0241	1.0489	1.0732	1.1899	1.3013	1.4096	1.6208	1.8279	2.033
	h	1215.7	1255.5	1289.9	1302.7	1315.2	1327.4	1339.3	1351.1	1407.7	1462.5	1516.7	1625.5	1736.3	1850.0
	s	1.4586	1.4990	1.5323	1.5443	1.5558	1.5667	1.5773	1.5875	1.6343	1.6762	1.7147	1.7846	1.8476	1.9056
700 (503.10)	v		0.7277	0.7934	0.8177	0.8411	0.8639	0.8860	0.9077	1.0108	1.1082	1.2024	1.3853	1.5641	1.7405
	h		1243.2	1280.6	1294.3	1307.5	1320.3	1332.8	1345.0	1403.2	1459.0	1515.9	1623.5	1734.8	1848.8
	s		1.4722	1.5084	1.5212	1.5333	1.5449	1.5559	1.5665	1.6147	1.6573	1.6963	1.7666	1.8299	1.8881
800 (518.23)	v		0.6154	0.6779	0.7006	0.7223	0.7433	0.7635	0.7833	0.8763	0.9633	1.0470	1.2088	1.3662	1.5214
	h		1229.8	1270.7	1285.4	1299.4	1312.9	1325.9	1338.6	1398.6	1455.4	1511.0	1621.4	1733.2	1847.5
	s		1.4467	1.4863	1.5000	1.5129	1.5250	1.5366	1.5476	1.5972	1.6407	1.6801	1.7510	1.8146	1.8729
900 (531.98)	v		0.5264	0.5873	0.6089	0.6294	0.6491	0.6680	0.6863	0.7716	0.8506	0.9262	1.0714	1.2124	1.3509
	h		1215.0	1260.1	1275.9	1290.9	1305.1	1318.8	1332.1	1393.9	1451.8	1508.1	1619.3	1731.6	1846.3
	s		1.4216	1.4653	1.4800	1.4938	1.5066	1.5187	1.5303	1.5814	1.6257	1.6656	1.7371	1.8009	1.8595
1000 (544.61)	v		0.4533	0.5140	0.5350	0.5546	0.5733	0.5912	0.6084	0.6878	0.7604	0.8294	0.9615	1.0893	1.2146
	h		1198.3	1248.8	1265.9	1281.9	1297.0	1311.4	1325.3	1389.2	1448.2	1505.1	1617.3	1730.0	1845.0
	s		1.3961	1.4450	1.4610	1.4757	1.4893	1.5021	1.5141	1.5670	1.6121	1.6525	1.7245	1.7886	1.8474
1100 (556.31)	v			0.4532	0.4738	0.4929	0.5110	0.5281	0.5445	0.6191	0.6866	0.7503	0.8716	0.9885	1.1031
	h			1236.7	1255.3	1272.4	1288.5	1303.7	1318.3	1384.3	1444.5	1502.2	1615.2	1728.4	1843.8
	s			1.4251	1.4425	1.4583	1.4728	1.4862	1.4989	1.5535	1.5995	1.6405	1.7130	1.7775	1.8363
1200 (567.22)	v			0.4016	0.4222	0.4410	0.4586	0.4752	0.4909	0.5617	0.6250	0.6843	0.7967	0.9046	1.0101
	h			1223.5	1243.9	1262.4	1279.6	1295.7	1311.0	1379.3	1440.7	1499.2	1613.1	1726.9	1842.5
	s			1.4052	1.4243	1.4413	1.4568	1.4710	1.4843	1.5409	1.5879	1.6293	1.7025	1.7672	1.8263
1400 (587.10)	v			0.3174	0.3390	0.3580	0.3753	0.3912	0.4062	0.4714	0.5281	0.5805	0.6789	0.7727	0.8640
	h			1193.0	1218.4	1240.4	1260.3	1278.5	1295.5	1369.1	1433.1	1493.2	1608.9	1723.7	1840.0
	s			1.3639	1.3877	1.4079	1.4258	1.4419	1.4567	1.5177	1.5666	1.6093	1.6836	1.7489	1.8083
1600 (604.90)	v				0.2733	0.2936	0.3112	0.3271	0.3417	0.4034	0.4553	0.5027	0.5906	0.6738	0.7545
	h				1187.8	1215.2	1238.7	1259.6	1278.7	1358.4	1425.3	1487.0	1604.6	1720.5	1837.5
	s				1.3489	1.3741	1.3952	1.4137	1.4303	1.4964	1.5476	1.5914	1.6669	1.7328	1.7926
1800 (621.03)	v				0.2407	0.2597	0.2760	0.2907	0.3042	0.3502	0.3986	0.4421	0.5218	0.5968	0.6693
	h				1185.1	1214.0	1238.5	1260.3	1347.2	1417.4	1480.8	1600.4	1717.3	1835.0
	s				1.3377	1.3638	1.3855	1.4044	1.4765	1.5301	1.5752	1.6520	1.7185	1.7786
2000 (635.82)	v				0.1936	0.2161	0.2337	0.2489	0.3074	0.3532	0.3935	0.4668	0.5352	0.6011
	h				1145.6	1184.9	1214.8	1240.0	1335.5	1409.2	1474.5	1596.1	1714.1	1832.5
	s				1.2945	1.3300	1.3564	1.3783	1.4576	1.5139	1.5603	1.6384	1.7055	1.7660
2500 (668.13)	v						0.1484	0.1686	0.2294	0.2710	0.3061	0.3678	0.4244	0.4784
	h						1132.3	1176.8	1303.6	1387.8	1458.4	1585.3	1706.1	1826.2
	s						1.2687	1.3073	1.4127	1.4772	1.5273	1.6088	1.6775	1.7389
3000 (695.36)	v							0.0984	0.1760	0.2159	0.2476	0.3018	0.3505	0.3966
	h							1060.7	1292.2	1365.0	1441.8	1574.3	1698.0	1819.9
	s							1.1966	1.3690	1.4439	1.4984	1.5837	1.6540	1.7163
3206.2 (705.40)	v								0.1583	0.1981	0.2288	0.2806	0.3267	0.3703
	h								1250.5	1355.2	1434.7	1569.8	1694.6	1817.2
	s								1.3508	1.4309	1.4874	1.5742	1.6452	1.7080
3500	v								0.0306	0.1364	0.1762	0.2058	0.2546	0.2977	0.3381
	h								780.5	1224.9	1340.7	1424.5	1563.3	1689.8	1813.6
	s								0.9515	1.3241	1.4127	1.4723	1.5615	1.6336	1.6968
4000	v								0.0287	0.1052	0.1462	0.1743	0.2192	0.2581	0.2943
	h								763.8	1174.8	1314.4	1406.8	1552.1	1681.7	1807.2
	s								0.9347	1.2757	1.3827	1.4482	1.5417	1.6154	1.6795
4500	v								0.0276	0.0798	0.1226	0.1500	0.1917	0.2273	0.2602
	h								753.5	1113.9	1286.5	1388.4	1540.8	1673.5	1800.9
	s								0.9235	1.2204	1.3529	1.4253	1.5235	1.5990	1.6640
5000	v								0.0268	0.0593	0.1036	0.1303	0.1696	0.2027	0.2329
	h								746.4	1047.1	1256.5	1369.5	1529.5	1665.3	1704.5
	s								0.9152	1.1622	1.3231	1.4034	1.5066	1.5839	1.6499
5500	v								0.0262	0.0463	0.0880	0.1143	0.1516	0.1825	0.2106
	h								741.3	985.0	1224.1	1349.3	1518.2	1657.0	1788.1
	s								0.9090	1.1093	1.2930	1.3821	1.4908	1.5699	1.6360

* Abridged from "Thermodynamic Properties of Steam," by Joseph H. Keenan and Frederick G. Keyes, John Wiley & Sons, Inc., New York, 1937.

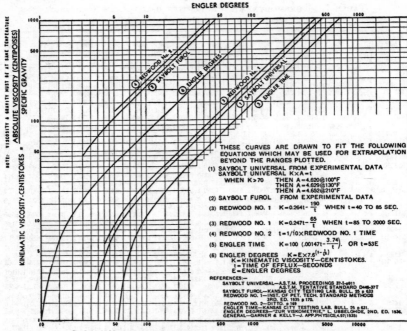

Fig. 13a. Viscosity conversion chart.

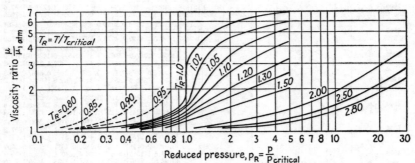

Fig. 13b. Viscosity correction chart for gases at different pressures. [Comings and Egly, *Ind. Eng. Chem.*, **32**, 715 (1940).]

APPENDIX OF CALCULATION DATA 821

Viscosities of Petroleum Fractions
For temperature ranges employed in the text
Coordinates to be used with Fig. 14

	X	Y
76°API natural gasoline	14.4	6.4
56°API gasoline	14.0	10.5
42°API kerosene	11.6	16.0
35°API distillate	10.0	20.0
34°API mid-continent crude	10.3	21.3
28°API gas oil	10.0	23.6

Viscosities of Animal and Vegetable Oils*

	Acid No.	Sp gr, 20/4°C	X	Y
Almond	2.85	0.9188	6.9	28.2
Coconut	0.01	0.9226	6.9	26.9
Cod liver	....	0.9138	7.7	27.7
Cottonseed	14.24	0.9187	7.0	28.0
Lard	3.39	0.9138	7.0	28.2
Linseed	3.42	0.9297	6.8	27.5
Mustard	....	0.9237	7.0	28.5
Neatsfoot	13.35	0.9158	6.5	28.0
Olive	....	0.9158	6.6	28.3
Palm kernel	9.0	0.9190	7.0	26.9
Perilla, raw	1.36	0.9297	8.1	27.2
Rapeseed	0.34	0.9114	7.0	28.8
Sardine	0.57	0.9384	7.7	27.3
Soybean	3.50	0.9228	8.3	27.5
Sperm	0.80	0.8829	7.7	26.3
Sunflower	2.76	0.9207	7.5	27.6
Whale, refined	0.73	0.9227	7.5	27.5

* Based on data at 100 and 210°F of A. R. Rescorla and F. L. Carnahan, *Ind. Eng. Chem.*, **28**, 1212–1213 (1936).

Viscosities of Commercial Fatty Acids*
250 to 400°F

	Sp gr at 300°F	X	Y
Lauric	0.792	10.1	23.1
Oleic	0.799	10.0	25.2
Palmitic	0.786	9.2	25.9
Stearic	0.789	10.5	25.5

* From data of D. Q. Kern and W. Van Nostrand, *Ind. Eng. Chem.*, **41**, 2209 (1949).

Viscosities of Liquids*
Coordinates to be used with Fig. 14

Liquid	X	Y	Liquid	X	Y
Acetaldehyde	15.2	4.8	Freon-21	15.7	7.5
Acetic acid, 100%	12.1	14.2	Freon-22	17.2	4.7
Acetic acid, 70%	9.5	17.0	Freon-113	12.5	11.4
Acetic anhydride	12.7	12.8	Freon-114	14.6	8.3
Acetone, 100%	14.5	7.2	Glycerol, 100%	2.0	30.0
Acetone, 35%	7.9	15.0	Glycerol, 50%	6.9	19.6
Allyl alcohol	10.2	14.3	Heptane	14.1	8.4
Ammonia, 100%	12.6	2.0	Hexane	14.7	7.0
Ammonia, 26%	10.1	13.9	Hydrochloric acid, 31.5%	13.0	16.6
Amyl acetate	11.8	12.5	Isobutyl alcohol	7.1	18.0
Amyl alcohol	7.5	18.4	Isobutyric acid	12.2	14.4
Aniline	8.1	18.7	Isopropyl alcohol	8.2	16.0
Anisole	12.3	13.5	Mercury	18.4	16.4
Arsenic trichloride	13.9	14.5	Methanol, 100%	12.4	10.5
Benzene	12.5	10.9	Methanol, 90%	12.3	11.8
Brine, CaCl$_2$, 25%	6.6	15.9	Methanol, 40%	7.8	15.5
Brine, NaCl, 25%	10.2	16.6	Methyl acetate	14.2	8.2
Bromine	14.2	13.2	Methyl chloride	15.0	3.8
Bromotoluene	20.0	15.9	Methyl ethyl ketone	13.9	8.6
n-Butane	15.3	3.3	Naphthalene	7.9	18.1
Isobutane	14.5	3.7	Nitric acid, 95%	12.8	13.8
Butyl acetate	12.3	11.0	Nitric acid, 60%	10.8	17.0
Butyl alcohol	8.6	17.2	Nitrobenzene	10.6	16.2
Butyric acid	12.1	15.3	Nitrotoluene	11.0	17.0
Carbon dioxide	11.6	0.3	Octane	13.7	10.0
Carbon disulfide	16.1	7.5	Octyl alcohol	6.6	21.1
Carbon tetrachloride	12.7	13.1	Pentachloroethane	10.9	17.3
Chlorobenzene	12.3	12.4	Pentane	14.9	5.2
Chloroform	14.4	10.2	Phenol	6.9	20.8
Chlorosulfonic acid	11.2	18.1	Phosphorus tribromide	13.8	16.7
Chlorotoluene, ortho	13.0	13.3	Phosphorus trichloride	16.2	10.9
Chlorotoluene, meta	13.3	12.5	Propane	15.3	1.0
Chlorotoluene, para	13.3	12.5	Propionic acid	12.8	13.8
Cresol, meta	2.5	20.8	Propyl alcohol	9.1	16.5
Cyclohexanol	2.9	24.3	Propyl bromide	14.5	9.6
Dibromoethane	12.7	15.8	Propyl chloride	14.4	7.5
Dichloroethane	13.2	12.2	Propyl iodide	14.1	11.6
Dichloromethane	14.6	8.9	Sodium	16.4	13.9
Diethyl oxalate	11.0	16.4	Sodium hydroxide, 50%	3.2	25.8
Dimethyl oxalate	12.3	15.8	Stannic chloride	13.5	12.8
Diphenyl	12.0	18.3	Sulfur dioxide	15.2	7.1
Dipropyl oxalate	10.3	17.7	Sulfuric acid, 110%	7.2	27.4
Ethyl acetate	13.7	9.1	Sulfuric acid, 98%	7.0	24.8
Ethyl alcohol, 100%	10.5	13.8	Sulfuric acid, 60%	10.2	21.3
Ethyl alcohol, 95%	9.8	14.3	Sulfuryl chloride	15.2	12.4
Ethyl alcohol, 40%	6.5	16.6	Tetrachloroethane	11.9	15.7
Ethyl benzene	13.2	11.5	Tetrachloroethylene	14.2	12.7
Ethyl bromide	14.5	8.1	Titanium tetrachloride	14.4	12.3
Ethyl chloride	14.8	6.0	Toluene	13.7	10.4
Ethyl ether	14.5	5.3	Trichloroethylene	14.8	10.5
Ethyl formate	14.2	8.4	Turpentine	11.5	14.9
Ethyl iodide	14.7	10.3	Vinyl acetate	14.0	8.8
Ethylene glycol	6.0	23.6	Water	10.2	13.0
Formic acid	10.7	15.8	Xylene, ortho	13.5	12.1
Freon-11	14.4	9.0	Xylene, meta	13.9	10.6
Freon-12	16.8	5.6	Xylene, para	13.9	10.9

* From Perry, J. H., "Chemical Engineers' Handbook," 3d ed., McGraw-Hill Book Company, Inc., New York, 1950.

APPENDIX OF CALCULATION DATA 823

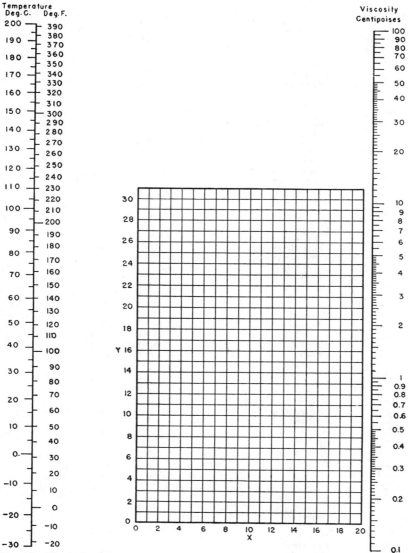

FIG. 14. Viscosities of liquids. (Perry, "*Chemical Engineers' Handbook*," 3d ed., McGraw Hill Book Company, Inc., New York, 1950.)

Viscosities of Gases*
Coordinates to be used with Fig. 15

Gas	X	Y
Acetic acid	7.7	14.3
Acetone	8.9	13.0
Acetylene	9.8	14.9
Air	11.0	20.0
Ammonia	8.4	16.0
Argon	10.5	22.4
Benzene	8.5	13.2
Bromine	8.9	19.2
Butene	9.2	13.7
Butylene	8.9	13.0
Carbon dioxide	9.5	18.7
Carbon disulfide	8.0	16.0
Carbon monoxide	11.0	20.0
Chlorine	9.0	18.4
Chloroform	8.9	15.7
Cyanogen	9.2	15.2
Cyclohexane	9.2	12.0
Ethane	9.1	14.5
Ethyl acetate	8.5	13.2
Ethyl alcohol	9.2	14.2
Ethyl chloride	8.5	15.6
Ethyl ether	8.9	13.0
Ethylene	9.5	15.1
Fluorine	7.3	23.8
Freon-11	10.6	15.1
Freon-12	11.1	16.0
Freon-21	10.8	15.3
Freon-22	10.1	17.0
Freon-113	11.3	14.0
Helium	10.9	20.5
Hexane	8.6	11.8
Hydrogen	11.2	12.4
$3H_2 + 1N_2$	11.2	17.2
Hydrogen bromide	8.8	20.9
Hydrogen chloride	8.8	18.7
Hydrogen cyanide	9.8	14.9
Hydrogen iodide	9.0	21.3
Hydrogen sulfide	8.6	18.0
Iodine	9.0	18.4
Mercury	5.3	22.9
Methane	9.9	15.5
Methyl alcohol	8.5	15.6
Nitric oxide	10.9	20.5
Nitrogen	10.6	20.0
Nitrosyl chloride	8.0	17.6
Nitrous oxide	8.8	19.0
Oxygen	11.0	21.3
Pentane	7.0	12.8
Propane	9.7	12.9
Propyl alcohol	8.4	13.4
Propylene	9.0	13.8
Sulfur dioxide	9.6	17.0
Toluene	8.6	12.4
2, 3, 3-Trimethylbutane	9.5	10.5
Water	8.0	16.0
Xenon	9.3	23.0

* From Perry, J. H., "Chemical Engineers' Handbook," 3d ed., McGraw-Hill Book Company, Inc., New York, 1950.

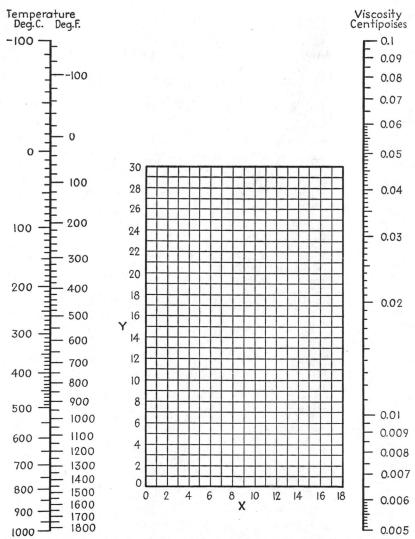

FIG. 15. Viscosities of gases. (Perry, "Chemical Engineers' Handbook," 3d ed., McGraw-Hill Book Company, Inc., New York, 1950.)

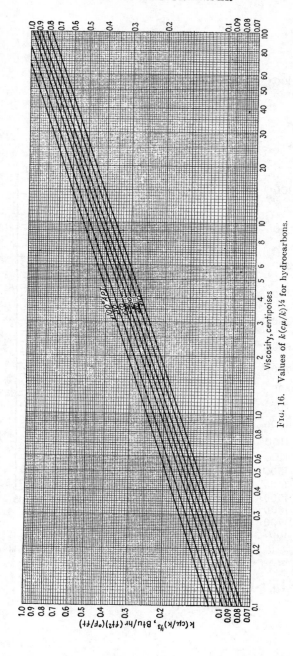

Fig. 16. Values of $k(c\mu/k)^{1/3}$ for hydrocarbons.

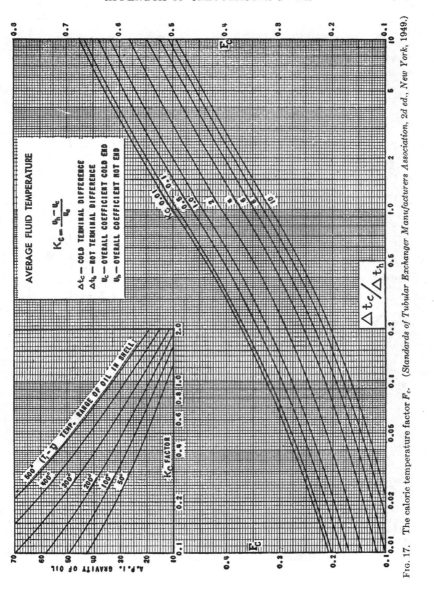

Fig. 17. The caloric temperature factor F_c. (*Standards of Tubular Exchanger Manufacturers Association*, 2d ed., New York, 1949.)

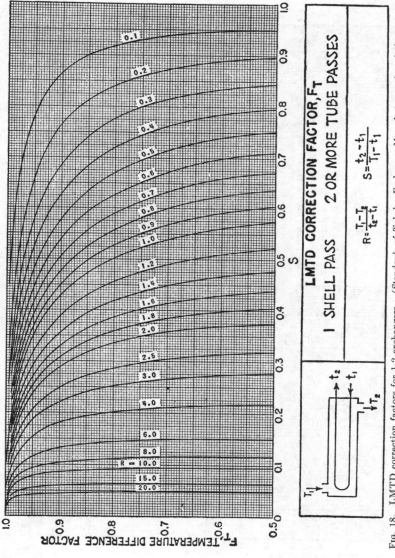

Fig. 18. LMTD correction factors for 1-2 exchangers. (*Standards of Tubular Exchanger Manufacturers Association*, 2d ed., New York, 1949.)

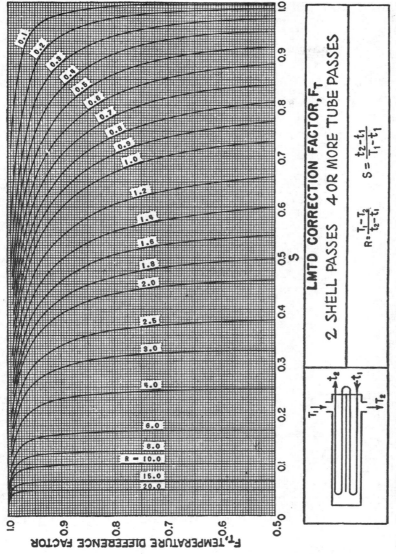

Fig. 19. LMTD correction factors for 2-4 exchangers. (*Standards of Tubular Exchanger Manufacturers Association* 2d ed., New York, 1949.)

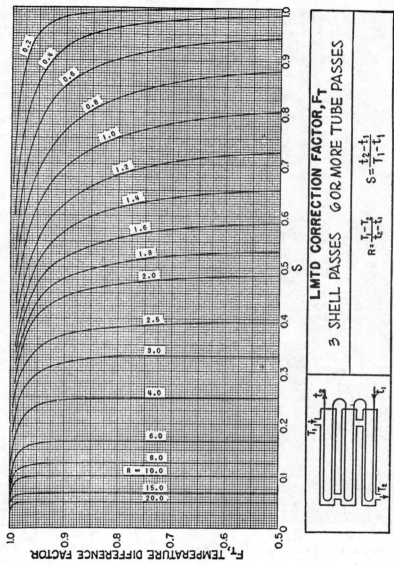

Fig. 20. LMTD correction factors for 3-6 exchangers. (*Standards of Tubular Exchanger Manufacturers Association*, 2d ed., New York, 1949.)

APPENDIX OF CALCULATION DATA

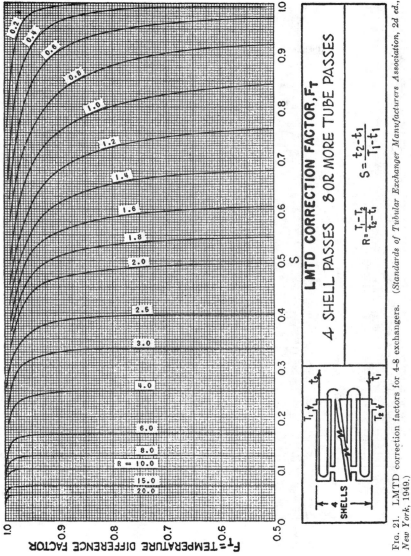

Fig. 21. LMTD correction factors for 4-8 exchangers. (*Standards of Tubular Exchanger Manufacturers Association*, 2d ed., New York, 1949.)

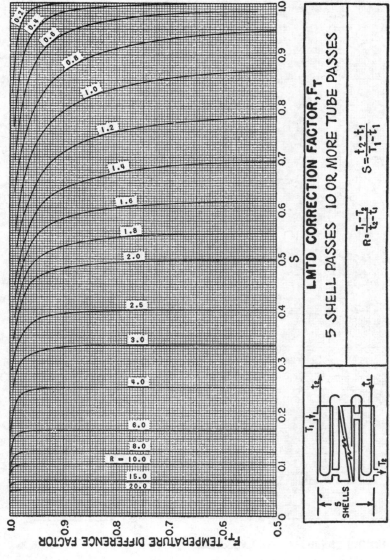

Fig. 22. LMTD correction factors for 5-10 exchangers. (*Standards of Tubular Exchanger Manufacturers Association*, 2d ed., New York, 1949.)

APPENDIX OF CALCULATION DATA 833

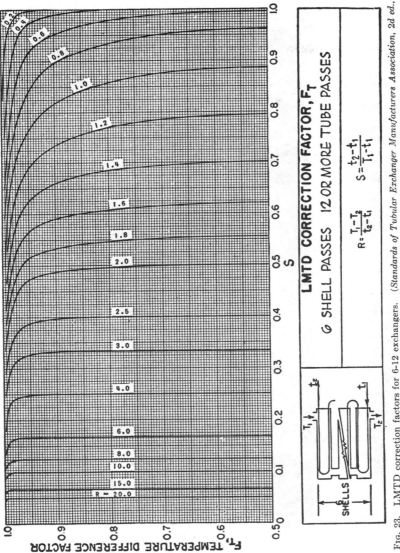

Fig. 23. LMTD correction factors for 6-12 exchangers. (*Standards of Tubular Exchanger Manufacturers Association*, 2d ed., New York, 1949.)

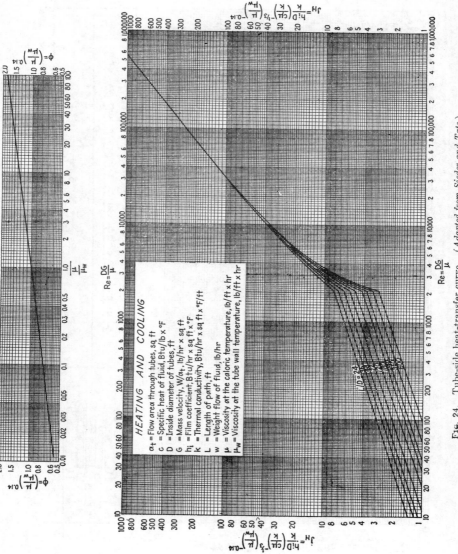

Fig. 24. Tube-side heat-transfer curve. (Adapted from Sieder and Tate.)

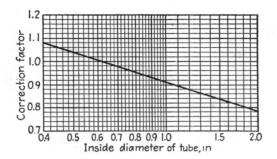
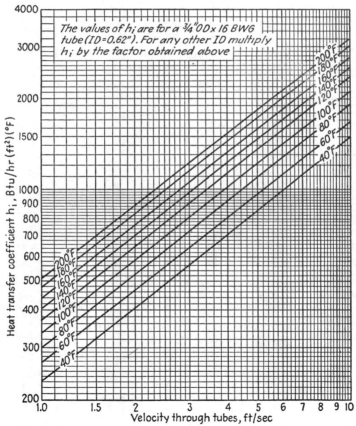

Fig. 25. Tube-side water-heat-transfer curve. [*Adapted from Eagle and Ferguson, Proc. Roy. Soc.*, **A127**, 540 (1930).]

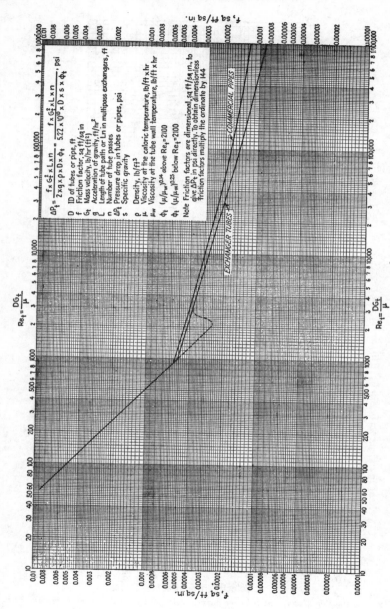

Fig. 26. Tube-side friction factors. (*Standards of Tubular Exchanger Manufacturers Association*, 2d ed., New York, 1949.)

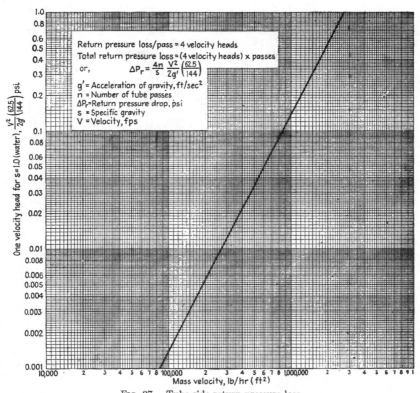

Fig. 27. Tube-side return pressure loss.

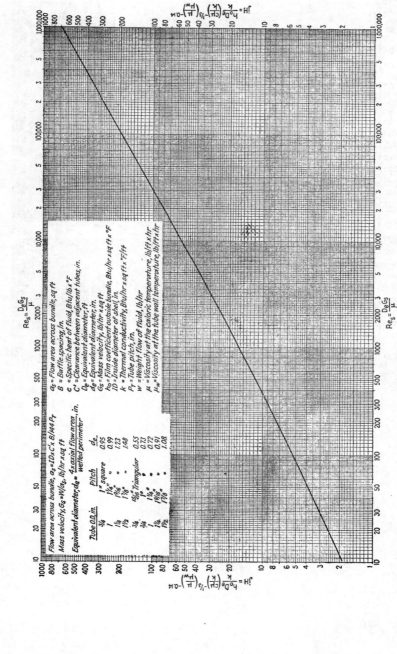

Fig. 28. Shell-side heat-transfer curve for bundles with 25% cut segmental baffles.

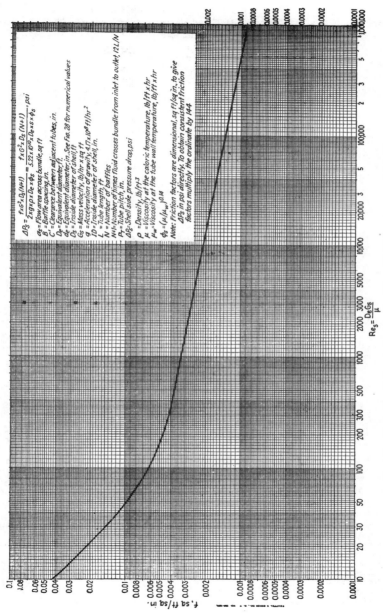

Fig. 29. Shell-side friction factors for bundles with 25% cut segmental baffles.

TABLE 8. APPROXIMATE OVERALL DESIGN COEFFICIENTS
Values include total dirt factors of 0.003 and allowable pressure drops of 5 to 10 psi on the controlling stream

Coolers

Hot fluid	Cold fluid	Overall U_D
Water	Water	250–500§
Methanol	Water	250–500§
Ammonia	Water	250–500§
Aqueous solutions	Water	250–500§
Light organics*	Water	75–150
Medium organics†	Water	50–125
Heavy organics‡	Water	5–75 ‖
Gases	Water	2–50 ¶
Water	Brine	100–200
Light organics	Brine	40–100

Heaters

Hot fluid	Cold fluid	Overall U_D
Steam	Water	200–700§
Steam	Methanol	200–700§
Steam	Ammonia	200–700§
Steam	Aqueous solutions:	
Steam	Less than 2.0 cp	200–700
Steam	More than 2.0 cp	100–500§
Steam	Light organics	100–200
Steam	Medium organics	50–100
Steam	Heavy organics	6–60
Steam	Gases	5–50 ¶

Exchangers

Hot fluid	Cold fluid	Overall U_D
Water	Water	250–500§
Aqueous solutions	Aqueous solutions	250–500§
Light organics	Light organics	40–75
Medium organics	Medium organics	20–60
Heavy organics	Heavy organics	10–40
Heavy organics	Light organics	30–60
Light organics	Heavy organics	10–40

* *Light organics* are fluids with viscosities of less than 0.5 centipoise and include benzene, toluene, acetone, ethanol, methyl ethyl ketone, gasoline, light kerosene, and naphtha.

† *Medium organics* have viscosities of 0.5 to 1.0 centipoise and include kerosene, straw oil, hot gas oil, hot absorber oil, and some crudes.

‡ *Heavy organics* have viscosities above 1.0 centipose and include cold gas oil, lube oils, fuel oils, reduced crude oils, tars, and asphalts.

§ Dirt factor 0.001.

‖ **Pressure** drop 20 to 30 psi.

¶ **These rates are greatly influenced by the operating pressure.**

TABLE 9. TUBE-SHEET LAYOUTS (TUBE COUNTS)
Square Pitch

Shell ID, in.	1-P	2-P	4-P	6-P	8-P	Shell ID, in.	1-P	2-P	4-P	6-P	8-P
\multicolumn{6}{c}{¾ in. OD tubes on 1-in. square pitch}	\multicolumn{6}{c}{1 in. OD tubes on 1¼-in. square pitch}
8	32	26	20	20		8	21	16	14		
10	52	52	40	36		10	32	32	26	24	
12	81	76	68	68	60	12	48	45	40	38	36
13¼	97	90	82	76	70	13¼	61	56	52	48	44
15¼	137	124	116	108	108	15¼	81	76	68	68	64
17¼	177	166	158	150	142	17¼	112	112	96	90	82
19¼	224	220	204	192	188	19¼	138	132	128	122	116
21¼	277	270	246	240	234	21¼	177	166	158	152	148
23¼	341	324	308	302	292	23¼	213	208	192	184	184
25	413	394	370	356	346	25	260	252	238	226	222
27	481	460	432	420	408	27	300	288	278	268	260
29	553	526	480	468	456	29	341	326	300	294	286
31	657	640	600	580	560	31	406	398	380	368	358
33	749	718	688	676	648	33	465	460	432	420	414
35	845	824	780	766	748	35	522	518	488	484	472
37	934	914	886	866	838	37	596	574	562	544	532
39	1049	1024	982	968	948	39	665	644	624	612	600
\multicolumn{6}{c}{1¼ in. OD tubes on 1 9/16-in. square pitch}	\multicolumn{6}{c}{1½ in. OD tubes on 1⅞-in. square pitch}
10	16	12	10								
12	30	24	22	16	16	12	16	16	12	12	
13¼	32	30	30	22	22	13¼	22	22	16	16	
15¼	44	40	37	35	31	15¼	29	29	25	24	22
17¼	56	53	51	48	44	17¼	39	39	34	32	29
19¼	78	73	71	64	56	19¼	50	48	45	43	39
21¼	96	90	86	82	78	21¼	62	60	57	54	50
23¼	127	112	106	102	96	23¼	78	74	70	66	62
25	140	135	127	123	115	25	94	90	86	84	78
27	166	160	151	146	140	27	112	108	102	98	94
29	193	188	178	174	166	29	131	127	120	116	112
31	226	220	209	202	193	31	151	146	141	138	131
33	258	252	244	238	226	33	176	170	164	160	151
35	293	287	275	268	258	35	202	196	188	182	176
37	334	322	311	304	293	37	224	220	217	210	202
39	370	362	348	342	336	39	252	246	237	230	224

TABLE 9. TUBE-SHEET LAYOUTS (TUBE COUNTS).—*(Continued)*
Triangular Pitch

¾ in. OD tubes on 15/16-in. triangular pitch						¾ in. OD tubes on 1-in. triangular pitch					
Shell ID, in.	1-P	2-P	4-P	6-P	8-P	Shell ID, in.	1-P	2-P	4-P	6-P	8-P
8	36	32	26	24	18	8	37	30	24	24	
10	62	56	47	42	36	10	61	52	40	36	
12	109	98	86	82	78	12	92	82	76	74	70
13¼	127	114	96	90	86	13¼	109	106	86	82	74
15¼	170	160	140	136	128	15¼	151	138	122	118	110
17¼	239	224	194	188	178	17¼	203	196	178	172	166
19¼	301	282	252	244	234	19¼	262	250	226	216	210
21¼	361	342	314	306	290	21¼	316	302	278	272	260
23¼	442	420	386	378	364	23¼	384	376	352	342	328
25	532	506	468	446	434	25	470	452	422	394	382
27	637	602	550	536	524	27	559	534	488	474	464
29	721	692	640	620	594	29	630	604	556	538	508
31	847	822	766	722	720	31	745	728	678	666	640
33	974	938	878	852	826	33	856	830	774	760	732
35	1102	1068	1004	988	958	35	970	938	882	864	848
37	1240	1200	1144	1104	1072	37	1074	1044	1012	986	870
39	1377	1330	1258	1248	1212	39	1206	1176	1128	1100	1078

1 in. OD tubes on 1¼-in. triangular pitch						1¼ in. OD tubes on 19/16-in. triangular pitch					
8	21	16	16	14							
10	32	32	26	24		10	20	18	14		
12	55	52	48	46	44	12	32	30	26	22	20
13¼	68	66	58	54	50	13¼	38	36	32	28	26
15¼	91	86	80	74	72	15¼	54	51	45	42	38
17¼	131	118	106	104	94	17¼	69	66	62	58	54
19¼	163	152	140	136	128	19¼	95	91	86	78	69
21¼	199	188	170	164	160	21¼	117	112	105	101	95
23¼	241	232	212	212	202	23¼	140	136	130	123	117
25	294	282	256	252	242	25	170	164	155	150	140
27	349	334	302	296	286	27	202	196	185	179	170
29	397	376	338	334	316	29	235	228	217	212	202
31	472	454	430	424	400	31	275	270	255	245	235
33	538	522	486	470	454	33	315	305	297	288	275
35	608	592	562	546	532	35	357	348	335	327	315
37	674	664	632	614	598	37	407	390	380	374	357
39	766	736	700	688	672	39	449	436	425	419	407

1½ in. OD tubes on 1⅞-in. triangular pitch					
12	18	14	14	12	12
13¼	27	22	18	16	14
15¼	36	34	32	30	27
17¼	48	44	42	38	36
19¼	61	58	55	51	48
21¼	76	72	70	66	61
23¼	95	91	86	80	76
25	115	110	105	98	95
27	136	131	125	118	115
29	160	154	147	141	136
31	184	177	172	165	160
33	215	206	200	190	184
35	246	238	230	220	215
37	275	268	260	252	246
39	307	299	290	284	275

APPENDIX OF CALCULATION DATA 843

TABLE 10. HEAT EXCHANGER AND CONDENSER TUBE DATA

Tube OD, in.	BWG	Wall thickness, in.	ID, in.	Flow area per tube, in.^{2}	Surface per lin ft, ft^2		Weight per lin ft, lb steel
					Outside	Inside	
½	12	0.109	0.282	0.0625	0.1309	0.0748	0.493
	14	0.083	0.334	0.0876		0.0874	0.403
	16	0.065	0.370	0.1076		0.0969	0.329
	18	0.049	0.402	0.127		0.1052	0.258
	20	0.035	0.430	0.145		0.1125	0.190
¾	10	0.134	0.482	0.182	0.1963	0.1263	0.965
	11	0.120	0.510	0.204		0.1335	0.884
	12	0.109	0.532	0.223		0.1393	0.817
	13	0.095	0.560	0.247		0.1466	0.727
	14	0.083	0.584	0.268		0.1529	0.647
	15	0.072	0.606	0.289		0.1587	0.571
	16	0.065	0.620	0.302		0.1623	0.520
	17	0.058	0.634	0.314		0.1660	0.469
	18	0.049	0.652	0.334		0.1707	0.401
1	8	0.165	0.670	0.355	0.2618	0.1754	1.61
	9	0.148	0.704	0.389		0.1843	1.47
	10	0.134	0.732	0.421		0.1916	1.36
	11	0.120	0.760	0.455		0.1990	1.23
	12	0.109	0.782	0.479		0.2048	1.14
	13	0.095	0.810	0.515		0.2121	1.00
	14	0.083	0.834	0.546		0.2183	0.890
	15	0.072	0.856	0.576		0.2241	0.781
	16	0.065	0.870	0.594		0.2277	0.710
	17	0.058	0.884	0.613		0.2314	0.639
	18	0.049	0.902	0.639		0.2361	0.545
1¼	8	0.165	0.920	0.665	0.3271	0.2409	2.09
	9	0.148	0.954	0.714		0.2498	1.91
	10	0.134	0.982	0.757		0.2572	1.75
	11	0.120	1.01	0.800		0.2644	1.58
	12	0.109	1.03	0.836		0.2701	1.45
	13	0.095	1.06	0.884		0.2775	1.28
	14	0.083	1.08	0.923		0.2839	1.13
	15	0.072	1.11	0.960		0.2896	0.991
	16	0.065	1.12	0.985		0.2932	0.900
	17	0.058	1.13	1.01		0.2969	0.808
	18	0.049	1.15	1.04		0.3015	0.688
1½	8	0.165	1.17	1.075	0.3925	0.3063	2.57
	9	0.148	1.20	1.14		0.3152	2.34
	10	0.134	1.23	1.19		0.3225	2.14
	11	0.120	1.26	1.25		0.3299	1.98
	12	0.109	1.28	1.29		0.3356	1.77
	13	0.095	1.31	1.35		0.3430	1.56
	14	0.083	1.33	1.40		0.3492	1.37
	15	0.072	1.36	1.44		0.3555	1.20
	16	0.065	1.37	1.47		0.3587	1.09
	17	0.058	1.38	1.50		0.3623	0.978
	18	0.049	1.40	1.54		0.3670	0.831

TABLE 11. DIMENSIONS OF STEEL PIPE (IPS)

Nominal pipe size, IPS, in.	OD, in.	Schedule No.	ID, in.	Flow area per pipe, in.^{2}	Surface per lin ft, ft.^{2}/ft.		Weight per lin ft, lb steel
					Outside	Inside	
⅛	0.405	40* 80†	0.269 0.215	0.058 0.036	0.106	0.070 0.056	0.25 0.32
¼	0.540	40* 80†	0.364 0.302	0.104 0.072	0.141	0.095 0.079	0.43 0.54
⅜	0.675	40* 80†	0.493 0.423	0.192 0.141	0.177	0.129 0.111	0.57 0.74
½	0.840	40* 80†	0.622 0.546	0.304 0.235	0.220	0.163 0.143	0.85 1.09
¾	1.05	40* 80†	0.824 0.742	0.534 0.432	0.275	0.216 0.194	1.13 1.48
1	1.32	40* 80†	1.049 0.957	0.864 0.718	0.344	0.274 0.250	1.68 2.17
1¼	1.66	40* 80†	1.380 1.278	1.50 1.28	0.435	0.362 0.335	2.28 3.00
1½	1.90	40* 80†	1.610 1.500	2.04 1.76	0.498	0.422 0.393	2.72 3.64
2	2.38	40* 80†	2.067 1.939	3.35 2.95	0.622	0.542 0.508	3.66 5.03
2½	2.88	40* 80†	2.469 2.323	4.79 4.23	0.753	0.647 0.609	5.80 7.67
3	3.50	40* 80†	3.068 2.900	7.38 6.61	0.917	0.804 0.760	7.58 10.3
4	4.50	40* 80†	4.026 3.826	12.7 11.5	1.178	1.055 1.002	10.8 15.0
6	6.625	40* 80†	6.065 5.761	28.9 26.1	1.734	1.590 1.510	19.0 28.6
8	8.625	40* 80†	7.981 7.625	50.0 45.7	2.258	2.090 2.000	28.6 43.4
10	10.75	40* 60	10.02 9.75	78.8 74.6	2.814	2.62 2.55	40.5 54.8
12	12.75	30	12.09	115	3.338	3.17	43.8
14	14.0	30	13.25	138	3.665	3.47	54.6
16	16.0	30	15.25	183	4.189	4.00	62.6
18	18.0	20‡	17.25	234	4.712	4.52	72.7
20	20.0	20	19.25	291	5.236	5.05	78.6
22	22.0	20‡	21.25	355	5.747	5.56	84.0
24	24.0	20	23.25	425	6.283	6.09	94.7

* Commonly known as standard.
† Commonly known as extra heavy.
‡ Approximately.

APPENDIX OF CALCULATION DATA

TABLE 12. FOULING FACTORS*

Temperature of heating medium	Up to 240°F		240–400°F†	
Temperature of water	125°F or less		Over 125°F	
Water	Water velocity, fps		Water velocity, fps	
	3 ft and less	Over 3 ft	3 ft and less	Over 3 ft
Sea water	0.0005	0.0005	0.001	0.001
Brackish water	0.002	0.001	0.003	0.002
Cooling tower and artificial spray pond:				
Treated make-up	0.001	0.001	0.002	0.002
Untreated	0.003	0.003	0.005	0.004
City or well water (such as Great Lakes)	0.001	0.001	0.002	0.002
Great Lakes	0.001	0.001	0.002	0.002
River water:				
Minimum	0.002	0.001	0.003	0.022
Mississippi	0.003	0.002	0.004	0.003
Delaware, Schuylkill	0.003	0.002	0.004	0.003
East River and New York Bay	0.003	0.002	0.004	0.003
Chicago sanitary canal	0.008	0.006	0.010	0.008
Muddy or silty	0.003	0.002	0.004	0.003
Hard (over 15 grains/gal)	0.003	0.003	0.005	0.005
Engine jacket	0.001	0.001	0.001	0.001
Distilled	0.0005	0.0005	0.0005	0.0005
Treated boiler feedwater	0.001	0.0005	0.001	0.001
Boiler blowdown	0.002	0.002	0.002	0.002

† Ratings in the last two columns are based on a temperature of the heating medium of 240 to 400°F. If the heating medium temperature is over 400°F, and the cooling medium is known to scale these ratings should be modified accordingly.

Petroleum Fractions

Oils (industrial):
  Fuel oil.................... 0.005
  Clean recirculating oil........ 0.001
  Machinery and transformer oils 0.001
  Quenching oil................ 0.004
  Vegetable oils............... 0.003
Gases, vapors (industrial):
  Coke-oven gas, manufactured gas.................... 0.01
  Diesel-engine exhaust gas..... 0.01
  Organic vapors.............. 0.0005
  Steam (non-oil bearing)...... 0.0
  Alcohol vapors.............. 0.0
  Steam, exhaust (oil bearing from reciprocating engines) 0.001
  Refrigerating vapors (condensing from reciprocating compressors).................. 0.002
  Air........................ 0.002

Liquids (industrial):
  Organic.................... 0.001
  Refrigerating liquids, heating, cooling, or evaporating..... 0.001
  Brine (cooling)............. 0.001
Atmospheric distillation units:
  Residual bottoms, less than 25°API.................... 0.005
  Distillate bottoms, 25°API or above..................... 0.002
Atmospheric distillation units:
  Overhead untreated vapors... 0.0013
  Overhead treated vapors..... 0.003
  Side-stream cuts............ 0.0013
Vacuum distillation units:
  Overhead vapors to oil:
    From bubble tower (partial condenser).............. 0.001
    From flash pot (no appreciable reflux)............. 0.003

* Standards of Tubular Exchanger Manufacturers Association, 2d ed., New York, 1949.

TABLE 12. FOULING FACTORS.*—(Continued)

Item	Value
Overhead vapors in water-cooled condensers:	
From bubble tower (final condenser)	0.001
From flash pot	0.04
Side stream:	
To oil	0.001
To water	0.002
Residual bottoms, less than 20°API	0.005
Distillate bottoms, over 20°API	0.002
Natural gasoline stabilizer units:	
Feed	0.0005
O.H. vapors	0.0005
Product coolers and exchangers	0.0005
Product reboilers	0.001
H_2S Removal Units:	
For overhead vapors	0.001
Solution exchanger coolers	0.0016
Reboiler	0.0016
Cracking units:	
Gas oil feed:	
Under 500°F	0.002
500°F and over	0.003
Naphtha feed:	
Under 500°F	0.002
500°F and over	0.004
Separator vapors (vapors from separator, flash pot, and vaporizer)	0.006
Bubble-tower vapors	0.002
Residuum	0.010
Absorption units:	
Gas	0.002
Fat oil	0.002
Lean oil	0.002
Overhead vapors	0.001
Gasoline	0.0005
Debutanizer, Depropanizer, Depentanizer, and Alkylation Units:	
Feed	0.001
Overhead vapors	0.001
Product coolers	0.001
Product reboilers	0.002
Reactor feed	0.002
Lube treating units:	
Solvent oil mixed feed	0.002
Overhead vapors	0.001
Refined oil	0.001
Refined oil heat exchangers water cooled‡	0.003
Gums and tars:	
Oil-cooled and steam generators	0.005
Water-cooled	0.003
Solvent	0.001
Deasphaltizing units:	
Feed oil	0.002
Solvent	0.001
Asphalt and resin:	
Oil-cooled and steam generators	0.005
Water-cooled	0.003
Solvent vapors	0.001
Refined oil	0.001
Refined oil water cooled	0.003
Dewaxing units:	
Lube oil	0.001
Solvent	0.001
Oil wax mix heating	0.001
Oil wax mix cooling‡	0.003

‡ Precautions must be taken against deposition of wax.

Crude Oil Streams

	0–199°F			200–299°F			300–499°F			500°F and over		
	Velocity, fps											
	Under 2 ft	2–4 ft	4 ft and over	Under 2 ft	2–4 ft.	4 ft and over	Under 2 ft	2–4 ft	4 ft and over	Under 2 ft	2–4 ft	4 ft and over
Dry	0.003	0.002	0.002	0.003	0.002	0.002	0.004	0.003	0.002	0.005	0.004	0.003
Salt§	0.003	0.002	0.002	0.005	0.004	0.004	0.006	0.005	0.004	0.007	0.006	0.005

§ Refers to a wet crude—any crude that has not been dehydrated.

AUTHOR INDEX

A

Adams, L. H., 645
Alves, G., 206
Anderson, E. D., 526
Arnold, J. H., 576
Awbery, J., 21

B

Badger, W. L., 493
Bailey, A., 18
Baker, E. M., 338
Bates, O. K., 7
Bays, G. S., Jr., 728, 746, 747
Beiswenger, G. A., 353
Boelter, L. M. K., 206, 605, 640
Bolland, J. L., 7
Bonilla, C. F., 412, 521, 553
Boussinesq, J., 202
Bowman, R. A., 140, 547, 625, 727
Breidenbach, E. P., 137
Brenner, E., 718
Bridgman, P. W., 8, 30, 38
Brinn, M. S., 748
Brown, G. G., 356, 806, 807
Buckingham, E., 38

C

Carnahan, F. L., 821
Carpenter, C. L., 246, 283
Carslaw, H. S., 512, 640
Cartinhour, J., 554
Chaddock, R. E., 625
Cherry, V. H., 640
Chilton, T. H., 340, 574, 576, 717, 718, 722, 804, 815
Colburn, A. P., 94–96, 99, 137, 159, 266, 270, 318, 340, 564, 565, 570, 574, 576, 598, 716, 721, 804, 815
Comings, E. W., 190, 820
Considine, D. M., 766
Craig, H. L., 206
Cryder, D. S. 378

D

De Lorenzo, B., 526
Dodd, G. D., 409
Dodge, B. F., 170, 664
Donohue, D. A., 130

Douglass, R. D., 539
Drew, T. B., 40, 53, 253, 319, 377, 717, 718, 722, 728, 746, 747
Dunn, W. E., 553
Dusinberre, G. M., 640
Dwyer, J. B., 674

E

Eagle, A., 155, 836
Egbert, R. B., 691, 692
Egly, R. S., 190, 820
Evans, J. E., 697, 706

F

Ferguson, R. M., 155, 836
Finalborgo, A. C., 378
Fischer, F. K., 178
Fishenden, M., 77
Fisher, R. C., 625
Fourier, J. B., 11, 37, 86, 106, 107, 110, 144, 177, 201, 205, 639, 641
Friedman, S. J., 748
Friend, L., 601
Fritz, W., 375–377
Furnas, C. C., 669

G

Gaffert, G. A., 675
Gardner, K. A., 515, 538, 740
Gibbs, J. W., 252, 313
Gilliland, E. R., 344, 493
Goldschmidt, H., 640
Graetz, L., 40, 201, 202, 746
Grober, H., 640, 647, 648, 666
Guckert, F. A., 748
Gunter, A. Y., 554
Gurney, H. P., 651

H

Haigler, E. D., 769
Hashmall, F., 601
Hawley, R. W., 668
Hazelton, R., 338
Heilgenstadt, W., 665
Heilman, R. H., 18
Hilpert, R., 575
Holcomb, D. E., 806, 807

Hottel, H. C., 80, 674, 684, 685, 687, 689, 691, 692, 696, 697, 706
Hougen, O. A., 340
Hurd, N. L., 740
Hutchinson, E., 7

I

Ingersoll, A. C., 640
Ingersoll, L. R., 640

J

Jackson, D. H., 397
Jaeger, J. C., 512, 640
Jakob, M., 375–378, 512, 544, 640
Jameson, S. L., 554
Jebens, R. H., 717, 718, 722
Jenny, F. J., 810, 812–814
Jeschke, D., 721
Johnson, H. A., 640
Josefowitz, S., 190

K

Kallam, F. L., 731, 733
Kalustian, P., 443
Katz, D. L., 356
Kayan, C. F., 527
Keenan, J. H., 584, 816–819
Kelley, K. K., 799
Kern, D. Q., 206, 246, 283, 427, 821
Keyes, F. G., 584, 816–819
Koo, E. C., 53

L

Lansing, N. E., 206
Lewis, W. K., 493, 575
Lichtmann, R. S., 722
Lindseth, E. L., 705
Lobo, W. E., 601, 674, 697, 706, 707
Lof, G. O., 668
London, A. L., 605
Lund, G., 664
Lurie, J., 651
Lyell, N. C., 18

M

McAdams, W. H., 53, 137, 215, 225, 264, 270, 281, 376, 377, 493, 593, 599, 640, 674, 721, 728, 746, 747
McCabe, W. L., 493, 496
Mack, D. E., 717
Mahony, L. H., 722
Marks, L. S., 795–799
Martinelli, R. C., 206, 640
Mason, W. E., 605
Meisler, J., 409, 427
Melville, H. W., 7
Merkel, F., 587
Milne, W. E., 615

Monrad, C. C., 708
Morris, F. H., 46, 59, 93, 103
Morrison, M. S., 718
Morrow, D. G., 705
Mueller, A. C., 140, 377, 547, 625, 727
Mullikin, H. F., 687, 697, 706
Murphy, G. B., 5, 360
Murray, W. M., 515

N

Nagle, W. M., 140, 253, 547, 625, 727
Nelson, E. F., 5, 360
Nelson, W. L., 733
Newman, A. B., 649, 650, 654, 656
Nusselt, W., 253, 256, 270, 272, 318, 330, 339, 547

O

O'Connell, H. E., 137
Orrok, G. A., 697
Othmer, D. F., 190, 206, 302, 339, 399, 400

P

Packie, J. W., 353, 355–357
Parekh, M. D., 599
Partridge, E. P., 640
Perry, J. H., 493, 564, 674, 689, 716, 721, 766, 787, 800–802, 804, 805, 822–825
Phillips, G. A., 718
Pigford, R. L., 565, 748
Piroomov, R. S., 353
Porter, R. W., 766
Prandtl, L., 56, 57

R

Ray, B. B., 214
Reed, C. E., 539, 615, 640, 657, 660
Rescorla, A. R., 821
Reynolds, O., 29, 54, 56
Richter, G. A., 721
Robinson, C. S., 493
Rushton, J. H., 722

S

Saha, M. N., 7
Saunders, M. T., 625
Saunders, O. A., 77
Schack, A., 640, 641, 664
Scheibel, E. G., 810, 812–814
Schmidt, E., 544, 657, 660
Schofield, F., 21
Schumann, T. E., 668
Seltzer, M., 53
Shaw, W. A., 554
Sherwood, T. K., 344, 539, 565, 576, 598, 599, 614, 640, 657, 660
Short, B. E., 137
Sieder, E. N., 60, 99, 103, 223, 228, 229, 576, 835
Simpson, W. M., 598, 615

Smith, D. M., 547
Smith, J. F. D., 8
Smith, R. L., 191
Smith, W. Q., 253
Southwell, C. J., 206
Srivastava, B. N., 7
Stanton, T. E., 56
Sutherland, W., 9

T

Tate, G. E., 60, 99, 103, 554, 576, 835
Taylor, G. I., 56
Ten Broeck, H., 169, 178, 635
Thiele, E. W., 496

U

Uhl, V. W., 717
Underwood, A. J. V., 140, 145

V

Van Nostrand, W., 821
Vernon, H. C., 804, 815

W

Walker, W. H., 493
Watson, K. M., 5, 191, 360
Weinberg, E. B., 206
White, A. M., 718
Whitman, W. G., 46, 59, 93, 103
Williamson, E. D., 645
Wilson, D. W., 697, 706
Wilson, R. E., 53, 309
Wohlenberg, W. J., 697, 705, 706

Z

Zenz, F., 601
Zobel, O. J., 640

SUBJECT INDEX

(The principal tables, graphs, and equations are shown in boldface type)

A

Absolute pressure in surface condensers, 305
Absolute temperature, 3, 5, 74, 192
Absolute viscosity, 28
Absorbents, 222, 341
Absorbers, 222
Absorption, 341
  in gasoline recovery, 222
  of luminous flames, 690
Absorption oil, 4, 235
Absorptivity, definition of, 67
Acceleration head loss, 486
Acetic acid, 185
Acetone, 185, 470
Adiabatic compression, 192, 196, 197
Adiabatic humidification, 571–573
Adiabatic temperature, 192
Adsorptive processes, 747
After condensers, 394, 395, 397
  jet, 394, 564
  surface, 395, 397
Aftercoolers and aftercooling, 192–195
Agitation in vessels, effect of, on coefficient, **717–719**, 721–733
  on transfer time, **725–733**
Agitators, paddle, 717, 722
  turbine, 722
Air, 191, 196
  atmospheric, reasonable maximum temperatures of, **596**
  summer wet-bulb temperatures of, **594–596**
  compression of, 192, 195
  dissolved in water, 154, 302
  enthalpies for, 585
  over finned strip heaters, **757**
  over finned tubes, 535, 543
  friction factors for, 191, **836, 839**
  heat-transfer coefficients in, 192
  humid heat of, 588
  humidities of, **585, 586**
  as steam contaminant, 302, 397
  over strip heaters, **756**
  (*See also* Gas; Gases)
Air compressors, 195
Air conditioning, 544, 611
Air gap, resistance of, 15
Air jet pump, 305, 306, 394
  (*See also* Ejector)
Air leakage in surface condensers, 306, 397
Air preheaters, **678, 680, 701**

Alcohol heater, calculation of 241
Algae, 154, 724
Allowable pressure drop, in condensers, **272**
  in exchangers, 108
Alloys, properties of, **799**
American Petroleum Institute (API), degrees API, definition of, 4
American Society for Testing Materials (ASTM), 353
Ammonia, 193, 619
Analogies, between electrical and heat conduction, 12, 527
  between heat transfer, and fluid friction, 54–57, 564, 574
  between heat transfer and mass transfer, 340, 564, 574
Analysis of performance, in cooling towers, **602**
  in double pipe exchangers, **110**
  in 1–2 exchangers, **148**
  in 2–4 exchangers, **181**
Aniline, 217
Annealing, 645
Annual fixed charges, 21, 158, 159, 222, 224
  in distillation, 325
  in evaporators, 386, 389
Annual operating hours, 158
Annuli, in bayonets, 745
  in condensing films, 264
  in core tubes, 210
  in double pipes, **104–106**
API (*see* American Petroleum Institute)
Apparent overall coefficients, 399
Apparent temperature drop, 399
Approach temperature, in barometric condensers, 396
  in cooling towers, 594
  in exchangers, 85
  in parallel flow–counterflow exchangers, 146
Aqueous solutions, 398, 453
  heat-transfer coefficients of, in pipes and tubes, 103
  properties of, **161**
  (*See also* specific chemical names)
Area, crossflow, in shell-and-tube exchangers, 138
  effective, in radiant heat transfer, 78–82, 683, 684, 686
  for heat transfer in cylindrical resistances, 16
  in tubular equipment, 129, 306
Asphalt, 739
ASTM distillation, 353
  calculations using, 356–367

Atmospheric coolers, 581, 594, **730–738**
Atomic energy, 515
Atomic volumes, table of, **344**
Automatic control, 766
Automatic-control mechanisms, 771
Automatic reset, 770
Average fluid temperature (*see* Caloric temperature)

B

Back-pressure regulators, 767, 776
Baffle pitch, **129**, 227
Baffle spacers, 129
Baffles, 129-131
  dam, 294, 331
  disc and doughnut, 130
  helical, in jackets, 717
  longitudinal, in divided flow exchangers, 478
    in 2-4 exchangers, 175, 176, 179
  orifice, 131
  segmental, 130
    horizontally and vertically cut, 180
    maximum and minimum spacing for, 129, **226**, 227
    for side-to-side flow, 130
  split segmental, 180
  for up-and-down flow, 130
Balanced pressure drop, 280, 294
Bare pipes, heat losses from, 17–21
Bare tube coolers, 581, 594, **730–738**
Barometric condensers, 394, 396, 425, 427
  water requirements in, 396
Batch cooling and heating, of liquids, 625, 659
  agitated, calculation of time for, 635–636
    coefficients for, 627
    by coils, 626–628
    by counterflow exchangers, 628–631
    by 1-2 exchangers, 631–633
    by 2-4 exchangers, 633
    by jackets, 626–628
  without agitation, calculation of time for, 636–637
    by coils, 633
    by counterflow exchangers, 633–635
    by 1-2 exchangers, 635
    by 2-4 exchangers, 635
    by jackets, 633
  of solids (*see* Conduction, unsteady)
Batch processes, 624
  vs. continuous processes, 624
  instrumentation for, 787
Batch stills, 637
Bayonet exchangers, 738–746
Beam length, 689
Beattie-Bridgman equation, 191
Benzene, 113, 338, 502, 635
Berl saddles, humidification characteristics of, 600
Bessel functions, 539
Biot number, 37
Birmingham wire gage (BWG), 128
Black ash, 426
Black body, 66

Black liquor, 426
Blanketing in vaporization, 376
Blast furnace, 664
Blast-furnace charge, heat transfer to, 670
Bleed heater, 388
Bleed steam, 387, 417
Blowdown, 381, 384, 393, 454
Bohr, 63
Boil-up rate, 298
Boiler feed water, 302
Boilers, definition of, 378
  steam-generating, 675
    efficiency of, 678, 701
    fire-tube, 675, 695
    water-tube, 675
Boiling, 375
  film and nuclear, 377
  (*See also* Evaporation; Vaporization)
Boiling point vs. bubble point, 322
Boiling-point elevation, 400
Boiling-point rise (BPR), 399, 400
Boiling range in multicomponent mixtures, 322, 468
Boltzmann, 3, 74
Booster-ejectors, 393–395, 443
  steam consumption in, graphs, 444
Bottoms product, 255
Boundary layer, 54–57
Breathing in reboilers, 485
Brine, 433, 473
  specific gravity of, 808
  specific heat of, 804
  thermal conductivity of, 800
  viscosities of, 822
Brix, definition of, 418
Bubble caps, 254
Bubble point, vs. boiling point, 322
  definition of, 322
Bubble-point mixtures, multicomponent, calculation of, 322–325
  using equilibrium constants, 324
  using relative volatilities, 325
Buffer layer, 57
Building materials, thermal conductivities of, **795**
Bulk temperature, 206, 215
Bundle, 135, 217, 229
Butane, 285, 316, 317, 331, 464
Butane vapor-pressure curve, 321
By-passing, 772, 773
By-product power, 391

C

Calcium scale, 380
Calcium sulfite, 426
Calculation, of desuperheater-condensers, 283–289
  of double pipe exchangers, **110**
  of double pipe extended-surface exchangers, **530**
  of 1-2 exchangers, **148**
    in series, 184
  of 2-4 exchangers, 181
  of 1-2 extended-surface exchangers, 534

SUBJECT INDEX

Calculation, of 1-2 horizontal condensers, 274–277
  of surface condensers, 307
  trial-and-error, 19, 231
  of 1-2 vertical condensers, 277–280
  (*See also* Index to Principal Apparatus Calculations)
Calming sections, 44
Caloric fraction, 96
Caloric temperature, **93–97**
  calculation of, 97
Caloric-temperature factor, 96, **827**
Caramalizing, 420
Carbon dioxide, 344, 690
Carbon disulfide, 299
Carnot cycle, 301
Cartridge heaters, 754
Cascade coolers, 727
Catalysts, fixed and moving bed, 624, **668**
Catalytic processes, 725, 747, 769
Catchalls, 403, 425
Caustic embrittlement, 433
Caustic soda (*see* Sodium hydroxide)
Cell liquor, 438
Centipoise, 28
Centistokes, 28
Centrifugal pumps, flow control in, **772**
Ceramics, 727
Channels, 129, 133, 305
  flow area in, 202
Characterization factor, 5
Chemical evaporators, 378, 400–449
Chemical furnaces, 674
Chemical reaction, 298
Chiller, 459, 472–475
  instrumentation for, 786
Chlorbenzene, 338
Clausius-Clapeyron equation, 400
Clean overall coefficient, 106
Cleaning lanes, 128
Cleaning transfer surface, 108
Cleanliness factor, 307
Clearance between tubes, 128
Coal, as fuel, 675, 678, 689, 705
  heat transfer to, 670
Coefficient of heat transfer (*see* Heat-transfer coefficient)
Coils, pancake, 721
  pipe, submerged, 723–725
    vertical, 722
  in tanks, time for heating with, 626–628
  tube, helical, 720
    calculation of, 723
Coke, heat transfer to, 670
Coke deposits, 675, 679
Colloidal dispersions, properties of, 161
Combustion, 689–692
Components in phase rule, 315
Composition of vapor mixture, 313
  calculation of, 323–327
  between dew point and bubble point, 327, 362
Compression, adiabatic, 192
  polytropic, 192

Compressor aftercoolers, 192–195
  temperature control for, **776**
Compressors, single and multistage, 192, 195
Condensate, pumped return of, 273
Condensate depression in surface condensers, 305
Condensate loading (*see* Condensation)
Condensate return pump, 273
Condensation, 252–312, 313–374
  dropwise, 252, 338
  film, 252
    in bayonet exchangers, 738
    coefficients of, graph, **267**
    with condensable impurities, 281
    dimensional analysis of, 253
    of exhaust steam, 268, 397
    graphical analysis of, 308
    heat load vs. temperature in, 313
    in horizontal tubes, condensate loading of, 269
    on horizontal tubes, 261
      coefficients of, equaton, **265**
      condensate loading in, 265, 266
    horizontal vs. vertical, 268
    on inclined surfaces, 260
    isothermal, 252
    with noncondensable impurities, 281, 397
    nonisothermal, 281
    Nusselt's theory of, 256–263
    recommended curves for, **267**, 270
    recommended equation, **266**, 267
    of single vapors, 252–312
    of steam, equations, **266**
      presence of air in, 281
      in surface condensers, **301–309**
      nomenclature, 305
    of superheated vapor, 281
    vapor-loading, 265
    variation of, with temperature, 313
    on vertical surfaces, 257–260
      coefficients of, equation, **265**
      condensate loading in, 265
      flow transition height of, 270
      working equations for, **264–266**
    in vertical tubes, semiempirical coefficients of, **270**
    zone of, 282, 289
Condensation of mixed vapors, 313–370
  binary mixture, 318
  condensing curve for, 319
    calculation of, 329
  condensing range of, by phase rule, 317
  differential, 328
  immiscibles from noncondensable gas in, 352–372
  miscibles from immiscible in, 337
  multicomponent mixture in, 319–337
  from noncondensable gas, 313, 317, 564
  phase-rule applications in, 315
  of two miscible vapors, 317, 318
  of steam from air, 343
  from noncondensable gas, 339, 367
    calculation of, 346–**351**

Condenser tubes, 128
  dimensions of, **843**
Condenser vacuum, in surface condensers, 305
Condensers, air-cooled, 545
  barometric, 394, 396, 425, 427, 432
  for batch distillation, 639
  calculation types of, 271
  definition of, 102
  and dephlegmators, 280
  evaporative, 732, 738
  instrumentation for, 780–783
  low-level, 396
  operating pressure of, 325, **577**
  partial, 280, 318
    instrumentation for, 783
  reflux, 288
  shell-and-tube, condenser-subcooler, calculation of, 290–293
    1-1, 282, 289, 299
    1-2, 271, **274**, 277, 280, 281, 283, 285, 289, 290, 295, 325, 332, 345
    desuperheater, calculation of, 285–289
    divided-flow, 246
    horizontal, 268, 269
      with gravity reflux, 268
      n-propanol-water, calculation of, 274–277
      split-flow, 274
    horizontal vs. vertical, 269
    knock-back, 298
    multicomponent-mixture, calculation of, 335
    partial, 280
    reflux-type, 298
      calculation of, 299–301
    steam-CO_2, calculation of, 346
    surface, for steam, 255, 301–309, 427, 433
      calculation of, 308
      crossflow in, 304
      dissolved air, in, 302
      pressure drop in, 302
    vertical, 268
      condenser-subcooler, calculation of, 289–293
      n-propanol-water, calculation of, 277–280
      with single tube pass, 271
    vacuum, 274, 738
  (*See also* Aftercondensers; Condensation; Desuperheater-condensers; Intercondensers)
Condensing curve, 319, 329
  influence of components on, 352
  for multicomponent mixture, calculation of, 332, 366
Condensing range for multicomponent mixture, calculation of, 332
Condensing zone, 282, 289, 295, 331
  for mixed vapors, 331
Conditioning of air, 544, 611
Conductance, definition of, 6
  unit, definition of, 17
Conduction, Fourier's general equation of, 11
  periodic, 624, 662–664
  steady, 2, 6–23
    through composite wall, 14
    through composite pipe wall, 16

Conduction, steady, through condensate film, 263
  in extended surfaces (*see* Extended surfaces)
  in fins (*see* Extended surfaces)
  through fluids, 8
  through granular materials, 747
  graphical solution of, 21
  through insulation, 17, 20
  through metals, 12
  through nonmetal solids, **7**
  through pipe wall, 15
  through resistances in series, 14
  with ribs, 22
  through solids, 8
  temperature gradient in, 2, 10
  three-dimensional, 11
  through wall, 13
  unsteady, 624, 639–670
    annealing, 645
    with constant environment temperature, in brick shape, 653, 656–657
    with contact resistance, 641, 647–657
    without contact resistance, 640–647
    in cube, 645, 646, 651
    in cylinders, 645, 646, 651
    with constant environment temperature, in finitely thick wall, 644, 647, 651, 657
    graphical solution of, 657–662
    Gurney-Lurie charts for, **652–655**
    in infinitely thick wall, 641
    in infinitely wide slab, 645, 646, 651
    Newman's method for, 653–657
    in rectangular shapes, 648
    Schack charts for, **649, 650**
    Schmidt method for, 657–662
    in spheres, 645, 646, 651
    in square bars, 645, 646, 651
    time-temperature-distance in, 641, 657–662
    and Fränkl regenerators, 664
    to granular materials, 668–670
    with heat-source variations, 624
    heat-storage factor in, 666
    in infinitely thick wall, calculation of, 643
    with periodic temperatures, 624, 662–664
    in regenerators, 625, 664–668
    in reversing exchangers, 515, 664
Conductivity (*see* Thermal conductivity; Electrical conductivity)
Conductors, definition of, 8
Configuration factor, 555
Consistent units, 37
Constant boiling mixture, 281, 318, 319, 638
Constant pressure in condensers, 316, 322
Constants, 794
Contact resistance, 9, 640, 647
Continuity equation, 31, 567
Continuous vs. batch processes, 624
Contraction head loss, 108, 112, 148, 273, 487
Control process, automatic, 766
  manual, 772
Controllers, pilot-acting, 767
  self-acting, 766
Controlling film coefficient, 87, 96

SUBJECT INDEX 855

Convection, 2, 25–27
  in cooling towers, 586–590
  forced, 2, 25–27
    in annuli, double pipe exchangers, 105
      with extended surface, 524
    in batch cooling and heating, 626–637
    in batch operations, 625–659
    in condensate films, 270
    in pipes and tubes, correlation of, 46–51
      experimental determination of, 43–46
      for liquids, 103
      recommended equations, **103**
      for solutions, 166
      for water, 155, **835**
    in shells, 136–140
      for gases, 192
      for liquids, 136, 166, **838**
      recommended equation, **137, 838**
      for solutions, 166
      for water, 155
    and simultaneous diffusion, 586–590
  free, 3
    from coils, 721
    dimensional analysis of, 206
    in evaporators, 375, 400
    for gases, 215
    heater calculation of, 217
    in jacketed vessels, 716
    in pipes and tubes, horizontal, 205
      vertical, 206
    outside pipes and tubes, 214
      recommended equation, **215**
    from planes, 215
    and radiation, 18, 218
    recommended equations, **216, 217**
    and streamline flow, 206
  in furnaces, 675
  natural (*see* Convection, free)
Conversion factors, 793, 794
  force to mass, 34
  kinetic energy to heat, 34
Coolers, 102
  atmospheric, 581, 594, **730–738**
  cascade, 727
  direct-contact, 614–621
  drip, 727
  horizontal-film, 727
  instrumentation for, **775**
  open, 730
  S-type, 727
  serpentine, 727
  spray, 730
  submerged-pipe coil, 723–**727**
  trickle, 727
  trombone, 727–730
    temperature difference in, 727–728
  (*See also* Exchangers)
Cooling towers, 563, 570, 576–611
  air enthalpies for, **585**
  air loading in, 584, 601
  approach temperature in, 594, 606

Cooling towers, atmospheric, 577, 578
  wind velocities for, **580**
  calculation of diffusion units of, 591, 602–605, 609–611
  crossflow in, 579
  as dehumidifiers, 611, 613, 614
  depreciation in, 602
  design conditions for, 594–598, 601, 602, 607
  diffusion unit in, 591, 602–604, 612
  diffusion units vs. performance, 591
  drift loss in, 578, 583
  droplet surface in, 582
  elevation of, above sea level of, 606
  enthalpies for, **585**
    above sea level, **586**
  enthalpy-temperature plot for, 591
  equations for, **588**
  fans for, 609, 613
  fill in, 581–583, 590
  film surface in, 582
  flooding in, 601
  fogging from, 598
  forced draft in, 577
  graphical integration in, 591
  guarantees for, 602, 605, 608–611
  heat balance in, 583–585, 614
  induced draft in, 577
  liquid film resistance in, 593
  loading in, 584, 601
  loading variations in, 609
  log mean enthalpy difference in, 592, 603
  make-up in, 584
  mechanical draft in, 577
  natural circulation in, 577, 578–581
  natural draft in, 577, 579
  number of diffusion units in, 590
  numerical integration in, 591
  operating range in, 607
  performance calculations of, **602–605**
  performance characteristics of, 599
  performance curves of, 609–611
  pressure drop in, 601
  process conditions for, 594–598
  process conditions affecting size of, 605–608
  recirculation in, 578
  seasonal variations in, 608
  without simplified Lewis numbers, 614
  staging in, 608
  summer wet-bulb, **595**
  unsaturation of inlet air, 607
Cooling-tower water in evaporator-condensers, 390
Core tubes, 209
Correlation of experimental data on fluid friction, 51–54
  on heat transfer, to extended surfaces, 526
  in pipes, 46–51
Cosine law, 79
Cosmic rays, 62
Cost, of cooling water, 154, 159
  of exchangers, installed, 223
    minimum annual, 225
    in relation to size, 224
    tube diameter, 228

Cost, minimum, of insulation, 21
  of steam, 165
Countercurrent (*see* Counterflow)
Counterflow, definition of, 85
  heat recovery in, 92
  methods of approximating, 184
  temperature difference in, 87
Cox chart, 356
Cracking, thermal, 678, 679
Critical radius, 20
Critical temperature difference, 376, 377, 400
Critical velocity, 53
Crossflow, 545
  in surface condensers, 304, 307
  temperature difference in, 546–553, 727–728
Crossflow area in shell-and-tube exchangers, 138
Crude oil, 4, 121, 151, 203
  utilization of, 678
Cube, conduction in, 645, 646, 651
Curvilinear squares, 22
Cycle efficiency, 303
Cycles, Carnot, 301
  power, 301, 388, 391
Cylinders, free convection from, 214, **216**
  radiation from, 77
  unsteady conduction in, 645

D

Dalton's law, 320
Dam baffle, 294, 331
Deaeration of feedwater, 302
Defecation of sugar, 418
Degree of freedom, 315
Dehumidification, 576, 611, 613, 614
Dephlegmator, 280
Derivative function, 770
Design, thermal, of exchangers,
    method for, 225–229
  outline of, **229–231**
Design overall coefficient, 106
Desuperheater-condensers, 282–285
Desuperheaters, 198
Desuperheating, 271, 281, 454
Desuperheating zone, 282, 331
Dew point, 195, 316, 339
  of multicomponent mixtures, 316, 322
    calculation of, 322–325
      using equilibrium constants, 324
      using relative volatilities, 326
  of vapor-gas mixtures, 195
    calculation of, 196, 361
Dew-point vs. wet-bulb temperature, 573
Dew points, systems with two, 352
Diameter, equivalent,
  in annuli, **104**, 745
  in core tubes, 210
  in shells, **138**, 167, 535, **838**
  spherical, for solids, 669
Diameters, of pipes, **844**
  of tubes, **843**
Diaphragm valves, 768
Differential condensation, 328

Diffusion, 253, 281, 318, 339, 367, 564–570, 613
  film thickness in, 576
  height of transfer unit of, 570
  rate of, 341, 343
  and simultaneous convection, 586–590
  transfer unit for, 570
  two-film theory of, 565
  of vapor through gas, 303, 339
  (*See also* Cooling towers; Direct-contact transfer)
Diffusion coefficient, 341, 572
Diffusion unit, 591, 602–604, 612
Diffusional heat transfer, 340
Diffusivity, 343, 567
  calculation of, 344
  (*See also* Thermal diffusivity)
Digesters, 426
Dimensional analysis, 30–41
  of condensation, 253
  evaluation of equations from, 44
  of fluid friction, 34
  of forced convection, streamline flow, 40
    turbulent flow, 35
  of free convection, 206
  by Pi theorem, 38–40
  problems in, 57
  theory of models in, 38
  of unsteady conduction, 643
Dimensionless constants, 31
Dimensionless groups, 36
  names of common, 37
Dimensions, of common variables, table, **33**
  consistent, 37
  derived, 30
  fundamental, 30
  six-dimension system of, 32
Direct-contact transfer, 564–621
  calculation of gas cooler for, 615–619, 620
  channeling in, 600
  for cooling gases, 612–614
  dehumidifiers in, 611, 613, 614
  flooding in, 601
  heat balances in, 614
  height of transfer unit in, 570
  humidification characteristics of packings in, 599, **600**
  with liquid film resistances, 593
  without mass transfer, 619
  packed towers for, 598–601, 619
    humidification coefficients, **598–601**
  sensible-heat transfer in, 619–621
  theory of, 565–570
  transfer unit in, 570
  (*See also* Cooling towers)
Direct-contact vs. tubular equipment, 563
Direct-fired kettle, 708
Direct-fired still, 637
Direct-fired tank, 708
Direct-fired tubular furnace, 678
Dirt factors (*see* Fouling factors)
Dirty overall coefficient, 106

SUBJECT INDEX 857

Disengagement, in chemical evaporators, 404
  in power-plant evaporators, 379, 380, 454
  in vaporizers, 454
Disengaging drums, 455, 456
Distillate, 255
Distillation, batch, 624, 637–639
  instrumentation for, 787
  Rayleigh equation for, 638
  continuous, 253–255, 281, 289, 338, 353, 453, **491–502,** 777
  ASTM, 353
  of crudes, 678
  flash reference line for, 354
  in gasoline recovery, 222
  instrumentation for, 788
  minimum reflux for, 499
  nomenclature for, 254
  operating pressure of, 325
  processes of, **491–502**
  reference line for, 354
  reflux control of, 772
  steam, 222, 337, 352
    instrumentation for, 789
  theoretical plate, 498
  trapout, 457
  true boiling point (TBP), 353
Distillery waste, 418
Distilling columns, 222
  condenser arrangements for, 268, 271, 274, 290
  operating pressure of, 325
Divided flow, 246, 274, 478
Double pipe exchangers (see Exchangers)
Downcomer, 255
Drag in fluid flow, 55
Drip coolers, 727
Dropwise condensation, 252
Dry gases, coolers for, 192
Duhring's rule, 399

E

Economic (see Optimum)
Economizers, 192, 544, 554, 677, 678
EF curve, 355
  (see also Equilibrium flash)
Effective temperature difference (see Temperature difference)
Efficiency of exchanger, 170
Ejectors, 305, 306, 394, 443, 457
  calculation of, 445
  steam consumption in boosters, graph, 444
Elapsed-time controller, 788
Electric resistance heating, 753–763
Electrical conductivity, 12
Electrolytic decomposition, 433
Electrolytic solutions, 320
Electromagnetic theory, 62
Emissive power, 65
Emissivity, 3, 68
  of diatomic gases, 690
  of metals and their oxides, **70–72**
  of miscellaneous materials, **72**
Emissivity factor, 82, 687

Emulsions, properties of, 161
Endothermic reaction, 679
Energy, conversion factors for, **793**
  spectral, 64
Engines, internal-combustion, 731
  steam, 165, 301, 390, 397
Enlargement head loss, 112, 148, 273, 487
Enthalpy, of air, saturated, **585, 586**
  of light hydrocarbons, **813**
  of petroleum fractions, **814**
  of pure hydrocarbons, **812**
  of steam, **816–819**
Enthalpy change, in turbines, 301
Entrance head loss, 112, 148, 273, 487
Equilibrium, in phase rule, 314
Equilibrium constants, 321, **810**
Equilibrium flash, 354
Equilibrium temperature, 252
Equivalent diameters, in annuli, **104, 745**
  in core tubes, 210
  shell-side crossflow, **138, 838**
  spherical, for solids, 669
  volumetric, in crossflow, 555
Ethyl alcohol, 241, 470
Evaporation, 375–449
  blanketing in, 377, 378
  bubble formation in, 375
  chemical, 378, 400–409
    flashing in forward-feed, 408, 411
    free, 409
    mixed feed in, 408
    parallel feed in, 408
    solids distribution in, 427
    (See also Evaporators)
  critical temperature difference in, 376
  effect of pressure and properties on, 377
  factors affecting, 377
  heat flux in, allowable, 384
    definition of, 377
  Kelvin theory of, 375
  maximum attainable flux in, 377
  mechanism of, 375–377
  multiple-effect, 382–390, 408
    with parallel feed, 385, 389, 408
    from pools, 375–398, 459
  power-plant processes of, 386–393
  pressure correction for, 377
  for pure water 375–397
  roughness of surface in, 375
  surface tension in, 375
  vacuum, 390
  vapor binding in, 377, 378
  (See also Vaporization)
Evaporative condensers, 738
Evaporator-condensers, 387, 388
Evaporators, chemical, absence of blowdown in, 398
  accepted heat fluxes for sugar solutions in, 421, 425
  accepted overall coefficients in, 379, 412, 414, 453
  for soda waste liquor, 429
  for sugar solutions, 421, 425

858 PROCESS HEAT TRANSFER

Evaporators, chemical, algebraic calculations of, 409–413
  apparent temperature difference in, 383
  apparent temperature drop in, 399
  backward-feed, 399, 409
    outline of, 411–412
    operation difficulty in, 416
  boiling-point rise in, 399, 400
  calculation of, 409–449
  calculation of sugar, 418–426
  calculation of triple-effect backward-feed, 414, 416
  calculation of triple-effect forward-feed, 412, 414
  catchalls for, 403, 425
  crystallization in, 401
  economy in, definition of, 413
  extrapolated evaporation in, 422
  forced-circulation, 400, 405, 406, 457
    caustic soda in, calculation of, 433–442
    1-2 exchangers as, 470
    inside vertical element in, 405
    outside horizontal element in, 406
    outside vertical element in, 406
    velocities in, 440
  forward-feed, 398, 408
    outline of, 409–411
  hydrostatic head in, 417, 426
  industrial problems of, 418–449
  minimum initial cost of, 412
  minimum surface for, 412
  multiple-effect, 398, 408
  natural-circulation, 400–405
    basket-type, 401, 403, 453
    calandria vertical-tube, 401–403
    horizontal-tube, 401, 453
    long-tube vertical, 401–404, 425, 426, 459, 486, 491
    soda liquor in, calculation of, 426–433
    standard, 403
    sugar, calculation of, 418–426, 447–449
    (See also Reboilers, natural-circulation, vertical thermosyphon)
  nonalgebraic calculation of, 413–449
  optimum number of effects for, 416
  parallel feed in, 408
  pressure distribution in, 413
  quadruple-effect, 398
  sextuple-effect, 427
  single-effect, 400
  standard number of effects for, 416
  steam cost in, 409
  steam economy in, 413
  supersaturation in, 408
  temperature distribution in, 415, 421, 427
  thermocompression, 418, 442–447
    sugar solutions in, calculation of, 447–449
  (See also Evaporation, chemical)
  definition of, 102, 378, 453
  design methods in, 379
  disengagement in, 379
  function of, 378
  instrumentation for, 787

Evaporators, power-plant, 378–397, 453
  bent-tube, 382
  blowdown in, 381
  descaling, 381
  distillers in, 386
  double-effect, 384, 386, 388, 389
  evaporator-condensers with, 387, 388
  heat flux in, 384
  heat head in, 383
  heat transformers in, 386, 390–392
  make-up of, 386–389
  multiple-effect, 384–386, 389
    pressure distribution in, 390
    vacuum in, 389
  overall coefficients for, 382, 383
  process for, 386, 389
  quadruple-effect, 389
  reducing valve in, 392
  salt-water distillers for, 386, 392
  separators for, 380
  serpentine-tube, 382
  shocking, 382
  single-effect, 386
  temperature head of, 383
  triple-effect, 386, 388, 389
  types of scale in, 381
  water treatment for, 380–382
Exchanger performance, 148
Exchanger rating, 148
Exchangers, approach temperature in, 85, 146, 175
  bayonet, 738–746
  cleaning of, 108
  definition of, 102
  double pipe, 102–123
    calculation for series, 113–115
    calculation outline for, 110–112
    calculation for series-parallel, 121–123
    condensation in, 269
    with extended surfaces (see Extended surfaces)
    flow areas and equivalent diameters of, 110
    series-parallel arrangements in, 115
    series-parallel calculation of, 121
    standard fittings for, 103
    true temperature difference in, 120
    with viscosity correction, 120
  efficiency of, 170
  guarantees for, 775
  instrumentation for, 777
  with longitudinal fins, 523, 534
  shell-and-tube, annual fixed charges for, 222, 224
    cost of, in relation to size, 224
    in relation to tube diameter, 228
    in relation to tube length, 229
  1-1, 187
    for batch heating, 628–632
    with extended surfaces, 534, 544
    as vaporizers, 457, 488
  1-2, aqueous solution in, calculation of, 161–163, 167
    without baffles, 166

SUBJECT INDEX 859

Exchangers, shell-and-tube, 1-2, for batch heating, 626
  calculation of, 151–153
  calculation outline of, **149–151**
  condensation in shells of, 266
  as condensers, 271
  as coolers, 154, 161, 238
  core tubes for, 206
  definition of, 145
  double-tube sheets, 135
  efficiency of, 170
  with extended surfaces, 534
  with fixed tube sheets, 132
  with floating head, 133
  as gas aftercooler, calculation of, 193–195
  as heaters, 164, 241
    calculation of, 167
  oil-to-oil, calculation of, 151–153, 231–235
  outlet temperatures of clean, 170
  with packed floating head, 135
  with pull-through floating head, 132
  in series, 176, 184
    calculation of, 184–187
  shell-side flow area of, 138
  shell-side friction factors, **838**
  shell-side mass velocity, 137
  shell-side pressure drop, 147, **838**
  with split-ring floating head, 133
  with streamline flow, 203, 207
  U-bend, 135
  as vaporizers (*see* Vaporizers)
  water-to-water, calculation of, 155–158
2-2, 176
2-4, for batch heating, 626
  calculation of, **181**
  definition of, 175
  with extended surfaces, 534
  oil-to-oil, calculation of, 235
  oil-to-water, calculation of, 181–184
  in series, calculation of, 235
  with removable-baffle floating head, 179
  with welded-baffle floating head, 179
3-6, 177, 235
4-8, 178, 184
  oil-to-oil, calculation of, 235
  with extended surfaces (*see* Extended surfaces, shell-and-tube exchangers)
  falling-film, 746
  optimum, 224
  optimum velocities in, **227**
  mass velocities in, 227
  rating of, 148, 149, 225
  return pressure loss in, **148**
  shell-side pressure drop for, 147, **838**
  split-flow, 245
    calculation of, 246–250
  with stationary tube sheets, 129, 131
  with tube coils, 720
  tube-side pressure drop in, 148, **836**
  with solutions, 160
  with streamline flow, 202
  temperature cross in, 146, 175
  temperature-cross increase in, 175

Exchangers, with transverse fins, 544
  vaporizing, 378. 453, 458
Exhaust steam, 165
Exhaust temperature, 301
Exit head loss, 112, 148, 273, 487
Exothermic reactions, heat removal, 245
Expansion head loss, 112, 148, 273, 487
Expansion joints, 131
Experimental evaluation of convection coefficients, 43–51
Extended surfaces, 192, 512–562
  air-cooled, 545, 731
  double pipe exchangers, 514, 523
    calculation of, **530–534**
    common sizes of, **524**
    efficiency curves for, **524**
    heat-transfer coefficients for, **525**
    pressure drops in, **525**
    temperature difference in, 524
  fins, classification of, 514
    discontinuous, 515, 554
    effectiveness of, 519
    ideal, 544
    longitudinal, 514, 538
      efficiency calculation of, 522
      efficiency curves of, **543**
      efficiency derivation of, 515–519, 538, 541
      weighted-efficiency calculation of, 522
    transverse, 514, 538–546, 553–562
      disc, 514, 554
      efficiency curves for, **542**
      efficiency derivation of, 538–541
      heat-transfer coefficients for, 553, **555**
      helical, 514, 554
      pressure drop for, 555
      star, 514
      weighted efficiency of, 541
    wall temperature of, 527–530
    weighted efficiency of, 519, 524
  pegs, 515, 559
    efficiency curves of, **543**
  shell-and-tube exchangers, 514, 534
    calculation of, 535–538
  spines, 515, 541, 559
    efficiency curves of, **543**
Extrapolated evaporation, 422

F

Falling-film exchangers, 746
Fanning equation, 52
  for annuli, 104
  friction factors for, **53**
Fatty acids, properties of (*see* Liquids)
  specific gravity of, 821
  viscosity of, 821
Feed water, deaeration of, 302
Fermentation of sulfite liquors, 427
Ferrules, 127, 402
Fictitious compounds, 356
Field tubes, 738
Film boiling, 377

Film coefficients, 25, 26
  control of, 87
  (*See also* Heat-transfer coefficients)
Film resistance, 25
Film temperature, in condensation, 260
  in forced convection, 99
  in free convection, 215
Film theory, 26
Filmwise condensation, 252
Finned strip heaters, 754, 756, 757, 762
Fins (*see* Extended surfaces)
Firebrick, heat transfer to, 670
Fixed charges, 21, 158, 159, 222, 224
Fixed-tube-sheet exchangers, 129, 131
Flared nozzle, 271
Flash evaporation, 409, 416, 435
Flash reference line, 354
Flash tanks, 387
Flash vaporization curve, 355
Floating control, 770
Floating head, 132
  flow area in, 202
  split, 180
Floating-head cover, 132
Floating-tube sheet, 132
Flow, side-to-side, in condensers, 272
  in exchangers, 130
Flow control, 771
Flow of fluids (*see* Fluid flow)
Flue gas, 246
Fluid flow, conversion factors for, 793
  dimensional analysis of, 34
  drag in, 55
  pressure drop in, 34, 109
  streamline, 29, 40, 201
  transition velocity in, 53
  turbulent, 29, 35
  velocity distribution in, 29
  (*See also* Friction factors; Pressure drop)
Fluid friction, 34
  (*See also* Friction factors; Pressure drop)
Fluid shear, 27
Foaming, 381, 400
  in soda process, 426
Forced circulation, in evaporators, 405, 406
  in vaporizers, 454
Forced convection (*see* Convection)
Fouling factors, 106, **845**
  in water-to-water exchangers, 155
Fourier equation, 87, 106, 107, 110, 201, 225
  for parallel-counterflow exchangers, 144, 177
Fourier general equation, 11, 86
Fourier number, 37
Fourier series, 645, 663
Fractional distillation, 493–502
Fractions from crude oil, 4
Free convection (*see* Convection)
  and radiation, 18, 218
  and streamline flow, 206
Frequency of radiation, 62
Friction, analogy of, to heat transfer, 54–57, 564, 574
Friction factors, 52

Friction factors, for Fanning equation, **53**
  for fluids, in shells, **839**
    in tubes, **836**
    baffled, **839**
    unbaffled, 167
  (*See also* Pressure drop)
Fronts, in multicomponent mixtures, 329, 333
Fuel oil, 4, 675, 678, 689
Fugacity, 320
Furnaces, air preheaters for, 678, 680, 701
  atomizing steam for, 698, 711
  average flux in, 687, 696
  blast, 664
  calculations of, Lobo and Evans method, 697, 698–705
    Orrok-Hudson equation, 697, 706, 707
    Wilson, Lobo, and Hottel method, 697, 705
    Wohlenberg simplified method, 697, 705
  chemical, 674
  coking in, 675, 679
  conductivity factor for, 687
  design methods for, 696–708
  with double tube rows, **687**
  dry ash in, 689
  effective refractory surface for, 701
  effectiveness factors for, 685, 687
  emissivity factors for, 82, 687
  enclosing surfaces for, 684, 692–696
  equivalent cold plane, 685
  excess air in, 689, 690, 701, 706, 711
  fuels for, 675
  gas vs. coal-firing, 689
  heat balances in, 698
  heat sinks for, 684–689
  heat sources for, 684, 689–692
  heat-treating, 674
  kilns, 674
  maximum flux in, 687
  mean beam length, **690–692**
  oil vs. coal-firing, 689
  open-hearth, 664
  Orsat analysis of, 712
  ovens for, 674
  permissible flux in, 674, 702, 712
  petroleum-refinery, 678–683
    "A" frame, 682
    average flux in, 686, 696
    box-type, 678
    bridgewalls for, 678
    convection surface in, 678
    De Florez, 680
    double-radiant box-type, 680
    down-draft convection-bank, 682
    efficiency of, 678, 701
    maximum flux in, 686
    multiple radiant-section, 682
    overhead convection-bank, 680
    polymers in, 679
    radiant surface in, 678
    soaking section in, 679
    tube spacing in, 686
  pulverized-coal-firing, 689, 706
  radiant-rate distribution, 685

SUBJECT INDEX 861

Furnaces, regenerative, 624
  reradiation in, 685
  shield tubes in, 707
  slag factor, 687, 688
  slagging in, 674, 687
  stoker-firing, 689, 706
  tube cleaning in, 711
  tube-skin temperature, 713
  (*See also* Boilers, steam-generating; Radiation)

G

Gases, absorption of, 222
  compression of, 192, 195
  cooling by direct contact, 614–621
  corrosive, 727
  direct-contact transfer, 564
  in exchangers, **190–198**, 245, 246–250
    ammonia cooler for, calculation of, 193–195
    coolers for, 192, 246–250
      wet, calculation of, 197
  flue, 246
  free convection to, **216**
  friction factors for, 191, **836, 839**
  as fuels, 675, 678, 689
  heat-transfer coefficients of, 192
  lean, 222
  natural, 222
  noncondensable, in vapor mixtures, 302, 313, 339–352
  nonluminous, 674, 689
  Prandtl numbers for, **191**
  properties of, compared with liquids, 190
  radiating, 690
  reactor, 285
  rich, 222
  specific heat of, **805**
  thermal conductivity of, **801**
  under vacuum, 191, 195
  viscosity of, **825**
  viscosity correction for, **820**
  wet, coolers for, 195
Gas-turbine cycles, 515
Gas oil, 4, 46, 209, 482, 530
Gasoline, 4, 222, 476
Gauss's error integral, 641
Gibbs' phase rule, 313–318
Gilliland formula, 344, 568
Graetz number, 37, 202, 748
Granular materials, in beds, 625, 668–670
  in tubes, 747–752
Graphical solution, of conduction, 21
  of process temperatures, 224
  of steam-condensing coefficients, 309
Grashof number, 37, 206, 215, 400, 405
Gravel, heat transfer to, 670
Gurney-Lurie charts, **651–655**

H

h, individual heat-transfer coefficient, definition of, 26
Hairpin, 102, 523
HDU, height of diffusion unit, 591

Heat, mechanical equivalent of, **34**
  of reaction, 679
  of solution, 625, 630, 631
  total (*see* Enthalpy)
  of vaporization, **815**
Heat capacity (*see* Specific heat)
Heat economizers, 192
Heat-exchanger tubes, 128, **843**
Heat exchangers (*see* Exchangers)
Heat flow (*see* Conduction)
Heat head in surface condensers, 305
Heat interchangers (*see* Exchangers)
Heat loss from pipe, 17
  to air, coefficients of, **18**
Heat recovery, in counterflow, 92
  in 1-2 exchangers, 169, **170**, 177
  in 2-4 exchangers, 177, 178, **179**
  instrumentation for, 777
  lack of, in exchangers, 175
  in parallel flow, 93
  of waste heat, 192
Heat regenerators, 625, 664–668
Heat transfer, definition of, 1
Heat-transfer coefficients, in annuli, **105**
  with extended surfaces, **525**
  for atmospheric coolers, **734, 736**
  for bayonet exchangers, 745
  for coils, **718**, 721–723
  condensation, of binary mixtures, 319
    graph of, **267**
    of miscible-immiscible mixtures, **338**
    of miscibles and immiscibles from noncondensable, 352, 367
    of multicomponent mixtures, 330
    of steam, **266**
      in surface condensers, **306**
    in steam distillation, 338
    in tube, horizontal, 269
      vertical, equation, **265**
    on tube, horizontal, equation, **265**
    on tubes, horizontal, equation, **266**
      vertical, equation, **265**
  controlling, 87
  conversion factors for, **793**
  definition of, 3, 18
  desuperheating, calculation of, 286
  evaporation, accepted overall, 379, 412, 414 453
    apparent overall, 399
    effect of pressure on, 378
    of soda waste liquor, 429
    of sodium hydroxide, 441, 442
    of sugar solutions, 421
    of water from pools, **382**
  extended surfaces, double pipe exchangers, **526**
    longitudinal fins, **525**, 535
    transverse fins, **555**
  in falling-film exchangers, 747
  forced convection, for gases (vapors), 192
    in pipes and tubes, **834**
    in shells, **838**
  for liquids, in pipes and tubes, **103, 834**
    in shells, **838**

Heat-transfer coefficients, forced convection, for
water in pipes and tubes, **835**
free convection, from miscellaneous shapes, **215**
from pipes, to gases, **215, 216**
to liquids, **216, 217**
for granular materials in beds, 670
individual, 26
in surface condensers, 308
in jacketed vessels, **716–719**
overall, 86
approximate, **840**
arithmetic mean, 94
clean, 106
definition of, 106
dirty, design, 106
in evaporators, power-plant, **382**, 383
in surface condensers, **306**
true mean, 94
in vapor-noncondensable systems, 343
periodic, 665
from pipes to air, **18**
from pipes to gases, 214, **216**
in pipes and tubes, **834**
water, **835**
for radiation, **77**
for reboilers, 474
in shells, baffled, 136–139, **838**
unbaffled, **166**
for slurries, 725
for solids, 670
for solids in rodlike flow, 749
for steam as heating medium, 164
effect of dissolved air on, 302
for steam from noncondensables, 367
for suspensions, 725
in tubes, 103
gases and liquids, **834**
water, **835**
for unsteady state, 717, 722
vaporization, forced-circulation, isothermal boiling, 461
maximum allowable, 460
nonisothermal boiling, 469
in shells, 461
in tubes, 470
natural-circulation, 474
in chillers, 472, 474
in horizontal thermosyphons, 479
in kettle reboilers, 474
maximum allowable, **460**
in vertical thermosyphons, 488
weighted, 480
of vapors from noncondensables, 367
for vertical condenser-subcoolers, 289
volumetric, 668, 669
for water, in shells, **137**
in tubes, **155**
weighted clean overall, in desuperheater-condensers, 284
for mixed vapors, 331
for vapor from noncondensable gas, calculation of, 351
weighted trial overall, 285

Heat-transfer factor j_H, 50, 104, 111
Heat-transfer factor j_h, 192
Heat-treating furnaces, 674
Heaters, bleed, 388
definition of, 102
instrumentation for, 778–780
stage, 388
sugar-juice, 422, 425
(*See also* Exchangers)
Height of diffusion unit (HDU), 591
Height of transfer unit (HTU), 570, 590
Henry's law, 569, 638
n-Heptane, 282
i-Hexane, 355
n-Hexane, 322, 331
vapor-pressure curve of, 321
Holdup, in batch processes, 624
in distillation, 478
Horizontal condensers (*see* Condensers, shell-and-tube)
Horizontal condenser-subcooler, 294
calculation of, 295–297
Horizontal film coolers, 727
Horizontal vs. vertical condenser-subcoolers, 298
Hot well, 305
HTU, height of transfer unit, 570, 590
Humid heat, 588
Humidification, 563, 570–577, 613, 614
Humidification coefficients, 598–601
Humidifiers, 589, 614
Hydraulic radius, 104, 138
in condensate films, 264
Hydrocarbon gases and vapors, specific heat of, **807**
thermal conductivity of, **801**
Hydrocarbon liquids, equilibrium constants for, **810**
molecular weights of, **360**
Prandtl numbers for, **826**
specific gravity of, **809**
specific heat of, **806**
thermal conductivity of, **803**
vapor pressures of, **811**
Hydrocarbons, heat-transfer (*see* Heat-transfer coefficients)
light, enthalpies of, **813**
pure, enthalpies of, **812**
Hydrogen, 190, 690
Hydrogen chloride, 727
Hydrostatic head, in evaporators, **408**
for reflux return, 272
Hyperbolic tangents, **521**

I

IBP (*see* Initial boiling point)
Ideal solutions, 319, 495
Ideality, criteria of, 320
Individual film coefficients (*see* Heat-transfer coefficients)
Ilmenite, 749
Immersion heaters, 754, 755, 758
Immiscible mixtures, 314

Incinerators, rotary, 426
Infinite series, 747
Infrared, near and far, 65
Initial boiling point (IBP), 4, 353
Instrument Society of America, 773
Instrumentation symbols, 773
Instruments, 767
Insulating materials, thermal conductivity of, **795**
Insulation, maximum heat loss through pipe, 20
  optimum thickness of, 21
Insulator, definition of, 2, 8
Intensity of radiation, 64
Intercondensers, 397
Intercoolers, 192, 195, 197, 198, 339
  calculation of heat load in, 196
Ionic solutions, 320
IPS (*see* Iron pipe size)
Iron balls, heat transfer to, 670
Iron ores, heat transfer to, 670
Iron pipe dimensions, **844**
Iron pipe size (IPS), **843**

J

j_H, j_h (*see* Heat-transfer factor)
Jacket water cooling, 731
Jacketed vessels, 625, 716
  heat-transfer coefficients for, **716–719**
  time for heating, 626–628
Jet condenser, 394, 564
Jet propulsion, 515
Joule's law, 754

K

Kelvin evaporation theory, 375
Kerosene, 4, 151, 207
Kettle, direct-fired, 708
Kettle reboilers (*see* Reboilers, natural circulation)
Kilns, 674
Kinematic viscosity, 28
Kirchhoff's law, 67
Kleinschmidt still, 442
Knock-back condensers, 298
Knockout drums, 776

L

Laminar flow (*see* Streamline flow)
Laminar layer, 54–57
Latent heat of vaporization, **815**
Lean oil, 222, 235
Level control, 766, 776
Lewis number, 574, 612
Ligament between tubes, 128
Lignin, 426
Lime, 418, 433
Lime-soda process, 381
Limestone, heat transfer to, 670
Liquid-level control, 766, 776
Liquids, cooling and heating (*see* Heat-transfer coefficients)
  specific gravity of, **808**

Liquids, specific heat of, **804**
  thermal conductivity of, **800**
  viscosity of, **823**
LMTD (*see* Logarithmic mean temperature difference)
Loading, in cooling towers, 584, 601
  in surface condensers, 306
Logarithmic mean temperature difference (LMTD), 42
Loop seal, on horizontal condenser-subcoolers, 294, 331
  on vertical condenser-subcoolers, 289
Low-level condensers, 396
Lube oil, 4, 121

M

Magnesium bisulfite, 427
Magnesium scale, 380
Magnesium sulfite, 426
Mass-diffusion coefficient, 341
Mass transfer, 339, 564, 612, 614
Mass velocity, 35
  in shells, 137
  in tube bundles, 150
Material transfer (*see* Mass transfer)
Maxwell's electromagnetic theory, 62
Mean beam length, **690–692**
Mean temperature, 201
Mechanical equivalent of heat, 34
Metals, properties of, **799**
Methyl alcohol, 619
Micron, definition of, 62
Mineral oils (*see* Hydrocarbons, Petroleum)
Minimum annual cost of exchanger, 225
Miscible mixtures, 314
Mixing, in streamline flow, 202
  in temperature-difference assumptions, 140, 547
Models, theory of, 38
Mol fraction, 320
Molasses, 739
Molecular volume, 344
Momentum transfer, 54, 342
Monochromatic properties, 65
Multicomponent mixtures, 319–329
  empirical solution of, 353
  fictitious compounds of, 356
  modified by immiscibles and noncondensables, 352
  reboiler duty for, 505
Multipass exchangers (*see* Exchangers, 1–2)
Multiple-effect evaporation, 384, 389, 408

N

Naphtha, 4, 231, 482
Natural circulation, in evaporators, 400–405
  in vaporizers, 454
Natural convection (*see* Convection, free)
Natural gas, 222
Natural gasoline industry, 731
Newton's law of cooling, 3
Newton's rule, 27

Newton's second law, 30
Nickel, 438
Nitric oxide, 727
Nitrogen, 615, 690
Nomenclature (*see* conclusion of each chapter)
Noncondensable gas (*see* Gases)
Noncondensing turbines and engines, 165
Nonideal solutions, 320
Nonluminous gases, 674
Normal total emisivity, table, **70–73**
Nozzle, flared, 271
Nozzle entry allowances, **134**
Nuclear boiling, 377
Number of transfer units, 570, 590
Nusselt number, 37
Nusselt's theory, 256–263

O

Ohm's law, 6, 12, 754
Oils, as fuels, 675, 678, 689
  petroleum, classification of, 4
    degree API of, 4
    enthalpies of, **812–814**
    equilibrium constants of, **810**
    friction factors of (*see* Friction factors; Pressure drop)
    heat transfer to (*see* Heat-transfer coefficients)
    molecular weights of, **360**
    pressure drop in (*see* Friction factors; Pressure drop)
    specify gravity of liquid, **809**
    specific heat, of liquid, **806**
      of vapors from, **807**
    thermal conductivity, of liquid, **803**
      of vapors from, **801**
    vapor pressures of, **811**
    viscosity of, **821**
  (*See also* Specific names of oils)
On-and-off control, 770
Open coolers, 730
Open-hearth furnace, 664
Open-and-shut control, 770
Operating pressure, of condenser, 325, 577
  influence of, on gas heaters and coolers, 191
Optical glass, annealing, 645
Optimum atmospheric-cooler temperature, 734
Optimum effects in evaporation, 416
Optimum efficiency of cycles, 387
Optimum exchanger, 224
Optimum outlet-water temperature, 158
Optimum pressure in distillation, 325
Optimum process conditions, 221
Optimum recovery temperature, 224
Optimum reflux ratio, 499
Optimum thermal fin, 544
Optimum thickness of insulation, 21
Optimum use of exhaust and process steam, 165
Optimum velocities, 225
Organic fluids, heat transfer of, 103, 840
  (*See also* Heat-transfer coefficients)

Organic solutions, properties of, 161
Ottawa sand, 749
Outer tube limit, 134
Ovens, 674
Overall coefficients of heat transfer (*see* Heat-transfer coefficients)
  approximate values of, 840
  graphical analysis of, 308
Overhead product, 255
Oxygen, 535, 690
  dissolved in water, 306

P

Packed floating-head exchangers, 135
Packed towers (*see* Direct-contact transfer)
Paper-pulp processes, 426
Paper-pulp waste liquors, 426
Parabolic velocity distribution, 40, 201
Parallel-counterflow, 139
Parallel feed in evaporators, 384
Parallel flow, 85
  heat recovery in, 93
  temperature difference in, 90
Parallel flow-counterflow, 139
Partial condensers, 280, 318
Partial phase change, 283
Partial pressure, 321
Pebble heaters, 624
Peclet number, 37
i-Pentane, 290
n-Pentane, 290, 314, 316, 317, 331
  vapor pressure curve of, 321
Perfect gas law, 191, 192
Performance, of cooling tower, 602
  of exchanger, 148
Periodic heat-transfer coefficient, 665
Petroleum coke, as fuel, 678
Petroleum fractions, classification of, 4
  light ends, 4
Petroleum industry, 4, 353
Petroleum oils (*see* Oils)
Petroleum-refinery furnaces (*see* Furnaces)
Phase, definition of, 313
Phase changes, partial, 283
Phase equilibrium, 314, 321
Phase rule, 252, 313–318, 564
Phosphate solution cooler, 161
Pi theorem, 38
  forced convection by, 39
Pipe stills (*see* Furnaces, petroleum-refinery)
Pipes, dimensions of, **843**
  friction in (*see* Friction factors; Pressure drop)
  heat losses from, 17–21
Pipe-wall temperature, 97
Pitch, baffle, 129
  tube, 128
Planck, 63
Planck's law, 66, 69
Plate in distillation, 254
Poise, 28
Poiseuille, 28

Polytropic compression, 192
Potential, definition of, 6
Power, by-product of, 391
 conversion factors for, 793
Power cycles, 301
Power-plant evaporators, 378–397
 (*See also* Evaporators)
Power series, 31
Prandtl analogy, 57
Prandtl number, 37, 50, 215
Prandtl numbers for gases, table, **191**
Precondensers, 395, 397
Preheaters, in batch distillation, 639
 in continuous distillation, 221
 with make-up evaporators, 387
Pressure, constant, in distillation, 271
 conversion factors of, **794**
 effect of, on boiling coefficient, 378, 382
 on thermal conductivity, 8
 on viscosity, **820**
Pressure booster, 390
Pressure-control valve, 767, 776
Pressure distribution in evaporators, 413
Pressure drop, 34
 allowable, in compressor intercoolers, 195
 in condensers, 272
 in cooling towers, 601
 in exchangers, 108
 in annuli, 109
 balanced, in subcoolers, 280, 294
 in chillers, 475
 of condensing vapors, 273
 in double pipe exchangers, entrance and exit losses, 109
 on extended surfaces, longitudinal fins, **525**
 transverse fins, **555**
 in horizontal thermosyphons, 480
 instrumentation for low, 777
 in kettle reboilers, 475
 in partial condensers, 280
 in pipes and tubes, 148, 202, **836**
 of condensing steam, 165
 of condensing vapors, 273
 deviation of, 34
 equations for, **52**
 friction factors of, **53, 836**
 of gases, **191**
 of liquids, **148**
 over pipes and tubes, in crossflow, 554
 in reboilers, natural-circulation, 469
 in relation to heat transfer, 54–57
 in shells, **147, 839**
 balanced, in condensation, 280, 294
 of condensing steam, **165**
 of condensing vapors, 273
 of gases, **191**
 split-flow, 245
 of steam, 165
 unbaffled, 166
 in surface condensers, 307
 in tubes, friction factors for, **836**
 return losses for, 148, **837**
 Williams and Hazen equation, 307

Pressure drop, in vaporizers, natural-circulation, 461, 469
Pressure-drop allowance, 108
 in compressor intercoolers, 195
 in condensers, 272
Pressure gradient, 34
Pressure-volume-temperature relationships, 320
Priming in evaporators, 381
Probability integral, 641
Process conditions, 110, 229
 optimum, 221
Process heat transfer, 3
Process lags, 768
Process steam, 165
 vs. exhaust steam, 165
Process temperatures, 85
Process variables, 765
Program controllers, 788
n-Propanol, 274
Properties, of liquid solutions, 161
 of saturated steam, **816, 817**
 of superheated steam, **818, 819**
Proportional control, 770
 and reset, 770
Propyl alcohol, 274
Psychrometric charts, 584
Pull-through floating-head exchangers, 132
Pumps, selection of, 774

Q

Quantum theory, 63
Quenching, 641, 647, 653

R

Radiant energy (*see* Radiation)
Radiant-heat flux, 674
Radiant-heat transfer, 78–82
 (*See also* Radiation)
Radiant-heat-transfer factors, 683–696
Radiation, 62–84, 674–715
 absorptivity of, 67
 between any source and receiver, 78
 black-body, 66
 characteristics of, 62
 coefficient of heat-transfer, 77
 to completely absorbing receiver, 77
 between concentric cylinders, 77
 between concentric spheres, 77
 constant, 74
 cosine law of, 74
 definition of, 3, 62
 distribution of, 64
 emission of, 63
 emissive power of, 65
 influence of temperature on, 69
 emissivity of, 3, 68
 experimental determination of, 68
 normal total, 69
 table of, **70–73**
 and free convection, 18, 218
 frequency of, 62

866   *PROCESS HEAT TRANSFER*

Radiation, of furnaces (*see* Furnaces)
  geometric factor of, 80
  heat-transfer coefficient of, 77
  incidence of, 66
  intensity of, 64
  Kirchhoff's law of, 67
  from luminous flames, 689
  monochromatic properties of, 65
  from nonluminous gases, 674, 689
    carbon dioxide, 690, 693
    carbon monoxide, 690
    sulfur dioxide, 690
    temperature over bridgewall, 699
    water-vapor, 690, 694
  origin of, 63
  oscillating field of, 62
  oscillator, 63, 64
  overall heat-exchange factor of, 700
    between parallel planes, 74
    between perpendicular planes, 81
    between pipe and duct, 82
    Planck equation for, 66
    between planes of different emissivity, 74
    between plate and plane, 81
    quantum theory of, 63, 66
    reflectivity of, 67
    Stefan-Boltzmann law of, 69
    and temperature of sun, 66
    total emissive power of, 65
    transmissivity of, 67
    ultraviolet, 66
    velocity of propagation of, 62
    wave velocity of, 62
    wave length of, 62
    Wien's displacement law of, 66
Radio waves, 62
Raoult's law, 320, 638
Raschig rings, humidification characteristics of, 600
Rating an exchanger, 148
  outline of, **149**
Rayleigh equation, 638
Reaction time, 624
Reboilers, 102, 255, 378, 453, 454, 459
  for batch stills, 637
  divided-flow, 246
  effect of column pressure on, 325
  forced-circulation, for aqueous solutions, 470
    film coefficients for, 461
    instrumentation for, 784
    maximum allowable flux for, **459**
    maximum coefficient for, **460**
    pressure drop in, 461–463
    pump-through, 457, 458, 464–470
    specific gravities in, 461–463
    with vapors in tubes, 470, 471
  heat balances for, 491
  heat-flux limitations in, **459**
  influence of feed on, 500
  natural-circulation, baffled thermosyphon, 485
    bundle-in-column, 459, 478
    horizontal thermosyphon, 459, 478–482, 485
      calculation of, 482–485

Reboilers, natural-circulation, instrumentation for, 786
  kettle, 459, 471
    calculation of, 475–477
    maximum allowable flux for, **459**
    maximum coefficient for, **460**
  once-through, 458, 479, 480
  overdesign in, 485
  recirculating, 458, 479, 480, 482
  vertical thermosyphon, 459, 482, 486–488
    calculation of, 488–491
  temperature-difference limitations of, **459**
  vaporization curves of, 469
Receiver, definition of, 1
Reciprocating pumps, flow control with, 772
Recirculation rate, 482
Recirculation ratio, 482, 486–488
Recorders, 766, 772
Recovery of gasoline from natural gas, 222
  (*See also* Heat recovery)
Reference line, distillation, 354
Reflectivity, 67
Reflux, 255
  flow control of, 772
  gravity return of, 268, 271, 272
  instrumentation for return of, 781
  (*See also* Distillation)
Reflux condensers, 298
Refrigerants, 627
Refrigeration cycles, 472
Regenerators, 625, 664–668
Regulators, 767
Reheater, 392
Reheating, 176
Relative volatilities, 325
Relief valves, 789
Removable bundle exchangers, 132
Resin, 426
Resistance, contact, 9, 640
  definition of, 6
  to electricity flow, 6, 13
  to fluid flow, 6
  to heat flow, 6
  unit, definition of, 18
Resistance heating, 753–763
Resistances in series, 14
Resistivity, electrical, 12
Return pressure drop, 148, 202, **837**
Reversing exchanger, 515, 664
Reynolds analogy, 54–57
  applied, to diffusion, 342
  to distillation, 343
Reynolds number, 29
  critical value, 53
  modified for agitation, 718
  variation of, in streamline flow, 202
Ribs, conduction in, 22
Rich oil, 222, 235
Rodlike flow, in fluids (*see* Streamline flow)
  in solids, 747
Rotated square pitch, 128
Roughness, effect of, on fluid friction, 53
  on heat transfer, 104

SUBJECT INDEX

S

S-type coolers, 727
Safety valves, 789
Salt solutions, properties, 161
Saturation temperature, 252
Saybolt Seconds Universal (SSU), 29
  conversion factors for, **820**
Saybolt viscometer, 29
Scale, calcium and magnesium, 380
  in evaporators, 379–382, 400
  impurities causing, 301
  removal of, by shock, 382, 733
Scale factors (*see* Fouling factors)
Schack charts for rectangular shapes, 649, 650
Schmidt method, 657–662
Schmidt number, 37, 342
Sea water, 392
Self-diffusion, 564
Semiplastics, 739
Separators, 380, 454, 457
  steam, 380, 455
Series, Fourier, 645, 663
  infinite, 747
  power, 31
Serpentine coolers, 727
Shear stress in fluids, 27
Shell-and-tube condensers (*see* Condensers)
Shell-and-tube exchangers (*see* Exchangers)
Shell-side film coefficients (see Heat-transfer coefficients)
Shell-side equivalent diameters, crossflow, 137, **838**
Shell-side flow area, 138
Shell-side mass velocity, 137
Shell sizes, **129**
Shocking for scale removal, 382
Sieder-Tate equation, **103**, 206
Sieder-Tate heat-transfer factor, 104
Skin friction, 55, 342
Slabs, unsteady conduction in, 645, 657
Slag in boilers, 674
Sludge as fuel, 675
Slurries, 725–727
Soaking in cracking, 679, 682
Soap in water conditioning, 380
Soda ash, 433
Soda process, 426
Soda waste liquor, BPR of, 430
  evaporation of, 426–433
  specific heat of, 429
Sodium carbonate, 426
Sodium chloride, 398
  number of effects for, 416
Sodium hydroxide, 161, 398, 399, 426
  boiling point vs. pressure relations of, 435
  cooler for, calculation of, 238
  corrosion by, 438
  evaporation of, 433–442
  number of effects for, 416
  relative heat contents of, 438
  specific-gravity curve of, 436
  specific-heat curve of, 437
Sodium sulfide, 426

Sodium zeolite, 381
Solids, specific heat of, **799**
  thermal conductivity of, **795–799**
Solids in beds, 668–670
Solution exchangers, 160
Solutions, ideal, 319, 495
  nonideal, 320, 324
Source, definition of, 1
Specific gravity, of alloys, **799**
  of hydrocarbons, **809**
  of liquids, **808**
  of petroleum fractions, **809**
  of metals, **799**
Specific heat, of alloys, **799**
  of gases, 190, **805**
    correction for pressure, 191
  of granular solids, 750
  of hydrocarbon liquids, **806**
  of hydrocarbon vapors, **807**
  of liquids, **804**
  of liquid solutions, 161
  of metals, **799**
  of petroleum fractions, 806
  of vapors, **807**
Spectral energy, 64
Spheres, radiation from, 77
  unsteady conduction in, 645
Split flow, 245, 274
Split-ring assembly, 133
Split-ring floating-head exchangers, 133
Spray coolers, 730
Spray driers, 576
Spray ponds, 576, 579, 581
Spray towers, 576, 579, 581, 600
Spray washers, 570, 611
Square pitch, 128
  rotated, 128
SSU (*see* Saybolt Seconds Universal)
Stage heater, 388
Stanton number, 37
Stationary tube-sheet exchangers, 129, 131
Steady state, definition of, 11
Steam, bleed, 387
  condensation of, 253
  economy of, in evaporators, 384, 413, 416
  as heating medium, 163, 167, 203, 207, 211, 242, 464
  optimum use of exhaust and process, 165
  superheated, 386
    as heating medium, 164, 198
Steam boilers, 675
Steam distillation, 222, 337, 352
Steam engines, 165, 301, 390, 397
Steam generator, unfired, 457, 486
Steam heater, calculation of, 167
Steam load in surface condensers, 305
Steam loading in surface condensers, 306
Steam motive in ejectors, 395, 446
Steam pressure drop in shells and tubes, 165
Steam tables, **816–819**
Steam turbines, 165, 301, 386, 390, 397
Steel industry, 624, 664
Steel pipe dimensions, **843**

Stefan-Boltzmann constant, 74
Stefan-Boltzmann law, 69
Stokes, 28
Storage tanks, free-convection heaters for, 217
Straw oil, 4, 46, 231
Streamline flow, in condensate films, 256
  transition from, 270
  definition of, 29
  dimensional analysis of, 40
  in 1-2 exchangers, mixed, calculation of, 203
    unmixed, calculation of, 207
  fouling factors for, 214
  parabolic velocity distribution in, 40, 201
  in pipes and tubes, core tubes for, 209
    correction of, for free convection, 206
    heat-transfer equation for, **103**
    highest attainable temperature in, 201
  in shells, 214
  variation of Reynolds number in, 202
Streamline flow and free convection, 206
Streamline shapes, 605
Strip heaters, 754–756, 759
Stripping (*see* Distillation, steam)
Subcooling, 269, 280, 282, 289–298
  in surface condensers, 305
Subcooling zone, horizontal, 294, 331
  vertical, 289
Submerged pipe coil, 723–725
Submergence in condenser-subcoolers, 289, 294
Sucrose (*see* Sugar)
Suction pressure, 305
Suction tank heater, 738
Sugar, 398, 401, 417–426
Sugar solution heater, 167
Sugar solutions, BPR for, 420
  evaporation of, 418–427, 447–449
    number of effects of, 416
  specific heats of, 420
Sulfate process, 426
Sulfite liquor fermentation, 427
Sufite process, 426
Sulfur dioxide, 690, 729
Superheat, 252
  in distilling column, 454
  in steam vs. organics, 281
Superheated steam, 281, **818**
Superheated vapor, 281, 283, 454
Superheaters, in furnaces, 675
Surface, effective, in radiant-heat transfer, 78–82, 683, 684, 686
  in tubular equipment, 129, 306
Surface coefficient of heat transfer, 18
Surface condensers, 301–309, 427, 433
  coefficients for, **306**
  dual-bank, 304
Surface tension of organics and water, 376, 381
Suspensions, 725
Sutherland constant, 9

T

Table salt, 398
  production of, number of effects for, 416

Tails in multicomponent mixtures, 329, 333
Tank suction heater, 738
TBP (*see* True boiling point)
Temperature, absolute, 3, 5, 74, 192
  in adiabatic compression, 192
  arithmetic mean, 46
  bulk, 206, 215
  caloric, 93–97
  conversion factors for, 794
  cut-off, 639
  dry-bulb, 571
  film, in condensation, 260
    in forced convection, 99
    in free convection, 215
  mean, 201
  periodically varying, 625
  pipe-wall, **97**
  tube-skin, 713
  wet-bulb, 570–573
Temperature approach, in barometric condensers, 396
  in cooling towers, 594, 607
  in exchangers, 146
Temperature control, 765–789
Temperature cross, 146, 175
  in desuperheater-condensers, 284
  increase of, in exchangers, 175
Temperature difference, 85
  arithmetic mean, 201
  critical, in evaporation, 376, 377
    in vaporization, 460
  in desuperheating, 282
  effective (*see* Temperature difference, true, weighted)
  between fluid and pipe, 41
  logarithmic mean, 42
    calculation of, 90–93
    correction factors in, for 1-2 exchangers, **828**
    derivation of, 144
    correction in, for 2-4 exchangers, **829**
      for 3-6 exchangers, **830**
      for 4-8 exchangers, **831**
      for 5-10 exchangers, **832**
      for 6-12 exchangers, **833**
      for R greater than 20, 729
    counterflow derivation of, 87
    parallel-flow derivation of, 90
  mean, 201
  optimum, in coolers, 158–160
    in heaters, 166
  with partial phase changes (*see* Temperature difference, weighted)
  in power-plant evaporators, 383
  in submerged pipe coils, 724
  true (*see* Temperature difference, weighted)
    from ASTM curve, 354
    in atmospheric coolers, 736
    in bayonet exchangers, 740–745
    comparison of, between 1-2 and 2-4 exchangers, 177
    in condensation, of binary mixtures, 319
      of vapor from noncondensable gas, **343**
    in condenser-subcoolers, vertical, 295

SUBJECT INDEX 869

Temperature difference, true, in condensing mixtures, 314
  in counterflow, 97, 553
  in crossflow, 546–553, 727–728
  crossflow correction curves in, 549
  crossflow vs. parallel-counterflow, 547, 553
  in desuperheater-condensers, 283
  in divided flow, 479
  in 1-2 exchangers, 139–147, **828**
  in 2-4 exchangers, 176–179, **829**
  in horizontal thermosyphons, 479
  for R greater than 20, 729
  in series-parallel counterflow, 117–120
  in split-flow exchangers, 245
  from TBP curve, 354
  in trombone coolers, 727–728
  vs. tube length, 86
  weighted, in condensation, of miscibles and immiscibles from noncondensable, 356
    of multicomponent mixture, 329, 332
  in condenser-subcoolers, horizontal, 295
    vertical, 290, 291
  in condensers, batch, 639
  in desuperheater-condensers, 283
    calculation of, 286
  in preheater-vaporizers, 461, 465
Temperature-difference correction for R greater than 20, 729
Temperature distribution, vs. heat load, in condensation, 282, 313
  in counterflow, 42
  in moving fluids, 202, 747–751
  in moving solids, 748
  vs. tube length, in 1-2 condensers, isothermal condensation of, 283
    in counterflow, 86
    in 1-2 desuperheater-condensers, 283
    in 1-2 exchangers, 140, 145, 176
    in 2-4 exchangers, 176
    in parallel flow, 86
    in split flow, 245
Temperature effect on thermal conductivity, 8
Temperature-entropy plot, 574
Temperature gradient, 2
  in petroleum soaking, 679, 682
  in streamline flow, 202
Temperature range, 85
  in condensation, 281, 285, 313, 314
  in desuperheating, 285
Temperature rise in surface condensers, 305
Tempering coils, 544
Terminal-temperature difference, 306
Theory of models, 38
Thermal conductivity, of alloys, **799**
  of building and insulating materials, **795–798**
  conversion factors for, **794**
  from electrical conductivities, 12
  experimental determination of, for liquids and gases, 8
  for nonmetal solids, 7
  of gases, 190
    and vapors, **801**

Thermal conductivity, of granular solids, 749
  of hydrocarbon liquids, **803**
  influence of temperature and pressure on, 8
  of liquid solutions, 161
  of liquids, **800**
  of metals, **799**
  of petroleum fractions, **803**
  units of, 7
Thermal cracking, 678, 679
Thermal diffusivity, 11, 641, 662
Thermal-expansion elements, 767
Thermal shock for scale removal. 382, 733
Thermocompression, 418, 442–447
  steam economy of, 444
  steam requirement for, graphs, **443**
  sugar evaporation by, 447–451
Thermocompression evaporation, 447–449
Thermocouples, 767
Thermodynamics, definition of, 1
Thermometer wells, 775
Thermosyphons (see Reboilers, natural-circulation)
Time of cooling and heating batches, calculation of, 635–637
  with coil-in-tank, 625–628
  with counterflow exchangers, 625, 628–631, 633–635
  with 1-2 exchangers, 626, 631–633, 635
  with 2-4 exchangers, 633, 635
  with jacketed vessels, 625–628
Time of cooling and heating solids, 639–662
  (See also Conduction, unsteady)
Time of reaction, 624
Toluene, 113, 338, 502
Total emissive power, 65
Towers (see Direct-contact transfer; Distilling column)
Transfer unit, 570, 590
Transition region, 52, 103
Transition velocity, 53
Trapout, 457
Treating, chemical, 298
Triangular pitch, 128
Trickle coolers, 727
Trombone coolers, 727–730
  temperature difference in, 727–728
Trouton's rule, 495
True boiling point (TBP), 353
True temperature difference (see Temperature difference)
Tube bank (see Tube bundle)
Tube bundle, 135, 217, 229
Tube coils, 625
Tube counts, 134, 226, **840, 841**
  reduction of, for extra passes, 248
Tube diameters, common, 228
Tube expanding, 127
Tube lengths, common, 228
Tube passes, 131, 227
Tube pitch, common, 128
  radial, 305, 307
  staggered, 272
Tube resistance to heat transfer, 309

Tube rolling, 127
Tube-sheet layouts, 128, 133, 840, 841
Tube sheets, 127
Tube standards, 228
Tube stills (*see* Furnaces, petroleum-refinery)
Tubes, bent, 382
 common sizes of, 128
 field, 738
 serpentine, 382
Tubular equipment (*see* Exchangers, shell-and-tube)
Tubular Exchanger Manufacturers Association, Standards of the, 137, 148, 828–834, 837, 845
Turbines, 165, 301, 386, 390
Turbulent flow, in condensate films, 270
 definition of, 29
Two-position control, 770

U

U, overall coefficient of heat transfer, 86
U-bend exchangers, 135
 double-tube sheet, 135
 as suction heaters, 740
Ultraviolet, 66
Unit resistance, definition of, 18
Units, consistent, 37
Unsteady conduction (*see* Conduction)
Unsteady state, 624

V

Vacuum, in evaporation processes, 393
 gases under, 191
 in surface condensers, 305
Vacuum processes, 390
 water temperatures for, 594
Valves, regulating, 767
Van der Waals' equation, 191
Vapor, saturated, 282, 289
 superheated, 281
Vapor belt, 271
Vapor binding, 376, 378
Vapor-diffusion (*see* Diffusion)
Vapor-liquid equilibria in phase rule, 313
Vapor-liquid relationships, of binary mixtures, 318
 of multicomponent mixture, 319–329
 of vapor and noncondensable gas, 339
Vapor mixtures (*see* Condensation of mixed vapors; Vaporization)
 common types of, 315
 compositions of, 313
Vapor pressure, of hydrocarbons, **811**
 of water, **816**
Vapor-recovery system, 222
Vaporization, 252, 322, 326, 375, 400
 blanketing in, 458
 from clean surfaces, 459
 maximum coefficients for, **459**
 from pools, 459
 (*See also* Evaporation; Evaporators; Vaporizers; Reboilers)

Vaporizers, 102, 378, 453, 458, 459
 1-2 exchangers as, 459, 461
 forced-circulation, 454
  calculation of, 464–468
  as evaporators, 470
  film coefficients for, 461
  isothermal boiling, 461
  maximum allowable flux, **459**
  maximum coefficient, **460**
  nonisothermal boiling, 468
  pressure drop in, 461–463
  pump problems in, 457
  specific gravities in, 461–463
  vapors in tubes of, 470, 471
 forced- vs. natural-circulation, 457
 fouling in, 455
 natural-circulation, 456
  hydrostatic head in, 456
  maximum allowable flux in, **459**
  preheating in, 461
  recirculation in, 455, 456
Vaporizing equipment, classification of, 378
Vaporizing exchangers, 378, 453, 458
Vaporizing processes, 453, 465
Velocity, of drainage, 256
 of water, in coolers, 155
 in surface condensers, 306
Velocity distribution, parabolic, 40, 201
Velocity gradient, 27
Vertical condensers (*see* Condensers, shell-and-tube)
Vertical condenser-subcooler, 289
 calculation of, 290–293
Vessels, with coils, time for heating, 626–628
 direct-fired, 709
 jacketed, heat-transfer coefficients for, 716–719
  time for heating, 626–628
Viscosity, 27
 of animal and vegetable oils, **821**
 conversion chart for, **820**
 conversion factors for, **794**
 of fatty acids, **821**
 of gases, 190, **824, 825**
 of liquids, **822, 823**
 of liquid solutions, 161
 of petroleum fractions, **821**
 pound-force, 28, 258
 pound-mass, 28, 258
Viscosity correction, for gases, **820**
 for nonisothermal flow, 99, 120
Viscous flow (*see* Streamline flow)
Volumetric equivalent diameter, 555
Volumetric heat-transfer coefficient, 668, 669

W

Waste heat recovery, 192
Waste liquor, sulfate, 426
Water, algae in, 154, 724
 as cooling medium, 154, 161, 193, 594
 distilled, 155, 384, 443
 evaporation of (*see* Evaporation)

Water, hard and soft, 380
  optimum outlet temperature of, in exchangers, 158
    calculation of, 159
  in power cycles, 301
  raw, 155
  sea, 392
  in shells, heat-transfer coefficients of, 137
  treatment of boiler-feed, 380–382
  in tubes, heat-transfer coefficients of, 155, **835**
  velocity of, in coolers, 155
    in surface condensers, 306
  well, 285
Water conditioning, 380–382
Water boxes, 129–133, 305

Water reuse, 577
Wavelength, 4
Weighted temperature difference (*see* Temperature difference)
Well water, 285
Wet-bulb temperature, 570–573
  vs. dew-point, 573
Wet gases, coolers for, 195
Wien's displacement law, 66
Wind, maximum velocities of, 597
  summer data on, 582
Wood as fuel, 675

Z

Zeolite process, 381